P9-DDI-976

Molecular Pathology

Author

Antoni Horst, M.D., Ph.D.
Professor Emeritus
Experimental Pathology and Human Genetics
Institute of Human Genetics
Polish Academy of Sciences
Poznań, Poland

CRC Press
Boca Raton Ann Arbor Boston

Library of Congress Cataloging-in-Publication Data

Horst, Antoni.
Molecular pathology / author, Antoni Horst.
p. cm.
Includes bibliographical references and index.
ISBN 0-8493-6088-9
1. Pathology, Molecular. I. Title.
[DNLM: 1. Molecular Biology. 2. Pathology. QZ H819m]
RB113.h65 1991
616.07—dc20
DNLM/DLC
for Library of Congress 90-15127
CIP

Direct all inquiries to CRC Press, Inc., 2000 Corporate Blvd., N.W., Boca Raton, Florida 33431.

International Standard Book Number 0-8493-6088-9

Library of Congress Card Number 90-15127
Printed in the United States

PREFACE

Molecular biology is based on the elaboration of totally new methods of analysis of body constituents i.e., genes (nucleic acids) and their products: proteins, including enzymes responsible for the physiological functions of the organism and for the synthesis of the remaining body constituents (sugars, lipids), and structural proteins which form the cyto-skeleton and other cellular organelles. Using the classical methods of isolation and structure determination of active polypeptides (e.g., insulin) or other active substances present in particular organs or tissues in mili- or micro-grams, these substances had to be extracted from many hundreds or thousands of kilograms of crude material, a process several decades, whereas by modern techniques of molecular biology it becomes possible to determine the structures and functions of active substances present in particular tissues or organs in nano- or pico-grams by the elaboration of minimal crude material in a few years. For example, the atrial natriuretic factor, which for several years was postulated as a renal factor, was localized in 1981 in the cardiac atrial extract. Using modern biomedical methods, it was possible to determine its structure within 2 years, and in the next 2 years was performed its synthesis and release, and elucidation of its function and therapeutic use, expecially the homeostatic regulation of fluid, salt, and blood pressure. Similar results have been obtained in relation to other body constitutents.

My intention in writing this book was not to describe all the known molecular abnormalities, but to present a new conception of the pathogenesis of diseases based on molecular abnormalitites not explicable by previously held views. Typical examples are disorders caused by globin abnormalities (sickle cell anemia, unstable hemoglobins, hemoglobins with partial or total loss of oxygen transport, increased or decreased oxygen affinity, thalassemias caused by deletion or point mutations, hereditary persistence of fetal homoglobin, and others) which could not be explained by morphological methods.

The most astounding factors are pathological events caused by transformation of body constituents into substances of totally new (sometimes opposite) functions. The best known example is α_1-antitrypsin Pittsburg. The primary function of α_1-antitrypsin is inhibition of neutrophil elastase, a protease capable of hydrolyzing most connective tissue components. Hereditary deficiency of α_1-antytrypsin results in early onset of emphysema caused by degradation of eleastic fibers of the lungs.

Another example is α_2-antiplasmin Enschede (after a city in the Netherlands). Normally, α_2-antiplasmin inhibits the action of plasmin, a clot-dissolving enzyme. The insertion of an alanine residue near its active site converts α_2-antiplasmin Enschede into a substrate for plasmin (instead of its inhibitor), resulting in a serious blooding disorder because of the lack of normal α_2-antiplasmin activity which degrades the clot-dissolving enzyme plasmin.

Point mutations of oncogenic proteins are futher examples of pathogenic function of mutated proteins. The plasma membrane protein p21 binding GTP or GDP, analogous to G-proteins, is known to transduce signals from various cell-surface receptors to activate adenyl cyclase. GTP bound to this protein denotes the excited state of this protein, and its dissociation to GDP the relaxed state. The p21 proteins encoded by the H-*ras* gene (Harvey or Kirsten sarcoma virus) demonstrate an amino acid substitution at residue 12. The mutated protein p21 binds GTP effectively, but its hydrolysis to GDP is greatly reduced. In consequence the mutated p21 remains excited, permanently inducing adenyl cyclase activity which seems to be the major contributing factor for urine bladder cancer.

These few examples indicate that mutation of proteins may totally change their functions as a consequence of the loss of their specific functions. The best known are the mutation consequences of the globin molecule that comprise several hundreds of different mutations with tens of changed functions. Without knowing the enormous number of proteins of a living organism (probably some hundred thousands) with all their mutations causing changes in their functions, we cannot know the crucial role these functional changes play in health and disease. The molecular era is just in its beginning stages, and incessantly new discoveries modify the results we have had and the views we have had hitherto. Therefore, our present knowledge of molecular mechanisms and their role in pathogenesis of diseases is incomplete and will probablly require a long time and thousands of experiments to be complete and will clarify probably all the unknown pathogenic mechanisms of contemporary medicine.

Molecular pathology is a new branch of knowlege on the pathogenesis of idiopathic diseases. Molecular biology created totally new views about living matter, and abolished the strong differences between living and dead matter. Molecular biologists are able to create "living" viruses, normal genes able to replace abnormal ones, etc. These great progress made must surely have their reflection on medical sciences especially because gene therapy has already been done in some monogenetic diseases.

The plan of this book comprises a description of the known molecular bases of various diseases. The best known example of various diseases caused by pathological variants of hemoglobin are blood disorders/hemolytical anemias, e.g., sicle cell anemia, methemoglobinemias, thalassemias, erytremia caused by variants with increased O_2 affinity, decreased malaria morbidity in cases of mild hemoglobinopathies, or G-6-PD deficiency.

Molecular pathology of blood disorders is best known because of the accessibility of hemoglobin. But hemoglobin is only one protein, and there are some hundred thousands of them. According to genetic rules, every gene may undergo mutations with a specified frequency. Therefore, mutations must be common in all proteins and are not known solely because of their inaccessibility. As in blood disorders, there must exist protein variants responsible for the function or organs, and consequently, various diseases of these organs must occur. Thus, it is clear that the future of medicine lies in the expanding knowledge of the molecular mechanisms of disease and in therapy based on these events.

Antoni Horst

THE AUTHOR

Antoni Horst, M.D., Ph.D., is Professor Emeritus of General and Experimental Pathology and Human Genetics at the Institute of Human Genetics of the Polish Academy of Sciences, and Doctor honoris causa of the Medical Academy (School) of Poznań.

Dr. Horst graduated from the Medical Faculty of Poznań University, was Associate Professor of Internal Medicine (1945 to 1950), Professor and Head of the Department of General and Experimental Pathology (1950 to 1974) and the Department of Human Genetics (1963 to 1974) in the Medical Academy (School) of Poznań. He is founder and former Director of the Institute of Human Genetics of the Polish Academy of Sciences (1974 to 1985).

Among other honors, Dr. Horst received the award of the Medical Department of the Polish Academy of Sciences, and several awards of the Minister of Health and Social Welfare of the People's Republic of Poland. He was fellow of the Rockefeller Foundation in 1958/9.

Dr. Horst is member of the Polish Academy of Sciences Medical Department, President of the Committee of Cell Pathophysiology of the Polish Academy of Sciences, Honorary Member of the Polish Genetical Society, Polish Immunological Society, Polish Society of Histochemists and Cytochemists, Poznań Society of Friends of Sciences, and Member of the Polish Biochemical Society, Polish Physiological Society and many others.

Dr. Horst has published more than 200 research and scientific papters, and ten scientific books, among them the *Textbook Phyiological Pathology* (9 editions). His current major research interests include gene regulation in eukaryotes, and molecular pathology as a general concept of pathogenesis of diseases.

TABLE OF CONTENTS

Chapter 1

INTRODUCTION

The origin of disease has aroused remarkable human interest from the dawn of history. Primitive man believed that illnesses are caused by environmental pathogenic factors. To the primitive man it was difficult to believe that a disease could be caused by idiopathic somatic disturbances, and therefore he ascribed its origins to various insignificant environmental factors, such as cold, fatigue, famine, troubles, etc. or even to the action of supernatural factors, such as malicious pagan gods. On the other hand, he believed that good health was a gift of good-natured gods.

In general, the more sober were views upon life at a given era, the more rational were views about the causes of disease. Therefore, the reasonable ancient thinkers noticed some kind of somatic causes of disease: Hippocrates, in his humoral theory of disease, taught that an unbalanced composition of body fluids (dyscrasia) was the main cause of disease, and Asklepiades, in his atomic theory of disease, viewed the main cause as the malformed structure of body components (atoms).

Because of thc mystical views about life in the Middle Ages, beliefs that disease is caused by God as a kind of punishment for sins, etc. re-emerged.

In the following centuries, when strict thinking became dominant, living organisms were considered as a kind of complicated machine, composed of relatively simple components, which principally did not differ from artificially constructed machines. After the discovery in the early 19th century (first in plants and later in animals) that cells were the principal body constituents of all living organisms, Virchow, in the middle of the 19th century established his well-known cellular theory of disease. This theory was the prevailing theory up to the middle of the 20th century.

Soon, however, it became obvious that this theory could not explain many aspects of life processes and, consequently, various illnesses. In fact, this theory was nothing more than a theory which transposed the living phenomena from the whole organism to the same living phenomena restricted to the cells. Consequently, all incomprehensible living phenomena were related to a mysterious vital function. In other words, the ancient vitalistic view of life processes was revived.

The vitalistic character of the cellular theory, although not able to provide a final resolution of vital processes, enabled man to acquire an enormous amount of knowledge in the field of many physiological and pathological processes.

In the period of the "cellular theory" many extraorganismic causes of disease were recognized. The most important findings in this field were the great microbial discoveries at the end of the 19th century, when recognition of pathogenic microbes seemed to be identical with the recognition of causes of the particular disease. Consequently, an increasing number of erroneous beliefs proposed that illnesses were principally caused by noxious environmental factors.

There were many objections to this doctrine. For instance, René Dubos named this trend as "a doctrine of specific etiology of diseases" which does not take into account the specificity of living organism, although it was well known that not all species suffered from the same infectious diseases, and to find animals susceptible to some infectious diseases which occur only in man was very difficult and sometimes impossible.

The conclusion was obvious that, for the occurrence of a particular disease, an adequate susceptible living organism was necessary. Unfortunately, the general knowledge of living processes was, at that time, so unsatisfactory that it was impossible to establish a substantial theory about causes of disease (considering the extraorganismic noxious factors and the intraorganismic factors which make the organism susceptible to the disease). Therefore, despite the noted progress in many branches of medical science, the fundamental resolution of the essence of the vital processes also was impossible.

Correct views of the life processes did not occur prior to discovery of the molecular bases of heredity and knowledge of the mechanisms of protein biosynthesis and the function of protein molecules. With these discoveries it was proven, for the first time, that specific characteristics are inherited through a chemical substance, i.e., deoxyribonucleic acid (DNA) which determines (codes) the sequence of amino acids in synthesizing protein molecules that are responsible for all vital functions. These findings made it possible to consider the dependence of disease on the presence of specific active substances in living organisms, whose normal structure and function are responsible for health, and abnormal structure and function are responsible for disease. This produced a radical change of view about the causes of disease from mainly extraorganismic to the organismic ones and, as a matter of fact, that diseases are not caused by deviation of mysterious vital processes of the whole organism but by molecules, the function of which is exclusively based on well-known physicochemical reactions.

In aspects of molecular biology, living processes occur when reacting molecules appear at an adequate place and time in the organism. Usually, many kinds of interacting molecules are necessary to initiate and maintain vital functions. At the moment when the stimulus for the life function disappears, the life function must also expire. This is accomplished by inactivation of the reacting molecules, usually by decay of these molecules. Therefore, vital function arises usually by the synthesis or activation of preexisting molecules specific for the given reaction and is terminated by inactivation of the reacting molecules. Very often the synthesized active molecules are so labile that they disintegrate almost immediately after their synthesis; in other cases their decay is caused by specific enzymes. In this aspect a vital function comes into existence by the permanent synthesis of active molecules which, almost immediately after their synthesis, disintegrate.

The permanent synthesis and decay of protein molecules (the average half-life time of proteins is 3 days) require a mechanism which always ensures the synthesis of the same molecules, otherwise continued life of a given organism would be impossible. The permanent synthesis of the same protein molecules is always defined by the genetic code. Therefore, we come to the conclusion that a normal genetic code and efficient mechanisms of synthesis of protein and other components of an organism are essential processes for life.

The purpose of this book is to present evidence about the molecular mechanisms essential for the occurrence of disease. Molecular events happen in living organisms, therefore the molecular mechanisms of disease turn from the environment to the organism, lending itself (or in combination with environmental factors) to the production of disease. It must be emphasized that the environmental factors must not be pathogenic per se, but factors which in combination with molecular defects of the organism become pathogenic. That is, in molecular biology, the causes of illnesses are to be found principally inside the living organism, and not so much in environmental pathogenic factors.

This change in concept of disease causes requires a thorough knowledge of the elements of molecular biology which enables one to interpret events leading to the occurrence of disease. These elements comprise knowledge of specific structure, synthesis, and function of molecules reacting in the living organisms. To these elements belong structure and function of genes, regulation of gene expression, mechanisms of protein biosynthesis, molecular bases of cell differentiation, mechanisms of mutations, pathological variants of proteins with disturbances of stability and function, etc.

Molecular disturbances as causes of disease usually are recognized in heritable diseases. This seems inappropriate because all protein molecules of living organisms depend on the genetic code and, as is well known, every protein molecule may undergo mutations some more often and others less often. At present our knowledge about variants of protein molecules is very limited. The best known are globin molecules, principally because of their accessibility and appearance in the erythrocytes in an almost pure state. Thanks to this example, we know many pathological health disturbances are caused by point mutation with the consequence of a single substitution of only one amino acid (e.g., sickle cell anemia is caused by the instability of the globin beta chain [erythremia] caused by increased oxygen affinity, methemoglobinopathies being unable to transport oxygen, etc.) or disturbances of globin synthesis (e.g., various kinds of thalassemias caused by gene deletion or disturbances of the particular steps of protein synthesis).

The question arises if other diseases may be caused by pathological variants of proteins. Besides pathological variants of globin, we also know some other protein variants, although these have not been studied as well (e.g., pathological variants or lack of fibrinogen and other blood clotting factors, and other blood plasma components). Variants of tissue proteins are very little known, principally because of their inaccessibility. According to the genetic rule that every gene may undergo mutations we must accept the idea that tissue proteins also are composed of an enormous number of protein variants which differ in a better and more efficient, or a less and even totally inefficient function. Only in this way can it be explained why one person is a very efficient runner and the other is not; why one person does not suffer from hypoxia in high mountains and the other does; why one person dies in his early youth because of heart failure and the other lives to old age; why one person is susceptible to an infectious disease and the other is not. There are countless such examples. The only explanation we can assume is that the cause of these differences lies in the varied quality of molecules responsible for the particular functions. Since defective functions are the essential cause of disease, it must be accepted that defective variants of protein components of living organisms are the main causes of all idiopathic diseases.

Chapter 2

THE MOLECULE: ANATOMY AND PHYSIOLOGY

I. THE STRUCTURE OF MOLECULES

A molecule is an aggregate of atoms held together by chemical bonds (strong covalent bonds and weak chemical bonds or interactions) of a definite size, structure, and function. Cells are composed of molecules which form organelles (being aggregations of molecules of specific function). The distribution of molecules inside the cell, as well as in its organelles, is not random but well-defined, which makes the very different functions of different cells possible. The arrangement of different molecules inside the cell and its organelles is controlled by weak chemical bonds. Therefore, the knowledge of these bonds is very important in understanding the role of molecules in performing vital functions.

A chemical bond is an attractive force that holds atoms together. We distinguish strong covalent bonds and weak chemical bonds. The most characteristic feature of bonds is their strength. Strong bonds never dissociate at physiological temperatures. Atoms bound by covalent bonds belong to the same molecule. Covalent bonds hold atoms closer than weak chemical bonds. For example, two hydrogen atoms bound covalently are 0.074 nm (0.74 Å) apart, while bound by weak van der Waals' contacts are 0.12 nm (1.2 Å) apart. Atoms bound by covalent bonds are capable of weak interactions with neighboring atoms. In living cells, covalent bonds can be formed or dissociated only by the action of appropriate enzymes.

Weak bonds are very important for the function of molecules. The most important are van der Waals' bonds, hydrogen bonds, and ionic bonds. Weak chemical bonds, also called "secondary bonds," exist not only between atoms of the same molecule, but also between atoms of different neighboring molecules. Weak bonds can be broken easily at normal physiological temperatures, which enables some flexibility of molecules. Single weak bonds are not strong enough to bind two atoms together when present singly, but a group of weak bonds can bind atoms and even molecules also for relatively longer periods of time. The most important feature of weak bonds is that they allow movement of atoms inside the molecule. The number of covalent bonds of a given molecule is limited by its valence, while the number of weak bonds is limited only by the number of atoms which simultaneously touch each other. Single covalent bonds permit free rotation of bound atoms, while two or more bonds delimit more or less the movements of the bound atoms. In contrast, weak bonds (van der Waals' ionic) do not restrict bound atoms.

The nature of bonds can be explained by quantum mechanics which states that all bonds, independently of their type, are based on electrostatic forces. The spontaneous formation of a bond involves release of internal energy of the unbound atoms, which is converted into another form of energy. Thus, the reaction between two atoms can be described by the formula

$$A + B \rightarrow AB + \text{energy}$$

where A and B represent the particular atoms, and AB the molecule. The rate of the reaction depends on the frequency of collisions between the reacting atoms. The stronger the bond, the greater amount of energy is released. The most common unit to measure energy is the calorie, and since thousands of calories are necessary to break a mole of bonds, the chemical energy for this purpose is expressed in kilocalories (kcal) or joules (J) and kilojoules (kJ) per mole (mol).

Weak chemical bonds also can break. The most important forces which cause dissociation of chemical bonds come from heat energy. Because of the kinetic energy, collisions with fast moving atoms or molecules may push apart the bound atoms. The higher the temperature, the faster the molecules are moving and the greater is the probability that in consequence of the collisions the bonds will break. Therefore, with an increase of temperature, the stability of molecules decreases. The breaking of a molecule may be described by the formula

$$AB + \text{energy} \rightarrow A + B$$

which denotes that energy necessary to break a bond is equal to the amount of energy released during formation of the molecule. This is in accordance with the first law of thermodynamics that energy can be neither made nor destroyed.

Very important for the living organisms is the equilibrium between formation and breaking of chemical bonds, i.e., that the amount of forming and breaking bonds is equal.

Biologically important is the so-called "free energy" of chemical reactions because it enables metabolic reactions. In biological systems free energy is the energy that has the ability to do work, represented by the symbol ΔG. According to the second law of thermodynamics, during all spontaneous reactions a decrease of free energy occurs (ΔG is negative). After reaching equilibrium no change in the amount of free energy occurs ($\Delta G = 0$). Therefore in a state of equilibrium a closed number of atoms possesses the lowest amount of free energy.

Free energy may be transformed into heat or it increases the amount of entropy (in a simplified form entropy may be described as the amount of disorder, i.e., structure, of a molecule). Therefore, the greater the disorder (i.e., the more complicated the structure of the molecule), the greater is the amount of entropy. On the contrary, crystallization is the state where atoms are bound in a strict order. The increase of entropy during a reaction denotes that the given reaction does not liberate heat.

II. THE POLARITY OF MOLECULES

Molecules in which atoms are symmetrically distributed are nonpolar molecules, and molecules in which atoms are nonsymetrically distributed are polar molecules. The nonpolar molecules are uncharged. Electronegative atoms have the tendency to gain electrons, and electropositive atoms have the tendency to give up electrons. In a polar molecule, e.g., a water molecule (H_2O) the two hydrogen atoms are electropositive, and the one oxygen atom is electronegative. The center of the positive charge is on the one side of the molecule and the negative on the other side of the molecule. Separated negative and positive charges create an electric dipole moment. A dipole is characteristic for polar molecules (see Figure 1).

The distribution of charges inside the molecule also can be influenced by neighboring molecules. Nonpolar molecules are especially affected by polar molecules. In this case a fluctuating charge distribution occurs and the nonpolar molecule acquires a small charge; consequently, its chemical bonds become weaker.

A. VAN DER WAALS' FORCES

Weak van der Waals' forces originate when two neighboring atoms lie near enough to each other that an induced fluctuating charge, and consequently, permanent charge distribution, occurs. These forces can be present among all types of molecules, both polar and nonpolar ones. Van der Waals' bonds are strictly dependent upon the distance between the reacting molecules, being inversely proportional to the sixth power of the distance. Therefore, too great a distance abolished the van der Waals' bonds, and too close a distance causes repulsion of the atoms instead of attraction. The attractive and repulsive van der Waals'

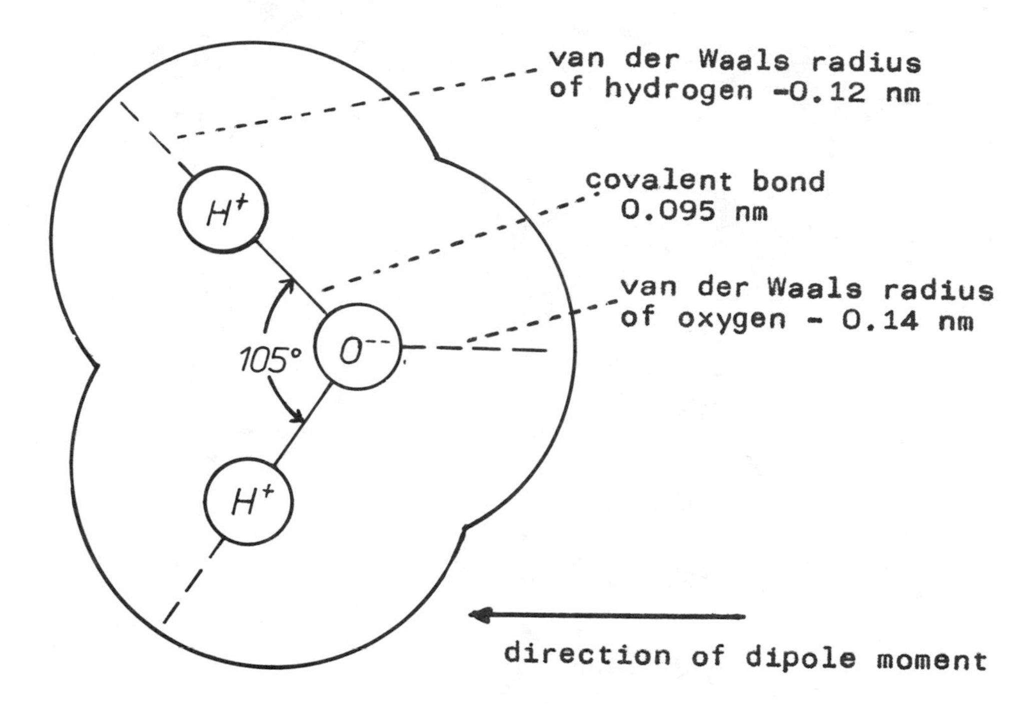

FIGURE 1. Model of a polar molecule (water molecule).

forces hold atoms in a balanced position dependent on the interaction of these forces (see Figure 2).

The distance between two atoms bonded by the van der Waals' forces represents the so-called van der Waals' radius. The bonding energy of the van der Waals' forces increases with the size of the bonded atoms. Van der Waals' forces belong to the weakest chemical bonds. When two atoms lie at a distance of the sum of their van der Waals' radii the average van der Waals' force is only about 4 kJ/mol (1 kcal/mol). This is only slightly more than the thermal energy of molecules at room temperature, i.e., 2.5 kJ/mol (0.6 kcal/mol). Therefore, it is understandable that even slight elevation of the temperature dissociates the van der Waals' bonds.

The distance between atoms bound by the van der Waals' forces is decisive for the binding of the reacting molecules, e.g., binding of antigens by antibodies. In order that several atoms within the surfaces of the reacting molecules interact effectively, the sum of their van der Waals' radii cannot be in excess positively or negatively. In the case that this distance is greater, the binding forces are too small. On the contrary, when the distance is too small, repulsive forces instead of attractive forces arise. One of the reacting molecules usually has a cavity into which an adequate group of atoms of the other reacting molecule protrude. In this case the binding forces can be 40 kJ/mol (10 kcal per mole).

B. HYDROGEN BONDS

Bonds between covalently bounded positively charged hydrogen atoms with covalently bounded negatively charged acceptor atoms are defined as hydrogen bonds. Hydrogen bonds may join single atoms as well as groups of atoms. For example, the hydrogen atom of the imino group (N–H) bound with negatively charged keto oxygen atom (C=O), as well as positively charged group NH_3^+ bound with negatively charged group COO^-. Hydrogen bonds are stronger than the van der Waals' forces and range between 13 to 29 kJ/mol (3 to 7 kcal/mol).

Contrary to the van der Waals' bonds, hydrogen bonds are highly directional. The

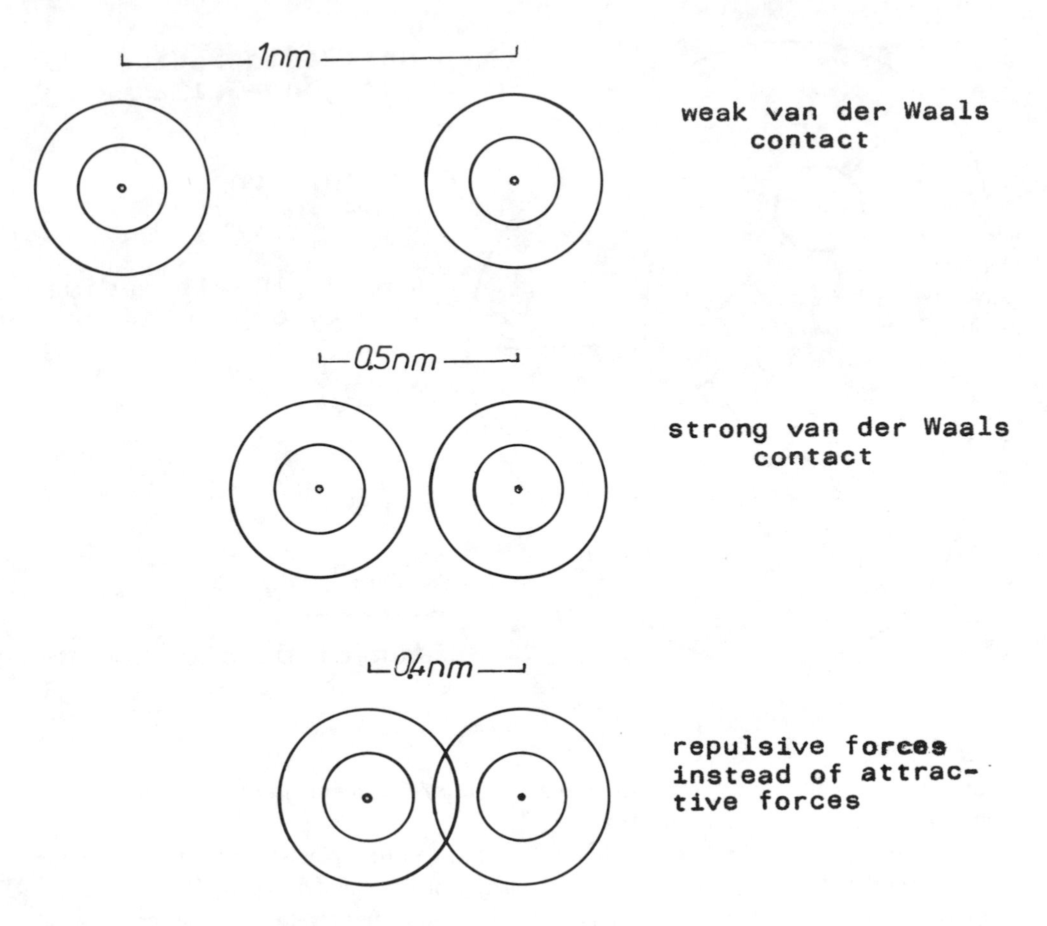

FIGURE 2. Model of van der Waals' contacts.

hydrogen atom points directly to the acceptor atom. Therefore the hydrogen bonds are more specific than the van der Waals' bonds since the donor hydrogen atom must have a complementary acceptor atom.

The most biologically important hydrogen bonds after J. Watson are between a hydroxyl group of serine and a peptide group.

between peptide groups

$$\mathrm{R-\underset{\underset{H}{|}}{\overset{\overset{H}{|}}{C}}-\underset{\underset{\underset{H}{|}}{N-}}{\overset{}{C}}=O\;\cdots\cdots\;HN-\overset{\overset{|}{C=O}}{\underset{}{CH}}-R}$$

between two hydroxyl groups

$$\mathrm{C=\overset{H}{O}\;\cdots\cdots\;H-O-\underset{\underset{H}{|}}{\overset{|}{C}}-}$$

between charged carboxyl group and a hydroxyl group of tyrosine

$$\mathrm{-C_6H_4-O-H\;\cdots\cdots\;O^-\!\!\setminus C=O}$$

between a charged amino group and a charged carboxyl group $_2HN — H \cdots\cdots O^-$ \ $C = O$ |

C. IONIC BONDS

Charged positive or negative groups inside the molecules usually are neutralized by neighboring oppositely charged groups. These bonds are defined as ionic bonds. For example, phosphate groups of negatively charged mononucleotides exhibit a positively charged amino group (NH_3^+), and a negatively charged carboxyl group (COO^-). The average energy of the ionic bonds is about 21 kJ/mol (5 kcal/mol).

In water both the hydrogen and oxygen atoms form strong hydrogen bonds. Water is a polar molecule (H–O–H) in which the oxygen atom can form a bond with the two external hydrogen atoms, and each hydrogen atom can form a bond with one neighboring oxygen atom. In temperature below 0°C (in ice) these bonds are very rigid which make the arrangement of water molecules fixed. In higher temperatures (above 0°C) the thermal collisions cause disruption of the hydrogen bonds which allow the water molecules to change the neighboring water molecules.

D. WEAK BONDS IN WATER SOLUTION

In the living organism the molecules exist dissolved in water which enables their activity, according to the old sentence: ''corpora non agunt nisi soluta.'' In aqueous solution the molecules move until they find another molecule with which they can form the strongest secondary bonds. Because of the disruptive influence of heat, the arrangement of molecules in aqueous solution changes from one form to another. In living organisms metabolic processes continually transform the molecules with simultaneous changes in the nature of secondary bonds. Hence, in living organisms the bonding of molecules is influenced by the motion of dissolved molecules, as also be metabolic transformations of the molecules. Because the hydrogen bonds are stronger than the van der Waals' forces, molecules that can form hydrogen bonds will form them in preference to van der Waals' bonds. Therefore, when we put molecules which cannot form hydrogen bonds into water, e.g., benzene which cannot form hydrogen bonds, the water molecules will immediately form hydrogen bonds and the benzene molecules will attach to each other by van der Waals' contacts. Consequently, the water molecules and the benzene molecules will separate from each other. Conversely, molecules that can form hydrogen bonds are soluble in water, e.g., glucose and other sugars. In other words, they are hydrophilic. In contrast, molecules that cannot form hydrogen bonds are hydrophobic.

III. THE CONFORMATION OF MOLECULES AND ITS ROLE IN MOLECULAR-MOLECULAR INTERACTIONS

Interactions of molecules are performed almost purely by means of weak bonds and as the weak bonds depend on the distance between the reacting atoms, the surfaces of the interacting molecules, composed of many atoms, must correspond exactly to one another. Even in the case when only one atom protrudes, it may hinder the formation of chemical bonds because of the repulsive forces, and when the distance is too great between the reacting surfaces, the bonding forces are too weak which makes it impossible to bind the two molecules together.

As explained earlier, weak bonds can bind atoms together only when they are present in sufficient number. The greater the number of weak bonds, the stronger the interactions between the reacting molecules. This alone makes the specificity of reactions between the appropriate molecules which enables them to perform the complex vital functions. For example, metabolism of living organisms belongs to the most significant attributes of life. Metabolism of living organisms would be impossible without the action of enzymes which enables the performance of the metabolic reactions by lowering the amount of necessary energy of the reactions. Through the action of enzymes, reactions that would be performed during years, are performed in seconds. The second characteristic feature of enzymes is that they always react only with molecules (named substrates) specific for the given enzyme and not with just any molecules. The specificity of binding the appropriate substrates by enzymes depends on the adequate surfaces of the molecules enabling them to form weak chemical bonds. The presence of weak bonds of interacting molecules causes not only the specificity of the reacting molecules, but also the velocity of the reactions. Enzymes catalyze both the processes of synthesis and dissociation. In this aspect the weak bonding forces which enable easy conversion of synthesis to dissociation and vice versa, are very important for maintaining the necessary equilibrium between these two reactions. The ease of formation and breakage of weak chemical bonds enables an extremely rapid course of the reactions. It has been calculated that 10^6 enzyme-substrate reactions can be performed during one second.

As a rule, both the strong covalent bonds as well as the weak chemical bonds participate in forming the definite conformation of molecules. The conformation of a molecule depends primarily on covalent bonds but, in formation of the definite conformation, weak bonds play an important role. Very important in this aspect is the flexibility of atoms or groups (e.g., methyl groups, COOH, CH_2OH, etc.) bonded by single covalent bonds. Atoms or groups bonded by two or more covalent bonds cannot rotate and are inflexible. As a rule, in living organisms molecules exist in an aqueous environment. Therefore, the formation of weak chemical bonds depends on the simultaneous possibility of the formation of stronger hydrogen bonds between water molecules. The conformation of polymeric molecules usually depends on the presence of a linear backbone with regular side groups held together by weak bonds. In the case of regular side groups, the most appropriate is the helical configuration because it allows each group to be placed in an identical orientation inside the molecule. This also is the most stable molecular configuration. Since most biopolymers (proteins) have irregular side groups, the helical structure of these molecules is usually interrupted by nonhelical structures. In this case, a sheet-like structure, also called beta structures, arises in which a compromise exists between the tendency of the regular backbone to form a helix and of the irregular side groups to take such a position in which the strength of their secondary bonds is minimal. The majority of proteins has irregular side groups (various amino acids). Therefore, the conformation of proteins in which the helical structure exists alternately with the nonhelical usually is irregular. Contrary to the proteins, nucleic acids have regular side groups, therefore, their structure is regularly helical.

The next essential feature of molecules that enables vital functions is the possibility of movements. Among the most longstanding and primitive definitions of life is the one which states that, "life is all that is spontaneously moving." Recent research has shown that this attribute is among the basic properties of macromolecules, or biopolymers. The essence of this motion may be explained by means of the simplest possible example: the contraction of polyacrylic ester, or polyacrylate. The structure of polyacrylate is as simple as can be; it is made up of a chain of unlimited length (see Figure 3) with carboxyl groups (COO^-) ionized in an alkaline medium (hence negatively charged). These negative charges repel each other causing the polyacrylate chain to accrete. When such a chain is placed in an acidic medium (with a surplus of positive hydrogen ions—H^+), there is no dissociation of the slightly acid COOH groups into COO^-. This causes the negative charges of the COO^- groups of polyacrylate to disappear and the chain becomes shorter. Repeating this at regular intervals, we obtain a rhythmic expansion and contraction of the fiber. How great a force

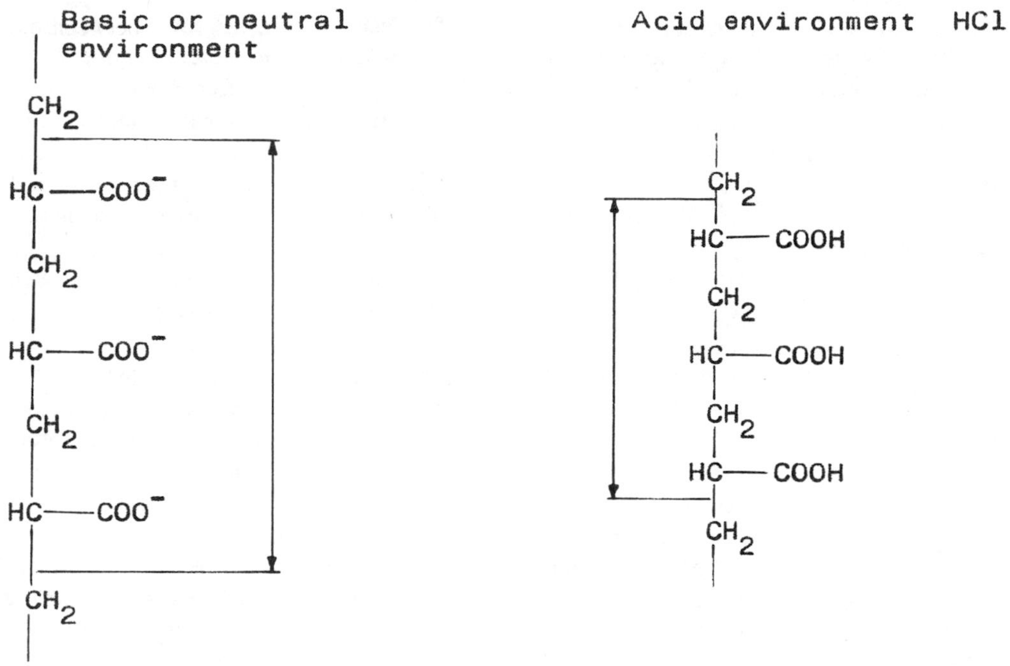

Basic or neutral environment

CH_2 — HC—COO^- — CH_2 — HC—COO^- — CH_2 — HC—COO^- — CH_2

Acid environment HCl

CH_2 — HC—COOH — CH_2 — HC—COOH — CH_2 — HC—COOH — CH_2

FIGURE 3. Contraction and expansion of polyacrylate chain.

can be engendered in this way was demonstrated at Expo 1958 in Brussels, where an artificial muscle based on this principle was exhibited. A suitably large number of polyacrylate fibers was made to lift a weight of 50 kg. As long as the fibers remained in an alkaline or neutral medium, they expanded and the weight rested on its base. With a change of the environment from basic to acidic (by the addition of hydrochloric acid), the fibers contracted at once, lifting the 50-kg weight several meters high. When the acid was removed, the fibers expanded again and the weight dropped back on the base. This simple demonstration is the best proof of the ability of macromolecules to perform a motion, depending on the electrostatic charge of the molecule's component elements.

The ability of molecules to change conformations is probably indispensable for their role in performing life processes. This is best seen in the case of enzymes changing conformation during the course of reactions. It is generally accepted that conformational changes of enzymes are necessary for performing catalytic reactions. In this respect, interaction of a given enzyme with its appropriate substrate changes of enzyme conformation occurs, which enables them to perform the reaction so that the positions of the reacting atoms become nearer to one another. This enables transfer of charges and formation of new bonds until the reaction is finished. Finally, the enzymes disconnect from the transformed substrates and their conformations revert to the previous ones which enable them to react with the substrate's next molecules.

Usually enzymes have an active center (active site) located in a hollow (fissure or pocket) in which the substrate fits. The tight-fitting substrate makes a closed apolar hydrophobic environment in which by lowering the dielectric constant, the enzymatic reaction is maximally facilitated. The hydrophobic environment works in this aspect as some kind of "organic solvent" which enables the easy performance of the reactions. For example, a hydrous solution of acetic and propionic acids being side chains of the aspartic and glutamic acids of the active site of chymotrypsin practically do not hydrolyze chitin at room temperature, whereas in the hydrophobic active center of chymotrypsin they hydrolyze chitin during some milliseconds.

Circumstantial evidence for active movements of molecules during the course of reactions catalyzed by enzymes may be lowering of temperature which arrests these reactions. Lowering of temperature, decreasing thermal collisions of molecules, strengthens the weak chemical bonds of molecules (principally the most important van der Waals' forces) which make the movements of molecules impossible, thus stopping the reactions. This "cold phenomenon" is widely used for preparation of biological materials containing enzymes and their substrates; for examination of these materials all laboratory work must be done at low temperature.

A very convincing example is the possibility of freezing of living organisms (e.g., rats). Using an appropriate procedure it is possible to freeze an animal to a lump of ice with no signs of life, and after appropriate thawing the animal revives without any signs of disturbance of its functions or changes of its organs. This is understandable only in aspect of the stoppage of all enzymatically catalyzed metabolic reactions responsible for life processes. Because of the instantaneous cessation of all reactions, any negative consequences, such as consumption of substrates or production of toxic metabolites, cannot take place. Therefore, after thawing, all metabolic reactions necessary for the vital functions, stopped in the state of congealment, can be continued (start again) which is synonymous with life.

The function of proteins is realized by their tertiary structure. The primary protein structure denotes the sequence of amino acid residues and number of chains forming the protein molecule depending directly on the genetic code. Several polypeptide chains are held together by secondary bonds (weak chemical bonds, as also by disulfide bonds). Sometimes protein molecules also have nonprotein (prosthetic) groups of vital importance to them (e.g., heme group in the hemoglobin molecule). Prosthetic groups, combined with proteins, acquire new properties. For example, heme combines with oxygen irreversibly but in combination with globin heme combines with oxygen reversibly.

The secondary structure of proteins may be helical in the case of regular side groups or beta structure (sheetlike structure) in the case of irregular side groups.

The tertiary or tridimensional structure of proteins is responsible for their specific functions. The tridimensional protein structure is defined by the backbone chain composed of amino acid residues bound by covalent peptide bonds and secondary bonds that force the molecule to form the final conformation. Because of the different and specific amino acid residual sequences in proteins their definite conformation is extremely irregular. This is dependent on the diverse chemical nature of the amino acid side groups. The disulfide bonds that stabilize the protein molecules also are very important. This is best seen in the case of protein denaturation, when the denaturing agent causes breakage of these bonds and after removal of the denaturing agent the disulfide bonds reappear at the same places, causing the renaturation of proteins with the reappearance of the previous conformations.

IV. THE FUNCTION OF MOLECULES DEPENDENT ON THEIR STRUCTURE

Earlier, molecules were considered to be relatively static structures. In the case of enzymes the decisive role in their functions has been ascribed to a motionless active center. Contrary to this, modern views ascribe to molecules totally new functions. First of all, e.g., in enzyme molecules, the active centers play not only the decisive role in their functions, but the entire molecules. This is best seen on the well-described molecule, i.e., hemoglobin molecule which can be described in somewhat vitalistic fashion: the hemoglobin molecule (due to the increased oxygen affinity in the lungs) "catches" oxygen from the air, then (due to the high oxygen affinity) transports oxygen to all tissues, and finally in the tissues (due to the diminished oxygen affinity) the hemoglobin molecule "pushes" the oxygen molecules into the tissues. Returning to the lungs the cycle of action of the hemoglobin molecule, i.e.,

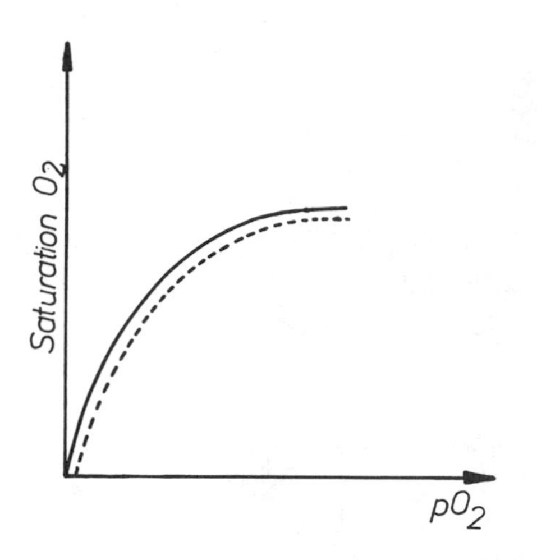

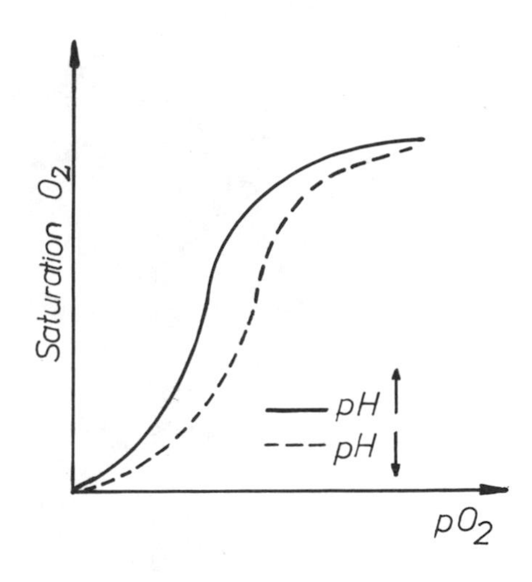

FIGURE 4. Oxygen saturation curves of myoglobin or hemoglobin (left), and hemoglobin A (right). The oxygen saturation curve of myoglobin or hemoglobin H is of exponential type; Bohr's effect is absent. The oxygen saturation curve of hemoglobin A is of the sigmoid type, resulting from heme-heme interaction (Hill's constant); Bohr's effect is present (change of affinity of hemoglobin to oxygen effected by pH). A lowered affinity to oxygen causes shift of the curve to the right.

taking oxygen in the lungs and giving up in the tissues recurs. Simultaneously, to some extent, the oxygen molecule replaces in the tissues oxygen for carbon dioxide (CO_2) and in the lungs replaces carbon dioxide for oxygen. It is particularly noteworthy that both binding of oxygen in the lungs or giving up oxygen to the tissues are active forms of hemoglobin functions.

The hemoglobin molecule consists of four polypeptide chains. In normal hemoglobin A there are two alpha and two beta chains (α_2 β_2). Each of the polypeptide chains is coiled and folded in a characteristic way so that the whole molecule demonstrates a three-dimensional complex structure. The prosthetic group in the hemoglobin molecule is heme. Four heme groups form the porphyrin ring with an iron atom in the center. In the three-dimensional structure composed of four globin chains, the heme groups lie in separate pockets formed by the globin chains. The heme groups are bound with the polypeptide globin chains by a single coordinate bond between the iron atom and particular histidine residues of the particular globin chains. Combination of oxygen with the iron atoms of hemoglobin is associated with changes of the three-dimensional conformation of the whole hemoglobin molecule.

The reversible combination of oxygen with the hemoglobin molecule is dependent on temperature, ionic strength (electrolyte composition), pCO_2, and pH of the reaction medium. The pCO_2-pH influence on the reaction is known as the Bohr effect (see Figure 4). The higher the pCO_2 (hydrogen ion concentration) and lower the pH, the lower is the affinity of the hemoglobin molecule for oxygen (the lower the oxygen dissociation curve in Figure 4).

The function of the hemoglobin molecule as oxygen transporter from the air to the tissues depends on its complex structure, which is well documented thanks to the knowledge of functional disturbances of pathological forms of hemoglobin. The normal hemoglobin molecule consists of four polypeptide chains, two alpha and two beta chains which differ in number of amino acid residues. The alpha chain is composed of 141 amino acid residues and the beta chain of 146 amino acid residues, although there is a basic similarity among the two chains. The polypeptide chains are held together by a number of weak chemical bonds: about 60 contacts which bind each heme in the particular heme pockets of the globin chains, about 80 contacts between α_1-β_2 chains and about 110 contacts between α_1-β_1 chains (see Figure 5). The bonds are so weak that the hemoglobin tetramer easily tends to dissociate

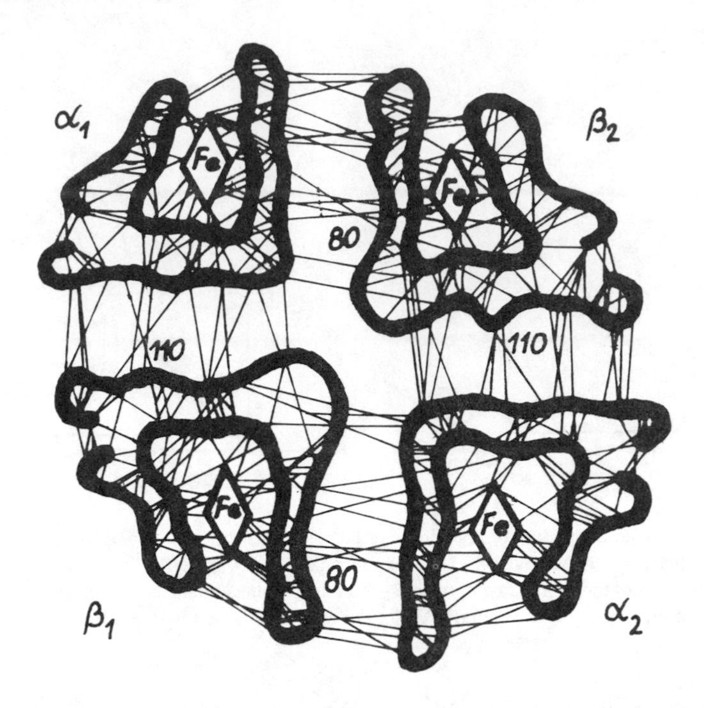

FIGURE 5. Diagrammatic representation of a hemoglobin molecule which represents a tetramer comprising two alpha chains and two beta chains as well as four iron-containing hemes built into each of the globin chains. The numbers stand for weak bonds which join individual elements of the tetramer into the whole.

into separate chains. The weak binding forces make it possible to break one of the bonds and to form new ones which permits the motion of the particular parts of the molecule.

The performance of the hemoglobin function is possible only through movements of the particular parts of the molecule. The combination of oxygen with hemoglobin molecules causes a heme-heme interaction in which the ferric atom in the beta chains approximate from 4.03 nm (40.3 Å) to 3.44 nm (34.4 Å), with simultaneous change of the shape of the molecule (see Figure 6) which in the oxygenated state is more spherical, and in the deoxygenated state is more elongate.

Change of conformation of proteins when performing their functions seems to be a regular feature. For example, after interaction of the active site of the enzyme carboxypeptidase A with its substrate (glycine-L-tyrosine group), tyrosine 248 changes its position almost 180°, enabling in this way near contact with the imino group of the substrate. Simultaneously, arginine 145 combines with the carboxy group of the substrate, and glutamic acid 270 with the free amino group of the substrate. All these conformational changes bring nearer the reacting groups enabling in this way the performance of the catalytic reaction of carboxypeptidase A (through transfer of electrons).

The role of the complex asymmetric structure of the hemoglobin molecule on the highly specific transport of oxygen is best evident if compared with binding of oxygen by myoglobin or hemoglobin H. Myoglobin, contrary to hemoglobin, is composed of a single polypeptide chain. Myoglobin occurs in muscles where it binds oxygen, although in an entirely different way from hemoglobin. Myoglobin has a greater affinity to oxygen than hemoglobin, it has a greater facility of binding oxygen and is not as prone to release it. Moreover, the saturation curve of myoglobin is of the exponential type, and that of hemoglobin is of the sigmoid type. Finally, myoglobin does not possess Bohr's effect, under which hemoglobin varies its affinity to oxygen depending on its alkalinity or acidity (see Figure 4). All the peculiar properties of hemoglobin which bear on its combination with oxygen are of great importance

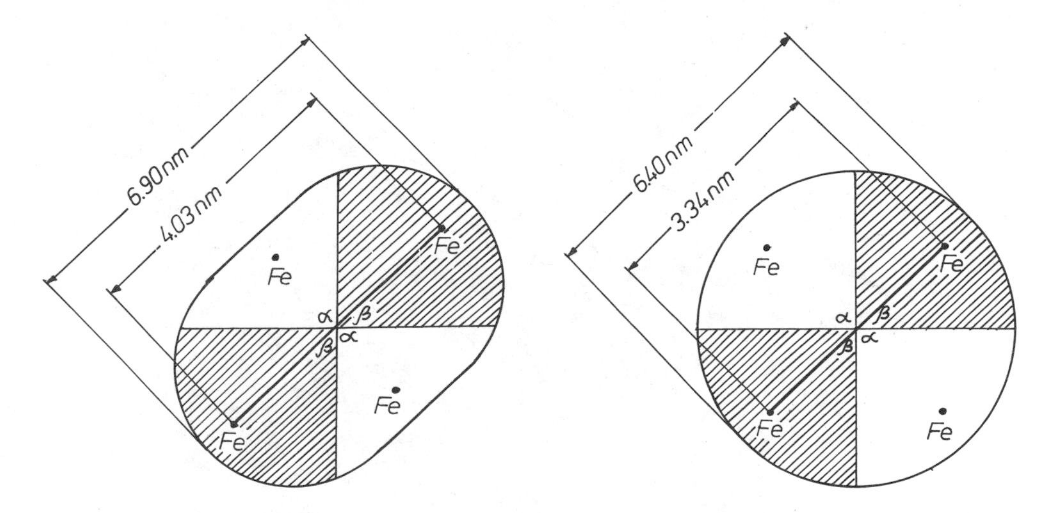

FIGURE 6. The respiratory movement of the hemoglobin molecule. The combining of O_2 causes (through the heme-heme interaction) a change in conformation which consists in an approximation of ferrum atoms in the beta chains from 4.03 nm to 3.44 nm, with simultaneous change of the shape of the molecule. The alpha chains do not engage in such movement.

for the normal functioning of hemoglobin: its excessive affinity to oxygen would prevent a rapid release of oxygen to the tissues; after an initial period of reduced affinity to oxygen, the sigmoid curve enables quick saturation; with a change of the milieu in the tissues (to a more acidic milieu which reduces the affinity of hemoglobin to oxygen) the Bohr effect enables a rapid release of oxygen in the tissues. Let us state right away that all these properties of hemoglobin ensue from the fact that it constitutes a complex asymmetric tetramer, whereas myoglobin is a monomer. The dependence of all these properties on the asymmetry of the molecule is proved by the fact that hemoglobin H (which is composed of four identical beta chains, being a tetramer of four beta chains) behaves in exactly the same way as myoglobin. The asymmetry of the hemoglobin molecule enables also the heme-heme interaction which consists in the mutual approximation of the iron atoms of the beta chains affected by combination of the oxygen molecule with hemoglobin, and their recession under the influence of giving up the oxygen molecule. This combination and giving up of oxygen causes the hemoglobin molecule to perform rhythmic movements known as respiratory movements.

The structure of the β globin chain is presented after Perutz in Figure 7. The helical segments are denoted by letters A to H, and the nonhelical segments by combined letters AB, CD, etc. which denote the position of the nonhelical segment between the helical segments. NA denotes the nonhelical amino end of the chain, and HC the carboxyl end of the chain. According to Perutz, residues within each segment are numbered from the amino end, e.g., F8 denotes the eighth position of the amino acid residue of the helical segment F; NA2 the second position of the nonhelical segment of the amino end of the chain.

Molecular changes caused by oxygenation of the hemoglobin molecule had been discovered in 1938 when Haurowitz proved that crystals of deoxyhemoglobin break after their oxygenation.

Oxygenation of the whole hemoglobin molecule is a very complex process. Oxygenation of the particular hemoglobin subunits depends on their conformation. For example; in the case of two hemoglobin molecules, when three hemes of the one hemoglobin molecule are oxygenated and any one heme of the other hemoglobin molecule is oxygenated, there is a 70% probability the fourth heme of the first hemoglobin molecule will be oxygenated before any one of the hemes of the nonoxygenated hemoglobin molecule will be oxygenated. This depends on different conformations of oxygenated and deoxygenated hemoglobin molecules.

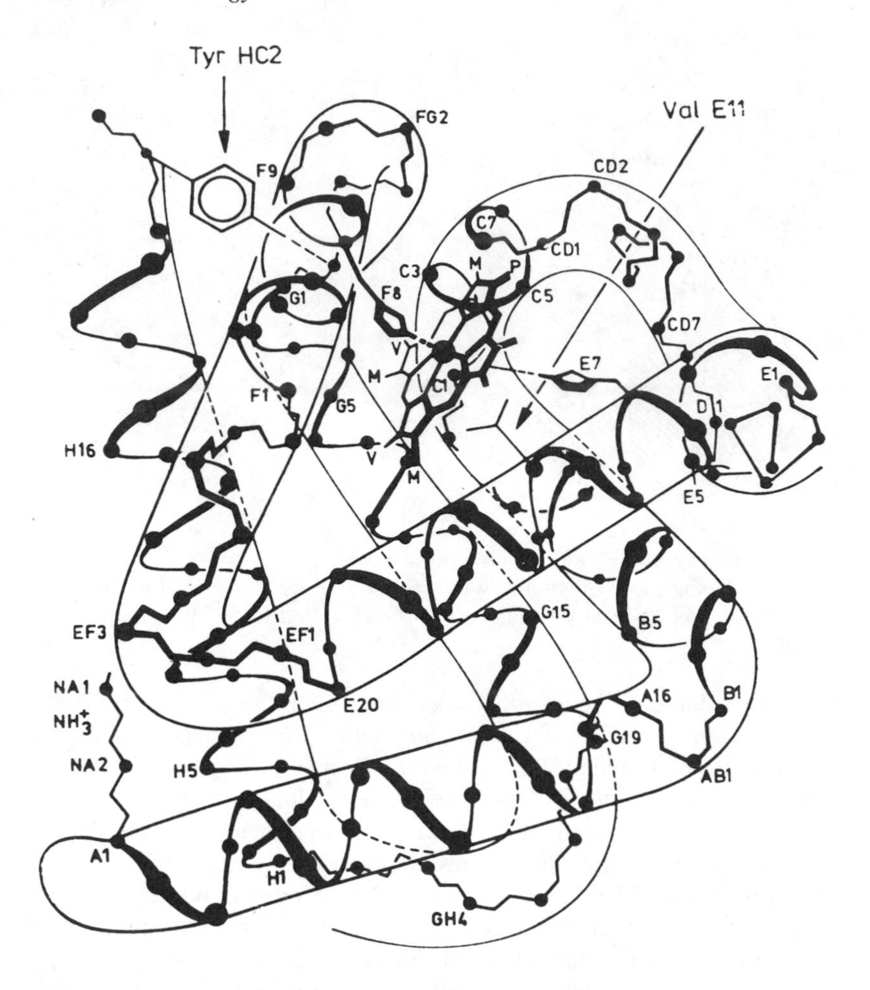

FIGURE 7. Model of β-globin chain. Helical segments are signed by single letters A to H, nonhelical segments by double letters (e.g., NA, AB, BC until HC). (Redrawn from Reference 4.)

The oxygen affinity of the oxygenated hemoglobin molecule is greater than that of non-oxygenated hemoglobin molecule.

The physiological role of heme-heme interaction consists not in an increase of oxygen affinity of the hemoglobin molecule but on its decrease which enables dissociation and the giving up of oxygen to the tissues. When hemoglobin has had a hyperbolic curve of oxygen saturation then it could give up only a small amount of oxygen molecules to the tissues and suffocation would occur. The importance of physiological mechanisms decreasing oxygen affinity of hemoglobin are proved by the existence of supplementary mechanisms, i.e., the Bohr effect and 2,3-diphosphoglycerate decreasing oxygen affinity of hemoglobin.

Change of the oxygen affinity of hemes during oxygenation and deoxygenation is dependent on spin changes that accompany reactions with ligands. The heme-heme interaction arises from reversible transition of tertiary and quaternary conformations. In the oxy form the iron is low spin, and the heme group is approximately planar, and the heme linked to the nitrogen of histidine. F8 lies at about 0.2 nm from the plane of the porphyrin ring. Contrary to this, in the deoxy form the iron is high spin which causes lengthening of the iron nitrogen bonds and in consequence displacement of the iron atom and the heme-linked histidine from the plane of the porphyrin ring by about 0.075 nm to 0.29 nm, respectively (Perutz, 1970). Because of lengthening of the iron nitrogen bonds there are five coordinate bonds in deoxyhemoglobin and six coordinate bonds in oxyhemoglobin. The quaternary

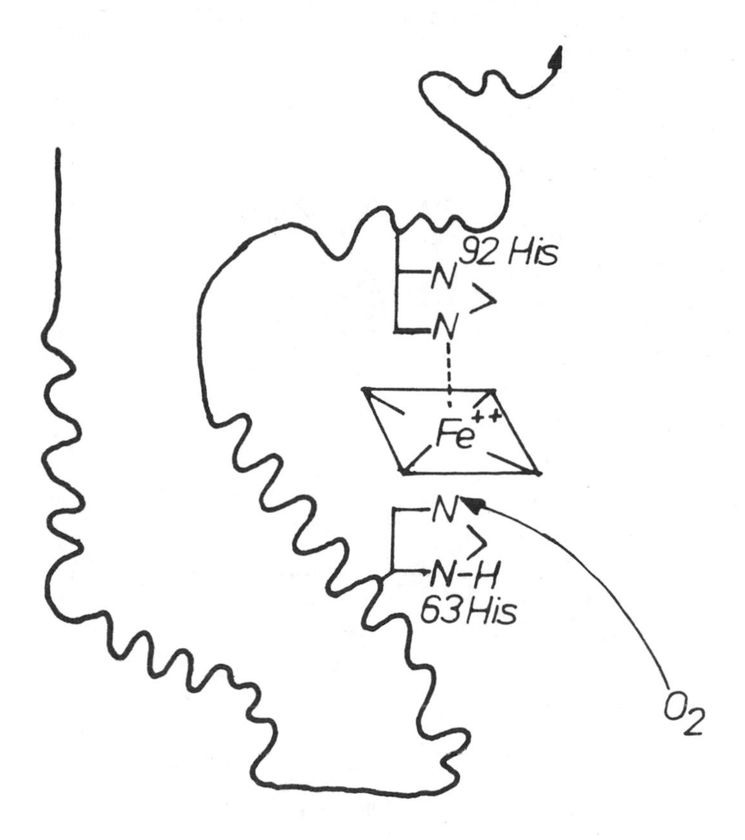

FIGURE 8. Schematic representation of β-globin chain. The ferrous atom is bound by one coordinated bond with proximal histidine (position 92). The oxygen molecule enters during oxygenation between the ferrous atom of porphyrin and distal histidine (position 63).

structure of hemoglobin in mammals is dependent exclusively on the number of coordinate bonds and not on the valency or spin of the iron atom. Displacement of the iron atom in the porphyrin ring results from the proximal histidine. Oxygenation of the iron atom causes its movement to the plane of the porphyrin ring.

A. THE HEME POCKET

The heme is covalently bound with histidine F8; on the other side there are distal histidine E7 and valine E11. The ligand enters between these residues and the iron atom (see Figure 8). Besides, about 60 van der Waals' contacts exist between the atoms of the porphyrin ring and the globin chain. In the α chain pockets there is enough room for the ligand. In the β chain it depends on the oxy or deoxy state. In β chains the distance between the iron atom of the porphyrin ring and valine E11 shrinks by about 0.1 nm in going from the oxy to the deoxy state so that there is no room for even the smallest ligand (including the oxygen molecule). Consequently, in deoxy forms of the β chains the valines E11 must move to make room for ligands. Summarizing, the width of the heme pocket of the α chains has room for ligands in both forms, i.e., in deoxy and oxy forms, while in the deoxy form of β chain the heme pocket must widen for a place allowing the binding of ligands.

B. THE C-TERMINAL RESIDUES

The C-terminal residues of both chains have in the oxy form total freedom of rotation, as well as the penultimate tyrosines which lie between helices F and H. Contrary to this, in the deoxy form they are doubly anchored by salt bridges between helices F and G, and

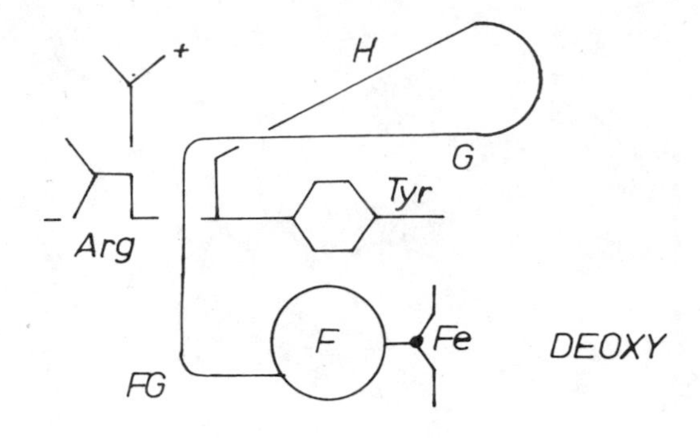

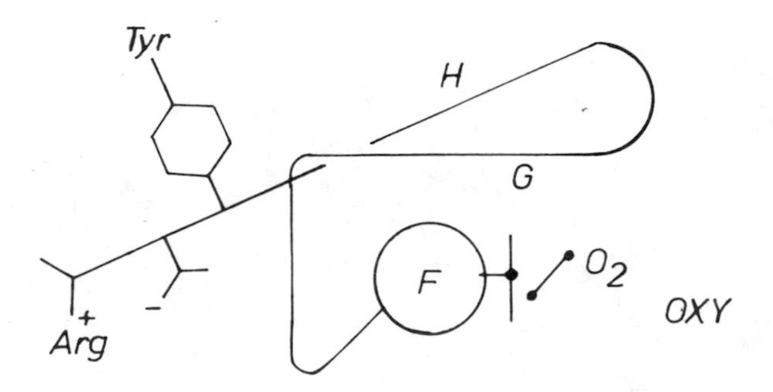

FIGURE 9. Schematic representation of the tertiary structure of hemoglobin subunits during interaction with ligands. In the deoxy conformation penultimate tyrosine (Tyr) lies between helices F and H. During oxygenation the ferrous atom moves about 0.08 nm toward the plane of the porphyrin ring (from the site of proximal histidine) which causes movement of helix F toward the helix H, protruding in this way the penultimate tyrosine from its pocket between helices F and H, and liberating the so-called Bohr protons. (Redrawn from Reference 4.)

cannot be displaced from the pockets without breaking of the salt bridges (see Figure 9). According to Perutz, the movements of the penultimate tyrosines (C-terminals) are of vital importance to the function of the hemoglobin molecule. This is best documented by enzymatic removal of the C-terminal residues which affects the heme-heme interaction and the Bohr effect. These effects are probably caused by their importance for constraining crosslinks (salt bridges) between the subunits in deoxyhemoglobin that makes the quaternary structure of deoxyhemoglobin unstable. Removal of C-terminal residues causes increased oxygen affinity. Therefore, it can be concluded that the C-terminal residues stabilize the deoxy form of hemoglobin.

C. CONFORMATIONAL CHANGES OF α AND β CHAINS

Tertiary changes caused by binding of ligands with the iron atoms of hemes precede transition of the deoxy form of hemoglobin to the oxy form. The main changes concern width of the penultimate tyrosine pockets, heme pockets, and reciprocal movements of subunits toward each other. Under the influence of the binding of ligands the width of the penultimate tyrosine pockets becomes narrower—about 0.13 nm in the α chains and about 0.2 nm in the β chains. There are also possible other changes of the tertiary structure of the particular subunits (at this point not precisely defined).

Changes of quaternary structure of the hemoglobin molecule are very significant. Replacement of α_1 β_1 and α_2 β_2 subunits is minimal and extends about 0.1 nm, while that of α_1 β_2 subunits is considerable and extends about 0.7 nm. Because of the displacement of

α_1 β_2 subunits during oxygenation, breakage of the ionic bond between histidine HC3β_1 and C5α_2 occurs, and in addition some new van der Waals' contacts arise, while the number of hydrogen bonds probably does not change.

Immediate contacts between β chains do not exist. Contacts between these subunits are realized by 2,3-diphosphoglycerate (2,3-DPG) which in the deoxy form of hemoglobin enters between the β chains.

Changes of quaternary structure of the hemoglobin molecules occur mainly within contacts α_1 β_2 (contacts α_1 β_1 practically do not change). Within contacts α_1 β_2 the CD region of the one chain fits in the form of a dovetail into the FG region of the other chain, which during change of the quaternary structure enables rotation relative to each other of about 13.5°.

During oxygenation of the hemoglobin molecule the particular subunits pass from the deoxy form to the oxy form. The hemoglobin molecule as a whole remains in the deoxy form until sufficient subunits pass to the oxy form due to oxygenation. Therefore, oxy structures of particular subunits may exist within the quaternary deoxy structure, and conversely, deoxy subunits may exist in the quaternary oxy structure.

D. THE MECHANISMS OF THE CONFORMATIONAL CHANGES OF THE HEMOGLOBIN MOLECULE

The distances between the heme groups are too long (0.25 to 3.7 nm) for electromagnetic interactions to arise, therefore only stereochemical effects are possible. The heme pockets of α chains are permanently "open" and do not change significantly during transition from the deoxy to the oxy form of hemoglobin, therefore these subunits are probably oxygenated at first. Transition from the deoxy to the oxy form causes conformational changes of these chains: binding of heme iron with ligand [O_2] causes its movement toward the plane of the porphyrin ring and simultaneous pull with it the proximal histidine toward the porphyrin ring about 0.075 to 0.095 nm. Consequently, the segment F moves toward the center of the molecule which extrudes the penultimate tyrosine HC2 from its pocket between helices F and H. The extruded tyrosine pulls the ultimate arginine HC3 with it, thus breaking its salt bridges with the opposite α chain and releasing Bohr protons.

In the β chains the changes are more complex. Prior to the binding of ligand (O_2) opening of the heme pocket must occur (thermal vibrations provide activation energy for this purpose). After binding of the ligand (O_2) with the heme iron, changes of conformation such as those in α chains occur. The iron moves into the plane of the porphyrin ring which involves simultaneous movement of the proximal histidine toward the porphyrin ring. Consequently, segment F moves toward the center of the molecule which extrudes the penultimate tyrosine HC2 from its pocket between helices F and H. The extruded tyrosine pulls the ultimate histidine HC3 with it, thus breaking salt bridges with aspartate FG1 of the β chain and lysine C5 of the α chain, and releasing Bohr protons.

The probable sequence of conformational changes during oxygenation of a hemoglobin molecule is presented in Figure 10. It is probable that one of the α subunits reacts with the ligand at first because of the permanent opening of the heme pocket. Besides, the affinity to oxygen of β subunits is lower and several conformational changes must occur until reaction with the ligand is possible. Therefore, probably the first to be oxygenated would be the iron atom of the first α subunit with consequent conformational changes described above, i.e., breaking of salt bridges with the opposite α subunit and release of Bohr protons (see Figure 10[2]). The next reacting subunit may be the second α subunit with the same conformational changes and release of the next Bohr protons (see Figure 10[3]).

At this step of the reaction, four of the six ionic bonds responsible for the deoxy state of the hemoglobin molecule break, which makes it possible to transform the hemoglobin molecule from the deoxy state to the oxy state. Environmental changes, such as change of

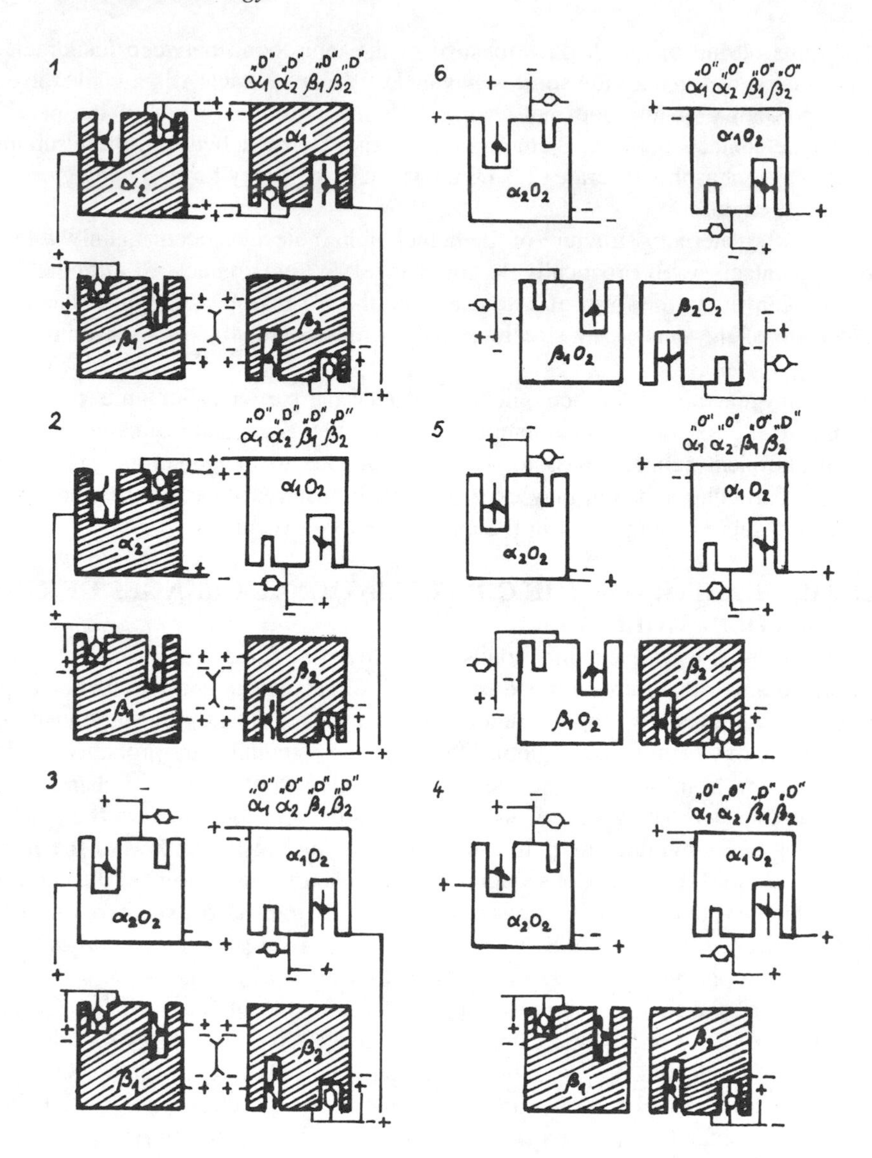

FIGURE 10. Schematic representation of oxygenation of hemoglobin molecule. 1-deoxy hemoglobin: the C-terminal ends are bound by ionic cross-bindings of α chain arginine with β chain histidine and with one molecule of 2,3-DPG between β chains. 1-2 and 2-3—oxygenation of the α subunits: the penultimate tyrosines are expelled from their pockets between helices F and H which become narrower; the salt bridges with the partner α chains are disrupted. 3-4—the quaternary structure clicks to the oxy form, 2,3-DPG is expelled and salt bridges between subunits α_1 β_2 and α_2 β_1 are disrupted. 4-5 and 5-6—oxygenation of β subunits: the heme pockets enlarge, the penultimate tyrosines are expelled from their pockets which become narrower and the salt bridges of the C-terminal histidines with asparagine residues are disrupted. (Redrawn from Reference 4.)

pH, concentration of CO_2, or phosphates, contribute to the transition from the deoxy to the oxy form of hemoglobin. Loosening of the α_1 β_2 and α_2 β_1 contacts enables click of the deoxy to the oxy conformation and break of the remaining salt bridges constraining the deoxy conformation of hemoglobin. Because of rotation of the subunits the distance between the β subunits and 2.3-DPG becomes so long that it causes breakage of the bonds between 2.3-DPG and the β chains (liberation of 2.3-DPG from the β subunits) without release of Bohr protons, (see Figure 10[4]). At this step of the reaction all the salt bridges between the α and β subunits are broken, which diminishes the amount of energy necessary for the extrusion of the penultimate tyrosines from their pockets. Simultaneously, enlargement of the heme pockets occurs which enables binding of iron atoms with ligands (O_2) with the consequent conformational changes of the β subunits, i.e., extrusion of the penultimate tyrosines from their pockets between segments F and H, and liberation of Bohr protons (Figure 10[5] and [6]). In this way the number of liberated Bohr protons per one hemoglobin molecule corresponds to the number of oxygen atoms taken up.

α_1 β_1 dimers, with α_1 β_2 contacts open, can bind oxygen because their deoxy form is not constrained by interchain bonds $\alpha_1\beta_2$. Conversely, the important function of hemoglobin to give up oxygen to the tissues is possible only in the tetrameric form because of bonds stabilizing the deoxy conformation of the hemoglobin molecule.

The heme-heme interaction consists (Perutz, 1972) of a change of tension at the hemes, caused by transition between the two alternative quaternary structures. In the quaternary oxy conformation the heme structure is relaxed, which causes its high oxygen affinity, and in the deoxy conformation the heme structure is constrained by the contacts between the subunits, which causes its low oxygen affinity. In the tense state of heme there is high iron spin, and in the relaxed state low iron spin.

Further influence on the reactivity of hemoglobin molecules have allosteric effects caused by hydrogen ion concentration and 2,3-diphosphoglycerate (2,3-DPG). Hydrogen ions strengthen the ionic bonds which support tense form of hemes by an increase of positively charged molecules. 2,3-DPG causes an increase of free energy of interactions because of introducing additional ionic bonds, specific for the tense form of heme with low oxygen affinity. Therefore, either hydrogen ions or 2,3-DPG lower oxygen affinity of the quaternary structure of hemoglobin.

E. THE BOHR EFFECT

The Bohr effect depends on liberation of three protons (so-called Bohr protons) from each oxygenated hemoglobin subunit because of breaking of ionic bonds of the C-terminal amino acid residues. These protons pass in the lungs to the blood serum where they facilitate removal of CO_2 to the air. In tissues, when hemoglobin dissociates and oxygen passes to the tissues, the Bohr protons return to the hemoglobin molecules facilitating in this way binding of CO_2 by the blood serum. These reactions are best demonstrated as follows:

$$\begin{bmatrix}\text{Hb-Bohr}\\ \text{protons}\end{bmatrix} \xrightarrow[\text{in lungs}]{+\ O_2} HbO_2 + \begin{matrix}\text{Bohr protons}\\ \text{in serum}\end{matrix} \xrightarrow[\text{in tissues}]{-\ O_2} \begin{bmatrix}\text{Hb-Bohr}\\ \text{protons}\end{bmatrix} + \begin{matrix}CO_2\\ \text{in}\\ \text{serum}\end{matrix}$$

The importance of the Bohr effect lies in its physiological role as a regulatory mechanism of oxygen transport and exchange with carbon dioxide. This is also a good model for studying the effects of hydrogen ion concentrations on functional properties of protein molecules.

In physiological conditions only, an alkaline Bohr effect can occur. It neutralizes the protons released in the blood and returns to the hemoglobin molecules during deoxygenation which facilitates uptake of CO_2 from the tissues and its transport to the lungs. Acid medium (lactic acid, bicarbonates) lowers affinity of hemoglobin to oxygen and facilitates release of oxygen. Strong acid medium (pH below 6.0) causes an inversed Bohr effect—protons are liberated during release of ligands. The involvement of the C-terminal arginines in the Bohr effect was confirmed through the enzymatic removal of the C-terminal residues which diminish significantly the alkaline Bohr effect.

Great importance in maintaining the proper structure of molecules depends on free SH groups. The reactivity of the SH group at cysteine (position 93) increases proportionally to the increase of oxygenation of the hemoglobin molecule which reflects on conformational changes of the C-terminal residues of the β chain. In the oxy form of hemoglobin the SH groups are accessible and in the deoxy form they restrict access to them. The expulsion of penultimate tyrosines from their pockets and breakage of restricting salt bridges during oxygenation of the β chains, secure the proportional course of conformational changes of the C-terminal residues with simultaneously increased oxygenation. A similar mechanism is bound with spin changes of cysteine 93. In deoxyhemoglobin the tyrosine pocket is occupied, and consequently the spin is free, and in oxyhemoglobin the tyrosine pocket is open and the spin may be immobilized. The transition from the free to the immobilized conformation is therefore proportional to the number of oxgenated β chains.

F. BINDING OF 2,3-Diphosphoglycerate (DPG)

2,3-DPG is present in red blood cells of many species. It lowers the oxygen affinity of hemoglobin. 2,3-DPG augments free energy of the heme-heme interactions, without release of Bohr protons. In deoxyhemoglobin 2,3-DPG binds preferentially to the β chains at a ratio of 1 mol/1 mol at neutral pH and a physiological salt concentration of 1.3×10^5, which corresponds to free binding energy of 29 kJ/mol (7 kcal/mol). The binding of 2,3-DPG to hemoglobin is totally inhibited at higher salt concentrations which denotes that this bond is of electrostatic character.

During oxygenation the distance between the α amino groups, with which 2,3-DPG is bonded, increase from 1.6 nm to 2 nm (16 Å to 20 Å) which make bonds between these groups and 2,3-DPG impossible, and because of narrowing of the distance between the β subunits 2,3-DPG is expelled from the central cavity of the molecule. 2,3-DPG can decrease the oxygen affinity of hemoglobin about 30 times.

It is interesting that hemoglobin of some species (e.g., sheep hemoglobin) do not demonstrate affinity for 2,3-DPG. This is caused by shortage of their β chains more than 0.3 nm (3 Å) which increases the distance between the two amino groups of the C-terminals to about 2.2 nm (22 Å) and makes bonds between the two phosphate groups of 2,3-DPG with these amino groups impossible.

Other phosphates like ATP, ADP, and AMP also decrease the oxygen affinity of hemoglobin although their binding activity and lowering of oxygen affinity of hemoglobin are lower than that of 2,3-DPG. Pyridoxal phosphate binds with the β chain of hemoglobin at the same site and ratio as 2,3-DPG, thus inhibiting binding of 2,3-DPG.

Much data about the function of particular segments and even of single amino acid residues of the hemoglobin molecule have been recognized on the basis of functional disturbances of pathological hemoglobins. Differences between the function of normal and pathological hemoglobin explain many obscure functions of the hemoglobin molecule. Despite all the known details about function of the hemoglobin molecule, we must be aware that our knowledge in this field is far from complete even in the case of the hemoglobin molecule, and much less about the functions of the many thousands of molecules, components of living organisms.

As seen in the case of hemoglobin molecules, their functions can be subordinated to regulatory mechanisms which are common with all vital functions. 2,3-DPG stabilizes the quaternary structure of deoxyhemoglobin by four cross-links (without influence on the tertiary structure). In turn, hydrogen ions alter the tertiary structure of oxyhemoglobin and stabilize its quaternary structure. In this way (other mechanisms cannot be excluded) the function of hemoglobin molecule is regulated in a very efficient dynamic manner, changing completely from the deoxy to the oxy state, and inversely, dependent on change of the hydrogen ions which is manifested by the Bohr effect, change of 2,3-DPG concentration and pCO_2. As seen from this one relatively well-known molecule, many vital functions can be explained by its function, and it seems probable that after learning all the functions of all molecules

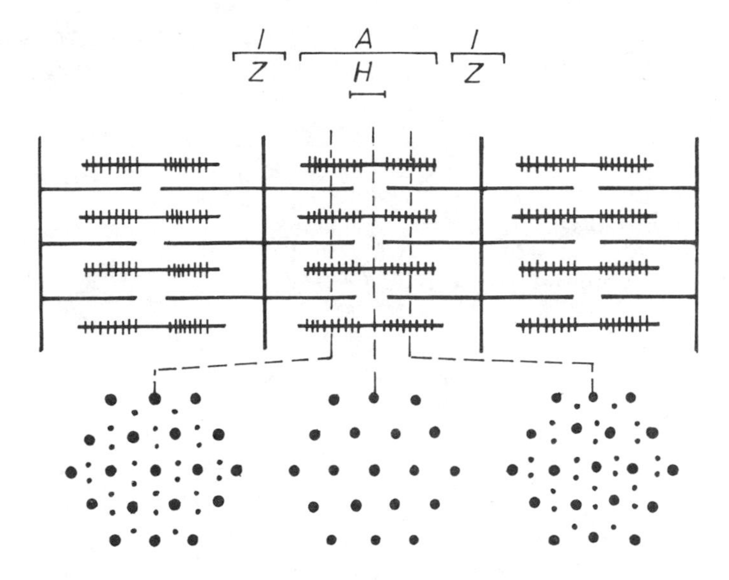

FIGURE 11. Schematic structure of a striated muscle. Thick filaments of myosin with regular heads of ATP-ase activity. Z—line where thin filaments are attached. A—anisotropic bands, containing only thin filaments. In the middle of the A band is the H zone which increases or decreases in width according to the resting or contracted muscle (Redrawn from Huxley.)

(probably several hundred thousand) of the living organism we shall be able to explain all vital functions of living organisms by the functions of their molecules.

G. THE COOPERATION OF MOLECULES IN PERFORMING COMPLEX VITAL FUNCTIONS

Despite the great complexity of the function of single molecules, as seen on the hemoglobin molecule, it cannot explain complex vital processes of a living organism, although reminding one of illusorily vital processes. Only interactions between particular molecules can fulfill this requirement.

A simple model of interaction between molecules is the enzyme-substrate interaction, leading to degradation or synthesis of further molecules. For vital processes usually interaction of many molecules is necessary in which final effects are vital functions. As an example of this kind of interaction, the interaction of muscle proteins leading to muscle contraction will be presented.

The muscle contraction arises by cooperation of four major muscle proteins: myosin, actin, tropomyosin, and troponin. Besides those proteins, there are a number of further proteins in muscles only slightly known. Their function and localization are uncertain. All the major muscle proteins can be easily isolated and examined separately, but their function depend on the high-ordered structure they form in the muscle cell and their cooperative biochemical interactions.

The scheme of the ordered structure in the form of thick and thin filaments is presented in Figure 11, as viewed through an electron micrograph of high resolution.

The thick filaments contain myosin molecules (molecular weight about 500,000) composed of two symmetrical polypeptide chains in the form of a long rod, coiled in the form of a simple α helix and next a super helix. At one end of the myosin molecules there are two small globular heads, composed of two proper heads with ATP-ase activity and two light chains (Figure 12). The light chains can be easily dissociated from the remaining myosin molecules. The part of myosin molecules between their heads and rod-like structures exhibit great flexibility.

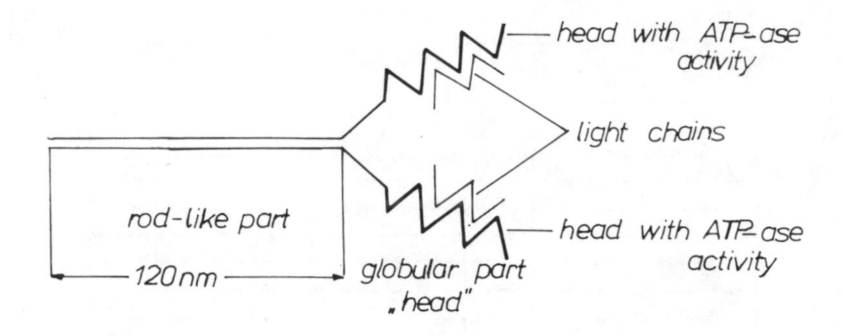

FIGURE 12. Diagrammatic presentation of a myosin molecule.

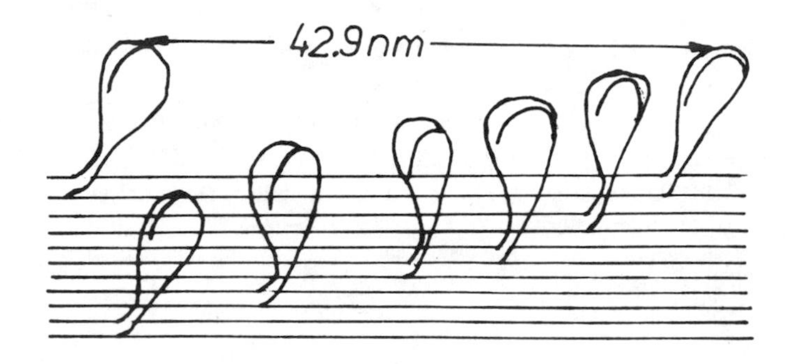

FIGURE 13. Scheme of a myosin ''thick'' filament, composed of aggregated myosin molecules.

In rabbit myosin three kinds of light chains have been demonstrated. The first one of molecular weight 18,000, present in the amount of 2 mol/1 mol of myosin. Its deprivation does not abolish ATP-ase activity of myosin. The second light chain of molecular weight 25,000 and the third of molecular weight 16,000 are necessary for the ATP-ase activity—their deprivation abolishes ATP-ase activity of the myosin heads.

The myosin molecules exhibit a great tendency to aggregate. In consequence they form thick filaments, composed of many myosin molecules in this way that the rod-like parts of them are situated side by side with the globular heads protruding outside the filaments, visible through the electron microscope. Earlier these globular heads have been denominated ''bridges''—now heads with ATP-ase activity (see Figure 13). These heads are present on the surface of thick filaments at regular intervals of 14.3 nm and according to the turns of the helix every 42.9 nm.

In the myosin filament there is another globular protein M (molecular weight about 88,000) present which acts probably as the aggregating factor of single myosin filaments into a thick filament. Finally, protein C is present in the myosin filament, but its function is unknown.

The thin filaments comprise the remaining major muscle proteins, i.e., monomers of G-actin, tropomyosin, and troponin. The actin molecules are small globular proteins of molecular weight of about 45,000 combined with ATP and calcium ions. G-actin polymerizes into F-actin filaments with dissociation of inorganic P from ATP, bound with G-actin. F-actin molecules are arranged into filaments twisted in the form of a double helix, resembling two strings of pearls twisted in relation to each other at distances 35.5 nm/1 turn (see Figure 14). In addition, the thin filaments contain two other proteins, i.e., tropomyosin and troponin (Figure 14).

Tropomyosin belongs to the fibrillar proteins composed of two dextrorotatory α chains,

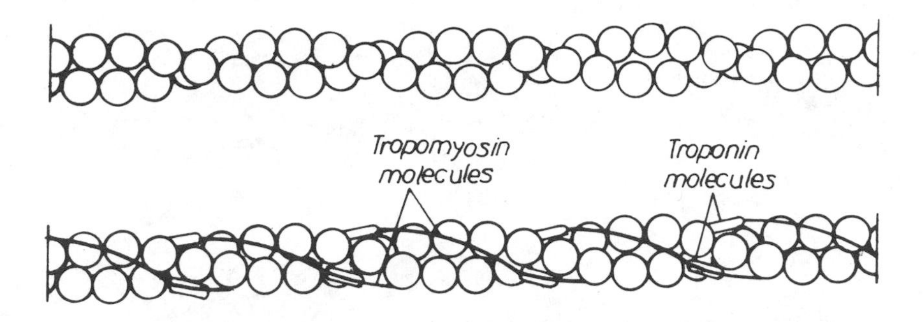

FIGURE 14. "Thin" filament of actin molecules (at top) with interaction of troponin and tropomyosin (at bottom).

twisted into a superhelix 40.0 nm long, 2.0 nm thick, and a molecular weight of 70,000. The joining ends of tropomyosin lie in the groove of F-actin. One molecule of tropomyosin extends over seven actin molecules.

Troponin belongs to the globular proteins and is affixed near one end of each tropomyosin molecule (Figure 14). Troponin is composed of three subunits: TnT of molecular weight 40,200, and is bound very tightly with tropomyosin; TnI of molecular weight 23,500 is bound with actin, and TnC of molecular weight 18,300 is bound with calcium ions. Troponin is a protein of regulatory character.

Line Z (compare with Figure 11) is composed of α actin molecules (a dimer composed of two monomers of molecular weight 94,000). The α actin molecules convert F-actin into gel with regular paracrystals on the whole length of F-actin. Tropomyosin is hardly competitive to α actinin, and extending over the whole F-actin molecules, it inhibits interaction of α actinin with F-actin over the whole F-actin filament. Only the ends of F-actin are free which interact very little with α actinin resulting in attachment of the thin filaments to the line Z.

The first step in explaining the nature of muscle contraction was the demonstration of Huxley and collaborators that during muscle contraction what occurs is not shortening of muscle fibers, a displacement of filaments to each other. Comparing contracted and relaxed muscles Huxley et al. demonstrated that in the relaxed muscles all the myosin heads are oriented perpendicularly to the axis of myosin filaments while during contraction they decline toward the axis of these filaments. Therefore, the conclusion was that muscle contraction is caused through a joining of myosin heads in a perpendicular position with the thin filaments of actin and because of a hinge movement of myosin heads the displacement of actin filaments occurs, which causes shortening of the muscle as a whole, in other words contraction of the muscle (see Figure 15).

In the final phase of this movement disconnection of the link between the myosin heads with actin molecules occurs, resulting in reversion to the relaxed state of the muscle and enabling a new cycle of events leading to muscle contraction. The energy for these movements arises from ATP degradation; for every myosin head, degradation of one molecule of ATP per one cycle.

The described interactions of myosin with actin may occur as long as ATP is available which enables relaxation of muscles. With depletion of ATP, relaxation of muscles becomes impossible. A typical example of this state is rigor mortis.

The decisive role in initiation of muscle contraction is played by calcium ions. Inside the muscle fiber, in addition to sarcoplasmic reticulum, there are also transversal channels which arose through invagination of sarcolemma. Ebashi demonstrated that the concentration of ATP is regulated by the concentration of calcium ions (concentration 10^{-7} to 10^{-5} causes 20 times augmentation of ATP-ase activity). In the resting muscle calcium ions are stored

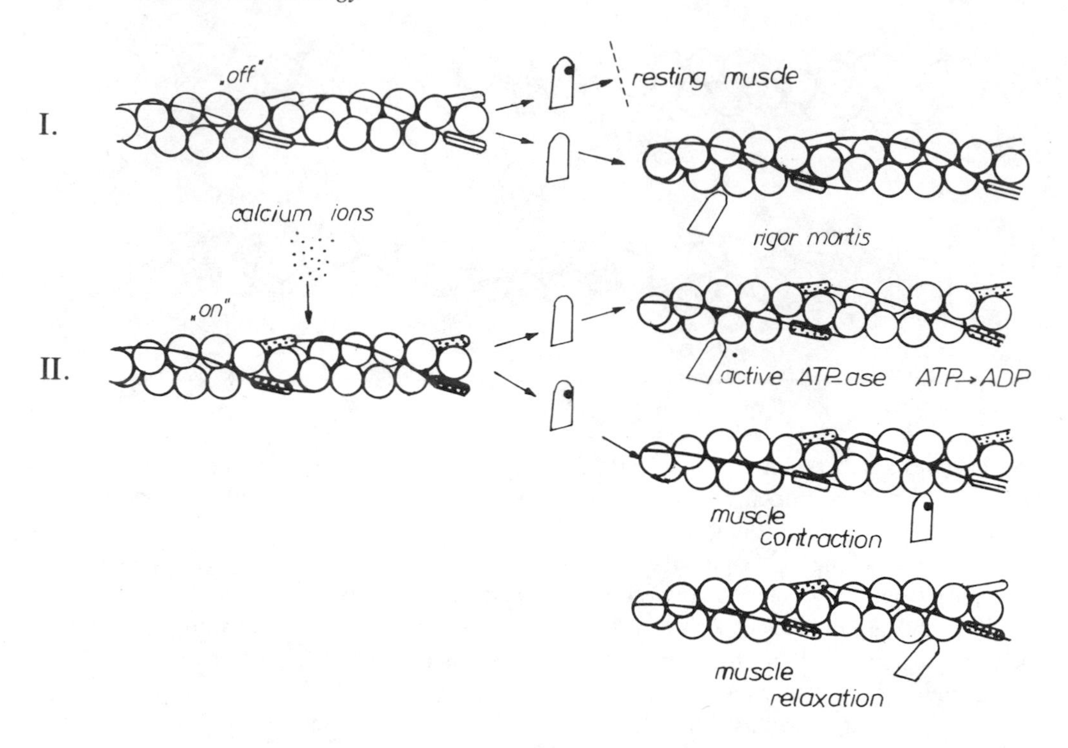

FIGURE 15. Regulation of muscle contraction: I—thin filament in ''off'' position (sites of interaction of myosin heads with actin molecules are inaccessible). (Top): myosin head contains ATP—resting muscle. (Bottom): myosin head depleted of ATP—rigor mortis. II—thin filament in ''on'' position (sites of interaction of myosin heads with actin molecules are accessible); (Top): active ATP-ase hydrolyses ATP to ADP. (Bottom): normal muscle contraction occurs. (Modified from Reference 2)

in the sarcoplasmic reticulum. During the muscle contraction calcium ions pass to the myosin filaments where they activate ATP-ase which results in muscle contraction, after which they return to the sarcoplasmic reticulum. Consequently another cycle of events necessary for muscle contraction can occur.

Explanation of the mechanism of the action of calcium ions in muscle contraction became possible only after discovery of the role of regulatory proteins: tropomyosin and troponin. A hypothetical model of the action of regulatory proteins combined with the influx of calcium ions was presented by C. Cohen (see Figure 16). According to this model: in the resting muscle troponin I (TnI) makes strong bonds with actin which immobilizes tropomyosin molecules and inhibits binding of actin with the heads of myosin. Consequently, actin is in an ''off'' position. Under the influence of calcium ions binding with troponin C (TnC) strong interaction between troponin subunits occur. Simultaneously, bonds between troponin I (TnI) became weaker which enabled movement of tropomyosin from the blocking position nearer the actin groove. Consequently, actin passes to the ''on'' position which enables binding of myosin heads with actin molecules resulting in activation of ATP-ase, degradation of ATP and muscle contraction.

Under the influence of calcium ions strengthening of bonds between troponin subunits occurs with simultaneous weakening of interaction between troponin and actin which allows the ''on'' position of actin. Diminution of calcium ions concentration resulting in weakening of the interaction between the troponin subunits enables troponin subunit I (TnI) to make a strong bond with actin which causes movement of tropomyosin and the ''off'' position of actin, inhibiting binding of myosin heads. The role of tropomyosin in this process consists on the transfer of the unblocking or blocking effect simultaneously on seven molecules of actin.

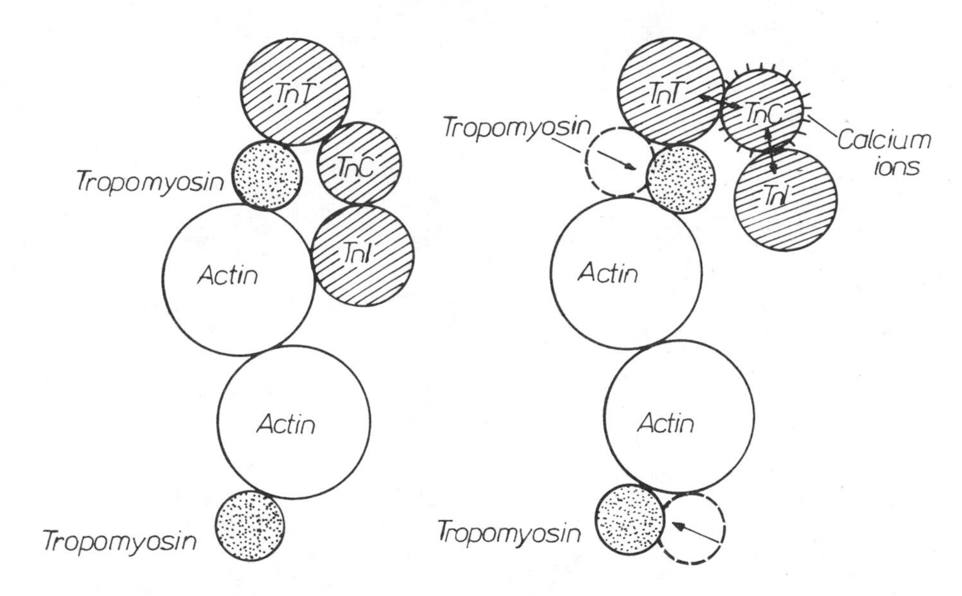

FIGURE 16. Interaction of muscle proteins: actin, tropomyosin and subunits of troponin (TnT—subunit binding with tropomyosin, TnC—subunit binding calcium ions, and TnI—inhibitory subunit, binding with actin). In the resting muscle (left), interaction between troponin subunits is weak. TnI binds actin, making interaction with myosin heads impossible. Under the influence of calcium (right), the interaction between the particular subunits of troponin becomes stronger, troponin moves deeper to the groove of actin and exposes sites of interaction with myosin heads, which activates the ATP-ase and enables muscle contraction. (Redrawn from Reference 1.)

REFERENCES

1. **Cohen, C.,** The protein switch of muscle contraction, *Sci. Am.,* 233, 36, 1975.
2. **Murray, J. M. and Weber, A.,** The cooperative action of muscle proteins, *Sci. Am.,* 230, 58, 1974.
3. **Perutz, M. F., Muirhead, H., Cox, J. M., and Goaman, L. C. G.,** Three-dimensional Fourier synthesis of horse oxyhaemoglobin at 2.8 Å resolution: the atomic model, *Nature (London),* 219, 131, 1968.
4. **Perutz, M. F.,** Stereochemistry of cooperative effects in haemoglobin, *Nature (London),* 228, 726, 1970.
5. **Perutz, M. F.,** The Bohr effect and combination with organic phosphates, *Nature (London),* 228, 734, 1970.
6. **Watson, J. D.,** *Molecular Biology of the Gene,* W. A. Benjamin, Menlo Park, CA, 1970.

Chapter 3

THE GENOME IN EUKARYOTES AND ITS FUNCTION

I. INTRODUCTION

Genetics is a relatively young discipline. The first scientific bases derived from experiments of Gregor Mendel in the nineteenth century. Mendel, on the basis of experiments on pea flowers, described "hereditary factors" as responsible for the traits of living organisms. These factors have been called "genes" by the Danish biologist Johannsen. Morgan et al.,[42] on the basis of experiments performed on *Drosophila,* created the so-called chromosomal or classical theory of heredity in which genes have been located on chromosomes.

The new molecular era of genetics and biology arose from the experiments of F. Griffith[43] who demonstrated that transformation of nonpathogenic pneumococci to pathogenic ones may be caused by culture of living nonpathogenic pneumococci with heat-killed pathogenic pneumococci in which the factor (gene) responsible for pathogenicity was transferred from the pathogenic pneumococci to the nonpathogenic ones. This is the first evidence that nonliving material may be responsible for the arising of traits in living organisms. Soon this material was recognized by O. T. Avery, et al.[44] as nucleic acid.

The first description of nucleic acids and nucleoproteins derives from F. Miescher.[45] The description of DNA structure with some implications to its function (J. D. Watson, et al.)[46] was the most important for further investigations in the field of genetics.

The genetic information in cells is stored on deoxyribonucleic acid (DNA), transcribed into messenger RNA (mRNA) and then translated into protein. In viruses, genetic information may be stored in either DNA or RNA. A molecule of the nucleic acid DNA is a long, thin double-helix, each strand of which is a chain of nucleotides. Each nucleotide is composed of three components: a chemical group called a base, a sugar (deoxyribose in DNA, and ribose in RNA), and a phosphate group. The linked phosphates and sugars form the backbone of the molecule. The genetic information is encoded in particular sequences of bases. In DNA there are four bases: purines, adenine (A) and guanine (G); and pyrimidines thymine (T) and cytosine (C). In RNA uracil (U) takes the place of thymine. The bases are complementary and they form pairs in accordance with the possibility to form hydrogen bonds: two hydrogen bonds between adenine and thymine, and three hydrogen bonds between guanine and cytosine (see Figure 1). The complementarity of bases enables correct transcription and replication of the DNA.

The sequence of bases, which represents the genetic information, arises on the binding of nucleotides through phosphate groups into a long chain, as seen in Figure 2. Since each amino acid is encoded by a three-nucleotide codon and a protein is on an average composed of 300 amino acids, about 900 base pairs are usually necessary to encode one protein molecule. Therefore, the more complex the living organism is, the longer must be the DNA coding for all the necessary proteins. Simple prokaryotic organisms, such as *E. coli* have a single chromosome with about three million base pairs, enough for coding about 3000 different protein molecules.

In eukaryotic cells the total amount of DNA generally increases with the complexity of the organism. In man the total DNA is three billion to four billion bases long, enough for coding about three to four million protein molecules. Since the total number of proteins synthesized in human cells is estimated to be from 30,000 to 150,000, the number of base pairs of human DNA is too large. It is accepted that only a little part of the DNA base pairs encode proteins, and the other part serves a regulatory purpose of protein synthesis.

Genetic information of prokaryotes and eukaryotes differ in more than their length. In

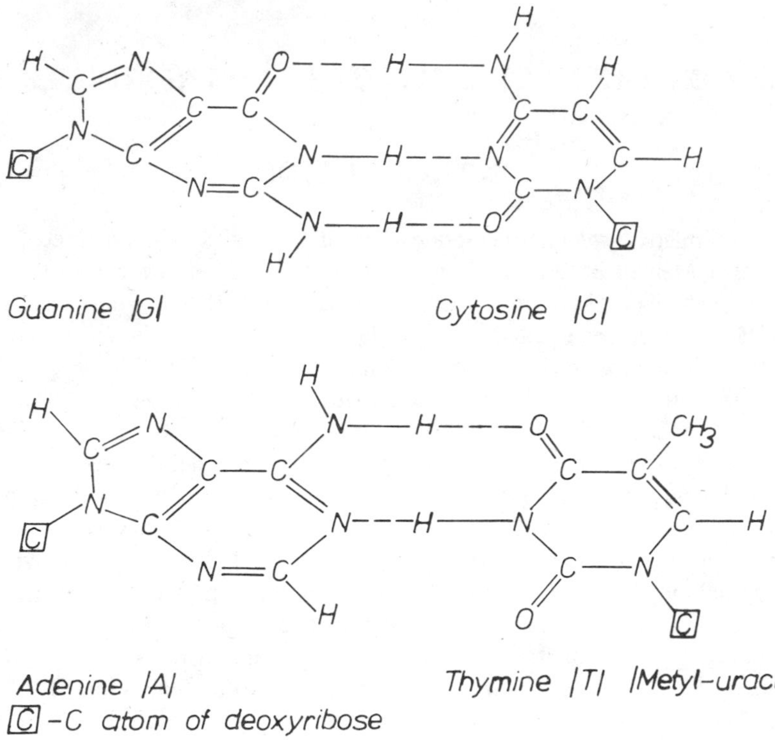

Guanine /G/ Cytosine /C/

Adenine /A/ Thymine /T/ /Metyl-uracil/

[C] -C atom of deoxyribose

FIGURE 1. Base-pairs: Guanine–Cytosine (three hydrogen bonds); Adenine–Thymine (two hydrogen bonds).

prokaryotes the DNA molecule appears as a closed loop, tightly folded to fit inside the bacterial cell. In eukaryotes the DNA is complexed with nuclear proteins and creates a specific cell organ—the nucleus—separated from the cytoplasm by the nuclear membrane.

The DNA molecule is composed of two nucleotide chains coiled around one another to form a right-handed double helix (see Figure 3). The two chains run in opposite directions and are held together by hydrogen bonds between the complementary bases A-T and C-G. Since pairing of the bases is obligatory the parallel strands of the DNA molecule are complementary to one another. In higher organisms the ratio of pairs A-T:C-G is about 1.4 to 1.0.

For approximately 25 years only the DNA model of Watson and Crick has been available. The prevalence of this model exists in the possibility of establishing very probable mechanisms of DNA replication and transfer of genetic information from the cell nuclei to the cytoplasm where proteins are synthesized. In the 1970s many other models of DNA structure were constructed, the most important one being the left-handed model discovered by the group of A. Rich et al.[47] and independently by R. Dickerson.[48] The latter is called Z-DNA because the chains run in a zigzag line instead of a regular line as in the right-handed model of Watson and Crick. Some authors suggest that the Z-DNA may be the active form of DNA. With the description of different DNA structures it is generally accepted that the DNA structure in cells, contrary to previous views, is a flexible, changeable structure and not a rigid, nonchangeable one. It is accepted that specific forms of DNA may be responsible for different functional possibilities of the DNA molecule.

To fit inside a cell nucleus, the enormous long DNA molecule must bend and kink at many sites. The basic structure of DNA in higher organisms is the elementary fiber, 11 nm in diameter, composed of repeating units called nucleosomes. The core of nucleosomes is

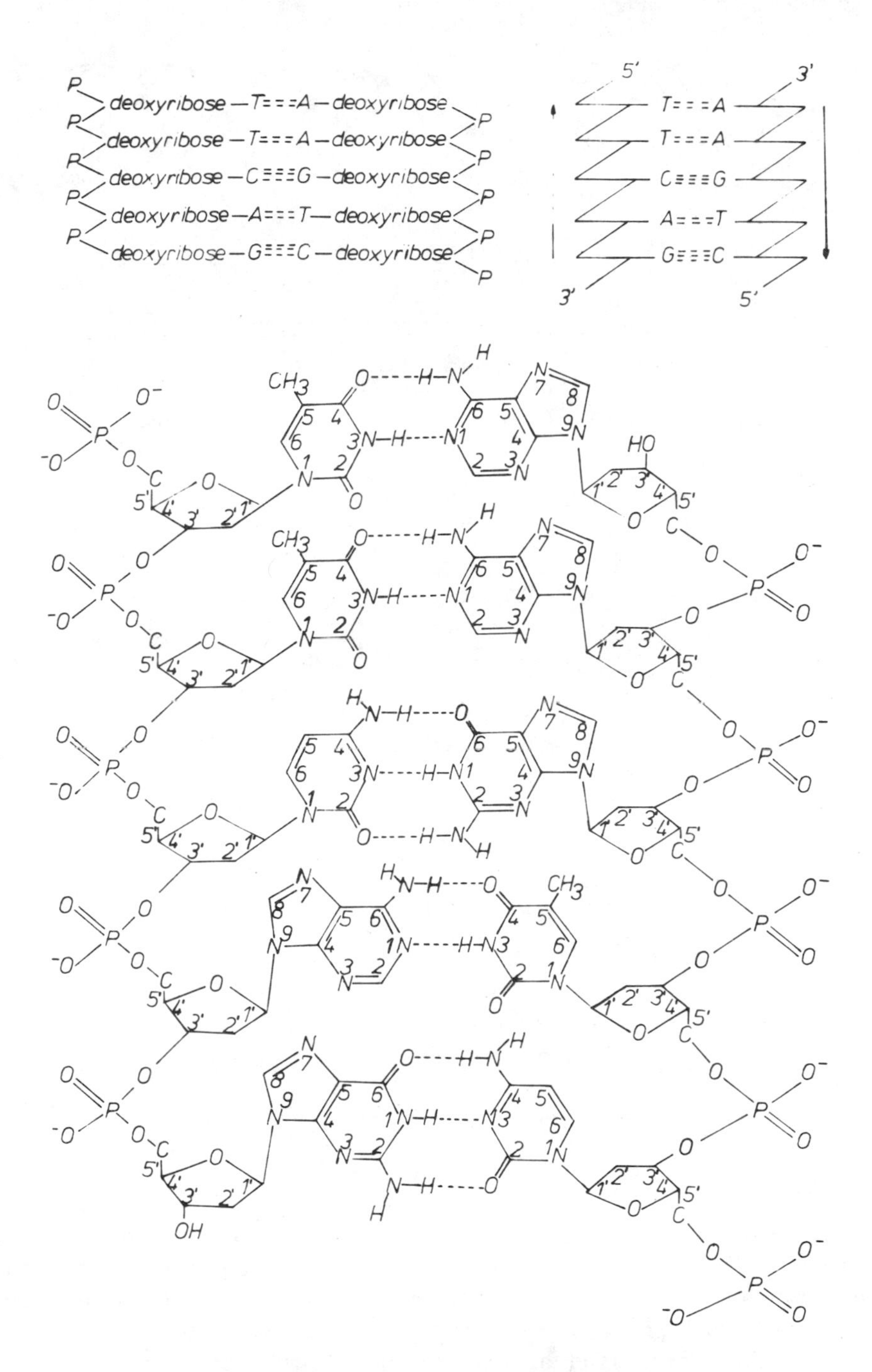

FIGURE 2. Three variants of description of the same DNA fragment: top left—simplified description, T—thymine, C—cytosine, A—adenine, G—guanine: top right—description by symbols, the two chains running in opposite directions. Bottom—structural pattern of the same DNA fragment.

composed of a double set of histones (H2A, H2B, H3, and H4) around which the DNA molecule is coiled $1^3/_4$ times (see Figure 4). Another histone molecule (H1) binds the nucleosomes to each other. These histone molecules appear to be responsible for contraction of the DNA molecule. In nucleated chicken erythrocytes histone H1 is replaced by histone H5.

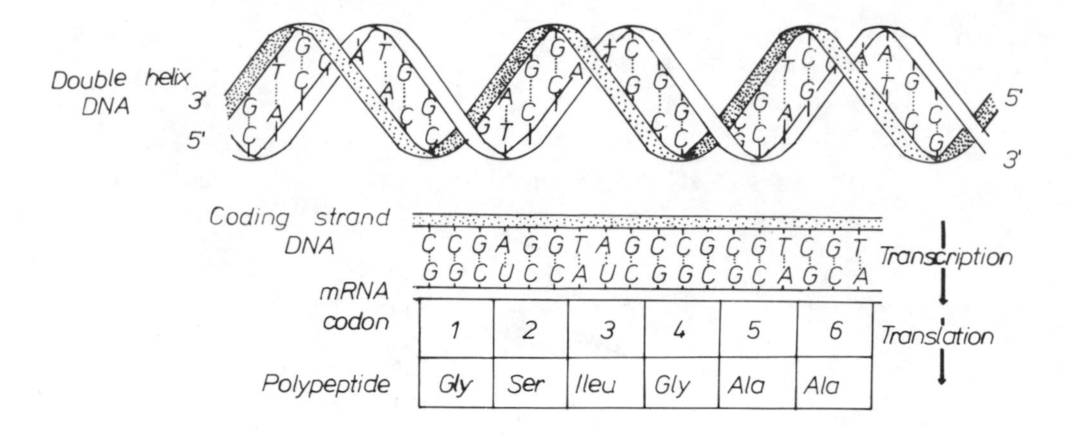

FIGURE 3. Model of a DNA fragment—transcription—translation.

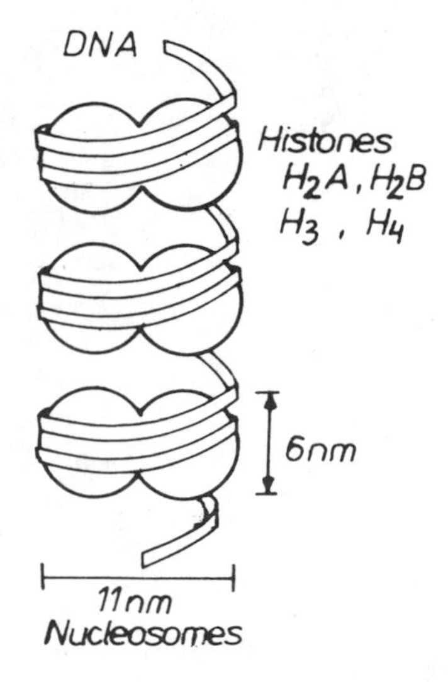

FIGURE 4. The structure of nucleosomes.

The elementary fiber of linked nucleosomes is coiled in chromatin fibers in the form of a solenoid with the DNA outside, 36 nm in diameter, which may be visualized in the electron microscope (see Figure 5). The next step of DNA condensation includes the chromosomes. In metaphase chromosomes, to the central acidic scaffold protein, chromatin fibers are attached at repeated sequences forming the chromosome body, about 0.6 μ in diameter, composed of chromatin fibers in the form of loops of about 200,000 base pairs each (so-called Laemli loops) radiating from the scaffold protein see Figure 6. This kind of DNA compaction (magnification approximately × 40) is probably the most adequate for doubling during cell division. DNA in metaphase chromosomes is condensed about 10,000 times which is reached by forming a supersolenoid in the form of an empty cylinder about 400 nm in diameter which consists of chromatin fibers of about 36 nm. In this way, in spite of contraction the DNA is accessible from all sides of the supersolenoid.

A question arises about mechanisms causing various forms of structure, and the degree and sites of the bending and kinking of the DNA molecule. On a nucleosome, DNA is bent through 360° in the space of about 80 base pairs (bp). A question arises if the sites of bending are random or caused by specific base sequences and proteins bound with the DNA

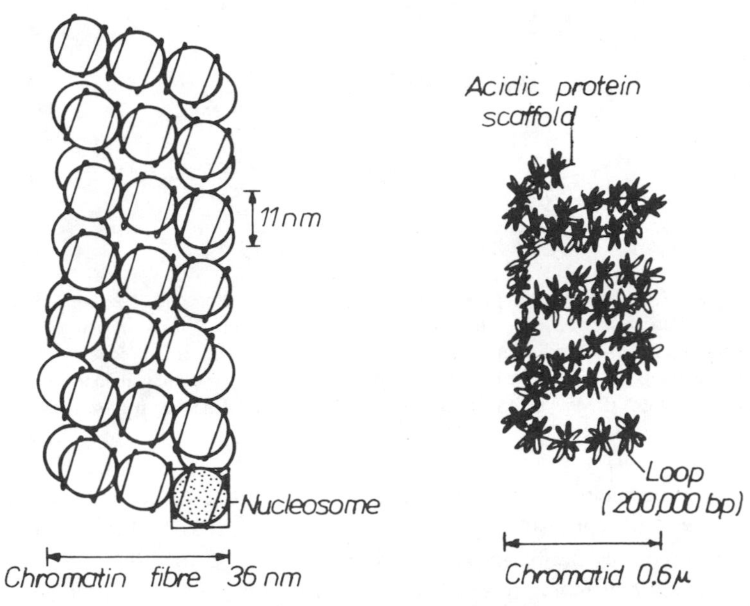

FIGURE 5. The structure of a chromatin fiber.

FIGURE 6. The structure of a chromatid.

molecule. New studies demonstrate that the structure and function of the DNA molecule can be affected in an important way both by the DNA sequence itself and by the interaction of specific proteins with the DNA. It is suggested that altered DNA conformations play an important role in DNA packaging and in consequent specific recognition of DNA sequences by proteins.

II. THE NUCLEAR PROTEINS

Nuclear proteins can be divided into two main groups: histone and nonhistone proteins.

Histones belong to the most conserved proteins. They contain basic amino acids which cause their basic character. H2A (molecular weight 14,500) and H2B (molecular weight 13,700) histones, rich in lysine and H3 (molecular weight 15,300), and H_4 (molecular weight 11,300) histones, rich in arginine are constituents of the core of nucleosomes. Histone H1 (molecular weight 23,000), rich in lysine binds the particular nucleosomes into the elementary chromatin fiber. Some specific histones have been described in spermatozoa.

The original investigations on histone proteins were initiated by Bonner et al.[49] who demonstrated that addition of histones inhibited the protein-synthesizing system *in vitro* because of inhibition of the template activity of DNA. Contrary to this, removal of histones increased template activity. Next Alfrey and Mirsky[50] demonstrated that removal of histones from isolated cell nuclei enabled RNA synthesis under adequate experimental conditions. Histones are synthesized only during DNA synthesis and because of their basic character they bind immediately with the new synthesized DNA. On the basis of these experiments it was concluded that histones may be the suppressors of noncontrolled template activity of DNA.

Histones do not contain tryptophan. The content of basic amino acids in histone molecules seems to be responsible for condensation of DNA. H1 contains about 30% of basic amino acids, while H5 about 37% of basic amino acids, principally lysine, which may be the reason that DNA in tissues where H1 is replaced by H5 is more condensed. The basic parts of histones (mainly lysine residues) are symmetrically distributed on both ends of the histone

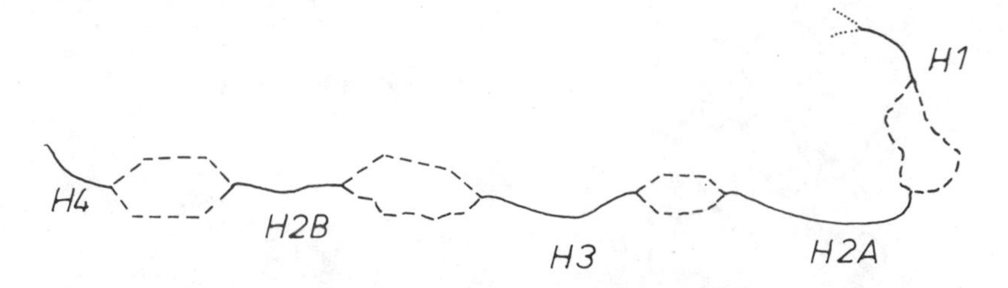

FIGURE 7. Electron-micrograph of histone genes, succeeding: H4, H2B, H3, H2A, and H1. Spacers between the particular genes, rich in AT sequences in chosen temperature, are melted (seen as single chains) while the particular genes rich in CG sequences (more stable at the same temperature because of three hydrogen bonds) still remain in the double-helix structure. (Redrawn from Portmann, Schaffner, and Birnstiel, *Nature,* 264, 31, 1976).

molecules, while the middle of the molecules is apolar. The apolar part of the molecules seems to be responsible for easy aggregation of these molecules, while binding with DNA is executed principally by the basic parts. Histone DNA ratio in cells is generally accepted at a ratio 1:1, which means that every DNA molecule is bound with one histone molecule.

Because of the role of histones as general repressors of DNA, their repressory function must be neutralized in the case of transcription of DNA. The amount of histones in proportion to DNA does not change during transcription. Therefore, the only possibility is the modification of histones which may loosen binding of histones with DNA. According to many experiments this is done by postranslational modifications of histones, i.e., their acetylation, phosphorylation and methylation. Experiments on *Tetrahymena pyriformis* in which only DNA in the macronucleus, where histones bound with the DNA of the macronucleus are acetylated, is there active transcription while in the micronucleus, where histones bound with DNA are not acetylated, DNA is not transcribed. It seems that acetylation of histones may be characteristic for ''actively transcribed chromatin''.

Histones are synthesized in the cytoplasm of cells where they undergo some modifications. In cytoplasm, serine of N-terminal H4 is reversibly phosphorylated and acetylated. After transfer to the nuclei, additional amino acid residues, mainly lysine, become acetylated which neutralizes the amino groups. Preferentially acetylated are lysine residues with direct contacts with DNA. Histones H1 and H3 are phosphorylated also in the cytoplasm, but dephosphorylated during transfer from the cytoplasm to the nuclei and again phosphorylated in the nuclei. During the cell cycle, histones H1 and H3 are reversibly phosphorylated. During G_2 and mitosis the amount of phosphorylated nuclear proteins increases and during G_1 decreases. During various periods of development the intensity of histone synthesis differs.

Histone genes are organized in tandemly repeated highly conserved quintets of five genes, encoding the major histone proteins in the following order: H4, H2B, H3, H2A, and H1 (see Figure 7). From earlier studies it has been accepted that histones, irrespective from which tissue they are derived are almost all identical. Experiments performed in mammals revealed that the numbers of reiterated histone genes are not too high, and presently there are five classes of histone genes (Zweidler, after Old and Woodland):[1]

1. Replication-dependent variants, synthesized from the start to the end of DNA synthesis.
2. Partially replication-dependent variants, synthesized from the start of DNA synthesis, but not completely repressed beyond the S phase.
3. Replication-independent variants, synthesized throughout the cell cycle and in nondividing cells, also called basic histones.
4. Minor histones, synthesized in small amounts in nondividing somatic cells. These replication-independent histones may be incorporated into chromatin in place of pre-

viously synthesized replication-dependent histones. Therefore they are also called replacement histones.

5. Tissue-specific histones, e.g., H5 (closely related to H1) in erythrocytes of birds instead of H1. Unusual histones are present above all in spermatocytes; in fish, protamines are not related to histones. In vertebrates there have been many descriptions of atypical histones (in *Xenopus* H3 and H4 are combined with an H1-like sperm basic protein) and several smaller, testis specific basic proteins. In mammals there are also variants of H1, H2A, H2B, and H3 histones, typical for mature sperm.

The role of histone variants is obscure. Because of higher condensation of DNA in erythrocytes of birds it is accepted that H5 causes higher condensation of DNA than H1. The presence of histone variants if sperm cells may speak for their role in the differentiation process of these cells. Speaking against a major role of histone variants in cell differentiation is the absence of these variants in *Drosophila* and yeast. Yeast contains two nonfunctional H2B mutants differing by four amino acids. Despite this, the organism is viable.

Finally it may be accepted that histones protect the genetic material (DNA), take part in chromosome structure, and act as gene regulators, mainly suppressors.

III. NONHISTONE CHROMATIN PROTEINS

Contrary to the histones which represent a narrow, well-defined group of basic proteins, the nonhistone chromatin proteins represent a highly variable group of acidic proteins of various functions. The molecular weight of these proteins is very different, from 10,000 and less to 200,000 and more. The exact number of nonhistone chromatin proteins is presently not known—some authors claim that there anywhere from hundreds to thousands, while others estimate hundreds of thousands of them. Among these proteins are a great number of enzymes and factors active in replication of DNA and transcription processes, as well as enzymes modifying DNA and nuclear proteins. Contractible cell elements (actin) also belong to these proteins. Finally, it is suggested that a small group of nonhistone chromatin proteins may be of regulatory character for gene expression.

Contrary to histones whose composition in various differentiated tissues is the same, the nonhistone chromatin proteins differ in various differentiated tissues. In support of a regulatory role of the nonhistone chromatin proteins one sees: quick metabolic turnover, quantitative and qualitative tissue specificity, accumulation in tissues of active transcription processes and augmentation in tissues with induced gene activities (e.g., in corticosterone- or other corticoid-dependent tissues after administration of these compounds) and diminution or even absence in tissues with total gene repression.

The main property of a gene-inducing agent must be its potential to bind with specific DNA sequences, otherwise this agent would induce any gene. This would be incompatible with the specific adequate function of the particular genes and such an organism could not exist. At first, Kleinsmith et al.[36] using affinity chromatography demonstrated that a small fraction of nonhistone chromatin proteins from rat liver binds specifically with DNA from rat liver. In a column with DNA extracted from rat liver a small quantity of nonhistone chromatin proteins from rat liver has been retained, while the same proteins have not been retained in a column containing DNA from bacteria or DNA from sperm of salmon.

Paul and Gilmour[37] proved the influence of nonhistone chromatin proteins on gene activation in eukaryotes. This conclusion was drawn from experiments upon template activity measured by the amount of synthesized RNA in experiments with native chromatin and purified DNA, extracted from this chromatin. The amount of RNA synthesized in native chromatin was only 5 to 10% in relation to the amount of DNA, while in purified DNA it was 40 to 50%. Besides, RNA synthesized on the basis of native chromatin was organ-specific, while on purified DNA it was organ-nonspecific. Experiments with reconstituted chromatin (i.e., chromatin separated into DNA, histones and nonhistones, and then through adequate procedures reconstituted) were very impressive. In these experiments RNA synthesized on chromatin reconstituted with DNA, histones and nonhistones from the same organ, was specific for this organ, e.g., chromatin reconstituted with DNA and histones from thymus and bone marrow but nonhistone proteins only from thymus synthesized RNA specific for the thymus (despite presence of DNA and histones from bone marrow). The same fractions of DNA and histones from thymus and bone marrow but supplemented only with nonhistones from bone marrow, synthesized RNA specific only for bone marrow.

The method of hybridization of gene products with its cDNA (DNA received by treatment of gene product—mRNA with revertase, i.e., an enzyme synthesizing DNA complementary to RNA, in this case mRNA) enables the detection of the smallest specific gene products (mRNAs). Using this method Paul and Gilmour[37] examined chromatin from fetal liver which synthesizes globin mRNA and brain cells which do not synthesize globin mRNA. Chromatin from these organs was dissociated into fractions containing DNA, histones and nonhistones and then DNA + histones from both fetal liver and brain cells reconstituted with nonhistones from the fetal liver synthesized globin mRNA, while when reconstituted with nonhistones from brain cells did not synthesize globin mRNA.

In experiments performed by other authors the results have been similar. Axel et al.[51] using cDNA for mRNA of duck globin mRNA showed that on the basis of isolated chromatin of reticulocytes globin mRNA was synthesized, while on liver chromatin, or chromatin from reticulocytes depleted of nonhistones and histones, synthesis of globin mRNA did not occur.

Histones, synthesized usually only during the S phase of the cell cycle, are synthesized also during G_1 when nonhistone proteins isolated from the S phase have been added during G_1.[52]

On the basis of these experiments it has been suggested that nonhistone chromatin proteins may play the role of gene regulators, but there is a very serious obstacle. Nonhistone chromatin proteins must be synthesized on the basis of their genetic codes and every time synthesis of these proteins occurs, activation of their genetic codes must also occur. Therefore, the nonhistone proteins per se cannot be the primary gene regulators. Probably a small percentage of them is necessary for initiation of transcription. Their role probably depends on recognition of specific DNA sequences, while the main bulk of nonhistone proteins is composed of enzymes and other factors necessary for the transcription process. In this respect the nonhistone chromatin proteins are probably responsible for the proper conformation of chromatin in differentiated cells. These proteins react on intracellular or extracellular stimuli by a chain reaction with induction of genes synthesizing proper proteins, among them also nonhistone proteins necessary for recognition of the proper DNA sequences and proteins necessary for the transcription process. Therefore, the essential regulatory signals for gene activation must be signals on the DNA.

IV. THE GENETIC CODE

The main task after discovery of DNA as the genetic material was to elucidate in which way DNA codes biosynthesis of proteins. Nirenberg and Matthaei,[53] using a synthetic polynucleotide, composed of only uracils (poly-U) in an *in vitro* system—after adding adequate enzymes and an activated pool of amino acids—resulted in the synthesis of a polypeptide composed only of phenylalanines (poly-Phe). This experiment proved that a triplet of uracils –U–U–U– causes addition of amino acid phenylalanine to the synthesized polypeptide (protein) chain. These and other experiments allowed in a short time the elucidation of the entire genetic code (see Table 1).

The genetic code is ternary, which denotes three bases code for one amino acid. The

TABLE 1
The Genetic Code

First position (5′ end)	Second position U	C	A	G	Third position (3′ end)
U	Phe	Ser	Tyr	Cys	U
	Phe	Ser	Tyr	Cys	C
	Leu	Ser	Term[a]	Term	A
	Leu	Ser	Term	Trp	G
C	Leu	Pro	His	Arg	U
	Leu	Pro	His	Arg	C
	Leu	Pro	Glu-N	Arg	A
	Leu	Pro	Glu-N	Arg	G
A	Ileu	Thr	Asp-N	Ser	U
	Ileu	Thr	Asp-N	Ser	C
	Ileu	Thr	Lys	Arg	A
	Meth	Thr	Lys	Arg	G
G	Val	Ala	Asp	Gly	U
	Val	Ala	Asp	Gly	C
	Val	Ala	Glu	Gly	A
	Val	Ala	Glu	Gly	G

[a] Chain terminating codon (formerly "nonsense codon").

three bases of the genetic code are not equivalent. Some amino acids (e.g., serine) can be selected independently of the third base that can be A, G, C, or U, while the first two bases remain the same. In the case of leucine the first base can also change. The fact that some amino acids can be coded by more than one codon is called codon degeneracy. It was accepted that at least 61 different tRNAs would exist, plus three chain-terminating codons. This has been calculated as follows: 4 bases $\times$ 3 = 4^3 = 64, minus 3 terminating codons = 61. After discovery of known tRNA sequence, however, it became obvious that the same tRNA can recognize several codons. Another case was the presence of inosine in some anticodons, derived from adenine whose 6-amino group through enzymatic deamination became modified into the 6-keto group of inosine (I). So, the wobble theory was established which states that the base at the 5′ end allows the formation of hydrogen bonds with most of the bases at the 3′ end of the codon. The following base pairing combinations are possible according to the wobble theory:

Base in codon	Base in anticodon
U or C	G
G	C
U	A
A or G	U
A, U, or C	I

According to the wobble theory any single tRNA cannot recognize four different codons, and three codons can be recognized only when I is on the third position. The wobble theory predicted correctly that for the six serine codons there exist three different tRNAs, as well as for the other two amino acids (leucine and methionine) whose codons differ in one of the first two positions and have different tRNAs.

AUG and GUG have been recognized as initiation codons. Some complications arose with the discovery of tRNA specific for *N*-formyl methionine which usually initiates translation of bacterial proteins. Further experiments, however, revealed that both methionine and *N*-formyl methionine have the same anticodon. Therefore, it must be accepted that signals for the starting amino acids must be much more complex than signals for all other amino acids present in the polypeptide chain. Further complication arises from experiments *in vitro* that the tRNA for *N*-formyl methionine recognizes AUG as well as GUG codons (GUG normally recognizes valine). Therefore, it must be accepted that a very specific wobble must exist for amino acids which initiate protein synthesis.

Termination of translation occurs at codons UAA, UAG, and UGA, known as chain terminating codons (formerly known as "nonsense" codons), which are not recognized by specific tRNAs but specific proteins ("release" factors). Each of the two known release factors recognizes two termination codons. This is in accordance with the presence of two chain-terminating codons UAA UAG, terminating the code for the coat protein of phage R17. Whether this is true for other proteins is not known.

Another characteristic feature of the genetic code is that the code is not overlapping. This denotes that the base pairs of the genetic code always code only one protein. This rule in the case of bacterias appears to be untrue. After DNA sequencing of phage ø174 it became clear that the same fragment of DNA codes for two or even three different proteins (Barrel, G. et al.,[38] Sanger, F. et al.[39]). This is possible on the basis of frameshift reading of the code. Frameshift reading can be explained by the following example. Any code of one's choice, e.g., TAGATGCGCA can be read beginning with the first base: TAG ATG CGC; beginning with the second base: AGA TGC GCA; and beginning with the third base: GAT GCG and so on. As seen from this example the same code may be read in three versions. Whether this is true only for bacteria and not for higher organisms is unclear, but such a possibility exists.

This is not all, some proteins have part of the code common with the other protein as follows: when the code for the first protein is terminated, the code for the other protein continues its work until the second protein is synthesized to its end. These data suggest that coding of particular proteins undergoes precise regulatory mechanisms selecting the proper code for every synthesized protein, independently of whether the code is overlapped or not.

The next characteristic feature of the genetic code is its colinearity which denotes that the sequence of bases of the genetic code (DNA) corresponds with the sequence of amino acid residues of the synthesized protein. This is unquestionably true for the sequences of bases in mRNA and amino acid residues in the synthesized protein, but sequences in DNA do not precisely correspond to those in the synthesized protein. At first, Hozumi and Tonegawa[40] showed that the sequence of amino acid residues in the immunoglobulin chain do not correspond to its DNA. Next, Breathnach et al.[41] revealed that the sequence of bases in the ovalbumin gene do not correspond to the sequence of amino acid residues of this protein. This leads to the second revolution in molecular biology and genetics caused by sequencing of bases of DNA. It became obvious that the primary idea upon gene structure as a simple fragment of the DNA chain was a very simplified one. Today we know that the gene structure is very complicated and comprises coding as well as noncoding DNA sequences.[2]

In higher organisms most of the genes are probably discontinuous, comprising coding sequences, so-called exons, and noncoding sequences, so-called introns. The number of introns and exons varies from one protein to another, likewise the number of bases in introns and exons. For example the β globin chain gene comprises two introns and three exons, while the ovalbumin gene comprises eight exons and seven introns. At the same time the total number of bases of the ovalbumin gene is 7700 bps, from which coding there are only 1872 base pairs (see Figure 8).

The next characteristic feature of the genetic code is its universality which denotes that the genetic code was the same in all living organisms (from the lowest to the highest). Newer

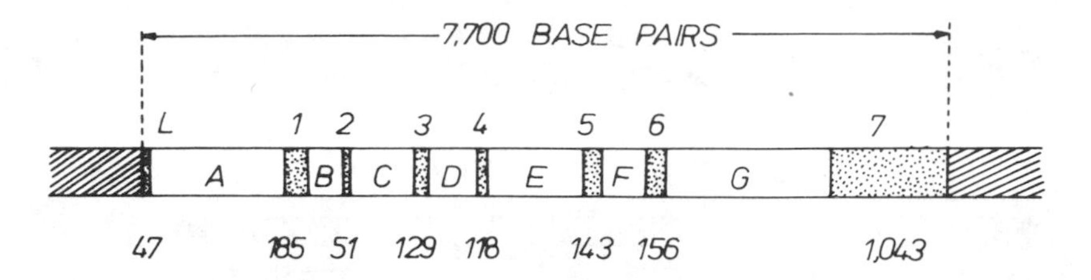

FIGURE 8. Scheme of the eukaryotic gene for ovalbumin, as an example of a mosaic or split gene. The total number of base pairs is 7700, there are eight exons (L,1—7), and seven introns (A—G) at the bottom the numbers of base pairs of the particular exons (L-leader sequences, responsible for the transport of the synthesized protein inside the cell compartments).

investigations, however, revealed that this is not always true. In mammalian mitochondria AGA and AGG are "stop" signals rather than codons for arginine, AUA codes for methionine rather than isoleucine, AUA and AUU may be start signals instead of AUG. Contrary to this in *Neurospora* and *Aspergillus* AGA, ACG and AUA code according to the universal code. In yeast AUA codes for methionine (as in mammals), but the four codons beginning with CU code for threonine rather than for leucine. In the mitochondria of mammals, yeast, *Neurospora* and *Aspergillus* tryptophan is coded by UGA and UGG, but in maize by CGG (Grivell).[3]

In recent times a new characteristic feature of the genetic code arose, i.e., movable elements, also called transposable genetic elements. These elements may influence gene expression both in physiological and in pathological conditions. Movable elements have been best described in prokaryota (bacteria), where two kinds of movable elements are present, the insertion sequences (IS) and transposons (Tn) (see Figure 9).

The insertion sequences belong to the simplest known movable elements of the genome. They have from 700 to 1500 bp and usually code only proteins causing their transposition. On both ends (5′ and 3′) there are so-called insertion sequences of inverted axis of symmetry, complementary to each other (TG......CA / AC......GT—"inverted repeat" orientation). During insertion of these elements into a new place of the genome a short preceding segment of the host genome and sequences at the end of the insertion undergo duplication, being of the same orientation—so-called "direct repeat" (e.g., AATTG........AATTG / TTAAC........TTAAC). The insertion sequences, because of their ability to move from place to place in the genome, can cause deletions, inversions, and translocations of the genome. Insertion sequences that contain promoter sequences, after insertion at the 5′ ends of the host genes (operons) may cause their activation. In the case of insertion into the host genome in an opposite direction to transcription, they may cause suppression of genes (operons).

Bacterial transposons are elements of at least thousands of base pairs which include codes for proteins responsible for their transposition (transposases) and some other genes, such as genes for antibiotic resistance (enzymes degrading or modifying antibiotics) (see Figure 9). At both ends of the transposons there are insertion sequences. Transposons are analogous to the insertion sequences, due to their ability to translocate fragments of the genome, and may activate or suppress particular genes (operons).

Typical movable elements similar to the transposons of bacteria have been described in yeast and *Drosophila* (Figure 9). The function of movable elements in eukaryotes is not known, nor is it not known if these elements code for proteins. The movable elements in *Drosophila* are called copia because of their repeated reiteration. In the germ line of *Drosophila* the copia elements appear in approximately 30 families of different structures. In one family the copia element is reiterated 20 to 40 times in the genome which makes about 5 to 10% of the whole genomic sequences. During differentiation the number of copia

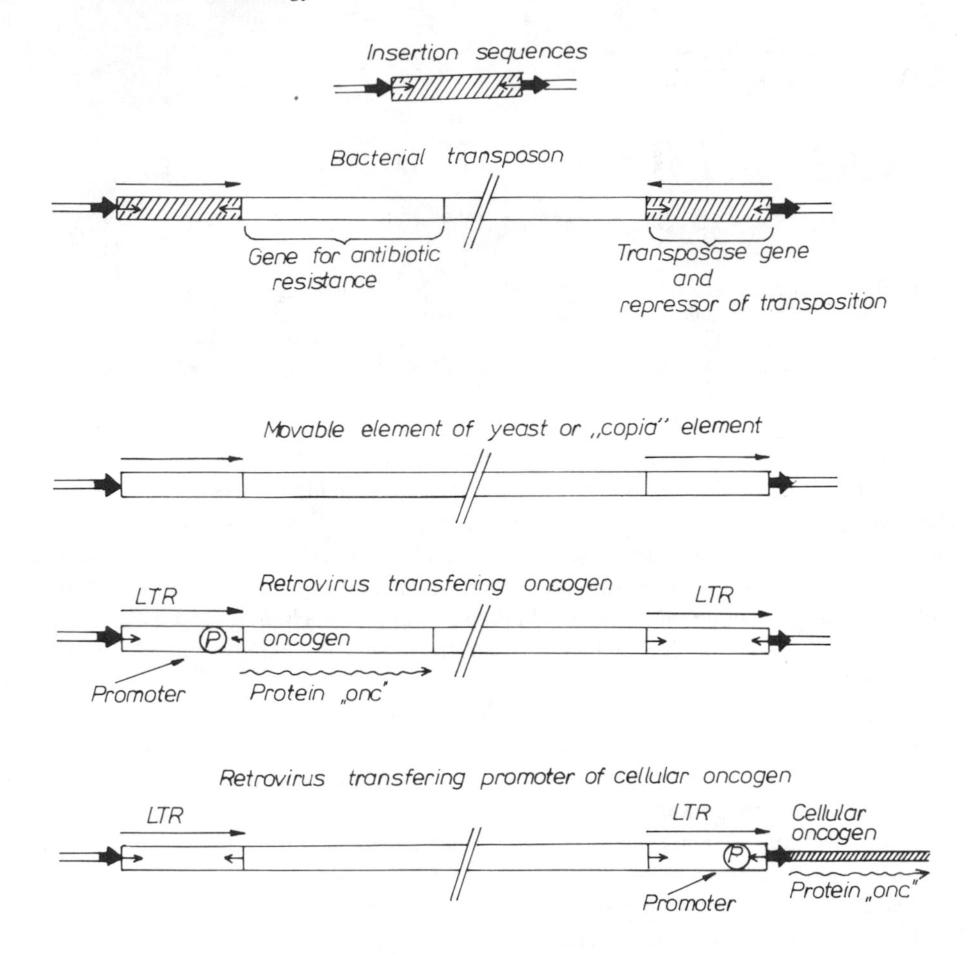

FIGURE 9. Structure of movable elements of the genome: insertion sequences, bacterial transposon, movable element of yeast or "copia" element of *Drosophila* and retroviruses. Thick arrows denote duplicated sequences of the host genome, thin arrows denote the ends of the insertion sequences: $^{\text{TG}}_{\text{AC}}$ at the 5′ and $^{\text{CA}}_{\text{GT}}$ at the 3′ end are in "inverted repeat" orientation.

elements in one family increases to about 250 with simultaneous translocations within the genome. As a consequence of copia element translocation into the promoter region of genes, they can be activated or inactivated.

In higher organisms typical movable elements have not been described. It is suggested that movable character in the genome of higher eukaryotes may have reiterating "Alu" sequences present in many recombination loci. Mobile character also have genes coding the variable regions of immunoglobulin genes, translocated during differentiation of immunocompetent cells and formation of immunoglobulin genes.

Similar to the structure and function of movable elements are proviruses of retroviruses (Figure 9). Retroviruses are viruses containing two identical RNA copies which replicate after infection of cells into DNA by use of the enzyme reverse transcriptase. After synthesis of the second DNA chain, a circular form of DNA originates which is inserted into the host's genome. This is the "provirus" phase of retroviruses. The structure of the proviruses resembles movable genomic elements. On both of their ends there are identical copies of reiterated sequences, the so-called Long Terminal Repeats (LTR). All have on their ends a short stretch of sequences of an inverted axis of symmetry—the so-called inverted repeats. During insertion of proviruses into the genome of eukaryotes a short preceding segment of the host genome and sequences on both ends of the provirus will be duplicated, being of

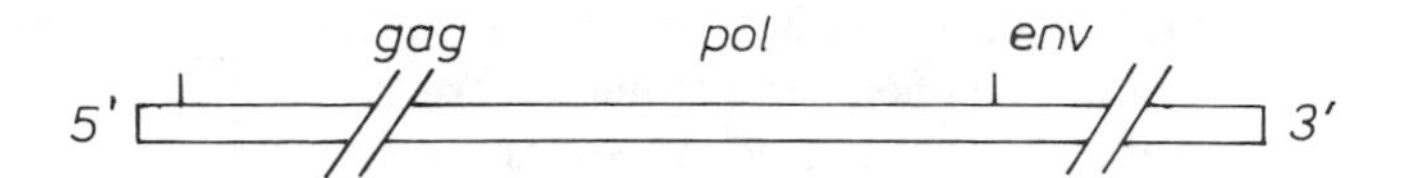

FIGURE 10. Scheme of the genome of retroviruses: *gag*—sequences coding for core; *pol*—sequences coding for the reverting enzyme (DNA-polymerase RNA-dependent) and endonuclease; *env*—seqeunces coding for envelope proteins.

the same orientation, so-called direct repeats (like insertion sequences, bacterial transposons, movable elements of yeast and copia elements of *Drosophila*). The structure of proviruses of retroviruses demonstrates all characteristic features of movable elements. The function of retroviruses shows the same characteristic features as movable bacterial elements, i.e., transfer of their own genetic information to other organisms. In the case of promoter transfer they may activate genes of host eukaryotes, and in the case of insertion into promoters of eukaryotic genes they may block these genes.

Along with virus genes that code proteins, realizing their replication and transcription, some of the retroviruses transfer the so-called oncogenes. Oncogenes have their analogues in genomes of normal cells of mammals and rodents. It is suggested that retroviruses in distant evolutional time did not receive oncogenes, and during transfer from the lysogenic into lytic phase, when the virus was excised from the host's genome, it took a fragment of the host's genome with it which contained the oncogene (so-called proto-oncogene). In normal cells proto-oncogenes are transcribed under the control of normal regulatory mechanisms and do not initiate malignant transformation of cells. It is suggested that proto-oncogenes may be normal genes which code proteins (factors) responsible for growth and differentiation of cells. Proto-oncogenes transposed together with retroviruses into the genome of cells, and transcribed free of normal cellular control mechanisms cause abnormal increase of growth factors which causes pathological growth of cells (tissues), the most significant feature of neoplasms.

Transposition of retroviruses is accomplished by a complex mechanism in which participating proteins are encoded by the retrovirus itself. A typical retrovirus contains in its genome a set of genes: ''gag'' coding for core, ''pol'' coding for reverse transcriptase, an endonuclease and some other proteins at the 3′ end, and ''env'' coding for protein of the viral envelope (see Figure 10). The biological role of the endonuclease probably consists of cutting chromosomal DNA to yield the four to six virus-specific base-pair duplications of host sequences that flank proviral DNA, and on cutting viral DNA in such a manner that proviruses can be joined with the host DNA at sites two bases from the ends of viral LTRs (long terminal repeats), containing inverted repeats within their LTRs. In this way during infection and integration, the endonuclease probably recognizes palindromic sequences necessary for circle junctions by the inverted repetitions at the end of the unintegrated linear DNA to join and form circular DNA. For example, insertion of a small circle junction fragment into spleen necrosis virus vector yields a new integration site within the viral genome (integrity of the inverted repeats is necessary for integration).

Transposable elements of *Drosophila* (copia-like elements) and *Saccharomycetes* (Ty elements) are organized in the same virus-like integrative mechanism, dependent upon the putative endonuclease action. Recently it has been shown that Ty elements transpose in a similar fashion as retroviral proviruses, i.e., via RNA intermediates. It is also suggested that copia-like elements of *Drosophila* and retrovirus-like elements transpose and amplify through RNA intermediates. Accepting this idea raised an evolutionary question: are retroviral proviruses of vertebrates with their capacity to spread from cell to cell and animal to animal advanced elements of eukaryotes? Or are such elements only remnants of ancient retroviral infections able to mediate intracellular transpositions?[4]

Another new feature has been detected while studying translocations between various kinds of cellular DNA. Several authors have found that, relatively frequently, translocations of the DNA from chloroplast genomes into mitochondrial genomes occurs in the plants as well as from chloroplast and mitochondrial genomes into the nuclear genomes. For this kind of DNA the term promiscuous DNA has been introduced. In higher organisms translocations of DNA from the mitochondrial genome into the nuclear genome have been observed. Whether nuclear DNA also can be translocated into the genomes of these organelles has not yet been determined. The mechanism of these translocations where there may be several mechanisms is also not yet clear.

A. PSEUDOGENES

The term pseudogene denotes defective sequences related to the normal gene. Typical examples are "globin pseudogenes" related to the particular productive genes (α pseudogene to α globin gene, β pseudoglobin gene to β globin gene). These genes may be transcribed but not translated (either α- or β-globin-chain is synthesized). Several pseudogenes in humans and other higher organisms have been described: globin pseudogenes, immunoglobulin variable pseudogenes in humans and mice, human ε heavy-chain and λ light-chain constant-region immunoglobulin pseudogenes, human snRNA pseudogenes, human β tubulin pseudogenes, human actin pseudogenes, human dihydrofolate reductase pseudogenes, human metallothionein pseudogenes, and human c-*ras* oncogene pseudogenes (Vanin).[5] Vanin enumerates two categories of pseudogenes: those which retain the intron-exon arrangement of their productive counterparts, and those which lack the intervening sequences of their productive counterparts, termed "processed pseudogenes".

Analysis of different pseudogenes revealed that they may arise as a result of multiple genetic lesions, such as insertions and deletions inside the gene sequences which may cause change in the reading frame or premature termination of translation resulting in an absence of the normal functional gene product (protein encoded by normal counterpart of the pseudogene). Other (processed) pseudogenes are characterized by the total absence of intervening sequences (introns) or presence of an oligo (A) tract at the 3′ end—these pseudogenes seem to derive from mRNAs by the action of the enzyme reverse transcriptase (DNA-polymerase RNA-dependent) which causes their transcription from an mRNA of the normal gene into an abnormal gene. Because mRNA comes into existence through its "processing", these pseudogenes are termed "processed pseudogenes"

Pseudogenes can also arise as a result of mutations of the flanking sequences, so-called direct repeats on both ends (5′ and 3′). Because flanking sequences are necessary for normal gene function, mutations of them may result in inability of these genes to serve as normal templates for the synthesis of adequate proteins. Different polyadenylation has also been found in some human Alu I family sequences. Many of the processed pseudogenes are flanked by direct repeats (9 to 14 bps) which occur immediately from 3′ to the oligo (A) tract and immediately at the 5′ end. These sequences have been found flanking the human Alu I family sequences. The direct repeats flanking various pseudogene sequences may be the result of a mechanism by which they are integrated into the genome.

V. TRANSCRIPTION

Genetic information stored in DNA, for the generation of its function, must be transformed into another form of genetic information. Usually DNA is transformed into RNA, and RNA into protein. This is the basic dogma of molecule biology, formulated by Crick.[54] In 1970 another flux of information was discovered (Temin)[55] i.e., transformation of RNA into DNA. Transformations of DNA into RNA, or DNA into DNA are usually termed transcription, and synthesis of polypeptide chains of proteins on the basis of RNAs is termed translation. The terms transcription and translation refer to the term "code", which is transcribed into another code (RNA) and then translated into protein.

Besides transcription of DNA into RNA, another kind of transcription is also possible, i.e., transformation of DNA into DNA, called replication of DNA, which yields a double amount of DNA necessary to provide dividing cells with an identical amount of DNA as in the mother cell. The same is possible in the case of RNA which can replicate into two RNA molecules which yields a double amount of RNA in multiplying RNA-viruses. Multiplication of RNA also occurs sometimes in the case of mRNAs which yields increased amounts of mRNAs for the translation of adequate proteins.

DNA transcription is a very complicated process, different in prokaryotes and eukaryotes. Transcription is realized by various enzymes, known under the comprehensive term RNA-polymerase DNA dependent (E.C. 2776, Nucleosidotriphosphate: RNA nucleotidyltransferase). There is only one known bacterial RNA-polymerase, while in eukaryotes there are known at least three RNA-polymerases, I, II, and III (termed also A, B, and C), responsible for transcription of various types of DNA sequences.

Bacterial RNA-polymerase is a complex enzyme, composed of four subunits α, β, β' and σ in molar ratio 2:1:1:1. The enzyme deprived of the σ subunit, preserves its catalytic activity, but loses its specificity. This indicates that the σ subunit is responsible for the search for the proper site on the DNA chain, while the core of the enzyme (α, β, β^1) is responsible for elongation of the RNA chain. The characteristic feature of bacterial RNA-polymerase is its sensitivity to rifampicin, which probably binds with the β subunit inhibiting its activity. Products of bacterial RNA-polymerase are mRNA, pre-rRNA and pre-tRNA.

The RNA-polymerases of eukaryotes can be divided into three main classes: RNA-polymerase I is located in the nucleoli, where it catalyses transcription of DNA for ribosomal RNAs (rRNAs). Molecular weights of the most common RNA-polymerases I are approximately 2×10^5 and 1.2×10^5, usually present in a ration of 1:1.

RNA-polymerase II is located in the nucleoplasm and is responsible for transcription of unique DNA sequences (mainly mRNAs for particular proteins). RNA-polymerase II is composed of eight subunits and its molecular weight is about 4.5×10^5 to about 5×10^5. The characteristic feature of RNA-polymerase II is its sensitivity to α-amanitin which causes 100% inhibition in molar concentration 10^{-8} M/l.

RNA-polymerase III is the largest of the all known RNA-polymerases, it is composed of ten subunits and its molecular weight is about 6.3×10^6. RNA-polymerase III is located in nucleoplasm and is responsible for the transcription of tRNA genes and 5S RNA genes. RNA-polymerase III is inhibited by α-amanitin in a concentration 10^{-5} M/l. The product of eukaryotic RNA-polymerase I is pre-rRNA, of RNA-polymerase II pre-mRNA and of RNA-polymerase III pre-tRNA and 5S RNA.

In addition, in cells of eukaryotic organelles there are: mitochondrial RNA-polymerase (located in mitochondria, responsible for transcription of mitochondrial DNA yielding mitochondrial RNA), and chloroplast RNA-polymerase located in chloroplasts (responsible for transcription of DNA of chloroplasts, yielding chloroplast RNA).

The first step in transcription is the binding of RNA-polymerase at an adequate site of the genome (DNA) with subsequent activation of the RNA-polymerase.

The second step is initiation of transcription (usually the first bond arises between a purine nucleotide and the first nucleotide of the DNA template).

The third step is elongation of the polynucleotide (during this step the synthesized polynucleotide RNA-chain is joined with the DNA template).

The fourth and last step of transcription is its termination and release of the synthesized polynucleotide RNA-chain from the DNA-template.

For better understanding of the particular processes of eukaryotic gene function, before I go into further detail, study the entire process as presented in Figure 11. As a typical

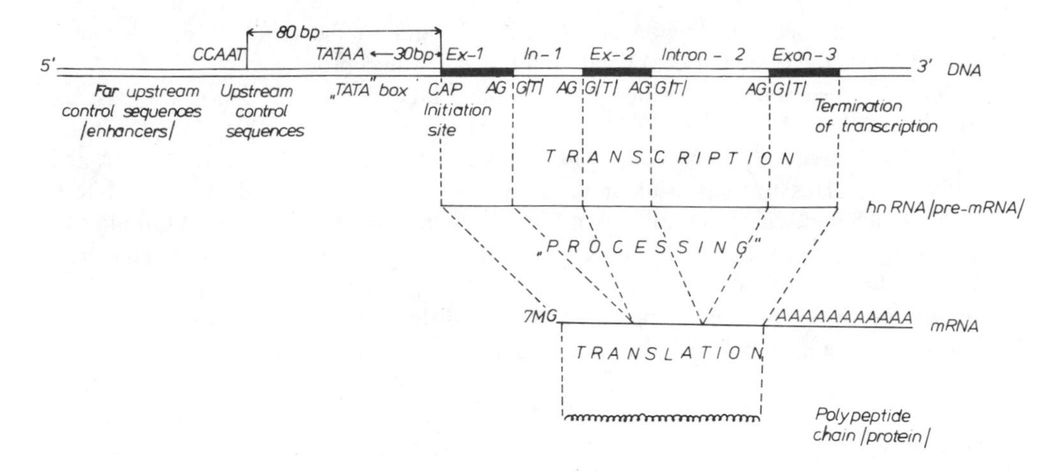

FIGURE 11. Scheme of a eukaryotic gene (the globin gene is represented). The gene is composed of coding and noncoding sequences (termed exons and introns, respectively). In addition, important for the function of the gene are the upstream and downstream flanking sequences. To the upstream sequences belong the TATA box, preceding by about 25 to 30 bp the initiation site joined with CAP (catabolite activator protein); CCAAT upstream control sequences, preceding by about 80 bp the initiation site, and far upstream are control sequences (enhancers) preceding the initiation site by some hundreds to thousands of bp. At the 5′ and 3′ ends of exons and introns there are signal sequences for splicing. In active genes, exons and introns first are transcribed into hnRNA (pre-mRNA). Its processing yields mature mRNA which is translated into a polypeptide (protein).

eukaryotic gene, the globin gene is presented with two introns and three exons, as well as the flanking sequences of basic importance for gene regulation.

For transcription of proper DNA sequences, the RNA-polymerases must be specialized by binding with specific factors which ensure correct recognition of the proper genes. To these factors belong σ subunits of RNA-polymerases, present in large numbers in cells. It seems that transcription of specific genes is connected with change of the σ factors. Several authors have found that heat-shock response of *Escherichia coli* caused by a sudden increase of temperature stops the transcription of vegetative genes with the simultaneous start of transcription of 15 heat-shock genes which requires the htpR gene product for their expression (Travers).[6] This product has a strong homology with the σ factor possessing two domains, one binding with the DNA and the other binding with the core of RNA-polymerase. In other experiments it has been shown that the htpR gene product acts *in vitro* as a σ factor binding to the core RNA-polymerase and activating in this way the heat-shock promoter, which is nonactive in vegetative bacteria.

Independently of the σ factors there are many other factors which cooperate in the process of transcription initiation. In mammalian cells there have been described transcription factors isolated from cell extracts, fractionated into fractions A, B, C, and D. Fraction C revealed RNA-polymerase I activity, and could be replaced by purified RNA-polymerase I; fraction B inhibited random initiation of transcription and could be replaced by poly (ADP-ribose) polymerase; fraction A enhanced the transcription and fraction D demonstrated species-dependency of transcription (Sommerville).[7]

RNA-polymerase II transcribes a great number of unique genes (in contrast to RNA-polymerases I and III which transcribe single genes). Therefore, the cognition of particular unique genes presents a peculiar task. Upstream of the initiation site (see Figure 11) there are several noncoding sequences recognized by the RNA-polymerase, termed the promoter. The promoter is composed of several consensus sequences. RNA-polymerase binds with these sequences and going along the DNA chain starts the transcription at the initiation site. The nearest consensus sequence of approximately 8 bp TATAAATA occurs about 25 to 30 bp upstream of the initiation site. There are many variants of the consensus sequence. The most conserved are the first four nucleotides, therefore this site is usually termed "TATA box", or Goldberg box who described this sequence in 1979.[56] An analogous sequence in procaryotes was described approximately 10 bp upstream of the start of transcription, termed "Pribnow box".

Recognition of these sequences is realized by specific proteins. Several of them have been described. For example, *Drosophila* RNA-polymerase II requires two distinct transcription factors (A and B) from which factor B specifically binds to a region 65 bp upstream of the start point of transcription (Parker and Topol).[8] This factor binds with the same DNA sequences also in absence of RNA-polymerase II. The same authors[9] described also a *Drosophila* RNA-polymerase II transcription factor product of the heat-shock 70 gene (70 kilodalton [kDa]) which binds to the upstream regulatory site 55 bp upstream of the TATA box. Sequences located 30 to 100 bp upstream of the starting point of transcription are involved in activation of transcription of heat shock genes in *Drosophila,* metallothionein genes, and also viral genes. Proteins that bind with these sequences may act similarly to the CAP (catabolite activator protein of *E. coli*) (see Figure 11). Bram and Kornberg[10] described specific proteins binding to far upstream activating sequences in RNA-polymerase II promoters. There can be distinguished two classes of regions that potentiate transcription *in vitro*. One (principally TATA box), not too far from the initiation site, and the second some hundreds to thousands of base pairs from the initiation site. The regions are termed upstream activation sequences, in the case of immunoglobulin genes—enhancers. In the case of *Saccharomyces cerevisiae*, where several GAL genes (genes which encode galactose utilization pathway) produce proteins that activate transcription of adequate genes through binding with the upstream activation sequences, while others inhibit this process.

According to modern views, recognition of adequate genes by RNA-polymerase II is realized through binding of specific proteins with upstream activating sequences. Among other proteins such a human promoter-specific transcription factor interacts with DNA sequences in a region of 21 bp direct repeats, 48 to 100 bp from the starting point of transcription for early proteins of the Simian Virus 40. This factor has been found also in noninfected cells which indicates its role in expression of cellular as well as viral genes (Gidoni et al.).[11]

Catabolite gene activator protein (CAP) (termed also cAMP receptor protein), binds with a 14 bp consensus sequence with either adenine or thymine at the end. This protein is necessary to form the initiation complex of transcription. It is probable, that CAP + RNA-polymerase + gene inductor cause transformation of inactive chromatin into active chromatin, which enables the process of transcription.

The next stage of transcription is elongation of the RNA chain. Its synthesis occurs always from the 5′ to the 3′ end, therefore exonucleases which cleave nucleic acids from the 3′ cannot cleave the originating RNA. After the RNA is synthesized, its 5′ end undergo modifications which inhibit the action of exonucleases.

The last step of transcription is termination and simultaneous release of the synthesized RNA from its template-DNA. Termination of transcription requires an adequate signal on the DNA, so-called chain terminating codons (earlier nonsense codons) UAG and UAA. The terminating signal is recognized by the (protein) ρ. Lack of factor ρ, or its mutation, causes elongation of transcription beyond the gene. On the other hand mutations (so-called nonsense mutations) of normal codons which yield a nonsense codon cause premature termination of transcription that results in the absence of proper gene products (the synthesized polypeptide chain is shorter because of premature termination of transcription and therefore the nonfunctional product).

RNA-polymerase III transcribes tRNA and 5S RNA genes. For accurate transcription several specific factors are necessary. Investigations of these factors initiated discovery of antibodies from patients suffering from autoimmune diseases, particularly systemic lupus erythematosus and Sjögren's syndrome, directed against ribonucleoprotein complexes of these diseases. Sera from these patients inhibited RNA-polymerase III. Two sera have been described, first—anti-SS-B (or La) directed against a 50 kDa polypeptide, part of a larger ribonucleoprotein complex, and second—anti-SpNo directed against a 100 kDa polypeptide. Both sera depleted transcription factors from the probe, which are necessary for the action of RNA-polymerase III. Extract S100 from HeLa cells restored the transcription ability of RNA-polymerase III.[12]

Further experiments upon transcription realized by RNA-polymerase revealed some rules for all the RNA-polymerases (I, II, and III) of eukaryotes. For accurate initiation of transcription, an initiation complex must be formed by the interaction of a 40 kDa protein (40 kd factor; TFIIIA) with a 50 nucleotide region in the center of the 5S ribosomal RNA gene. The complex is essential for the correct starting site of transcription of the gene. In contrast to this, termination of transcription occurs when the RNA-polymerase recognize a simple consensus at the end of the gene. The 40 kDa factor binds specifically to the 5S RNA genes which is essential for formation of the stable transcription complex. Limited proteolysis of the 40 kDa factor revealed that the one end of this protein is essential for binding with the essential internal control region of the 5S RNA gene and the other for efficient transcription.[13]

A not yet strictly defined role in transcription is played by a major component of mammalian short-length, middle-repetitive DNA, the Alu-equivalent (non-Alu) family which consists of 10^5 closely related sequences comprising 3 to 5% of the total DNA. Most of them are interspersed (not clustered) throughout the genome, frequently flanked by direct repeats which demonstrate RNA-polymerase III promoter activity. It is suggested that these sequences play a role in gene expression, transcription and genetic recombination. Alu-like middle repetitive DNA has been found in an unusual tandem arrangement with rat growth hormone which is of interest because RNA-polymerase III promoters occur as tandemly arranged tRNA genes (from which only the first is transcribed). Besides, sequences of these repeat elements have been found in some unique genes of the rat brain.[14]

Several authors have described different repressors of transcription which may be of interest for interpretation of some diseases. A bacterial repressor protein that blocks upstream activation of a yeast gene was described by Brent and Ptashne.[15] Enhancers usually activate gene expression, however, in recent times Borelli et al.[16] described repression of enhancer if induced stimulation of transcription by adenovirus-2-early region products which repress transcription induced by Simian Virus 40, polyoma virus, and adenovirus-2 enhancers. Repression is probably realized through interaction of adenovirus-2 products with the enhancer elements.

VI. PROCESSING OF RNA

The products of DNA transcription are RNA molecules. The main kinds of RNAs are: messenger RNA (mRNA), the protein-coding RNA; transfer RNA (tRNA), earlier termed soluble RNA (sRNA), which binds the particular amino acids and places them into protein synthesizing machinery of cells according to the complementary rule of the codon on the mRNA chain and the anticodon on the tRNA chain. Thus, each amino acid has its own tRNA, ribosomal RNA (rRNA), a component of the ribosome, the cytoplasmic organelle that acts as the protein synthesizing machinery. All these RNAs are much larger than the end-products participating in the protein synthesis. The process which yields mature RNA products is termed RNA processing.

Examination of RNA products preceding rRNA (immediately after synthesis) revealed a sedimentation value of 45S. It contains three spacers and three RNAs: 18S, 5.8S, and 28S (Figure 12). Shortly after synthesis the precursor of ribosomal RNA is cleaved into two rRNA molecules: a larger 28S and a smaller 18S; and beyond 5.8S RNA, tRNA is also realized by cleavage of a molecule approximately 20 to 30 nucleotides larger.

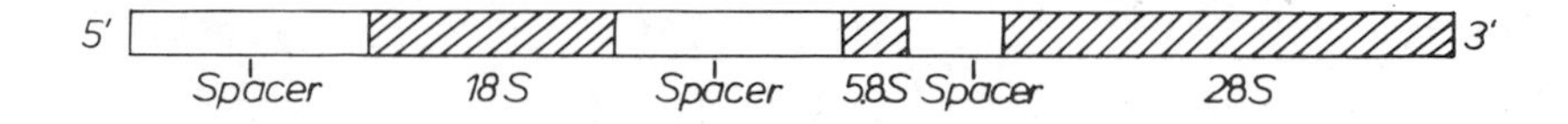

FIGURE 12. Precursor of rRNA (ribosomal RNA) of eukaryotes—spacers are removed during processing.

Another problem arose when it became obvious that the genes of the eukaryote genes do not represent a continuous stretch of DNA chain but that they may be (and probably are) a stretch of DNA interrupted by coding exon and noncoding intron sequences. This product was termed heterogenous nuclear RNA (hnRNA) or premature pre-mRNA. This discovery raised the following questions: by what mechanism are introns removed (this process is termed "splicing") from the pre-mRNA and what is the origin and function of introns? The removal of introns from transfer RNA revealed that introns themselves can catalyze their removal in the following way: the exons are folded in such a manner that nuclease and ligase can excise the introns and accurately join the exon ends. Moreover, in some classes of introns the excision and splicing can be performed by the intron itself. In that case RNA plays an enzymatic role in splicing. In this reaction in some lower organisms introns of mitochondrial messenger RNA and ribosomal RNA fold the molecule into a configuration which enables contact of the two ends of the exons and subsequent catalysis of the splicing reaction.[17]

More complicated is the splicing of mRNAs. For this purpose a splicing complex (so-called "spliceosome") must be made.[18] In yeast cells this is a complex of about 40S and in mammalian cells about 60S (for comparison the larger 30S ribosomal subunit contains 21 protein molecules and rRNA about 1500 nucleotides). The enormous task to be done by splicing is best illustrated by the fact that some protein genes may contain about 50 introns or more, and thousands of nucleotides each, and the pre-mRNA may be 200,000 nucleotides long or more. It is suggested that ribonucleoprotein particles (larger hnRNP—heterogenous ribonucleoprotein particles, and smaller snRNP—small ribonucleoprotein particles) may be the products of cleavage of the spliceosomes. Of particular interest is the ubiquitous presence in mammals of the small U1 RNA which has already been shown to be involved in splicing processes and which binds only with the 5′ and not 3′ splice sites (Lewin),[17] and antibodies against these particles inhibit the *in vitro* splicing.

Splicing is performed in two steps: first, cleavage of the 5′ site occurs, and at the second step the intron folds back and forms a lariat structure which brings together the two ends of the exons and both ends of the intron with subsequent ligation and removal of the lariat (Padgett et al.,[19] Konarska et al.,[20] Keller[21]) (see Figure 13).

Very important for accurate splicing are signals at their ends: a conserved sequence of nine nucleotides termed the 5′ splice end (with the first nucleotides G [T] which hybridizes with the 3′ splice end that is about an 11 nucleotides pyrimidine-rich region with the last nucleotides AG (see Figure 11). A conserved internal signal in yeast, near the 3′ end of the intron TACTAAC, has been discovered. This conserved sequence was found in sea urchin, mouse, rat, *Drosophila,* duck, chicken, and human genes. This signal is required for accurate splicing. This signal is termed the branch sequence and it is necessary for forming the lariat (see Figure 13). Mutations of this sequence disturb normal process of splicing or even inhibit it completely.[22]

Processing of pre-mRNA, besides splicing, includes cap structure, polyadenylation of the 3′ end, internal methylation and addition of specific nucleotides.

The cap structure consists of adding the nuclei into the 5′ end of the nascent pre-mRNAs a guanine molecule methylated in position 7—$m^7G(5')ppp(5')N$ (N = nondefined nucleotide). This common modification facilitates mRNA translation, stabilizes the structure of pre-mRNA by protecting against 5′ exonucleases, and contributes to the splicing process.

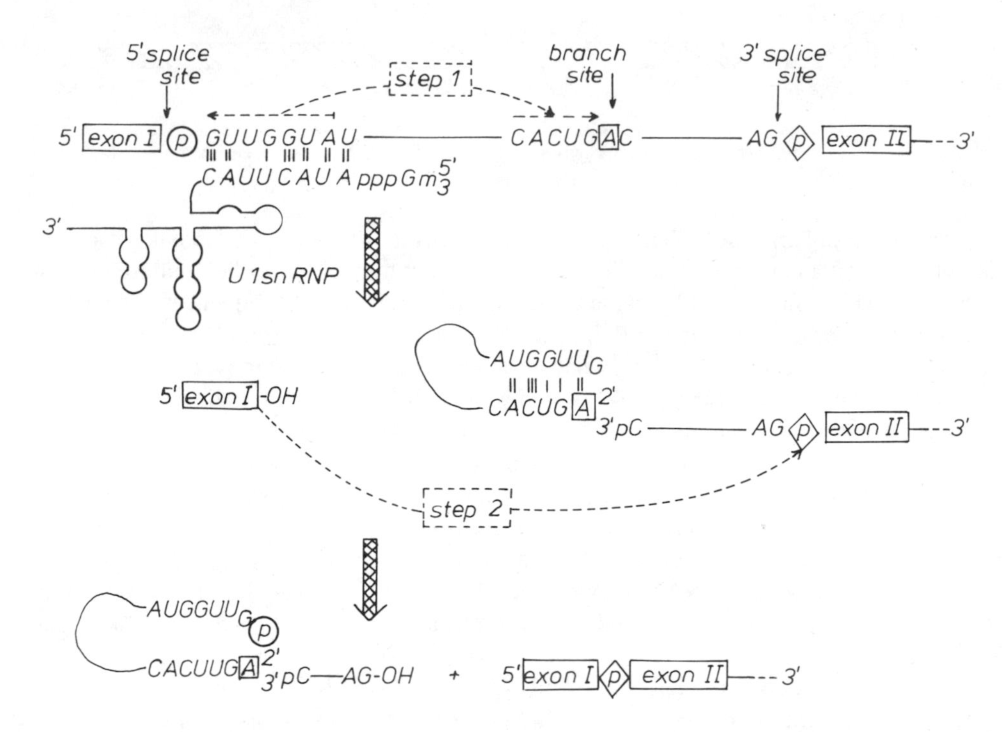

FIGURE 13. Scheme of the supposed splicing reactions. Step 1: recognition of splice site of the 3′ end of exon I, probably mediated by U1 snRNP; after cleavage the first sequences of the intron hybridize (in inverted orientation) with the branch site of the intron—consequently a lariat is formed. Step 2: exon I and II are ligated, the lariat is rejected. (Modified from Reference 21.)

The role of cap in splicing probably consists in formation of the specific ribonucleoprotein complex necessary for splicing.[23]

On the 3′ end of mRNAs usually poly(A) addition occurs. After the poly(A) addition signal AAUAAA, a number of about 50 to 200 and more adenine (A) nucleotides, catalyzed by the polyadenylate nucleotidyl transferase are added. Some mRNA species do not demonstrate polyadenylation. Polyadenylation protects the 3′ ends of mRNAs against exonucleases. Therefore, mRNAs deprived of poly(A) are usually short-lived, contrary to the polyadenylated which are long-lived.

Contrary to the previous views, termination of transcription is performed by a complex process, and does not occur simply by falling off the gene template from the terminating codons. Citron et al.[24] demonstrated that transcription termination occurs within 700 to 2000 nucleotides downstream from the poly(A) addition site of the mouse globin (major) gene.

The previously recognized signal for only poly(A) addition site AAUAAA serves also as an essential part of the recognition site for 3′ end processing of the pre-mRNA. It was shown that mutation of nucleotides of this signal causes transcription beyond the normal 3′ end of the gene. These longer than normal transcripts are unstable and degrade easily causing deficiency of gene products. In the case of a longer globin mRNA which degrades, thalassemia may occur. Experiments on transcription of sea urchin histone genes into *Xenopus* oocytes revealed that additional sea urchin extracts are necessary for accurate transcription and termination of transcription. These and other experiments suggest that specific post-transcriptional processing of RNA-polymerase II products must also include termination of transcription in which AAUAAA forms a part of the recognition signal for the endonucleolytic cleavage of the pre-mRNA, and besides which other factors also are necessary which form a termination complex for termination of transcription, analogical to the initiation complex for initiation of transcription (Proudfoot).[25] It seems that into this process some snRNP particles are involved, other than in the 5′ end recognition.

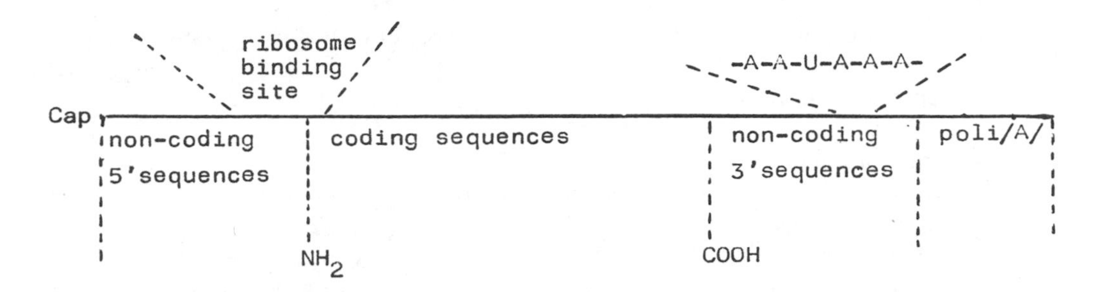

FIGURE 14. Scheme of eukaryotic mRNA. Cap—nucleotides protecting 5′ end (7-methylguanine) (Modified from Proudfoot and Brownlee, *Nature,* 263, 211, 1976.)

Processing of pre-mRNAs also includes modifications of nucleotides. The most important of these is methylation of bases which always occurs at the same sites and appears important in regulation of gene expression.

Processing must not always occur in the same way. Alternative splicing may be the cause of various mRNAs. Schwarzbauer et al.[26] demonstrated three different fibronectin mRNAs by alternative splicing of the same coding region (gene) which yielded various molecules of fibronectin, i.e., large glycoproteins which are important in cellular adhesion and morphology, cytoskeletal organization, cell migration, hemostasis and thrombosis, and malignant transformation.

The discovery of introns raised some questions about their origin and role in evolution since genomes of primitive organisms (bacteria) do not reveal those. It has been hypothesized that exons encode the particular functional domains of proteins (discovered in immunoglobulins). According to this statement it was suggested that during evolution different genomes of primitive organisms were joined in a single split gene in which the particular primitive genes were divided by introns. It was also proposed that introns participated in the joining process of these genes. When a greater number of genes had been sequenced, however, it became obvious that introns do not always divide exons with specific functions. Therefore, this question must remain unsolved. In this aspect, exon shuffling and intron insertion has been investigated in serine protease genes.[27] Comparing genes of various serine proteases: human clotting factor IX, human thrombin, human tissue plasminogen activator, pig urokinase, rat trypsin, kallikrein, nerve growth factor, rat chymotrypsin, human complement factor B, and human haptoglobin, it seems probable that most of the introns have been inserted into genes and have nothing to do with the particular domains of proteins. Rather, they may have some influence on gene expression.

A. mRNA

mRNA is a heterogenous nucleic acid molecule, which encodes protein molecules. At the 5′ end there is the cap structure and at the 3′ end, there usually is poly(A). The mRNA molecules represent the minority of RNA molecules (about 5% of the total RNA). The mRNA molecule is unstable, its life in bacteria is only some minutes, in higher organisms some hours or even days (the last property is caused by the cap on the 5′ end and polyadenylation of the 3′ end). After joining with ribosome, mRNA is translated into protein molecules. In addition to introns, which are excised during processing, not all the bases code for proteins. As seen in Figure 14, at the 5′ end and 3′ end there are bases which do not encode any protein. The role of the noncoding sequences is not yet known, but it is supposed that these sequences play a regulatory role.

1. Pathology of mRNA

Disturbances of the complicated process of mRNA synthesis may cause various diseases. Classical examples are represented by thalassemias caused by decreased synthesis of the particular α or β globin chains (α^+ resp. β^+ thalassemias), or total absence of α or β globin chains (α^0 or β^0 thalassemias).

The cause of β^0 thalassemia may be:

1. Deletion of the proper β globin gene
2. Presence of genes which do not undergo transcription (caused by inhibitory factors—Ferrara type of thalassemia)
3. Presence of changed mRNAs for globin chains (Ferrara type of thalassemia)
4. Presence of normal mRNA (hybridizing with cDNA globin) which does not undergo translation (probably lack of translation factors)

VII. TRANSLATION

Translation is the final step in gene expression. During transcription the genetic information, stored in DNA, is transformed into complementary mRNA, which during translation is transformed into its complementary polypeptide chain. The translation process is performed on ribosomes, present amply in cytoplasm of cells. For further recognition of the genetic information of mRNAs, Crick[54] suggested the presence of specific adapter molecules which from the one side could form hydrogen bonds with the codon nucleotides of mRNAs, and from the other side bind specific amino acids. Soon such molecules had been discovered in the form of tRNA (transfer RNA).

At the first step of the protein synthesizing process, amino acids which are inactive molecules, must be activated (their hydroxyl groups [OH] must bind with 3′ hydroxyl groups of tRNA). Every tRNA is highly specific for only one amino acid which determines high specificity of the translation process. Like other RNAs, tRNA is synthesized on the basis of the genetic code (in other words on DNA) in the form of a larger molecule which is processed to the normal size.

Because of the high specificity of particular tRNAs in relation to the particular amino acids, the total number of various tRNAs must be at least about 20, i.e., the number of amino acids present in proteins. In reality the number of tRNAs is higher because some amino acids have more than one kind of tRNA. For example, in bacteria there are two tRNAs for methionine, formyl-methionine tRNA ($tRNA_f^{Met}$) for initiation of translation, and methionine tRNA ($tRNA_m^{Met}$) for elongation of the synthesized polypeptide chain.

tRNA molecules contain approximately 77 to 85 nucleotides. All tRNAs reveal some common features. At the 5′ end there is always guanine, at the 3′ end the following sequence pCpCpA of nucleotides. The terminal adenine reveals a free hydroxyl group (OH) at the 3′ position of ribose, which binds the activated amino acid. This reaction is catalyzed by specific enzymes aminoacyl-tRNA synthetases, which form energy-rich acyl bonds. The pCpCpA sequence of nucleotides is the acceptor site for the activated amino acid. The aminoacyl-tRNA synthetases recognize both the amino acid and its proper tRNA. The reaction is performed in two steps.

1. Amino acid + ATP + enzyme ⇄ aminoacyl-AMP-enzyme complex + PP
2. Aminoacyl-AMP-enzyme + tRNA ⇄ amino acid-tRNA + enzyme + AMP

The structure of tRNA resembles a cloverleaf the stem of which represents the acceptor site and both arms complementary regions which form loops. The opposite loop of the acceptor site represents the anticodon, i.e., a triplet of nucleotides which form the codon of mRNA. Recognition of the particular amino acids for the sequence of the synthesizing polypeptide chain is possible due to the complementarity of codons within the mRNA with the anticodons of the tRNA. The anticodon appears between the 30th and 40th nucleotide

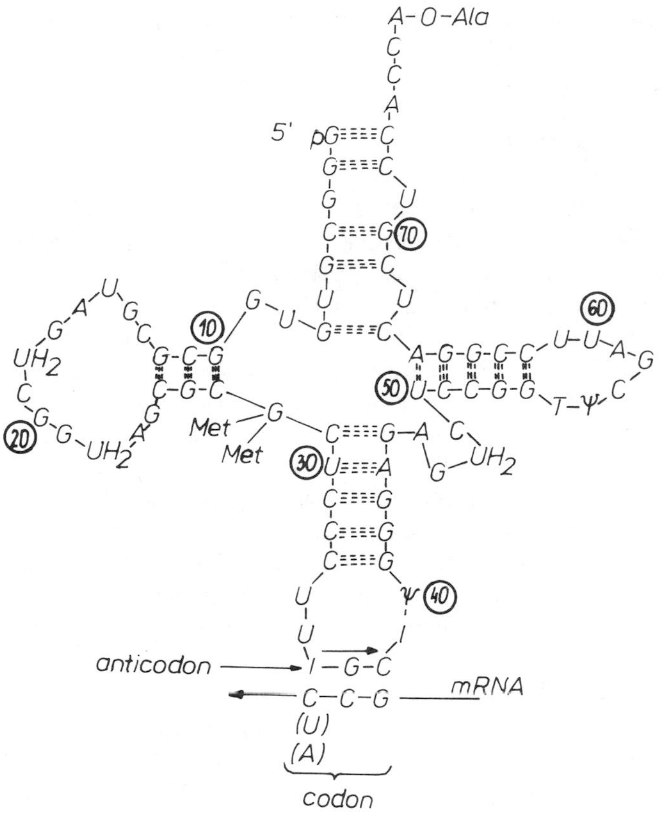

FIGURE 15. Nucleotide sequences and secondary structure of tRNA for alanine with a fragment of mRNA, coding for this tRNA.

from the 5′ end of the tRNA. In all tRNAs the anticodon is preceded by a uracil nucleotide and ends with an atypical nucleotide (see Figure 15). Characteristic features of tRNA are atypical nucleotides such as: ribosylthymine (rT), pseudouridine (ψU), inosine (I), 2′-methylcytosine (2′MeC), 1-methyladenosine (MeA), 2′-methylguanosine (2′MeG), and others.

A. rRNA

Ribosomal RNA represents the third kind of RNA engaged in gene expression. rRNAs are essential components of ribosomes, which contain about 60% rRNA and about 40% proteins. Ribosomes are present in cell cytoplasm as free or bound with the endoplasmic reticulum. Ribosomes are present in all living organisms. They are composed of two subunits, a smaller one about 30S in prokaryotes and about 40S in eukaryotes, and a larger one about 50S in prokaryotes and about 60S in eukaryotes. Ribosomes of most bacteria demonstrate a sedimentation coefficient of about 70S, and that of eukaryotes of about 80S (ribosomes of yeasts, plants, and animals demonstrate somewhat lesser amounts of rRNA).

The best known are ribosomes of *E. coli* with a sedimentation coefficient of 70S. In a medium with a lower concentration of Mg^{++} they degrade into two subunits: 30S and 50S. The subunit 50S contains a core and easily split proteins. The core is composed of 23S and 5S RNAs, and proteins. The 30S subunit exhibits the same structure, only the core rRNA molecule demonstrates a sedimentation coefficient of 16S. The core of eukaryotic 60S subunit contains 28S, 5.8S and 5S rRNAs, and proteins, and the 40S subunit 18S rRNA and proteins. Ribosomes are formed in the nucleolar region. At first, a large molecule of pre-rRNA is synthesized, which is methylated and then dissociates into 28S and 16S rRNAs, which become surrounded with proteins and form the ribosomal subunits 30S and 50S. After transfer to the cytoplasm the subunits join into ribosomes of 70S. The 30S subunits are synthesized in some minutes, and the 50S subunits a little longer.

B. RIBOSOMAL PROTEINS

The 30S ribosomal subunit of *E. coli* contains 21 protein molecules per each 16 rRNA molecule, termed S1 to S21, and the 50S subunit 34 protein molecules, termed L1 to L34.

Binding of mRNA with the 30S subunit by protein S1 begins formation of the initiation complex. Usually the methionine codon AUG initiates the synthesis of a polypeptide (in prokaryotes GUG also may serve as an initiation codon, which codes for valine inside the chain). The codons AUG and GUG are recognized by the initiator tRNA for methionine, different from the elongator tRNA, active in elongation of the chain. In prokaryotes the initiator tRNA is formylated by the enzyme transformylase, absent in eukaryotes. Therefore the initiator tRNA in eukaryotes is not formylated. Both, the initiator $tRNA_f^{Met}$ and the elongator $tRNA_m^{Met}$ are acylated by the same synthetase methionyl-tRNA, formylated became only the amino group of the initiator $tRNA_f^{Met}$, while the elongator $tRNA_m^{Met}$ is not formylated by this enzyme. In eukaryotes there are also two tRNAs for methionine, one corresponds with the bacterial $tRNA_f^{Met}$ and is termed the initiator tRNA ($tRNA_i^{Met}$), and the other is the elongator tRNA ($tRNA_m^{Met}$). Structural differences between the initiator and elongator tRNAs determine that the initiator tRNA after binding with protein initiation factors binds with the initiation site (P) on the ribosome, and the elongator tRNA after binding with protein elongation factors binds with the acceptor site (A) on the ribosome.

During its function, the ribosomes undergo several conformational changes, which result from interactions of ribosomes with mRNA, tRNA, and initiation, elongation and termination factors. Every factor participating in the translation process is strictly localized on the surface of ribosomes. Two sites seem to be crucial, P (from peptidyl) where during initiation of translation the initiator tRNA is bound and during elongation of translation the same site binds peptidyl-tRNA. The second site A (from aminoacyl) binds aminoacyl-tRNA (with activated amino acid bound with its acceptor site). Closely located to the A and P sites is the enzymatic center for the peptidyl transferase which catalyses transfer of the peptidyl residue from the peptidyl-tRNA on the amino group of the aminoacyl-tRNA. The peptidyl transferase is localized on the larger ribosome subunit, together with the center for GTPase, which hydrolyses GTP.

The crucial step in protein synthesis is initiation of translation in which several initiation factors participate. Those factors facilitate binding of initiator tRNA (formylmethionyl-tRNA in prokaryotes and methionyl-tRNA in eukaryotes) with the ribosomal small subunit which forms the 30S initiator complex in prokaryotes, and the 40S initiator complex in eukaryotes. After forming this complex the initiation factors disjoin from the small ribosomal subunit and bind with the large ribosomal subunit forming the 70S initiation complex in prokaryotes or 80S initiation complex in eukaryotes.

In the initiation process of polypeptide synthesis of prokaryotes, three initiation factors and GTP participate. In eukaryotes, in the same process, however, eight to nine initiation factors besides GTP and ATP participate. Eukaryotic initiation factors and their functions are presented in Table 2. The reaction is initiated by binding of methionyl-tRNA with the small ribosomal subunit 40S. In this reaction the initiation factor eIF-2 which is composed of three subunits participates. eIF-2 and GTP form the ternary complex which reacts with the 40S ribosomal subunit. One of the eIF-2 subunits may be phosphorylated by a specific protein kinase which is very important to the regulation of translation. Some of the initiation factors are very complex, e.g., eIF-3 is composed of 8 to 10 subunits, the roles of which are unknown. ATP, which participates in binding of mRNA with the initiation complex probably is hydrolyzed by eIF-4B.

TABLE 2
Initiation, Elongation, and Termination Factors of Translation in Eukaryotes

	Da	
eIF-1[a]	15,000	"Pleotropic" - necessary for all reactions forming the initiation complex
eIF-2	125,000	Formation of ternary complex (with Met-$tRNA_f$ and GTP)
eIF-2A	65,000	Binding of tRNA with 40S ribosomal subunit in the presence of AUG
eIF-3	500,000	Binding of mRNA with 40S ribosomal subunits—dissociation of 80S ribosomes into subunits (anti-association activity)
eIF-4A	50,000	Binding of mRNA with ribosomes
eIF-4B	80,000	Recognition of (Cap) mRNA
eIF-4C	17,000	Association of subunits
eIF-4D	15,000	Association of subunits
eIF-5	150,000	Ribosome-dependent GTPase; liberation of eIF from 40S complex
EF-1[b]	50,000	Association of aa-tRNA to the ribosome
EF-$1_{\beta,\gamma}$	?	Regeneration of EF-1-GTP complex
EF-2	97,000	Catalysis of translocation
RF[c]	56,500	Recognition of termination codons UAA, UAG and UGA

[a] eIF - eukaryotic initiation factor.
[b] EF - elongation factor.
[c] RF - release factor.

C. ELONGATION OF THE POLYPEPTIDE CHAIN

During elongation of the polypeptide chain the ribosome moves along the mRNA from the 5′ end, and catalyzes binding of subsequent amino acids to the amino end of the polypeptide. This process is initiated by binding of aminoacyl-tRNA to the A site of the initiation complex. The elongation reaction is composed of binding of the aminoacyl-tRNA to the A site, forming peptide bonds (transfer of the growing peptidyl group from the CCA terminus of tRNA to the aminoacyl group attached to the CCA terminus of the next incoming tRNA molecule) with translocation of mRNA to the next codon. During this reaction the free tRNA is released from the P site, and replaced by a new peptidyl-tRNA molecule from the A site.

In the elongation reaction protein elongation factors (see Table 2) Tu,Ts and G in prokaryotes, and EF-1 and EF-2 in eukaryotes participate. Factor Tu forms a ternary complex with GTP and aminoacyl-tRNA. After the joining of this complex with the ribosome, GTP is hydrolyzed and Tu joined with GDP is released from the ribosome. Ts in prokaryotes and EF-1 in eukaryotes causes regeneration of the GTP complex which enables binding with the next aminoacyl-tRNA. Elongation factor G in prokaryotes (EF-2 in eukaryotes) catalyzes translocation of the ribosome along the mRNA molecule.

D. TERMINATION OF TRANSLATION

Termination of translation occurs when the elongation reaches the termination codons (UAA, UAG or UGA), which are recognized by the terminating factors (three release factors in prokaryotes RF-1, RF-2, and RF-3, and one release factor in eukaryotes—RF). The release of the synthesized polypeptide is catalyzed by the peptidyl transferase. Simultaneously with release of the peptide chain, mRNA is also released from the ribosome, and the ribosome dissociates into subunits. The energy for the release processes originates from GTP which dissociates into GDP.

E. SYNTHESIS OF SECRETORY PROTEINS

Proteins are synthesized on free polyribosomes or bound with membranes of the endoplasmic reticulum. Proteins synthesized on free polyribosomes in cells are used by them-

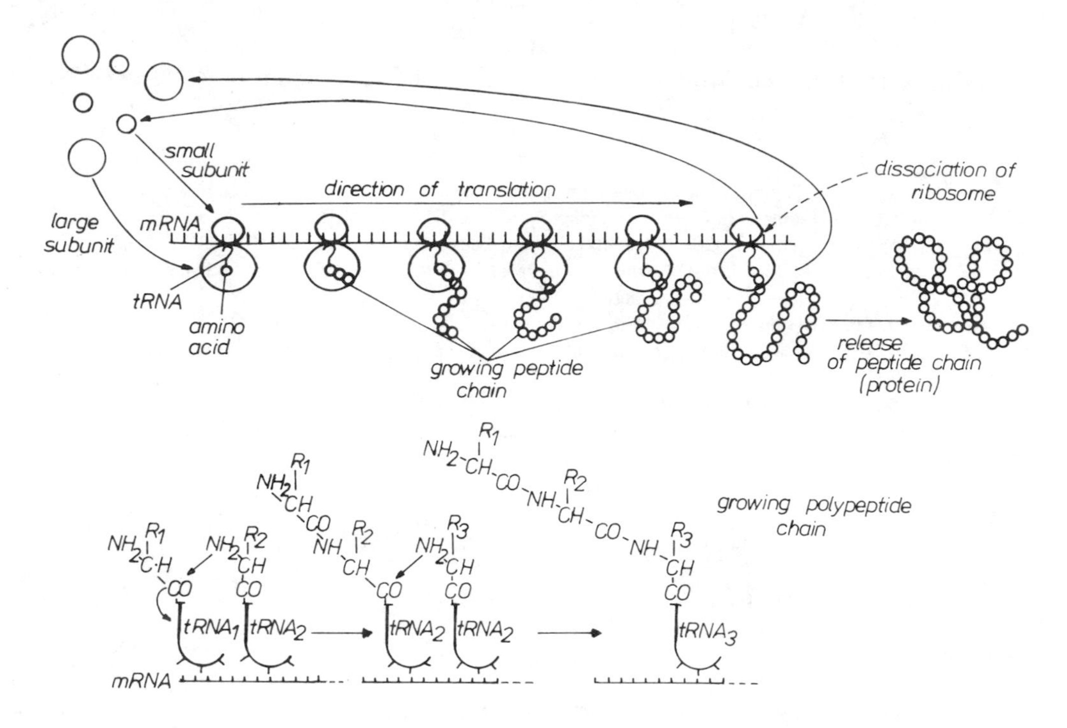

FIGURE 16. (Top): role of ribosomes in biosynthesis of proteins (translation of the genetic code). (Bottom): formation of peptide bonds.

selves, while those which are synthesized on polyribosomes bound with the endoplasmic reticulum are secreted for use of other cells of the multicellular organism. Typical secretory proteins are hormones, synthesized as pre-prohormones and pre-proenzymes. The mechanism of synthesis and secretion of secretory proteins has been explained by Blobel and Dobberstein (Figure 16).[28]

According to the theory of Blobel and Dobberstein secretory proteins are synthesized with an additional piece on the N-end of the polypeptide, which is composed principally of hydrophobic amino acids. This piece gives the name "pre" to the synthesized polypeptide (e.g., pre-proinsulin), and is cut from the polypeptide by a peptidase during passage through membranes. The hydrophobic piece of the polypeptide is termed "signal", because it indicates the way by which the polypeptide passes through the membrane. The hydrophobic character of the signal peptide enables penetration of the synthesized polypeptide into the hydrophobic part of membranes and formation of a kind of tunnel through which the hydrophilic polypeptide chain can pass. After passage of the polypeptide chain through or into the membrane the hydrophobic signal piece is cut by a peptidase present in the membrane (see Figure 17).

The hydrophobic signal piece of the polypeptide chain is encoded by an appropriate fragment of DNA, preceding the nucleotide pairs of the proper gene.

Besides the signal hydrophobic piece, the synthesized polypeptide in the case of active protein molecules (enzymes, hormones) usually is comprised of the next piece which is responsible for inactivation of the synthesized molecule by obturating its active site. The synthesized enzymes (proenzymes) and hormones (prohormones) of protein character do not become active until the obturating piece is removed (usually by a specific enzyme), which causes conversion of the inactive form of proenzymes or prohormones into active ones. For example: conversion of protrypsin into trypsin or proinsulin into insulin.

After the synthesis of a polypeptide is terminated, the ribosomes dissociate from the endoplasmic reticulum and their subunits can be used for formation of further ribosomes.

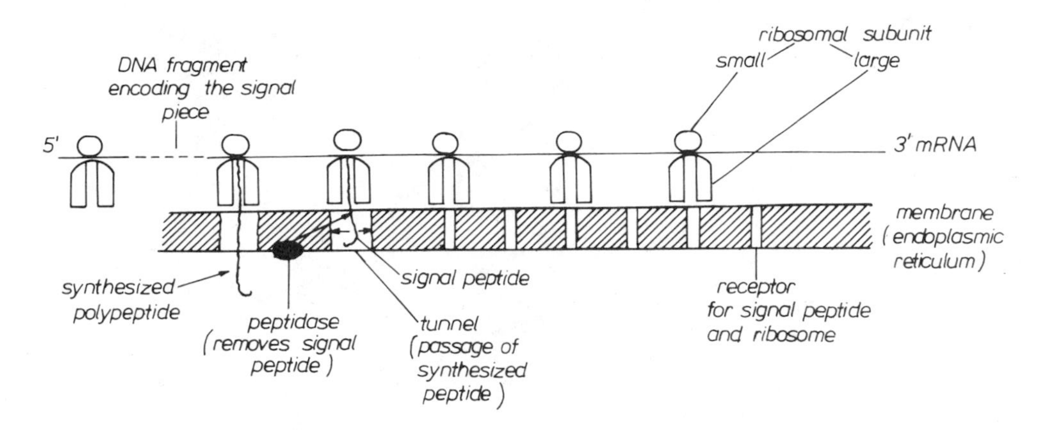

FIGURE 17. Scheme of the signal theory of Blobel for passage of proteins through membranes.

F. POST-TRANSLATIONAL MODIFICATIONS OF PROTEINS

The synthesized protein molecules can undergo various modifications. The most frequent modification is glycosylation, i.e., binding of proteins by covalent bonds with various sugars or phosphorylation by binding with phosphates. Glycosylation is usually performed in the Golgi apparatus, where adequate enzymes are present.

In the case of abnormally high glucose levels in diabetics, pathological glycosylation of hemoglobin may occur which may disturb the physiological function of this protein.

VIII. REPLICATION OF DNA

In dividing cells the amount of DNA must be duplicated to provide daughter cells with the same amount of identical DNA copies as was present in the progenitor cell. For this purpose replication always precedes cell divisions.

Replication of DNA is a very complicated process in which many proteins participate mediating this basic genetic process. In general, DNA replication is performed in a semi-conservative manner in which every DNA strand serves as a template for construction of a new, identical strand of DNA (see Figure 18).

The structure of the DNA molecule seems to predestine the mode of its replication. After breaking hydrogen bonds which join the two DNA strands, each of them may serve as a template for the synthesis of a new identical strand. In this way two new strands originate. Consequently, the double DNA helix is always made up of one new and one old strand. In the place where the DNA synthesis occurs the two strands unwind and a so-called replication fork is formed. The DNA polymerase simultaneously synthetizes two DNA strands, always in the direction from the 5′ end to the 3′ end. This predicts that the two DNA strands cannot be continuous (Okazaki et al.).[29] The authors showed that DNA replication occurs on one so-called leading strand, and on the other so-called lagging strand. On the lagging strand the new DNA is synthesized always in short fragments (termed "Okazaki fragments") as soon as the template of the lagging strand is accessible to the DNA polymerase. Synthesis of short Okazaki fragments is initiated by short RNA fragments which play the role of primers. They recognize the starting points for DNA polymerases and are responsible for correct joining of the Okazaki fragments. After their joining the RNA primers are excised.

As the replication process is a very complicated one, a series of proteins are necessary for its execution. The basic replication apparatus consists of a moving complex of several polypeptides (different in various organisms) in which the entire replication fork is embedded. In the T4 bacteriophage, these proteins have a total mass of 900,000 Da, plus a helix

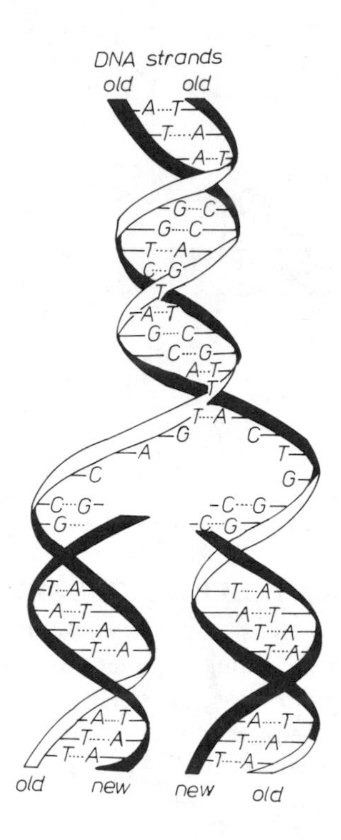

FIGURE 18. A generalized scheme of semiconservative replication of DNA.

destabilizing protein bound to the single stranded DNA. These proteins proceed unidirectionally along the DNA strands about 500 nucleotides per second.[30]

The main work in DNA replication is done by DNA polymerase. In eukaryotes three DNA polymerases in nuclei and one in mitochondrias have been described. DNA polymerase α (about 15.5×10^4 Da) seems to be the main DNA polymerase in DNA replication which works as well *in vitro* as *in vivo*. DNA polymerase α is inhibited by salt concentration above 25 m*M*/l NaCl, as well as by ethidium bromide and *N*-ethyl malonate. DNA polymerase β (about 4×10^4 Da) is present mainly in the nuclei at the same level during the whole cell cycle, therefore it is supposed that this polymerase is engaged mainly in DNA repair. DNA polymerase β is not inhibited by the typical inhibitors like ethidium bromide and *N*-ethyl malonate. DNA polymerase γ (about 20×10^4 Da) is characterized by its ability to replicate native and artificial, single and double stranded DNA. DNA polymerase γ cannot replicate RNA. This polymerase is inhibited by ethidium bromide and ethyl malonate. Mitochondrial DNA polymerase is probably identical with DNA polymerase γ.

Besides DNA polymerases many other protein molecules are necessary for the replication process. Among them are ligases which join the short Okazaki fragments into one long strand; peptides which make the double DNA structure accessible to the replication enzymes. To those peptides belong: "unwinding proteins" which bind with single-stranded DNA facilitating their unwinding and in this way their replication (those proteins protect the single-stranded DNA structures against association into double-stranded structures, and inhibit action of endonucleases and random replication); helix relaxation proteins (helicases) which cause unwinding of DNA superhelices (these proteins make incisions of strands and after rotation of the strands catalyze their joining); gyrases which cause winding of circular DNA into negative superhelices (under the action of those enzymes double-stranded circular DNA is wound around this enzyme molecule, which results in abatement of forces binding the

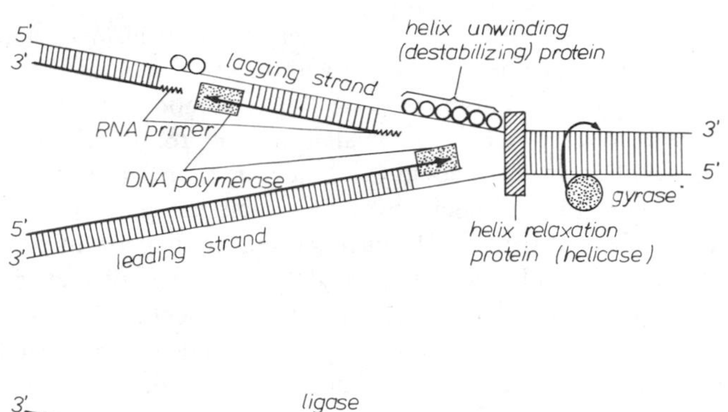

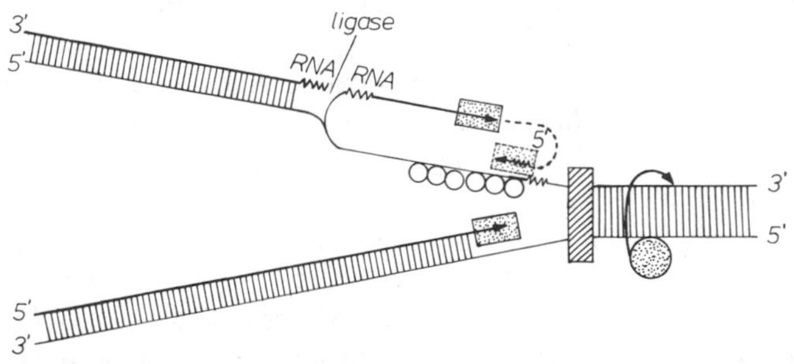

FIGURE 19. Scheme of the replication fork (description in text).

two DNA strands together); nucleases which make incisions of double-stranded DNA at starting points of replication, excise RNA primer fragments and erroneous DNA sequences which make the normal replication process impossible, and liberate nucleotides for renewed utilization in DNA synthesis.

Energy for DNA replication is supplied by hydrolysis of ATP. Some of the enzymes active in DNA replication exhibit ATPase properties which yield the necessary energy for this process.

In prokaryotes there is a single starting point from which replication is performed in both directions of the DNA molecule. In eukaryotes there are more starting points, which occur always at the same region of the given DNA molecule.

The main events on the replication fork are presented in Figure 19. At constant starting points, several enzymes prepare the double helix for replication. Those enzymes are helix unwinding (destabilizing) protein, helix relaxation protein (helicase), gyrase, and some factors using energy from ATP hydrolysis which dissociate the double-stranded DNA into two separate strands. To each single strand joins a DNA polymerase which synthesizes two new strands of DNA, always from the 5′ end to the 3′ end of the molecule. The leading strand is synthesized continuously from the 5′ end to the 3′ end, while the lagging strand is synthesized in the form of interrupted Okazaki fragments (synthesized also from the 5′ end to the 3′ end). After synthesis of about a 1500 nucleotide long Okazaki fragment including the RNA primer fragment, the Okazaki fragment forms a loop which turns this fragment approximately 180°, so that the DNA polymerase is put to a new starting point for synthesis of a new Okazaki fragment. After excision of the RNA primers, the two ends of the newly synthesized DNA strand are joined by an appropriate ligase. Thanks to this maneuver, the DNA polymerase becomes translocated from the lagging strand to the starting position and can serve for the synthesis of another Okazaki fragment.

The double structure of DNA is stabilized by histone proteins which are synthesized at the same time during the S phase of the cell cycle in eukaryotes. The presence of histones during DNA replication is indispensable (inhibition of their synthesis arrests the replication process). During replication, DNA probably is liberated from histones and the new DNA molecule, immediately after synthesis is bound with newly synthesized histone proteins.

Very important are mechanisms governing the replication process which makes an enormous task for each dividing cell. In mammalian cells approximately 3×10^9 nucleotides must replicate. The DNA molecule is about 1 m long, due to the chromosomal organization, is divided into at least 10,000 smaller replicating subunits, termed "replicons", which replicate in a time of about 8 h. It has been found that replication of the whole genomic DNA follows a distinct temporal order which might be the result of sequential activation of initiation sites. On the basis of their own investigations Calza et al.[31] suggest that the gene's position in the chromosomes rather than the sequence of nucleotides, determines the time of replication. These positions may be associated with proximity to sites that control the temporal order of replication.

In chromosomes of eukaryotes there are several initiation sites. The sites are always the same, therefore initiation of replication of particular chromosomes (or their parts) presents a characteristic feature of them. For example: late replication is characteristic for the inactive X chromosome.

In general DNA replication in mammalian cells is temporally bimodal. Active genes replicate usually in the first stage of the S phase of cell growth. According to this rule all "housekeeping" genes (active in all cells, necessary for maintenance of life processes) and tissue specific genes replicate at first, and the nonactive genes in the second stage of the S phase.[32]

IX. THE ENZYMATIC FUNCTION OF POLYRIBONUCLEIC ACIDS

An important mechanism of splicing was discovered in 1981. Cech and coworkers found that excision of an intervening sequence of *Tetrahymena thermophila* (intron 413 nucleotides long) occurred through transesterification without any protein enzyme, only after addition of micromolar quantities of GTP (or other guanosine derivatives). The 3′ hydroxyl group of guanosine attacks the phosphodiester bond between the 3′ end of the preceding exon and the first nucleotide of the following intron. As a result, the guanosine residue is attached to the 5′ end of the intron. At the next step of the reaction, the liberated 3′ end of the exon attacks the distant end of the intron which enables a splice to be made between the two exons (Figure 20).

The liberated intron undergoes a cascade of spontaneous reactions. At first, the 3′ end of the polyribonucleotide attacks the phosphate bond between the 15th and 16th nucleotide from the 5′ end of the liberated intron (in this way the liberated intron is shortened from 413 to 399 nucleotides). The remaining polynucleotide (399 nucleotides) forms a circular structure. When this circular structure is exposed to change of pH (from 7.5 to 9.0), this structure opens at the same site where it had been formed. This linear polynucleotide has enzymic properties; its active 3′ end attacks again and ejects four nucleotides from the 5′ end, and the remaining polynucleotide forms once again a circular structure (395 nucleotides—Figure 20).

The discovery of polyribonucleic acids as enzymes is revolutionary in the field of catalysis. Until that point, protein enzymes seemed to be obligatory for every catalyzed reaction.[34] The discovered enzymatic reaction of polyribonucleic acids does not play a role only in excision of introns, but also in reactions performed on ribosomes, where most enzymes exhibit ribonucleoprotein character. In this case the ribonucleic part of the enzyme is probably the catalytic site of the enzyme.

The discovery of the enzymatic character of polyribonucleotides also has a great influence on views about the beginning of life. Because of their self-replication and catalytic properties,

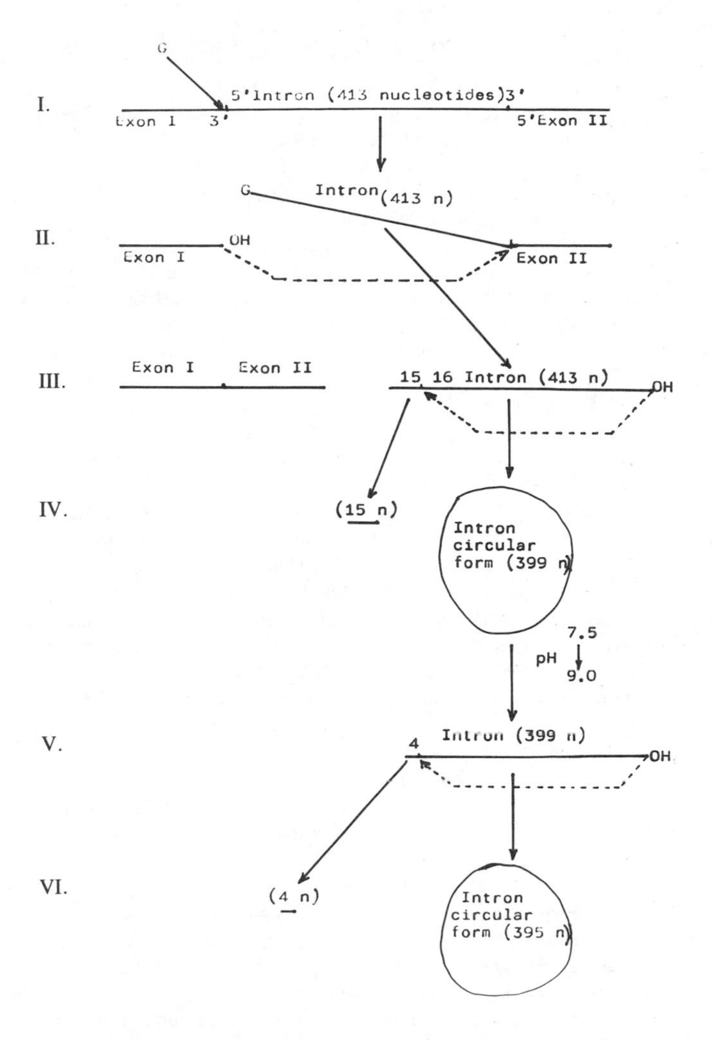

FIGURE 20. Scheme of enzymatic function of polyribonucleic acids as enzymes: I. Addition of guanosine (or its derivatives) initiates the reaction. II. 3′ hydroxyl group of guanosine attacks the phosphodiester bond between Exon I and Intron. III. Exon I and II splice; 3′ hydroxyl group of the liberated intron attacks between the 15th and 16th nucleotide from the 5′ end. IV. A 15 polynucleotide is ejected; the remaining 399 nucleotides form a circular structure. V. Change of pH from 7.5 to 9.0 opens the circular structure, and a circular structure is formed from the remaining 399 nucleotides. VI. A tetramer nucleotide is ejected; the remaining 395 nucleotides form a circular structure.

these molecules appear to be the best candidates for the primitive molecules of life. The self-splicing intron can splice itself out of an RNA molecule, or back into the same or other appropriate nucleotide sequence. This indicates that RNA may be a self-sufficient molecule for all functions of the genetic apparatus; it may be the source of genetic information (e.g., RNA-viruses); it may transform immature genetic information into functional information (e.g., splicing of pre-mRNA by rejection of introns or introduction of new nucleotides into existing polynucleotides); it can serve as a template for the synthesis of proteins (translation of mRNA); and through reverse transcription it can transform into DNA molecules. It is very probable that RNA initiated the most primitive life processes,[35] and during evolution these molecules acquired increasingly larger properties until reaching the level they presently exhibit.

REFERENCES

1. **Old, R. W. and Woodland, H. R.,** Histone genes: not so simple after all, *Cell,* 38, 624, 1984.
2. **Chambon, P.,** Split genes, *Sci. Am.,* 244 (5), 48, 1981.
3. **Grivell, L.,** Mitochondrial DNA, *Sci. Am.,* 248 (3), 78, 1983.
4. **Varmus, H. E.,** Reverse transcriptase rides again, *Nature (London),* 314, 583, 1985.
5. **Vanin, E. F.,** Processed pseudogenes, characteristics and evolution, *B.B.A. Libr.,* 782, 231, 1984.
6. **Travers, A.,** Sigma factors in multitude, *Nature (London),* 313, 15, 1985.
7. **Sommerville, J.,** RNA polymerase I promoters and transcription factors, *Nature (London),* 310, 189, 1984.
8. **Parker, C. S. and Topol, J.,** A drosophila RNA polymerase II transcription factor contains a promoter-region-specific DNA-binding activity, *Cell,* 36, 357, 1984.
9. **Parker, C. S., and Topol, J.,** A drosophila RNA polymerase II transcription factor binds to the regulatory site of an hsp 70 gene, *Cell,* 37, 273, 1984.
10. **Bram, R. J. and Kornberg, R. D.,** Specific protein binding to far upstream activating sequences in polymerase II promoters, *Proc. Natl. Acad. Sci. U.S.A.,* 82, 43, 1985.
11. **Gidoni, D., Dynan, W. S., and Tjian, R.,** Multiple specific contacts between a mammalian transcription factor and its cognate promoters, *Nature (London),* 312, 409, 1984.
12. **Gottesfeld, J. M., Andrews, D. L., and Hoch, S.,** Association of an RNA polymerase transcription factor with a ribonucleoprotein complex recognized by autoimmune sera, *Nucl. Acids Res.,* 12, 3185, 1984.
13. **Smith, D. R., Jackson, I. J., and Brown, D. D.,** Domains of the positive transcription factor specific for the Xenopus 5S RNA gene, *Cell,* 37, 645, 1984.
14. **Guttierez-Hartmann, A., Lieberburg, I., Gardner, D., Baxter, J. D., and Cathala, G. G.,** Transcription of two classes of rat growth hormone gene-associated repetitive DNA: differences in activity and effects of tandem repeat structure, *Nucl. Acids Res.,* 12, 7153, 1984.
15. **Brent, R. and Ptashne, M.,** A bacterial repressor protein or a yeast transcriptional terminator can block upstream activation of a yeast gene, *Nature (London),* 312, 612, 1984.
16. **Borelli, E., Hen, R., and Chambon, P.,** Adenovirus-2 E1A products repress enhancer-induced stimulation of transcription, *Nature (London),* 312, 608, 1984.
17. **Lewin, R.,** More progress in messenger RNA splicing, *Science,* 228, 977, 1985.
18. **Brody, E. and Abelson, J.,** The "spliceosome": yeast pre-messenger RNA associates with a 40S complex in a splicing-dependent reaction, *Science,* 228, 963, 1985.
19. **Padgett, R. A., Konarska, M. M., Grabowski, P. J., Hardy, S. F., and Sharp, P. A.,** Lariat RNA's as intermediates and products in the splicing of messenger RNA precursors, *Science,* 225, 898, 1984.
20. **Konarska, M. M., Grabowski, P. J., Padgett, R. A., and Sharp, P. A.,** Characterization of the branch site in lariat RNAs produced by splicing of mRNA precursors, *Nature (London),* 313, 552, 1985.
21. **Keller, W.,** The RNA lariat: a new ring to the splicing of mRNA precursors, *Cell,* 39, 423, 1984.
22. **Keller, E. B. and Noon, W. A.,** Intron splicing: a conserved internal signal in introns of animal pre-mRNAs, *Proc. Natl. Acad. Sci. U.S.A.,* 81, 7417, 1984.
23. **Konarska, M. M., Padgett, R. A., and Sharp, P. A.,** Recognition of cap structure in splicing *in vitro* of mRNA precursors, *Cell,* 38, 731, 1984.
24. **Citron, B., Falck-Pedersen, E., Saditt-Georgieff, M., and Darnell, J. E., Jr.,** Transcription termination occurs within a 1000 base pair region downstream from the poly(A) site of the mouse β-globin (major) gene, *Nucl. Acids Res.,* 12, 8723, 1984.
25. **Proudfoot, N.,** The end of the message and beyond, *Nature (London),* 307, 412, 1984.
26. **Schwarzbauer, J. E., Tamkun, J. W., Lemischka, I. R. and Hynes, R. O.,** Three different fibronection mRNAs arise by alternative splicing within the encoding region, *Cell,* 35, 421, 1983.
27. **Rogers, J.,** Exon shuffling and intron insertion in serine protease genes, *Nature (London),* 315, 458, 1985.
28. **Blobel, G. and Dobberstein, B.,** Transfer of proteins across membranes, *J. Cell Biol.,* 67, 835, 1975.
29. **Okazaki, R., Okazaki, T., Sakabe, K., Sugimoto, K., Kainuma, R., Sugino, A., and Iwatsuki, N.,** *In vivo* mechanism of DNA chain growth, *Cold Spring Harbor Symp. Quant. Biol.,* 33, 129, 1969.
30. **Alberts, B. M.,** Protein machines mediate the basic genetic processes, *Trends Genet.,* 1, 26, 1985.
31. **Calza, R. E., Eckhardt, L. A., DelGuidice, T., and Schildkraut, C. L.,** Changes in gene position are accompanied by a change in time of replication, *Cell,* 36, 689, 1984.
32. **Goldman, M. A., Holmquist, G. P., Gray, M. C., Caston, L. A., Nag, A.,** Replication timing of genes and middle repetitive sequences, *Science,* 224, 686, 1984.
33. **Zaug, A. J. and Cech, T. R.,** The intervening sequence RNA of Tetrahymena is an enzyme, *Science,* 231, 470, 1986.
34. **Westheimer, F. H.,** Polyribonucleic acids as enzymes, *Nature (London),* 319, 534, 1986.
35. **Gilbert, W.,** The RNA world, *Nature (London),* 319, 618, 1986.

For References 36 to 56 see page 87.

Chapter 4

GENE REGULATION

I. GENE REGULATION IN PROKARYOTES

As a rule proteins can be divided into two: (1) constitutive synthesized by cells in fixed amounts, independent of need, and (2) inducible or repressible whose synthesis is dependent on the regulatory mechanisms of adequate genes. Generally, the first group is necessary for basic life processes, and the second group for specific functions, e.g., enzymes necessary for degradation of lactose by *E. coli* are useless in a medium without this sugar, but in a medium containing lactose (without these enzymes lactose cannot be utilized).

First, Jacob and Monod[1] performed experiments upon inducible and repressible enzyme systems in bacteria. These systems were well-known for over 60 years when it was observed that in the presence of a specific substrate (e.g., lactose) enzymes necessary for their degradation were synthesized, and in the absence of the substrate, these enzymes disappeared. This effect, named ''enzymatic adaptation'' was the subject of many experiments and speculations until the experiments of Jacob and Monod threw some light upon the essence of this phenomenon.

The main idea of these experiments was the operon model comprising enzymes which catalyze the given biochemical reaction, directed by a specific regulatory mechanism. Enzymes are synthesized on the basis of genes which determine their structure, therefore named ''structural genes''. The regulatory mechanism is composed of specific regulatory genes, preceding the structural genes. The structural genes and the regulatory genes compose a unit, named ''operon''. The regulatory genes occur normally far from the structural genes.

The regulatory genes make some kind of switching mechanism which turns on and off a whole set of structural genes of the given operon. Turning ''on'' causes transcription of the structural genes, and turning ''off'' stops their transcription. Transcription of structural genes in a bacterial operon follows in an uninterrupted manner, and a polycistronic mRNA is synthesized, i.e., a long RNA molecule containing mRNAs for the entire set of enzymes, necessary to perform the given biochemical reaction. The polycistronic character of the synthesized mRNA tells us: mutation in the regulatory region causes partial or total inhibition of expression of all enzymes in the operon; mutation in the early region of transcription causes inhibition of transcription of all the following (located downstream) genes of the operon; transcription of all the structural genes is coordinated.

The regulatory system of the operon works in a typical ''feedback'' manner of reaction. The regulatory genes are synthesizing proteins—cytoplasmic repressors—which through interaction with the ''operator'' (region of the DNA upstream of the proper gene which interacts with the cytoplasmic repressor) may switch ''on'' or ''off'' transcription of the structural genes in the operon. Negative or positive types of these reactions exist.

A. NEGATIVE REGULATORY SYSTEM

As a typical negative regulatory reaction, the lactose operon in *E. coli* can be considered. This operon is composed of three structural genes coding for beta-galactosidase (which converts beta-D-galactosides into alcohol and beta-D-galactose), lactose permease and acetyltransferase, preceded by the regulatory region (see Figure 1). The regulatory region is composed of a regulator gene, a promoter (the region which binds with the RNA polymerase DNA dependent), and an operator (the region which binds the cytoplasmic repressor). Binding of the operator with the cytoplasmic repressor causes inhibition of the whole operon that synthesizes mRNAs for the synthesis of the three enzymes necessary for degradation

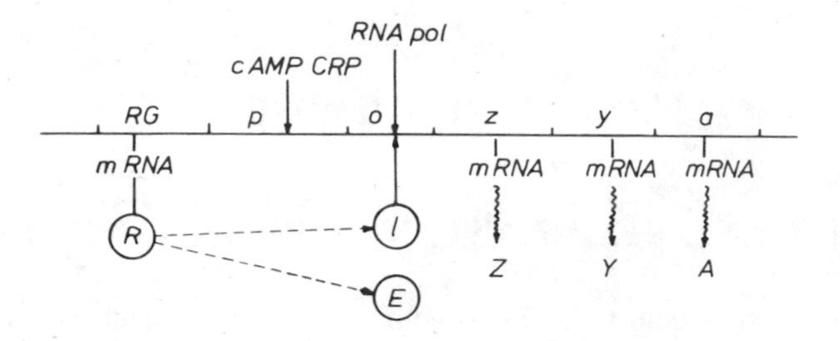

FIGURE 1. The "lac" operon in *E. coli*. RG—regulatory gene, p—promoter, o—operator, z, y, a—beta-galactosidase, permease, acetyltransferase structural genes (respectively), cAMP—cyclic AMP, CRP—catabolite repression protein, RNA pol—RNA polymerase, R—cytoplasmic repressor, I—inducer, E—effector (e.g., lactose).

of lactose. In contrast, adding lactose to the medium as the only source of carbon, causes binding of lactose (allo-lactose) with the cytoplasmic receptor which abolishes its affinity to the cytoplasmic repressor, hindering its binding with the operator. Consequently, synthesis of mRNAs for lactose degrading enzymes occurs. Further investigations revealed that the cytoplasmic repressor does not bind immediately to the operator, but with the inducer, inhibiting binding of the inducer with the operator, blocking in this way the whole operon. The presence of lactose or other substances with affinity to the cytoplasmic repressor forms repressor-inducer complexes which makes binding of the cytoplasmic repressor with the inducer impossible. Consequently, the operon is not blocked and synthesis of enzymes may occur. Substances which bind with the cytoplasmic repressor are named effectors.

$LacI^-$ mutants cause constitutive synthesis of the lac operon, i.e., synthesis of the lac enzymes is not permanently regulated because of inability to synthesize active repressor molecules. $LacI^s$ mutants synthesize repressor molecules which are unable to bind with the effector molecules binding only with the inducers, and therefore synthesis of the enzymes does not occur even in the presence of effector molecules. Mutations of the operator may cause the same effects, as mutations of the regulatory genes. Promoter sequences do not synthesize any protein molecules; therefore, mutations in this region do not cause regulatory disturbances, but only decreased or increased affinity to the RNA polymerase, and consequently decreased or increased synthesis of enzymes by hindering or facilitating binding of RNA polymerase with the promoter.

From earlier experiments it is known that in the presence of glucose the addition of lactose does not cause expression of the lac operon. This is caused by another regulatory system, named "catabolite repression". This is a more general system which causes preferential catabolism of glucose. This system is regulated by a protein molecule, the "catabolite repression protein" (CRP) which is inactivated after binding with cyclic adenosine monophosphate (cAMP). CRP forms complexes with cAMP (CRP-cAMP). CRP-cAMP complexes bind with an adequate region of the promoter which enables initiation of transcription by binding RNA polymerase. In the presence of glucose the amount of cAMP is diminished which causes simultaneous diminution of CRP-cAMP complexes. Consequently, the amount of CRP-cAMP complexes is insufficient for induction of the operon expression.

The nucleotide sequences of the promoter revealed two parts: one of 46 nucleotide pairs to which RNA polymerase binds, and a second rich in AT sequences, flanked by CG sequences. AT sequences reveal a more flexible structure and CG sequences a more stable structure. It is suggested that the AT region is protected by the flanking CG sequences.

To summarize: in the absence of glucose, CRP-cAMP complexes bind to the CG promoter region which destabilizes the DNA structure and makes the AT promoter region accessible

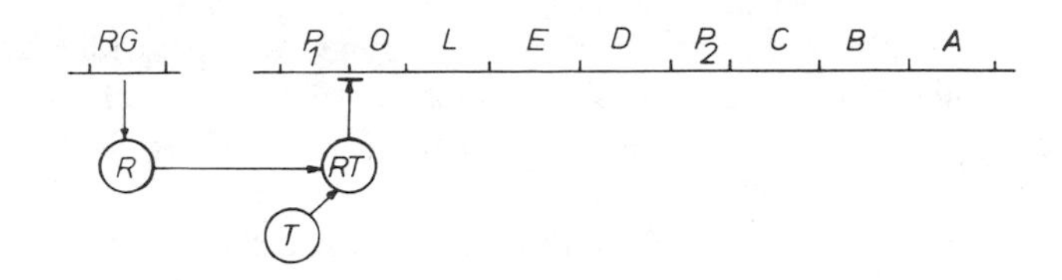

FIGURE 2. Tryptophan operon in *E. coli*. A, B, C, D, E—structural genes coding for particular enzymes of tryptophan synthesis pathway, P_1,P_2—promoters, L—leader sequences, RG—regulatory gene, R—repressor, T—tryptophan, RT—repressor-tryptophan complex.

to the RNA polymerase. In the presence of effector molecules (lactose) the inducer is not bound with the cytoplasmic repressor which enables induction of the whole lac operon, synthesis of mRNAs and afterward synthesis of enzymes degrading lactose.

B. POSITIVE REGULATORY SYSTEM

The positive regulatory system is best known in the case of the degradation of arabinose. The regulatory region of arabinose is composed of two parts: araI and araO. In the absence of arabinose the repressor protein (inactive inducer) binds with araO. In this position the repressor protein inhibits synthesis of enzymes degrading arabinose. In the presence of arabinose the repressor protein (i.e., inactive inducer) changes its affinity and binds with the araI region. Consequently, RNA polymerase can bind with the araI region which causes expression of the ara operon and synthesis of enzymes degrading arabinose (i.e., the inactive inducer becomes transformed into an active one).

The ara operon is also glucose dependent. Addition of glucose to the medium causes, as in the case of lac operon, diminution of about 50% of the catabolite activity of the ara operon. Diminution of the catabolite activity is caused by preferential catabolism of glucose also. Similarly, as in the case of the lac operon, CRP binds with cAMP, which destabilizes the DNA structure, enables binding with RNA polymerase, and consequently causes expression of the whole operon. In the presence of glucose, the amount of cAMP became diminished which causes diminution of CRP-cAMP complexes to a level insufficient for induction of the operon.

Comparing the expression of the described lac and ara operons, the differences between negative and positive mechanisms are evident: in the negative mechanism, the binding of the cytoplasmic repressor with the effector inactivates the repressor which enables expression; and in the positive mechanism, the binding of the regulatory gene's product (inactive inducer) activates the inducer and enables expression of the whole operon. Many experiments have proved that both mechanisms may play roles in adequate regulation of the gene expression, which enables very precise adaptation to the changing environmental conditions.

C. GENE REGULATION THROUGH TERMINATION OF TRANSCRIPTION

The described regulatory mechanisms are active in catabolic processes and refer to the initiation of transcription. Besides, there are also regulatory mechanisms which refer to the termination of transcription, principally concerning anabolic processes, e.g., synthesis of amino acids.[2] The best known is the operon of tryptophan synthesis, composed of five structural genes coding for five enzymes A, B, C, D, E, synthesizing tryptophan from its precursor chorismic acid through intermediates: Anthranilic acid, *N*-5′-(Phosphoribosyl)-Anthranilic acid,1-Co-Carboxyphenylamino 1′-Desoxyribulose-5-phosphate,Indole-3-Glycerolphosphate and Indole into tryptophan. Besides, the tryptophan operon comprises two promoter regions, operator and leader sequences (see Figure 2). The affinity of RNA polymerase to the P_1 promoter is greater than that to the P_2 promoter. Independently of the tryptophan operon, there is a tryptophan regulatory gene, which synthesizes an inactive repressor protein which after binding with tryptophan (tryptophan acts as a corepressor) inhibits the binding of RNA polymerase to the operator and thus is also an expression of the operon.

Measuring the amount of operon transcripts it was found that there are ten times more short transcripts (comprising sequences corresponding to genes which are adjacent to the first promoter, i.e., genes E and D) than sequences corresponding to the whole operon. This was explained by the greater affinity of the first promoter to RNA polymerase than that of the second. In the case of the shorter transcripts, transcription stopped at the region of the second promoter and passed only 1:10 along the second part of the operon transcribing the whole operon. Such premature stoppage transcription is named "attenuation", while the first part comprises leader sequences. This denotes that there are two regulatory systems, the first concerning initiation of transcription, and the second concerning termination of transcription. Both systems work independently, and what is highly important is that this kind of gene's regulation seems to be very effective.

The described kind of regulation which governs anabolic processes appears to prevail in the synthesis of all amino acids. Well-known is the histidine operon, which for a long time was estimated as controlled by a negative system, but now it is evident that the histidine operon is regulated similarly to the tryptophan operon. In this case the modified histidine-tRNA complex plays the role of corepressor (modification concerns introduction of two pseudouridine nucleotides into the anticodon loop). The nonmodified histidine-tRNA complex, although active in protein synthesis, is inactive as corepressor.

D. TRANSLATIONAL CONTROL OF TRANSCRIPTION

The best known are tryptophan, histidine, and phenylalanine operons in *E. coli* with about 160, 210, and 170 nucleotides, respectively. The common characteristic feature shared by them is the presence in the leader sequences of multiple codons for the synthesized amino acid: 2/14 codons of the tryptophan operon, 7/14 codons of the histidine operon and 7/14 codons of the phenylalanine operon. The presence of these codons (triplets) makes it possible to form secondary structures, and suggests that gene expression may be regulated through attenuation. The rate of translation depends on the concentration of adequate amino-acyl-tRNA complexes, thus also on the concentration of adequate amino acids. Complementary sequences of the transcripts (i.e., triplets coding the same amino acids) may form secondary structures, and thus influence termination of transcription. In this way translation of the leader sequences may determine transcription.

E. SELF-GENERATED CONTROL OF GENE EXPRESSION

The mechanism of control of the regulatory genes represents an important problem. If every gene were to have its own regulatory protein, an enormous number of genes should be necessary to control expression of these genes, and if this were so, another enormous number of genes would be necessary to control them, and so on, to an indefinite number of genes. As a way to escape this impasse, a model of self-generated control of the regulatory genes was established.[3] In this model products of gene expression may play the role of inducers or repressors. There also may exist negative or positive mechanisms of gene expression. Besides, some of the regulatory genes may be expressed constitutively.

Undoubtedly, apart from the described regulatory systems which seem to work principally at the local level, there also must be superior systems which work independently from the local systems in the matter of the whole living organism. To such systems belongs the already described catabolic repression which works independently of the other systems, and can be observed as a superior system for all catabolic reactions. This system works on the basis of cAMP, synthesized inside the cells from ATP catalyzed by adenyl cyclase, and turned into AMP by cAMP phosphodiesterase. The addition of glucose to the medium causes immediate diminution of cAMP concentration in the cells. For a long time it was not known if glucose inhibited synthesis or degradation of cAMP. At present it is known that the activity of adenyl cyclase and transport of glucose into the cell is subordinate to the same regulatory system. cAMP forms complexes with CRP (catabolite repression protein) which are necessary to initiate transcription by facilitating binding of RNA polymerase with the promoter region. The cAMP-CRP complexes do not influence the transcription process, but suppress only the catabolic reactions which yield carbon sources for the metabolic process.

Besides the described system of catabolic repression of carbon, an analogous system for catabolic processes yielding nitrogen compounds (e.g., ammonium ions) has been detected. The repression of nitrogen catabolic processes is mediated by glutamine synthetase. In the case of nitrogen shortage, the amount of non-adenylated glutamine synthetase increases, which plays the role of gene activator.

II. GENE REGULATION IN EUKARYOTES

Eukaryotes represent a very large family of organisms ranging from relatively simple unicellular organisms (e.g., protozoa) to very complicated organisms including mammals and human beings. The regulatory mechanisms of gene expression in prokaryotes are relatively well-known, although study of the structure of their chromatin is far from complete. In unicellular eukaryotes chromatin for the first time occurs in a specialized cellular organelle—the nucleus. In multicellular organisms a new event occurs, i.e., differentiation of cells for realization of their specific functions (liver, kidneys, brain, and cells of other organs). Because each cell of a given multicellular organism has the same genetic information, the performance of their specific functions can be realized only through gene regulatory mechanisms which from the enormous amount of genes in every cell must turn on genes responsible for the given function, or turn off genes whose products are useless or even depressant to this function.

Chromosomal DNA from either unicellular or multicellular eukaryote organisms is supercoiled to about the same extent. Because of the complexity of the chromatin structure, different regulatory mechanisms of gene expression must occur dependent on the kind of the performed function. It is generally accepted that metabolic pathways, the crucial necessities of life, are regulated either by positive or negative feedback regulatory mechanisms. Per analogy, genes responsible for the synthesis of particular enzymes also are produced by positive or negative regulatory mechanisms.

Consider an example of a positive gene's regulatory mechanism—inducible enzymes, synthesized by inducible genes. These enzymes are not synthesized in absence of their substrates which act as inducers. The presence of inducers in cells increases synthesis of these genes' products (proteins, enzymes), dependent to a large measure, on cellular environment. Most inducible genes are involved in degradation of substances derived from the environment and utilized by the organism either as a source of energy or as a supply of molecular fragments for the synthesis of complex substances. Usually inducible genes are active in catabolic processes of living organisms.

Now consider an example of a negative gene's regulatory mechanism—repressible enzymes, synthesized by repressible genes. These enzymes are normally present in cells, and their synthesis ceases when their intracellular concentration reaches a certain level. Usually these genes synthesize enzymes active in metabolic pathways synthesizing end products (e.g., tryptophan from its precursors). Usually repressible genes are active in anabolic processes of living organisms.

Besides positive and negative gene regulatory mechanisms there also exists another type of gene regulation, i.e., constitutive. In this case the gene's expression is permanent and independent of need, in contrast to inducible or repressible genes. Consequently, products of these genes are synthesized in fixed amounts. These genes appear to be responsible for the basic life processes in cells independent of their specific functions.

The described types of gene control do not explain the mechanisms of gene expression which must be studied on the genes themselves. The most important question is: which of the known steps of gene expression (transcription, pre-mRNA processing, translation, post-translational modification of synthesized protein molecules) can serve as control mechanisms? The answer to this question seems to be conclusively explained in that transcriptional control dominates over other possible mechanisms, including the processing of RNA. For elucidation of this question Darnell[4] and coworkers extracted all the mRNA from mouse liver cells, and using reverse transcriptase copied it into DNA. The DNA was cloned in *E. coli* to obtain sufficient amount of identical liver mouse genes. Exposing this DNA to total mRNAs from various organs they could identify genes coding for specific liver proteins which hybridized with mRNA from liver cells only, and genes coding for common proteins which hybridized with mRNA from cells of other organs. This experiment demonstrated that liver specific mRNAs were synthesized in liver cells only and not in cells of other organs.

To exclude the possibility of inappropriate processing of properly transcribed pre-mRNAs the authors collected nuclei from brain, kidney, and liver cells and allowed them to synthesize RNA from radioactively labeled precursors for a brief period (10 min). Under these conditions all the newly synthesized and nonprocessed transcripts were labeled. When the transcripts were exposed to the various cloned DNAs, the RNA from brain and kidney hybridized with DNAs encoding common proteins only and not with liver specific ones. The result confirms that genes for liver specific proteins are transcribed in the liver only and not in other organs.

There are some hints also that processing of pre-mRNA may play a role in gene regulation. The best example is differential processing which yields two different mRNAs from the same primary transcript. Until the discovery of introns (1977) it was generally accepted that a gene represents a stretch of DNA encoding a single protein. The discovery of introns raises the question: in which way can processing of pre-mRNA serve as a mechanism for regulation of gene expression? Crucial may be the presence of more than one (poly-A) signals on the same transcript which could cut the primary transcript into more than one mRNA with different 3′ ends. In consequence different protein molecules would then be synthesized with the same amino acid sequences at the beginning, and different terminal amino acid sequences.

Recently, several viral and cellular genes have been described which encode more than one protein. Such genes are known under the name "complex transcription units." A typical example is the gene encoding calcitonin in the thyroid gland and calcitonin-gene-related protein synthesized in the pituitary gland. As seen in Figure 3, the DNA stretch is the same in the thyroid and pituitary gland. There are four introns with two poly-A sites one at the 3′ end of the fourth exon, and the other on the 3′ end of the sixth exon. In the thyroid gland transcription is terminated at the 3′ end of the fourth exon, yielding a pre-mRNA which is processed into mRNA encoding the hormone calcitonin. In the hypophysis transcription is terminated at the 3′ end of the sixth exon yielding a pre-mRNA which is processed into mRNA encoding the hormone calcitonin-gene-related protein (the physiological function of this hormone is not exactly known.

As can be seen from the general scheme of an eukaryotic gene (Chapter 3, Figure 11) the regulatory signals for transcription are present on the DNA stretch itself. These are the TATA box, the upstream element and the distant upstream elements (also known as enhancers). In this aspect the DNA molecule could function as a self-governing unit, but the question arises as to which way those units are controlled? If those sequences were the sole control elements, all genes equipped with them should be expressed, which is not what occurs. The differentiated cells in particular organs synthesize many kinds of different

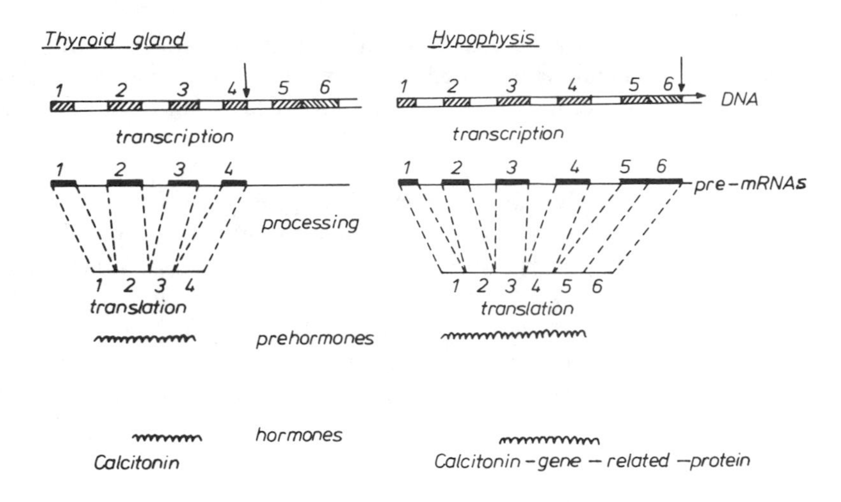

FIGURE 3. Different processing yielding two different mRNAs, encoding calcitonin in the thyroid gland and calcitonin-gene-related-protein in the hypophysis. Description in text.

proteins, specific for those organs, not synthesized in other organs despite the presence of the same nucleotide sequences. Therefore, in cells in which the same DNA is transcribed, specific regulatory mechanisms must exist which enable transcription, while in cells in which the same DNA sequences are not transcribed, those mechanisms must be absent or specific repressory mechanisms must exist.

Based on hybridization experiments a differentiated cell contains a minimum of 10,000 tissue-specific genes. In cells of particular organs only a small number of genes is expressed, while most of them are inactive. The inactivity of those genes is probably due to a potent repression system, which may be specific or general. There are several well-documented examples of specific trans-acting repressors, e.g., a fibroblast chromosome that represses specific liver functions, or adenovirus E1a gene product which represses the SV40 and polyoma enhancers.[5]

A. THE ROLE OF CHROMATIN IN GENE REGULATION

Chromatin is composed of DNA, histones, nonhistone chromatin proteins, and a small amount of RNA. It is generally accepted that the presence of histones represses gene activation. This effect is caused by the folding of chromosomes into a higher order of conformations, which makes DNA inaccessible to the transcription factors. Cytogenetic studies of heterochromatin, where genes are inactive, and studies on X-chromosome inactivation, where condensed chromatin causes its genetic inactivity, support these views. In this aspect DNA-ases I or II preferentially cleave active DNA sequences, which enables subsequent removal of nonactive sequences by Mg^{2+} precipitation. The result of DNA-ase digestion of chromatin must be interpreted so that the active chromatin has a conformation which is accessible either to transcription factors or nucleases.

These experiments paid attention to the structure of chromatin as the most probable point of different expression of specific genes. Earlier studies about removal of particular chromatin components (histones, nonhistones or DNA), and reconstruction experiments showed that removal of histones makes DNA accessible to nonspecific agents and an increase of transcription, while addition of histones causes diminution of transcription. Naked DNA (deprived of histones and nonhistones) showed an increase of nonspecific transcription. The most important for specific transcription were the nonhistone chromatin proteins. Addition of these proteins was conclusive for the expression of specific genes, irrespective of the origin of the examined genes. So, addition of nonhistone proteins extracted from liver cells and reconstituted with brain chromatin caused transcription of genes characteristic for liver cells and not for brain cells, despite the presence of DNA and histones extracted from the brain cells.

DNA in chromatin is packaged in histone-containing nucleosomes. The higher order structure of nucleosomes probably plays the main role in gene repression. Among histones principally responsible for the repressed chromatin structure is the noncore histone H1, which in contrast to other histone proteins is highly modified and polymorphic. Structural analysis of chromatin has shown that histone H1 forms crosslinks which bind together inactive nucleosomes in the inactive state of chromatin. In the active chromatin these bonds are loose which enables access of transcription factors (Weintraub).[5] These views have been supported by experiments with inactive oocyte genes S5, which after extraction of histones H1 become active.

A further argument for the role of chromatin structure in regulation of gene expression is experimentation with specific genes introduced into other cells (in cell or tissue culture). In this case genes present in the transduced chromatin became active under specific inducers, while the inactive genes of the host cells remained inactive.

The repression of inactive genes in eukaryotes is very efficient. Comparative studies of globin genes transduced into fibroblasts demonstrated that they synthesize about 10 to 100 mRNA molecules encoding globin per template—this is about three orders lower than synthesis of globin mRNAs in erythrocytes. These experiments prove that for the diminution of globin mRNA synthesis in fibroblasts is not lack of the transcription signals responsible, but repression of transcription and probably lack of specific positive control mechanisms for those genes present in erythrocytes and absent in fibroblasts. The replication system propagates during replication. This is evidenced by the fact that transfected genes remain active through many generations, while endogenous genes remain inactive.

Weintraub[5] suggests that repression is the basic state of chromatin, and activation is performed by specific activation elements which antagonize the generalized repression system. It has been described that the known transcription factor (5S-specific transcription factor—TFIIIA, encoding 5S RNA) promotes the active chromatin assembly by preventing conversion of this assembly into inactive chromatin.

Further experiments revealed that the repression state may be removed by several factors. The most important during cell differentiation seems to be the transient presence of histone $H1^0$—its replacement of histone H1 may cause transient loosening of the repression state of chromatin. Loosening of repression in differentiated cells may be caused by the high mobility group proteins 14 and 17 (HMG 14 and HMG 17) by histone ubiquitination or acetylation changing the character of histone H1 or its exchange with other proteins (for review see Weintraub[5]).

Investigations upon X-chromosome inactivation demonstrate that the heterochromatin structure of inactive X-chromosomes in mammals is heritable. There are also data that methylation is responsible for its repression which is documented by experiments that inhibition of DNA methylation by 5-azacytidine causes activation of some genes located on the X-chromosome which primarily were inactivated (e.g., glucose-6-phosphate dehydrogenase). A similar observation has been recorded in the case of beta-thalassemia treatment with the same drug. In beta-thalassemia the synthesis of beta-globin chain is diminished because of deletion of beta-globin gene and the synthesis of gamma-globin chains is absent because of repression of these genes. Treatment with 5-azacytidine causes demethylation of the gamma-genes with subsequent transient synthesis of gamma-globin chains resulting in the synthesis of fetal hemoglobin ($\alpha2\ \gamma2$). Recent data about the role of DNA methylation in DNA suppression suggest this to be rather secondary character because DNA extracted from inactive X-chromosomes from extraembryonic tissues does not demonstrate covalent modifications causing inactivation of genes.

The stable propagation of active and inactive X-chromosomes in mammalians and absence of methylation in *Drosophila* X-chromosomes suggest that there two separate mechanisms may exist: one responsible for gene expression, and the other for replication of that mechanism. Experiments performed to examine if the induced hypersensitive sites remain on chromatin in the absence of their inducers demonstrated that some hypersensitive sites were present in progenitor cells despite cessation of their transcription (e.g., beta-globin genes in fibroblasts transformed by insertion of Rous sarcoma virus secreted beta-globin chains. After removal of the viral gene responsible for this transformation the fibroblasts ceased to secrete beta-globin chains, but the hypersensitive sites of chromatin remained). These results demonstrate that hypersensitive sites for nucleases remain after removal of inducers, and their presence is needed although not indispensable for transcription. There are also other examples that induction of hypersensitive sites causes stable hypersensitive sites in spite of removal of the adequate inducer. Typical examples are the hypersensitive sites induced by some steroid hormones (estrogen) which also remain after removal of those hormones. These observations suggest that some inducers are able to induce stable chromatin modifications which also remain in the absence of these inducers.

There are additional examples that when specific inducers necessary for induction of gene expression are removed, the expression of those genes remains. For example, consider immunoglobulin genes. For high-level production of immunoglobulin genes in B-cells, enhancer sequences are necessary, but after DNA rearrangements the enhancer sequences are removed and the high level of immunoglobulin gene expression remains. This result suggests that the enhancer sequences are necessary for modification of DNA, but not for its maintenance.

B. THE ROLE OF DNA STRUCTURE IN GENE EXPRESSION

DNA supercoiling also seems to be relevant for gene expression. Coiling of the DNA is fundamental for replication, recombination, and transcription in bacteria.[6] Supercoiling in bacterial DNA depends upon two enzymes of antagonistic function: DNA gyrase which supercoils the DNA, and DNA topoisomerase I which removes supercoils of the DNA. Supercoiling of the DNA depends upon the dynamic balance between the action of those two enzymes. It is also interesting that expression of the DNA gyrase depends upon DNA supercoiling. Several experiments demonstrated that the template activity of DNA is dependent upon its conformation. Relaxation of the DNA configuration leads to a tenfold increase of gyrase synthesis.

In contrast to the bacterial topoisomerases the eukaryotic topoisomerases have not been so intensively investigated. With increasing knowledge that the expression of repression of genes must depend upon different configuration of the DNA, interest arose concerning enzymes which are able to induce changes of DNA configurations. Similarly, as in bacteria, the eukaryotic topoisomerases which cause different DNA configurations without its nucleotide sequences belong to those enzymes. In other words these enzymes may be responsible for creating "open" DNA domains that seem to be necessary for gene expression in eukaryotes.[7]

The best known topoisomerase is the bacterial gyrase which alters the torsional twist in a reverse sense to that of the DNA helix. The process defined as negative supercoiling is possible due to transient breaks of both strands of the DNA. This topoisomerase is termed TOP II (TOP 2). Several experiments demonstrated that yeast TOP 2 is essential during mitosis, but not in other cell cycles of yeast. Mutants of TOP 2 were unable to terminate DNA replication. In contrast to mutations of TOP 2, mutation of TOP 1 did not cause any phenotypic effect which suggested that TOP 1 is essential at all stages in the cell cycle. TOP 1 makes only cuts on one of the two strands of DNA.

Experiments performed with inhibitors of bacterial gyrase and topoisomerase II in cultured *Drosophila* cells exposed to heat shock confirmed views about the role of DNA structure in gene regulation. Heat shock is a generalized reaction to elevated temperature from which ensues a series of specific changes in gene expression. Several so-called shock genes become induced with alterations to their chromatin structure. This reaction can be blocked by previous addition of novobiocin, an inhibitor of bacterial gyrase and *Drosophila* TOP 2. Novobiocin added after induction of heat shock genes inhibited their transcription, but not the establishment of "active" chromatin structure. The experiment suggests that *Drosophila* TOP 2 is necessary both for transcription and creation of "active" chromatin configuration.

Search for sites of TOP 2 action on DNA revealed that they correspond with both flanking sequences of heat shock genes, hypersensitive to nuclease digestion. The same was reported with SV40 infected monkey cells.

The question arises as to how the supercoiled DNA form is necessary for its activation. There are some possibilities: recently it was found that DNA-ase I hypersensitive sites, usually 200 to 400 base pairs long, are sensitive to nucleases (especially S1 nuclease and to chemicals characterizing the non-B structure of chromatin). In this respect, supercoiling of DNA caused by topoisomerases changing the DNA configuration enables transcription factors to form an "active" chromatin structure. It is suggested that transcription factors may have higher affinity to DNA regions when supercoiling does not have the normal B-form. Another possibility is that repression factors (e.g., histone H1) have lower affinity to the supercoiled DNA. Supercoiling may be also necessary to transduce signals from distant parts of the DNA which through supercoiling are placed near the mRNA starting site.

The role of topoisomerases in making up the chromosome scaffold was documented by experiments in which after the nuclease digested mitotic chromosomes, and after the extraction of all soluble proteins, the remaining isomerases were localized at the bases of the radical loop domains of mitotic chromosomes.[7] If these loops represent the active domains of DNA, then this location of topoisomerases supports the suggestion of a regulatory role of topoisomerases in eukaryotic gene regulation.

The importance of DNA configuration to regulation of gene expression attracts the most attention concerning the Z-DNA form, discovered some years ago by Rich (see Figure 4). The left-handed Z-DNA form is totally different from the usual right-handed B-DNA form of Watson and Crick. In the B-DNA helix there are regularly about 10 base pairs per turn. The B-DNA molecule reveals two grooves, a major and a minor one. There are also other right-handed DNA forms, e.g., the A-form in which the bases are shifted away from the axis of the helix and are inclined with respect to that axis.[8]

In contrast to the naturally occurring right-handed A, B, C DNA forms, the left-handed Z-DNA form was obtained artificially. The Z-DNA form is an unstable molecule with only one groove and a zigzag backbone with 12 base pairs for a helical turn. The first question was in which way the Z-DNA form can be stabilized. The instability of the Z-DNA is caused by large sequences containing alternatively purines and pyrimidines. Consequently, water molecules are disordered in the deep groove around stretches of adenine-thymine base pairs.[9,10] Therefore, the Z-structure can easily change into the B-structure which requires less energy. Methylation of the fifth position of cytosine stabilizes the Z-structure. Binding of B-DNA with certain proteins may cause change of the B-structure to the Z-structure. Negative supercoiling catalyzed by topoisomerases also stabilizes the Z-structure of DNA. The ease with which the DNA structure is changed and its stability being dependent upon factors active in gene regulation, gave hints that this may be one of the bases of gene regulatory processes, but at present direct evidence is lacking.

An important role in gene expression seems to be exerted by transposable elements which cause rearrangements and genome instability in eukaryotes. Transposable elements in eukaryotes are represented by the Ty1 element in yeast, copia elements in *Drosophila*, Alu family in humans, Alu-like family in rodents and some other mammals, and immu-

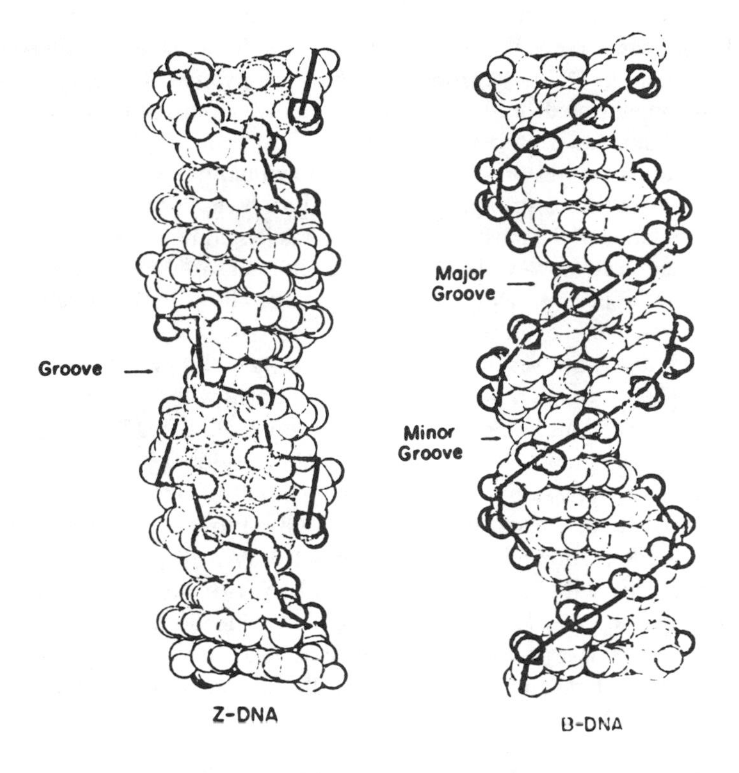

FIGURE 4. Two structural forms of DNA, on the left: Z-DNA, on the right: B-DNA. Description in text.

noglobulin genes in mammals.[11] Their role in gene expression is not well-defined, but there is a possibility that movable DNA elements can introduce transcription signals into regions of silent genes which in this way may be activated. Investigations upon DNA inversions in prokaryotes (instability of flagellar antigen in *Salmonella typhimurium,* variation in the host range of *E. coli*-bacteriophage Mu, invertible element in an ultraviolet-excisable segment of the *E. coli* chromosome) demonstrate that movable elements of DNA can switch "on" and "off" adequate genes.[12]

It is also noteworthy that movable elements can control expression of their own protein at the translational level.[13]

A very important mechanism of gene activation is represented by viruses, mainly retroviruses, which insert their DNA into host's genomes. Transcription signals for retroviruses are present in their long terminal repeat (LTR) sequences, flanking the inserted proviruses. In the case of retrovirus infection the LTR of those viruses may be inserted into the host's genome in such a manner that it causes permanent activation of one or more genes. In the case that the abnormally activated gene causes activation of other genes (e.g., genes of growth factors, like epidermal growth factor, nerve growth factor, platelet derived growth factor, etc.) abnormal proliferative processes may occur, typical for neoplasms.

C. THE ROLE OF DNA METHYLATION IN GENE EXPRESSION

DNA methylation is well known from the discovery of the DNA structure and its role in genetic information. The biological role of DNA methylation is highly controversial. DNA methylation is a post-replicative process performed by cell methylases. The distribution of methyl groups along the DNA is specific to the cell type. In some organisms all methylatable sites are methylated *(E. coli),* in other (eukaryotes) some methylatable sites remain unmethylated.[14] Eukaryotic DNA contains cytosine methylated at the fifth position as the sole modified base, exclusively at CpG sequences. Methylation patterns of eukaryotic DNA are heritable features. The DNA in sperm cells is fully methylated. Tissue DNA of expressed genes is often hypomethylated. It is suggested that methyl groups in those tissues have been lost during the differentiation process. Methylated and hypomethylated sites along a gene sequence constitute the methylation pattern of an individual gene in a given tissue.

Because of the specific distribution of methyl groups at specific gene sequences and extreme hypomethylation of the most active genes, for a long time there has been the question of whether methylation is the key element in the regulation of eukaryotic gene expression. One attractive hypothesis suggests that methylation stabilizes the inactive state of a gene and hypomethylation stabilizes its active state.[15]

Actual transcription, however, depends on transacting factors which interact with tissue-specific proteins. There are experiments that cells transfected with both a methylated gamma-globin gene (methylated at all its CpG sequences) and unmethylated beta-globin gene, were able to express the beta-globin gene but not the gamma-globin gene. Thus, methylation appears to inhibit gene expression.[16] In contrast to this, it also has been shown that complete DNA methylation does not prevent polyoma and SV40 virus early gene expression.[17]

There are also observations that a gene inactivated by methylation with a mammalian methylase can be reactivated by 5 azacytidine which inhibits methylation at the fifth position of cytidine. Changes in gene activation are often heritable for many generations (in the absence of the further 5-azacytidine treatment). Simultaneously, the activation frequencies may be 5 to 6 orders of magnitude greater than those observed for mutagenic agents.[18] Finally, 5-azacytidine also may activate silent genes in a previously noninducible cell type.[19]

Dosage compensation in X-chromosome-linked genes in females seems to be dependent on their methylation. Studies of DNA from males and females with active or inactive chromosomes revealed that remarkably dense clusters of CpG dinucleotides in the 3′ coding sequences in active glucose-6-phosphate dehydrogenase genes are highly hypomethylated, and highly methylated in inactive ones.[20] Hypoxanthine phosphoribosyltransferase (HPRT) genes on inactive X-chromosomes can be reactivated by the use of 5-azacytidine. In this case a diminution of methylation at the gene locus was observed.[21]

Immunoglobulin genes in their germline are transcriptionally silent, becoming transcriptionally active after their rearrangement. The constant genes for the kappa light chain immunoglobulin (C_k genes) are transcriptionally active even in the unrearranged (or germline) configuration, but the V_k (variant) genes are transcriptionally inactive and become transcriptionally active only after joining with the C_k genes. After fusion the V_k genes become hypomethylated and DNA-ase I sensitive extending several kilobases upstream of the transcription site of initiation, but not extending to the adjacent genes.[22]

As seen from this short review, the results concerning the role of DNA methylation in gene regulation are controversial. It seems that this is caused by the multiplicity of factors engaged in gene regulation. Among them are chromatin structure, sensitivity to nucleases, differential binding of chromosomal proteins, hypomethylation, and probably a number of other factors, all of which are not yet known. It is probable that gene activation cannot take place unless several of these factors cooperate. Therefore, gene activation can be performed by hypomethylation in those cases in which all the other factors necessary for gene activation are present and only methylation inhibited transcription. In other cases methylation does not inhibit transcription when the other factors of gene regulation are more important and may cause transcription in spite of methylation of the adequate genes.

D. THE ROLE OF CHROMOSOMAL PROTEINS IN GENE REGULATION

The first irrefutable arguments that different functions of specialized cells in higher organisms depend upon gene regulatory mechanisms was confirmed by the experiments of J. B. Gurdon who transplanted nuclei from differentiated intestinal frog cells into enucleated frog eggs and received a percent of normal frogs. These experiments demonstrated that nuclei of intestinal cells contain all the frog's genetic information to form normal frog organisms and that in the intestinal cells only a part of the genetic information necessary to intestinal function is expressed. What is more, these experiments revealed that expression of differentiated cells can be modified when resting cells are activated, e.g., stimulated by specific hormones. In this case a complex change of cell metabolism occurred by switching on an adequate cell's genetic program. It is generally accepted that less than 10% of the total genetic information is expressed at any one time. Specific regulatory mechanisms activate and inactivate appropriate genes for transcription according to the needs of specific cells.

Usually, chromosomal proteins are divided into histones and nonhistone proteins.

1. The Histones

Histones are basic proteins with a positive charge which enables them to bind nucleic acids. Histones had been discovered at the end of the nineteenth century by Kossel. The first indications about the regulatory function of histones derived from the experiences of E. Stedmans (1943) who found that growing tissues contained fewer histones than non-growing ones.

The first appropriate experiments upon the role of histones in gene regulatory processes were performed by Bonner et al. (1963) who found that the addition of histones to the synthesizing proteins systems *in vitro* caused dimunition of the template activity of DNA, and removal of histones caused an increase of protein synthesis. Next, Allfrey and Mirsky demonstrated that depletion of histones of the isolated cell's nuclei caused adequate RNA synthesis *in vitro*.

At present a threefold function of histones is accepted: (1) protection of the genetic material, (2) participation in chromatin structure, (3) regulation of gene expression.

Histones belong to a relatively homogeneous group of proteins which has changed insignificantly during evolution. Therefore, histones from various sources (e.g., histones from calf thymus, *Neurospora crassa*, or *Chlorella*) are very similar. This does not mean that these proteins did not undergo mutations. On the contrary, they must have undergone numerous mutations as did all other proteins. Perhaps the only organisms with mutated histones mostly died out because of the importance of the normal function of histones to the vital processes of those organisms. Among other features of histones is the lack of tryptophan, probably caused by the absence of this amino acid at the time when histones originated during the evolution of living organisms. Mostly, histones do not have poly(A) sequences at the 3′ end of their mRNAs which is because they are short-lived molecules. The biosynthesis of histones is coordinated with the synthesis of DNA at the cell cycle (phase S). This is guaranteed by the short-lived mRNA molecules which disintegrate simultaneously with cessation of DNA synthesis.

Histones are divided into five main classes: H1—rich in lysine, alanine and proline; H2A—rich in leucine, alanine and lysine; H2B—rich in lysine, alanine and serine; H3—rich in alanine, arginine and glutamic acid; and H4—rich in glycine, arginine and lysine. One end of histones H1, H3, and H4 is basic and the other is acid or neutral, while both ends of H2A and H2B are basic. The middle of the molecules is acidic or neutral. The basic ends of the molecules exhibit mostly nonhelical structure, while the remaining parts possess a distinct helical structure. It is accepted that only the basic, nonhelical ends bind with DNA.

In the past there have been described also other histones, e.g., F1 and F2b appearing only in sexually mature testes. It is supposed that these histones may play a role in gene regulation in highly differentiated organs of higher organisms.

Eight histone molecules (in groups of two H2A, H2B, H3, and H4) make the core of smaller subunits of chromosomes, the so-called nucleosomes (see Chapter 3, Figure 4), to which the double DNA strand is wound (from 165 nucleotides in fungi to 210 in chicken). The bulk of the chromatin is condensed to 25 to 30 nm thick fibers which are supercoiled to a solenoid. The presence of histone H1 is probably necessary for the formation and stabilization of this structure. It is possible that the solenoid structure becomes unfolded when a gene becomes transcriptionally active or is committed to activity. Since histone H1 is essential for the fiber formation of chromatin, its presence in active chromatin appears to be greatly depleted. It is also possible that nonhistone proteins of the "high mobility group" may replace the H1 histones in actively transcribed chromatin.[23]

The highest level structure of chromatin appears to be the folding of the solenoids into loops and domains of chromatin, which are suggested to be supercoiled stretches of chromatin comprising from 35 to 100 kbp of DNA. These structures are anchored by specific nonhistone proteins located at the base of these loops which form the matrix, nuclear membrane, or scaffold in the case of metaphase chromosomes (therefore, the structure of metaphase chromosomes becomes destroyed only after extraction of nonhistone proteins). The less compacted interphase chromatin is probably organized into the same supercoiled domains as the loops in metaphase chromosomes.[23] Recent investigations support the views that supercoiled chromatin domains may be the essential elements of the eukaryotic chromatin structure as replication and transcription units.[23]

a. Nuclease Digestion of Chromatin

Chromatin domains of active genes are preferentially digested by various endonucleases. This argues that the active genes are packaged in an altered nucleosome form, making them more accessible to enzymes. Weintraub and Groudine[54] first demonstrated that the globin gene is sensitive to DNAase I in chick erythrocyte nuclei, but not in oviduct cells where it is inactive. In contrast, the ovalbumin gene is preferentially sensitive in the oviduct cells where it is active, but not in erythrocytes where it is inactive (for a review see Reference 23). Although the experiments appear to be very convincing it is not clear whether the result is caused solely by the altered conformation of the individual nucleosomes or by a higher order chromatin structure, and whether the nuclease-sensitive region is confined to the transcribed region of the active gene, and reversible after transition of the active gene to an inactive one. Further experiments revealed that the sensitive region extends far upstream and downstream including nontranscribed sequences surrounding the coding sequences. These results suggest that sequences far removed from the coding sequences themselves may be important to the control of the transcription of the particular genes. Besides, it was found that the active genes become sensitive immediately after their replication preceding their transcription already in the development.

Further experiments revealed that the region of the gene itself was highly sensitive while the adjacent 20 to 50 kilobase-pairs were moderately sensitive. This suggests a higher-order chromatin structure in the region of the transcribed genes.

Another type of chromatin hypersensitivity to nucleases was discovered in the heat shock genes. These genes become transcriptionally active as a result of heat or other stresses. In the vicinity of the heat shock genes (usually near the 5′ end) sites hypersensitive to DNAase I have been discovered. These sites were one order of magnitude more sensitive than usually active genes and two orders of magnitude more sensitive than inactive chromatin.

Hypersensitive areas of chromatin to nucleases usually include hypomethylation of the DNA, hyperacetylation of histones and the presence of nucleosomes combined with the high-mobility group of nonhistone proteins.

2. Enhancer Elements

Enhancer or activator sequences are promoter elements associated with genes transcribed by RNA polymerase II. Most of them are located 5′ upstream of the site of transcription initiation -20 to -110 bp, usually two groups of nucleotides, the TATA (or Goldberg-Hogness) box, and the upstream CCAAT sequence. Enhancers can function upstream, downstream, or within genes. Sometimes they act over considerable distances—probably more than 10 kbp. These sequences are probably free of nucleosomes. Enhancers are usually present as short, tandemly repeated, *cis*-acting regulatory sequences that increase transcriptional efficiency, independently of the position of, and orientation to, the activated gene. Enhancers can influence more than one promoter.

The mechanism of enhancer action is unclear. Reeder suggests that the enhancers may act by attracting a transcription factor or factors whose function is to bind them to the gene's promoter and activate it for transcription.[24] This hypothesis is based on observation that enhancers cloned on a plasmid with no promoter adjacent will still effectively compete against a promoter on a separate plasmid, and a plasmid with many enhancers is more effective than a plasmid with few enhancers.[24] Once the proposed factor is bound with enhancers it causes a stable activation of the promoter. Studies *in vitro* demonstrate that promoter recognition of a tight binding complex by transcription factors is distinct from the RNA polymerase itself.[24] The enhancer sequences can function also when removed from their original location to distant promoter elements, even when inserted in an inverted orientation.

Enhancers from several viruses, especially papova- and retroviruses cause an increase of transcription of different eukaryotic genes: conalbumin, chicken lysozyme, mouse metalothionein, and p21 transforming protein of the myc oncogene. Enhancer like elements partially homologous to those found in viruses have been identified in various human cells, suggesting that these sequences may represent the LTRs of the endogenous retroviruses.[23] Just this finding prompted the view that the pathological function of enhancers may activate pathological cellular proliferation and cancer.

Enhancer sequences 5′ upstream of the insulin and chymotrypsin genes reveal highly conserved consensus sequences for viral enhancers. This observation suggests that cell-type-specific enhancer elements may participate in regulation of specific genes during cellular differentiation. Gene specific enhancer elements are activated also by binding steroid hormone-receptor complexes, especially the glucocorticoid receptor. Steroid hormones modulate gene expression by interaction with intracellular protein receptors. The subsequent allosteric alteration of the hormone-receptor complexes is the cause of their association with specific genomic loci.[25] *In vivo*, glucocorticoid-receptor complexes bind to multiple DNA regions of the mouse mammary tumor virus including sequences within the long terminal repeat (LTR) upstream of the transcription starting site.[25] There is a strong suggestion that specific glucocorticoid receptor-DNA interactions alter the conformation of the DNA (or chromatin) in the neighborhood of the binding sites creating in this way an active transcriptional enhancer.[25]

Of particular interest is the mechanism of immunoglobulin gene regulation which presents the best evidence that the cellular genomes may contain internal enhancer sequences. The regulatory elements, like the immunoglobulin genes themselves, are widely separated in the germ-line genome and joined by somatic rearrangement in B-lymphoid cells. There is also evidence that the function of these elements may depend on the interaction with a transacting factor which is active in a different way during lymphocyte maturation.[26] The main DNA segments that form an immunoglobulin gene are the constant (C) and the variable (V) segments, separated in the germ-line DNA. The V segments are transcriptionally silent, and become active only through their rearrangement. The constant segments (constant and joining portions of these genes) are transcriptionally active in the germ-line as well as in the rearranged form. The transcriptional ability of the constant fragments is dependent on the presence of enhancer elements between the joining and constant segments. These enhancer elements are preferentially active in lymphoid cells and extend about 200 nucleotides.

Recently, besides the TATA box, a highly conserved octanucleotide (ATTTGCAT) was described (located about 60 to 80 bp upstream of the initiation site).[26] Gene transfer experiments demonstrated that this octanucleotide is necessary for effective transcription of the immunoglobulin genes. This observation demonstrates that the enhancer element seems to require a tissue-specific *trans*-acting factor induced by external stimuli during the early maturation phases and which become constitutive during the later phase of the immune reaction (B-cell and plasma-cell). The promoter also seems to be induced by *trans*-acting factors.

a. Tissue-Specific Expression of Transferred Genes

In a broad formulation, probably all living organisms may be transgenic in the sense that they have stably integrated foreign DNA into their genomes, especially DNA of viral origin. Usually, this definition is limited to those animals that have integrated foreign DNA into their germ-line as a consequence of experimental introduction of foreign DNA.[27]

The aim of studies involving generation of transgenic mice was to investigate developmental processes. The principal question raised by introducing foreign genes, which are usually expressed in a tissue-specific manner, was whether *cis*-acting DNA involved in developmental programming of gene expression would be present in transgenic animals. A series of experiments performed by various authors confirmed that *trans*-specific gene expression occurs in transgenic mice. The most convincing result was expression of the rat elastase gene in the acinar cells of the pancreas, in which the level of expression was 10,000-fold greater than in other tissues. Human beta-globin and kappa light chain genes were expressed in erythroid cells, and a rearranged immunoglobulin gene was equally expressed in B- and T-lymphocytes (for a review see Reference 27).

The greatest progress in precise localization of the cis-acting elements for tissue-specific gene expression has been made on the rat elastase gene to a 213 bp region contiguous with the promoter, and for the human beta-globin gene where the tissue-specific elements seem to lie within the gene or at the 3′ end of the structural gene (for review see Reference 27).

Gene expression in transgenic mice also raise other questions concerning this problem. Introduction of metalothionein-growth hormone fusion genes from rat into mice may cause different results: (1) the offspring is of normal size (the gene did not integrate into the mouse genome); (2) the gene integrated, but did not express—the mouse is of normal size; (3) the gene integrated, but is not fully expressed—the mouse is only a little bigger; and (4) the gene integrated and is fully expressed—the mouse has grown to become a giant mouse. In the transgenic mice the growth hormone is synthesized mainly in the liver under the control of constitutively expressed metalothionein gene promoter. In this way, the normal feedback control mechanisms from the pituitary gland are abolished (the somatotroph cells that normally synthesize growth hormone decline) and the growth hormone is synthesized constitutively (without feedback control of the pituitary gland). In addition, the sexual differentiation of the liver functions become abnormal and the fertility of females is impaired.[27]

Transgenic mice also were used to solve the question concerning regulation of the immune system. According to the ‘‘allelic exclusion’’ theory, the activated B-lymphocytes always synthesize only one functional immunoglobulin. Introduction of a rearranged kappa light chain gene in a transgenic mouse should solve the question if allelic exclusion is a regulated or a random process. The performed experiments showed that the production of functional immunoglobulin prevents further rearrangement, which suggests a regulated process in allelic exclusion.[27] In hybridomas, in which the foreign kappa gene is inactive or unassociated with a heavy chain, the endogenous kappa genes rearrange and form functional immuoglobulins.

b. Repressive Action of Enhancers

The E1A early antigen of human adenovirus, together with the viral gene E1B, cause transformation of rat cells to grow indefinitely *in vitro*. The E1A is a transcriptional regulator that can activate or repress transcription from various promoters. It seems that in undifferentiated embryonal carcinoma cells there is present a regulatory activity which is lost during differentiation. Repression by E1A involves enhancers, regulatory *cis*-acting upstream or downstream DNA elements of viral or cellular promoters which stimulate initiation of transcription independently of their position and orientation.[28] The repressive function of E1A protein on transcription has been shown on several early viral proteins, immunoglobulin genes, and human major heat shock genes.

Many of the enhancers that are repressed by E1A protein are unable to activate efficiently undifferentiated embryonal carcinoma cells. The cellular factor that represses these repressors is homologous to E1A. Adenovirus mutants with defective function of E1A replicate poorly in differentiated cells because products of the early genes necessary for their replication are poorly synthesized. In enbryonal carcinoma cells these mutants replicate normally which prove that these cells contain an E1A-like activity, enabling transcription of the early genes in the absence of E1A protein. The result suggests that the activating and repressing activities of E1A and embryonal carcinoma cells may be the same.[28]

E. ACTIVATION OF HEAT SHOCK GENES

When cells, both of prokaryotes and eukaryotes are exposed to heat (some degrees above their normal growth temperature) they exhibit synthesis of proteins, termed "heat shock proteins". This denotes that the synthesis of heat shock proteins is a generalized reaction of all living organisms. The synthesis of these proteins can be induced by a variety of agents other than heat (heavy metal ions, amino acid analogues incorporated into proteins, hydrogen peroxide, ethanol, arsenite, hypoxia, glucose starvation and refeeding, and others) causing damage of cellular proteins. The general function of these proteins seems to be protection of cells from lethal effects of heat and other toxic agents. The mechanism of action of the heat shock proteins seems to depend on facilitation of repair of reversible damage caused by heat or other stress factors.

The best known heat shock proteins are: heat shock protein 70 (hsp 70), produced in small amounts also in unstressed cells and bound to the cytoskeleton—in stressed cells its synthesis increases and it is bound with the nuclei with transient concentration in nucleoli. Its function seems to facilitate reassembly of damaged ribonuclear proteins (RNPs). Hsp 90 is a soluble cytoplasmic protein—its function seems to lie in the recognition of hydrophobic surfaces of denatured proteins and binding to them, preventing them from forming insoluble damaging precipitates.

Most important is the mechanism of induction of these proteins because of dramatic events at all levels of gene expression. The first step is activation of transcription initiated from a highly conserved promoter (a hyphenated dyad C-GAA-TTC-G—termed heat-shock-element) situated about 20 bp upstream of the TATA box in all eukaryotic cells.[29] The heat-shock-element is hypersensitive to DNAase I. During activation of heat shock genes, both the heat-shock-element and the TATA box sequences are nuclease resistant. This indicates that both are bound with proteins protecting them from digestion. In unstressed cells the TATA box is protected, but not the heat-shock-element. Furthermore, the heat-shock-element can be made nonsensitive by adding a protein-sensitive extract from heat-shocked nuclei. This denotes that during the activation of the heat shock genes a protein must be bound to the heat-shock-element.[29] Another protein is bound with the TATA box so that the two proteins together cover about 130 bp of the heat shock promoter. The heat shock binding protein can be extracted from normal and heat-shocked tissue. It is noteworthy that the extract from heat-shocked tissue is more active than that from normal cells.

In response to stress the heat shock genes usually are expressed coordinately. Sometimes they also are activated in normal conditions (a minimal amount of heat shock proteins is present in normal cells, especially the heat shock protein 83 (hsp 83) in *Drosophila*). Besides, genes of the small heat shock proteins in *Drosophila* can be activated by the juvenile hormone ecdysone. This suggests that the heat shock genes may have an additional promoter element or factor activating them in normal conditions. Therefore, it may be suggested that during stress the two promoters, one activated specifically by heat or stress and bound with the heat-shock-element of the DNA, and the other nonspecific promoter bound with the TATA box, become complexed. It is suggested that the complexed promoters synthesize a factor which promotes transcription of the heat shock genes—termed ''heat shock transcription factor''.[29] At that moment a dramatic change in maximal expression of the heat shock genes occur. It seems probable that the heat transcription factor may be a specifically modified sigma factor which belongs to the necessary transcription factors.[30]

Analysis of the agents activating the heat shock genes demonstrated that accumulation of denatured or damaged protein molecules inside the cell act as heat-shock-gene inducers. To these agents belong factors causing translation errors and, consequently, synthesis of abnormal proteins. Among these agents are ethanol, amino acid analogues, and puromycin. To a group of factors denaturing or damaging existing proteins belong the following: heat—causes unfolding of proteins; heavy metals—bind to sulfhydryl groups of proteins, causing conformational changes in them (e.g., copper-chelating compounds, arsenite, iodoacetamide, *p*-chloromercuribenzoate); hydrogen peroxide, superoxide ions, and other free radicals: cause fragmentation of protein molecules; ammonium chloride: causes inhibition of proteolysis; antimycin, azide, dinitrophenol, rotenone, ionophores, and others: inhibit oxidative phosphorylation with changes in the redox state and covalent modifications of proteins; hydroxylamine: cleaves asparagine-glycine bonds in proteins.[31]

Heat shock proteins decay *in vivo* in a relatively short time. They are not degraded by specific enzyme systems but show a slow proteolytic activity upon themselves. *In vitro*, heat shock proteins usually degrade spontaneously. This speaks for an ancient origin of these proteins which enables termination of the heat shock reaction. Heat shock genes also may be inactivated by methylation of adenine in the DNA. In adequate experiments it has been shown that after administration of 5-methylthioadenosine in *Drosophila* the heat-shock reaction was inhibited.

Activation of heat shock genes confirms the suppressive role of histones in gene regulation. With increasing transcription of the heat shock genes they first become depleted of H1 and then of the remaining histone proteins.[32] The absence of histones on the coding region of heat shock genes demonstrate a dramatic reconstruction of inactive chromatin into an active form bringing to light some characteristic features of transcribed chromatin. Because of the disappearance of histones H2A, H2B, H3, and H4 being decisive for formation of nucleosomes, the transcribed DNA stretch cannot represent a constricted, nucleosomal structure, becoming accessible to transcription factors and RNA polymerase which is decisive for initiation and performance of transcription.

Although our knowledge of the activation of heat shock genes clarify many obscure details of gene regulation, many unanswered questions remain. The exact mechanism of heat shock gene activation and heat shock proteins function are not completely known as yet. Moreover, heat shock has profound effects on mRNA stability and translation processes.[29] Nevertheless, activation of heat shock genes seems to be a good model for studying gene regulation in eukaryotes.

F. THE NONHISTONE CHROMATIN PROTEINS (NHP)

Nonhistone chromatin proteins constitute a large group of highly heterogenous, functional nuclear proteins: enzymes (DNA polymerase, RNA polymerases I, II, and II, topoisomerases, etc.), transcription factors, and proteins or protein complexes regulating gene expression. It has been demonstrated that an extremely small NHP fraction specifically recognizes DNA sequences of a given organ. For example, a very small fraction of NHP extracted from rat liver binds only with the DNA extracted from rat liver, and not with DNA extracted from other rat organs or DNA of other species. Reconstitution experiments of chromatin with various nonhistone chromatin protein fractions extracted from different organs always exhibited synthesis of RNA specific for the organ from which they have been extracted. For example, DNA and histones extracted from embryonic rat liver, and reconstituted with NHP from brain did not synthesize mRNA for globin chains, but DNA and histones extracted from brain and reconstituted with NHP from rat embryonic liver synthesized mRNAs for globin chains although the brain cells never synthesized globin mRNAs, and embryonic liver cells synthesize them.

In spite of these observations the nonhistone chromatin proteins per se cannot be the sole gene regulators because their synthesis needs similar regulatory mechanisms, etc. in an unfinished series of reactions. Therefore, it seems that they contain regulatory factors which originate from specific organismic functions, e.g., from the endocrine system in the form of specific hormonal receptors which specifically induce adequate genes. Nevertheless, modifications of nonhistone chromatin proteins seem to be necessary for performance of the complicated and manifold process of gene regulation.

G. MODIFICATIONS OF CHROMOSOMAL PROTEINS

There are many conflicting views upon the role of protein modifications in gene regulation. Preferential binding of specific proteins to the particular gene regions have been demonstrated. Particularly noteworthy is the "high mobility group" (HMG) of nonhistone chromatin proteins binding preferentially to transcriptionally active chromatin.

The HMG proteins demonstrate unusual solubility being extractable in 0.35 *M* NaCl and nonprecipitable in 2% trichloracetic acid. They can be postsynthetically modified by acetylation, methylation, phosphorylation, ADP-ribosylation, and glycosylation.[23] The best known HMG proteins are 1 and 2 (M_r 28,000), and 14 and 17 (M_r 10,000 to 12,000). *In vitro*, HMGs 1 and 2 seem to destabilize and unwind (maybe also by induction of superhelicity) the DNA of circular plasmids. HMGs 14 and 17 inhibit the histone deacetylase enzymes. In some experiments binding of HMGs with DNA did not result in loosening but rather in stabilizing of chromatin. HMG proteins bind preferentially with single-stranded DNA (which usually is the region of nuclease hypersensitivity). Potential Z-DNA structures are protected from digestion by HMG1 and HMG2. After the removal of HMG 14 and 17 proteins, chromatin loses its activity (examined by hypersensitivity to DNAase I). Chromatin reconstituted with these proteins regains its hypersensitivity to nucleases which denotes that these proteins are necessary for the "active" chromatin conformation. For this function glycosylation of HMG 14 and 17 proteins seems to be responsible.[23]

1. Ubiquitination and Poly(ADP)-Ribosylation of Chromatin Proteins in Gene Regulation

Ubiquitin forms in chromatin specific adducts with 5 to 15% of histones H2A (uH2A), and in minor amounts with H2B (uH2B). Usually only one H2A molecule per nucleosome is converted to uH2A. Recently, a striking amount of uH2A histones was detected in transcribed copia element of *Drosophila* and heat shock genes, and its absence in nontranscribed satellite DNA; an augmentation of uH2A was also observed in nucleosomes of the transcriptionally active dihydrofolate reductase during experiments with amplification of dihydrofolate reductase genes.[23] It is suggested that ubiquitation, which normally facilitates degradation of proteins, prevents the formation of higher order chromosomal structure.

Poly(ADP)-ribosylation was described in most eukaryotic cells. It was suggested that poly(ADP)-ribosylation of nucleosomal and H1 histones facilitates transcription by relaxing chromatin structure and preventing formation of higher order chromatin conformation.[23]

Acetylation of histones in eukaryotic gene control has been discussed for a long time. Some controversial views were put forward that acetylation of core histones may be an important but not the sole mechanism of gene control. Acetylation of core histones mainly involves the basic terminal fragments which bind with the acid DNA. In this way, hyperacetylation by weakening of the ionic binding forces would destabilize the chromatin solenoid, facilitating its unwinding. The cation-condensed hyperacetylated chromatin is 0 to 30% longer than unacetylated solenoids. It is generally accepted that histone hyperacetylation is an important, but not sufficient mechanism for decondensing highly compacted chromatin. Chromatin acetylation occurs probably not only during the process of chromatin activation, but also during other processes like replication of the DNA. The short chain fatty acid *n*-butyrate induces reversibly hyperacetylation of histones (H3 and H4) by inhibition of the deacetylase enzyme. Nuclease sensitivity of the hyperacetylated butyrate-treated chromatin resembles that of transcriptionally active chromatin. Besides, butyrate induces phosphorylation of the HMG proteins, and also stimulates nuclear poly(ADP-ribose) polymerase resulting in poly(ADP)-ribosylation of histones, both enriched in transcriptionally active chromatin. Butyrate seems to have an effect on the cytoskeleton with consequent loosening of the nuclear structure. Finally, butyrate although inducing some genes also has an inhibiting influence on the glucocorticoid induction of specific genes in hormone-responsive cells.[23]

H. THE ROLE OF THE NUCLEAR MATRIX ON GENE REGULATION

The supercoiled chromatin is probably anchored to a scaffolding structure of mitotic chromosomes, or to the nuclear matrix membrane mediated by specific nonhistone chromatin proteins and specific DNA sequences. It has been demonstrated that both replicating and transcribing DNA sequences are preferentially associated with the nuclear matrix. In the absence of transcription the gene sequences are no longer associated with the nuclear matrix.[23]

On the basis of these findings it is probable that two different levels of transcription control may be present in eukaryotic cells: first, the genetic factors involved immediately with the gene expression control (promoter, terminator, and enhancer sequences of the DNA), and second, the differentiated cellular organization established during development of these cells, which creates the potential possibility of those cells to transcribe. This is the position effect of transcribable DNA sequences. In *Drosophila* genetic rearrangements have been found that link a region of uncondensed transcriptionally active euchromatin to a highly condensed transcriptionally inactive heterochromatin.

I. THE ROLE OF HORMONES ON GENE REGULATION

In 1968 Karlson discovered that administration of a series of hormones caused synthesis of RNA in target cells for these hormones. The result indicated that the nonprotein hormones act through synthesis of specific effector proteins, mainly enzymes activating the reactions induced by the hormones. The next step in elucidating the mechanism of action of these hormones was the discovery of specific hormone receptors which interact with nuclear proteins or even with the DNA sequences in search of the proper gene. It was demonstrated very early that the isolated estrogen receptor of uterus mucosa from castrated mice inhibits the normal activity of uterus mucosa. This inhibitory effect disappeared after previous injection of estrogen to those mice.

Edelman[33] demonstrated that the mechanism of aldosteron action through the isolated frog urinary bladder depends on penetration of the aldosteron molecules into the bladder cell nuclei with subsequent synthesis of mRNAs and proteins—probably enzymes involved in the metabolic processes necessary to perform the active transport of Na^+ through the bladder wall. The transport of sodium ions can be blocked either by actinomycin D (inhibitor

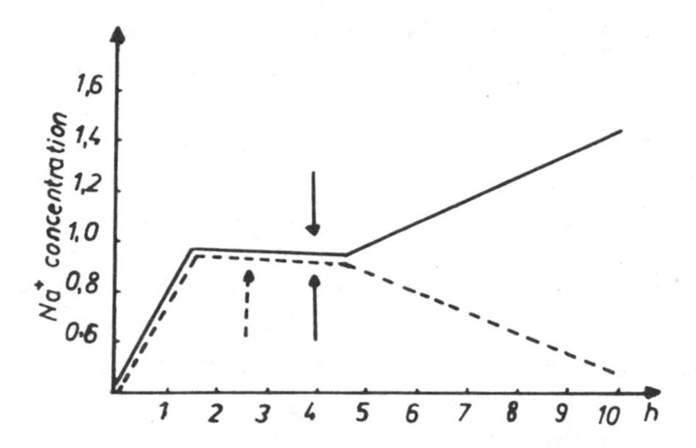

FIGURE 5. The action of aldosteron on isolated frog urinary bladder, preincubated 14 h in a medium containing Na^+. Solid arrows denote in both cases addition of aldosteron. In the first experiment (solid line), after addition of aldosterone, occurs a latency of 60 to 90 min active of transport of sodium through the sensitive membrane; in the second experiment (broken line) 30 min before addition of aldosterone, actinomycin-D which blocks transcription or puromycin which blocks translation, was added (broken arrow)—there is no response on addition of aldosterone which denotes that aldosterone acts through induction of genes which synthesize protein(s) necessary for realization of aldosterone action. (Modified from Edelman et al., *Proc. Natl. Acad. Sci. U.S.A.*, 50, 1169, 1963.)

of mRNA synthesis) or puromycin (inhibitor of translation of mRNAs into proteins) (see Figure 5).

The next investigations were concerned with the action of steroid hormones on the genome of their target cells. Any agent can react specifically with cells when they exhibit adequate receptors, usually proteins (glycoproteins, lipoproteins) but also glycolipids. Membranal receptors are floating in the phospholipid layer of the cell membrane. After binding with their specific ligands, they change their conformation, agglomerate, and become internalized into the cell, where they activate specific metabolic pathways or induce specific genes with subsequent synthesis of effector substances. In the case of hydrophobic substances, which can pass through the cell membrane without additional mechanisms, cytoplasmic receptors may occur instead of membranal receptors.

The mechanism of steroid hormone receptors is best known in the case of estrogen. Labeled estradiol, during the first 15 to 20 min after administration, preferentially (70 to 80%) becomes bound with the nuclei of target cells. Hydrophobic steroid hormones pass the cell membrane without any additional mechanism.

Estrogen receptors were first described by Toft and Gorski in 1966 in the rat uterine cytosol. Two years later Jensen et al. demonstrated that one hour after hormone administration the receptor accumulates in the target cell nuclei. These and other findings led to the two-step hypothesis of hormone action in which it was accepted that estrogen receptors are cytoplasmic proteins which, after binding with the hormones, form complexes, acquiring the ability to pass to the nuclei, where they bind with nonhistone proteins or DNA sequences of genes and activate them. Over the years, some experiments have been performed which suggest that the two-step model may be false. In the case of glucocorticoids this question has been analyzed by the isolation of variant mouse lymphoma cell lines abnormally unresponsive to glucocorticoids. Three variants have been described: deficient in the cytoplasmic receptor (r^-), defective in nuclear transport (nt^-) and a variant with increased nuclear transfer (nt^i).[35]

An argument suggesting direct binding of the glucocorticoid receptor with specific DNA sequences was based on the discovery that the activated glucocorticoid receptor recognizes DNA sequences upstream of the main promoter of the LTR region of the mouse mammary tumor virus (MTV).[37,38] Glucocorticoids enhance the transcription of MTV by direct interaction with the sites in a glucocorticoid regulatory element (located −72 and −192 bp upstream of the initiation transcription site within the proviral LTR region). These binding sites represent the hexanucleotide $\begin{smallmatrix}\text{5'-T-G-T-T-C-T-3'}\\ \text{3'-A-C-A-A-G-A-5'}\end{smallmatrix}$. Further experiments demonstrated that the glucocorticoid and progesterone receptors bind to the same DNA sites.[39] This denotes that the same DNA sites may act as regulatory elements for various gene inducers (e.g., hormones, viruses, etc.).

The chicken oviduct progesterone receptor consists of two subunits A and B with functional binding sites. The subunit A (M_r 79,000) contains a DNA-binding site which recognizes specific DNA sequences upstream of the target gene, probably being the gene ''activitor'' protein. The subunit B (M_r 108,000 ± 5,000) is structurally similar to subunit A and binds to chromatin at physiological ionic strength in preference to DNA. The B subunit seems to be the ''specifier protein'' which directs the activator (subunit A) to the site of action by modification of chromatin proteins.[38] What is more, the progesteron receptor gene may be a target gene for estrogen because the level of progesteron receptors depends on the level of estrogen in the organism.[39]

According to Chambon et al.,[40] the chick oviduct is a particularly adequate system for studying molecular mechanisms regulating gene expression in eukaryotes. The synthesis of the major egg white proteins in this organ can be modulated by administration or removal of a series of hormones (estrogens, progestins, glucocorticoids and androgens), each acting through specific receptors. In immature chicks estrogen causes proliferation and differentiation of tubular gland cells which synthesize the major egg white proteins (ovalbumin, conalbumin, ovomucoid, and lysozyme) with accumulation of their mRNAs (increased synthesis and decreased degradation). Removal of estrogen causes cessation of synthesis of all the egg white proteins, which can be restimulated to synthesis by any one of the above-mentioned steroids.

The exact mechanisms by which the steroid hormones induce adequate genes is not known. It is probable that the steroid hormone-receptor complexes bind the regulatory sequences of the induced gene. Studies on the molecular level involving transfer of genes to hormone-responsive cells demonstrated that the DNA sequences containing the binding sites are decisive for positive and negative control of transcription by glucocorticoids and progesterone.[40] Chambon et al.[40] showed that in the absence of steroid hormone the ovalbumin promoter is negatively regulated by a control element (between −295 and −420 bp upstream of the initiation-site of transcription), and that the repression can be abolished by administration of estradiol or progesterone.

The ovalbumin gene promoter is active only in differentiated tubular gland cells of the oviduct activated by steroid hormones. It is noteworthy that the ovalbumin gene is never expressed in the chicken liver cells despite the presence of the estradiol receptor.

On the basis of their experiments, Chambon et al.[40] explain the function of steroid hormones on the ovalbumin gene promoter in chick oviduct tubular gland cells as follows: chicken hepatocytes contain cell-specific factor(s) which cause constitutive expression of the microinjected ovalbumin promoter in the absence of steroid hormones (the control element is localized upstream −132 bp). This indicates that hepatocytes are able to express the ovalbumin gene in the situation when foreign promoter sequences are introduced. A similar cell-specific factor is present in oviduct tubular gland cells which permits the constitutive expression of the ovalbumin promoter (the control element is localized upstream −295 bp). In nondifferentiated cells this cell-specific factor is absent, e.g., in fibroblasts. In oviduct tubular gland cells, but not in hepatocytes, there is present a negative control element on the activity of ovalbumin promoter (localized between −295 and −420 bp). The repressive activity of this element in oviduct tubular gland cells is abolished by steroid hormones (estradiol, progesteron, and probably also by glucocorticoids) which enables synthesis of ovalbumin despite the repressor element.

These experiments proved that the inactivity of the ovalbumin gene in chicken tubular gland cells in the absence of steroid hormones is caused by the existence of the negative regulatory element (blocker), located −295 and −420 bp upstream of the cap-site. The blocker is active only in the presence of a *trans*-acting repressor. Therefore, in hepatocytes the active promoter is not expressed, not because of the presence of the blocker, but because of *cis*-acting negative mechanism (e.g., by an "inactive" chromatin conformation such as hypermethylation of the DNA).[40] The discovery of a negative blocker element enriches our knowledge on regulation of gene expression in eukaryotes simply because the presence of enhancers and blockers enables precise differentiated regulation of gene expression indispensable in multicellular organisms.

Induction of genes in target cells of the endocrine system is not limited to the steroid hormones. Several experiments demonstrate regulatory interactions by gene induction between hypothalamus and pituitary gland. For example, growth hormone-releasing factor caused a 2- to 2.5-fold increase of growth hormone mRNA levels in monolayer cultures of rat anterior pituitary cells.[41] So it seems that the hypothalamic releasing factors may independently either stimulate synthesis of the pituitary hormones and/or regulate their release. Similarly, pituitary hormones may control the transcription of the polypeptide hormones belonging to the peripheral glands of the endocrine system. For example, thyrotropin (TSH) stimulates thyroid follicular cells to synthesize and secrete thyroid hormones. Secretion is achieved by the lysosomal digestion of thyroglobulin. Although the main effect of TSH seems to be mediated by cyclic AMP-dependent mechanism, in recent times it was discovered that TSH also controls transcription of the thyroglobulin gene.[42]

J. GENE REGULATION BY INTERACTION OF mRNAs

Not so rarely, genes of higher eukaryotes appear in the genome in many copies and clusters. Typical examples are clusters of alpha and beta globin genes which switch from the embryonic to adult ones. The question arises: does there exist transcriptional interference between the duplicated identical alpha globin genes? Using molecular types of experiments, Proudfoot[43] proved that transcriptional interference causes substantial inhibition of the downstream alpha gene by transcription of the upstream alpha gene. Further experiments revealed that this inhibition can be moderated by placing transcriptional termination signals between the two alpha genes. This observation suggests that transcriptional termination signals may be important for preventing interference between adjacent genes.

Expression of eukaryotic genes transcribed by RNA polymerase II belongs to the most important unresolved problems of modern molecular biology and medicine. This process is controlled by a variety of *cis*-acting genetic elements (promoters, enhancers, blockers) and other responsive elements of the genome. These *cis*-acting elements are binding sites for *trans*-acting regulatory proteins. The question arises in which way transcription of genes can be controlled by regulatory proteins that bind to sites on the DNA either nearby or at a remote distance. According to Ptashne[44] several possibilities exist: looping, in which the interaction of adjacent DNA-bound proteins occurs by passing of the RNA polymerase over DNA loops until the two interacting proteins lie nearby; twisting, through a similar although not identical process; and sliding, in which a regulatory protein bound with RNA polymerase recognizes a specific site of DNA and then moves along the DNA until it reaches another specific protein with which it interacts and consequently initiates transcription.

K. MULTIPLE LEVELS OF GENE CONTROL IN EUKARYOTIC CELLS

Gene expression can be controlled on the transcriptional level which includes transcription of the chosen region of DNA, and mRNA processing which includes modifications of the 5′ and 3′ ends, splicing and modifications (mainly methylation) of the particular nucleotides, transport of the processed mRNA from the nuclei to the cytoplasm, translation of mRNAs into polypeptide chains, and finally post-translational modifications of proteins. Each of these steps may act as a control element, however, it is generally accepted that regulation of transcription is the primary and most important controlling element of eukaryotic gene expression. In many cases the differential expression of eukaryotic genes seems to be tissue-dependent. Only in this way can it be explained, why in the chicken the ovalbumin gene became activated by estrogen in the oviduct cells, in which the vitellogenin gene is permanently non-expressed, in contrast to the hepatocytes where the vitellogenin gene is permanently activated by estrogens while the ovalbumin gene is permanently nonexpressed. The chicken lysozyme gene can be activated by estrogens, progesteron, glucocorticoids, and testosterone in the oviduct but not in the fibroblasts despite presence of glucocorticoid receptors.[45]

1. The Role of Translation in Regulating Synthesis of Proteins

Different mRNAs are usually translated in eukaryotic cells at different rates which is best evidenced in β and δ globin genes. Because β globin genes are predominantly expressed while delta genes are expressed very little, synthesis of adult Hb A hemoglobin composed of β globin chains prevails over the synthesis of Hb A_2 hemoglobin composed of δ globin chains. As a result the adult hemoglobin contains 92% of Hb A and only 2.5% of Hb A_2. A satisfactory explanation of this occurrence is lacking. On the basis of experiments on viruses it was discovered that mRNAs must compete for a limiting message-discriminating factor in order to be translated and that competitive inhibition of translation of one mRNA by other mRNAs may be an important factor in regulating their initiation rate of translation.[46]

Another mechanism of translational control lies in the modifications of translation factors. The best known is the eukaryotic initiation factor 2 where phosphorylation disturbs the formation of the initiation complex. The first step of the polypeptide chain initiation is the formation of a ternary complex containing the eukaryotic initiation factor (eIF-2), GTP, and eukaryotic initiator—methionyl-tRNA (Met-$tRNA_i$). This complex binds with the 40S ribosomal subunit, forming the pre-initiation complex. Afterwards, the 80S initiation complex is formed, containing 60S ribosomal subunit, mRNA, additional factors, and ATP, which enables polypeptide chain elongation. Phosphorylation of the alpha subunit of eIF-2, upon activation of the heme-controlled translational inhibitor or the double-stranded RNA-activated inhibitor, interferes with the ability of the GDP-GTP exchange factor to remove eIF-2 bound GDP, and blocks the first step of protein synthesis initiation.[47] In the case of hemoglobin synthesis the limiting factor is the presence of heme.

Disturbances of translation can also be caused by extrasomatic agents. The best known is diphtheria toxin and toxin A from *Pseudomonas aeruginosa*. These toxins inactivate the elongation factor 2 (EF-2) which catalyzes the translocation of the growing polypeptide chain from the amino-acyl-tRNA site to the peptidyl-tRNA site on the ribosome, where an additional amino acid is added to the growing polypeptide chain. The mechanism of this disturbance lies in ADP-ribosylation of the EF-2.

2. Post-translational Modifications of Proteins

Only 20 amino acids are used in the protein synthesis by translation of mRNAs, but about 140 are found in proteins as the result of post-translational modifications.[48] The main mechanisms of these modifications are phosphorylation, acetylation, and methylation. Phosphorylation and acetylation may be decisive for activation or inactivation of the given enzymes. Another example of modification is gamma-carboxyglutamine which is an important chelate compound enabling binding of calcium ions to prothrombin and other coagulation factors of the prothrombin family. Disturbances of this mechanism, in the absence of vitamin K, causes severe bleeding disease.

III. REGULATION OF MESSENGER RNA TURNOVER IN EUKARYOTES

Recently, it has been discovered that the synthesis rate of many proteins may be influenced by how rapidly their mRNAs degrade, than by how rapidly it is synthesized.[49,50] Its durability may be regulated by the 5′ and 3′ nontranslated regions (UTR) of mRNA. The 5′ UTR of the c-*myc* mRNA, in case of its alteration, can cause prolongation of its half-life and consequent overproduction of c-*myc* protein and lymphnode cells, becoming cancerous. Several experiments have demonstrated that the 3′ UTR of mRNA may be important for the persistence of certain mRNAs. For example, the delta-globin mRNA is degraded four times faster than the beta-globin mRNA, and the only difference between the two are different 3′ UTRs.

A. POST-TRANSCRIPTIONAL GENE REGULATION

The molecular basis for understanding the post-transcriptional mechanism of gene regulation can be best explained by the coordinate but simultaneously opposite regulation of ferritin and transferrin receptor (TfR) genes. The uptake of iron by cells when it is needed, and detoxification when surplus, is executed by two different proteins: ferritin which binds iron in a neutral form, and TfR which enables its endocytosis.

The expression of both ferritin and TfR genes is regulated by the amount of iron. Limiting iron results in an increase of TfR and a decrease of ferritin, and in the reverse situation of plentiful iron with an increase of ferritin and a decrease of TfR. TfR and ferritin mRNAs contain *cis*-acting iron-responsive elements (IRE), which interact with a common IRE-binding protein (IRE-BP).[51] When cellular iron becomes limiting, the affinity of IRE-BP to IRE of the 5′ untranslated region (UTR) of ferritin mRNA increases, acting as a repressor of translation, resulting in a decrease of ferritin synthesis. Low affinity of IRE-BP to the 5′ UTR of ferritin mRNA prevents block of mRNA translation, resulting in ferritin biosynthesis. The same IRE-BP with high affinity to 3′ UTR of TfR mRNA prevents degradation of its mRNA, resulting in an increase of TfR synthesis and supplying cells with iron. The absence of IRE-BP binding to the 3′ UTR of TfR mRNA causes degradation of TfR mRNA, resulting in a decrease of TfR synthesis. In this way, the regulated binding of IRE-BP can coordinately regulate a decrease of ferritin biosynthesis (by translational repression) and an increase of TfR synthesis (inhibiting mRNA degradation),[51] according to the request of the cells. Crucial for the described post-transcriptional gene regulation is high or low affinity of IRE-BP to the 5′ UTR or 3′ UTR of mRNAs. Unfortunately, we do not know the exact mechanism of this change of affinity. But what is important is that the same protein, changing its features, can regulate genes expression without regulating of its own synthesis, which should also be regulated.

REFERENCES

1. **Jacob, F. and Monod, J.,** Genetic regulatory mechanisms in the synthesis of proteins, *J. Mol. Biol.*, 3, 318, 1961.
2. **Umbarger, H. E.,** Amino acid biosynthesis and its regulation, *Ann. Rev. Biochem.*, 47, 533, 1972.
3. **Goldberger, R. F.,** Autogeneous regulation of gene expression, *Science,* 183, 810, 1974.
4. **Darnell, J. E.,** RNA, *Sci. Amer.*, 253(4), 68, 1985.
5. **Weintraub, H.,** Assembly and propagation of repressed and derepressed chromosomal states, *Cell,* 42, 705, 1985.
6. **Fisher, L. M.,** DNA supercoiling and gene expression, *Nature (London),* 307, 686, 1984.
7. **North, G.,** Eukaryotic topoisomerases come into limelight, *Nature (London),* 316, 394, 1985.

8. **Felsenfeld, G.,** DNA, *Sci. Amer.*, 253(4), 58, 1985.
9. **Kolata, G.,** Z-DNA moves toward "real biology", *Science,* 222, 495, 1983.
10. **Marx, J. L.,** Z-DNA: still searching for a function, *Science,* 230, 794, 1985.
11. **Rakowicz-Szulczynska, E. M.,** Movable DNA sequences, *Acta Anthropogenetica,* 7, 265, 1983.
12. **Plasterk, R. H. A. and van de Putte, P.,** Genetic switches by DNA inversions in prokaryotes, *B.B.A.*, 782, 111, 1984.
13. **Simons, R. W. and Kleckner, N.,** Translational control of IS10 transposition, *Cell,* 34, 683, 1983.
14. **Razin, A. and Szyf, M.,** DNA methylation patterns—formation and function, *B.B.A.*, 782, 331, 1984.
15. **Jaenisch, R. and Jähner, D.,** Methylation, expression and chromosomal position of genes in mammals, *B.B.A.*, 782, 1, 1984.
16. **Bird, P. A.,** DNA methylation—how important in gene control? *Nature (London),* 307, 503, 1984.
17. **Graessmann, M., Graessmann, A., Wagner, H., Werner, E., and Simon, D.,** Complete DNA methylation does not prevent polyoma and simian virus 40 early gene expression, *Proc. Natl. Acad. Sci. U.S.A.*, 80, 6470, 1983.
18. **Jones, P. A.,** Altering gene expression with 5-azacytidine, *Cell,* 40, 485, 1985.
19. **Chiu, C. P. and Blau, H. M.,** 5-azacytidine permits gene activation in a previously noninducible cell type, *Cell,* 40, 417, 1985.
20. **Wolf, S. F., Dintzis, S., Toniolo, D., Persico, G., Lunnen, K. D., Axelman, J., and Migeon, B.,** Complete concordance between glucose-6-phosphate dehydrogenase activity and hypomethylation of 3′ CpG clusters: implications for X chromosome dosage compensation, *Nucleic Acid Res.*, 12, 9333, 1984.
21. **Wolf, S. F., Jolly, D. J., Lunnen, K. D., Friedmann, T., and Migeon, B.,** *Proc. Natl. Acad. Sci. U.S.A.*, 81, 2806, 1984.
22. **Mather, E. and Perry, R. P.,** Methylation status and DNAse I sensitivity of immunoglobulin genes: changes associated with rearrangement, *Proc. Natl. Acad. Sci. U.S.A.*, 80, 4689, 1983.
23. **Reeves, R.,** Transcriptionally active chromatin, *B.B.A.*, 782, 343, 1984.
24. **Reeder, R. H.,** Enhancers and ribosomal gene spacers, *Cell,* 38, 349, 1984.
25. **Zaret, K. S. and Yamamoto, K. R.,** Reversible and persistent changes in chromatin structure accompany activation of a glucocorticoid-dependent enhancer element, *Cell,* 38, 29, 1984.
26. **Perry, R. P.,** What controls the transcription of immunoglobulin genes?, *Nature (London),* 310, 14, 1984.
27. **Palmiter, R. D. and Brinster, R. L.,** Transgenic mice, *Cell,* 41, 343, 1985.
28. **Jones, N. C.,** Negative regulation of enhancers, *Nature,* 321, 202, 1986.
29. **Pelham, H.,** Activation of heat-shock genes in eukaryotes, *Trends Genet.*, 1, 31, 1985.
30. **Travers, A.,** Sigma factors in multitude, *Nature,* 313, 15, 1985.
31. **Ananthan, J., Goldberg, A. L., and Voellmy, R.,** Abnormal proteins serve as eukaryotic stress signals and trigger the activation of heat shock genes, *Science,* 232, 522, 1986.
32. **Karpov, V. L., Preobrazhenskaya, O., and Mirzabekov, A. D.,** Chromatin structure of hsp 70 genes, activated by heat shock: selective removal of histones from the coding region and their absence from the 5′ region, *Cell,* 36, 423, 1984.
33. **Edelman, I., Bogoroch, R., and Porter, G. A.,** On the mechanism of action of aldosteron on sodium transport, the role of protein synthesis, *Proc. Natl. Acad. Sci. U.S.A.*, 50, 1169, 1963.
34. **Schrader, W. T.,** New model for steroid hormone receptors? *Nature (London),* 308, 17, 1984.
35. **King, R. J. B.,** Enlightenment and confusion over steroid hormone receptors, *Nature (London),* 322, 701, 1984.
36. **Scheidereit, C. and Beato, M.,** Contacts between hormone receptor and DNA double helix within a glucocorticoid regulatory element of mouse mammary tumor virus, *Proc. Natl. Acad. Sci. U.S.A.*, 81, 3029, 1984.
37. **Groner, B., Kennedy, N., Skroch, P., Hynes, N. E., and Ponta, H.,** DNA sequences involved in the regulatory gene expression by glucocorticoid hormones, *B.B.A.*, 781, 1, 1984.
38. **Zarucki-Schulz, T., Kulomaa, M. S., Headon, R., Weigel, N. L., Baez, M., Edwards, D. P., McGuire, W. L., Schrader, W. T., and O'Malley, B. W.,** Molecular cloning of cDNA for the chicken progesterone receptor B antigen, *Proc. Natl. Acad. Sci. U.S.A.*, 81, 6358, 1984.
39. **von der Ahe, D., Janich, S., Scheidereit, C., Renkawitz, R., Schütz, G., and Beato, M.,** Glucocorticoid and progesteron receptors bind to the same sites in two hormonally regulated promoters, *Nature (London),* 313, 706, 1985.
40. **Chambon, P., Gaub, M. P., Lepennec, J. P., Dierich, A., and Astinott, D.,** Steroid hormones relieve repression of the ovalbumin gene promoter in chick oviduct tubular gland cells, in *Endocrinology,* Proc. 7th Int. Congress, Quebec 1984, Labrie and Proulx, Eds., Excerpta Medica, Amsterdam, 1984, 3.
41. **Barinaga, M., Bilezikjian, L. M., Vale, W. W., Rosenfeld, M. G., and Evans, R. M.,** Independent effects of growth hormone releasing factor on growth hormone release and gene transcription, *Nature,* 314, 279, 1985.
42. **van Heuverswyn, B., Streydio, C., Brocas, H., Refetoff, S., Dumont, J., and Vassart, G.,** Thyrotropin controls transcription of the thyroglobulin gene, *Proc. Natl. Acad. Sci. U.S.A.*, 81, 5941, 1984.

43. **Proudfoot, N. J.,** Transcriptional interference and termination between duplicated alpha-globin gene constructs suggests a novel mechanism for gene regulation, *Nature (London),* 322, 562, 1986.
44. **Ptashne, M.,** Gene regulation by proteins acting nearby and at a distance, *Nature (London),* 322, 697, 1986.
45. **North, G.,** Multiple levels of gene control in eukaryotic cells, *Nature (London),* 312, 308, 1984.
46. **Ray, B. K., Brendler, T. G., Adya, S., Daniels-Mc-Queen, S., Miller, J. K., Hershey, J. W. B., Grifo, J. A., Merrick, W. C., and Trach, R. E.,** Role of mRNA competition in regulating translation: further characterization of mRNA discriminatory initiation factors, *Proc. Natl. Acad. Sci. U.S.A.,* 80, 663, 1983.
47. **Siekierka, J., Manne, V., and Ochoa, S.,** Mechanism of translational control of the alpha subunit of eukaryotic initiation factor 2, *Proc. Natl. Acad. Sci. U.S.A.,* 81, 352, 1984.
48. **Uy, R. and Wold, F.,** Posttranslational covalent modification of proteins, *Science,* 198, 890, 1978.
49. **Raghow, R.,** Regulation of messenger RNA turnover in eukaryotes, *TIBS,* 12, 358, 1987.
50. **Ross, J.,** The turnover of messenger RNA, *Sci. Am.,* 4, 48, 1989.
51. **Klausner, R. D. and Harford, J. B.,** *Cis-trans* models for post-transcriptional gene regulation, *Science,* 246, 870, 1989.

The following are additions to References in Chapter 3:

36. **Kleinsmith et al.,** *Nature,* 226, p. 1925, 1970.
37. **Paul and Gilmour,** *J. Mol. Biol.,* 34, 305, 1968.
68. **Barrel, G. et al.,** *Nature,* 264, 34, 1976.
39. **Sanger, F. et al.,** *Nature,* 265, 687, 1977.
40. **Hozumi and Tonegawa,** *Proc. Natl. Acad. Sci. U.S.A.,* 73, 3628, 1976.
41. **Breathnach et al.,** *Nature,* 270, 314, 1977.
42. **Johannsen Morgan et al.**
43. **Griffith, F.,** 1928.
44. **Avery, O. T., MacLeod, C. M., and MacCarty, M.,** 1944.
45. **Miescher, F.,** 1971.
46. **Watson, J. D., Crick, F. H., and Wilkins, M.,** 1953.
47. **Rich, A. et al.**
48. **Dickerson, R.**
49. **Bonner et al.,** 1963.
50. **Alfrey and Mirsky.**
51. **Axel et al.**
52. **Park et al.,** 1976.
53. **Nirenberg and Matthaei,** 1961.
54. **Crick, F. H.,** 1958.
55. **Temin,** 1970.
56. **Goldberg,** 1979.

Chapter 5

THE PROTEINS

I. INTRODUCTION

Proteins are products of gene activation. The primary structure of proteins (sequence of amino acids) depends on sequence of nucleotides at the genetic code. Next, the secondary and tertiary structure depends directly on the presence of the primary structure. Besides, all proteins are built on the common foundation: the amino group (NH_2) and carboxylic group (COOH) of amino acids, both attached to the central carbon (called the alpha carbon). A hydrogen atom and a side chain are also attached to the alpha carbon. The nature of the side chains decide that amino acids differ from one another. The backbone of a protein is built by linking amino acids by peptide bonds (the peptide bond –CO–NH– is performed by linking of the amino and carboxylic groups with simultaneous removal of a water molecule). The properties of the peptide bond enable almost entire rotations of the alpha-carbon bonds. Therefore, the main influence on protein folding comes from the properties of the side chains. Interactions of side chains and molecules in the medium can force the polypeptide chain to fold into a compact globule with a specific shape, but then came a surprising discovery which shook these views.

Researchers in biotechnology tried to produce hormones by inserting genes of the particular hormones into bacteria to receive great quantities of protein-hormones. To their surprise, instead of a soluble protein synthesized on the basis of the inserted gene, they produced a clumpy mess.[1] The result proved that besides the known mechanism of protein synthesis by simple translation of mRNA, there must exist mechanisms which influence, or are responsible for the definitive shape (the tertiary structure) decisive for the function of the synthesized proteins.

Most of the proteins do not exhibit a regular alpha helix (e.g., the backbone of the protein coiled into a tight helix stabilized by hydrogen bonds), or β sheet (the polypeptide chains situated adjacent to one another and running either parallel or antiparallel with hydrogen bonds connecting the adjacent chains). Proteins can be composed solely of an α helix or a β sheet, but usually globular proteins are composed of a bundle of β strands inside the molecule covered on their surface with α helices.

Recently, an intermediate protein structure has been described, i.e., domains composed of two β strands connected by a segment of α helix arranged at particular angles. These β-α-β domains are important because usually such domains lying next to each other create a crevice which serves often as the binding site for this protein. The geometric arrangement of the domains constitutes the tertiary structure of these proteins.

Several proteins, beyond peptide chains, may also have nonpeptide components, e.g., metal ions which usually are essential to the activity of these proteins, mostly enzymes. To such proteins belong hemoglobins and chlorophyll with a porphyrin ring, glycoproteins with sugar molecules on their surfaces, and other features.

II. THE DYNAMIC STATE OF PROTEINS

X-ray crystallography yielded dramatic advances to the knowledge of protein structure, but also led to mistakes. The structure of proteins in a crystal is determined from the way the crystal diffracts, or scatters, a beam of X-rays. The mistake of X-ray diffraction lie in the misconception that the atoms in a protein are fixed in their positions and the entire protein molecule is rigid. By this method only gross changes could be demonstrated, e.g., a hemoglobin molecule changing its shape during binding or release of oxygen.

In recent years, views on the static character of protein structure have been radically revised. It is now recognized that the atoms in a protein molecule are in constant motion causing a dynamic state of the whole molecule. This causes the protein molecule to change its conformation constantly. A crystal of a protein generally consists of about 10^{20} protein molecules, and it is highly improbable that at any time even one individual protein molecule exhibits the average position as determined by X-ray crystallography.

The description of the dynamic state of protein molecules came from many sources, the most important were: experimental techniques, X-ray crystallography and computer simulations. The latter approach showed that in many cases activity of the given protein would be impossible when the molecule had a rigid structure. The dynamic state of proteins is particularly important in the case of enzymes which all are proteins that catalyze the speed of all metabolic reactions essential for living processes, and proteins that transport small molecules, electrons, and energy to all parts of the organisms.

To understand the role of atomic fluctuations in protein molecules, it is necessary to know their magnitude (how large they are), their probability (how often they occur), and their time scale (how long they take). The most satisfactory approach to these problems in a protein molecule is to treat them as particles responding to forces in accord to Newton's equations of motion.[2]

At present the dynamic state of only a few proteins has been analyzed. One of the best-known examples is myoglobin, the protein that stores oxygen in muscle tissue. In myoglobin the oxygen molecule is reversibly bound to the iron atom of heme which lies deep in the myoglobin molecule. The surrounding globular protein protects the iron atom from water (water would oxidize the iron atom from its ferrous state, Fe^{+2}, to the ferric state, Fe^{+3}, that cannot bind oxygen). If the myoglobin molecule were rigid, as stated in the X-ray crystallographic structure, it would be useless for the organism because binding or release of a single oxygen molecule would last years. It was calculated that because of the energy barriers, at least 100 kcal would be necessary for the entry or exit of a single oxygen molecule. This denotes that the time needed for this process would be billions of years. But, due to the fluctuations of atoms composing the myoglobin molecule, a series of possible situations may occur by which the energy barrier of the well would be lowered, or the oxygen molecule would be energized by collisions with atoms of the protein, or both would happen simultaneously (the most likely possibility). The oxygen atom would then move rapidly over the barrier and into the next well where the process would be repeated. The number of possible paths make it likely that the motion of the oxygen atoms becomes diffuse in character.[2] The analysis of myoglobin suggests that other proteins may function in a similar manner.

Several important steps in the action of proteins take place in time periods from nanoseconds (10^{-9}) to milliseconds (10^{-3}). In some enzymes the reaction time may be consumed for rearrangement of subunits or groups of atoms inside the protein molecule. Liver alcohol dehydrogenase, the ethyl alcohol oxidizing enzyme, may serve as an example where large-scale motions are involved in its activation. Ethyl alcohol dehydrogenase consists of two identical monomers, or subunits. Each subunit has two lobelike domains connected by strands of polypeptide chain, which may act as hinges. The catalytic region of this enzyme lies between two globular domains. The access to the catalytic site depends on hinge-bending deformations of the whole molecule. The motions of the globular domains may separate or join the lobes. The velocity of the reaction, the rate at which substrates are bound or products released, depends on the dynamics of the domains. X-ray crystallography has demonstrated that each domain has two configurations. The apo-enzyme has an open structure in which the lobes are spread apart, and the holo-enzyme with a closed structure in which the lobes are pressed together with the coenzyme NADH bound to one of the lobes. The open structure enables binding of the coenzyme, and the closed structure creates an environment in which the reaction can proceed. After the performance of the reaction an opening fluctuation moves the domains apart, the reaction product can escape, and the enzyme is ready for another reaction cycle.

Further experiments have revealed that the positions of atoms in the apo and holo configurations differ by large amounts. The catalytic domain (the domain opposed to the coenzyme binding domain) in the apo-enzyme is rotated so that the hinge between the lobes closes by about 10 degrees, and most of the atoms in the apo-enzyme are superposable on the equivalent atoms of the corresponding domain in the holo-enzyme.[2]

Similarly, as in the case of myoglobin there also exists in the case of alcohol dehydrogenase a barrier which prevents any rotation, because the holo-like configuration produced by rotation of the apo catalytic domain has an energy several thousand kcal per mol greater than that of the apo configuration. This barrier is also annihilated by atomic fluctuations which allows a hinge-bending motion. This is achieved by positional fluctuations of atoms in the hinge region and in areas of contact between the two domains which minimizes the energy of the rotated structure.

Enzymes play a decisive role in the metabolism of living organisms, mainly by reducing the energy necessary for the performance of chemical reactions. Enzymes act by binding the reactants, they bring them together, and transfer charges associated with the side chains of amino acids which enable the performance of the reaction. It is accepted that enzymes can perform their functions thanks to their flexibility and the movements occurring during the reaction. The motions of atoms composing the enzyme, the reactants, and the solvent water molecules at and near the active site (in a reaction in which water acts as one of the reagents) influence the rate of the reaction catalyzed by the given enzyme. In the enzyme catalysis the atomic motions play an essential role. The most important is the ''reaction zone'' which comprises the active site and its neighborhood, the reactants, and the solvent molecules, although the remaining part of the molecule cannot be neglected.

III. THE EVOLUTION OF PROTEINS

Important to the understanding of physiology and especially pathology of some proteins is the knowledge of their evolutionary origin. There are two methods of studying protein evolution. One is examination of amino acid sequences of the same protein in various organisms from the lowest prokaryotes to the highest eukaryotes, e.g., hemoglobin, cytochrome c, and other proteins (from bacteria to humans). The other is comparing structures of various proteins within a single species.[3]

The primary mechanism of protein evolution is probably gene duplication. In this case the original unchanged gene secures the normal function of this gene, while the redundant copies can mutate without harm to the organism. Most mutations produce useless or even harmful products, but occasionally a protein molecule with improved or even with a totally new function may originate. In this case the organism with the improved enzymatic function or organism with the new function may be better adapted to its environment and predominate over its predecessor.

Comparing various proteins within a single species, a series of proteins with similar fragments can be found; usually such proteins form families. To such a broad family belong proteases with serine in their active site. To this family belong trypsin, chymotrypsin, elastase, thrombin, fibrinolysin, collagenases, lysosomal proteases, and some components of the complement system. Various hemoglobin chains and myoglobins show clear similarities.

On the basis of the similarity of domains in different proteins which differ greatly in function but exhibit the common ability to bind coenzymes, e.g., nicotinamide adenine dinucleotide (NAD), flavin mononucleotide (FMN), or adenosine monophosphate (AMP), a hypothesis was established that all these proteins include a common domain with a binding site for mononucleotides. It was proposed that this domain was incorporated into several enzymes from a prototype enzyme that emerged in the first living systems,[3] and in a broader hypothesis that various enzymes originated by the shuffling of multiple domains. These enzymes may exhibit common amino acid sequences, as seen in blood coagulation factor X, epidermal growth factor precursor, low density lipoprotein receptor, urokinase, and complement component 9.[3]

IV. THE PATHOLOGICAL TRANSFORMATION OF PROTEINS

The concept of the common origin of proteins may be important to the explanation of certain pathological disturbances. A typical case represents the mutation of α_1-antitrypsin to antithrombin.[4] In a 14-year-old boy who died from a bleeding disorder, a variant of α_1-antitrypsin was identified in which methionine at position 358 was substituted by arginine (α_1-antitrypsin Pittsburgh variant). This mutation converted the elastase inhibitor α_1-antitrypsin into an inhibitor of thrombin. The protein retained its inhibitory activity against trypsin, lost its activity to inhibit pancreatic elastase, and demonstrated a 4000-fold increase in thrombin inhibiting activity.[5] The thrombin inhibiting activity of the α_1-antitrypsin molecule substituted at position Met 358 → Arg was confirmed on α_1-antitrypsin.

To study the possibility of modulating the biological properties of α_1-antitrypsin, Courtney et al.[6] introduced selected sequence modifications at the reactive site by *in vitro* mutation of a cloned α_1-antitrypsin complementary DNA. After insertion of the cDNA in *E. coli* they received two different variants, one α_1-antitrypsin Met 358 → Arg, identical to the α_1-antitrypsin Pittsburgh variant, which lost its elastase-inhibiting activity, and gained thrombin-inhibiting activity, and the other α_1-antitrypsin Met 358 → Val variant, which exhibited decreased elastase-inhibiting activity and become resistant to oxidative inactivation. The normal α_1-antitrypsin activity in experiments *in vitro* was partially inactivated by the oxidants in tobacco smoke and in the lungs in individuals smoking cigarettes.

On the basis of extensive homologies, Hunt and Dayhoff[7] suggested that α_1-antitrypsin, antithrombin III and ovalbumin (a protein without known inhibitor function) form a family that may have diverged from a common ancestral protease inhibitor 500 million years ago. Recent data suggest that α_2-plasmin inhibitor belongs to the same family.[8]

V. PROTEIN DEGRADATION

Various proteins demonstrate enormous differences in their lifetime in cells. Some of them survive as long as the cell does, while others survive only for seconds or even shorter. The lifetime of proteins defines the duration and course of biochemical reactions, i.e., the lifetime of proteins regulates life processes dependent on biochemical reactions. Until now the duration and course of protein degradation was ascribed to the action of enzymes degrading particular proteins. Recently, it was discovered that the lifetime of particular protein molecules may be dependent upon the presence of specific amino acid residues. Metabolic instability of particular proteins may allow rapid changes of their intracellular concentrations through quick synthesis or degradation which adjust them to the current needs of the organism.

Recent biochemical and genetic data suggest that in eukaryotes covalent conjugation of ubiquitin to short-living intracellular proteins determines their selective degradation.[9] Similar to the signal sequences of proteins that designate their ability to enter into distinct cellular compartments, it is suggested that proteins contain specific amino acid residues which determine their lifetime *in vivo*.[9,10] A, the genetic code for methionine, is both the starting signal for protein synthesis and the terminator signal for cessation of transcription, and so all proteins have methionine at both ends. During post-translational modification enzymes split some parts of the amino acid chain at both ends which means that the amino acid residues of the protein molecules at their starting and termination points are not methionine.[11]

At first it was discovered that a mutant cell line did not grow and divide at elevated temperatures. The cause of this phenomenon was elucidated by detection that at elevated temperature the enzyme which split ubiquitin from the protein molecule became inactivated. This led to the statement that deubiquitination causes rapid degradation of the protein molecules. On the basis of this observation it was suggested that binding of ubiquitin to synthesized protein molecules designates their lifetime. Proteins which bind ubiquitin are assigned to rapid degradation.

The N-end rule of proteins denotes that the lifetime of protein molecules is dependent upon the last amino acid residue on their amino-ends. Protein molecules with methionine, serine, alanine, threonine, valine, and glycine at their N-terminal ends have half-lifetimes of more than 20 h; those with isoleucine and glutamic acid residues at their N-terminal ends have half-lifetimes of about half an hour; those with tyrosine and glutamic acid residues have about 10 min of half-lifetime; those with phenylalanine, leucine, aspartic acid, and lysine residues have half-lifetimes of about 3 min; and those with arginine residue a half-lifetime of about 2 min. On the basis of this rule it was possible to establish variants of beta-galactosidase molecules which, bound with ubiquitin, have had a lifetime dependent upon the last amino acid residue at its N-terminal end.[9]

Other authors established the so-called PEST hypothesis (PEST denotes amino acid residues: P—for proline, E—for glutamic acid, S—for serine, and T—for threonine).[10] According to these authors, ten proteins with intracellular half-lifetimes less than 2 h contain one or more PEST regions. Usually the PEST regions are flanked by clusters of several positively charged amino acid residues. In contrast to this, 35 intracellular proteins with half-lifetimes between 20 and 220 h demonstrated the PEST sequences only in three cases.

The mechanism which determines how much amino acid residues are cleaved at the N-terminal end of polypeptide chains in post-translational modification is unknown.[11] Ubiquitin bound to the N-terminal ends of protein molecules does not designate these proteins for degradation. It probably only destabilizes proteins that are already destined for degradation (it must be only bound to lysine side chains). The N-end rule explains why bacteria which do not have ubiquitin degrade protein molecules without disturbances.

REFERENCES

1. **Kolata, G.,** Trying to crack the second half of the genetic code, *Science,* 233, 1037, 1986.
2. **Karplus, M. and McCammon, J. A.,** The dynamics of proteins, *Sci. Amer.,* 254(4), 42, 1986.
3. **Doolittle, R. F.,** Proteins, *Sci. Am.,* 253(10), 88, 1985.
4. **Owen, M. C., Brennan, S. O., Jewis, J. H., and Carrel, R. W.,** Mutation of antitrypsin to antithrombin, $alpha_1$-antitrypsin Pittsburgh (Met 358 leads to Arg), a fatal bleeding disorder, *N. Engl. J. Med.,* 309, 694, 1983.
5. **Harpel, P. C.,** Protease inhibitors—a precarious balance, *N. Engl. J. Med.,* 309, 725, 1983.
6. **Courtney, M., Jallat, S., Tessier, L. H., Benavente, A., Crystal, R. G., and Lecocq, J. P.,** Synthesis in *E. coli* of $alpha_1$-antitrypsin variants of therapeutic potential for emphysema and thrombosis, *Nature (London),* 313, 149, 1985.
7. **Hunt, L. T. and Dayhoff, M. O.,** A surprising new superfamily containing ovalbumin, antithrombin III and $alpha_1$-proteinase inhibitor, *Biochem. Biophys. Res. Commun.,* 95, 864, 1982.
8. **Lijnen, H. R., Wiman, B., and Collen, D.,** Partial primary structure of human $alpha_2$-antiplasmin-homology with other plasma protease inhibitors, *Thromb. Haemost.,* 48, 311, 1982.
9. **Bachmair, A., Finley, D., and Varshavsky, A.,** *In vivo* half-life of a protein is a function of its amino-terminal residue, *Science,* 234, 179, 1986.

10. **Rogers, S., Wells, R., and Rechsteiner, M.,** Amino acid sequences common to rapidly degraded proteins: the PEST hypothesis, *Science,* 234, 364, 1986.
11. **Kolata, G.,** New rule proposed for protein degradation, *Science,* 234, 151, 1986.

Chapter 6

MOLECULAR BASES OF CELL DIFFERENTIATION

I. INTRODUCTION

Differentiation denotes passage of cells from a more primitive state to a higher level of vital functions and simultaneous acquirement of special functions. Simultaneously, however, the more primitive functions disappear, which deprives the differentiated cells of the ability of independent survival and forces them to live in complementary aggregates of cells within the multicellular organism. This state requires a constant collaboration of differentiated cells with preservation of their integrality and collaboration securing the survival not only of single cells, but also of the entire aggregation of differentiated cells, i.e., survival of the multicellular organism.

Evolutionarily, the development of multicellular organisms depended first and foremost on specialization, i.e., differentiation of primary totipotent cells to specialized differentiated cells. Embryonic cells undergo development into specialized cell types by a two-stage process: first, determination, which results in conversion of embryonic cells into lineages of stem cells. Then the stem cells proliferate and differentiate activating genes responsible for initiation of specialized functions of the differentiated cell.

At present it is generally accepted that differentiation occurs by determination and commitment, grounded on genetic mechanisms of repression of the one and expression of the other genes of specialized cells. The differentiated cells differ from ancestral cells by acquiring a new or modified biochemical reaction. A series of such modifications may establish a new biochemical pathway whose product may initiate a new function improving adaptation of such an organism to the environment and consequently its preferential survival. Cells that once acquired a new metabolic pathway which makes it possible to perform a special function, usually do not change this attribute. This denotes stability of the differentiated state.

Although it is widely accepted that cell differentiation is regulated by genetic processes, the exact mechanism which causes stable repression of the one and expression of the other genes is poorly understood, and based principally on known mechanisms of gene regulation in eukaryotes (see Chapter 4). Gene regulation as an aspect of cell differentiation usually is examined by two methods: one based on genomic organization during development, when a multicellular organism generates from a single cell (the zygote), and step-by-step, through proliferation, new generations of specialized cells are made. The second method is based on studies of cell differentiation in adults, when new generations of specialized cells originate from stem cells (e.g., red blood cells, plasma cells producing antibodies, etc.). In both cases it is certain that all cells of a multicellular organism retain the whole set of genes present in the zygote cell, independently of which kind of modification this cell underwent, to which organ this cell belongs, or which function is performed by this cell.

The essence of differentiation can be best illustrated by analysis of metabolic pathways in various organs with different functions in the multicellular organism, e.g., glucose metabolism in the liver, kidneys, and adipose tissue. The central position in glucose metabolism has glucose-6-phosphate (G-6-P) which, with the aid of glucose-6-phosphatase, can split off the phosphate group and pass to the blood circulation as free glucose (G); G-6-P may be synthesized to glycogen; glycogen by glycogenolysis gives again G-6-P; G-6-P is the primary substrate to the glycolytic reaction and for the pentose metabolic pathway; finally, G-6-P is the product of neoglucogenesis (i.e., synthesis of glucose from non-hydrocarbonic substrates) (see Figure 1).

Glycogen
G-6-Pase
G-1-P
G-6-PD
G
G-6-P
pentose path
F-6-P
F-1,6-Pase
F 1,6 P
Glycolysis

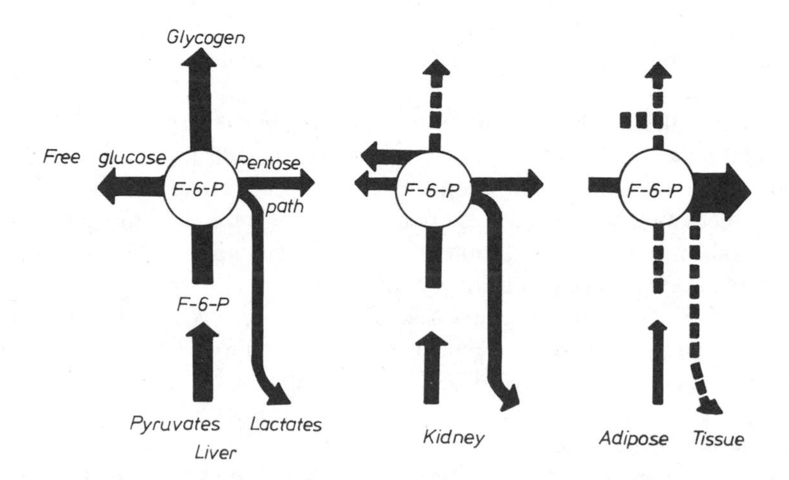

FIGURE 1. Comparison of the activities of glucose metabolic pathways in various tissues: liver, kidney, and adipose tissue. (Top): glucose metabolic pathways; (Bottom): glucose metabolic pathways in the liver, kidney, and adipose tissue. The thickness of the arrows indicates the intensity of metabolic reactions in relation to the liver standard at 100% activity for each pathway. G—glucose; G-1-P—glucose-1-phosphate; G-6-P—glucose-6-phosphate; G-6-Pase—glucose-6-phosphatase; G-6-PD—glucose-6-phosphate dehydrogenase; F-6-P—fructose-6-phosphate; F-1,6-P—fructose-1,6-diphosphate; F-1,6-Pase—fructose-1,6-diphosphatase.

In the liver, which is the central metabolic organ of the multicellular organism, all the glucose metabolic pathways are present, such as release of free glucose, glycogenogenesis, and glycogenolysis, the pentose pathway, and neoglucogenesis—although in different intensities (represented in Figure 1 by thick or thin arrows). In the kidneys the glycolytic pathway predominates as the energy-producing pathway necessary for renal function, and neoglucogenesis (production of glucose for energy supply). In the adipose tissue remains practically the only pentose metabolic pathway which provides the fatty acids pathway with reduced nicotinamide-adenine dinucleotide (NADH) and nicotinamide-adenine dinucleotide phosphate (NADPH) necessary for the synthesis of fatty acids. As seen from this example, the kind and intensity of the particular metabolic pathways of differentiated cells respond to the needs of specialized functions of the particular organs.

II. THE GENETIC MECHANISMS OF CELL DIFFERENTIATION

The expression of eukaryotic genes seems to be regulated principally at the transcriptional level. For this purpose, structural changes occur in the chromatin surrounding genes that must be expressed. These changes can be detected by increased sensitivity to digestion by DNAase I or II, and other nucleases. Recently, it has been suspected that some changes may also occur at the DNA molecule itself. Additionally, specific nucleotide sequences have been discovered on the DNA molecule—the TATA box, about 30 bp upstream of the initiation site; the upstream CCAAT control sequences, about 80 bp upstream of the initiation site; and enhancer or repressor sequences usually situated upstream of the initiation site, occasionally many hundreds or even thousands of sequences upstream, but sometimes also within the transcribed gene or even at sequences downstream of the gene sequences. It is noteworthy that hypersensitivity of the transcribed genes in differentiated cells can be detected before expression of those genes which denotes that the specific DNA or chromatin conformation also must exist before the gene is expressed, and is realized during the differentiation process of those cells.

Thus, the main question arises: what causes execution of the specific DNA conformation of the active genes in differentiated cells? Also, which are the agents or factors causing their expression? Recent investigations demonstrated that nuclear proteins are responsible for the DNA conformation, especially the nonhistone chromatin proteins with the high mobility group (HMG) of those proteins in the foreground, although recent investigations also have demonstrated that only in highly differentiated cells are specific histone proteins present which may be important for expression of specific genes in those cells.

Also important to differentiation may be the product of the transcribed genes, i.e., the mRNAs. In this aspect, transport of mRNAs to the cytoplasm may be controlled by interaction with other RNAs, but the most important function seems to be stabilization of the mRNA by ribonucleoprotein complexes in the cytoplasm. Alas, this problem remains very poorly understood.

III. DIFFERENTIATION OF STEM CELLS

Tissue specific phenotypes result from a series of developmental stages. The primary totipotent cells of the early embryo give rise to stem cells specific to three distinct layers; ectoderm, mesoderm, and entoderm, from which starts the development of all definitive tissues. Once a cell is determined, it undergoes specialization along a specific pathway, such as erythropoiesis, myogenesis, etc. The determined cell cannot generate other phenotypes, and its progeny always possesses the same limited potential. At particular points in development the determined cells always express the same phenotypes (by transcription of adequate genes).

In the course of differentiation, the cells do not lose either chromosomes or genes. This is best illustrated by the experiments performed by Gurdon, who transplanted specialized frog intestine nuclei into enucleated frog eggs and received in some cases entire frogs. In other experiments, generation of a diversity of normal tissue-specific cell types resulted from malignant neoplastic cells transplanted into early mouse embryos.[1]

Red blood cell differentiation represents a convenient model for examination of molecular mechanisms in coordinate regulation of gene expression. Red blood cells originate from multipotential precursor (stem) cells, which under the influence of various growth factors and environmental agents differentiate into early erythroid progenitor cells, which become sensitive to erythropoietin, probably by acquiring adequate receptors. Under the influence of erythropoietin the early erythroid progenitor cell synthesizes a series of proteins, among them proteins conditioning the main function of erythrocytes (i.e., transport of oxygen), i.e., hemoglobin, catalase, carbonic anhydrase, and a specific lipoxygenase; membrane proteins: spectrin, glycophorin, and transferrin receptor molecules; and enzymes of intermediate metabolism: pyruvate kinase, hexokinase, phosphofructokinase and glyceraldehyde-3-phosphate dehydrogenase.[2] The synthesis of those substances gives rise to morphological changes of the early erythroid progenitor cells which transform by successive stages into erythroblasts, reticulocytes, and erythrocytes.

The membranal protein, spectrin, is a typical component of cell membranes, present in most cells and is composed of two nonidentical subunits, α and β (spectrin has not been detected in smooth muscle, in microvilli of the intestine brush border nor in flagellae). The spectrin tetramer is bound to the plasma membrane by an intermediate protein—ankyrin—which in turn is bound with the major anion exchange protein. The α spectrin polypeptide interacts with calmodulin via another polypeptide. In various tissues, instead of the spectrin beta polypeptide chain there occurs different types of spectrin-like chains, e.g., fodrin in most nerve, liver, lens, fibroblast, and lymphoid cells. Binding of alpha-spectrin with calmodulin and actin suggests that spectrin regulates the deformability of membranes and mobility of membrane proteins.

Using monospecific antibodies, most of the red blood cell membranal proteins have been detected also in nonerythroid cells, e.g., ankyrin was detected in different normal cell types. These results suggest that the various components of the cytoskeletal network of red blood cells also may appear in and have similar functions in other cell types.[2]

Transferrin receptors have been found on all proliferating cells. In bone marrow, transferrin receptors are present in increasing amounts on the proerythroblasts and orthochromatic erythroblasts, on reticulocytes in very small amounts, and is absent on erythrocytes.

Because of the importance of iron for cell growth and the role of transferrin as the carrier protein in the mechanism of iron uptake and transport to adequate cells by the transferrin cycle, it was supposed that the transferrin cycle plays a regulatory role in erythroid cell differentiation.[3] The transferrin cycle involves binding of transferrin by specific transferrin receptors on the cell surface, and next, internalization via coated pits. In the acid environment of endocytic vesicles the ferric ions are released and transferred to the intracellular carrier protein, ferritin. Next, apotransferrin bound to its receptor is returned to the cell surface, where it can bind new ferric ions. Because of the role of heme in the control of hemoglobin synthesis, it seems probable that transferrin may play a coordinate role in the synthesis of erythrocyte proteins, i.e., hemin or hemoglobin.

Contrary to this, the other red blood cell proteins seem to be present only in erythrocytes. Glycophorins are synthesized mainly at the mid-erythroblast stage of differentiation, whereas their synthesis at later stages is very low.

Red blood cell lipoxygenase is synthesized only in peripheral rabbit reticulocytes and is present only in red blood cells. The enzyme is absent in early erythroblasts. In erythrocytes its mRNA is present in a form of inactive ribonucleoprotein complex (mRNP). Its physiological role seems to be lysis of mitochondria and inhibition of the respiratory chain.

The list of membranal and cytoskeletal proteins is incomplete. There are proteins of unknown function, for example a 19-kDa polypeptide.

IV. REGULATION OF GLOBIN GENE EXPRESSION

Globin genes are expressed only in erythroid cells. Exceptional globin gene transcripts have been detected also in nonerythroid cell lines exclusively from upstream initiation sites. Post-transcriptional control indicates that in chicken erythroblasts, transformed by avian erythroblastosis virus, the β-globin gene transcripts are normally processed, while the α-globin gene transcripts are not processed, and accumulate in the cell nuclei. This implies that the transport of these two transcripts to the cytoplasm is performed by two different mechanisms, which may act as the controlling mechanism of mRNA surplus. A question arises as to which way the non-globin mRNAs in reticulocytes become reduced. Two hypotheses have been established: one that non-globin mRNAs become destabilized during red blood cells differentiation, and the other that globin mRNA is more stable than non-globin mRNA. Some authors suggest that different expression of globin and non-globin mRNAs may be caused by binding with proteins. This suggestion is based on the detection of different globin mRNA complexes 20S mRNP and non-globin mRNA complexes 35S mRNP.[2]

Examination of chromatin changes during gene activation demonstrated that in chickens and humans both the embryonic/fetal and adult globin genes are hypersensitive to DNAase, although adult globin genes are not expressed during embryonic life. In adults the embryonic globin genes are not transcribed and become DNAase I resistant.

V. DNA METHYLATION DURING CELL DIFFERENTIATION

Many authors agree that DNA methylation is an important mechanism in gene regulation. A series of experimental results confirms these views. In humans the 6-week erythroblasts produce embryonic epsilon-chains; whereas the 12-week erythroblasts synthesize largely fetal gamma-chains, and the adult bone marrow erythroblasts almost exclusively synthesize adult β-chains. Strong correlation has been observed between synthesis of globin-chains at the particular developmental phases with hypomethylation in the close flanking sequences of the expressed globin genes.[4] This result suggests that modifications of the methylation patterns may be an important factor which regulates expression of human globin genes during embryonic → fetal → adult hemoglobin switches.[4] However, the mechanism initiating switch from embryonic to fetal and later to adult hemoglobin synthesis remains unknown. It is supposed that there must be either an internal "clock" mechanism, or stem cells become induced by environmental factors.[5]

DNA demethylation experiments by administration of 5-azacytidine confirm the role of methylation in cell differentiation. An induction of repressed human gamma-globin genes in adults was observed after administration of 5-azacytidine, evidenced by the synthesis of fetal hemoglobin. It is difficult to explain why "global" hypomethylation causes only derepression of human gamma-globin genes in a mouse erythroleukemia cell line (M11-X) which contains most of human chromosome 11 (human beta-like globin genes are located on chromosome 11).[6] In those experiments the genomic DNA was remethylated, but not sequences in the neighborhood of the human and mouse globin genes, which remained hypomethylated. The result suggests that remethylation of those regions is inhibited by an unknown mechanism.[6] In further experiments 5-azacytidine caused differentiation of the mouse mesodermal cell lineage C3H1OT1/2 clone 8 into skeletal muscle, chondrocyte, and adipocyte cell types, which proves that methylation and demethylation of the embryonic DNA may play an important role in cell differentiation.[7] It seems that methylation and demethylation may be important factors in cell differentiation, as one of the mechanisms which cause the particular genes to be expressed or not expressed, however, not as the guiding mechanism.

VI. *CIS*- AND *TRANS*-ACTING SUBSTANCES IN CELL DIFFERENTIATION

In Chapter 4, the induction of genes by hormones has been discussed. In cell differentiation during development, embryonic cell lineages of stem cells differentiate under the influence of various proliferation and differentiation factors. The problem can be studied best in short-lived hemopoetic cells which, for continuation of their necessary life functions, must permanently proliferate and differentiate. The initial cell of the hemopoetic system is the pluripotent hemopoetic bone marrow stem cell, which gives rise to the existence of all hemopoetic systems such as erythropoiesis, lymphopoiesis, megacariopoiesis, and granulopoiesis.

In the erythropoiesis system, the pluripotent hemopoetic stem cell transforms into early erythroid progenitor cell. Under the influence of the colony stimulating factor (produced by lymphocytes T) the early erythroid progenitor cells form early burst units in which proliferation and differentiation is controlled by the burst-promoting activity, synthesized also by T-lymphocytes. Under the influence of those substances, and next by erythropoietin (synthesized by the kidneys), the proerythroblast is generated, and it finally transforms into erythrocytes, mainly under the influence of erythropoietin. The molecular mechanism of action of these substances is not yet fully explained, but by comparison with the similar action of some hormones (glucocorticoids) it is very probable that they bind with specific receptors of cell nuclei and cause activation and expression of particular genes.

This explanation is suggested by the well-recognized *cis-* and *trans-*acting factors regulating globin gene expression. For effective transcription of the α- and β-globin genes, two conserved DNA sequences, the TATA box about 30 bp and the CCAAT box about 80 bp upstream of the initiation site of transcription are necessary. These sequences seem to be elements for recognition by the site of the initiation of transcription. Important differences in regulation of the α- and β-globin genes become evident by their responses to *cis-* or *trans-*acting viral regulator sequences *in vivo*. SV40 has a 72 bp enhancer sequence which increases transcription of SV40 and other viral genes. The enhancer sequences act only when covalently bound in cis-position and at a distance in either orientation. These enhancers are inactive *in vitro*. Therefore, it is supposed that they act by attachment of DNA to the nuclear matrix, or recognize chromatin structure, or DNA surrounding neighboring promoters.[2] The β-globin genes cloned into SV40 vectors need the presence of SV40 enhancer, while the α-globin genes are transcribed independently. This is explained by the fact that the β-globin genes have a sequence that interferes with the action of another enhancer, in this case probably the α-globin gene enhancer, whereas the stronger SV40 enhancer can overcome this interference.

The adenovirus regulatory protein (E1a) transcribes efficiently both α- and β-globin genes in the absence of SV40 enhancer. E1a protein synthesized by early adenovirus gene causes transcription of the late adenovirus genes by catalyzing transcriptional complexes. E1a stimulates all promoters that require enhancers for activity, probably by direct action on promoters. E1a protein switches transcription from the upstream starting position to the major cap site, which denotes that it interacts with the promoter sequences within the gene itself, rather than with the 5′ upstream flanking sequences.[2]

The differences in the action of α- and β-globin gene promoters are not fully explained, and future experiments must elucidate in which way α- and β-globin genes are coordinately regulated by *cis-* and *trans-*acting factors. One of the best methods for examination of *trans*-acting substances in somatic cells is transfer of genes by the cell-fusion method (transfer of purified genes by transfection of cloned DNA or single chromosomes). Harrison on the basis of experiments with three various mRNA fusions concludes that the three mRNAs are expressed or repressed in a coordinate manner in cell hybrids between erythroblasts and lymphoma cells or neuroblastoma cells respectively.[2] From further experiments he concluded that reduced expression of the α-globin gene in non-erythroid cells is *cis-*effective and it can be activated by a factor produced in tetraploid Friend cells. The *trans-*acting factors interact with DNAase I hypersensitive sites and they are related to erythroid nuclear proteins that bind specifically with DNAase I hypersensitive sites around coordinately regulated genes.[2]

VII. THE PLASTICITY OF DIFFERENTIATED CELLS

Contrary to a general suggestion that the differentiation state is very stable, Helen Blau et al.[1] demonstrated its plasticity. As a model for studying this problem, they used heterokaryons in which muscle genes can be stably expressed in nonmuscle cells. Using polyethylene alcohol they combined entire muscle cells (mouse muscle cell line C_2C_{12}) with cells of other phenotypes or species (usually of human origin). The fusion product is stable which enables monitoring for a relatively long period of time. In the heterokaryon cell division does not occur and the parental cell nuclei remain intact and retain their full complement of chromosomes. The activation of muscle genes in nonmuscle cells must be mediated by diffusible *trans*-acting substances that pass through the cytoplasm to the cell nuclei (the nuclei of the two cell types remain separate and distinct). The muscle genes in such specialized cells, such as skin, cartilage, lung, and liver, could be activated in heterokaryons by muscle genes regulatory factors.

On the basis of their experiments, Blau et al.[1] discuss four regulatory stages of muscle cell differentiation:

1. Transformation of totipotential stem cells to mesodermal stem cells
2. Transformation of the mesodermal stem cells to myoblasts by determination of adequate genes
3. Differentiation of myoblasts to fetal myotubes
4. Maturation of fetal myotubes to adult myotubes

The kinetics of human muscle gene expression in the experiments of Blau et al.[1] differed and resulted mainly from the kind of human cells used. The earliest and highest expression was observed in lung fibroblasts, the lower in kerationocytes and the lowest in hepatocytes. The next question was the mechanism of gene activation. The authors proved that *de novo* synthesis of DNA was not necessary and inhibition of DNA synthesis did not influence the level of muscle gene expression in heterokaryons.[8] Contrary to this in heterokaryons with HeLa cells (a malignant aneuploid cell type) activation of muscle genes occurred only after pretreatment of HeLa cells with 5-azacytidine. The result of this experiment suggests that in neoplastic cells the normal genes become hypermethylated, and only their hypomethylation makes it possible to activate them. In heterokaryons with human liver cells the nonmuscle genes have been also expressed, although to a lower extent than the muscle genes.

In our own experiments,[9] we found that myeloma cells synthesize immunoglobulins spontaneously (line RPC5 synthesizing gamma 2A, and line ABPC synthesizing IgM) any of the characteristic fractions of nonhistone chromatin proteins described by us in immunoglobulin producing spleen cells could not be detected. The result suggests that in neoplastic cells, normal genes responsible for immunoglobulin production have been repressed.

Another important finding in the experiments of Blau et al.[1] was the detection of regulatory circuits between nuclei in heterokaryons. For this purpose the authors examined whether the nonmuscle cell influences gene expression in the muscle cell. The experiments suggested that, the human gene (in this case α-cardiac actin gene) responds to mouse muscle cytoplasmic factors by producing transcripts with the time course and levels typical for human, but not for mouse muscle cell culture. The differences could arise in the *cis*-regulating regions of mouse and human α-cardiac actin genes. The result is compatible with the hypothesis that *trans*-acting mouse muscle factors activate muscle regulatory genes and therefore activation of human muscle genes occurred. The same phenomenon also causes domination of activated human phenotype over the mouse muscle genes transcripts of which are overridden by the presence of human muscle nuclei.

Besides the heterokaryon technique, transplantation of nuclei also proves that differentiation may be reversible. Nuclei from specialized frog cells introduced into enucleated frog eggs may be able to develop normal frogs to some percentage (experiments of Briggs and King, and Gurdon; for a review see Reference 10).

Most of the experiments of Blau et al.[1] were performed on mouse muscle cells in the differentiated stage (i.e., of muscle cell specialization) and the question arose if results would be similar in other developmental stages. For this purpose the authors determined fetal myosin and neonatal myosin izoenzymes using specific monoclonals. The experiments demonstrated that a progressive amount of neonatal myosin was observed over time in culture. The result suggests that neural contact and other activating substances are not necessary for certain steps in muscle maturation.

VIII. THE ACTIVATION OF DORMANT GENES

Activation of dormant genes in differentiated cells can be evoked in some lower vertebrates by the mechanisms of cellular metaplasia (or transdifferentiation). The best known examples are (1) transdifferentiation of pigment cells of the iris epithelium during regeneration of the lens (producing lens-fiber specific gamma-crystallins and other lens proteins) in several adult urodelen species; and (2) regeneration of the urodelen amphibian limbs after their amputation, including dedifferentiation of internal limb tissue cells (skeletal muscle, bone, cartilage, and connective tissue cells), their proliferation and redifferentiation (forming differentiated cells from related primordial origin cells).[10]

IX. CELL DIFFERENTIATION BY DNA REARRANGEMENTS

Cell differentiation also may occur by DNA rearrangements. This kind of cell differentiation is best known in lymphoid cell differentiation producing all kinds of cells participating in the immune reaction (T- and B-lymphocytes, T-helper and repressor cells, immunoglobulin producing plasma cells, macrophages, etc.). For details see Chapter 11.

Polypeptide growth factors play an important role in cell proliferation involved in many different physiological and pathological processes such as embryogenesis, growth and development (cell differentiation), selective cell survival, hemopoiesis, tissue repair, immune responses, atherosclerosis, and neoplasia. The best known are factors causing cell proliferation and differentiation in immune reactions (see Chapter 11).

Several authors suggest that cancer may be considered as a cell differentiation disease. According to this theory it was supposed that cancer cells may emerge from genetically misprogrammed normal cells. More recent experiments contradict these hypotheses. For example, human myeloid leukemia cells *in vitro*, and in mouse *in vivo*, can be reversed to normal granulocytes and macrophages after administration of MGI (macrophage and granulocyte inducer necessary for differentiation of normal myeloblasts).[11]

The molecular mechanisms of cell differentiation may be different including changes of chromatin structure caused by specific DNA-protein interactions (*cis*- and *trans*-acting substances), DNA methylation, DNA rearrangements, and loss of DNA fragments. Among all those mechanisms, only loss of DNA seems to be absolutely irreversible.

X. THE HOMEO BOX GENES

The genetic mechanisms that in the early embryonic development control temporal and spatial gene expression are of the greatest importance to the whole of biology and medicine. It is commonly accepted that development is based on differential expression of gene sets which program particular tissues and organs of the developing organism. The interdepencence of different parts of the developing organism excludes the possibility that each structural gene can be regulated independently. Therefore, a hierarchical control of groups of structural genes coding proteins necessary for the development of particular tissues and organs is necessary, and the recent discovery of the homeo box genes fulfill this requirement to a certain degree.

Genetic examinations of *Drosophila* mutants with disturbed morphogenesis made discovery of those genes possible. Two main types of mutations have been described: (1)

homeotic mutants, characterized by transversion of one body organ into another, e.g., antennae into legs, termed Antennapedia (ANT-C), or an abdominal segment into a thoracic segment, termed Bithorax (BX-C); (2) segmentation mutants characterized by disorder in the number of polarity of body segments, e.g., fushi tarazu (ftz) caused by disturbances of temporal and spatial expression of homeo box genes.[12] Homeosis refers to the replacement of one body structure by a homologous structure from another body segment.

Some years ago experiments performed on the transparent roundworm *Caenorhabditis elegans* and *Drosophila melanogaster* detected a group of genes responsible for the spatial organization in these organisms. The genes represent a common segment of DNA, termed homeo box which control activity of groups of other genes. Transcribed homeo box genes yield a protein which binds a fragment of DNA comprising a cluster of genes, turning them on or off. For example, an appropriate group of genes turned on in cells in the *Drosophila* embryo may produce a wing, while another group of genes in other cells may produce antennae. Similar groups of genes have been described in mice, humans, and other organisms. It is supposed that this discovery will be the key to understanding mechanisms responsible for the development of multicellular organisms from a zygote.

The worm *Caenorhabditis elegans* is specifically useful for these investigations because of the limited number of 959 cells in the adult worm. Each of them can be traced back to the fertilized egg. The cell lineages in this animal are specifically stable, unless they are mutated. In order to generate differentiated cells in adults, the cell lineages must branch off at exactly the right time. This is done by specific genes called chronogenes. Mutation of those genes causes disturbances in the development because the mutated genes cause branching off sooner or later of the subordinate genes. For example, mutation of the chronogene lin-14 causes the worm to undergo two extra molts which cause the cuticulum to lag behind (being larval) even though the animal is mature. This denotes that the lin-14 chronogene is responsible for the cuticle-forming cells to produce the adult cuticle at the right time.[13]

The segmentation disturbances are best studied in the development of *Drosophila*. The development of a *Drosophila* individual starts in the ovary of the female *Drosophila*. The germ cell divides four times, yielding 15 nurse cells and the oocyte, which transforms into the egg. At fertilization the egg and sperm join into the zygote, which divides. The daughter nuclei divide every 10 min. The daughter nuclei share a common cytoplasm and are scattered through the cytoplasm. After they reach the number of 256 nuclei (eight divisions), the nuclei migrate to the surface forming the cortex of the egg, forming a layer one nucleus thick. After 13 divisions, the cell membranes divide and a cellular monolayer is formed, called blastoderm. During the next hours interior cell layers are formed, and the embryo becomes divided into segments corresponding to the segments of the adult insect. There are three segments which later form the head (Md, Mx, and Lb), the three thoracic segments (T1 to T3) and the eight abdominal segments (A1 to A8).[13]

One day after fertilization the embryo hatches and becomes a larva, molts twice, pupates, and metamorphoses into the adult fly, preserving in all the development phases its segmental structure. Prior to the formation of the blastoderm the nuclei are totipotent, i.e., they can be predecessors of any organ of the adult fly. As soon as the blastoderm is formed, however, genetic determination of the cells begins. The fate of the particular cells can be examined easily by marking individual cells in the blastoderm (by use of controlled doses of X-rays which induce genetic, mainly chromosomal aberrations serving as markers of particular cells or nuclei). From now on, cells of the blastoderm are committed to becoming parts of particular segments. Thus, the segmental structure of *Drosophila* becomes established at the blastoderm stage, and particular cells are destined to one segment in the larval and adult body. Next, within the segments, cells become localized in the anterior or posterior parts of the particular segments. Following further development, adult structures become determined. At first the imaginal discs become distinguished; they contain small groups of genes, coding proteins for particular parts of the adult organism, e.g., one part of a leg, antennae, or other organs. The small groups of genes undergo a series of steps until the final days of the larva, when they transform under the influence of adequate hormones into their adult organs. The conclusion is that cells in the *Drosophila* embryo are determined in a series of progressively finer gradations that terminates at metamorphosis.[13]

XI. DEVELOPMENTAL MUTATIONS

Three kinds of developmental mutations have been distinguished: maternal effect mutations, segmentation mutations, and homeotic mutations.

Maternal effect mutations may disturb the spatial polarity of the embryo. A typical maternal mutation is the dicephalic (two-headed) fly. This mutation results from an abnormal distribution of nurse cells. Normally, nurse cells generate from the first four divisions, yielding 15 nurse cells and one oocyte. During the following steps of development the nurse cells accumulate at the anterior pole of the egg, but in the dicephalic mutant they accumulate at both poles. This results in embryos with two sets of anterior structures joined in the middle, and lack of posterior structures. Homozygotic female *Drosophila*, for the dicephalic mutations, produces aberrant follicles; thus, it produces aberrant embryos irrespective of the genotype of the father. Heterozygous mothers produce only normal eggs. The result demonstrates that the anterior-posterior polarity is formed in the ovary under the control of the maternal genome. Maternal effect mutants also may cause disturbances of the dorso-ventral polarity. This observation suggests that the spatial determination is dependent upon substances produced by the egg, which disappear after fertilization. A further suggestion is that some of those substances are produced from maternal mRNAs stored in the eggs.[13,14]

Usually segmentation mutations and homeotic mutations are expressed after the zygotic genome is activated to form the blastoderm.

Segmentation mutations interfere with the normal order division of the embryo into subunits. The best known is the fushi tarazu mutant with missing segments and the joining of incomplete segments. In fushi tarazu mutant the anterior segments Mx, T1, T3, A2, A4, A6, and A8 are fused with segments behind them, i.e., Lb, T2, A1, A3, A5, and A7. Consequently the embryo has seven segments instead of 14, and dies before hatching into larva.

A homeotic mutation causes transformation of one body part into another usually located on a different segment, e.g., legs became antennae, eyes became wings, etc. The mutant with legs on his head instead of antennae was termed Nasobemia, which belongs to a greater class of mutations called the Antennapedia complex (ANT-C). The ANT-C genes determine adult structures of the head and anterior thorax. The other complex, named Bithorax (Bx-C), includes genes that determine the posterior thorax and abdominal segments. Each of those complexes is determined by the combined activity of homeotic genes.[14] Since mutation causes disturbances in segmentation and proper localization of particular organs within the embryo, a hypothesis was established that normal homeotic genes must determine proper localization of organs in the developing organism. The hypothesis could not be proved before cloning of the homeotic genes in *Drosophila* was performed. Two homeotic clusters of genes have been cloned, i.e., the Bithorax complex and the Antennapedia complex. After their purification it was proven that the homeotic genes are very large and complex. For example, the Antennapedia complex is approximately 100,000 bp long, containing many exons separated by introns.[13]

The first step in isolating the Antennapedia complex genes was to assemble a set of slightly overlapping DNA fragments of the chromosomal region with the Antennapedia gene. This was performed by the method named "walking along the chromosome." The method is based on hybridization of overlapping fragments, beginning with a short fragment of DNA located near the gene being searched for, to which by means of hybridization of an unknown overlapping fragment become attached and so on until the whole gene is known.[13] Using these DNA sequences, a cDNA probe was constructed which hybridized with mRNA of the Antennapedia sequences, but also with the fushi tarazu locus, which lies near the Antennapedia locus and the Bithorax complex named also Ultrabithorax (Ubx-C).[13]

Using Southern blotting methods, the homeo box genes have been found also in other organisms including vertebrates and man.[14] The comparison of DNA sequences revealed homologies between the three main homeo boxes, ranging from 75 to 79%. This suggests that the homeo boxes code for a highly conserved protein with a large domain of basic amino acids (30% of all the residues are either arginine or lysine). The two *Xenopus* homeo boxes demonstrate a marked homology to the *Drosophila* sequences (59 out of 60 amino acid residues = 98% of homology). There is also great homology for the mouse homeo domain (73% and for the two human homeo domains (90%).[14]

The mechanism of action of the homeo complex genes probably lies in binding their product with adequate DNA sequences. This hypothesis suggests the presence of the highly conserved large domain composed of basic amino acid residues. Localization experiments of homeo box genes performed with radioactive cDNA revealed changes in the particular mutants. First, the fushi tarazu mutant, by using a radioactive fushi tarazu probe, revealed accumulation of dark grains (radioactive cDNA bound to mRNA) in the nuclei lined up in the cortical cytoplasm (before cell membranes are formed) at seven sections of the missing segments in the fushi tarazu mutant. The transcripts were no longer detected in the embryonic segments. This experiment proved that early in the enbryogenesis the normal fushi tarazu gene must be expressed in alternate sections so that the segmental plan can be realized correctly. This implies that the bare nuclei must have a ''sensor'' which enables them to determine the correct position in the cortical cytoplasm. To prove this hypothesis an artificial gene was constructed, including the fushi tarazu upstream control region and the bacterial gene for beta-galactosidase. The artificial gene was inserted into a *Drosophila* embryo and the expression of the beta-galactosidase gene was tested by a staining reaction. Beta-galactosidase had been found precisely at the pattern of the fushi tarazu transcripts. This denotes that after the nucleus is placed in the cortical cytoplasm, a substance present in this region must interact with the control region of the fushi tarazu gene turning it on or off, dependent on the position of the nucleus. Next, the product of the fushi tarazu gene must interact with a group of genes regulating adequate sets of genes responsible for the structural organization of the organism.[13,14]

Further experiments revealed that as soon as the fushi tarazu gene is turned on, the engrailed gene is also expressed in a pattern of 14 narrow sites corresponding to the posterior compartments of the segments. The ''en'' (engrailed) gene is involved in the spatial organization of the embryo, principally in subdivision of segments into anterior and posterior compartments.[14] Apparently, the embryo is first organized in segments, which divide next into compartments. Finally, the transcripts of various genes of Antennapedia complex and the Bithorax complex accumulate in a segment-specific fashion in the developing ventral nervous system of the embryo, which forms one ganglion per body segment in the embryonic development.[14,16]

The dependence of the position of the homeo box strongly support the hypothesis that these genes are involved in the spatial organization of the embryo. *In situ* hybridization of Antennapedia probes in embryos with Bithorax deletion complex revealed that the Antennapedia complex is regulated by the Bithorax complex.[14]

Beyond the main homeotic gene clusters, a relatively great number of homeotic genes had been discovered. Several murine and human homeo box sequences have been cloned and sequenced. Differential expression of homeo box regions in various adult tissues has been observed.[17]

The sites of transcript accumulation for six different homeotic loci of the Antennapedia and Bithorax gene complexes were identified. These transcripts were discovered in non-overlapping regions of the embryonic central nervous system.[18-20]

The next step of development after blastoderm formation is gastrulation. At this developmental phase the embryonic genome becomes completely activated and expressed which results in the appearance of new cell surface proteins, and new proteins (antigens detected by specific antibodies) on mesodermal, endodermal and ectodermal cells.[21]

A great surprise was the discovery that the products of two homeotic loci—the Notch locus in *Drosophila* and lin-12 locus in *Caenorhabditis elegans*—demonstrate homology with the epidermal growth factor. The Notch locus in *Drosophila* has been known since 1916. Mutants of this locus demonstrate unusual proliferation of neuronal precursor cells. On the basis of RNA analysis the amino acid composition of the Notch protein was predicted, including about 2700 amino acid residues, with the N-terminal composed of 36 repeats, each of 40 amino acid residues, which all demonstrate homology to EGF (epidermal growth factor).[22]

Mutations of lin-12 were recognized by their production of multiple abnormal vulvae in the hermaphrodite (dominant mutants). Thanks to these mutants, the lin-12 locus has been cloned and partially sequenced. The peptide synthesized on the basis of lin-12 locus contain 11 tandem repeats of about 38 amino acid residues. Each repeat copy is homologous to EGF. EGF is a 53 amino acid residues peptide, split off from a precursor 1200 amino acids long, containing nine other copies of EGF-like homology units, resembling the Notch and lin-12 proteins. EGF homology also was demonstrated in a family of transforming growth factors, vaccinia virus growth factor, blood clotting factors VIII and IX, tissue plasminogen activator, urokinase, and LDL receptor.[22] It is difficult to discover a possible common function joining such different protein molecules. The most probable seems to be a specific cell receptor function. In spite of a tremendous number of papers dealing with homeotic genes in connection with the development of multicellular organisms, the problem of genetic mechanisms directing the development of those organisms remains unresolved.[23] The mammalian homeo boxes are similar to those in *Drosophila*. It is supposed that proteins synthesized on the basis of the mammalian homeo box genes play the same role in mammals that proteins synthesized on the basis of *Drosophila* homeo box genes do in those insects. However, these substances are not yet known and it must be proved after their isolation, if they really play this role in mammals. This could be obtained by *in situ* localization of homeo box-related transcripts and proteins in the developing embryo using adequate probes. The molecular mechanism of homeo box proteins function should be elucidated not only in mammals, but also in *Drosophila*. There are an enormous number of events that control the development of a multicellular organism (transcription of an enormous number of genes, their translation resulting in generation of many unknown products with unknown function), and a sole mutation of a homeo box gene may cause disturbances in segmentation or disorder in the development or organs located at unnatural sites in the organism. Although it is very probable that the homeo box genes play an important, perhaps decisive, role in controlling development in mammals. The crucial evidence is still missing, and only time will confirm or disprove it.[24]

Summarizing, the differentiation process in higher eukaryotes requires multiple distinct DNA binding factors which interact with each other and modulate the transcription process for obtaining proper transcripts for the differentiated cells, tissues, and organs. Such multiple transcriptional factors have been discovered in herpes simplex virus-thymidine kinase, human metallothionein and human heat shock gene promoters. Some of those factors have been purified and characterized biochemically, e.g., the promoter-specific transcription factor (Sp1) from HeLa cells, composed of two polypeptides 105 and 95 kDa, recognizing and interacting specifically with the GC box promoter.[25] The same was found in viral infections of mammalian cells, where gene expression may be catalyzed initially by interaction of viral regulatory proteins with cellular transcription factors.[26] Therefore, it seems that transcriptional specificity and control in higher eukaryotes must be dependent on the combined action of multiple specific DNA binding factors interacting with distinct contra-elements in a selective manner to induce synthesis of proper RNAs, and in consequence synthesis of proteins necessary to perform the differentiated functions of a multicellular organism.

REFERENCES

1. **Blau, H., Paviath, G. K., Hardeman, E. C., Chiu, C. P., Silberstein, L., Webster, S. G., Miller, S. C., and Webster, C.**, Plasticity of the differentiated state, *Science,* 230, 759, 1985.
2. **Harrison, P. R.**, Molecular analysis of erythropoiesis, *Exp. Cell Res.*, 155, 321, 1984.
3. **Schmidt, J. A., Marshall, J., Hayman, M. J., Ponka, P., and Beug, H.**, Control of erythroid differentiation: possible role of transferrin cycle, *Cell,* 46, 41, 1986.
4. **Mavilio, F., Giampaolo, A., Carè, A., Migliaccio, G., Calandrini, M., Russo, G., Pagliardi, G. L., Mastroberardino, G., Marinucci, M., and Peschile, C.**, Molecular mechanisms of human hemoglobin switching: selective undermethylation and expression of globin genes in embryonic, fetal, and adult erythroblasts, *Proc. Natl. Acad. Sci. U.S.A.*, 80, 6907, 1983.
5. **Wood, W. G., Bunch, C., Kelly, S., Gunn, Y., and Breckon, G.**, Control of haemoglobin switching by a developmental clock? *Nature (London),* 313, 320, 1985.
6. **Ley, T. J., Chiang, Y. L., Haidaris, D., Anagnou, N. P., Wilson, V. L., and Anderson, W. F.**, DNA methylation and regulation of the human beta-globin-line genes in mouse erythroleukemia cells containing human chromosome 11, *Proc. Natl. Acad. Sci. U.S.A.*, 81, 6618, 1984.
7. **Konieczny, S. F. and Emerson, Jr, Ch.P.**, 5-azacytidine induction of stable mesodermal stem cell lineages from 10T1/2 cells: evidence for regulatory genes controlling determination, *Cell,* 38, 791, 1984.
8. **Chiu, C. P. and Blau, H. M.**, Reprogramming cell differentiation in the absence of DNA synthesis, *Cell,* 37, 879, 1984.
9. **Rakowicz-Szulczynska, E. M. and Horst, A.**, Synthesis of nonhistone chromatin proteins of mice in spleen cells and myeloma cells RPC 5 and ABPC 22, *Mol. Cell. Biochem.*, 33, 115, 1980.
10. **DiBernardino, M. A., Hoffner, N. J., and Etkin, L. D.**, Activation of dormant genes in specialized cells, *Science,* 224, 946, 1984.
11. **Nicola, N. A., Begley, C. G., and Metcalf, D.**, Identification of the human analogue that induces differentiation in murine leukaemic cells, *Nature,* 314, 625, 1985.
12. **Chisholm, R. L.**, Homeoboxes: what do they do? *Trends in Genet.*, 2, 224, 1986.
13. **Gehring, W. J.**, The molecular basis of development, *Sci. Am.*, 253(10), 152B, 1985.
14. **Gehring, W. J.**, The homeo box: a key to the understanding of development, *Cell,* 40, 3, 1985.
15. **Carrol, S. B., Winslow, G. M., Schüpbach, T., and Scott, M. P.**, Maternal control of *Drosophila* segmentation gene expression, *Nature,* 323, 278, 1986.
16. **Douarin, N. M.**, Cell line segregation during peripheral nervous system ontogeny, *Science,* 231, 1515, 1986.
17. **Colberg-Poley, A. M., Voss, S. D., Chowdhury, K., Stewart, C. L., Wagner, E. F., and Gruss, P.**, Clustered homeo boxes are differentially expressed during murine development, *Cell,* 43, 39, 1985.
18. **Harding, K., Wedeen, C., McGinnis, W., and Levine, M.**, Spatially regulated expression of homeotic genes in *Drosophila, Science,* 229, 1236, 1985.
19. **Coulter, D. and Wieschaus, E.**, Segmentation genes and the distributions of transcripts, *Nature,* 324, 472, 1986.
20. **Kilchherr, F., Baumgartner, S., Bopp, D., Frei, E., and Noll, M.**, Isolation of the paired gene of *Drosophila* and its spatial expression during early embryogenesis, *Nature,* 324, 493, 1986.
21. **McClay, D. R. and Wessel, G. M.**, The surface of the sea urchin embryo at gastrulation: a molecular mosaic, *Trends Genet.*, 1, 12, 1985.
22. **Bender, W.**, Homeotic gene products as growth factors, *Cell,* 43, 559, 1985.
23. **Marx, J. L.**, The continuing saga of "homeo-madness", *Science,* 232, 158, 1986.
24. **Manley, J. L., Levine, M. S.**, The homeo box and mammalian development, *Cell,* 43, 1985.
25. **Briggs, M. R., Kadonaga, J. T., Bell, S. O., and Tjian, R.**, Purification and biochemical characterization of the promoter-specific transcription factor, Sp1, *Science,* 234, 47, 1986.
26. **Coen, D. M., Weinheimer, S. P., and McKnight, S. L.**, A genetic approach to promoter recognition during trans induction of viral gene expression, *Science,* 234, 53, 1986.

Chapter 7

THE MOLECULAR STRUCTURES OF THE EUKARYOTIC CELL

I. THE CYTOSKELETON

According to earlier views the cytoplasm of a cell was a formless area in which the nucleus and the other organelles were dispersed. In contrast to this, recent investigations have revealed three kinds of filaments in the cytoplasm that considerably confine the displacement of the particular organelles in cells. The finest filaments are microfilaments, predominantly 6 nm in diameter, composed of the protein called actin. The largest filaments are microtubules, 22 nm in diameter, composed of the protein called tubulin. The third kind of filaments are intermediate filaments, 7 to 11 nm in diameter; the protein content of these filaments varies, dependent on the cell type. All the filaments form a network that stabilizes shape and cell structure, and is termed the cytoskeleton.

The microtubules radiate from the centrosome (also called cytocenter) to the cell membrane. The organization of microtubules changes during mitosis. The chromosomes in the nucleus condense and separate into two sets. Simultaneously, the microtubules degrade and tubulin becomes organized into the mitotic spindle. A network of microtubules extends from the poles of the dividing cell to the chromosome sets and attracts them to the two poles.

The intermediate filaments are differently organized in different tissues according to their constituent protein. In the epithelial cells they consist of keratin. Intermediate filaments usually are dispersed in the whole cytoplasm, but in several cells they appear in the form of bundles or individual filaments. In many cells they are combined with microtubules.

The basic protein of microfilaments—actin—represents a monomer, known as globular actin, termed G-actin. G-actin is stored in connection with profilin. In the presence of ATP G-actin polymerizes into F-actin—filamentous actin in the form of long double helices (the energy for this reaction is generated by degradation of ATP to ADP). The F-actin can polymerase (elongate) at the plus end, and depolymerase (shorten) at the minus end. F-actin in cells usually is bound with a large group of proteins called actin-binding proteins. Among these proteins are gelation factors which characteristically increase the viscosity of F-actin. F-actin forms flexible but tight bonds between the crisscrossed filaments. Other factors, so-called bundling factors, comprise a set of proteins that form dense bundles. Finally, rodlike spacing factors form bridges between parallel actin filaments at a distance of about 200 nm. Sequestering and stabilizing factors bind to single filaments influencing their length and stability. Under the influence of sequestering factors the viscosity of F-actin decreases dramatically. Contrary to this, the stabilizing factors, also named cofilamentous proteins, protect the filaments. The most important stabilizing factor in nonmuscle cells is the filamentous protein tropomyosin.

The assembly of F-actin fibers is governed by so-called capping factors. There is a permanent exchange of G-actin subunits in such a manner that new G-actin subunits are added to the plus end, and simultaneously at the minus end another subunit is released. This process is named treadmilling and is regulated by the capping factors. Thus, the capping factors stop addition and removal of G-actin, and stabilize the amount of polymerized F-actin. The equilibrium between G- and F-actin is secured additionally by the sequestering factors, which form complexes with G-actin, thus preventing them from polymerizing. These G-actin molecules can be used by the cell to form new filaments.

A question arises as to what kind of mechanism regulates the formation, linking, capping, and severing for microfilaments. Some hints to that question are that some of the severing, capping, and spacing proteins are regulated by calcium ions mediated by the calcium binding protein calmodulin. The sequestering proteins may be conditioned by some lipids.

Actin-binding proteins are responsible both for formation of the network, and for activity of these proteins. For example, myosin bound with F-actin causes cell movements by converting ATP into ADP through energy delivery for this reaction (myosin is an ATPase enzyme that converts ATP into ADP). In muscle cells myosin forms associated with F-actin contractile units named sarcomeres. In nonmuscle cells myosin is dispersed in the whole network of microfilaments.

Summarizing, the actin-binding proteins control the activity of actin: profilin, which acts as a sequestering factor that keeps G-actin in storage; capping factors, which regulate addition of G-actin to one end and simultaneously shedding G-actin molecules at the other end of F-actin strands; severing factors, which divide the completed strands; stabilizing factors, which counteract the severing factors by binding parallel to actin strands (thus protecting them); bundling, gelation, and spacing factors, which control the spatial organization of actin filaments; and contraction factors (myosin), which enable contraction of muscle cells by binding to actin filaments having opposite polarity and slides them along each other.[1]

Actin filaments may be the decisive elements that characterize specialized cells. For example, in red blood cells, just under the plasma membrane, there are very short α- and β-actin filaments (about 12 monomers) at the junctions of the network, interconnected by the spacing factor spectrin, which is bound to another protein—ankyrin. Ankyrin is connected to another protein embedded in the plasma membrane and protruding from the outer surface of the membrane. This protein is known as band III protein, according to its position on electrophoretic gels, or anion exchange protein. There are also stabilizing proteins that strengthen the structure of erythrocytes. The plasma membrane structure of erythrocytes enables them to pass through the finest capillaries.

The actin-binding proteins may play an important role in cell physiology. For example, the band III protein, or anion exchange protein, acts as a channel for ion exchange through the cell membrane, enabling the discharge of CO_2 in the lungs and consesquently the acceptance of oxygen. The same protein binds some glycolytic enzymes that enable degradation of glucose.

In other cells some protein kinases that cause phosphorylation of proteins may be present at the subsurface network. Phosphorylation of proteins causes disturbances in their functions. Some tumour viruses particularly may cause dramatic changes in the cytoskeleton by inducing synthesis of proteins encoded by cancer-inducing oncogenes. In some of these cases phosphorylation of specific cytoskeletal proteins has been demonstrated.

A very complicated network of actin was described in microvilli of the external surface of intestinal epithelial cells. These cells have approximately 1000 microvilli on their surface through which nutrients are absorbed. In each microvillus a bundle of F-actin is present. The bundling factors in these cells are proteins: villin and fimbrin. A series of bonds establish bridges between the actin bundle and plasma membrane of the microvillus. It is supposed that this is a transmembranal protein which joins the actin bundles with the surface membrane of the microvillus. In which way the upper end of the bundle is fastened is unknown. At the lower end the actin bundle is anchored in a network of a spectrin-like spacing factor, and next the actin fibers are mixed with the intermediate filaments of the cell—in this case keratin. The keratin filaments are anchored with the cell membrane at specialized regions—desmosomes—which join together epithelial cells.[1] The described structure of microvilli is accommodated specifically to the function of intestinal epithelial cells enabling both an increase of the absorbing surface, and movements that facilitate absorption of nutrients.

Microtubules, composed of another protein, tubulin, play an important role in intracellular organization, function, and shape of particular cells. Tubulin also is the main component of cilia and flagella in cells, and enables them to move. Tubulin is a globular protein that consists of two different polypeptide chains, α and β. In the presence of GTP (energy donator), the tubulin dimers composed of α- and β-tubulin chains polymerase, forming a tubelike structure (about 22 nm in diameter) with a hollow core (about 10 nm in diameter). Assembly of tubulin occurs after α- and β-tubulin have made 13 protofilaments arrayed around the hollow core. Microtubules are connected with other protein molecules, penetrating into the cytoplasm. Microtubules demonstrate the same polarity as microfilaments *in vitro*: one end becomes elongated by polymerization while the other end is shortened by the release of tubulin molecules. There also is a suggestion that microtubuli play a role in cell division, particularly in formation of the mitotic spindle.

Tubulin-associated proteins may be present only at specific stages or specific cell compartments. For example, a unique kind of protein is associated with tubulin in the mitotic spindle, in neurons other specific proteins are associated with tubulin in dendrites and axons, and another is associated in cilia and flagella where tubulin is bound with a flexible protein, dynein, which (like myosin) is an ATPase. Dynein or other myosin-like proteins, associated with tubulin in the cytoplasm, may enable cell movements, and by binding capping factors support shortening or lengthening of microtubules.

A specific mechanism of tubulin-associated protein has been described in transport of vesicles along the axons of nerve cells and movement of other organelles in the cytoplasm of these cells. *In vitro* experiments demonstrated that isolated vesicles move along the purified microtubules one micrometer per second (in the presence of ATP and a crude mixture of cellular proteins). It is supposed that within this crude mixture of proteins a specific translocator-protein may be present that translocates the vesicles using ATP energy.[1] The specific inhibitor of microtubules is colcemid. In cells treated with this drug the microtubules depolymerize and disappear. After removal of the drug, the microtubules reappear without any difference as compared with normal cell (within 75 min at a temperature of 37°). The microtubules reappear in a unidirectional manner, elongating outward from the centrosomes.

The microtubules also influence the shape and orientation of different cells. For example, fibroblasts in culture are usually asymmetric, demonstrating a ruffled leading edge and a slender tail. After addition of colcemid, which causes disappearance of microtubules, the cells become more symmetrical and their movement stops. Colcemid treatment of cells causes characteristic changes of intracellular organelles: thc movement of these organelles stops, the Golgi membranes become disordered and move to the centrosome region, and channels of the endoplasmic reticulum retract. Similar disturbances after colcemid treatment also demonstrate the intermediate filaments, they retract and coil around the cell nucleus. Taxol, a drug that promotes tubulin polymerization, causes the cell's microtubules to form bundles that no longer connect with the centrosome.[1]

In conclusion, it seems that microtubules play an important and specific role in highly asymmetric cells, particularly in neurons whose axons extend for several meters and without which guidance of the transport within the cell without the system of microtubules would be impossible.

The intermediate filaments are encoded by a family of genes of which various genes are expressed in various tissues, developmentally determined. There are five distinct kinds of intermediate filaments:

1. Epithelial keratin in epithelial cells
2. Neurofilaments in most neurons
3. Desmin in muscle cells
4. Glial filaments in glial cells
5. Vimentin filaments in connective tissue in blood and lymph vessels (in cells of mesenchymal origin)

There exist a great number of subtypes, e.g., approximately 20 different keratin molecules, that have been described in various kinds of epithelia.

The structural basis of intermediate filaments are rodlike central regions of invariable length and similarity (30 to 70% of identical amino acid residues). The intermediate filaments are about 10 nm in diameter (about 30 molecules in an interlocking arrangement). The assembly of intermediate filaments does not need ATP or GTP.

The terminal regions of intermediate filaments are not organized in filaments. They extend into the cytoplasm, where they bind with structural components. The intermediate filaments are combined with associated proteins, but only some of them are known. Inhibition of intermediate filaments does not disturb cell movements and cell division; therefore it is supposed that these filaments play a very subtle role, i.e., they participate in the stability and strength of the epithelial cells. This view is supported by the anchoring of keratin filaments at desmosomes that join neighboring cells together, due to the presence of glycoprotein molecules that interact with both desmosomes of the neighboring cells.

Most of the known structure and properties of the cytoskeleton proteins are based on experiments *in vitro*. Recent experiments *in vivo* demonstrated that five to seven long thin filaments (about 180 nm long) of purified spectrin tetramers are attached by their ends to junctions of short stubby actin filaments, generating a series of polygons (mostly hexagons) in which the sides are spectrin molecules and the vertices actin filaments.[2] The actin filaments are probably associated with additional protein molecules. Near the middle of the spectrin molecules there is present an ankyrin band III membrane attachment site for spectrin.

Recently, a dimeric calmodulin binding protein (M_r 97,000 to 103,000) has been discovered within the membrane skeleton, one dimer per filament of 16 actin monomers. Only a short actin filament containing six actin monomers is needed to bind six spectrin molecules. However, such a complex of spectrin, protein 4.1 and actin would not be self-regulating because of the tendency of actin to polymerize into long filaments. Therefore, the actin filament is bound with additional proteins that inhibit polymerization of actin which probably regulates the number of spectrin molecules attached to the junctions and specifies the regular lattice structure of the network.[2] An unexplained fact is that although *in vitro* actin filaments bind in equimolar amounts spectrin-protein 4.1, *in vivo* only six spectrin molecules are bound to any of the 13 to 15 monomer long filaments.[2]

Changes of shape and structure of erythrocytes may be caused by phosphorylation of the cytoskeleton proteins: protein 4.1, protein 4.9, and protein M_r 97,000 to 103,000 (calmodulin-binding dimer). These proteins can be phosphorylated by calcium ions, calmodulin-dependent protein, protein kinase C activity, and cyclic AMP-dependent kinase. It was demonstrated that kinase C and cyclic AMP-dependent kinase phosphorylate proteins at different sites. Phosphorylation of protein 4.1 by cyclic AMP-independent kinase in erythrocytes causes a decrease of binding affinity of protein 4.1 to spectrin and influences organization of the membrane skeleton.[2] It is supposed that phosphorylation by cyclic AMP kinase takes place at a spectrin-binding site.

The development of the cytoskeleton was examined in chick embryo erythroid cells which contain a membrane skeleton composed of the protein complex spectrin-band 4.1 and spectrin-binding protein ankyrin. Ankyrin binds spectrin with the integral protein band 3 (being the ultimate membrane skeletal receptor) that limits the amount of newly synthesized ankyrin. The spectrin beta-chain contains ankyrin binding site, consequently limiting beta-chain incorporation that also limits alpha-chain incorporation. According to this, the spectrin chains that are not incorporated into the skeleton become degraded.[3,4]

II. THE MOLECULAR PATHOLOGY OF THE SKELETON

Molecular disturbances of the skeleton are best visible on free movable cells in body fluids, in the first place in erythrocytes that change their shapes and become elliptic, spheric, etc. At present, our rudimentary knowledge about inherited disturbances of the particular proteins of the cytoskeleton is limited principally to disturbances of spectrin molecules, while that of the remaining cytoskeleton proteins is practically none. The best known pathophysiological disorders caused by structural changes of spectrin molecules are some congenital hemolytic anemias.[5]

A. HEREDITARY SPHEROCYTOSIS

Many abnormalities are known in hereditary spherocytosis including changes in shape, membrane cation permeability, disturbances of intracellular metabolism, and increased splenic entrapment with subsequent hemolysis. Not all the disturbances have been observed in all cases; therefore, it is possible that a molecular defect in cytoskeleton structure may not be the basic defect in that disease. There are some other possibilities: (1) spectrin deficiency has been shown in autosomal recessive spherocytosis in mouse mutants and partial spectrin deficiency in humans with hereditary spherocytosis; (2) a functional defect observed in spectrin purified from erythrocytes of patients with autosomal dominant hereditary spherocytosis: defective binding capacity for protein 4.1; (3) tighter binding of spectrin to the erythrocyte membrane.[6] These defects may contribute to the spherical shape of erythrocytes and their hemolysis. Alas, direct proof that pathophysiology of hereditary spherocytosis could be explained by a primary molecular defect of the cytoskeleton proteins is lacking.

B. HEREDITARY ELLIPTOCYTOSIS

Elliptocytosis is a clinically and biochemically heterogenous group of diseases characterized by elliptically shaped red blood cells. The disease is characterized by an autosomal dominant mode of inheritance. The principal molecular defect lies in disturbances of structure and properties of spectrin molecules.

Lawler et al.[7] described patients with elliptocytosis in which self-association of spectrin heterodimers to tetramers was defective, designated HE(SpD-SpD). Limited tryptic digestion of spectrin from those patients revealed diminution in the 80,000 Da domain of the alpha subunit, which is involved in spectrin dimer self-association. The decrease of this spectrin subunit was associated with an increase in atypical peptide fragments (74,000 Da fragment in 8 out of 9 families, 46,000 Da and 17,000 Da fragments in one family). The atypical spectrin fragments have been described in variants of pyropoikilocytosis, also with defective self-association of spectrin. The conclusion is that two distinct structural variants of spectrin may cause defective self-association of spectrin dimers. A similar case was described by Dhermy et al.[8] with diminution of the 80,000 Da dimer and presence of an atypical 74,000 Da spectrin fragment. This spectrin abnormality manifested clinically as a congenital hemolytic poikilocytic anemia. Lawler et al.[9] described also a third variant of spectrin α-chain in hereditary elliptocytosis, designated Sp α I/65 in three unrelated families. In all patients examined the percentage of SpD (in low ionic strength, O°C) in membrane extracts increased to 19% and 32%. Tryptic digestion revealed a decrease of α I domain and the presence of an atypical fragment 65,000 Da (isoelectric points 5.2 and 5.3). Derivation of this fragment from the α-chain was confirmed by staining specific antiserum to the spectrin α-chain.

Pyropoikilocytosis is a disease related to elliptocytosis. It is a rare hemolytic anemia characterized by erythrocyte budding and fragmentation. In pyropoikilocytosis the N-terminal domain of the spectrin α subunit demonstrates increased susceptibility to tryptic digestion. A 50,000-Da fragment is split off, probably because of the presence of a change in primary structure of the spectrin α I domain that alters conformation and creates a new, atypical cleavage site. This change impairs formation of spectrin oligomers *in vitro* consistent with the role of α I T80 in spectrin self-association.[10] Clinically this spectrin disturbance manifested in mild elliptocytosis without hemolysis to severe poikilocytic anemia resembling pyropoikilocytosis. Clinical disturbances were directly proportional to the amount of the atypical 50,000-Da fragment. In some cases changes in α II and III domains of the spectrin molecule were present.[10]

C. DISTURBANCES ASSOCIATED WITH CHANGES OF ERYTHROCYTE MEMBRANE PROTEINS

Reduction of approximately 30% of erythrocyte membrane protein band 4.1, dominantly inherited, does not cause clinical manifestations. Moderate clinical manifestation (mild elliptocytosis) was present in cases with recessive transmission. Deficiency of protein 4.1 seems to be relatively frequent and without clinical manifestations. In one heterozygous family an atypical protein band 4.1, shortened by about 8500 Da, involving both subcomponents 4.1a and 4.1b, without clinical manifestations was described.[11]

Self-digestion of erythrocyte membrane protein bands 2.1, 3, and 4.2 was faster in patients with chorea-acanthocytosis than in controls. Binding of spectrin with band 3 protein at the cytoplasmic side of the erythrocyte membrane revealed some conformational defects that may be responsible for the low fluidity in the interior erythrocyte membranes.[12] The question is open if these molecular defects are also present in other cell membranes.

D. THE ROLE OF CYTOSKELETON IN PARASITIC DISEASES

The malaria parasites can enter inside the erythrocytes only in the presence of ATP in the host cell membrane. In the absence of ATP, the malaria parasites bind to the cell receptors on the extracellular surface, but are not incorporated into the erythrocytes and remain at the surface of erythrocytes. The process of entry of the malaria parasites probably involves active participation of the cell membrane cytoskeleton.[13] The authors showed that in this case, ATP acts by the membrane-associated, cyclic AMP-independent kinase of erythrocytes that phosphorylates spectrin molecules. Therefore, the activated substrate in malaria parasites incorporation seems to be phosphorylation of spectrin molecules.

E. THE ROLE OF CYTOSKELETON IN CARCINOGENESIS

Cytoskeleton filaments are present in many tumors: keratins in carcinomas (tumors of epithelial origin), the glial-filament protein in gliomas, vimentin in lymphomas and non-muscle sarcomas, and protein of neurofilaments in tumors originating in the sympathetic nervous system. The presence of characteristic keratin variants in particular carcinomas (e.g., squamous-cell carcinomas and adenocarcinomas) may be used for clinical diagnosis of these carcinomas, including early lymph node or bone marrow metastases. Next, intermediate filament typing can be used for prenatal diagnosis, e.g., the presence of amniotic fluid cells containing glial filaments or neurofilaments may suggest a malformation of the central neural system. Abnormal intermediate filaments may be present in muscle cells in various muscle diseases and brain cells, probably in Alzheimer's disease and others.[1]

III. THE CELL MEMBRANE

From one side the cytoskeleton, and from the other side the cell membranes enable localization of cellular metabolic processes in such a manner that the particular metabolic reactions can go in the order necessary for realization of the particular metabolic pathways. Precisely this process has been realized *in vitro* in the procedure of immobilized enzymes, which bound with a stable substrate can govern and imitate the course of biochemical reactions in an order that exists in naturally occurring metabolic pathways. Besides, the cell membranes isolate cells from their environment and form compartments inside the cells which are necessary for the course of biochemical reactions organized in specific pathways. Finally, the cell membranes must be permeable to nutrients that enter into cells and their compartments, and simultaneously to waste materials that leave them. Biological membranes serve not only as dividing walls between cells and their organelles (and their environments), but they also serve as specific mediators in life processes based on their specific structure.

The cell membranes represent a thin stable film of lipids and proteins. For the different functions of the cell membranes, the protein molecules are most important, although the role of lipids should not be underestimated. A bilayer of phospholipids forms the basic framework of all membranes. Phospholipids have a hydrophylic head, predominantly choline, called phosphatidylcholine or sphingomyelin. Choline is bound to a glycerol group to which two hydrocarbon chains are attached. The hydrocarbon chains are hydrophobic, each of them is a fatty acid chain. Instead of choline the heads also may consist of ethanolamine in phosphatidylethanolamine, serine in phosphatidylserine, and inositol in phosphatidylinositol. The electric charge makes the head groups hydrophilic. In water the individual phospholipid molecules produce a bilayer in a form of closed spherical vesicles having two separated compartments, with fluid both inside and outside of the vesicle. This property makes the phospholipids particularly useful in performing biological membranes that are impermeable to most biological molecules (amino acids, sugars, proteins, nucleic acids), and to ions that are insoluble in hydrocarbon solvents and soluble in water.

The naturally existing phospholipid bilayer is a liquid that enables exchange of phospholipid molecules between neighboring molecules. In contrast to this, exchange of phospholipid molecules in opposite monolayers is extremely rare. Additionally, the local distribution of phospholipids is ordered in a manner that phosphatidylcholine and sphingomyelin are present mostly at the outer monolayer (in erythrocytes) while phosphatidylethanolamine and phosphatidylserine are at the cytoplasmic side of the bilayer (phosphatidylinositol probably is also at the cytoplasmic side of the bilayer).[14] The fluidity of the hydrocarbon interior of biological membranes is secured by the chaotic Brownian movements of the phospholipid molecules and random configuration of the tails. Therefore, the "fluid mosaic model" of biological membranes in which all the molecules that form the membrane are mobile, protein molecules included.

Two other lipids are present in animal cell membranes, i.e., glycolipids and cholesterol. Glycolipids are composed of a hydrophilic end of various sugars in the form of linear or branched structures, and a hydrophobic tail. Glycolipids appear only at the outer side of the monolayer and there are only a few of them inside the membrane. The physiological role of glycolipids is scarcely known (they probably serve as antigens that allow recognition of various kinds of cells by the immunological system of organisms).

Cholesterol is a major membrane lipid with one hydrophilic end and the rest hydrophobic. The hydrophobic part of cholesterol is embedded in the hydrophobic part of the plasma membrane (about equal numbers of cholesterol and phospholipids). The presence of the "rigid" cholesterol molecule within the phospholipid layer makes the membrane a little less flexible and less permeable. The cholesterol molecules keep the phospholipid tails relatively fixed and ordered close to the hydrophilic heads.

A. THE MEMBRANAL PROTEINS

The membranal proteins play the decisive role in the specific functions of cell membranes. They can be classified into two main groups according to their shapes: the rodlike proteins (proteins of α-helix structure), and globular proteins (proteins of a globular structure within the membrane's hydrophobic region).

A typical example of a α-helical protein is represented by glycophorin (the major membranal glycoprotein of erythrocytes). This protein spans the cell membrane in a manner, that a sequence of 26 hydrophobic amino acids lie inside the hydrophobic phospholipid bilayer, while on the other side both ends of this protein have a hydrophilic character. The function of this protein is not clearly known, it is supposed that the transmembranal localization of this protein argue for transmembranal transmission of signals from the outside to the inside of the cell.

Besides the proteins that span the phospholipid bilayer there are also proteins that are anchored to the cell surface by a short hydrophobic tail. Globular membranal proteins may contain more hydrophobic segments joined by hydrophilic ones. In this case the hydrophobic segments are embedded in the hydrocarbon core of the membrane forming a bundle bound by the hydrophilic segments. A typical example is bacteriorhodopsin with seven hydrophobic segments, joined by short hydrophilic ones. The hydrophobic segments are embedded in the hydrocarbon core of the membrane, bound from the outer and inner side by short hydrophilic segments (see Figure 1). To one of the hydrophobic segments a retinal molecule is attached that captures solar protons, triggering the protein to pump protons against the energy gradient across the membrane of certain salt-loving bacteria, which starts an unusual kind of photosynthesis.[15] The potential energy achieved by proton pumping is used for ATP synthesis which provides energy for biosynthetic pathways.

A very important and noteworthy protein of globular membranal structures (described in erythrocytes) is the anion exchange protein with several hydrophobic segments embedded in the hydrocarbon core, and hydrophilic tails, one bound inside the cell with ankyrin and then with spectrin, and the other at the outer side of the cell (this protein is also named band 3 of the erythrocyte plasma membrane). This protein forms a passageway for anions through the protein, which enables them to pass across the phospholipid bilayer of the cell membrane.

Most of the membranal proteins of eukaryotic cells have one or more oligosaccharide chain on their extracellular segments. The physiological role of these oligosaccharide chains is not clear, but probably they facilitate recognition of cells of the same character. The membrane proteins float freely in the lipid bilayer which enables them to diffuse sideways from one end of the cell to the other in only a few minutes. It has been calculated that membrane proteins can diffuse about 10 μm in 1 min.

Cells, especially the epithelial cells, usually make sheets one cell thick. The epithelial sheets in the gut have two surfaces, one faces the digestive tract (the apical one), the other faces the blood (the basolateral surface).

The position of epithelial cells in sheets is held by three kinds of immobilizing structures: the tight junction, desmosome, and gap junctions. The tight junction in the form of a circular belt lies nearest the apical surface of the cell. This structure prevents leaking and divides the plasma membrane into the apical and basolateral compartments, preventing mixing of membrane proteins of those two membranal compartments that secures specific functions of the apical and basolateral surfaces of cells. The next structure, desmosomes, occur near the basolateral surface and bind together two neighboring cells. Finally, there are gap junctions (see Figure 1). The gap junctions are composed of 12 subunits, six from each cell. Each group of the six subunits is arranged in a hexagon in the plasma membrane of apposed cells; the two hexagons lock onto each other and form a channel between the apposed cells. These channels can be open or closed. In opposition to the other cell membrane components, the gap junctions are relatively motionless. Small molecules (about 2 nm in diameter) such as ions, amino acids, oligosaccharides, and nucleotides can pass freely from the cytoplasm of one cell to the cytoplasm of opposed cells.

The tight junction of epithelial cells maintains the functional asymmetry necessary to transport materials only in one direction. For example, sodium is taken up from the gut on the apical surface of epithelial cells by specific proteins, and pumped on the basolateral surface into the blood by a different set of proteins which makes the transport of sodium extremely selective.[14] The question arises as to which way the epithelial cells separate their proteins, some to the apical surface and others to the basolateral surface. Sorting of membranal proteins is best documented in virus experiments. In a sheet of epithelial cells infected with influenza virus, the progeny viruses emerge only from the apical surface of the cells, while vesicular stomatitis virus (VSV) emerges only from the basolateral surface. Because viruses can leave host-cells coated only with proteins, it is concluded that the coat proteins of influenza viruses are directed to the apical surface, while the coat proteins of vesicular stomatitis virus are directed to the basolateral surface.

B. INTRACELLULAR ORGANELLES

Besides the cell membrane, there are numerous membranes in organelles of eukaryotic cells, e.g., endoplasmic reticulum, Golgi apparatus, lysosomes, etc. These intracellular organelles have their own membranes which play an important role in synthesis and transport of various substances. Some of the membrane components are synthesized in the endoplasmic reticulum, and oligosaccharides are added to the membrane proteins in the Golgi apparatus. There exists a continuous transfer of membranes from the endoplasmic reticulum to the Golgi apparatus, and from the Golgi apparatus to the cell membrane, probably mediated by phospholipid vesicles. It has been suggested that the transfer of membranes must be selective, but the mechanism of this transfer is unknown. Membranes of cellular organelles are composed of phospholipid monolayers instead of bilayers, as in cell membranes.

The intracellular pathway and assembly of newly formed variable surface glycoproteins is best known in African trypanosomes. These parasites have a continuous 12 to 15 nm thick glycoprotein coat (VSG—variable surface glycoproteins) that vary during different stages of infection, thus preventing the immunological attack. *Trypanosoma brucei* and *T. congolense* contain two classes of carbohydrate side chains. One is the internal side chain located in the COOH-terminal third of the glycoprotein molecule that contains mannose and *N*-acetylglucosamine. The other is localized at the COOH-terminal end and bound through an ethanolamine linkage to the α-carboxyl group, that contains mannose, *N*-acetylglucosamine, and galactose. This oligosaccharide chain remains stable on all variable antigens of *Trypanosoma brucei* and *T. congolense*. On the basis of pulse-chase experiments with ^{35}S-methionine it was found that variable surface glycoproteins are synthesized in the rough endoplasmic reticulum of *Trypanosoma* during 6 to 8 min, shuttled in the next 8 min to the *trans*-Golgi region (*cis*-Golgi region is not labeled).[16]

The Golgi apparatus, also called the Golgi complex, represents a series of membranal compartments through which membranal proteins, secretory vesicles, and lysosomes move sequentially. Different classes of proteins, destined for particular cell membranes, are sorted into different vesicles in the last Golgi compartment, named the *trans*-Golgi network, which corresponds to a tubular reticulum on the *trans*-side of the Golgi apparatus, previously called the Golgi endoplasmic reticulum lysosomes.[17]

The cells of a multicellular organism are surrounded by an aqueous medium that contains an enormous amount of ingredients derived from the blood usually at extremely low concentrations. These ingredients include nutrients (ions, amino acids, sugars, vitamins, etc.), intercellular signals (mainly hormones), and also toxic substances derived from cellular metabolic degrading processes. These substances must be taken up by selective mechanisms to the adequate cells (hormones to the particular target cells of the endocrine system, toxic substances to cells that degrade them, etc.).

IV. MECHANISMS THAT ENABLE CROSSING OF CELLULAR MEMBRANES

A. MEMBRANE CHANNELS

Membrane channels are composed of specific proteins that enable passage of small water soluble substances such as ions, amino acids, sugars, etc. A very typical channel represents the sodium channel that opens per depolarization, and closes in an inactivated state. The energy that transports sodium across membranes is obtained by degradation of ATP to ADP by the enzyme (Na^+ + K^+)ATPase. The passage of sodium ions is executed by exchange with potassium ions.

A very important mechanism for many cellular functions represents the calcium ions channel. Calcium ion concentration inside the cells can increase by releasing calcium ions from the intracellular calcium storage sites and/or by moving calcium ions from the extra-

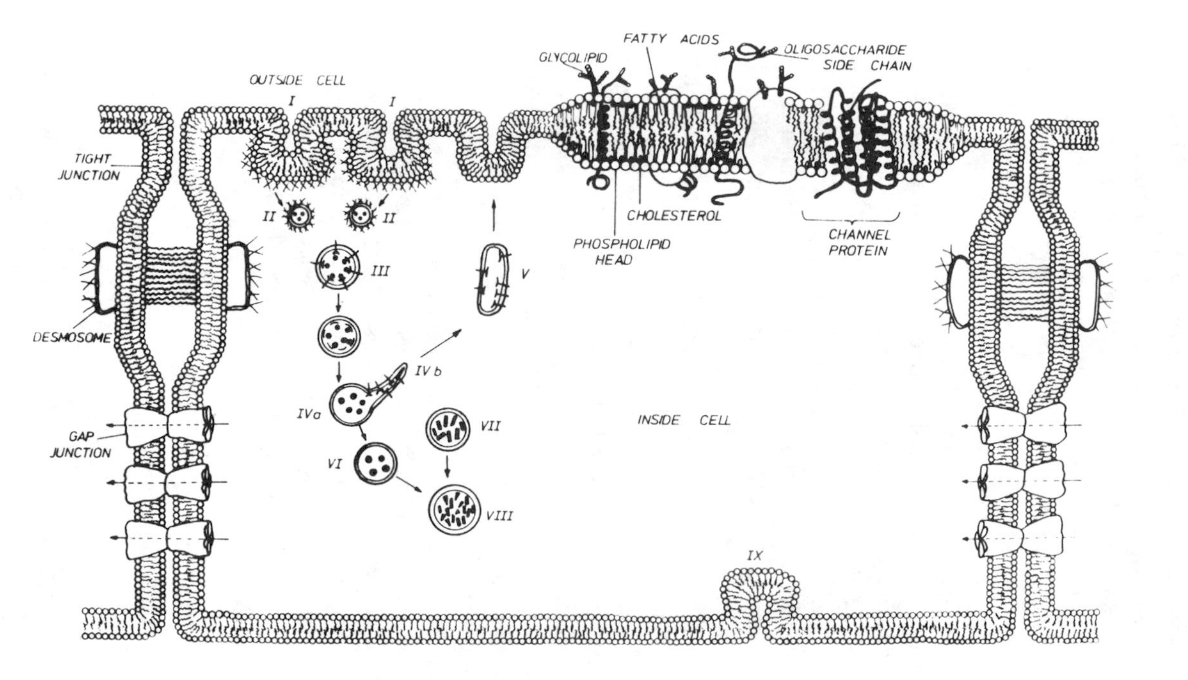

FIGURE 1. Schematic diagram of plasma membrane. Plasma membrane represents a phospholipid bilayer in which cholesterol and various proteins are embedded. The phospholipid bilayer is composed of phospholipid heads (round circles) and tails representing fatty acids of each phospholipid molecule. At top (right) a complex protein is presented that forms a channel enabling transport of ions, protein, sugars, etc. across the membrane. This protein is composed of hydrophobic segments embedded inside the membrane bilayer joint by hydrophilic fragments outside the membrane into one molecule. On the extreme left, there is an integral membrane protein with a single α-helix hydrophobic chain embedded inside the phospholipid bilayer, with hydrophilic fragments outside the bilayer—the hydrophilic fragments join the particular hydrophobic chains into one molecule. The integral membranal proteins have the N terminals outside the cell often with oligosaccharide side chains, and the C terminals inside the cell. Oligosaccharides are represented by small round circles. Finally, this fragment also possesses other globular integral proteins and cholesterol molecules between the fatty acids. On the adjacent cells there are tight junctions that separate the upper part of the membrane from the lower part, then the desmosome structure and gap junctions at the lowest part. On the left part of the cell the endocytosis mechanism is presented: I—coated pits, II—coated vesicles, III—endosome, IV CURL—(a) the vesicular, and (b) the tubular portion, V—tubular portion migrating to the cell surface with nondegraded receptors that recycle at the surface to coated pits, VI—the vesicular portion fuses with the lysosome VII → VIII, which causes degradation of the ligand. An analogous mechanism appears for exocytosis which transports cellular products outside the cell to the circulation, represented in this scheme by a single incomplete circle (IX) at the basolateral cell surface. More details in text.

cellular space through the calcium ion channels. Calcium channels are built by specific proteins, embedded in the lipid bilayer of the cell membrane. The cellular structure of calcium channels forming proteins is responsible for the selectivity of ions that pass those channels and for the membrane potential-dependent opening and closing of the channels. There are different calcium ion channels that influence the particular cell functions. Three different calcium ion channels in cultured sensory neurons[18] and two different types in mammalian cardiac cells[19] have been described. The particular calcium ion channels in sensory neurones differ by functional properties of individual channels: the L-type channels are characterized by repeating openings that produce long-lasting inward calcium ions (Ca^{2+}) current through the membrane, the T-type channel that opens at much more negative membrane potentials and produces transient inward membrane potential, and the N-type channel activated by large depolarization from a very negative membrane potential.[19] The L- and T-type channels also are present in cardiac muscle, while the N-type channels are present only in neuronal tissue.

The calcium channels are decisive in many cell functions. Their opening is essential for rhythmic firing of nerve cells, in neurosecretion, and in activation of contraction in cardiac and smooth muscles. The activity of L-type calcium channels in cardiac cells is modulated by neurotransmitters (catecholamine and acetylcholine) which probably causes AMP-dependent phosphorylation of the channel proteins. Modulation of calcium channels influences heart functions.[19] In sensory neurones noradrenaline modulates Ca^{2+} channels, independently of cyclic nucleotides. The T-type Ca^{2+} channels retain their function in isolated membranes, whereas L-type channels do not.[19]

Biochemically the Ca^{2+} channel may be a glycoprotein of about 200,000 M_r, consisting of three subunits.

Phosphatidylinositol is a minor component of mammalian cell membranes. Its role in activation of various surface receptors is not fully explained. It was thought that its breakdown by cytoplasmic phospholipase and resynthesis might control Ca^{2+} influx by opening and closing of calcium channels. Recent investigations suggest that, rather, breakdown of phosphatidylinositol may be caused by Ca^{2+} influx, and that phosphatidylinositol activates various protein kinases, and conversion of phosphatidylinositol to diaglycerol may produce a localized increase in membrane fluidity that could facilitate protein-protein interactions in the membrane.[20] Recent investigations revealed that neuronal calcium channels can be regulated by GTP-binding proteins, and activation of sodium proton exchange is a prerequisite for calcium ion mobilization in human blood platelets which is an essential step in the events that cause increase of free Ca^{2+} in platelets.

B. MEMBRANAL GLUCOSE TRANSPORTING PROTEIN

Glucose transport inside the cell, studied in erythrocytes, occurs by the mechanism of a facilitated diffusion carrier. It is an integral membrane glycoprotein (M_r approximately 55,000) that contains probably a single N-linked oligosaccharide. The purified protein exhibits specific D-glucose transport when reconstituted into lipid vesicles.[21] The uptake of glucose into muscle and fat cells increases after administration of insulin. It is suggested that insulin causes translocation of glucose transporting protein from the intracellular storage to the cell membrane that enables increasing uptake of glucose by these cells. The glucose transporting protein probably has 12 membrane-spanning domains. Most of them are probably amphipathic α-helices that contain several hydroxyl and amide side chains binding sugars or lining transmembrane pores through which the sugar passes. The amino terminus, carboxyl terminus, and a highly hydrophilic domain in the center of the protein lie probably on the cytoplasmatic face.[21]

C. ENDOCYTOSIS AND EXOCYTOSIS

Large molecules and material are brought into the cells by a process called endocytosis, and leave cells by an analogous process called exocytosis. In general, endocytosis represents a mechanism in which a patch of plasma membrane encloses the material to be taken up in the form of a vesicle that is incorporated into cells by specific mechanisms. These mechanisms are: phagocytosis, pinocytosis, and receptor-mediated endocytosis.[22]

D. PHAGOCYTOSIS

Very large particles trigger an expansion of the cell membrane around themselves that become incorporated into the cell in the form of a vesicle (several micrometers in diameter) composed of an invaginated patch of the membrane. In higher organisms phagocytosis occurs mainly in the immune system by which macrophages engulf bacteria and kill them.

E. PINOCYTOSIS

Pinocytosis refers to nonspecific uptake of fluids. Similarly, as in phagocytosis, a droplet of fluid becomes surrounded by a patch of invaginated plasma membrane and incorporated in the form of a vesicle (about 1 μm in diameter) ions and small molecules become incorporated into cells by this method.

F. RECEPTOR-MEDIATED ENDOCYTOSIS

In contrast to phagocytosis and pinocytosis, receptor-mediated endocytosis is extremely specific, warranted by specific cell receptors. The receptors are membrane proteins that have binding sites fitting the particular ligands (macromolecules like cholesterol, transferrin, proteins, etc.). The receptor binds usually one particle from the extracellular fluid (even when present in extremely low concentration) that triggers an invagination of the membrane and forms a vesicle enclosing the receptor and brings it into the cell.

The receptors that mediate endocytosis are amphiphatic; they have two hydrophilic regions (one outside and the other inside the cell) and a central hydrophobic region that binds the fatty acids that form the core of the membrane. The molecular structure of most receptors is unknown. One of the best known is the transferrin receptor, a glycoprotein with M_r about 180,000, composed of two identical polypeptide chains, each of about 800 amino acid residues, linked by a disulfide bond. Each polypeptide chain carries three carbohydrate chains and a fatty acid (palmitate) that help to anchor the receptor in the membrane. Each receptor binds two iron (ferric ions) loaded molecules of transferrin (this step of endocytosis does not require energy and is realized as well at low temperature) and is immediately internalized (this step requires energy and is blocked at low temperature). Transferrin has a very high affinity for its ligand—its dissociation constant is 5 nm (about 350 μg/l); it binds transferrin even when its concentration is about .00001 of the total concentration of proteins in the blood.[22]

Ligand-receptor complexes accumulate in so-called pits that represent a thick proteinaceous layer on the inner side of the plasma membrane. Under each pit, a fibrous protein—clathrin—is present. Coated pits furnished with ligand-receptor complexes invaginate and form vesicles that incorporate into the cells. Besides the membranaceous coat, the surfaces of the vesicles have a fibrous network composed of clathrin which forms pentagons and hexagons. Clathrin can be dissociated into three-armed subunits, called triskelions.

Deep in the cytoplasm the coated vesicles shed their clathrin coats and fuse into large, smooth-faced vesicles called endosomes. In the acidic environment of the endosomes the ligands are released from their receptors. The ligands and the receptors separate in the endosomes forming a structure with a vesicular portion on one end and a tubular portion on the other. The receptors and the ligands in the tubular portion accumulate in the vesicular portion. This structure is calleld CURL ("compartment of uncoupling of receptor and ligand"). The tubular portion then migrates to the cell surface receptors, thereby escaping degradation, whereas the vesicular portion fuses with lysosomes and the ligand particles become degraded, migrate into the cytoplasm, and are used as raw materials. The recycled receptors on the cell surface bind new ligands and internalize them in a cycle of about 10 to 15 min (see Figure 1).[22]

A question arises as to where the enormous number of different receptors that mediate endocytosis are sorted. Using immunoelectron microscopy, three different receptors have been analyzed: receptors for asialoglycoprotein (present in liver parenchymal cells, responsible for clearance of galactose terminal glycoproteins from the circulation); receptors for mannose-6-phosphate ligands (involved in targeting of newly synthesized lysosomal enzymes to primary lysosomes and binding mannose-6-phosphate residues present only on lysosomal enzymes); and receptors for polymeric IgA (known as the membrane secretory component that directs the transfer of IgA from the blood to the bile, and is responsible for transfer of immunoglobulins from the mother's blood to that of the newborn). Among these receptors, the asialoglycoprotein receptor is taken up in clathrin-coated pits and vesicles, uncoupled in CURL and retransported to the cell surface. The mannose-6-phosphate receptors bind mannose-6-phosphate molecules to lysosomal enzymes in the Golgi apparatus and then migrate probably to the cell membrane and through endocytosis to the lysosomes, where lysosomal enzymes separate. The IgA receptor binds polymeric immunoglobulins at the sinusoidal membrane surface. Then the receptor-ligand complexes are internalized through endocytosis. These complexes are directed to the bile canalicular membrane, where IgA is released and coupled to the extracytoplasmatic portion of the receptor. In contrast to the two other receptors, this receptor is not recycled.[23]

The physiological function of coated pits is diversified. They are present on the cell surface of almost all cells, except erythrocytes. Their role is to bring specific macromolecules into the cells by the specific receptor-mediated reaction. For example, oocytes have receptors for yolk proteins that bring these proteins to the eggs. Infant gut epithelial cells transfer the mother's immunoglobulins G to the infant's blood and provide them with passive immunity.

Coated pits may act as molecular filters; on fibroblasts they contain several different receptors: the LDL (low density lipoprotein) receptors, receptors for transferrin, transcobalamin, epidermal growth factor, alpha$_2$-macroglobulin, and proteins carrying mannose-6-phosphate residues.[14] Each pit has an area about 0.1 μm^2, and may contain 1000 different receptors.

The rate at which coated pits internalize has been described in various systems, e.g., the uptake of LDL by human fibroblasts, the uptake of asialoglycoproteins by hepatocytes was estimated at about 1 min. The receptor transit times (recycling) are different for various receptors, e.g., the transferrin receptors return to the cell surface in about 20 min, while the LDL-receptors in about $^1/_4$ min.[14]

G. CAPPING OF SURFACE ANTIGENS

In the endocytic cycle mediated by coated pits, the cells internalize specific receptors over the entire cell surface, but the receptors return to a particular cell region. The sites of endocytosis and exocytosis are not coincident, therefore the returning receptors must move from the sites of exocytosis to the sites of endocytosis. The flow of receptors and lipids may influence cell motion and may be an important factor that causes cell motion.[14]

Some of the receptors, e.g., the LDL-receptors, occur in the coated pits even in the absence of bound LDL particles, whereas transferrin receptors and galactose-terminal glycoproteins are usually distributed on the entire plasma membrane and accumulate in pits only when the ligand is bound and at a minimal temperature of 37°C. The LDL-receptors are cycling probably through human fibroblasts whether or not they have bound ligands, whereas receptors for transferrin and galactose-terminal glycoprotein are internalized only after binding with ligands.

Unlike other ligands, transferrin is neither degraded nor stored in cells after internalization—in an acidic environment it releases its iron ions and is secreted from the cells as iron-free apotransferrin to the cell surface and dissociates from its receptor. After binding with iron, transferrin undergoes another cycle of internalizing. The total time for one cycle is about 16 min.

Cells may internalize a large part of their plasma membrane for endocytosis, phagocytosis, and pinocytosis—in total 50% per h and 2% of the whole plasma membrane of cells growing in culture is taken up by deepening coated pits. They are generated not only by the plasma membrane, but also by membranes of intracellular organelles, e.g., membranes of the Golgi apparatus.

Protein transport across the membranes of endoplasmic reticulum. Proteins that are synthesized at ribosomes may remain in the same cellular compartment in which they have been synthesized, whereas others have to cross several cellular hydrophobic membranes to reach the intracellular compartment or extracellular site where they exert their function. Three distinct classes of proteins are translocated through the endoplasmic reticulum: secretory proteins, lysosomal proteins, and certain integral membrane proteins.

A very important element of the translocation is the signal recognition particle (SRP), an 11S cytoplasmic ribonucleoprotein consisting of nonidentical polypeptide chains and one molecule of RNA (about 300 nucleotides) called 7SL RNA.[24,25] Sequence analysis revealed that 7SL RNA is about 80% homologous to the Alu repeats at their ends.[25] The core part of the 7SL RNA does not show homology to Alu DNA and is present in the genome at middle repetitive frequency. Nucleolytic digestion of signal recognition particles causes their degradation and inactivation.

The next isolated component of the protein translocating machinery is the SRP-receptor, also called "docking protein".[24] The SRP-receptor is an integral membrane protein of the endoplasmic reticulum. Both the SRP and SRP-receptor require free sulfhydryl groups for their activation. Free SRP may exist (probably) in equilibrium with a membrane-bound form or ribosomebound form. Upon translation of the signal sequence of a nascent protein that must be translocated (secretory or lysosomal protein) the affinity of SRP to the translating ribosome dramatically increases, which simultaneously arrests elongation of the synthesized protein. It is supposed that in recognition and arrest of translation the 7SL RNA may be engaged, but this has not yet been experimentally documented. SRP-mediated elongation of protein synthesis may be an important regulator of protein synthesis at the translational level.[24] The arrest-releasing activity of the microsomal membrane fraction is ascribed to the SRP-receptor. With releasing the elongation arrest, the affinity of SRP to SRP-receptor decreases which enables the SRP and SRP-receptor to be recycled.

During translation/translocation several enzymes localized on the luminal site of the endoplasmic reticulum act on the nascent polypeptide. The signal peptides are removed by signal peptidase, mannose core oligosaccharides are translocated on asparagine residues, and cysteine residues are oxidized to form disulfide bonds. Translocation across the endoplasmic reticulum does not proceed spontaneously; even in the absence of protein synthesis, energy substrates are necessary for translocation.[26]

H. THE ARG-GLY-ASP RECOGNITION SIGNAL

Fibronectin, a large extracellular glycoprotein of approximately 2500 amino acid residues mediates adhesion of cells to the extracellular matrix. Crucial for the interaction is a cell surface receptor that recognizes the tripeptide Arg-Gly-Asp in the fibronectin polypeptide. Analysis of the cell attachment sites of several other proteins revealed the same amino acid triplet. The synthetic peptides containing Arg-Gly-Asp sequence on an insoluble substrate mediate cell attachment, whereas in solution they inhibit cell attachment to fibronectin. The same inhibition occurs to vitronectin, the active component of serum spreading factor and the S-protein of the complement membrane attack complex. Fibrinogen has two, and the von Willebrand factor one Arg-Gly-Asp sequence important for platelet adhesion of each of these proteins.[27] The fibronectin receptor for the Arg-Gly-Asp sequence consists of two subunits, each about 140 kDa, forming an integral membrane protein.

The vitronectin receptor for the Arg-Gly-Asp sequence also has two subunits with molecular masses 125 and 115 kDa. The vitronectin receptor binds also to Sepharose coupled with the Arg-Gly-Asp sequence containing hexapeptide (the fibronectin receptor does not), which suggests that the vitronectin receptor has a higher affinity for small peptides than does the fibronectin receptor. A third Arg-Gly-Asp sequence receptor was isolated from platelets which also is a heterodimer of subunits slightly smaller than those of the fibronectin and vitronectin receptors. The platelet receptor binds fibrinogen, fibronectin, vitronectin, and the von Willebrand factor.[27,28,30]

Several other Arg-Gly-Asp sequence receptors have been described, e.g., the fibronectin receptor of monocytic cells, a protein complex from chicken fibroblasts resembling the fibronectin receptor (it differs by recognizing laminin). To the same family may belong receptors for collagen, since various collagens contain Arg-Gly-Asp sequences and collagens exhibit cell-attachment activity. The epidermal growth factor (EGF) also has the Arg-Gly-Asp sequence, although it probably does not have physiological value. The "reverse" sequence of this tripeptide may inhibit cell attachment.[27] The question arises whether this sequence is always active. It is supposed that, apart from the binding site, local tertiary structure dependent on neighboring amino acid residues may be decisive for the binding reaction.

V. THE BASEMENT MEMBRANE

Basement membranes are ubiquitous extracellular structures at the boundary between cells and the connective tissue. Most of the basement membranes are 40 to 60 nm thick. The basement membranes contain several proteins. Some of them are present in all basement membranes, synthesized and secreted in cells resting on basement membranes (type IV collagen, laminin, heparan sulfate proteoglycan, and probably entactin), whereas the others are present only in some of the basement membranes synthesized and secreted in cells that do not rest on basement membranes, usually with a prominent filtering function (fibronectin, type V collagen, and chondroitin sulfate). Morphologically (in the electron microscope), the following can be distinguished, one after the other: lamina rara externa (the nearest to the epithelial cells), lamina densa, and lamina interna (the nearest to the connective tissue). Not all basement membranes exhibit all three laminae, sometimes there are more or less laminae.

Type IV collagen is present only in basement membranes. Each collagen molecule is composed of three α-chains (two α_1 and one α_2) with amino acid residues characteristic for each collagen type. The general formula for all α-chains is Gly-X-Y, where Gly stands for glycine, X is whichever amino acid is chosen, and Y is frequently hydroxyproline or hydroxylysine. Type IV collagen exhibits a characteristic amino acid sequence, its helical structure seems to be interrupted by segments other than Gly-X-Y which makes it susceptible to enzymes other than collagenase. It contains large amounts of 3-hydroxyproline (in other types of collagen, 4-hydroxyproline), is rich in carbohydrate side chains, and is incorporated into basement membranes without extracellular modifications (the procollagen amino- and carboxy-terminals are not released as in other types of collagen). Finally, this type of collagen contains a unique 7S region, relatively resistant to bacterial collagenase. This region probably represents the crosslinking site of type IV collagen.

Laminin is one of the best characterized noncollagenous components of basement membranes. Laminin is a glycoprotein (M_r approximately 900,000) composed of three subunits: the largest of M_r 400,000, the smaller subunits of M_r 200,000. It is thought that laminin interacts with collagen type IV. Laminin is susceptible to several proteases (trypsin, pepsin) and causes cell adhesion and attachment *in vivo* and *in vitro*.[28]

A. HEPARAN SULFATE PROTEOGLYCAN

This proteoglycan is composed of multiple glycosaminoglycan chains covalently bound to a core protein. The molecular weight for proteoglycan ranges between 160,000 to 750,000, and for glycosaminoglycan between 14,000 to 70,000. Heparan sulfate proteoglycan probably plays a major role in filtration and cell attachment.

Entactin has a molecular weight of approximately 150,000 and is highly sulfated. It is probably present in several basement membranes (this must be verified).

B. FIBRONECTIN

This protein is present in two forms: plasma and tissue fibronectin.[30] Their amino acid composition is similar and it is unclear whether these two forms represent posttranslational modifications or products of two different genes. The molecule is composed of two identical subunits, each of molecular weight 220,000, held together by disulfide bonds. Traces of fibronectin are present only in basement membranes with high filtration.

C. TYPE V COLLAGEN

This type of collagen is composed of three nonidentical type V alpha chains: α 1 (V), α 2 (V), and α 3 (V). Type V collagen is present in vascular adventitia, in stroma surrounding type I collagen bundles, and anchoring some basement membranes to surrounding tissues.

The components of basement membranes are synthesized by cells resting upon them. It denotes that epithelial, endothelial, muscular, and adipose tissue cells synthesize basement membrane components. Intracellular synthesis of these components was best studied in the murine parietal yolk sac cells that synthesize and secrete type IV collagen, laminin, entactin, and heparan sulfate proteoglycan. The basement membranes are characterized by marked heterogeneity caused by different ratios of synthesis of their particular components.

The physiological functions of basement membranes provide (1) physical strength to structures such as renal tubules, blood vessels, lens capsule, etc., (2) cell attachment, (3) ultrafiltration, particularly in capillaries and renal glomeruli. Crucial to filtration is the high anionic charge in basement membranes, due mainly to the presence of heparan sulfate. These three functions—physical support, cell attachment, and filtration—permit particular cells or groups of cells to generate and maintain their optimal environment.[29]

VI. TRAFFIC CONTROL IN THE EUKARYOTIC NUCLEUS

The nuclear envelope and double membrane provide a distinct specific macromolecular composition between the nucleus and cytoplasm. Molecules are transported across the nuclear envelope by proteinaceous channels, termed nuclear pore complexes. Their diameter is about 10 nm. Proteins forming the pore complexes are more concentrated in the nucleus than in the cytoplasm. These karyophilic proteins have a structure that is responsible for their accumulation in the nucleus. Molecules of small molecular weight (M_r 20,000 to 40,000) can probably pass pore complexes by passive diffusion and then by intranuclear binding. Larger molecules are transported probably by specific mediated-transport. Similar to the transport of proteins from the endoplasmic reticulum, the transport from cytoplasm to the nucleus probably requires specific signals that direct them into nuclei. Studies for this purpose have been performed on the pentameric nucleophilic protein—nucleoplasmin. On the basis of limited proteolytic digestion a small tail (M_r 10) has been found that mediates the selective uptake of this protein to the nucleus.[31] In the T-antigen of simian virus 40 (SV40) a short amino acid sequence (positions 126 to 132) was determined as responsible for nuclear transport of this protein. After attachment of these sequences to large, normally cytoplasmic proteins, they have been detected in the nuclei.[31]

Further experiments revealed that the mechanism that regulates import of nuclear proteins is dependent on U2 RNA (snRNP). Proteins complexed with those particles are transported into nuclei (after transcription of U2 RNA). Microinjection of U2 RNA into oocytes induces assembly of these U2-binding proteins that migrate into the nuclei.[31]

The inner surface of the nuclear envelope is lined by a nuclear assembly of "lamin" polypeptides called lamina. This structure probably anchors interphase chromosomes. Three (M_r 60 to 70) lamins A, B, C are present in typical mammalian cells. The lamins have about 300 amino acid residue sequences with striking homology to 300 amino acid residue sequences characteristic for all intermediate filament proteins.[31] During mitosis the lamins undergo total structural reorganization: they become reversibly depolymerized to 4S-5S form, probably by hyperphosphorylation. The reversible depolymerization probably enables the reconstruction of the nuclear envelope that occurs in a later stage of mitosis.

VII. THE BLOOD-BRAIN BARRIER

The maintenance of a constant internal milieu is an indispensable requirement of higher living organisms, and the internal milieu of the brain is the most important. The extracellular concentrations of hormones, amino acids, and ions (potassium, sodium, calcium, etc.) undergo continuous small fluctuations in the blood, particularly after stresses, meals, etc. If these fluctuations occurred in the brain it would result in an uncontrolled function, because some of the hormones and amino acids play an important role as neurotransmitters, and the ions, especially the potassium ion, are necessary for the function (firing) of the nerve cells. Therefore, the brain must be exactly isolated from changes in body fluids. The first experiments were those of Ehrlich et al., who injected various dyes intravenously or to the cerebrospinal fluid of small laboratory animals. They found that after intravenous injection, all peripheral organs had been stained with the injected dye, with one exception—the brain; and vice versa, after injection into cerebrospinal fluid, only the brain was stained but not the peripheral organs. On the basis of these experiments, the blood-brain barrier hypothesis has been established. Next, it was found that this effect is caused by a unique structure of brain capillaries that prevent the entry to the brain of many blood constituents except necessary nutrients (amino acids), some ions, etc. The question arose about the mechanism that enables nutrients to pass to the brain, and pumping of the surplus substances out of the brain.

The morphological structure of brain blood capillaries was elucidated by electron microscopy. The epithelial cells of blood capillaries form a tube of continuous tight junction (tight junction is present in all mammalian cells occupying only a small part of the cell that prevents leaks, even ions, and separates cells into two compartments of non-mixing components). In addition, the brain capillaries are lacking gaps and channels, present in body capillaries. Finally, the brain capillaries are surrounded by processes of astrocytes. Some brain areas (the pituitary gland, the pineal gland, and some parts of the hypothalamus) are deprived of blood-brain barrier capillaries. It is supposed that these regions may allow for transport of hormones and other substances entering the brain. Lipid soluble substances readily pass the blood-brain barrier and enter the brain.

Substances highly soluble in water, such as nutrients, mainly glucose and some amino acids enter the brain by specific transport systems. Experiments performed with a useful nutrient D-glucose and its useless stereoisomer L-glucose demonstrated that only D-glucose is extracted from the blood by brain tissue, whereas entry of the two isomers into other tissues is facilitated well. Experiments performed on isolated brain capillaries revealed that the transport systems for glucose are very similar to those of the erythrocytes. Glucose enters the epithelial cells of the capillaries at the luminal side and leaves them through the antiluminal (brain) side. Finally, it was demonstrated that glucose passes the capillaries by specific transporters (undoubtedly a protein of not yet defined composition) that span the membrane forming a kind of channel. The amount of these transporters is limited, and thus also limited is the amount of glucose that can pass from the blood to the brain. A similar situation occurs with the amino acids. The large neutral amino acids, required for the synthesis of neuro-transmitters and brain proteins possess specific transporters (about ten of them compete for the same transporter). The basic and small acidic amino acids possess other transporters specific for each of them. Small neutral amino acids that can be synthesized by the brain tissue are not transported from the blood to the brain. It is noteworthy that glycine, belonging to the small neutral amino acids, is a potent inhibitory neurotransmitter and a surplus of this amino acid could disturb the function of the brain. Further, it was discovered that glycine transport may be asymmetric, it can be taken up by the luminal side into the epithelial cells of capillaries, but not transported from these cells to the brain. Another asymmetric transport concerns the endothelial transport of ions, especially potassium and sodium. The transport of potassium is "active". The antiluminal membrane contains an overbalance of the sodium-potassium ATPase that simultaneously transports sodium out of the endothelium into the brain and potassium out of the brain into the endothelium.[32]

Another kind of blocking entry of substances into the brain is the "metabolic barrier". L-dopa, the amino acid of the neurotransmitters dopamine and noradrenaline, readily enters the endothelium, but then it is modified by specific enzymes to a form that cannot enter the brain.[32]

The next question arose about the mechanism that directs development of endothelial cells to form blood-brain barrier. This was explained in part by transplantation experiments. When brain tissue was transplanted to the gut, microvessels from the gut growing into the brain transplant developed capillaries characteristic for the blood-brain barrier; and vice versa, brain microvessels growing into the gut transplant lost their character of a blood-brain barrier.[32] These experiments suggest that the signal for adequate transformation of the capillaries generates in the brain, probably by induction of specific genes. Recent investigations revealed that astrocytes may induce development of blood-brain barrier in non-neural endothelial cells.[33]

Further experiments revealed that hyperosmotic sugar solution injected into the carotid artery causes transient increase of the permeability of the blood-brain barrier, evoked by temporary loosening of the tight junctions between the endothelial cells.

Receptor mediated transport of insulin across the endothelial cells is reminiscent of the transport of different substances through the blood-brain barrier. Insulin is transported from the blood to the target tissue through the endothelial cells. The endothelial cells can internalize insulin and release it promptly in a practically intact state. The transport of insulin through the endothelial cells is temperature sensitive. The uptake of labeled insulin can be inhibited by unlabeled insulin or by antibody specific for insulin receptor. These experiments suggest that insulin, and probably other hormones of a peptide character, are transported across the endothelial cells by a receptor-mediated mechanism.[34]

VIII. ENZYMATIC SYSTEMS OF CELL MEMBRANES

The cells of multicellular organisms are permanently submitted to a great diversity of signals. Their settlement is achieved on the basis of multiple specific cell receptors, but there is also need for a superior regulation that could coordinate the rival signals or those exceeding that known for surface receptors. The first investigation was on the family of GTP-binding proteins that mediates a great number of different fields: vision, olfaction, control of cell proliferation, ribosomal protein synthesis, and regulation of cellular functions by a great number of hormones and neurotransmitters.[35] The molecule that performs this task is a guanine nucleotide-binding protein that works in the following way: the nucleotide-binding protein stimulated by a receptor protein releases GDP and binds GTP; G-binding protein (GTP complex) regulates the effectors (enzymes or ion channels); hydrolysis of GTP terminates the regulatory effect of this complex. The GDP binding protein belongs to a family of G-proteins which comprises the *ras* proteins, i.e., elongation factors involved in protein synthesis machinery.

The G-binding protein binds the GDP molecule with four amino acid residues: Asp 80, which fixes the β phosphoryl via an interposed Mg^{2+} ion, Lys 24 which is also coordinated with the phosphoryls of GDP, and Asn 135 and Asp 138 that bind the keto and amino substituents of the guanine ring. Phosphorylation of GDP to GTP probably increases the distance between the phosphoryl bonds of GDP and guanine ring which enables binding of the G-protein with the multiple effectors. This mechanism is best explained in mutational events. Substitution of Gly 12 in p21 near the phosphoryls stabilizes the GTP-protein G complex.

A series of mutations support the hypothesis of G-protein function. Substitution of Gly 12 in p21 stabilizes GTP-protein G complex that results in malignant transformation. Similar effects may be caused by other substitutions of p21. Substitution of lysine by asparagine of bacterial elongation factor EF-Tu at the position corresponding to Lys 24 of p21 facilitates exchange of GDP- to GTP-protein G complex.[35]

G-proteins contain three distinct chains—an α-chain that binds GTP and is responsible for the specificity of detector and effector proteins, and β/γ complex that probably anchors the cytoplasmatic side of the plasma membrane. At present, eight different α-chains are known, among them two transducin α-chains, one expressed only in rods and responsible for night vision, and the other in cones, responsible for color vision. In both cases, activated transducin stimulates a cyclic GMP phosphodiesterase. Two different α-chains of G_s, a G-protein that stimulates adenyl cyclase activated by several hormones and neurotransmitters, among them GTP-dependent cAMP synthesis in olphactory epithelia. In patients with hereditary pseudo-hypoparathyroidism a partial deficiency of G_s was demonstrated, which explains why each of five examined patients were unable to identify common odorants.[35]

The α-chain G_1 mediates inhibition of adenyl cyclase by receptors for different hormones and neurotransmitters (e.g., somatostatin) and is a substrate for pertussis toxin-catalyzed ADP-ribosylation that uncouples G_1 from its receptors.

Besides the great number of α-chain variants, variants of the β and γ chains have been also detected, but their function is unknown.

The *ras* protein family comprises two *ras* proteins that control a cyclase in yeast, and three p21 proteins which are products of the three *ras*-protoncogenes Ha-Ki- and N-*ras*. Several variants of these proteins have been detected. Their function also is unknown. Among these variants, the N-*ras* protein probably couples receptors for activation of phospholipase C (PLC). Among the G-protein effector enzymes are: adenyl cyclase, phosphodiesterase, and phospholipase C.

A. PROTEIN KINASE C

The calcium phospholipid-dependent protein kinase (protein kinase C) emerges as a family of closely related structures. Protein kinase C, a serine-threonine protein kinase is dependent on three regulatory factors: phospholipids (especially phosphatidylserine), calcium ions, and diacylglycerols.[36]

Protein kinase C was discovered in 1977 as a proteolytically activated protein kinase. The enzyme is present in all tissues. In tissues other than brain its inactive form is present in soluble fraction, which after stimulation is translocated to membranes in a Ca^{2+}-dependent fashion. Its precise intracellular localization is only scarcely known. The enzyme is probably absent in the nucleus. Protein kinase C requires Ca^{2+} and phospholipid particularly phosphatidylserine for its activation. Diaglycerol dramatically increases the activity of protein kinase C, even in absence of a net increase of Ca^{2+} concentration. Thus, the enzyme is

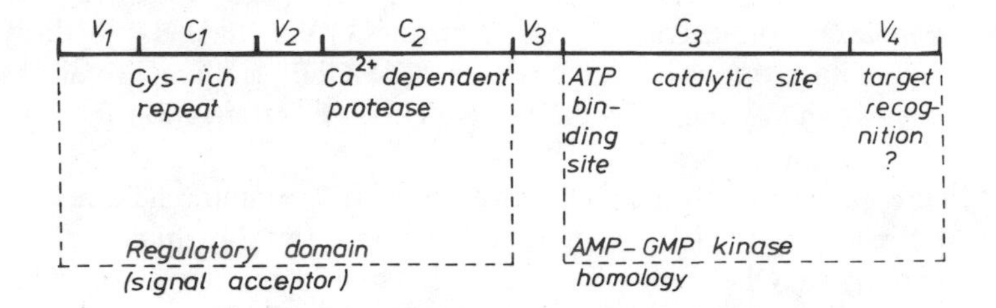

FIGURE 2. Schematic representation of protein kinase C. V_1, V_2, V_3 and V_4—variable regions, C_1—cysteine rich repeat, C_2—Ca^{2+} dependent protease, C_3—ATP binding site, C_3—catalytic binding site, V_4—probable target recognition site. V_1 to V_3—regulatory domain (signal receptor). V_3 to V_4—AMP and GMP dependent kinase homology. (Modified from Reference 38.)

biochemically dependent on Ca^{2+}, but sometimes also Ca^{2+} independent. Normally the enzyme is activated by an increase of Ca^{2+} concentration and formation of diacylglycerol.[37]

A family of three different types of rabbit protein kinase C, α, β, and γ, encoded on different chromosomes revealed a closely related sequence of amino acid residues (673, 671, and 672 amino acid residues, respectively). Two of them exhibited an identical N-terminal sequence of 621 amino acid residues encoded by a single gene. The C terminals of all kinases revealed typical kinase domains and were highly homologous to the other protein kinases. They have potential calcium binding sites, and demonstrate different tissue specificities. The γ type was present in various tissues, while the α and β types predominated in the brain.[38] Similar fragments exist in the sequences of all protein kinase C families. The molecule contains four variable (V1 to V4), and four interspersed conserved regions (C1 to C4). The conserved three and four region exhibits the greatest homology with the cyclic AMP and GMP-dependent kinases (see Figure 2).

Several hormones, neurotransmitters, and growth factors bind with membrane receptors and stimulate the phosphodiesteratic hydrolysis of phosphatidylinositol 4,5-biphosphate generating the mediators 1,4,5-triphosphate inositol (I 1,4,5-P_3) and diacylglycerol. Diacylglycerol stimulates protein kinase C, while inositol triphosphate binds specific receptors, both leading to release of Ca^{2+} from the endoplasmic reticulum.

Inositol triphosphate is recognized as an important second messenger in cellular signal transduction. Phosphatidylinositol 4,5-biphosphate belongs to the inositol lipids located at the inner side of the plasma membrane, and is formed by a two-stage phosphorylation of phosphatidylinositol.

The first indications that GTP-binding protein may function in signal transduction for calcium mobilizing receptors was the observation that GTP reduces the affinity of noradrenaline for α_1-receptors. Pertussis toxin reduces the modulatory effect of GTP on α_2 adrenoreceptors by inhibiting adenylate cylcase, but has no effect on calcium binding α_1 receptors—this suggested that the two receptor mechanisms are regulated by different GTP-binding proteins. The next observation was that stimulatory effects of guanine nucleotides introduced into different cells stimulate the hydrolysis of inositol lipids. Finally, the *ras* gene product, p21, a membrane-bound protein can bind and hydrolyze GTP, and function similarly as the GTP-binding protein to link receptors to phosphatidylinositol 4,5-diphosphate phosphodiesterase.[40]

The list of tissues that respond to external stimuli by a change in the hydrolysis of phosphatidylinositol 4,5-biphosphate is permanently growing. From a review of 1984 it now comprises[40] stimulation by acetylcholine in rabbit iris smooth muscle, brain, parotid, and pancreas; stimulation by adrenaline, ATP, vasopressin, angiotensin, or platelet-activating factor—liver; stimulation by ADP, thrombin, platelet-activating factor—blood platelets; stimulation by vasopressin—growth hormone, pituitary tumor cells; stimulation by angiotensin—adrenal cortex; stimulation by f-metionyl-leucyl-phenylalanine—HL leukemic cells; stimulation by *Pseudomonas* leukocidin—leukocytes; stimulation by phytohemagglutinin—T-lymphoblastoid cells; stimulation by PDGF (platelet derived growth factor), bombesin, or vasopressin—Swiss 3T3 cells; and stimulation by photons—photoreceptors, etc.

Since its discovery, it became clear that diacylglycerol, as the second messenger of protein kinase C, regulates several intracellular functions: regulation of cell growth and differentiation, secretion, activation of platelets and neutrophils, regulation of gene expression of cell surface receptors, and cellular metabolism.[41] Diacylglycerol is one of the second messengers produced by degradation of phosphatidylinositol 4,5-biphosphate (PIP_2) caused by a transmembrane signal; the other messenger, inositol triphosphate (IP_3) functions in mobilization of intracellular Ca^{2+}. Diacylglycerol and inositol triphosphate are rapidly degraded after stimulation.

The transmembrane signaling that causes diacylglycerol production is not yet sufficiently known. It is supposed that the receptor-mediated turnover of phosphatidylinositol 4,5-biphosphate (PIP_2) through a GTP-protein-dependent activation of phospholipase C causes degradation of phosphatidylserine phosphatidylinositol 4,5-biphosphate into diacylglycerol and inositol triphosphate (IP_3). IP_3 causes an increase of Ca^{2+} concentration. This mechanism is similar to that for transmembrane signaling coupled with cyclic AMP.

Studies about proteinase C activation using Triton X-100 mixed micelles revealed that:

1. Protein kinase C is fully activated by mixed micelles containing 8 mol% phosphatidylserine and 2.5 mol% diaglycerol (molc fractions similar to those of the plasma membrane during activation).
2. Activation does not require a phospholipid bilayer.
3. Activation of phosphatidylserine is highly cooperative and requires four or more molecules of phosphatidylserine.
4. Activation by diacylglycerol is noncooperative and requires only a single molecule of diacylglycerol.
5. Protein kinase C monomers bind to mixed micelles in a phosphatidylserine Ca^{2+}-dependent manner and are inactive in the absence of diacylglycerol.[41]

The minimum stechiometry of protein kinase C activation from these data is: a monomeric protein kinase C, one molecule of diacylglycerol, one molecule of Ca^{2+}, and four molecules of phosphatidylserine.

The list of protein kinase C activities in cellular responses is still growing and resembles that of phosphatidylserine 4,5-biphosphate and that of diacylglycerol. The possible roles are listed below, after Nishizuka.[37]

Endocrine system—Adrenal medulla—catecholamine secretion; adrenal cortex—aldosteron secretion, steroidogenesis; pancreatic islets—insulin release; pituitary cells—release of pituitary hormones (growth hormone, luteinizing hormone, prolactin, thyrotropin); parathyroid cells — parathormone release; tharoid cells—calcitonin release; Leydig cells—steroidogenesis.

Exocrine system—Pancreas—amylase secretion; parotid gland—amylase and mucin secretion; gastric glands—pepsinogen and gastric acid secretion; lung alveolar cells—surfactant secretion.

Nervous system—Caudate nucleus and ileal nerve endings—acetylcholine release; neuromuscular junction—transmitter release; neurone—dopamine release.

Muscles—Vascular smooth muscles—muscle contraction and relaxation.

Inflammation and immune reaction—Platelets—serotonin, lysosomal enzymes and arachidonate release, thromboxan synthesis; neutrophils—superoxide generation, lysosomal enzymes release, hexose transport; basophils and mast cells—histamine release; lymphocytes—T- and B-lymphocyte activation.

Metabolism of particular cells—Adipocytes—lipogenesis and glucose transport; hepatocytes—glycogenolysis; epidermal cells, fibroblasts, hepatocytes—inhibition of gap junctions.

Proposed substrate proteins of protein kinase C—Receptor proteins—epidermal growth factor (EGF), insulin, somatomedin C, transferrin, interleukin-2, nicotinic acetylcholine, β-adrenergic, and immunoglobulin E receptors.

- Membrane proteins—Ca^{2+}-Transport ATPase, Na^+/K^+-ATPase, Na^+ channel protein, Na^+/H^+ exchange system, glucose transporter, GTP-binding protein, HLA antigen, chromaffin granule binding protein, synaptic B50 protein.
- Contractile and cytoskeletal proteins—myosin light chain, troponin, vinculin, filamin, caldesmon, cardiac C protein, microtubule associated proteins.
- Enzymes—glycogen phosphorylase kinase, glycogen synthase, phosphofructokinase, β-hydroxy-β-methylgluratyl-CoA reductase, tyrosine hydroxylase, NADP oxidase, cytochrome P 450, guanylate cyclase, DNA methylase, myosin light chain kinase, initiation factor 2.
- Proteins—fibrinogen, retinoid-binding protein, ribosomal S6 protein, stress proteins, myelin basic protein, high-mobility group proteins, middle T-antigen, $pp60^{src}$ protein.

The polypeptide growth factor—Interleukin-2 (IL-2) stimulates proliferation and differentiation of T-lymphocytes. The receptor for interleukin-2 is synthesized by activated T lymphocytes. Analysis of the structure of IL-2 revealed that IL-2 does not have intrinsic signal transduction, unlike other growth factors. IL-2 with its receptor activates the calcium-phosphate dependent protein kinase C. Thus, IL-2 is the first recognized polypeptide growth factor activity which depends on GTP-binding protein.[42]

The cell signaling system seems to use two major receptor classes for transduction of information across the cell membrane; one that depends on generation of cAMP as second messenger, and another that depends on protein kinase C that causes turnover of inositol phospholipids and mobilization of Ca^{2+}. The latter causes release of arachidonate and an increase of cGMP. Thus, protein kinase C activation, Ca^{2+} mobilization, arachidonate release, and increase of cGMP can be integrated into a single receptor cascade.[37]

Tumor-promoting phorbol esters, especially 12-*O*-tetradecanoyl-phorbol-13-acetate (TRA) have a structure very similar to diacylglycerol and activate protein kinase C (*in vitro* and *in vivo*). The affinity of phorbol esters to protein kinase C sometimes exceeds that of the natural agent, i.e., diacylglycerol. Therefore, the phorbol esters can mimic the effect of diacylglycerol in enzyme activation.[43,44] The discovery of inositol triphosphate and an intracellular second messenger suggested that cell growth and development may depend on mobilization of calcium from intracellular stores. On the other hand, it also was known that growth factors cause changes in inositol lipid metabolism. Diacylglycerol acts within the plane of the membrane to stimulate protein kinase C, which is also the site of action of tumor promoting phorbol esters. The action of growth factors PDGF, bombesin, and vasopressin on Swiss 3T3 cells results in an increased formation of diacylglycerol and inositol triphosphate. Thus, hydrolysis of phosphatidylinositol 4,5-biphosphate results in the activation of two signals that regulate cell proliferation, i.e., increase of diacylglycerol that causes increase in pH, and inositol triphosphate that causes elevation of Ca^{2+} concentration.

The enzymatic reactions induced by oncogenes and growth factors on cell proliferation are presented in Figure 3. C-oncogenes (cellular oncogenes) are responsible for the synthesis of various growth factors: c-*sis* for PDGF that stimulates inositol metabolism; c-*erb* B for truncated EGF receptor that stimulates increase of Ca^{2+} in fibroblasts; and c-*src* and c-*ros* for kinases and receptors involved in inositol metabolism. The normal c-*ras* gene product binds and hydrolyzes GTP. Mutation of codon 12 of c-*ras* encodes an oncogenic protein

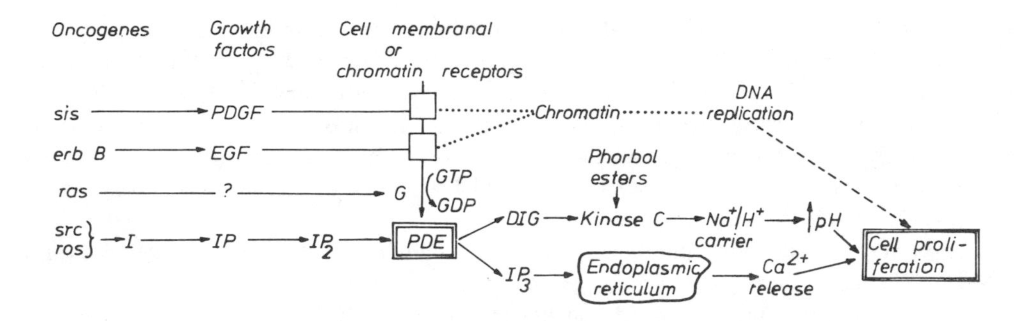

FIGURE 3. The proposed role of diacylglycerol (DIG) and inositol triphosphate (IP_3) as intracellular second messengers regulating cell proliferation. PDE—phospodiesterase; I—phosphatidylinositol; IP—phosphatidylinositol-4-phosphate; IP_2—phosphatidylinositol-4,5-biphosphate. (Modified from Reference 40.)

that binds GTP, but cannot hydrolyze it which results in permanent activation leading to formulation of inositol triphosphate and diacylglycerol that causes uncontrolled activation of growth factors and malignant transformation of cells. Recently, another mode of the action of growth factors has been recognized, namely indirect action on chromatin,[45] that may cause modifications of chromatin leading to uncontrolled growth of cells by a yet unknown method.

Activation of protein kinase C also affects the pH inside the cells by stimulation of the plasma membrane Na^+-H^+ exchanger which is responsible for the exchange of sodium ions for hydrogen ions across the outer membrane of cells resulting in an elevation of pH.[46] The fundamental acid-base problem lies in a permanent tendency toward intracellular acidosis, caused by influx of acids (H^+) or efflux of bases (HCO_3^-), or by metabolic generation of acids. To prevent this acidosis, cells use a transporter that exchanges external Na^+ for internal H^+. This mechanism can be modified by cAMP that stimulates the Na^+/HCO_3^--Cl^-/H^+ exchanger. The Na^+-H^+ exchanger may be involved in regulation of cell growth by extracellular stimuli. Quiescent Swiss 3T3 cells exposed to PDGF, vasopressin, or insulin cause a Na^+-dependent rise in pH. Increase of pH caused by EGF was inhibited by the Na^+-H^+ exchanger inhibitor. These results suggest that intracellular rise of pH induced by growth factors stimulates DNA synthesis which may be critical for cell proliferation.[46]

Inositol metabolism also is an important mechanism in movements of cells in early developmental stages. Ionic currents flowing through oocytes may be responsible for segregation of cytoplasmic components to form an anterior-posterior axis in early development. Muscarinic receptors, accumulated around the animal pole, respond to acetylcholine by inducing membrane depolimerization and opening the chloride channels regulated by calcium release from intracellular stores of inositol triphosphate. Eggs also display spontaneous fluctuations in chloride conductance that probably account for the chloride current that enters the animal pole.[40]

IX. THE MOLECULAR PATHOLOGY OF CELL MEMBRANES

The activities of membrane enzymes are dependent on their milieu, i.e., the membrane lipid bilayer which is the major factor that determines how membrane enzymes function. Most of these enzymes are inactive in an aqueous environment. Thus, the composition and state of membrane lipids must be taken into account when considering the role of cell membranes in life processes.

Unlike simple solvent systems, membranes are composed of various components. Most membrane components are in motion. The membranes form a fluid environment in which protein molecules (enzymes, transporter proteins) migrate slower than the lipid molecules.

In bilayers, composed of a single lipid species, lipid is of critical importance in determining the activity of membrane enzymes. The order of importance of bilayer features is as follows: lipid head group > lipid acyl chain length and saturation/unsaturation > lipid backbone > bilayer "fluidity".[47] Observations on a simple bilayer system of one lipid revealed that the activity of the reconstituted sugar transport is a complex function that may include bilayer surface potential, bilayer thickness, bilayer lipid backbone, and bilayer lipid acyl chain order.[47]

Cholesterol is the major lipid in most eukaryotic membranes. Cholesterol does not form bilayers, changing the physical properties of the bilayer. It suppresses bilayer phase transitions by modifying lipid packing. In sufficient amount cholesterol transforms the bilayer to an overall physical state between fully fluid and fully crystalline.

A. FREE RADICAL PATHOLOGY

Free radical pathology results from abnormal radical reactions occurring in cells. A free radical is any molecule that has an odd number of electrons. Most chemicals have their electron orbitals filled with paired electrons that spin in opposite directions. A free radical is a chemical that has an unpaired electron in an outer orbital. Free radicals can occur in living organisms in small quantities physiologically. Many components of living organisms can be made to have a free radical state induced by ionizing radiation (X-rays, ultraviolet light) or induced by chemicals that are already radicals, or which may be metabolized into radicals, or which may induce radical states in normal chemical constituents of cells. Substances that become free radicals change their properties and shapes. Lipid molecules in cell membranes and nucleic acids are highly susceptible to radical reactions. Alcoholic liver disease, exposure to certain chemical toxins, radiation pathology, and the interaction of chemical carcinogens with biomolecules can be recognized as molecular diseases.[48]

Radical reactions can occur in proteins, lipids, carbohydrates, nucleic acids, and electron transport factors. Water radicals ($OH^{\bullet}$, $H^{\bullet}$, $HOO^{\bullet}$ and others) formed directly in macromolecules and membranes play a remarkable role in free radical pathology.

Molecular oxygen can be written $O{=}O$ or $\dot{O}{-}\dot{O}$. Both of these structures are not completely accurate because the bond energy between the two oxygen atoms is stronger than just one covalent bond. Therefore $\dot{O}{-}\dot{O}$ is incorrect, but sometimes molecular oxygen can be diradical and therefore $O{=}O$ is also incorrect. This results from the ability of oxygen to behave like a diradical which endows it with powerful capacities for initiation and/or addition reactions. Molecular oxygen is the main source of free radicals in living organisms. The solubility of oxygen is very important to the free radical pathology of membranes. In a non-polar environment—and such an environment exists solely in cell membranes—oxygen is seven to eight times more soluble than in a polar environment. Thus, free radical pathology occurs predominantly in polyunsaturated fatty acids.

In membranes of cellular organelles the lipids do not form bilayers and are thought to be intertwined with proteins. The phospholipids of organelles contain more polyunsaturated fatty acids than cell membranes. Therefore, the membranes of organelles are more susceptible to peroxidation.

Unsaturated fatty acids are particularly liable to peroxidation because of the presence of double bonds that weaken the carbon-hydrogen bond of the carbon atom adjacent to a carbon with an unsaturated bond (see Figure 4). In the case of linoleic acid, there are two double bonds between atoms 9 and 10, and between 12 and 13 (as seen below):

```
     H     H     H     H     H     H
  —  C  —  C  =  C  —  C  —  C  =  C  —
     H                 H
     8     9    10    11    12    13
```

Carbon atoms 8 and 11 gave weak bonds to their hydrogen atoms. These carbons are "α methylene" carbons, and their hydrogens are "allylic". This means that less energy is

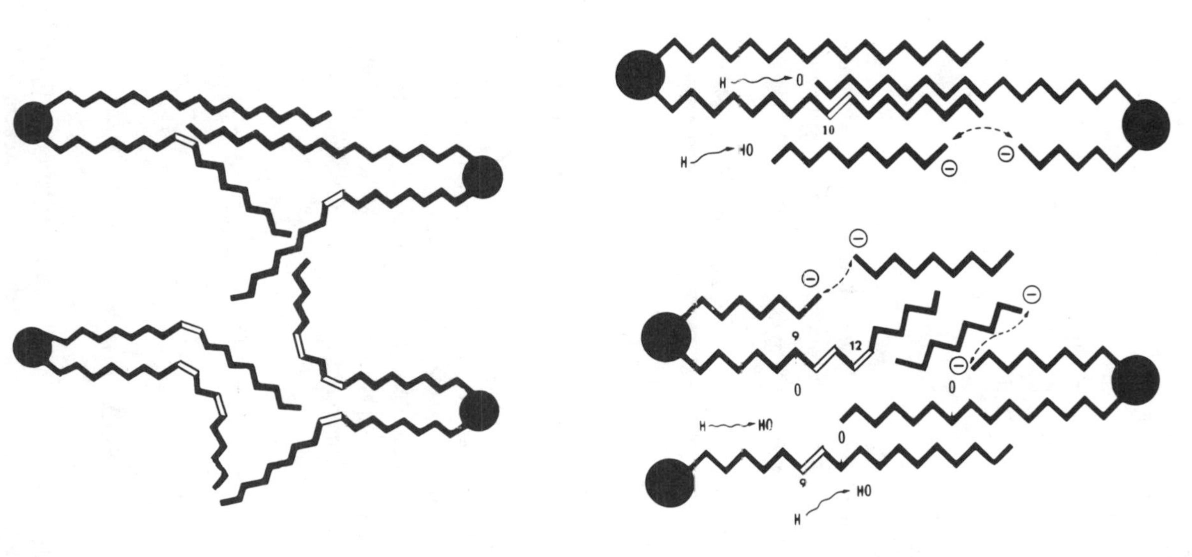

FIGURE 4. Left: schematic representation of the plasma membrane phospholipid bilayer with polar hydrophilic glycerophosphate heads and hydrophobic fatty acid tails. Unsaturated bonds (in white) are bent at an angle of 123° in the *cis*-isomeric conformation which forms spaces between for the steroids to intercalate into the fatty acids. (Right): scheme of structural consequences of lipid peroxidation of the hydrophobic part of the cell membrane. Most of the fatty acid tails are straightened. A saturated carbon no longer separates the carbons with unsaturated bonds (this is referred to as conjugation); RO˙ are present and form peroxides; mobile ˙OH radicals are shown as the result of hyperoxide fission; hydrogens are shown being abstracted by ˙OH (from adjacent lipid and protein molecules); abstracted hydrogens react with hydroxyls and form water in the hydrophobic part of the bilayer; fragmentation of fatty acid tails may generate negatively charged carboxylic acid groups (represented as minus). The numbers 9, 10, and 12 denote carbon atom numbers in the carbon chain of fatty acids. (Redrawn from Reference 48.)

required to start a reaction involving carbons 8 and 11 because they are partially "activated" and small amounts of initiators, or prooxidants, will result in complete activation. Thus, molecular oxygen and a minimal amount of certain metal complexes may initiate a reaction by abstracting an allylic hydrogen, which generates a radical center on the α-methylene carbon. This radical becomes highly reactive, and there may be addition of the perhydroxy radical, as shown below:

```
                                          alkyl radical                      hydroperoxide

   H     H     H     H ————————> H     H     H     H ————————>  H     H     H     H
 — C  —  C  =  C  —  C —       — C• —  C  =  C  —  C —        — C  —  C  =  C  —  C —
  (H)                H                             H      H     |                 H
   |                                                   O•— O    O— OH
   | +                                                 perhydroxy
   ↓                                                    radical
 O•— O•———————————————————> — H  —  H  —  H  —  H —
 peroxy                        C     C     C     C
 radical                             |     |
                                     O  —  O
```

The free radical series of reactions cause peroxidation of lipids with disturbances of structure and function of membranes (see Figure 4). Consequently, a dramatic change of functions of membrane proteins and membrane-attached structures (ribosomes, surface glycoproteins, various receptor sites, etc.) must occur.

Radical chain reactions might be initiated *in vivo*: (1) by molecule-induced hemolysis, (2) by redox reactions, (3) by radical production by radiation. The most important types of reactions are atom abstraction and addition, and electron transfer. Electron transfer is an important reaction in radiolysis, and several metal-catalyzed redox systems. Even relatively stable compounds may be oxidized at enhanced rates in the presence of readily oxidizable substances. This type of reaction is of great importance to the cell where readily oxidizable lipids may cause oxidation of relatively inert cellular compounds.

During hemolysis, radicals can be generated because of thermal hemolysis, by one-electron redox reactions and high energy radiation and photolysis.[49] Molecule-induced hemolysis occurs when bond-making accompanies bond-breaking, which lowers the activation energy of the overall process. For example, the normal dissociation of $Cl_2 \rightarrow 2Cl^{\cdot}$, but in a lipid environment a radical generates according to the equation $R{-}H + O_2 \rightarrow R^{\cdot} + HOO^{\cdot}$. Redox reaction that causes generation of free radicals is represented by the following equation:[49]

$$Fe^{2+} + ROOH \rightarrow Fe^{3+} + RO^{\cdot} + HO^{-}$$

$$Fe^{3+} + ROOH \rightarrow Fe^{2+} + ROO^{\cdot} + H^{+}$$

$$\text{Sum: } 2\ ROOH \rightarrow RO^{\cdot} + ROO^{\cdot} + H_2O$$

Radiolysis and photolysis generate radicals by absorption of high energy radiation or light. The radiolysis of water generates the cation-radical H_2O^{+} and an electron that reacts with water and each other with production of highly reactives $H^{\cdot}$ and $HO^{\cdot}$, an excited water molecule H_2O and H_2O_2. The hydrogen atom, the hydroxy radical and the solvated electron are capable of causing radiation damage to several molecules (DNA and proteins).[49]

1. Propagation Reactions: Hydrogen Transfer

These kinds of change are by atom transfer or group transfer, addition, atom abstraction, rearrangement, or beta scission.[49] β scission reaction is the reverse reaction of addition, it occurs in auto-oxidation chains. Rearrangement occurs very rarely in radical reactions. The addition reaction, like atom abstraction, is an important and frequent reaction of radicals. The best known reaction of greatest interest is the addition of a carbon-carbon double bond. The addition reaction is the basic reaction involved in vinyl polymerization used for plastics production. Hydrogen abstraction and addition to double bonds have similar rate constants, and often compete in particular systems, especially when olefin has reactive allylic hydrogens, which occurs in the auto-oxidation of most lipids.

The addition of radicals to aromatic rings is particularly important to the biology of nucleic acids because addition of radicals to the bases of DNA causes point mutations. Practically all radicals can be added to the DNA bases.

Electron transfer involves the transfer of an electron from a radical ion to a nonradical. The reaction is important in radiolysis and in several biochemical reactions (e.g., in flavin chemistry).[49] Lipids auto-oxidation occurs *in vivo* and *in vitro*. Peroxidized lipids are present in age pigments.

Lipid peroxidation *in vivo*, mainly polyunsaturated lipids, occurs in cellular mechanisms of aging; in air pollution oxidant damage to cells and to lungs; in atherosclerosis; in chlorinated hydrocarbon hepatotoxicity; in ethanol-induced liver injury; and in oxygen toxicity.[50] Lipid peroxidation is one of the most important damaging factors in human pathology. It involves the direct reaction of oxygen and lipid to form free radical intermediates and semistable peroxides. Proteins and enzymes subjected to lipid peroxidation undergo major reactions of polymerization, polypeptide scission, and chemical changes in individual amino acids. The consequence of polymerization or cross-linking of proteins causes an increase of the molecular weight to many multiples of that of the original protein. For example, human serum albumin becomes damaged on storage, and polymers of the cross-linked protein occur. The reaction mainly is possible because of peroxidation of unsaturated fatty acids bound to the protein molecule.

2. Damage to Membranes and Subcellular Organelles

Mitochondria and microsomal membranes contain relatively large amounts of polyunsaturated fatty acids in their phospholipids including 2, 4, 5, and 6 double bonds. Powerful catalysts are in close molecular proximity to these membranes. Thus, a free radical lipid reaction can propagate through the polyunsaturated fatty acids of the phospholipids. With progression of the chain reaction each completed peroxidation leaves a fatty acid hydroperoxide product. Peroxide decomposition accelerated by metal catalysts may lead to branching reactions in the lipid peroxidation process.[50]

Erythrocytes are very susceptible to lipid peroxidation because of their content of polyunsaturated acids and direct exposure to molecular oxygen. Lipid peroxidation causes hemolysis probably because of the damage to the lipid structure and inhibition of erythrocyte acetylcholinesterase activity by lipid peroxides.[50]

Lipid peroxidation damage to mitochondria may cause severe functional disturbances of the cellular metabolism. It can be initiated by redox agents (ferrous iron, ascorbic acid, glutathione). Finally, the mitochondria may disintegrate. As one of the evident symptoms of damage to mitochondria *in vivo* are disturbances in oxidative phosphorylation. In vitamin E deficiency disturbances of mitochondrial functions have been observed.

Microsomes also are very susceptible to lipid peroxidation, and consequently disturbances of their functions occur. Iron plus a reducing agent are powerful catalytic systems for lipid peroxidation of microsomes.

Experiments *in vitro* have demonstrated that lysosomes underwent lipid peroxidation less rapidly than mitochondria and microsomes, which corresponds with the smaller content of lipids in lysosomes.

Lipofuscin pigments accumulate in animal tissues as a function of age, especially in the brain and heart. It was discovered that lipofuscin and other age pigments are complexes of lipid-protein substances whose composition indicates that they were derived by lipid peroxidation of polyunsaturated lipids of subcellular membranes.[50]

3. Ethanol-Induced Fatty Liver

It is well known that chronic alcoholism causes fatty liver degeneration. The question was, if ethanol per se, or its metabolites, are responsible for these disturbances. The question was elucidated by administration of pyrazol, a competitive inhibitor of alcohol dehydrogenase. Administration of this inhibitor caused a prolonged and sustained alcohol concentration in the blood (also with intoxication), but without fatty liver development. This result suggests that ethanol metabolites are responsible for the fatty liver degeneration in alcoholics. The finding that ethanol must be metabolized to result in fatty liver degeneration suggests that the toxic effect depends on acetaldehyde and formation of free radicals and/or alteration in the redox potential of the hepatic cell due to increased generation of hydrogen equivalents resulting in alteration of NAD/NADH ratios.[51] Experiments with tetrachloride-injury support either the lipid peroxidation-antioxidant concept of ethanol-induced liver cell injury, or the importance of the dehydrogenation of ethanol as the molecular basis of ethanol-induced injury by hydrogen generating capacity of ethanol.[51]

4. Free Radical Damage of Nucleic Acids

Nucleic acid changes are the most noxious consequences of living organisms to ionizing radiation. These changes may occur either when radiation energy is absorbed by DNA molecules themselves or when it is absorbed by neighboring molecules, mainly water, and the reactive products diffuse to DNA and react with it. The principal lesions observed in DNA irradiated in aqueous media are breakage of hydrogen bonds, chain or single strand breaks, and degeneration of purine and pyrimidine bases. Hydrogen bond breaks are probably secondary reactions resulting from chain breaks or base degradations. Direct evidence of the formation of free radicals in nucleic acids per se by radiation have been well documented.[52] The only group in DNA that can form a free radical fragment -C(CH_3)-CH_2-5,6-dihydrothymidin-5-yl radical is the thymine moiety. Hydrogen atoms are produced by absorption of radiation energy in water and organic compounds. They are responsible for radiolytic processes in living organisms that react with organic compounds to give organic free radicals by abstraction of hydrogen atoms from saturated carbon atoms, and by addition to double and triple bonds.

Other nucleic acid bases with radical formation are methylcytosine that reacts with $H^{\cdot}$ to give a similar 5,6-dihydrothymidin-5-yl radical as does thymine; cytosine that by addition of $H^{\cdot}$ forms 5,6-dihydrocytosin-6-yl radical; 5-bromouracil that reacts with $H^{\cdot}$ to give 5,6-dihydrouracidyl radical; adenine, guanine, and xanthine react likewise with $H^{\cdot}$ giving analogous radicals.[52]

The initial reactions causing radical damage to nucleic acids consist of addition of $OH^{\cdot}$ to the bases and abstraction of hydrogen from pentoses to give organic free radicals. The extent of this reaction depends on the amount of the bases in nucleic acids. Other molecules in cells compete with nucleic acids for $OH^{\cdot}$, but the number of reaction sites on nucleic acids is so large that $OH^{\cdot}$ in the proximity of nucleic acid react with it. Nucleic acid radicals decay slowly which leads to permanent damage of nucleic acids. Unrepaired free radical damage to a base causes mutation, and damage to a pentose to a strand break. The attack of $H^{\cdot}$ on nucleic acids is similar to that of $OH^{\cdot}$, resulting in addition of $H^{\cdot}$ to the bases and abstraction of hydrogen from the pentose moieties. Studies on secondary reactions of nuclei acid radicals have revealed that oxygen and other small molecules react rapidly with the thymine-$OH^{\cdot}$ adduct and similar radicals. This led to a suggestion of chemical mechanisms by which 5-bromouracil, sulfhydryl compounds, and *N*-ethylmaleinide modify the response of nucleic acids to ionizing radiation.[52]

5. Tissue Injury Caused by Free Radicals

The biologic source of free radicals are byproducts of normal metabolism, caused by ionizing radiation, by drugs capable of redox cycling and by xenobiotics that can form free radical metabolites *in situ*. Several enzymes generate free radicals during their catalytic cycling. The best known is xanthine oxidase that generates $O_2^{\cdot}$ during reduction of oxygen to H_2O_2. Important sources that generate free radicals are subcellular organelles (mitochondria endoplasmic reticulum and nuclear membrane transport system, peroxisomes, and plasma membranes).

Peroxisomes are potent sources of cellular H_2O_2 because of their high concentration of oxidases. Peroxysomal H_2O_2 generating enzymes are D-amino acid oxidase, urate oxidase, L-α-hydroxyacid oxidase and fatty acyl-CoA oxidase.[53] Most of the H_2O_2 generated by peroxisomal oxidase is metabolized by peroxisomal catalase.

The plasma membrane is a critical site of free radical reactions. Extracellular-generated free radicals must cross the plasma membrane before they can react with cellular compounds. The unsaturated fatty acids of the plasma membrane (phospholipids, glycolipids, glycerides, and steroids) and the transmembrane proteins containing oxidizable amino acids are susceptible to free radical damage. Increased plasma membrane permeability caused by lipid peroxidation, or oxidation of structurally important proteins, can disturb its functions, e.g., abnormal transmembranal ion gradient, loss of secretory function, and disturbances of cellular metabolic processes. H_2O_2 can easily cross membranes, the charged $O_2^{\cdot}$ via transmembrane ion channel. The polyanionic cell surface attracts $H^{\cdot}$, resulting in low pH, which favors the formation of protonated form of $O_2^{\cdot}$, the perhydroxyl radical ($H^{\cdot} + O_2 \rightarrow HO^{\cdot}$) which is a stronger oxidant than $O_2^{\cdot}$ and causes toxic effects.[53]

The phagocytic cell plasma membrane NADPH oxidase-mediated production of free radicals is an important source of free radicals that can damage both the source cell and cells in close apposition to stimulated phagocytes.[53]

Free radical production of microsomal and plasma membrane-associated enzymes, lipoxygenase, and cyclooxygenase, is of great importance because of generating biologically potent arachidonic acid metabolites: prostaglandins, thromboxans, leukotriens, and slow reacting substances of anaphylaxis. The enzymatic oxidation of arachidonic acid by membrane-bound cyclooxygenase involves three radical intermediates. These free radical intermediates can cause irreversible oxidative deactivation of cyclooxygenase, and can be an important factor in feedback regulation of cyclooxygenase modulating both the rate and extent of prostaglandin biosynthesis, and cytotoxic effects caused by prostaglandins. Cyclooxygenase can metabolize xenobiotics to more toxic substances.[53]

Thromboxane synthesis in platelets is inhibited by the radical scavengers, imidazole and nordihydroguiaretic acid, suggesting that free radical reaction is caused by conversion of prostaglandin peroxide to thromboxanes. It seems that the biosynthesis of prostaglandins and thromboxanes results in hemoprotein-oxygen and carbon-centered free radicals capable of reacting with the biosynthetic enzymes themselves.[53]

The main free radical tissue injuries are consequences of protein cross-linking and denaturation, enzyme inhibition, mutations, permeability changes of plasma and organelles membranes, cell surface receptor changes, lipid cross-linking, cholesterol and fatty acid oxidation, decreased availability of nicotinamide- and flavin-containing cofactors, decreased neurotransmitter availability and activity especially of serotonin and adrenaline, and decreased availability of antioxidants (α-tocopherol, β-carotene).[53] The susceptibility of particular proteins to free radical damage depends on their amino acid composition. Due to the reactivity of unsaturated and sulfur-containing molecules with free radicals, proteins (enzymes) and other organic compounds containing the amino acids tryptophan, tyrosine, phenylalanine, histidine, methionine, and cysteine are particularly endangered by free radical-mediated amino acid modification. Cytoplasmic and membrane proteins can be cross-linked into dimers or larger aggregates after exposure to oxidizing agents, e.g., ozone and protoporphirin IX. Normally modification-resistant peptide bonds and amino acids (proline, lysine) may also be modified by the extreme reactivity of some free radicals, e.g., $OH^{\cdot}$.

Nucleic acids and DNA are equally susceptible to free radical injury. Ionizing radiation generates charged and neutral free radicals that injure DNA resulting in mutations—death from ionizing radiation is principally due to free radical reactions with DNA. Every kind of energy including ultraviolet and visible light, heat, gamma- and X-ray irradiations generate ions, free radicals and excited molecules that can injure several compounds of living organisms. Cytotoxicity of irradiations is mainly caused by chromosomal aberrations arising either from nucleic acid modifications or DNA strand scission. Cell death and mutations from free radicals generated by photochemical air pollutants have been ascribed to reactions with DNA.[53]

6. Membrane Injuries Caused by Free Radicals

The unsaturated fatty acids and cholesterol reacting with free radicals undergo peroxidation. This process, after initiation, can become autocatalytic and yield lipid peroxide, lipid alcohol, and aldehydic by-products. Peroxidation of fatty acids containing three or more double bonds will generate malondialdehyde. Plasma membrane and organelle lipid peroxidation is potentiated by the presence of metals which can serve as redox catalysts and catalyze the conversion of $O_2^{\cdot}$ and H_2O_2 to more potent oxidants. Peroxidation of membrane fatty acids generates shortened fatty acids containing R-OOH, R-COOH, R-CHO, and ROH groups that may impair membrane permeability and microviscosity. Oxidized phospholipids are split by phospholipase A as a step in prostaglandin synthesis.[53]

In the presence of so many sources of injurious free radicals without potent free radical scavengers, life probably would be impossible. Potent antioxidants are vitamin E (a series of isomers of tocopherol) that reduces $O_2^{\cdot}$, $OH^{\cdot}$, singlet oxygen, lipid peroxy radicals, and other free radicals. Another antioxidant is vitamin C (ascorbate) that has similar properties and probably is necessary to maintain vitamin E in the reduced active form. Beta-carotene is an efficient singlet oxygen scavenger that inhibits lipid peroxidation. Every molecule that reacts with free radical can be termed a "scavenger", and because unsaturated fatty acids, sugars, and amino acids containing sulfur can scavenge free radicals, they are in this sense free radical scavengers. The most important antioxidant of organismic origin is tryptophan which in cooperation with its reductant NADPH and enzymatic catalysts can reduce H_2O_2, lipid peroxides, disulfides, ascorbate, and free radicals. For example glutathion-peroxidase catalyzes peroxide reduction. The product of the reaction of glutathion with peroxides and disulfides is glutathion disulfide or a glutathion adduct of lipid or protein.[53]

X. MOLECULAR CHANGES OF THE BASEMENT MEMBRANE IN VARIOUS DISEASES

For a long time it has been known that the basement membrane becomes thicker in various diseases, especially in membraneous nephropathy and in diabetes mellitus.

Severe complication of diabetes mellitus is the generalized microangiopathy with neuropathy, retinopathy, and nephropathy, both in insulin-dependent and in insulin-independent diabetes. Basement thickening occurs in the capillaries of muscle, retina, skin, and kidney. A peculiar thickening is represented by thickening in the glomerular mesangium. The mesangial cell is a modified, smooth muscle cell and the mesangial matrix represents its basement membrane. Besides thickening, there are also severe functional changes, resulting in nephrotic syndrome (proteinuria, hypoproteinemia, and edema), retinal exudates, and microhemorrhages.[29]

The exact mechanism generating these changes is unknown. It has been postulated that diabetic microangiopathy is a separate, genetically determined disease. Against this hypothesis speak the following observations:

1. Diabetic patients who received renal transplants developed diabetic nephropathy in genetically unrelated transplanted kidney.
2. Patients either with hemochromatosis or growth hormone-producing pituitary adenomas may develop diabetes and microangiopathy.
3. In experimental diabetes rats developed thickening of the basal membrane.
4. After pancreatic islet transplantation microangiopathy in experimental diabetic animals disappeared.[29]

On the basis of these observations and experiments it is postulated that microangiopathy may develop because of nonenzymatic hyperglycosylation of plasma proteins. The best example is nonenzymatic glycosylation of hemoglobin and lens crystalline proteins causing diabetic or galactosemic cataracts. Hyperglycosylation with functional disturbances may also occur in other proteins. At present, the nonenzymatic hyperglycosylation as the sole cause of diabetic microangiopathy is inconclusive. Recently, it was proposed that in diabetic microangiopathy the membrane basement heparan sulfate proteoglycan synthesis is decreased, and thickening of the basement membrane may be the result of compensatory hypersecretion of type IV collagen and laminin.[29] This hypothesis is supported by studies on human kidneys which have demonstrated that in diabetic glomeruli an increase of hexuronic acid (heparan sulfate proteoglycan) with increased amounts of glycoprotein was observed. The loss of heparan sulfate proteoglycan explains the proteinuria in diabetic nephropathy.[29]

Lipid nephrosis is a glomerular disease characterized by proteinuria, hypoproteinemia, edema, and hyperlipidemia (nephrotic syndrome) without noticeable changes in the basement membrane. The etiology and pathogenesis of this disease is unknown. A similar syndrome can be produced in animals by injection of the aminonucleoside of puromycin. It is supposed that this syndrome is caused by the metabolic injury of glomerular podocytes with decreased synthesis and content of heparan sulfate proteoglycan in the basement membrane and with normal synthesis of laminin and collagen type IV. A general conclusion emerges that the decreased content of heparan sulfate proteoglycan may be involved in proteinuria of glomerular origin.

Hereditary progressive glomerulopathy (Alport's syndrome) is a hereditary, primary glomerular disease with persistent microscopic hematuria, recurrent gross hematuria, and progressive renal failure. The main ultrastructural abnormality is a marked attenuation of the glomerular basement membrane, often with discontinuity of the lamina densa. In more advanced cases the glomerular basement membrane becomes irregularly attenuated and thickened with characteristic splitting which causes a multilamellar appearance with interspersed granules and lipid droplets. Tubular basement membrane is also thickened and multilamellar, but to a lesser extent. Recent investigations have revealed that Alport's syndrome is an inherited, generalized metabolic disorder of basement membranes including epidermal and vascular membranes in the skin that demonstrates multilamellar morphology and abnormal distribution of laminin and type IV collagen (examined by specific antibodies).

A. HEREDITARY ONYCHO-OSTEODYSTROPHIA (NAIL-PATELLA SYNDROME)

The disease is manifested by dystrophic and hypoplastic nails, patellae, radial heads, iliac horns, and abnormal iridal pigmentation. Occasionally there is involvement of proteinuria. About 25% of patients develop renal failure. Morphologically, thickening of the glomerular basement membrane, segmental glomerulosclerosis, and cortical fibrosis occurs.

Ultrastructurally, the glomerular basement membrane is homogeneously thickened with typical cross-banded collagen fibers. The presence of cross-banded collagen fibers suggests that either there is a switch in the collagen synthesis (i.e., synthesis of another type of collagen, type IV), or there is extracellular cleavage of type IV collagen propeptides resulting in cross-banded collagen formation.[29]

B. IMMUNE-MEDIATED DISEASES

The most important are glomerular diseases caused by nonspecific entrapment of circulating antibodies in the glomerular filter resulting in complement fixation and chemotaxis which causes damage and thickening of the glomerular basement membrane. Goodpasture's syndrome was the first in which crescentic glomerulonephritis and hemorrhagic alveolitis with linear deposition of IgG in the glomerular and alveolar basement membranes have been demonstrated. The kidneys and lungs are edematous with hemorrhagic foci. Several antigens have been examined for elucidation of the nature of antibodies that cause the disease. Among several antigenic species, particular interest has been paid to hydroxylysine-galactose-glucose, a polypeptide specific for collagen. On the basis of these experiments, it was considered that collagen-like glycoproteins may be the antigens that are responsible for the synthesis of antibodies that entrap the glomerular membrane causing Goodpasture's syndrome.[29] Unfortunately, recent experiments did not support this hypothesis and it is supposed that a noncollagenous glycoprotein other than laminin may be the antigen that causes synthesis of antibodies in Goodpasture's syndrome.

REFERENCES

1. **Weber, K. and Osborn, M.,** The molecules of the cell matrix, *Sci. Am.,* 253(10), 110, 1985.
2. **Fowler, V. M.,** New views of the red cell network, *Nature (London),* 322, 777, 1986.
3. **Cohen, C. M.,** Origins of the cytoskeleton, *Nature (London),* 321, 18, 1986.
4. **Woods, C. M. and Lazarides, E.,** Spectrin assembly in avian erythroid development is determined by competing reactions of subunit homo- and hetero-oligomerization, *Nature (London),* 321, 85, 1986.
5. **Knowles, W. and Marchesi, S. L.,** Spectrin: structure, function, and abnormalities, *Semin. Hematol.,* 20(3), 159, 1983.
6. **Becker, P. S. and Lux, S. E.,** Hereditary spherocytosis and related disorders, *Clin. Hematol.,* 14, 15, 1985.
7. **Lawler, J., Liu, S. C., Palek, J., and Prchal, J.,** A molecular defect of spectrin in a subset of patients with hereditary elliptocytosis. Alterations in the alpha-subunit domain involved in spectrin self-association, *J. Clin. Invest.,* 73, 1688, 1984.
8. **Dhermy, D., Lecomte, M. C., Garbarz, M., Feo, C., Gautero, H., Bournier, O., Galand, C., Herrera, A., Gretillat, F., and Boivin, P.,** Molecular defect of spectrin in the family of a child with congenital hemolytic poikilocytic anemia, *Pediatr. Res.,* 18, 1005, 1984.
9. **Lawler, J., Coetzer, T. L., Palek, J., Jacob, H. S., and Luban, N.,** Sp alpha I/65: a new variant of the alpha subunit of spectrin in hereditary elliptocytosis, *Blood,* 66, 706, 1985.
10. **Marchesi, S. L., Knowles, W. J., Morrow, J. S., Bologna, M., and Marchesi, V. T.,** Abnormal spectrin in hereditary elliptocytosis, *Blood,* 67, 141, 1986.
11. **Alloisio, N., Dorleac, E., Morle, L., Girot, R., Galant, C., Boivin, P., and Delaunay, J.,** The genetic abnormalities involving red cell membrane protein 4.1 with or without elliptocytosis, *Biomed. Biochim. Acta,* 42, 538, 1983.
12. **Asano, K., Osawa, Y., Yanagisawa, N., Takahashi, Y., and Oshima, M.,** Erythrocyte membrane abnormalities in patients with amyotrophic chorea with acanthocytosis. Part 2. Abnormal degradation of membrane proteins, *J. Neurol. Sci.,* 68, 161, 1985.
13. **Rangachari, K., Dluzewski, A., Wilson, R. J. M., and Gratzer, W. B.,** Control of malarial invasion by phosphorylation of the host cell membrane cytoskeleton, *Nature,* 324, 364, 1986.
14. **Bretscher, M. S.,** The molecules of the cell membrane, *Sci. Am.,* 253(10), 100, 1985.

14a. **Bretscher, M. S.,** Endocytosis: relation to capping and cell locomotion, *Science,* 224, 681, 1984.

15. **Unwin, N. and Henderson, R.,** The structure of proteins in biological membranes, *Sci. Amer.,* 250(2), 78, 1984.
16. **Grab, D. J., Webster, P., and Verjee, Y.,** The intracellular pathway of newly formed variable surface glycoproteins of *Trypanosoma brucei, Proc. Natl. Acad. Sci. U.S.A.,* 81, 7703, 1984.
17. **Griffith, G. and Simons, K.,** The trans Golgi network: sorting at the exit of the Golgi complex, *Science,* 234, 438, 1986.
18. **Nowycky, M. C., Fox, A. P., and Tsien, R. W.,** Three types of neuronal calcium channel with different calcium agonist sensitivity, *Nature (London),* 316, 440, 1985.
19. **Reuter, H.,** A variety of calcium channels, *Nature (London),* 316, 391, 1985.
20. **Hawthorne, J. N.,** Is phosphatidylinositol now out of the calcium gate? *Nature (London),* 295, 281, 1982.
21. **Mueckler, M., Caruso, C., Baldwin, S. A., Panico, M., Blench, J., Morris, H. R., Allard, W. J., Lienhard, G. E., and Lodish, H. F.,** Sequence and structure of a human glucose transporter, *Science,* 229, 941, 1985.
22. **Dautry-Varsat, A. and Lodish, H. F.,** How receptors bring proteins and particles into cells, *Sci. Am.,* 250(5), 52, 1984.
23. **Geuze, H. J., Slot, J. W., Strous, G. J. A. M., Peppard, J., v. Figura, K., Hasilik, A., and Schwartz, A. L.,** Intracellular receptor sorting during endocytosis: comparative immunoelectron microscopy of multiple receptors in rat liver, *Cell,* 37, 195, 1984.
24. **Walter, P., Gilmore, R., and Blobel, G.,** Protein translocation across the endoplasmic reticulum, *Cell,* 38, 5, 1984.
25. **Robertson, M.,** Membrane traffic and the problem of protein secretion, *Nature,* 300, 594, 1984.
26. **Perara, E., Rothman, R. E., and Lingappa, V. R.,** Uncoupling translocation: implications for transport of proteins across membranes, *Science,* 232, 348, 1986.
27. **Ruoslahti, E. and Pierschbacher, M. D.,** Arg-Gly-Asp: a versatile cell recognition signal, *Cell,* 44, 517, 1986.
28. **Leptln, M.,** The fibronectin receptor tamıly, *Nature (London),* 321, 728, 1986.
29. **Martinez-Hernandez, A. and Amenta, A.,** The basement membrane in pathology, in *Advances in the Biology of Disease,* Vol. 1, Rubin, E. and Damjanov, I., Eds., Williams & Wilkins, Baltimore, 1984, 52.
30. **Mosesson, M. W. and Amrani, D. L.,** The structure of biologic activities of plasma fibronectin, *Blood,* 56, 145, 1980.
31. **Gerace, L.,** Traffic control and structural proteins in the eukaryotic nucleus, *Nature,* 318, 508, 1985.
32. **Goldstein, G. W. and Betz, A. L.,** The blood-brain barrier, *Sci. Am.,* 255, 74, 1986.
33. **Abbot, N. J.,** Glia and the blood-brain barrier, *Nature (London),* 325, 195, 1987.
34. **King, G. L. and Johnson, S. M.,** Receptor-mediated transport of insulin across endothelial cells, *Science,* 227, 1583, 1985.
35. **Bourne, H. R.,** One molecular machine can transduce diverse signals, *Nature (London),* 321, 814, 1986.
36. **Carpenter, D., Jackson, T., and Hanley, M. R.,** Coping with a growing family, *Nature (London),* 325, 107, 1987.
37. **Nishizuka, Y.,** Studies and perspectives of protein kinase C, *Science,* 233, 305, 1986.
38. **Ohno, S., Kawasaki, H., Imajoh, S., Suzuki, K., Inagaki, M., Yokokura, H., Sakoh, T., and Hidaka, H.,** Tissue-specific expression of three distinct types of rabbit protein kinase C, *Nature (London),* 325, 161, 1987.
39. **Coussens, L., Parker, P., Rhee, L., Yang-Feng, T. L., Waterfield, M. D., Francke, U., and Ullrich, A.,** Multiple distinct forms of bovine and human protein kinase C suggest diversity in cellular signaling pathways, *Science,* 233, 859, 1986.
40. **Berridge, M. J. and Irvine, R. F.,** Inositol triphosphate, a novel second messenger in cellular signal transduction, *Nature (London),* 312, 315, 1984.
41. **Bell, R. M.,** Protein kinase C activation by diacylglycerol second messengers, *Cell,* 45, 631, 1986.
42. **Evans, S. W., Beckner, S. K., and Farrar, W. L.,** Stimulation of specific GTP binding and hydrolysis activities in lymphocyte membrane by interleukin-2, *Nature (London),* 325, 166, 1987.
43. **May, W. S., Sahyoun, N., Wolf, M., and Cuatrecasas, P.,** Role of calcium mobilization in the regulation of protein kinase C-mediated membrane processes, *Nature (London),* 317, 549, 1985.
44. **Parker, P. J., Coussens, L., Totty, N., Rhee, L., Young, S., Chen, E., Stabel, S., Waterfield, M. D., and Ullrich, A.,** The complete primary structure of protein kinase C—the major phorbol ester receptor, *Science,* 233, 853, 1986.
45. **Rakowicz-Szulczynska, E., Rodeck, U., Harlyn, M., and Koprowski, H.,** Chromatin binding of epidermal growth factor, nerve growth factor, and platelet derived growth factor in cells bearing the appropriate surface receptors, *Proc. Natl. Acad. Sci. U.S.A.,* 83, 3728, 1986.
46. **Boron, W. F.,** The "basic" connection, *Nature,* 312, 312, 1984.
47. **Carruthers, A. and Melchior, D. L.,** How bilayer lipids affect membrane protein activity, *Trends Biochem. Sci.,* 11, 331, 1986.

48. **Demopoulos, H. B.,** The basis of free radical pathology, *Fed. Proc.,* 32, 1859, 1973.
49. **Pryor, W. A.,** Free radical reactions and their importance in biochemical systems, *Fed. Proc.,* 32, 1862, 1973.
50. **Tappel, A. L.,** Lipid peroxidation damage to cell components, *Fed. Proc.,* 32, 1870, 1973.
51. **Luzio, N. R.,** Antioxidants, lipid peroxidation and chemical-induced liver injury, *Fed. Proc.,* 32, 1875, 1973.
52. **Myers, Jr., L. S.,** Free radical damage of nucleic acids and their components by ionizing radiation, *Fed. Proc.,* 32, 1882, 1973.
53. **Freeman, B. A. and Crapo, J. D.,** Free radicals and tissue injury, in *Advances in the Biology of Disease,* Vol. 1,, Rubin, E. and Damjanov, I., Eds., Williams & Wilkins, Baltimore, 1984, 26.

Chapter 8

MUTATION

Mutation is any heritable change in the genetic material (not caused by genetic segregation or genetic recombination) which is transmitted to daughter cells and even to succeeding generations giving rise to mutant cells or mutant individuals (provided it does not act as a dominant lethal factor). Mutations in the germ line may be transmitted by the gametes to the next generation resulting in an individual with the mutant condition either in germ or somatic cells. Mutation usually affects the carrier of genetic information, i.e., DNA, affecting its chemical or physical constitution, mutability, replication, phenotypic function or recombination. The essence of mutation depends on nucleotides that may be added, deleted, substituted, inverted or transposed to other positions with or without inversion. Mutation may occur spontaneously or can be induced by mutagens. Spontaneous mutations are probably the result of several unknown and ubiquitous mutagens or mutagenic effects. A considerable number of spontaneous mutations seems to be the result of intracellular production of mutagens and antimutagens (e.g., free radicals or free radical scavengers). The number of spontaneous mutations is probably high, but usually they become repaired before they have time to be expressed.

Mutations can be divided into gene mutations involving only a single gene, and chromosomal mutations involving a part or whole chromosome with more than one gene. Since a functional gene consists of larger or smaller DNA fragments, and many sites may be involved in a single gene mutation, the consequences of a mutation vary dependent on the particular sites of mutation, e.g., mutations of the particular exons or introns, flanking sequences containing regulatory sequences like TATA box, upstream regulatory sequences, enhancers, etc.

On the molecular basis, gene mutations usually depend on substitution of a purine or pyrimidine base. Substitutions of a purine base on another purine base (adenine for guanine or vice versa) or a pyrimidine base on another pyrimidine base (thymine for cytosine or vice versa) are named transitions, and substitutions of a pyrimidine base on a purine base, and vice versa, are called transversions. Gene mutations also may be caused by deletion or insertion (addition) of a single or of multiple nucleotides.

The nucleotides are situated on the nucleic acid strands in a well-defined order necessary for the correct functioning of the genetic code. Therefore, every change of this order caused by substitutions, deletions, additions, inversions, or transpositions must cause disturbances in the correct reading of the genetic code, i.e., a mutation which results in synthesis of an abnormal structure of protein molecules and abnormal function of these proteins.

I. THE CONSEQUENCES OF GENE MUTATION

The consequences of gene mutations are

1. Atypical protein molecules caused by substitution of single amino acids; the best examples are various abnormal globin molecules which will be described later.
2. Nonsense mutation that denotes any mutation (base substitution or frameshift mutation) that converts a sense codon specifying any amino acid into a chain terminating codon (UAG, UAA, or UGA). Nonsense mutation causes premature termination of polypeptide chain synthesis at the site where the nonsense codon occurs in the mRNA.
3. Frameshift mutation represents a class of mutations that arise from insertion (+) or deletion (−) of a nucleotide, or any number of nucleotides other than three into DNA. Frameshift mutation displaces the starting point of genetic transcription of the genetic

code which results in "misreading" of mRNA from the point of nucleotide addition or deletion. Thus, once a frameshift mutation has been introduced into the gene, the reading frame becomes shifted so that all codons distal to the mutation are read out of phase of the reading frame.[1] Frameshift mutations can be caused by errors in genetic recombination due to unequal crossing over, and errors in DNA repair.

Chromosome mutations are structural changes of the particular chromosomes with gain, loss, or translocation of chromosome fragments. They may arise spontaneously (caused probably by intraorganismic production of mutagens, e.g., free radicals, and in normal cellular mechanisms, segregating chromosomes during meiosis or mitosis), or induced by chemical or physical mutagens.

Chromosome mutations are either intrachromosomal or interchromosomal. Intrachromosomal changes either involve one arm which is named paracentric, or both arms which are named pericentric mutations. Chromosomal structural changes that cause mutations involve deletions (loss of terminal or intercalary fragments of chromosomes), duplications (presence of additional fragment of a chromosome), inversions (chromosome fragment becomes inverted and reinserted into its original chromosomal position), and translocations (relocation of chromosome fragments within the same or among other chromosomes).

Chromosome mutations may result in:

1. Chromosome aberrations in which both chromatids are engaged in aberration formation at identical loci
2. Chromatid aberrations in which mutation concerns only one of the two chromatids of a single chromosome
3. Subchromatid aberration in which a fibrillar subunit (usually half of the chromatid) is involved in the aberration.

The chromosomal aberrations are the final results of complicated processes the nature of which are not yet completely clear.

Two main processes have been proposed: the breakage-reunion hypothesis and the exchange model. According to the breakage-reunion hypothesis, the primary lesions are breaks that cause discontinuities of chromosomes. These lesions occur spontaneously or may be caused by the action of mutagens. Most of those breaks (90 to 99%) become rejoined by "repair processes" of DNA. Restoration in these cases may be complete without any structural or functional changes, or intra- or inter-changes may occur because of failure to restore the original chromosome structure. The breakage surfaces may stabilize which causes further breaks producing fragments of chromosomes. The breakage may also reunite when two or more breaks interact and join in a new order producing a chromosomal structural change. According to the exchange hypothesis the primary lesions are only instabilities of the chromosomes (without discontinuity) which either immediately, or after a delay, undergo a mechanical exchange process. This process also may be repaired, but structural changes may occur by local instabilities, when pairwise cooperation of the instabilities causes exchange of the chromosomal segments and when the exchange process becomes analogous to crossing over.[1]

Mutation rate is the probability with which a particular mutational event takes place per biological entity (virus, cell, individual) per generation during the period required for the reproductive cycle. The mutation rate can be estimated from the mutation frequency. The mutation rate depends on (1) mutability of the particular gene, (2) the ability of mutation repair.

Mutation frequency is estimated on the frequency of particular mutants in a population.

Mutation pressure denotes the continuous recurrent production of a mutant allele which causes an increase of its frequency in the gene pool in a population. This should result in altered proportions of the normal and mutant allele which might ultimately cause the replacement of the normal allele by the mutant one. However, because of back mutations the population never becomes homozygous for one allele. An equilibrium will be established at that point where the number of mutations from normal to mutational and back from mutational to normal will be equal.

Back mutation is a heritable change in a mutant that results in a revertant that has regained the lost normal gene because of mutation. A true back mutation restores the original nucleotide sequences.

Mutability denotes an increased ability of a gene or genotype to undergo mutation. Mutability enables adaptation of gene pools to change in the environment. Since mutations are mostly deleterious, a high level of mutability would be unfavorable, and therefore a low level of mutability is usually advantageous for the population. The relative constancy of mutability is achieved by several mechanisms like hereditary changes that affect the nongenic parts of the cell causing resistance to, and neutralization of, mutagens, susceptibility to mutagens, and changes in repair mechanisms that neutralize a mutagen's action. Genes that cause a significant rise in the mutability of other genes are called mutator genes.

The hot spot of mutation is any region in a gene that mutates at very much higher frequencies than the neighboring regions of the same gene. The existence of hot spots explains why some kinds of mutations occur quite often, while others are very rare.

According to the present views, mutation is a constant process that occurs spontaneously at a defined number. The best argument for a spontaneous mechanism of mutations is that elevated temperature which causes acceleration of spontaneous biochemical reactions in living organisms simultaneously causes an increase of the number of mutations. This also explains the fate of the occurrence of mutations.

Genetic load denotes the proportion by which the fitness of the average or optimum genotype at the observed locus is decreased due to the presence of deleterious (lethal, sublethal, or subvital) genes. The genetic load denotes the genetic disability of an individual or population, and is expressed as the average number of potential genetic deaths per individual. The components of a genetic load are

1. Mutational loads due to deleterious mutations at loci that are homozygous for non-deleterious alleles (load maintained by recurrent mutations)
2. Segregational loads due to genes segregating from favored heterozygotes and giving rise to unfit homozygous (load maintained by segregation from advantageous heterozygotes)[1]

Minor components of genetic load are

1. Input loads, i.e., the load of inferior alleles in a gene pool caused by mutation and immigration;
2. Substitutional load—the cost to a population of replacing a particular allele by another in the course of evolutionary change;
3. Selection load—due to selection for an intermediate optimum of a quantitative character;
4. Interaction load—due to interaction of deleterious genes.

A genetic load is termed a balanced load where decreases in the overall fitness of a population due to segregation of inferior genotypes occur and whose component genes are maintained in the population because they improve fitness in different combinations, e.g., as heterozygotes.[1]

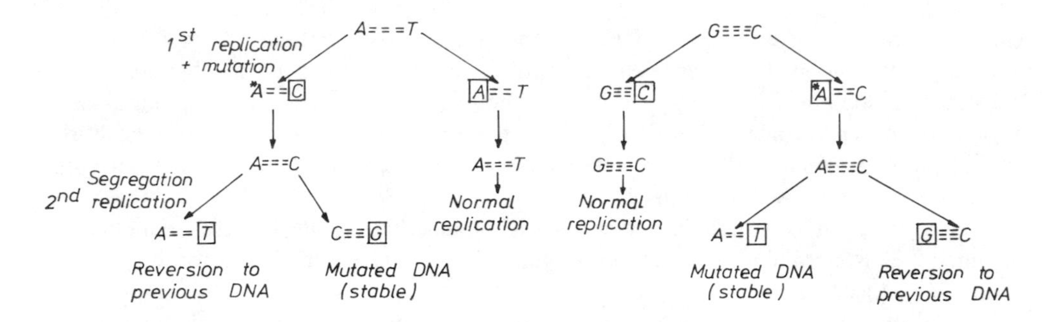

FIGURE 1. The effects of tautomerism on bases in subsequent DNA replication. The tautomeric bases are indicated by an asterisk, and bases incorporated from the environment are in squares. (Left): tautomerization of adenine present in the DNA; during the second replication reversion occurs to the previous state and gives rise to one mutated line of DNA (by transversion A → C). (Right): incorporation of tautomeric adenine from the environment; during the second replication reversion occurs to the previous state and gives rise to one mutated DNA line (by transition G → A).

II. THE MOLECULAR MECHANISMS OF MUTATION

The molecular mechanisms of mutations depend mainly on the intrinsic properties of nucleic acids. Double-stranded deoxyribonucleic acid occurs in a double helix with A (adenine) opposite T (thymine), and G (guanine) opposite C (cytosine) bound by weak hydrogen bonds. A change in one of the bases of the double-stranded chain may cause attraction of a wrong binding base altering all following events at this site. Such an abnormal attraction may be caused by tautomerism of the old base or by incorporation of a base analogue. Tautomerism denotes a rare event in which a rearrangement in the distribution of electrons and protons occurs that gives rise to a structure that is unable to make hydrogen bonds with its usual base pair and becomes able to form hydrogen bonds with atypical bases. A tautomer of A (adenine) binds C (cytosine) instead of T (thymine); a tautomer of C (cytosine) binds A (adenine) instead of G (guanine); a tautomer of G (guanine) binds T (thymine) instead of C (cytosine). All the tautomeric variants are able to make new erroneous purine-pyrimidine base pairs (see Figure 1).

Mutations of the genetic code also may be caused by different irradiations. The most common is ionizing radiation that deprives one proton from the first nitrogen atom of either thymine or guanine that generates an erroneous base pair thymine-guanine (T-G) or guanine-thymine (G-T). Further disturbances caused by the appearance of this erroneous base-pair are the same as presented in Figure 1.

Further changes of the DNA may be caused by incorporation of various structural analogs of purine or pyrimidine bases. The best known are changes caused by incorporation of 5-bromuracil (5BrU) instead of thymine (5-methyluracil). Bacteriophages cultivated in a medium containing 5-bromuracil incorporate at all positions 5-bromuracil instead of thymine. 5-bromuracil usually forms hydrogen bonds with adenine, however, it also sometimes bonds with guanine, which in the following replication generates an erroneous pair C-G with a mutated cell line as demonstrated in Figure 2.

Another kind of mutation may be caused by nitrous acid (HNO_2) which oxidizes the nitrous group (NH_2) to a hydroxyl group (OH). Deamination of cytosine generates uracil that forms pairs with adenine. Deamination of adenine generates hypoxanthine that forms hydrogen bonds with cytosine. Deamination of guanine that yields xanthine probably does not cause any pathological variant because xanthine forms hydrogen bonds with guanine. Mutagenic events caused by nitrous acid are shown in Figure 3.

Yet another kind of mutation is caused by radioactive ^{32}P. Phosphodiester cleavage by hydrolysis occurs in RNA, but not in DNA. DNA breaks occur usually as a consequence

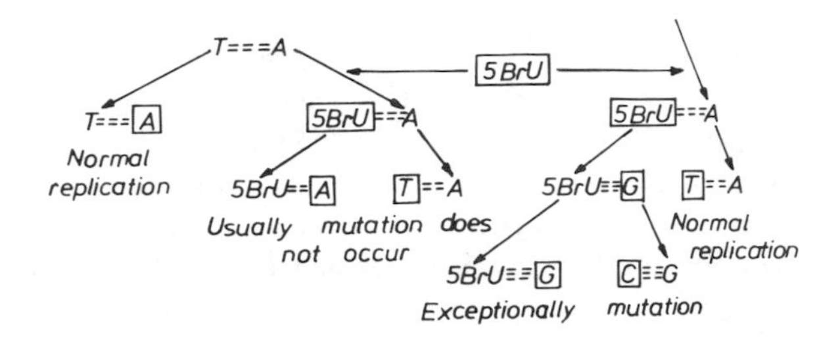

FIGURE 2. Mutagenic action of 5-bromuracil (5-BrU). Usually 5-BrU does not cause mutation because 5-BrU incorporates into DNA at the position of thymine and forms hydrogen bonds with adenine. In exceptional cases 5-BrU forms bonds with guanine which produces a persistent mutated line of DNA (AT → CG). Bases incorporated from the environment are in squares.

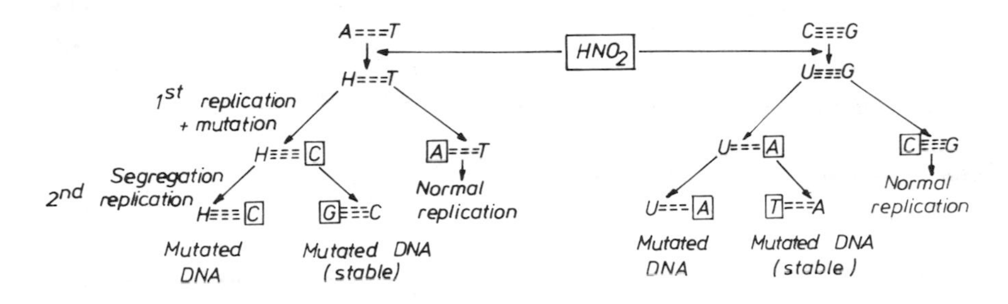

FIGURE 3. The mutagenic action of nitrous acid (NNO_2). (Left:) conversion of adenine into hypoxanthine (H). During the first replication H forms hydrogen bonds with C; during the second replication it produces a persistent pair CG; (transition T → C). (Right): conversion of cytosine into uracil. During the first replication uracil forms hydrogen bonds with adenine, during the second replication it forms a persistent mutation (transversion of C into A). Bases incorporated from the environment are in squares.

of depurination of depyrimidination. Hydrolytic cleavage of a base results in deoxyribose residues, that partially exist in aldehyde form. The rate constant for a hydrolytic break of the phosphodiester bond at an apurinic site is about 10^4 faster than depurination. Thus, all apurinic sites can be sites of breakage, even in the absence of endonucleases.[2] The estimated rate constants in double-stranded DNA are on the order of 10^{-11}, which nonenzymatic hydrolysis makes very improbable. However, radioactive disintegration of DNA may occur through incorporation of a considerable amount of ^{32}P into DNA which may cause a "suicidal" decay of the DNA by deprivation of phosphodiester bonds.

III. THE MUTAGENS

A mutagen is any physical or chemical agent that causes a significant increase of mutation rates above the spontaneous background level. Such mutations are called "induced mutations." The most widespread mutagens are ionizing radiation, ultraviolet rays, base analogs and alkylating agents.

A. ALKYLATING AGENTS

Several mutagens and carcinogens are mono- or bi-functional alkylating agents. The initial observation was that mustard gas becomes covalently bound with nucleic acids, particularly to the N^7 of guanine. Many alkyl derivatives in nucleic acids could not be detected before use of radiolabeled alkylating agents. The definite sites of modifications of nucleosides have been clarified with monofunctional aliphatic methylating and ethylating agents. Bifunctional agents act at the same sites, but they can also cyclize and form crosslinks.

Dimethylsulfate and methylmethanesulfonate react with N^7 of guanosine, N^1 and N^3 of adenine, and N^3 of cytosine. Their ethylating analogues are less reactive, although diethylsulfate and ethylmethanesulfonate react additionally with O^6 of guanine, N^6 and N^7 of adenine, and O^2 and N^4 of cytosine. The number of these reactions is considered small.

Dialkylation occurs only with alkyl iodides, probably as the result of alkaline reaction condition. This leads to imidazole ring-opened 7- and 1,7-alkyl guanosines.[2]

Alkylation of synthetic polynucleotides revealed differences between alkylation of nucleosides and polynucleotides. The differences are caused by the presence of phosphate in the phosphodiester bonds that can form phosphotriesters. The major sites of alkylation of single-stranded polynucleotides are the same as for nucleosides, only the extent of the reaction is lower.

The most studied alkylating agents in eukaryotes *in vivo* are alkyl sulfates and *N*-nitroso-compounds. Ethylating agents have greater affinity for modifying oxygens, than less reactive methylating agents. The reactive sites in single-stranded nucleic acids (RNA or DNA) are the same as in double-stranded DNA. All simple, direct-acting alkylating agents react with nucleic acids *in vivo* and *in vitro* at the same sites, and for each alkylating agent the amount and distribution of alkyl groups is characteristic.[2]

1. Monofunctional Alkylating Agents

a. Bisulfite

HSO_3^- is produced endogenously in mammals as an intermediate of sulfur containing amino acids. Another source of HSO_3^- is environmental exposure to sulfur dioxide, which in water becomes hydrated and forms H_2SO_3 that dissociates into bisulfite and sulfite. Bisulfite undergoes several reactions with nucleic acids from which the most important is deamination of cytosine into uracil. The consequence of that reaction is the same as that of nitrous acid.

5-Methylcytosine is also deaminated by bisulfite, but at a rate two orders of magnitude slower than cytosine. However, deamination of 5-methylcytosine converts it to thymine which is not repaired (as is uracil by a uracil glycosylase) causing transition of G-C into A-T, and thus can represent ''hot spots'' for spontaneous transition mutations.[2]

Cytosine deamination by bisulfite requires a high concentration of it (1 *M*) at pH 5.6. Different genetic effects have been observed also in neutral pH and lower concentrations which suggests that these effects could be caused by side reactions or reactions with proteins.[2] Finally, during these reactions free radicals may be also formed.

b. Hydroxylamine and Methoxyamine

Hydroxylamine and methoxyamine are mutagens that can cause an immediate change in base pairing. Hydroxylamine can react with adenine, cytosine, and uracil. The reaction with uracil is very slow and practically ineffective (it is formed only at pH 10). Guanine seems to be nonreactive to hydroxylamine. The reaction of cytidine with hydroxylamine and methoxyamine at pH 5.6 leads to replacement of the amino group by a hydroxylamino or methoxyamino group. Direct replacement of the amino group occurs most readily at high temperatures (40°C) and low concentration. The reaction that causes replacement of the adenosine amino group is extremely slow. Substitution of the N^4 or N^6 of adenosine by hydroxylamino or methoxyamino groups shifts the tautomeric equilibrium toward the imino form that can make an erroneous pair (N^4-methoxy cytosine for uracil).[2] Reactivity of hydroxylamine and methoxyamine is decreased in double-stranded nucleic acids, and mutations probably only occur with single-stranded fragments. Hydroxylamine at concentrations of 10^{-3} and 10^{-4} may undergo oxidative conversion with production of free radicals.

c. Hydrazine

Hydrazine is a weak mutagen *in vivo* and probably a human carcinogen. The major reaction of hydrazine is with cytidine that transforms N^4-cytidine into N^4-aminocytidine. That reaction is inhibited by protonation of hydrazine. At pH 6.0 only 0.4% of hydrazine is nonprotonated, and only little reaction can occur in aqueous solution. N^4-aminocytidine is the main compound that may cause mispairing of nucleic acid. Experiments with hydrazine reactions revealed that the pyrimidine ring of uracil and of thymine can be easily degraded, which may generate apyrimidinic sites in DNA.

d. Formaldehyde

Formaldehyde reacts reversibly with unbase-paired amino groups of nucleic acids. In single-stranded nucleic acids, hydroxymethyl derivatives of adenine, guanine, and cytosine have been found. A second, slower reaction results in $-CH_2=$ crosslinks formation described in formaldehyde-treated nucleic acids. However, biological effects of formaldehyde due to these reactions is doubtful, rather they are caused by secondary effects on cellular components. It is well-documented that mutations may occur through the formation of organic peroxides or peroxidic radicals from formaldehyde, aided by formaldehyde-induced catalase inhibition.

2. Bifunctional Alkylating Agents

Several alkylating agents are bifunctional carrying a halogen and a nitroso group. The best known are cyclic compounds: epoxides (chloroethylene oxide, ethylene oxide), lactones (β-propiolactone), *S*-mustards (mustard gas), *N*-mustards, and glicylaldehyde. The chemotherapeutic effect of the latter compounds correlate with the toxicity caused by their cross-linking activity, whereas their depurination reactions and chain breakages causes mutagenic and carcinogenic effects. Therefore, highly toxic levels are used in chemotherapy of Hodgkin's disease, some other lymphomas, and solid tumors.[2]

The cyclic alkylating agents owe their alkylating ability to reactive unstable ring structures. These reactive groups can form intra- and inter-strand crosslinks and cyclic derivatives.

a. Mustards

To that group belongs the known chemotherapeutic agent in cancer therapy—cyclophosphamide. Nitrogen and sulfur mustards were the first known chemical mutagens that reacted with the N^7 of guanine and formed monoadducts and cross-linked adducts to DNA. Further reactions at the N^1 and N^3 of adenine, and N^3 of cytosine were detected. The product of metabolism of *N*-mustard (known as cyclophosphamide) is phosphoramide which reacts with guanosine forming a highly unstable N^7 derivative that is probably the active agent of cyclophosphamide used in cancer therapy.

b. Lactones

The first known carcinogenic agent of lactones was β-propiolactone. Being easily decomposed in water, this compound is a relatively poor carcinogen, although it is an initiator and promoter of tumors in mouse skin. β-propiolactone as a typical alkylating agent reacts with polynucleotides and DNA at the N^7 of guanine and N^1 of adenine. It also performs a cyclizing reaction with adenosine. There also is indirect evidence that β-propiolactone can crosslink protein and nucleic acid.[2]

c. Epoxides

These, being three-membered rings, are highly reactive alkylating agents. The simplest epoxides, ethylene oxide and propylene oxide are weak mutagens. These compounds, being typical alkylating agents react with DNA and RNA at the N^7 of guanine, and N^1 and N^3 of adenine, forming hydroxyethyl or hydroxypropyl derivatives. Diepoxides, e.g., diepoxybutane, reacts at the N^7 of guanine. Crosslinking between the N^7 of two guanines on two strands of double helix occurs with D and L isomers. The diepoxide 1,2:5,6-dianhydrogalactitol, that preferentially reacts with Yoshida sarcoma in rats, yields three kinds of reaction-products on the N^7 of guanine (mono- and di-guanyl derivatives).

d. Vinyl Halides

Vinyl chloride is carcinogenic to humans and experimental animals, usually causing angiosarcoma of the liver, but also of the brain, lung, and hematolymphopoietic system. The biological effect of these compounds depend on metabolic activation by the microsomal P-450 monooxygenase system. Vinyl chloride forms a highly reactive metabolite—chloroethylene oxide—an epoxide that rapidly rearranges nonenzymatically to chloracetaldehyde. Guanine reacts directly with chloracetaldehyde to form N^2,3-ethenoguanine. Bromoacetaldehyde reacts with adenosine-5-phosphate and cytidine-5-phosphate forming etheno-derivatives (the reaction rates are tenfold faster than that with chloroacetylaldehyde).

Chloracetaldehyde modifies adenine and cytosine residues in single-stranded ribo- and deoxyribonucleosides (cytosine residues are about tenfold more reactive than adenosine residues). Vinyl chloride and vinyl bromide react with DNA and RNA—the earliest product is 1,N^6-ethenocytosine. After activation (*in vivo* or *in vitro*) of vinyl chloride with rat liver microsomes and an NADPH-regenerating system, 7-(2-hydroxyethyl)guanine is the major product of the reaction of vinyl halide with DNA.

e. Alkyl Carbamates

Ethyl carbamate (urethan) is carcinogenic in several species. It is supposed that this compound may be metabolically activated by dehydrogenation to vinyl carbamate that is next epoxidated, and is a more potent carcinogen than ethyl carbamate. Vinyl carbamate epoxide probably reacts with adenosine and cytidine forming etheno derivatives.[12]

f. Metabolically Activated Carcinogens

Polycyclic aromatic hydrocarbons formed mainly through combustion and pyrolysis of fuels and organic materials are widespread environmental pollutants of the atmosphere, water, and food chain, and may be a significant factor in the incidence of human mutations and cancer. Polycyclic aromatic hydrocarbons in cigarette smoke are powerful carcinogens in experimental animals and probably play an important role in the etiology of human lung cancer. Most of the polycyclic aromatic hydrocarbons to become mutagenic are metabolized by enzyme systems localized mainly in the liver, which although it serves mainly in detoxification of these compounds, may also produce much more potent mutagens and carcinogens than the initial material. The most studied polycyclic aromatic hydrocarbon is benzo(*a*)pyrene, detected by Kenneway in 1938 as the carcinogenic agent in coal tar.

A group of enzymes known as the cytochrome P-450-mediated mixed-function oxidases or monooxygenases found in microsomes of endoplasmic reticulum, but also in mitochondrial and nuclear fraction of cells, represents the most important enzyme system. The reactions catalyzed by these enzymes are hydroxylation reactions. Therefore, they are termed arylhydrocarbon hydroxylases (AHH). AHH is present in 90% of human tissues (liver, alveolar macrophages of lung, gastrointestinal tract, kidney, placenta, lymphocytes, and monocytes). The membrane-bound multicomponent system includes two proteins: cytochrome P-450 (a family of different proteins) and a protein NADPH-cytochrome P-450 reductase, as well as molecular oxygen and NADPH. For full activity the reconstituted enzyme needs more phospholipid phosphatidylcholine which would facilitate interaction between the cytochrome P-450 and the NADPH-cytochrome P-450 reductase.

Reactions of metabolically activated carcinogens with nucleic acids—The first interactions of chemical carcinogens (mutagens) with macromolecules were described about 30 years ago and concerned aminoazodyes and benzo[*a*]pyrene bound to proteins of organisms in which they produced tumors. Next, covalent binding of residues of these compounds with nucleic acids and proteins was described. The extent of binding of these residues to nucleic acids, but not to proteins, demonstrated a correlation with the carcinogenic potency, although this was not always true. On the basis of these observations a hypothesis has been established that the active carcinogenic agents are highly reactive electrophils that form covalent bonds with nucleophilic residues in macromolecules. Compounds that are not electrophilic per se undergo metabolism by mixed-function oxidase system to carcinogenic forms.[2]

B. POLYCYCLIC AROMATIC HYDROCARBONS

1. Benzo[*a*]pyrene (BP)

BP was the first polycyclic aromatic hydrocarbon that was covalently bound to DNA, RNA, and proteins. It was soon discovered that 7,8-diol metabolite of BP was bound to DNA to a greater extent than BP or other phenol metabolite. This enabled a hypothesis that the diol was converted to a diol epoxide which was the active form binding to DNA.[2] Direct evidence that the 7,8-diol metabolite converted to the diol epoxide comes from nucleic acid binding studies which confirmed that the major nucleoside products formed from reactions of the diol epoxide with polydeoxyguanosine were identical to those formed in DNA as the result of metabolic activation of BP by several cellular systems (human lungs or human epidermal keratinocytes). It also was suggested that the diol epoxide may react with the N^7 of guanine. The results obtained when binding BP metabolites to DNA support the view that BP-diol epoxides are the major metabolites involved in binding to DNA which results in mutagenesis as well as in carcinogenesis. These results do not exclude the possibility that other BP metabolites binding to DNA may be potent mutagens.

2. Benz[*a*]anthracene (BA)

The 3,4-diol-1,2 epoxide of BA is a potent mutagen and carcinogen that forms two adducts to DNA. Analysis of the structure of these adducts demonstrates that the hydrocarbon becomes attached to the exocyclic amino group of guanine in a similar way as described for BP-7,8-diol-9,10 epoxide.

3. Methylbenz[*a*]anthracene (MBA)

The metabolism of MBA in liver homogenate yields five dihydrodiols. Unlike DNA adducts with BA, in MBA the 8,9-dihydrodiol-10-11-oxide does not seem to be involved, and rather the 3,4-diol-1,2-epoxide region forms adducts with DNA.

4. 7,12-Dimethylbenz[*a*]anthracene (DMBA)

The major hydrocarbon-deoxyribonucleoside adduct present in the hydrolysate of DNA isolated from animals treated with DMBA revealed anthracene-like and not phenanthrene-like fluorescence spectra. That result suggests activated 1, 2, 3, 4 positions of DMBA are involved in reaction with DNA. DMBA represents a typical compound the mutagenic action of which, due to cellular metabolism, is different in microsomal systems and in whole cell systems.

5. 3-Methylcholanthrene (3MC)

Metabolic activation of 3MC yields not only diol epoxides, but also several hydroxylated derivatives. Thus, a relatively great number of different adducts occur: eight major and one minor deoxyribonucleoside adducts. Metabolic activation of 3MC occurs through the formation of diol epoxide 3MC-9,10-diol-7,8-oxide and other derivatives that retain the anthracene ring system. Some of the adducts with DNA are formed by the reaction of this diol epoxide, and the others by the reaction of the hydroxylated derivatives of this diol epoxide with DNA.

C. *N*-SUBSTITUTED AROMATIC COMPOUNDS

1. *N*-2-Acetylaminofluorene (AAF)

The best known interactions of mutagens (carcinogens) with macromolecules is the covalent attachment of the reactive metabolites of *N*-substituted aromatic compounds to nucleic acids, particularly the nonenzymatic binding of AAF with nucleic acids. After administration of the mutagen (carcinogen) *N*-OH-AAF to rats, three kinds of DNA adducts have been detected: the major product (80%) was *N*-(deoxyguanosin-8-yl)AAF, the other 3-(deoxyguanosin-N^2-yl)AAF, and *N*-(deoxyguanosin-8-yl)AAF.

2. *N*-Methyl-4-aminobenzene (MAB)

MAB belongs to the best studied azo dye compounds. After interaction of the synthetic ester *N*-benzoyloxy-MAB with DNA (*in vitro*), 6 MAB adducts were detected. The main product was formed by binding of the MAB residue to C-8 of guanosine. The next minor adduct detected in rat liver was 3-(deoxyguanosin-N^2-yl)MAB and the third 3-(deoxy-adenosin-N^6-yl)MAB.

3. 2-Naphthylamine (2NA)

This compound was the first recognized causative agent in human bladder cancer which occurred in workers exposed to 2NA, but also in cigarette smokers due to the presence of nanogram quantities of 2NA in cigarette smoke. 2NA is metabolically activated to *N*-OH-arylamines and forms *N*-OH-2-NA which becomes converted to arylnitrenium ion-carbocation electrophile that is able to bind covalently with cellular macromolecules (DNA included) by forming 1-(deoxyguanosin-N^2-yl)-2NA,1-(deoxyadenosin-N^6-*yl)2NA and a purine ring opened derivative of* *N*-(deoxyguanosin-8-yl)2NA. The same adducts were found in liver and urothelium of male dogs after 2NA administration (fourfold higher in the target tissue—urothelium after 2 d and eightfold after 7 d than in the nontarget tissue—liver.

4. 1-Naphthylamine (1NA)

The mutagenic (carcinogenic) effect of 1NA is doubtful, although reactions are known *in vitro* that yielded two kinds of DNA-adducts: *N*-(deoxyguanosin-O^6-yl)-1NA and 2-(deoxyguanosin-O^6-yl)1NA.

5. Benzidine (BZ)

BZ has been identified as a human urinary bladder carcinogen. It is supposed that acetylation may be necessary for its activation. Reactions that activate BZ to active compounds forming DNA adducts are not fully clear.

6. 4-Nitroquinoline-1-oxide (4NQO)

4NQO is a potent mutagen. It requires metabolic activation that enables binding to DNA. 4NQO is converted in liver, kidney, spleen and lung to 4-hydroxyaminoquinoline-1-oxide ($NADH_2$ and $NADPH_2$ serve as hydrogen donors in this reaction). It is suggested that both seryl- and prolyl-tRNA synthetases may participate in its activation. 4-Hydroxyamino-quinoline-1-oxide serves as an intermediate in the formation of DNA-bound derivatives.[2] Two esters of 4-hydroxyaminoquinoline-1-oxide and the acetyl derivatives react with purine nucleosides yielding five adducts, two with adenine and three with guanine residues.

7. Heterocyclic Amines from Amino Acid and Protein Pyrolysates

A number of mutagens have been identified in cooked foods in Japan (by bacterial tests). The mutagenic activity was dependent on the presence of postmitochondrial supernatant from homogenized livers of rats pretreated with Aroclor. This denotes that the mutagenic substances needed metabolic activation. Singer and Grunberger[2] isolated several mutagenic compounds from the pyrolysates of amino acids, broiled fish and beef. The mutagenic products from tryptophan pyrolysis were amino-q-carbolines: 3-amino-1,4-dimethyl-5H-pyridol[4,3-b]indole and 3-amino-1-methyl-5H-pyridol[4,3-b]indole; that of glutamic acid pyrolysis: 2-amino-6-methyl-dipyridol[1,2-a:3′,2′-a[imidazole and 2-amino-dipyridol]1,2-a:3′2′-a]imidazole; that of phenylalanine pyrolysis: 2-amino-5-phenylpyridine; and that of lysine pyrolysis: 3,4-cyclopentenopyridol[3,2-a]carbasole.

Further mutagenic compounds have been identified from pyrolysis of soybean globulin as 2-amino-9H-pyridol[2,3-b]indole and 2-amino-3-methyl-α-H-pyridol[2,3-b]indole; from boiled sardines: 2-amino-3-methylimidazo[4,5-f]quinoline and 2-amino-3,4-dimethylimidazo[4,5-f]quinoline; from fried beef: 2-amino-3,8-dimethylimidazo[4,5-f]quinoxaline. It was demonstrated that these compounds are formed in normally cooked foods, especially pyrolysates of casein, glutenglobulin, albumin, chicken meat, horse, and mackerel.

Most of these substances also reveal carcinogenic activity. All the pyrolysis products require metabolic activation by the cytochrome P-450-mediated monooxygenase system to be transformed into highly mutagenic intermediates. So far, binding to DNA C-8 guanine adducts were proved of 2-amino-6-methyldipyridol[1,2-a: 3′,2′-d]imidazole and 3-amino-1-methyl-5H-pirydol[4,3-b]indole; binding occurred only in the presence of microsomal fractions.

D. NATURALLY OCCURRING MUTAGENS (CARCINOGENS)

1. Aflatoxins

In the 1960s an epidemic killing of chicks, ducks, and turkeys occurred in England with acute hepatic necrosis, and trout with liver cancer after feeding these animals with nuts from Brazil and Africa infected by *Aspergillus flavus*. Soon, substances containing contaminating nuts had been isolated and termed aflatoxins B or G (dependent on blue or green fluorescence). Aflatoxin promptly induced hepatomas in trout. Simultaneously, it was stated that the frequency of hepatomas in humans in Africa and Southeast Asia correlated with the content of aflatoxin in the food.

Soon after, it was discovered that the mutagenic and carcinogenic activity of aflatoxins requires metabolic activation by the mixed-function oxidase system. Aflatoxin B_1 binds covalently to cellular macromolecules after activation by liver microsomes. The main aflatoxin B_1 metabolite that binds DNA is the aflatoxin B_1-2,3-oxide and part of the total DNA binding is caused by activation of its hydroxylated metabolites. The experiments showed that the N^7 of guanyl residues is the main target in DNA for metabolically activated aflatoxin B_1, or its metabolites, and the reaction leading to formation of DNA-bound forms is epoxidation of the vinyl bond in the terminal furan ring.

It is noteworthy that mouse liver forms very low levels of DNA adducts, which may be caused by limited ability for activation processes in mouse liver, or a very efficient system for removal of DNA adducts. The main pathological effect in the mouse occurs in the kidney, whereas the liver is insensitive, but in comparison to rat liver the amount of adducts in mouse kidney is about 100-fold lower than that observed in rat liver. In conclusion, this result supports the view that tissue specificity for carcinogenicity or mutagenicity depends on the total amount of activated aflatoxin B_1 residues bound to DNA.

2. Sterigmatocystin (ST)

This mycotoxin is produced by *Aspergillus*, *Penicillium*, and *Bipolaris*. ST causes hepatocellular hepatomas in rats. Its carcinogenic potency is about $^1/_{10}$ of the aflatoxin B_1 potency. ST binding to DNA occurs after induction of phenobarbital-induced rat liver microsomes with DNA and NADPH-generating systems. The generated adduct is 1,2-dihydro-2,2(N^7-guanyl)-1-hydroxysterigmatocystin (the metabolite that reacts with DNA is the ST-1,2-oxide). Since the same mechanism forms adducts in aflatoxin B_1 and ST, the differences in potencies seems to be caused either in the rate of performing metabolites or removal of reaction products.

Safrole (4-allyl-1,2-methylenedioxybenzene) is a natural plant product used in cosmetics and as a food flavoring supplement. It exhibits weak hepatocarcinogenic activity. The carcinogenic activity of its metabolite 1-hydroxysafrole is greater than that of safrole. The sulfuric ester of 1-hydroxysafrole seems to be the principal ester formed *in vivo*. Further metabolites are formed *in vivo*: one-1′-hydroxysafrole-2′,3′-oxide and the other safrole-2,3′-oxide formed in a NADPH-dependent reaction by liver microsomes. The 1′-hydroxysafrole causes formation of DNA-bound adducts. The cancerogen/mutagen activity becomes covalently bound to the exocyclic groups of guanine and adenine residues.

Estragole (1-allyl-4-methoxybenzene) is a naturally occurring flavoring agent of plant origin (chervil, sweet goldenrod, Mexican avocado leaves). Estragole is a weak hepatocarcinogenic agent. In the liver microsomal system estragole plus NADPH is metabolized to its 2′,3′-oxide. Further metabolic reactions generate: (1) a sulfuric acid ester, catalyzed by cytosolic sulfotransferase; (2) epoxidation of the 2′,3′ double bond to 1′-hydroxyestragole-2′,3′-oxide; (3) oxidation of the 1′ position to 1-oxoestragole. The cancerogen/mutagen activity becomes covalently bound to exocyclic groups of guanine and adenine residues (similar to safrole). Binding of estragole to N^2-guanine is similar to that of other carcinogens/mutagens like benzo[*a*]pyrene, *N*-2-acetyl-aminofluorene, and *N*-methyl-4-aminoazobenzene. Most of the N^2-guanine and N^6-adenine adducts of 1′-hydroxyestragole are removed from mouse liver very quickly.

To the different DNA damaging agents undoubtedly belongs also cigarette smoke which contains about 300 mutagenic compounds.[4] Examination of DNA demonstrated a spectrum of base adducts that can illustrate the intensity of exposure of humans to these agents. Such covalent DNA adducts have been detected even in human placenta.[5] Since several of these adducts can be converted into apurinic sites, they represent a potential source of mutations that in the future can lead to carcinogenesis.[4]

IV. IONIZING RADIATION

There are two types of energy-rich radiation: electromagnetic waves and corpuscular radiation. The biologic activity of electromagnetic waves depends on wavelength and energy. For a mutagenic effect the energy should at least be sufficient to lift an electron from an inner to an outer shell that makes the atom unstable and more prone to chemical reactions. The shorter the wavelength of radiation, the higher the radiation energy. The energy of infrared and visible light is insufficient to produce mutations, only being able to produce thermal effects. The energy of ultraviolet light (UV) can produce mutations provided it reaches the DNA. The most frequent chemical reaction (change) caused by UV radiation is linking of two adjacent thymine molecules which prevents them from pairing with adenine. UV-radiation is easily absorbed by the epidermis and therefore it cannot reach human germ cells. However, UV radiation can induce somatic mutations—usually skin cancer.

Radiation of higher energy (X-rays, gamma rays) are able to push electrons out of the internal shell so that the atom becomes a positive ion. The liberated electrons can react with other atoms converting them into negatively charged ions. These ions, together with free radicals, are capable of secondary chemical reactions. The energy needed for ionization of biological material is about 32 eV (X-rays) and the energy for atomic reactions (very hard rays) is about 3 MeV.

Corpuscular radiation consists not of photon energy, but of particles. They may be charged like electrons and protons, or uncharged like neutrons. Their effects depend on their kinetic energy. Ionization produced by neutrons is densely concentrated around the particle, whereas electromagnetic waves produce looser ionization. Irradiation may cause indirect reactions. For example, energy-rich neutrons may be accepted into atom nuclei or they can transfer their kinetic energy to hydrogen nuclei, i.e., protons which as accelerated particles may undergo several secondary reactions with other molecules.

For a long time it was unclear if ionizing radiation can induce gene (point) mutations. Experiments performed on *Neurospora crassa* revealed that among induced mutations there were 42% transitions, 37% insertions or deletions of single bases, and 21% of different origin mutations, including transversions.

A. RADIOCHEMISTRY OF RADIATION

The effect of radiation depends not only on its action on cells but also on tissue water. High energy radiation causes water dissociation into ions H^+ and H^- which join with neutral water producing H_3O^+ and OH^- ions. These ions disintegrate into free ions H^+ and OH^- and the radical $OH^{\cdot}$ and hydrogen atoms. Most of these elements are unstable and form new water molecules with liberation of free energy. In the presence of oxygen, which is permanently present in tissues, the free radical $HO_2^{\cdot}$ is formed which converts in tissues into injurious H_2O_2. Sulfur compounds, especially cysteine, inhibit these reactions. The biological effect on tissues depends mostly on their oxygenation that promotes, and presence of sulfur compounds that inhibit, production of toxic and mutagenic compounds arising during irradiation.

V. THE FIDELITY OF NUCLEIC ACID SYNTHESIS

The fidelity of genetic information transfer relies on the ability to form specific base pairs with the appropriate number of hydrogen bonds. This is the central postulate for formation of base pairs according to the Watson-Crick model of nucleic acid as the substrate of transmission of genetic information. There exists, however, an enormous number of possibilities that can change the original information such as tautomerism, sugar conformation, base orientation, steric hindrance, appropriate enzyme activity and participation of neighboring groups[2] that can increase or decrease fidelity of nucleic acid synthesis, as described above.

The principal elements that enables the necessary fidelity of DNA replication and transcription are polymerizing enzymes: DNA polymerase and RNA polymerase DNA-dependent. Without the polymerizing enzymes, complementary and noncomplementary nucleotides could form wrong base pairs at every 10 to 100 nucleotides (calculated from the free energy differences). Such a great error would make the genetic information of DNA unserviceable. Though the use of DNA polymerase *in vitro* one misincorporation may happen per 10^7 nucleotides, and *in vivo* in the range of 10^7 to 10^{11} nucleotides. Contrary to this, examining the fidelity of DNA polymerase copying synthetic nucleotides, the frequency of errors probably is much greater because of slippage of the primer along the synthetic template, or because of changes of the secondary structure and stacking interactions that differ from the original DNA. As a rule, prokaryotic polymerases are more accurate than eukaryotic ones.

Types of DNA damage:

1. Insertion of an erroneous nucleotide into the DNA may result because of the presence of tautomeric bases in the cell.
2. Substitution of nucleotide analogues, e.g., 5-bromuracil or 2-aminopurine.
3. Loss of a single base may occur spontaneously at a rate of 10^{-9} per cell cycle.
4. Cyclobutane pyrimidine dimers and (6-4) photoproducts formation may be caused by UV irradiation.
5. Cross-links may be formed by bifunctional alkylating agents that form networks between complementary DNA strands and between DNA and proteins.

6. One-strand breakages result by ionizing radiation (X-rays, corpuscular irradiation, radioactive ^{32}P) or loss of bases and SOS DNA replication.
7. Double-strand breakages may result by the same agents that cause one-strand breakages.

The DNA damage may be repaired which may restore the original DNA structure, or a mutation may be generated when repair is incomplete. Inactivation or cell death may occur in total absence of repair.

Because of the multiple possibilities that can change the structure of DNA (tautomerism, base rotation and protonation, base cleavage, deamination, and chain breakage) the associated proofreading exonucleotic activity of polymerases must contribute to error prevention. *In vivo*, a specific type of proofreading mismatch-specific endonucleases excise noncomplementary bases postsynthetically. By this way the nonspecific base insertions opposite apurinic-apyrimidinic sites become repaired. However, polymerases themselves may be the origins of inappropriate base pairing, as found in polymerases isolated from carcinogen-treated animals or in leukemic lymphocytes. Metals present in cells may also decrease fidelity of DNA replication or transcription probably because of interaction with the template. Modifications of triphosphates also may be an origin of errors, e.g., when the nucleoside triphosphate is an analogue it can cause a transition. Changes in the intracellular pool of nucleotides occurring in some diseases due to lack of specific kinases may also cause errors.

At present it is acknowledged that polymerases that distinguish normal nucleoside triphosphates from the erroneous ones are more responsible for the correct synthesis of DNA or RNA than complementarity of hydrogen-bonding of the particular nucleotides. Finally, the replication of the genetic code is performed through base pairing compatible with stereochemical rules of the Watson-Crick helix. When a wrong base can be fitted to the geometry of the helix and can be stabilized within the helix structure through staking or adequate hydrogen-bonding, strengthened by the nearest neighboring conditions—both transition and transversion may occur. Formation of wrong base-pairs of nucleotides usually leads to mutation that can be manifested as clinical symptoms, but may also be silent.

A. TRANSCRIPTION AND REPLICATION OF MODIFIED TEMPLATES

Several types of experiments have been constructed to analyze the mechanism of transcription and replication in modified templates which involve translation of messengers, codon-anticodon interaction, transcription, and translation using natural and synthetic templates. When the modified nucleoside can form at least two hydrogen bonds it can be used for transcription and replication, and the performed templates and mRNAs are active. In contrast, changes in the secondary structure of nucleosides may inhibit base pairing, e.g., attachment of a sugar molecule to a carbon replacing N^9 of adenine inhibits base pairing with poly(U).

The question arises about the influence of nucleic acid derivatives on transcription. Mutagenic modification of a polynucleotide usually causes production of several products. Transcription is usually stopped at the sites bound with adducts of metabolites of benzo[*a*]pyrene and acetylaminofluorene. Ribopolynucleotides of a single type of modified bases were examined for *in vitro* and *in vivo* transcription and translation. The rate and extent of transcription and translation were generally inhibited, irrespective of an increase of errors. The extent of total change or misincorporation in transcription is approximately equal to the amount of modified nucleosides in the polymer. Therefore, misincorporations are rare events. Prokaryotic DNA polymerases (*E. coli* DNA polymerase I) are particularly inhibited by modified bases in deoxypolynucleotide templates so that only a few errors can be made.

Important sites of modifications of nucleic acids by mutagens and carcinogens are the phosphodiesters. In ribopolymers the phosphodiesters are unstable, while in deoxypolymers they are extremely stable and resistant to the action of enzymes. Isomers with poly(dA-dT) containing triesters reveal decrease in rates of synthesis which suggests that the orientation of the ethyl group to the rest of the template backbone may be important in inhibiting the synthesis, probably by unfavorable steric interaction with the polymerase. The replacement of a negative charge by the ethyl group does not inhibit DNA synthesis, only the strength of the hydrogen bonding is weaker than that of normal nucleotides.

In general, the aromatic substituents of nucleic acids, irrespective of their position on a nucleoside, will interfere with replication simply because of size. The rapid depurination of these products may cause a mutagenic effect through error-prone repair of the apurinic site—the same may occur because of depurination of N^3 and N^7 alkyl purines. Smaller unrepaired alkyl derivatives affecting hydrogen-bonding may cause misincorporation without forming hydrogen bonds (if they can be placed in the helix by other forces—stacking or other interactions).

Mutations by nonalkylating agents may be generated usually because of deamination of bases or preferred tautomers in the modified derivative.

Various reactions of cytidine exemplify all possible changes in base pairing. Tautomerism, whether intrinsic or chemically caused, leads to uracil, as does deamination. Deaminated cytosine (i.e., uracil) can exhibit tautomerism and act as true cytosine. Finally, ambiguity transcription of cytosine may be caused by ionization, depyrimidinization or change to the imino tautomer resulting from chemical modification.[2]

VI. DNA REPAIR

Considering the great number of agents that can change the structure of DNA making it unusable as the normal template for the enormous number of synthesized life-giving macromolecules, life probably would be impossible if an equally great number of DNA repair mechanisms did not exist. This is best documented by the fact that some well-known human diseases exist which are caused by the deficiency of the DNA repair mechanisms.

DNA repair is a general term for the removal of a modified base or nucleotide or a modied group from DNA. After removal of the erroneous base or group the error-free base or group becomes inserted through repair mechanisms. Most of the repair enzymes have been isolated from prokaryotes, especially from *E. coli*. These include enzymes repairing UV dimers, apurinic and apyrimidinic sites, glycosylases acting on 3-alkyl adenosine, 7-alkyl guanine, ring-opened 7-alkyl guanine, deaminated adenine and cytosine, fragments of ring-opened pyrimidines and thymine glycol. The adequate substrate for these repair mechanisms is the double-stranded DNA. It is supposed that there may exist an enzyme that can directly insert purines into apurinic sites of the DNA, but there is not yet conclusive evidence for such an ''insertase''. Finally, there exist enzymes dealkylating O^4-methyl thymine and glycosylases acting on O^2-methyl thymine and on O^2-methyl cytosine.[2]

There are several known repair mechanisms: (1) photoenzymatic DNA repair; (2) an excision repair mechanism; (3) a postreplicative mechanism.

A. PHOTOENZYMATIC DNA REPAIR

Photoreactivation is the simplest DNA repair mechanism. The reaction is catalyzed by a photoreactive enzyme, photolyase (M_r 40,000). At the first step the enzyme recognizes cyclobutane pyrimidine dimers to which the enzyme becomes attached. This step of the reaction can be performed even in the dark. In the next step, visible light (300 to 600 nm) is necessary, which activates the enzyme and the energy of which is used for cleavage of the cyclobutane ring. Next, the original DNA structure becomes reconstructed, and the enzyme becomes liberated. The photoreaction occurs in double- and single-stranded DNA, but not in RNA.

B. DNA EXCISION REPAIR MECHANISM

Cyclobutane pyrimidine dimers can be also removed by enzyme-mediated excision, followed by resynthesis of the excised area. The typical sequence of base excision repair of DNA containing modified bases is executed by the following enzymes:

1. DNA glycosylase cleaves the base sugar-bond, releasing the free base and generating a free apurinic/apyrimidinic site.
2. The apurinic/apyrimidinic site is recognized by an endonuclease that catalyzes a chain break at the 5′ side of the lesion.
3. An exonuclease removes the deoxyribosephosphate moiety as a part of small oligonucleotide comprising the erroneous base.
4. Repair synthesis by a DNA polymerase occurs.
5. Ligation by DNA ligase joins the new synthesized oligonucleotide with the old DNA chain.

The progress in studies about DNA structure urged that the old scheme of ultraviolet light damage and its enzymatic repair should be reexamined.[3] DNA absorbs light in the range of 240 to 300 nm which results in excited energy state of the bases. The main DNA changes causing lethality and mutagenesis are cyclobutane pyrimidine dimers formed between adjacent pyrimidines. Other products of lesser value also are formed. The cyclobutane pyrimidine dimers are the most prevalent lesions caused by UV irradiation; the mutants defective for repair demonstrate increased sensitivity to lethal and mutagenic defects; photoreversal of cyclobutane pyrimidine dimers by visible light repairs most of the UV-irradiation-induced DNA damage.

At first two different UV-specific endonucleases have been isolated from *Micrococcus luteus* and bacteriophage T4. The first surprise was the incision site of these enzymes. The 5′-end scission sequences obtained by these enzymes were longer than those obtained by breakage of the DNA. Both enzymes demonstrate two different activities: *N*-glycosylase activity that cleaves the *N*-glycosyl bond between the 5′ pyrimidine of a dimer and the corresponding sugar, and an apurinic/apyrimidinic (AP) endonuclease activity that cleaves a phosphodiester bond 3′ to the newly created apurinic/apyrimidinic site. As the result of this reaction a 3′ phosphosugar remains at one terminus which is not a substrate for the DNA polymerase, and therefore this terminus must be removed with excision of the dimer attached to the 5′ end of the nick. Finally, resynthesis follows by DNA ligation.

The UV-specific endonucleases responsible for this reaction are not yet sufficiently characterized. They usually recognize a very limited range of base modifications in which cyclobutane pyrimidine dimers are the only known substrates. The major excision pathways of *E. coli*, encoded by the uvrA, uvrB, and uvrC genes are capable of repair of bulky adducts. The same reaction was seen in humans where the greater repair of DNA lesions is governed by multiple loci that constitute the xeroderma pigmentosum phenotype. For DNA strand incision three functional gene products uvrA (M_r 114,000), uvrB (M_r 84,000), uvrC (M_r 74,000), plus ATP and magnesium are required. The sites of incision are the eight phosphodiester bond 5′ hydrolized and the fourth (sometimes the fifth) phosphodiester bond 3′ to the dimer (see Figure 4). By that method the excised DNA fragment that includes the site of damage is 12 to 13 nucleotides long. The enzyme leaves a 3′ hydroxyl end adequate for elongation by the DNA polymerase I to the 5′ side of the damaged base, and a phosphoryl group suitable for ligation reactions to the 3′ site. The involvement of the three uvrABC products in the incision mechanism is explained by the requirement of all three proteins.

The above described two cut mechanisms may be useful for repair of DNA damage caused by cross-linking agents in which cuts on both sides of the crosslink might provide a substrate for recombination repair, and also for bypass DNA synthesis. The described

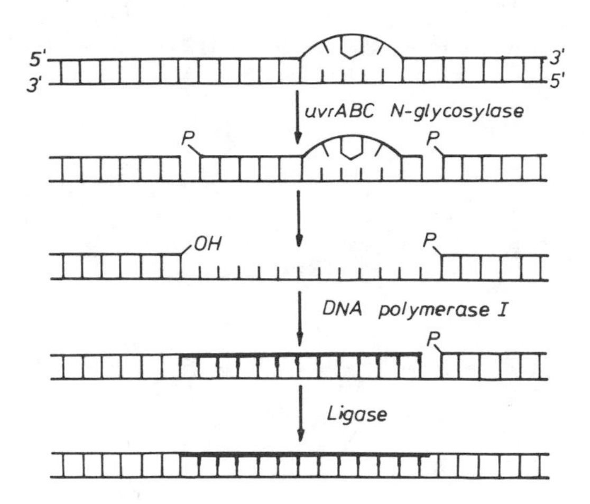

FIGURE 4. Scheme of repair mechanism of UV-damage of DNA (of cyclobutan pyrimidine dimers). The mechanism of incision at DNA damage sites by uvrABC exonuclease of *E. coli*. (Redrawn from Reference 3.)

mechanisms in *E. coli* are very similar to those seen in human cells and the genetic loci characteristic for DNA repair of xeroderma pigmentosum are very similar to uvrABC loci in *E. coli*.

Removal of apurinic sites is very important because they are the most frequent spontaneous alterations in DNA causing approximately 10^4 depurinations per mammalian cell per day. An apurinic site originates when the glycosylic bond connecting the purine base to deoxyribose is cleaved and the DNA phosphodiester backbone remains intact. Depurination can be caused by spontaneous hydrolysis, by enzymatic removal of changed bases by specific glycosylases, and by chemical modifications of bases that weaken the glycosylic bond and during DNA replication may lead to nontemplate incorporation of erroneous nucleotides by DNA polymerase.[4]

Apurinic sites inhibit chain elongation by DNA polymerases *in vitro*, although bypass of these sites can occur also with incorporation of nucleotides noncomplementary to the DNA template. The frequency of misincorporation is proportional to the extent of depurination. The possibility that apurinic sites may generate mutations was proved by many observations. In the case of various chemical mutagens, the apurinic sites seem to be intermediate events that finally lead to mutation. Several adducts produced by mutagens cause modifications of the N^3 and N^7 positions in purines that weaken the glycosylic bonds. Transversions have been observed with high prevalence in cells treated with certain mutagens, in which DNA SOS synthesis occurs, i.e., synthesis of the single DNA strand independent of the template strand which is damaged. The resultant altered DNA replication is the source of multiple mutation sites.[4]

C. (6-4) PHOTOPRODUCTS

Recently, another form of UV-light-induced DNA damage has been discovered. Cyclobutane pyrimidine dimers occur with greatest efficiency at adjacent thymidine residues, while chain terminating mutations occurred most frequently at TC sequences, and at some CC sequences, infrequent breaks at some TT sequences, but not at CT sites.[3] These photoproducts have been identified as Thy(6-4)pyo. This product generates only in previously UV irradiated DNA treated with alkali. This type of lesion is produced by reaction of the 5,6 bond of a 5′ thymidine with the 4 exocyclic group of the 3′ pyrimidine (amino or keto group of cytosine and thymine, respectively). Finally, a stable bond is formed between the 6 position and the

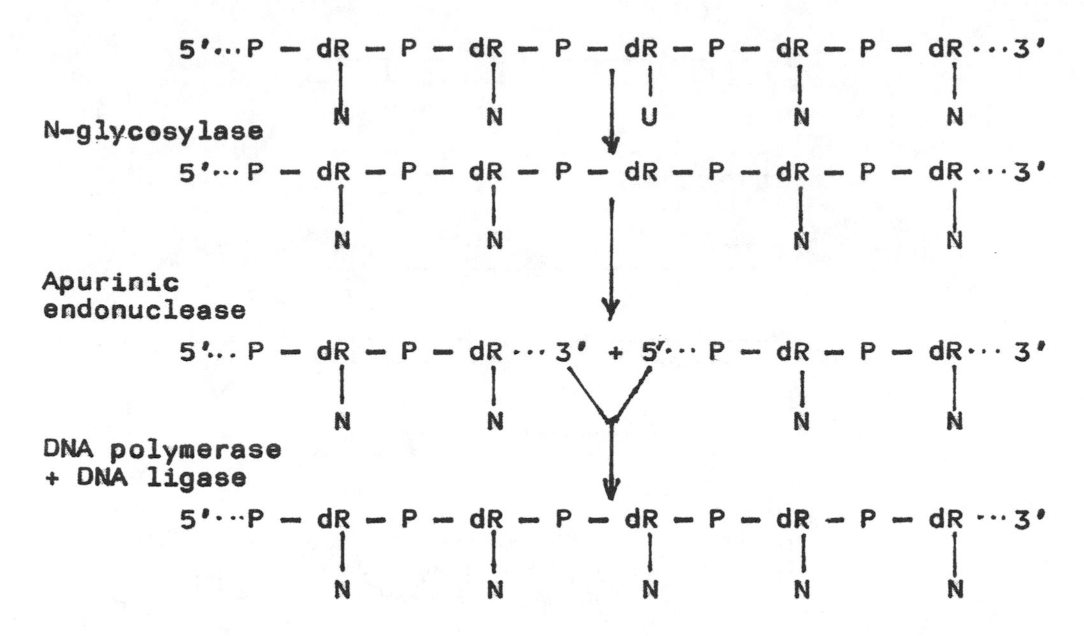

FIGURE 5. Scheme of DNA repair because of deamination of cytosine into uracil. dR = deoxyribose; N = whichever of the four bases: A,G,C,T; U = uracil.

4 position of the 3′ pyrimidine. The product of this reaction has been termed (6-4) photoproduct. The biological importance of these products is 10 times lower than that of cyclobutane pyrimidine dimers. The (6-4) photoproducts may be mutagenic. They can be removed by the repair system uvrABC in *E. coli* and a similar system in humans.

D. POSTREPLICATIVE DNA REPAIR

A typical example of postreplicative DNA repair caused by substitution of thymine by uracil is represented in Figure 5.

Ethyl and methyl base or phosphate-alkylated derivatives present in mammalian cells after treatment with alkylating agents disappear in a short time. Phosphodiesters are relatively stable in human fibroblasts, their half-life is more than 8 to 12 d. Ethyl triesters have a half-life of about 7 d, but after 10 weeks they are still present. The liver has the highest capacity to repair DNA damage. The DNA repair ability also depends on the amount of administered mutagen; the greater the dose the lower the repair possibility which can be explained by saturation of the constitutive enzymes. Finally, the methyl derivatives seem to be more easily repaired than the ethyl derivatives and higher homologues.

Some mutagens/carcinogens demonstrate organotropic specificity. This was first observed in N-nitroso compounds that have the potential to mispair with thymine, and therefore would cause the initiating events. On the other side, the labile alkyl purines generated by weak mutagens like dimethylsulfate would not persist and apurinic sites are repaired by the normal repair mechanism.

E. DNA REPAIR MECHANISMS OF ALKYL DERIVATIVES IN EUKARYOTES

The first discovered enzyme was the glycosylase activity that removes 3- and 7-alkyl purines that are present in human lymphoblasts by cleavage of glycosyl bonds of methyl and ethyl purines. The reaction was more efficient in methyl derivatives than in ethyl ones. In the rat liver 7-methylguanosine glycosylase has been discovered that removes *in vivo* 7-methyl guanosine. There probably exist two glycosylases from which the 3-methyl adenine glycosylase has been purified.

Methyltransferase represents an enzyme that transfers methyl or ethyl groups from O^6-alkyl guanine (a potent carcinogen) to the acceptor *S*-alkylcysteine. The enzyme is present in various organs of mammals. The human liver enzyme is about ten times more active in catalyzing demethylation of O^6-methylguanine than that of rat liver.

The question is open whether the repair enzymes—glycosylases and methyltransferases—are constitutive repair enzymes, or if they can be induced to produce more enzyme.[2] Some of the experiments performed in HeLa cells support the possibility of induction, while others, especially performed in xeroderma pigmentosum-derived human lymphoblastoid cell lines are unable to produce more methyltransferase activity. Partial hepatectomy increases the amount of methyltransferase activity but not of glycosylases which are synthesized parallel to the amount of DNA polymerase.

F. *IN VIVO* REMOVAL OF AROMATIC DERIVATIVES

The repair mechanism of DNA modified by aromatic compounds, mainly polycyclic aromatic hydrocarbons, is only slightly understood. The difficulty lies in the fact that these types of mutagens are metabolized to many active intermediates that can bind to different bases of DNA and form many types of adducts. The problem has been studied by comparison of the effects of various compounds on normal human cells and on repair-deficient human cell lines, e.g., xeroderma pigmentosum cell line XP 12BE. The cytotoxic effect of polycyclic aromatic hydrocarbons (benzo[*a*]pyrene, benzo[*a*]anthracene, dimethylbenzo[*a*]anthracene) was several times higher in XP 12BE cells than in normal human skin fibroblasts. Similar results have been obtained after administration of benzo[*a*]pyrene diol-epoxides whose adducts are repaired in normal cells by the excision repair mechanism, whereas in xeroderma pigmentosum cells they are not. The xeroderma pigmentosum cells perform less than 1% of normal cell repair of thymine dimers after ultraviolet irradiation.

Nondividing human lymphocytes react with 7-methylbromobenzo[*a*]anthracene by forming three types of adducts, all on excyclic amino groups. All the adducts are repaired in normal cells, the adenine adducts more quickly than those of guanine. Treatment of mammal cells with *N*-2-acetylaminofluorene results in formation of three types of adducts, one deacylated. The major adduct with a half-life of about 7 d disappeared quickly from liver DNA, but minor adducts remained in DNA up to eight weeks. Xeroderma pigmentosum cells are also more sensitive to toxic effects of these compounds, probably because of inability to repair the DNA damage.

Formation and removal of aflatoxin B_1 adducts have been studied in a human epithelial lung cell line. After 28 h exposure, 60% of the total adducts had been removed following 24 h of incubation.

Detailed information about mutagens can be found in Singer and Grunberger.[2]

VII. MOLECULAR PATHOLOGY OF DNA REPAIR

There are several heritable diseases caused by disturbances in DNA repair. The classical disease of this group is xeroderma pigmentosum.

A. XERODERMA PIGMENTOSUM

The main clinical symptoms can be divided into three groups: (1) cutaneous symptoms—erythema and bullae because of hypersensitivity to exposure to the sun, xerosis, hypopigmentation, teleangiectasiae, and atrophy; (2) ocular symptoms—blepharitis, erythema of the lids, pigmentation and keratosis, atrophy of lids leading to entropion, ectropion, loss of cilia and loss of lower lids, conjunctivitis with photophobia, pigmentation, dryness, symblepharon, exposure keratitis with edema, cellular invasion and vascularization of the cornea, ulceration of the cornea, iritis with synechiae and atrophy, and (3) neurological symptoms—

microcephaly, progressive mental deterioration, choreoathetosis, ataxia, spasticity, deafness, hyporeflexia, and areflexia. Several kinds of neoplasia are common to all surfaces following exposure to sun: actinic keratosis, basal and squamous cell carcinoma of skin, malignant melanoma, angioma, fibroma, sarcoma, and ocular neoplasmas: papillomas, apitheliomas, intraepithelial epithelioma of conjunctivae, and others.

The pathogenesis of xeroderma pigmentosum lies in hypersensitivity of skin and eyes to sunlight that causes damage to DNA with formation of cyclobutane pyrimidine dimers and (6-4) photoproducts that cannot be properly corrected because of molecular defects of the cellular DNA repair system. Several kinds of DNA repair systems have been detected in xeroderma pigmentosum patients: different degrees of reduced pyrimidine dimer excision, reduced levels of repair DNA replication causing reduced insertion of new bases after dimer excision. Xeroderma pigmentosum patients demonstrate levels of repair replication from 0 to 90% of normal. The reduction of DNA repair was discovered in all tissues examined (skin cells, peripheral lymphocytes, fibroblast and liver cell cultures). Common to all types of xeroderma pigmentosum are chromosomal aberrations. At least six different DNA repair mechanisms (variants), not yet fully characterized, have been detected.

Different agents causing DNA damage that are defectively repaired belong to: UV-irradiation, methoxypsoralen adducts, 4-nitroquinoline-1-oxide, bromobenzo[a]anthracene, benzo[a]anthracene epoxide, 1-nitropyridine-1-oxide, acetylaminofluorene, and others.

On the other hand agents causing DNA damage that is normally repaired in xeroderma pigmentosum belong to: X-rays, bromouracil photoproducts, methylmethane sulfonate, *N*-methyl-*N'*-nitro-*N*-nitrosoguanidine, methylnitrosourea, and others. At present the explanation, for example, of why UV-damage is not repaired in xeroderma patients and X-ray damage is corrected, is lacking.

B. BLOOM'S SYNDROME

The syndrome is characterized by extreme chromosomal fragility and numerous sister chromatid exchanges. The number of sister chromatid exchanges in bromodeoxyuracil substituted chromosomes is several times greater in Bloom's syndrome as compared to normal. Phytohemagglutinin stimulated lymphocytes in culture demonstrated about 90 sister chromatid exchanges as compared to 10 or fewer in normal persons. Interchanges between homologous chromosomes at homologous sites in Bloom's syndrome cells are ten times more frequent than in normal cells. These findings proved disturbances in DNA replication in Bloom's syndrome cells: the rate of replicon-fork progression was about 25% slower in fibroblasts and T-lymphocytes from Bloom's syndrome patients as compared to normal persons.[6] Clinically, Bloom's syndrome is characterized by dwarfness (height about 150 cm) with teleangiectasiae and erythema of lips, café au lait spots on the skin, susceptibility to infections and hypochromic anemias, and increased number of metaphases in lymphocyte and fibroblast cultures (10 to 20%).

Recent investigations have proved DNA ligase I deficiency in Bloom's syndrome.[7,8] Mammalian cells have two DNA ligases. The larger ligase I is induced during cell proliferation and probably serves in chromosomal replication, whereas DNA ligase II is permanently present in cell nuclei. DNA ligase I catalyzes the joining of blunt-ended DNA fragments, whereas DNA ligase II oligo(dT)molecules annealed to poly(rA).

Patients with Bloom's syndrome are hypersensitive to UV-irradiation. The carcinogenic risk of patients with Bloom's syndrome is about 5 to 10% of patients. Further investigations revealed that B-lymphoblastoid cells with a high level of sister chromatid exchanges are highly susceptible to the action of various carcinogens. This observation suggests that carcinogens are capable of transforming only those cells that have a critical level of sister chromatid exchanges (140 per cell) and not cells with only mildly elevated level (13 per cell).[9]

C. FANCONI ANEMIA

Characteristic features of Fanconi anemia are pancytopenic anemia with skeletal defects and dwarfness with hypo- and hyper-pigmentation of skin. Also frequent are kidney defects (renal hypoplasia, horseshoe kidney, double ureters), rotation of intestines, ocular disturbances (palpebral ptosis, nystagmus, strabismus, microphthalmus), deafness, congenital heart defects, and mental retardation in about 20% of patients. Acute leukemia may occur in older patients. The most common bone defects are radial abnormalities with a hypoplastic, absent, digitalized or supernumerary thumb, hypoplasia of the first metacarpal, hypoplasia or absence of the radius, and other defects.

The characteristic feature of cultured lymphocytes and fibroblasts are chromatid breaks, gaps, exchange figures and endoreduplication. The Fanconi anemia cells exhibit increased sensitivity to X-rays. It is suggested that defects of either exonucleases or DNA ligases of the DNA repair system may be the cause of Fanconi anemia, but persuasive arguments are lacking.

D. ATAXIA-TELEANGIECTASIA

This disorder starts early in infancy and leads to complete disability in the first decade of life. The main symptoms are: incoordination of limbs, tremor, and hyperkinesis. The muscles are hypotonic, tendon reflexes are lacking. Dysarthric speech, grimacing, and jerky eye movements appear during the course of the disease. Intellectual disturbances are frequently present at about 10 years of age. Teleangiectatic lesions appear at three to four years of age or later, mainly in the conjunctiva, ears, and neck. Finally, a profound immunological defect exists. All the patients reveal a cell-mediated immunity defect and often deficiency of IgA and IgE. The deficiency of these immunoglobulins probably results from their deficient synthesis caused by a defect in B-cell maturation. Because DNA sequences for IgA synthesis are present in ataxia-teleangiectasia patients, the deficient synthesis of this immunoglobulin probably results from defects in rearrangement and splicing processes in these patients.

Ataxia-teleangiectasia cells are extremely sensitive to ionizing radiation and show an increased susceptibility to cancer. Ataxia-teleangiectasia patients demonstrate chromosome instability, a defect of DNA repair, premature aging and a high level of serum alpha-fetoprotein.[10]

The neurological symptoms cannot be readily joined with the biochemical or immune defects. It is supposed that they may be the result of deficient differentiation of several tissues, especially the central nervous system, liver, and gonads. Experiments performed on ataxia-teleangiectasia cells suggest two different types of molecular events: a failure that inhibits DNA replication after gamma- or X-ray irradiation, and a defect in one or more DNA repair pathways.[10] The defect of DNA repair mechanism was evidenced by long-lived unrepaired lesions after X-ray irradiation. It is difficult to interpret the ascertained DNA repair defect with lack of the normal post-irradiation mitotic delay, caused by normal DNA replication in spite of X-ray irradiation.

E. COCKAYNE SYNDROME

This syndrome is characterized by dwarfness with progressive psychomotor deficiency. The disease usually starts at the second year of life with progressive deficiency in development and thriftness, photodermatitis with hyperpigmentation of scars, progeria and mental retardation. Patients with Cockayne syndrome demonstrate hypersensitivity to UV-irradiation. Defects of DNA repair are not sufficiently documented.[14]

F. PROGERIA (WERNER SYNDROME, HUTCHINSON-GILFORD SYNDROME)

Premature aging starts at infancy exhibiting dwarfness, alopecia, premature atherosclerosis with hyperlipidemia, osteoporosis, osteolysis, and cachexy. Hypersensitivity to UV-irradiation also exists. Cultured progeroid cells do not have the ability to rejoin DNA strand breakages as do normal cells which indicates a probable DNA repair defect (not fully recognized).[11]

REFERENCES

1. **Rieger, R., Michaelis, A., and Green, M. M.,** *Glossary of Genetics and Cytogenetics,* Springer-Verlag, New York, 1976.
2. **Singer, B. and Grunberger, D.,** *Molecular Biology of Mutagens and Carcinogens,* Plenum Press, New York, 1983.
3. **Haseltine, W.,** Ultraviolet light repair and mutagenesis revisited, *Cell,* 33, 13, 1983.
4. **Loeb, L. A.,** Apurinic sites as mutagenic intermediates, *Cell,* 40, 483, 1985.
5. **Everson, R. B., Randerath, E., Santella, R. M., Cefalo, R. C., Avitts, T. A., and Randerath, K.,** Detection of smoking-related covalent DNA adducts in human placenta, *Science,* 231, 54, 1986.
6. **Spanos, A., Holliday, R., and German, J.,** DNA-polymerase activity of cultured lymphoblastoid cells— Bloom's syndrome, *Hum. Genet.,* 73, 119, 1986.
7. **Willis, A. E. and Lindahl, T.,** DNA ligase I deficiency in Bloom's syndrome, *Nature (London),* 325, 355, 1987.
8. **Chan, J. Y. H., Becker, F. F., German, J., and Ray, J. H.,** Altered DNA ligase I activity in Bloom's syndrome cells, *Nature (London),* 325, 357, 1987.
9. **Shiraishi, Y., Yosida, T. H., and Sandberg, A. A.,** Malignant transformation of Bloom syndrome B-lymphoblastoid cell lines by carcinogens, *Proc. Natl. Acad. Sci. U.S.A.,* 82, 5102, 1985.
10. **Bridges, B. A., and Harnden, D.,** Untangling ataxia-teleangiectasia, *Nature (London),* 289, 222, 1981.
11. **Passarge, E.,** *Elemente der Klinischen Genetik,* Gustav Fischer, Stuttgart, 1979.

Chapter 9

THE PATHOLOGY OF GLOBIN MOLECULES

I. INTRODUCTION

Hemoglobin was the first subject that opened new perspectives in the elucidation of pathological events at the molecular level, and although there are now many examples of human pathology caused by mutation of single molecules, hemoglobin remains the most extensive and richest example in this field. The principal causes of any disease are disturbances of function of any organ or tissue. In the background of all functions of living organisms are molecules with their specific features that enable them to perform the most complicated vital functions, and precisely in this aspect hemoglobin is the most satisfactory example.

Hemoglobin is an example of a molecule that is perfectly adapted to its specialized role as a vehicle for the transport of oxygen within living organisms. Hemoglobin actively accepts oxygen molecules in the lung, transports them through blood circulation to the tissues, and in the tissues again actively, but opposite the active function in the lung, discharges oxygen to the tissues. The opposing function of hemoglobin in the lung and tissues is executed by modifications of the hemoglobin molecule in these two different environments, with the result that in the lung the oxygen affinity of hemoglobin increases and remaining high, and becomes a good transporter of oxygen in the blood circulation. In tissues the oxygen affinity of hemoglobin decreases, which causes discharge of oxygen. Besides, in the tissues, after discharge of oxygen, hemoglobin can accept metabolic intermediates from the tissues, mainly carbon dioxide which becomes discharged in the lung into the atmosphere. Without these specialized functions of hemoglobin, life of multicellular organisms would be impossible.

In the background of the specialized functions of hemoglobin lies its structure, encoded by specific genes, which according to the general rule of universal genetic polymorphism must be regularly and simultaneously represented by different variants in the population. Some of the variants may be neutral, i.e., dependent on a gene action that is roughly neutral in its effect on the action of the variant, and on the survival rate of the genotype in which it is contained. There are also, however, variants that disturb the normal function of a molecule or even make its function impossible. These variants may disturb vital functions of living organisms which then appear as a disease or may even cause death of these organisms.

Polymorphic variants may be a stable component of particular populations that were generated in ancient times and remained in the populations usually because their phenotypes did not differ from those of the normal population. There also may exist variants of regional or geographic character. These variants may be caused by selection which results in accumulation of variants better-adapted to the specific environmental conditions. In this case so-called negative selection may occur which prefers individuals with pathological molecules. Such a case represents accumulation of mild hemoglobin variants (sickle cell hemoglobin) in regions where malaria occurred endemically before the historical period. Finally, variants may be generated by fresh mutations. Usually these mutations appear in single cases because mutations are only very rarely advantageous to the organisms causing death of the cell in somatic cell mutations or death of the organism in germ cell mutations.

The primary effect of a gene mutation is a change in the amino acid sequence of a polypeptide molecule, the primary product of a structural gene. The sequence of events in mutation are the following: the mutated gene produces a mutated mRNA and a mutated mRNA synthesizes a single amino acid substitution in the synthesized protein. It may also cause other changes, e.g., premature termination of polypeptide synthesis (so-called "nonsense mutation") which results in synthesis of nonfunctional products.

At the same time with the development of knowledge of the mechanisms of gene functions, knowledge has developed of the mechanisms of molecular pathology caused by mutations and their consequences. To some extent mutations and disturbed functions of the particular molecules contributed to elucidation of mechanisms of their functions in the sense that mutations are a sort of experiments performed by nature itself. Therefore, knowledge of basic mechanisms and functions of molecular biology and genetics is necessary for understanding the essence of molecular pathology. This is best-visualized by the presence of mutations of a single polypeptide chain, the β-globin-chain where we know mutations of all steps in production of this polypeptide, including: impairment of transcription caused by mutations of the upstream regulatory elements (TATA box and others) which inhibit formation of the initiation complex; nonfunctional mRNA because of nonsense mutations, frameshift mutations, disturbances of RNA splicing, point mutations that cause replacements of particular amino acid residues because of a changed genetic code, and disturbances in chain termination.

The first molecular disease recognized as such by Pauling et al. was sickle cell anemia.[1] The logical reasons for the statement that sickle cell anemia must be caused by lesions of the hemoglobin molecule were: hemoglobin makes up nearly 100% of the protein content of red blood cells; sickling appears in deoxygenated state, being reversible in oxygenated state—therefore, changes in the shape of an erythrocyte should be caused by changes in physical state of oxygenated and deoxygenated hemoglobin, and cannot be caused by an inborn defect of an erythrocyte's cell structure. Soon this view was confirmed by chemical examinations. Ingram[2] demonstrated that the pathological variant of sickle cell hemoglobin was caused by replacement of glutamic acid by valine at the sixth position of the N-terminal end of the β-globin-chain. These findings were the start of a series of discoveries of pathological variants of hemoglobin, and in a short time some hundreds of pathological variants of hemoglobin molecule were described.

II. THE GLOBIN GENES

The globin genes are organized in clusters of α- and β-globin genes—the α-globin genes on the short arm of chromosome 16, and the β-globin genes on the short arm of chromosome 11. They are in the neighborhood of the insulin and c-Ha-*ras*-1 genes from the terminal end, and the parathyroid hormone gene from the centromere side of the chromosome.

The α-globin gene cluster is composed of two adult α_1 and α_2 (α_1, α_2) genes, one zeta (ζ) embryonic gene, two pseudognes zeta$_1$ ($\psi\zeta_1$) and alpha$_1$ ($\psi\alpha_1$). The beta globin gene cluster is composed of two adult, beta (β) and delta (δ) genes, two fetal gamma genes ($^G\gamma$ and $^A\gamma$), one embryonic gene epsilon (ε), and one pseudogene beta$_1$ ($\psi\beta_1$). The α globin genes are arranged in the order 5′ zeta, ψ zeta$_1$, ψ alpha$_1$, alpha$_2$ and alpha$_1$, and the beta globin genes in the order 5′ epsilon, ^{G}gamma, ^{A}gamma, ψ beta$_1$, delta and beta, 3′ end (see Figure 1). The order of both the α- and β-globin genes are in the same order in which they are expressed from the early embryonic to adult forms of hemoglobin during the development. The intriguing question is if the sequential arrangement of these genes comprises regulatory sequences in the intergenic fragments that could control the consecutive expression (switch) of each gene or gene group during embryonic and fetal development until expression of the adult globin genes. Pseudogenes are DNA sequences generated most probably by reverse transcription of mature mRNA, therefore these genes have no intervening sequences and cannot be transcribed. They are silent genes.

The structure of globin genes refers to the general mosaic organization of eukaryotic

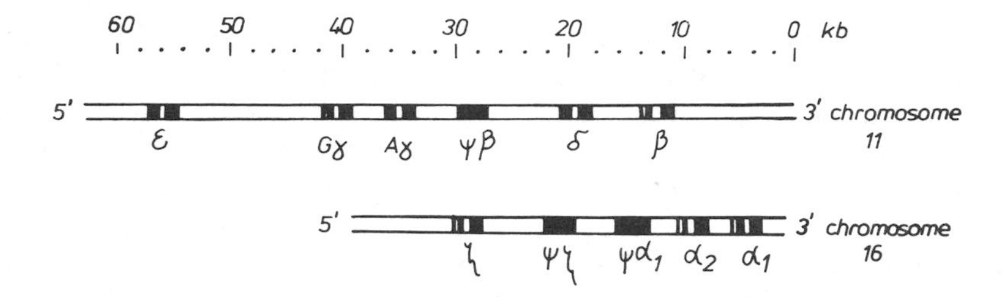

FIGURE 1. Linkage map of the human β-globin gene cluster (top) and α-globin gene cluster (bottom).

genes with exons and introns which are excised during mRNA processing (see Chapter 3, Figure 11). The majority of amino acid residues that are in contact with the heme group and amino acid residues that form the $\alpha_1\beta_2$ subunit interface are encoded by the central exon. Amino acid residues that form the $\alpha_1\beta_1$ bond are encoded mainly by the third exon. Finally, amino acid residues that contribute to the Bohr effect and 2,3-DPG binding are more randomly distributed.

A. REGULATION OF HEMOGLOBIN SYNTHESIS

The adjacent flanking sequences of globin genes are important for their transcription. Upstream of the coding sequences lies the "promoter" composed of the TATA box localized 28 to 30 bp upstream of the initiation codon, CCAAT box localized 70 to 80 bp upstream of the initiation codon. There are probably more regions important for transcription, localized in various regions of the genome, not yet sufficiently recognized. The promoter region serves as the recognition and binding site of RNA polymerase, DNA dependent.

At the 3′ end of the noncoding region of the globin genes within 15 to 19 bp 5′ from the codon terminating transcription is a highly conserved hexanucleotide sequence AATAAA that serves as the termination signal for RNA polymerase and other enzymes involved in polyadenylation of 3′ end of mRNA.

Mutations in the promoter region cause disturbances in transcription. The amount of mRNA molecules becomes significantly reduced. Mutations of the 3′ end cause disturbances in mRNA termination and polyadenylation, and in some cases prolongation of the synthesized globin chain.

In mammals hemoglobin is synthesized mainly in nucleated erythroblasts in bone marrow and to a lesser extent in reticulocytes. Mature erythrocytes do not synthesize hemoglobin. The synthesis of the hemoglobin molecule involves a group of biochemical pathways for formation of globin polypeptides and synthesis of porphyrines and heme. All these processes occur in a perfectly coordinated manner, synthesizing in adults equal quantities of complementary α-, β-, and δ-globin-chains and an adequate quantity of heme. The regulatory mechanism for coordinated synthesis of the globin-chains is not sufficiently elucidated. Opposed to this, the regulatory mechanism that functions on the translational level has been quite well recognized, particularly the one that coordinates the synthesis of hemoglobin dependent on the amount of heme.

Transcription of globin genes and processing of mRNA occurs according to the general rules described in Chapter 3. The translation process of globin mRNAs demonstrates its own mechanisms that will be described shortly. In the translation process of globin mRNAs the ribosomal subunits become attached to the mRNA near the 5′ end and form the initiation complex. Formation of the initiation complex requires coordinative action of several initiation factors and cofactors. The smaller ribosome subunit undergoes conversion to a native subunit $40S_N$ by association with the initiation factor eIF-3. Simultaneously, the ternary initiation complex is formed, containing initiation factor eIF-2, the nucleotide GTP, and $tRNA_f^{met}$

($tRNA_f^{met}$ is a methionyl initiator tRNA, containing the UAC anticodon complementary to the AUG initiation codon of mRNA). Next, the ternary complex binds to $40S_N$ subunit, and forms the preinitiation complex which binds with mRNA. The reaction requires ATP and initiation factors eIF-4A and eIF-4B. Finally, the initiation complex is formed by binding of the pre-initiation complex with the major 60S ribosomal subunit with the initiation factor eIF-5.

Elongation of the polypeptide chain requires two elongation factors EF-1 and EF-2. EF-1 becomes associated with the 80S ribosome (with GTP as cofactor of this reaction) and binds with the amino acyl tRNA of the next amino acid in the synthesized polypeptide. The ribosomal binding site of the newly attached tRNA (A, or amino acyl site) is adjacent to the methionyl initiator tRNA (P, or peptidyl site). Then transpeptidation occurs with formation of the peptide bond between the methionyl carboxyl group and the next α-amino group bound to its tRNA. Finally, the methionyl residue is released from its tRNA. Then, translocation of the ribosome to the position of the next mRNA codon occurs which requires GTP and elongation factor EF-2. Simultaneously a shift in the position of the bound peptidyl tRNA from the A to the P receptor site occurs with the expulsion of the deacylated initiator tRNA. The liberated A site can accept the next tRNA according to the mRNA codon, and the next cycle of reactions can follow. The amino-terminal methionyl residue is removed from the polypeptide chain probably by an aminopeptidase.

The termination of the globin polypeptide in reticulocytes occurs by a release factor (R factor) and GTP. After termination, the R factor with the termination codon hydrolyzes the peptidyl-tRNA complex followed by release of the completed polypeptide.

The porphirin ring of heme is synthesized by a series of biochemical reactions, catalyzed by mitochondrial and cytosolic enzymes. The biochemical pathway for heme synthesis is present in all mammalian cells, because of the requirement of heme as the prosthetic group in cytochromes and hemoprotein enzymes. In erythroid cells, regulatory mechanisms coordinate the heme synthesis with that of globin-chains. Heme regulates its own synthesis by a feedback inhibitory mechanism. An excess of protoporphirin in the cell inhibits its synthesis. There are also some suggestions that globin synthesis may influence heme synthesis.

The translation process has several possibilities to regulate synthesis of polypeptides. The greatest potential control of globin-chain synthesis is combined with the initiation of translation which has been shown to be rate-limiting. The role of heme as regulator of hemoglobin synthesis has been known for a fairly long time by demonstration that globin synthesis *in vitro* was stimulated by the addition of iron. Depletion of heme in an *in vitro* system synthesizing hemoglobin caused a significant decrease in globin synthesis in reticulocytes, and was associated with disaggregation of polyribosomes. The most important finding in this aspect was that cell-free lysates incubated without heme produce an inhibitor which, when added to globin synthesizing preparation, blocked the translational activity.[3] Addition of heme to the lysates prevented formation of this inhibitor, as the inhibitor was neutralized when heme was added shortly after its formation, whereas a prolonged incubation of lysates with the inhibitor could not reverse its function. The inhibitor termed "heme-controlled translational repressor" was recognized to be a cyclic AMP-independent protein kinase that specifically inactivates the initiation factor eIF-2 by its phosphorylation.[4]

α- and β-globin genes lie on different chromosomes and are independently transcribed. In spite of this, the amount of synthesized α- and β-globin-chains by bone marrow erythroblasts and reticulocytes in normal conditions is equal. Several regulatory mechanisms have been proposed, but none of them sufficiently explains this event. Most of these hypotheses suggested that the synthesis of the β-globin-chain is in some way suppressed by the synthesized α-globin-chain until balance between these two hemoglobin components is achieved. There have been several differences observed between the synthesis of α- and β-globin-chains. The most important are that the α chains are synthesized on smaller polyribosomes and the amount of ribosome-associated α-globin-chain mRNA is about 50% greater in reticulocytes than that of β-globin-chain mRNA, but the initiation rate of β-globin mRNA translation is about 40% greater, which equalizes the greater amount of α-globin-chain mRNAs.

Recently some differences in the promoters of the α- and β-globin genes have been discovered. β-globin genes cloned with SV40 require for their transcription the SV40 enhancer, but the α-globin genes do not. It is supposed that the enhancer of β-globin gene interferes with the α-globin chain enhancer, and only after the action of the strong SV40 enhancer can this inhibition be abolished. A recent suggestion that may contribute to compensation of unequal synthesis of two different constituents of the same protein may be the interaction of mRNAs. It was proved that transcriptional interference exists between the two α-globin genes. The downstream lying α-globin gene is inhibited by the upstream lying α_2-globulin gene. This inhibition can be moderated by insertion of transcriptional termination signals between the two α genes. This result suggests that intervening sequences between the adjacent genes may be important for their function (see Chapter 4).

The quantitative expression of β- and δ-globin genes is strikingly different. The amount of Hb A (α_2 β_2) is about 95% of normal adult hemoglobin and that of Hb A_2 ($\alpha_2\delta_2$) is only 2 to 3%. For an explanation of this difference several hypotheses have been established, but only after applying modern molecular biology methods of investigations could these hypotheses be proved. At present the most probable explanation of the small expression of δ-globin genes as compared to the β-globin genes are changes of the promoter regions. In place of the highly conserved CCAAT box about 80 bp upstream of the initiation site which is necessary for effective transcription, up-stream of the δ-globin gene, the CCAAC sequence was detected instead of CCAAT sequence,[5] which may be the cause of the poor transcription of the δ-globin gene.

In the past few years intensive study of human globin genes has provided important insights into normal gene structure and function and into the nature of molecular mechanisms leading to inherited diseases. The discovered lesions include practically all kinds of gene structure changes and all steps of gene function, i.e., promoter mutations, RNA processing defects caused by gene deletions and insertions, frameshift mutations, and point mutations involving transcription, mRNA processing, initiation, termination, poly(A)addition and globin-chain instability. Many forms of hereditary persistence of fetal hemoglobin (HPFH) caused by deletions of the β-like gene cluster involving regions that control the fetal to adult globin gene switch (regions responsible for this disturbance have not been identified with certainty).[6]

III. MUTATIONS OF GLOBIN GENES CAUSING STRUCTURAL GLOBIN-CHAIN ABNORMALITIES

A. SINGLE-POINT MUTATIONS

Single point mutations may occur at every place of the globin genes and there are no specific hypermutable hot spots that favor unusual mutation rates within the coding sequences.[7] Comparing the numbers of known mutations of α- and β-globin genes with that of δ- and γ-globin genes we can see that the distribution of known mutations differ significantly: approximately 60% are β-globin mutations, 30% α-globin mutations and 10% δ- and γ-globin mutations. This uneven distribution can be explained by the larger number of β-globin-chains because only one globin gene is expressed, contrary to α and γ-globin genes where two identical genes are expressed, and the δ-globin gene is expressed normally only to a lesser extent. Therefore, the β-globin chains represent the largest number of one gene products and the number of known mutations must be the largest in case of the β-globin chain.

Mutations that cause amino acid replacements at the same position are very informative. For example, the normal code for position β99 is GAT coding for aspartic acid. Mutation of the third base into A or G codes for glutamic acid. Because glutamic acid and aspartic acid carry an identical charge and have similar chemical and physical properties—this mutant may exhibit normal functional and electrophoretic properties and would not be detected by normal electrophoretic examination. Contrary to this, the other mutations in this position reveal characteristic functional and clinical manifestations in the form of erythrocytosis caused by the increased oxygen affinity of the mutant hemoglobins. To this group belong: Hb Ypsilanti β99 Asp → Tyr (GAT → TAT); Hb Yakima β99 Asp → His (GAT → CAT); Hb Kempsey β99 Asp → Asn (GAT → AAT); Hb Radcliffe β99 Asp → Ala (GAT → GCT); Hb Hôtel-Dieu β99 Asp → Gly (GAT → GGT); Hb Chemilly β99 Asp → Val (GAT → GTT).

Different mutations in the same position may cause also different functional and clinical manifestations. For example, mutations of position 101 of the β-globin-chain are known: Hb Rush β101 Glu → Gln (GAG → CAG); Hb British Columbia β101 Glu → Lys (GAG → AAG); Hb Alberta β101 Glu → Gly (GAG → GGG); Hb Potomac β101 Glu → Asp (GAG → GAC). In this group of hemoglobin mutations the functional properties vary and thus also the clinical manifestations: Hb Rush causes hemolytic anemia because of molecular instability of this mutant hemoglobin, while the remaining, i.e., Hb British Columbia, Hb Alberta and Hb Potomac cause erythrocytosis because of increased oxygen affinity of these mutant hemoglobins.

Among the more than 300 hemoglobin variants known with amino acid substitutions all except three are caused by single base changes in the affected codon.[7] These exceptions include Hb Bristol (β67 Val → Asp - GTG → GAT or GAC), Hb Edmonton (β50 Thr → Lys - ACT → AAA or AAG), Hb D Ouled Rabah (β19 Asn → Lys - AAC → AAA or AAG). The mechanism of double substitutions is not unmistakably clarified.

B. DOUBLE-POINT MUTATIONS

These mutations are characterized by the presence of two different amino acid substitutions within a single globin-chain. To these hemoglobin mutants belong:[7]

- Hb C Harlem (β6 Glu → Val and β73 Asp → Asn).
- Hb Arlington Park (β6 Glu → Lys and β95 Lys → Glu).
- Hb C Ziguinchor (β6 Glu → Val and β58 Pro → Arg).
- Hb S Travis (β6 Glu → Val and β142 Ala → Val).
- Hb Singapore (α78 Asn → Asp and α79 Ala → Gly).

The characteristic feature of this group of hemoglobin abnormalities is that most of them occur in populations where the particular single mutations composing the double ones are frequent in the given population. For example, Hb C Harlem includes the Hb S mutation (β6 Val → Glu) and the Korle Bu mutation (β73 Asp → Asn), both frequently occurring in populations of West Africa. Therefore, it is very probable that double-point mutations of distant positions of the same molecule may have been caused by recombination between the two different single point mutations.[7] This explanation does not concern the double-point mutation in Hb Singapore because of the adjacent point mutations.

C. NONSENSE MUTATIONS

There are only a few known cases of globin gene point mutations causing premature termination of mRNA translation resulting in a shortened polypeptide chain. One of these hemoglobins is Hb McKees Rocks β145 changing the normal code TAT for tyrosine into terminating codon TAA or TAG which shortens the β-globin-chain by two amino acid residues (tyrosine and histidine), important for oxygen affinity of the hemoglobin molecule. This results in high oxygen affinity of Hb McKees Rocks, causing erythrocytosis. A typical case of nonsense mutation represents the hemoglobin molecule in Ferrara type of β^{o} thalassemia caused by the β39 nonsense mutation.[8] The β39 nonsense mutation (change from normal code for Gln CAG → TAG) is widespread in the Mediterranean population with many variants. Generally, nonsense mutations are more frequent in thalassemias.

D. POINT MUTATIONS CAUSING PROLONGED GLOBIN CHAIN

There are two known β-globin-chain mutants at the N-terminal prolonged by nonremoved amino-terminal methionine. Normally, mRNA translation starts at the initiation codon AUG and anticodon of $tRNA_f^{met}$ bearing methionine. The amino-terminal methionyl residue is removed by an amino-terminal before termination of transcription. The two abnormal hemoglobins with a prolonged β-chain are: Hb Long Island—Marseille with a methionyl residue preceding the amino terminal valine and replacement at position β2 His → Pro (CAG → CCC), and Hb South Florida with a substitution of methionine for the N-terminal valine of the β-globin-chain.

Proteins, parts of which undergo removal, usually have serine, alanine, glycine, or valine as the NH_2-terminal residues. The first three residues favor acetylation. Proteins that retain the initiator methionine usually have a charged residue or methionine at the second position. About 20% of the Hb South Florida was acetylated at the NH_2-terminus of the β-globin-chain.[9]

In the α-globin-chain the mRNA termination codon UAA normally follows arginine codon 141. Several point mutations concern changes of this termination codon (UAA), changing this "stop" signal for an amino acid codon resulting in prolongation of the α-globin-chain until a new terminator codon UAA is reached. Because of this mutation up to 31 amino acid residues can be added. To these α-chain variants belong: Hb Constant Spring α142, termination codon TAA → CAA, encoding for glutamine (Gln), the α-chain has 172 amino acid residues instead of 141; Hb Icaria α142, mutation of termination codon TAA → AAA, encoding for lysine (Lys), the α-globin-chain has 172 amino acid residues instead of 141; Hb Koya Dora α142, mutation of termination codon TAA → TCA, encoding for serine (Ser), the α-globin-chain has also 172 amino acid residues; Hb Seal Rock α142, mutation of the termination codon TAA → GAA, encoding for glutamic acid (Glu), the α-chain has also 172 amino acid residues instead of 141. The cause of these mutations restricted to the α-globin-chain is not entirely clear.

E. DELETION AND INSERTION MUTATIONS — FRAMESHIFT MUTATIONS

Deletion of one or more amino acids (deletion of a code for one or more amino acids) usually causes unstable hemoglobin molecules resulting in hemolytic anemia dependent on position and number of deleted amino acids. Most of the deletions concern β-globin-chain. To the deletion mutations belong: Hb Leiden, del. β6 or 7 (GAG) glutamic acid (Glu), resulting in an unstable hemoglobin molecule which causes mild hemolysis; Hb Freiburg, del. β23 (GTT) valine (Val), resulting in increased oxygen affinity and a mild hemolytic anemia (which prevents erythrocytosis); Hb Niteroi, del. β42—44 (GAG-TCC-TTT)phenylalanine, glutamic acid and serine or del. β43—45 glutamic acid, serine and phenylalanine, resulting in an unstable hemoglobin molecule with hemolytic anemia and low oxygen affinity; Hb Tochigi, del. β56—59 (GGC-AAC-CCT-AAG) glycine, asparagine, proline, and lysine, resulting in an unstable hemoglobin molecule and hemolytic anemia; Hb St. Antoine, del. β74—75 (GGC-CTG) glycine, leucine, resulting in an unstable hemoglobin molecule and hemolytic anemia; Hb Vicksburg, del. β75 (CTG) leucine, causes β^{+} thalassemia; Hb Tours, del. β87(ACA) threonine, resulting in an unstable molecule and increased oxygen affinity, but causes no erythrocytosis because of hemolytic anemia; Hb

Gun Hill, del. β91-95 (CTG-CAC-TGT-GAC-AAC) leucine, histidine, cysteine, aspartic acid, and lysine, resulting in an unstable hemoglobin molecule which causes hemolytic anemia; Hb Coventry, del. β141 (CTG) leucine, results in an unstable hemoglobin molecule and hemolytic anemia; Hb Boyle Heights, del. α6 (GAC) aspartic acid, results in a slightly unstable hemoglobin molecule without clinical symptoms.

Quite different sequences occur when one or two bases become inserted or deleted. As compared with deletions of one or more amino acids (deletion of one or more coding triplets of bases) the consequences are more severe, resulting in frameshift mutations. In this kind of mutation the entire amino acid sequence of the carboxy-terminal fragment of the polypeptide chain would be changed beyond the site of mutation. All of these mutant chains are abnormally long, caused by changes of their termination codons. The following frameshift mutations are known:

Hb Wayne, α138 (TCC-TC) cytosine, the α-globin-chain is extended by five amino acid residues: α138 serine, 139 asparagine instead of lysine, 140 threonine instead of tyrosine, 141 valine instead of arginine and additionally 142 lysine, 143 leucine, 144 glycine, 145 proline, and 146 arginine—clinically without symptoms.

Hb Tak, β146 AC inserted after codon 146, additional insertion of 11 amino acid residues, resulting in increased oxygen affinity and clinically thalassemia-like symptoms.

Hb Cranston, β144 AG inserted after codon 144, additional insertion of 11 amino acid residues, resulting in addition of 11 amino acid residues causing an unstable hemoglobin molecule, clinically chronic hemolytic anemia.

Hb Saverne, β143, base deletion CAC-CC, resulting in addition of ten amino acid residues, resulting in unstable hemoglobin molecule with high oxygen affinity, clinically without erythrocytosis because of chronic hemolytic anemia.

F. FUSION-GENE MUTATIONS — Hb LEPORE

The name originates from an Italian-American family with clinical symptoms of β thalassemia in which this kind of pathological hemoglobin has been discovered. It contained δ-globin-chain sequences at its amino-terminal end, and β-globin-chain sequences at its carboxyl-terminal end. The elucidation of chain sequences suggested that this abnormal hemoglobin generated probably by nonhomologous recombination between the normal δ-globin-chains, because of extensive homology between these two β-globin-like genes. The result of mispairing is a deletion of about 7000 bp extending from the point where transcription of the δ-globin gene stops to the point where transcription of the β-globin gene starts. Consequently, chromosome bearing the δ-β hybrid gene is transcribed very slowly, which in the simultaneous absence of normal δ- and β-globin genes causes clinical symptoms of thalassemia (see Figure 2).

The nonhomologous gene recombination that generated the Lepore hemoglobins also explains generation of the reciprocal state, the anti-Lepore hemoglobins. In the anti-Lepore hemoglobin there is a 7000 bp insertion with formation of a β-δ hybrid gene flanked from both sides by β- and δ-globin gene sequences.

Because of the great homology of the β- and δ-globin genes (only ten amino acid residues are different) it is possible that the process of nonhomologous recombination occurs more frequently than the known Lepore and anti-Lepore hemoglobins, which because of the chain homology cannot be detected. For example, amino acid differences between δ- and β-globin-chains occur at the residues 87 and 116. If a crossing-over (nonhomologous recombination) occurred between these two positions, it could not be diagnosed because of identity of δ and β chain residues between these two positions. Therefore, it is no wonder that the most frequent Hb Lepore variants demonstrate δ-globin-chain residues until position 87, and the β-globin-chain residues start with the 116 position. Hb Lepore Hollandia has δ-chain residues from position 1 to 22, and β-chain residues from position 50 to 146. Hb Lepore Baltimore

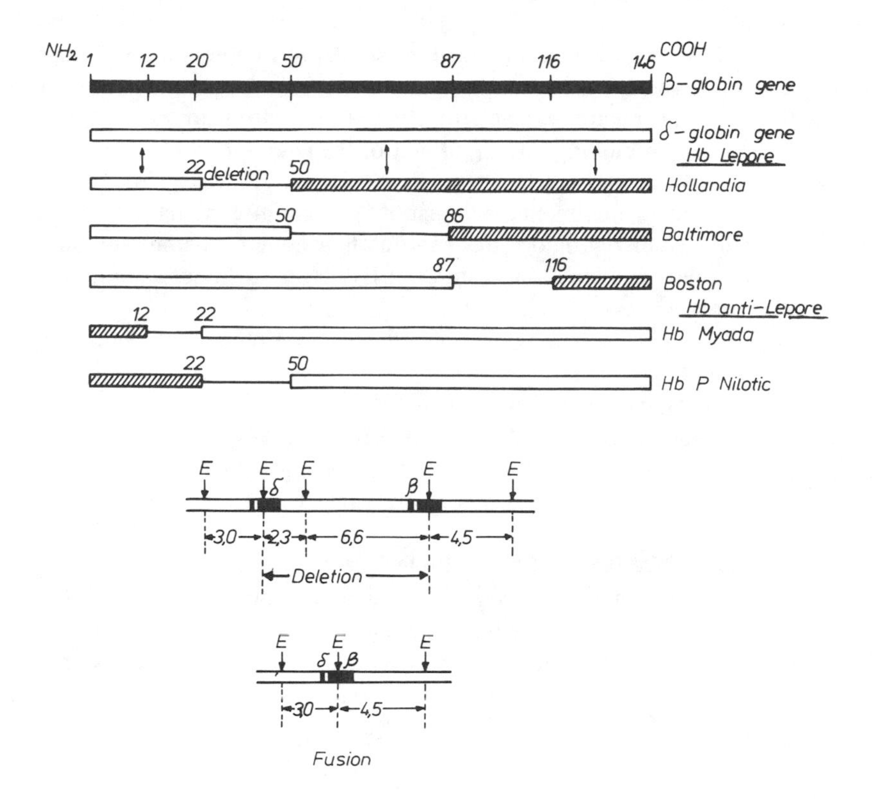

FIGURE 2. (Top): Hb Lepore (δ-globin/β-globin-gene hybrids) and anti-Lepore hemoglobins (β-globin/δ-globin gene hybrids). (Bottom): restriction enzyme analysis of Hb Lepore. E—site of EcoRI action.

demonstrates δ-chain residues from position 1 to 50, and β-chain residues from position 86 to 146. Hb Lepore Boston demonstrates δ-chain residues from position 1 to 87, and β-globin-chain residues from position 116 to 146.

Two anti-Lepore hemoglobins have been recognized: Hb Miyada with β-globin residues from positions 1 to 12 and δ-globin residues from positions 22 to 146, and Hb P Nilotic with β-globin residues from positions 1 to 22, and δ globin residues from position 50 to the end of the molecule, i.e., position 146.

The relative higher frequency of Lepore hemoglobins expression as compared to the anti-Lepore ones may be caused by negative selection forces in malaria regions. Besides, the synthesis of anti-Lepore hemoglobin variants is greatly decreased in reticulocytes as compared with bone marrow erythroid cells.

The Hb Parchman represents a specific type of δβ fusion hemoglobin. The polypeptide chain demonstrates δ-globin sequences from both ends, and β-globin sequences in the central region—this abnormal polypeptide chain particularly demonstrates δ-globin sequences at positions 1 to 12, then β-globin sequences at positions 22 to 50, and again δ globin sequences at positions 86 to 146. In Hb Parchman no deletion occurred as in other Lepore hemoglobins which may be explained by gene conversion between the δ- and β-globin genes. A gene conversion limited to the central part of the δ-globin gene would not produce gain or loss of nucleotides in the affected chromosome which is consistent with the Hb Parchman case[7] (the patient was not anemic and the total amount of abnormal hemoglobin was only 1.6% of the total amount of hemoglobin). Also known are globin hybrids involving γ with β-globin genes. A typical example represents Hb Kenya which represents a hybrid formed by ^{A}gamma globin (residues 1 to 81) and β-globin gene (residues 86 to 146). This γ/β hybrid probably arose by mispairing and recombination (similar to Lepore variants). In the case of Hb Kenya the deletion comprises part of the ^{A}gamma gene (from the second intron) including the remaining part of the ^{A}gamma gene (from the second intron to the end of this gene), the whole ^{G}gamma- and δ-globin genes, and part of the β-globin gene. The deletion in Hb Kenya comprises approximately 22,500 bp. Patients with Hb Kenya synthesize that hemoglobin approximately in amounts near to that of ^{G}gamma-globin-chains. This is probably caused by the influence of regulatory mechanisms that govern ^{G}gamma-globin synthesis. Patients with Hb Kenya usually exhibit features of HPFH, but without anemia, microcytosis and other thalassemia-like symptoms.

HbF Yamaguchi is a fusion hemoglobin of both gamma-globin genes containing ^{G}gamma sequences 1 to 75, and ^{A}gamma sequences 136 to 146. Mutation that causes this abnormality refers to the point mutation $^{A}\gamma$80,Asp → Asn (GAT → AAT). The fusion process results from a deletion of approximately 5000 bp. Similarly, as in Hb Kenya the synthesis rate is typical for ^{G}gamma regulatory mechanisms. (The ^{A}gamma globin gene which part is present in Hb F Yamaguchi is usually expressed at a lower rate.)

G. COMPLEX GLOBIN GENE MUTATION

Hb Lincoln Park is synthesized by a hybrid gene composed of part of the β-globin-chain (residues 1 to 22) and part of the δ-globin-chain (residues 50 to 146), and additionally deletion of valine (δ137 GTG). Thus, it represents a typical anti-Lepore fusion, caused by crossing-over recombination, and gene conversion resulting in the deletion of valine 137.

Hb Coventry (β141 del.Leu-CTG). There was a deletion of leucine at position 141, but otherwise a normal β-globin-chain sequence. The same patient revealed also the presence of Hb Sydney (β67, GTG → GCG, Val → Ala). This mutation results in an unstable hemoglobin molecule causing chronic hemolytic anemia. It is supposed that Hb Coventry might have arisen as part of a nonhomologous recombination between δ- and β-globin genes.

Hb Leiden (β6 or 7 del.GAG, glutamic acid). Clinically, because of the unstable Hb Leiden molecule a mild hemolytic anemia occurs. It is suggested that a β-δ hybrid has been formed by nonhomologous recombination between the β Leiden and a normal δ-globin gene with crossing over near the carboxyl terminal region. A gene conversion in connection with nonhomologous crossing-over was considered as another mechanism that might cause this abnormality.

IV. FUNCTIONAL DISTURBANCES OF THE HEMOGLOBIN MOLECULE CAUSED BY STRUCTURAL ABNORMALITIES

The main disorders caused by structural abnormalities of the hemoglobin molecule are:

- An unstable hemoglobin molecule
- A partial or total loss of oxygen transport
- An increased or decreased oxygen affinity

A. UNSTABLE HEMOGLOBIN

According to the atomic model of hemoglobin,[10-14] the heme lies in a nonpolar pocket of the globin chain with about 60, almost all nonpolar interactions and one covalent bond between the iron and the proximal histidine of the globin chain. The ability of ferrous ion (Fe^{2+}) to combine reversibly with oxygen is possible in its nonpolar environment.

Loss of globin-heme contacts may cause loss of the heme which deprives the particular globin chains of about 60 weak bonds. Consequently the hemoglobin molecule becomes unstable and precipitates. Clinically the precipitated hemoglobin forms insoluble protein inclusion bodies in the erythrocytes, the so-called "Heinz bodies". The formation of these precipitates leads to further reactions that damage the cell membrane resulting in hemolysis.

Typical examples of this kind of hemoglobin instability are represented by: Hb Torino, α43, replacement of Phe → Val removes two globin heme contacts of two CH groups, which makes the α subunit unstable; Hb Sydney, β67, replacement of Val → Ala, or Hb Santa Ana, β88, replacement of Leu → Pro causes loss of two methyl groups being in contact with the heme which causes the heme group to drop out of the heme pocket. In both cases unstable hemoglobin molecule generates.

Another example represents Hb Hammersmith, β42, replacement of Phe → Ser, which causes loss of two globin heme contacts with simultaneous introduction of two polar hydroxyl groups which causes the entry of water into the heme pocket causing the heme group to drop out of the heme pocket with degradation of the hemoglobin molecule. The same mechanism of hemoglobin instability by introducing polar groups into the heme pocket occurs in Hb Bristol, β67, replacement of Val → Asp.

To the same group of unstable hemoglobins causing defects of globin heme contacts belong: Hb Köln, β98, replacement of Val → Met, Hb Zürich, β63, replacement of His → Arg, Hb Genova, β28, replacement of Leu → Pro, and Hb Sabine, β91, replacement of Leu → Pro.

Another group of unstable hemoglobin molecules is caused by replacement of amino acid residues inside the molecule, e.g., Hb Sogn, β14, replacement of Leu → Arg. Replacement of leucine by arginine in Hb Sogn, although not introducing an internal polar group into the molecule, removes some nonpolar contacts inside the molecule which makes the β-globin-chain unstable.

Unstable molecules of hemoglobin may be caused by introduction of polar groups inside the molecule. For example, Hb Wien, β130, replacement of Tyr → Asp creates an extra positive ion which causes instability of the β-chain.

Another group of unstable hemoglobins is caused by introduction of proline into a helical region of the globin-chain. The presence of this amino acid residue because of its ring structure causes steric changes which makes it impossible to form regular helix structure of the globin-chain. To this group belong: Hb Genova, β10, replacement of Leu → Pro, Hb Sabine, β91, replacement of Leu → Pro, and others.

Deletion of one or more amino acid residues weakens the structure of the hemoglobin molecule resulting in its instability. Typical examples of such unstable hemoglobins present Hb Gun Hill, βdel. 91 to 95, deletion of the following amino acid residues: leucine, histidine, cysteine, aspartic acid and lysine, and Hb Freiburg, β del. 23, deletion of valine residue.

Instability of hemoglobin molecule also may be caused by replacements of amino acid residues on the surface of the molecule. Instability of such hemoglobin molecules occurs mainly in homozygotes, and very rarely in heterozygotes. The classical example represents Hb S, β6, replacement of glutamic acid by valine resulting in sickle cell anemia in homozygotes and Hb S trait in heterozygotes. To that group of unstable hemoglobins with amino acid replacements on the surface of the molecule belongs Hb Hofu, β126, caused by replacement of valine by glutamic acid, that makes the β-globin-chain unstable which in severe anoxemic conditions undergoes hemolysis. Hb S behaves similarly and in severe anoxemic environment undergoes hemolysis.

A separate group of unstable hemoglobins represent amino acid replacements at the subunit contacts, especially those of α_1 β_1 contacts. These replacements weaken bonds between the complementary chains which may result in dislocation of subunits, and because uncombined chains are very unstable, they precipitate easily with destruction of the hemoglobin molecule.

A common feature of all unstable hemoglobins is hemolysis accompanied by oxidative changes which may result in generation of superoxides and other products of oxidation (various types of free radicals—see in Chapter 6, the section on Free Radical Pathology). The principal clinical symptom of unstable hemoglobin are "Heinz bodies" inside the erythrocytes, in other words "inclusion bodies" which are generated by precipitation of free globin-chains. The precipitation process involves conversion of hemoglobin to hemichromes (hemoglobin intermediates formed by heme iron oxidation with internal bond formation) which undergo irreversible precipitation.[7] Erythrocytes containing Heinz bodies underwent preferential destruction in the spleen. The most common symptom of unstable hemoglobin is a mild chronic hemolytic anemia, usually compensated for by increased reticulocytosis.

Clinically the presence of an unstable hemoglobin also may be silent and brought to light only after exposure to sulfamides or other oxidant drugs. A typical example of such a case of unstable hemoglobin is Hb Zürich (β63, His → Arg) in which hemolysis occurs usually after treatment with sulfonamides. Hb Zürich also demonstrates a significant increase of carbon monoxide affinity. Therefore, in cigarette smokers the level of carbonmonoxy hemoglobin is three to four times higher than in nonsmoking individuals with Hb Zürich.[7]

Several thalassemic hemoglobins also belong to the unstable hemoglobins which will be discussed here.

1. Sickle Cell Anemia

Hb S (sickle hemoglobin) varies from normal adult hemoglobin (HbA) by the replacement of glutamic acid residue at the sixth position in the β-globin-chain by valine. Sickling was first described in 1910. In 1923 Huck demonstrated that sickling was reversible. In 1927 Hahn and Gillespie proved that sickling occurred in deoxygenated blood and was reversible after reoxygenation and correctly attributed the defect to the intracellular hemoglobin and not to the erythrocytes themselves. In 1940 Ham and Castle suggested the following pathophysiology of sickle cell anemia: the erythrocytes sickle in the peripheral blood circulation resulting in an increased blood viscosity which impairs blood passage through capillaries, decreasing further oxygen tension and by causing more sickling exacerbates the pathological symptoms of this disease. In 1949, Pauling et al.[7] demonstrated by electrophoretic examination a difference among the normal and sickle hemoglobin, and ascribed this phenomenon to change of electric charge of the globin-chain. Ingram[2] (1957) using the "fingerprinting method" demonstrated that sickle hemoglobin is caused by the substitution of the glutamic acid residue at the sixth position of the β-globin-chain by valine, which causes loss of two negative charges per one hemoglobin molecule.[15]

Hemoglobin S (Hb S), deoxygenated *in vitro*, becomes relatively insoluble as compared to normal hemoglobin A and aggregates into long polymers consisting of non-covalent linked arrays of deoxyhemoglobin S molecules. These polymers align themselves into paracrystalline gels. Further studies (polarizing microscopy, fiber X-ray diffraction, electron microscopy) revealed that in the polymer the hemoglobin tetramers are wound around a vertical axis resembling spirals of hemoglobin tetramers stacked upon one another. This tubelike structure is 20 to 22 nm in diameter. Recently, however, formation of true Hb S crystals has been documented.

All abnormal hemoglobin molecules with substitutions at the sixth position of the β-globin-chain of glutamic acid by valine demonstrate sickling independently of which other substitutions of amino acid residues are present on the same globin chain. This is documented by the following abnormal hemoglobins: Hb Ziguinchor β6,Glu → Val and β58,Pro → Arg; Hb Harlem β6,Glu → Val and β73,Asp → Asn; and Hb Travis β6,Glu → Val and β142,Ala → Val. However, substitutions of other amino acid residues at the same sixth position of the β-globin-chain do not exhibit sickling independently of which amino acid was substituted. Here we have the following abnormal hemoglobins: Hb C β6,Glu → Lys; Hb Arlington Park β6,Glu → Lys and β95,Lys → Glu; and Hb Leiden β6,Glu deleted. This observation led to the suggestion that valine substituted at the sixth position of the β-chain binds with valine at the first position of the β-globin-chain forming a ring structure which shortens the β-globin-chain to the same number of amino acid residues as in the α-globin-chain, i.e., 141. Because of this shortening of the β-globin-chain the homology between the α- and β-globin-chains increases enabling their aggregation.

The most physiologically important effector of Hb S polymerization is oxygen. Only the deoxygenated Hb S can gelate. All the liganded forms like oxy-, metoxy-, and carbomonoxy-Hb S aggregate normally. Thus, it is postulated that the deoxygenated Hb S aggregates into polymer because the deoxygenated conformation allows a sufficient number of intertetrameric contacts and bonds to supply energy enough to stabilize the polymer structure.[15] Anything that stabilizes the deoxy state, e.g., polymer Hb S formation, decreases oxygen affinity of hemoglobin, therefore, lowering of pH which decreases oxygen affinity of Hb S via the Bohr effect and promotes gelation. The increasing concentration of 2,3-DPG (2,3-diphosphoglycerate) that stabilizes the deoxy form of hemoglobin acts similarly.[15] Thus, a stay in rare air (high mountains, plane flight) enhances gelation of Hb S.

The inhibitory effect of Hb A and Hb F on Hb S gelation has been known for many years. The intact tetramers of deoxy-Hb A and Hb F as well as their hybrids ($\alpha_2\beta^S\beta^A$, $\alpha_2\beta^S\gamma$) enter the sickle polymer less easily than does deoxy-Hb S and therefore they inhibit gelation by a dilutional effect (diluted Hb S does not gelate). In this aspect, the gamma-containing tetramers more effectively inhibit gelation, probably because the gamma-globin-chains differ by 20 amino acid residues on their surfaces, while β-globin-chains differ only by one amino acid residue, i.e., valine instead of glutamic acid at the sixth position. These effects are probably the basis of the protective effect of heterozygosity for the sickle cell gene, and also the basis of high levels of fetal hemoglobin.[15]

Also important to Hb S sickling are changes of the cell membrane. When hemoglobin-free sickle cell membranes were mixed with normal hemoglobin (Hb A) and resealed, these hybrid cells did not sickle when deoxygenated. In contrast, when normal red-cell membranes were mixed with Hb S, and deoxygenated, sickling occurred. Investigations about membranal proteins of sickle erythrocytes revealed reduced binding of spectrin (the major skeletal protein) with ankyrin because its high affinity membrane binding site was highly reduced. These proteins influence cell shape, membrane flexibility, endocytosis, lipid organization, and lateral diffusion of integral membrane proteins. Thus, it was suggested that changes of these proteins may contribute to sickling of erythrocytes. Further investigations combined these changes with irreversible sickling, i.e., after reoxygenation sickling of erythrocytes did not disappear. Irreversibly sickled erythrocytes exhibit abnormal plastic deformation and reduced endocytosis. About 30% of the phosphatidyl etanolamine and phosphatidyl serine which are normally expelled from the cytoplasmic half of the membrane bilayer remain there, and are permanently exposed on the outer surface of the irreversibly sickled erythrocytes,[15,16] which influences receptor binding. The negative charge of sickled erythrocytes probably is caused by abnormal cluster formation of glycophorines.

The viscosity of sickle cell blood is increased primarily due to irreversibly sickled cells and increased gamma globulin levels. After deoxygenation, the viscosity increases further due to increased rigidity of sickled cells. This increases the exposure time of erythrocytes to the hypoxic environment and lower pH of tissues which promotes further sickling, resulting finally in occlusion of capillaries and arterioles, leading to infarction of surrounding tissues. Hemolysis probably occurs as a reaction to mechanical fragility of the deformed erythrocytes. The fibrous aggregates may exhibit a pattern of polymorphism in which the ratio of their helical pitch to their radius is approximately constant, but may influence the clinical picture of this disease.[17]

The clinical symptoms of sickling depend primarily on homozygous or heterozygous state. Heterozygotes are in normal oxygen conditions usually without pathological symptoms. The first pathological symptoms usually appear at the second half of the first year of life, when Hb S concentration in erythrocytes reaches a critical level. Hemolysis and a progressive hemolytic anemia with splenomegaly occurs. The requirement of folic acid is increased because of the increased rate of erythropoesis. The major dangerous disturbance for young children is the increased susceptibility to infections caused by *Streptococcus pneumoniae*, *Haemophilus influenzae*, *Salmonella*, *E. coli*, *Shigella*, and several other microbes. Bacterial infections of the lungs digestive and urinary tracts, meningitis, osteomyelitis and other disorders, including sepsis, are common among young children with sickle cell anemia. Factors that contribute to the severe infections are impaired antibody reaction, decreased opsonization, impaired complement activation (mainly in the alternative properdin pathway) and chemotaxis. Infection is the most frequent cause of death in young children with sickle cell anemia.

Because of increased blood viscosity vaso-occlusive crisis occurs in infancy usually with dactylitis, later involving the periosteum, bones and joints with subsequent infarction or thrombosis, osteomyelitis, septic arthritis, and pleural pain caused by vascular disturbances. Further clinical symptoms may include splenic sequestration, abdominal crises, cholelithiasis, hepatic infarcts with jaundice, hematuria, hypostenuria, skin ulcers, and other clinical manifestations.

Homozygous sickle hemoglobinopathy (Hb SS) causes the most severe form of sickle cell disease. Combination of Hb S with other hemoglobinopathies or thalassemias produces various clinical manifestations that are similar to that of Hb SS. Among the most severe sickling disorders caused by the doubly heterozygous combinations of Hb S are Hb S + Hb O Arab (β121Glu $\rightarrow$ Lys), Hb D Los Angeles (β121,Glu $\rightarrow$ Gln), Hb C Harlem (β6, Glu $\rightarrow$ Val, and β73,Asp $\rightarrow$ Asn). These abnormal hemoglobins very strongly promote the sickle transformation under conditions similar to those which cause sickle transformation of Hb SS erythrocytes. In particular, sickling promotion exerts doubly substituted Hb C Harlem with one substitution common with the β^S chain. The combination of Hb S with β^0 thalassemia causes similar severity in combination with heterozygous Hb S, although the intracellular concentration of β^S chains is lower than in Hb SS. In Hb S and δ/β thalassemia, and β^+ thalassemia sickling usually is less severe.

Hb SC diseases, although usually demonstrating less severe sickling, very frequently exhibit thromboembolic complications, renal papillary necrosis, aseptic necrosis of bones, and proliferative retinal vascular disorder.[7]

An interesting example of modified sickle cell disease represents Hb S in connection with Hb Memphis (α23,Glu $\rightarrow$ Gln). In this case Hb Memphis ameliorates the severity of sickle cell disease (the blood viscosity is lower than that in Hb SS patients), although it causes hemolytic disturbances similar to that in Hb SS. Similar amelioration of sickle cell disease manifestations also cause another abnormal hemoglobin: Hb Stanleyville (α78,Asn $\rightarrow$ Lys). Finally, an extremely mild course of sickle cell disease occurs in Hb H(4β-globin-chains) patients.

In heterozygotes (Hb $\alpha_2\beta^S\beta^A$) Hb S accounts for 32 to 45% of the hemoglobin in circulating erythrocytes. Pathological events caused in these patients are extremely rare, although some cryptic manifestations may be present. To these belong: isolated defects of kidneys with inability to produce normally concentrated urine, and very rarely hematuria. Severe complications may occur during severe hypoxia (flight at high altitude in an unpressurized plane) which may cause acute spleen infarction. The same disturbances may occur during surgical anesthesia associated with severe hypoxia. In general these disturbances are rather to very rare and individuals heterozygous for the sickle hemoglobin behave as healthy persons.

The severity of Hb SS manifestations are different in various populations. For example, in some rural parts of Africa (Kenya) about 100% of children with sickle cell anemia die within the first five years of life. On the contrary, in Saudi Arabia and in the Eastern Oasis population the Hb SS anemia is an extremely mild disease and seems to be compatible with normal survival. In Jamaica the mortality rate is about 10% of young children, thereafter many patients live to middle or old age. These differences are difficult to explain. The most important aspect seems to be the presence of other abnormal hemoglobins or other proteins that protect sickle hemoglobin from aggregation. To such protein belong undoubtedly fetal hemoglobin (gamma-globin-chains) which causes asymptomatic forms of Hb SS in Saudi Arabians in whom Hb F levels range from 20 to 30%, in individuals from Southern India whose Hb F levels are about 20% and in certain Afro-American families with increased levels of Hb F.[7]

Glucose-6-phosphate dehydrogenase (G-6-PD) deficiency is particularly frequent in the African population, which is explained, as in the similar case of Hb S, by negative selection caused by an increased resistance to *falciparum* malaria infection. Several investigations have been performed to determine if coexistence of these two disorders alleviates the negative symptoms of sickle cell disease. Some authors found that the presence of G-6-PD deficiency protects in some way Hb SS patients against sickling, while others did not confirm this statement.

Patients with hereditary spherocytosis, a predominantly inherited disorder of the red blood cell membrane, exhibit more severe hemolytic anemia with splenomegaly than Hb SS patients alone usually do, although the high levels of Hb F occurring in these patients (13 to 30%) may alleviate imminent severe Hb SS crises.

2. Intracellular Hemoglobin Crystallization

Hb C (β6,Glu $\rightarrow$ represents a typical example of intracellular hemoglobin crystallization. Hb C demonstrates decreased solubility in comparison to Hb A. Similarly as in Hb S, deoxygenation promotes crystallization of Hb C. Crystallization has been easily demonstrated in splenectomized Hb CC patients, but not in Hb CC patients with intact spleen. This observation suggests that in Hb CC patients crystallization of their hemoglobin is difficult to detect because Hb C crystals are removed by the spleen. Hb CC erythrocytes are more rigid than normal. Their shape is less deformed than that of Hb SS erythrocytes. The blood viscosity also is increased. The flow of Hb CC erythrocytes, thanks to the smaller diameter and size of these cells, is nearly normal.[7]

3. Unstable Hemoglobin

There have been approximately 70 unstable hemoglobin variants described, among them β-globin variants are about five times more frequent than α-globin variants. This disequilibrium is caused by different amounts of α-globin-chains (four normal genes) as compared with the β-globin-chain (only two genes). Therefore, when one of the four α genes undergoes mutation the remaining three can compensate for this disturbance. Thus, mutation of one of the β-globin genes modifies hemoglobin molecules to a greater extent than in the case of α gene mutation.

The instability of hemoglobin molecules results from various mutations that weaken the normal structure of the hemoglobin molecule. The structural changes that generate instability of the hemoglobin molecules are situated mainly at the amino acid residues in the vicinity of heme pockets, amino acid residues at the internal positions forming subunit contacts, and in the helical part of the molecule securing normal conformation of the subunits.[12]

B. HEMOGLOBINS WITH PARTIAL OR TOTAL LOSS OF OXYGEN TRANSPORT

1. The Hb M Globin Variants

Oxygen transport depends on the iron atom in ferrous form (Fe^{2+}) of hemoglobin. The heme irons have six coordination sites, four of them are occupied by the pyrrole ring tertiary nitrogen atoms of the heme, the fifth is occupied forming a bond with the proximal histidine (α87,β92), which lies at the opposite side of the heme and is free for binding oxygen (see

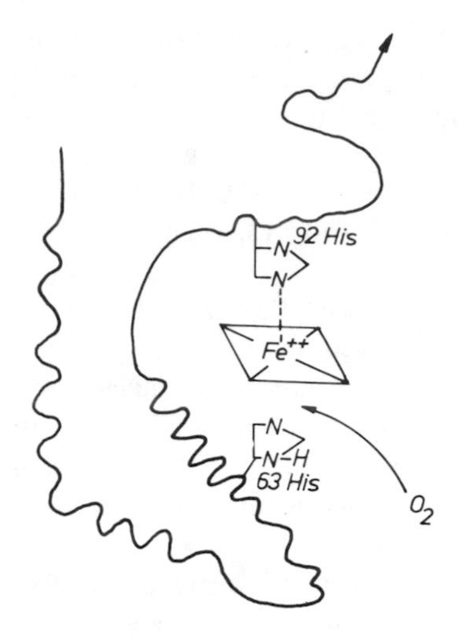

FIGURE 3. Scheme (simplified) of the β-globin chain. The heme iron has six coordination bonds, four with the pyrrole ring nitrogens, the fifth with proximal histidine (92 His) and the sixth (free) for oxygen binding (63 His).

Figure 3). Stabilization of the iron atom in ferric form (Fe^{3+}) is the main cause of hemoglobinopathies M (methemoglobinopathies). In Hb M Boston (α58,His → Tyr) and Hb M Saskatoon (β63,His → Tyr) the distal histidines are replaced by tyrosines. In Hb M Iwate (α87,His → Tyr) and Hb M Hyde Park (β92,His → Tyr) the proximal histidines are replaced by tyrosines. In all these cases the phenolic oxygen comes close (within 0.2 to 0.25 nm) to the iron atom. In these conditions an ionic bond is formed between the phenolic oxygen and the sixth site of the iron atom leaving no site free for a ligand (oxygen, CO) which results in inability of the hemoglobin molecule to bind reversibly with oxygen. These abnormal bonds have been proved by X-ray crystallography.

Hemoglobin M may also arise by replacement of other positions of the globin chain. For example, Hb M Milwaukee (β67,Val → Glu) with which glutamic acid makes a link, the iron atom stabilizing it is in the ferric form, and Hb Zürich (β63,His → Arg). In Hb Zürich the side chain of the substituted arginine is too long and protrudes from the heme pocket to the surface leaving the heme pocket empty, Hb Zürich binds easily with oxygen, but also becomes easily oxidized by sulfonamides or other oxidants resulting in generation of inclusion bodies in erythrocytes and hemolytic anemia after ingestion of sulfonamides.

Because of the impossibility to bind oxygen the Hb M hemoglobins reveal a brownish appearance even when well-oxygenated. The affected individual appears cyanotic. The gamma-chain mutant Hb F M Osaka (Gγ63,His → Tyr) is fully expressed at birth, whereas the others (Hb M Boston, Hb M Saskatoon, Hb M Hyde Park, and Hb M Iwate) at the age of 3 to 4 months old. In spite of their cyanotic appearance, individuals with methemoglobinopathy are not hypoxic, suffering only from mild hemolytic anemia.

Methemoglobinopathies must be differentiated from methemoglobinemias, caused by the deficiency of NADH dehydrogenase. Binding of oxygen by hemoglobin requires partial transfer of an electron from the heme iron to oxygen. Loss of this electron makes further binding by the oxidized ferric heme impossible, unless the electron is restored by some reducing mechanisms. The most important one requires NADH dehydrogenase. Inherited deficiency of methemoglobin reductase produces methemoglobinemia.

Cyanosis must be differentiated with other cyanotic symptoms in cardiac or pulmonary diseases. These diseases are easily distinguished from methemoglobinopathies and methemoglobinemias, although they reveal cyanosis, particularly at the insufficient state. Clinical symptoms in heterozygotes of methemoglobinopathies and methemoglobinemias (cyanosis) are reversible after administration of reducing agents in methemoglobinemias but not in methemoglobinopathies. Homozygotes of these disorders are incapable of life.

C. CHANGES IN OXYGEN AFFINITY OF THE HEMOGLOBIN MOLECULE

The physiological function of hemoglobin depends mainly on oxygen affinity changes in the lung and tissues, which is succored by so-called "respiratory movements" or heme-heme interaction of the hemoglobin molecule. The movements depend mainly on contacts between the subunits of the hemoglobin molecule.[13,14] Contacts between similar subunits are polar, between unlike ones predominantly non-polar. Dissociation of the hemoglobin tetramer into dimers usually takes place at the $\alpha_1\beta_2$ or $\alpha_2\beta_1$ contacts. Further dissociation into free α and β-globin-chain takes place at the $\alpha_1\beta_1$ and $\alpha_2\beta_2$ contacts.

Replacements of amino acid residues changing $\alpha_1\beta_1$ or $\alpha_2\beta_2$ contacts are not a predominate cause of greater clinical disturbances. On the contrary, replacements of amino acid residues influencing $\alpha_1\beta_2$ or $\alpha_2\beta_1$ contacts cause diminished heme-heme interactions resulting in increased oxygen affinity of the hemoglobin molecule. To this group belong about 80 different α- and β-globin-chains.

According to Honig and Adams[7] the following structural abnormalities of the hemoglobin molecule may cause increased oxygen affinity of the hemoglobin molecule:

Abnormalities of the heme pocket: Hb Zürich (β63,His $\rightarrow$ Arg), Hb Brisbane (β68,Leu $\rightarrow$ His).

Abnormalities affecting the Bohr effect and salt bond sites: Hb Rainier (β145,Tyr $\rightarrow$ Cys), Hb Hiroshima (β146,His $\rightarrow$ $\rightarrow$ Asp), Hb York (β146,His $\rightarrow$ Pro).

Abnormalities affecting 2.3-DPG binding: Hb Rahere (β82, Lys $\rightarrow$ Thr), Hb Toyoake (β142,Ala $\rightarrow$ Pro), Hb Little Rock (β143,His $\rightarrow$ Gln).

Abnormality affecting the oxy-deoxy equilibrium of individual subunits: Hb Shepherds Bush (β74,Gly $\rightarrow$ Asp).

Abnormalities affecting equilibrium between quaternary structures: Hb Chesapeake (α92,Arg $\rightarrow$ Leu), Hb Crteil (β89, Ser $\rightarrow$ Asn), Hb Yakima (β99, Asp $\rightarrow$ His).

Abnormalities causing increased subunit dissociation: Hb (β37,Trp $\rightarrow$ Arg), Hb Hirose (β37, Trp $\rightarrow$ Ser).

Increased oxygen affinity with unstable hemoglobin molecule: Hb Palmerston North (β23,Val $\rightarrow$ Phe).

Increased oxygen affinity of the hemoglobin molecule may be caused by mutations of globin genes affecting various sites of the molecule. Mutations that cause amino acid replacements to loosen the joining of the $\alpha_1\beta_2$ subunits are among the most important in this field. The best example of this kind of disturbance represents Hb Hirose the deoxy form of which dissociates easily into subunits which are believed to be the primary pathogenic cause of increased oxygen affinity of this abnormal hemoglobin.

Increased oxygen affinity of the hemoglobin molecule may also be caused by mutations of non-globin genes. The typical example involves mutation of the red blood cell's enzyme diphosphoglycerate mutase which synthesizes 2,3-diphosphoglycerate (2,3-DPG). In this case persistent erythrocytosis occurs with other symptoms characteristic for abnormal hemoglobins with increased oxygen affinity.

As typical consequences of increased oxygen affinity, the abnormal hemoglobin molecules cannot overcome their physiological barrier for unloading of oxygen in the tissues which results in hypoxia of tissues. Consequently, induction of erythropoetin synthesis may occur which causes overproduction of erythrocytes (i.e., erythrocytosis).

The increased oxygen affinity of hemoglobin molecule may also influence other enzymatic functions of erythrocytes. For example, Hb Wood (β97,His $\rightarrow$ Leu) belongs to the group of hemoglobins with increased oxygen affinity. Methemoglobin reduction in intact red cells containing 50% of Hb Wood occurs at a slower rate than in normal red cells. In a patient bearing both mutations (Hb Wood and methemoglobin reductase deficiency) the level of methemoglobin reduction was lower than in single mutations, which suggests that both mutations cause an additive effect.[19]

D. HEMOGLOBINS WITH DECREASED OXYGEN AFFINITY

Abnormal hemoglobins with decreased oxygen affinity characterize the lowered saturation curve as compared to normal blood. In these hemoglobins the major arterial blood remains deoxygenated. Because of lowered blood oxygenation a characteristic bluish (cyanotic) color of skin and mucous membranes occurs. To the typical examples of hemoglobins with decreased oxygen affinity that exhibit clinically cyanosis belong: Hb Kansas, (β102, Asn $\rightarrow$ Thr) and Hb Beth Israel (β102,Asn $\rightarrow$ Ser). To hemoglobins with decreased oxygen affinity and hemolytic anemia belong: Hb Hazebrouk (β38,Thr $\rightarrow$ Pro), Hb Hammersmith (β52, Phe $\rightarrow$ Ser), Hb Louisville Bucuresti (β42,Phe $\rightarrow$ Leu), Hb Niteroi (β del.42 to 44 or 43 to 45, del. Phe-Glu-Ser). Hb Cheverly (β45,Phe $\rightarrow$ Ser), Hb Okaloosa (β48,Leu $\rightarrow$ Arg), Hb Seattle (β70,Ala $\rightarrow$ Asp). In all cases of decreased oxygen affinity and hemolytic anemia the cause of hemolytic anemia is instability of molecules of hemoglobin. To the hemoglobins with decreased oxygen affinity without clinical symptoms belong: Hb Bologna (β61,Lys $\rightarrow$ Met), Hb Vancouver (β73,Asp $\rightarrow$ Tyr), Hb Presbyterian (β108,Asn $\rightarrow$ Lys), and Hb Hope (β136,Gly $\rightarrow$ Asp). As seen from the above list, two of the abnormal hemoglobins with decreased oxygen affinity exhibit cyanotic symptoms, mild to more severe hemolytic anemia, and five are without distinct clinical symptoms. This observation demonstrates once more that molecular changes of the same molecule may cause very different clinical symptoms dependent upon the site and kind of change.

A typical case of decreased oxygen affinity is Hb Kansas. The decreased oxygen affinity of this hemoglobin, which causes low oxygen saturation in the lung, is compensated in tissues where Hb Kansas unloads oxygen very efficiently. Therefore, in spite of the very low oxygenation resulting in cyanosis, as a physiological end-effect Hb Kansas transports oxygen efficiently like normal hemoglobin. The molecular basis of decreased oxygen affinity of Hb Kansas caused by β102 replacement of asparagine by threonine lies in loss of a hydrogen bond of $\alpha_1\beta_2$ subunits binding forces that normally stabilizes the "oxy" quaternary structure. This shifts the hemoglobin structure to the "deoxy" quaternary structure of Hb Kansas which is responsible for the decreased oxygen affinity of this hemoglobin.

To the other molecular mechanisms that decrease oxygen affinity belong: changes in the heme pocket influencing the Bohr effect, and oxy-deoxy equilibria within the particular subunits and quaternary structure.[14]

V. THE THALASSEMIAS

Thalassemias represent an extremely heterogenous group of blood disorders caused by disturbances in gene expression. The main disturbance is the defective synthesis of particular globin-chains: α,β,δ in hemoglobins A (adult) $\alpha_2\beta_2$, hemoglobin $A_2\alpha_2\delta_2$, γ chains in Hb F (fetal hemoglobin) $\alpha_2\gamma_2$. The name of a particular thalassemia depends on the nonsynthesized chain; we distinguish $\alpha,\beta,\delta,\delta\beta$ and $\gamma\delta\beta$ thalassemias. The molecular changes responsible for defective synthesis of the particular globin chains are gene deletions, promoter mutations, disturbances of mRNA processing, nonsense mutations with premature chain termination, frameshift mutations, and unstable globin molecules caused by single or multiple amino acid residues deletions or point mutations.

The defective synthesis of particular globin-chains causes imbalance of the nonsynthesized chains and surplus of the synthesized ones. For example, in β thalassemia there is an excess of α-globin-chains. The free α-globin-chains precipitate in intramedullary bone, causing ineffective erythropoesis with intramedullary hemoglobin breakdown and defective maturation of erythrocytes. Besides, free globin chains precipitate also in extramedullary regions and form inclusion bodies. In more severe cases hemolysis occurs. Summarizing, the main clinical manifestation of all thalassemias is anemia.

Depending on the presence or total absence of mRNAs for the synthesis of particular globin-chains, the thalassemias may be divided into α^0 or β^0 thalassemias with no α or β globin chain synthesis, and α^+ or β^+ thalassemias with reduced synthesis of these chains.

In addition, there is a closely related state, called Hereditary Persistence of Fetal Hemoglobin (HPFH), in which fetal hemoglobin is synthesized for all of the life of the affected individual.

Discussing the molecular bases of thalassemias we must first know the organization of the globin genes (see Figure 1).

Thalassemia genes contain mutations that directly change gene structure resulting in disturbances in gene function. Principally, thalassemias must be considered as *cis*-acting mutations.[20] *Cis*-acting mutation affects only the gene on the chromosome on which the mutation occurs, a trans-acting mutation affects both genes at a chromosomal locus. As a general rule, *cis*-acting mutations affect gene structure directly and usually they do not alter molecules engaged in gene expression. All thalassemia genes that have been examined to date act directly on gene structure and its function.[20]

Thalassemias can be divided into two main groups. The first is characterized by deletions of globin gene segments, an entire globin gene or multiple globin genes. The second group, the so-called "non-deletion" group is characterized by mutations that involve single base mutations and frameshift mutations.

A. DELETION FORMS OF α THALASSEMIAS

Because of four normal α genes: two α_1 and two α_2 genes, deletion can comprise all four α genes which results in clinically severe form of α^0 thalassemia. Milder forms of α^+ thalassemias are characterized by only partial loss of α-globin genes. Loss of one α-globin gene is represented by the genotype $\alpha\alpha/\alpha$-, loss of two α-globin genes by the genotype $\alpha\alpha$/—, loss of three α-globin genes by the genotype α-/—. The genotype α^0 can be represented —/—. Usually, the greater the loss of α genes the more severe the clinical manifestations.

α^0-Thalassemia heterozygotes, are able to synthesize at birth (and as fetuses) 5 to 10% of Hb Bart's (hemoglobin molecule composed of four γ-globin-chains) as adults they synthesize the same amount of Hb H (hemoglobin molecule composed of four β-globin-chains. Homozygotes of α^0 thalassemia synthesize at birth Hb Bart's, usually they demonstrate hydrops fetalis and are incapable of surviving. Heterozygotes demonstrate clinical symptoms dependent on the loss of one, two or three α-globin genes—from minimal hematological changes to severe, and synthesis of small amounts of Hb Bart's at birth, and Hb H in adults.

Hb Bart's and hydrops fetalis are observed frequently in South Asia and less frequently in the Mediterranean population. It presents a severe obstetric problem. Frequent complications include toxemia, obstructed labor, and postpartum hemorrhages. Stillbirth occurs most frequently. Infants who survive die in a short time. They are anemic, the erythrocytes are characterized by hypochromia, poikilocytosis, reticulocytosis, and numerous nucleated red cell precursors. About 80% of the hemoglobin belongs to Hb Bart's and 20% to Hb Portland ($\zeta_2\gamma_2$), without Hb F or Hb A.

Hb H disease (hemoglobin composed of four β-globin-chains) demonstrates various degrees of anemia and splenomegaly. Individuals with Hb H disease demonstrate a mild hypochromic anemia which very easily passes during infections to acute hemolytic anemia. There exists an increased reticulocytosis. Incubation of the red cells with brilliant cresyl blue demonstrates many inclusion bodies because the redox potential of the dye precipitates

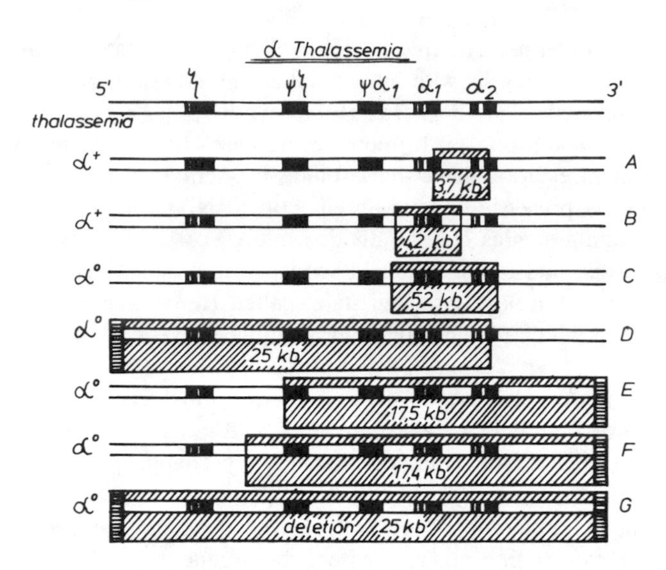

FIGURE 4. Deletion forms α-globin-chain the thalassemias. (First line): cluster of α-globin genes, A, B—common α^+ thalassemias, active one α-globin gene; C—rare Mediterranean thalassemia, synthesis of Hb Portland ($\zeta_2\gamma_2$); D—common thalassemia in south-east Asia, synthesis of Hb Portland; E, F, G various Mediterranean thalassemias.

Hb H. In splenectomized patients most of the erythrocytes incubated with methyl violet demonstrate large inclusion bodies. The hemoglobin pattern reveals Hb A and Hb H form 5 to 40% or more, Hb A_2 1.5 to 2% and variable amounts of Hb Bart's.

The deletion forms of α thalassemias are presented in Figure 4. Shown are two common α^+ thalassemias, resulting from the loss of 3.7 and 4.2 kb of the α gene cluster, leaving one α-globin gene intact and functional.[21] In Jamaica, about 35% of the entire population reveals α 3.7 deletion type thalassemia. The next two α^0 thalassemias are a reminder of the hereditary persistence of fetal hemoglobin (HPFH) because in these two cases the embryonic hemoglobin ($\zeta_2\gamma_2$) is synthesized (compare section V. devoted to HPFH). In both cases the deletion of 5.2 and 17.5 kb deprives both α-globin genes which results in the synthesis of the embryonic hemoglobin Hb Portland. In the next two α^0 thalassemias the deletion deprives the whole α-globin gene cluster which is incapable of synthesis of α-globin-chains, resulting in early death in homozygotes.

1. α-Globin-Chain Deletions Associated with α-Chain Structural Mutants

This kind of thalassemia is characterized by reduced production of α-globin-chains. There exists a remarkable heterogeneity of these thalassemias. To the structural hemoglobin mutants existing in linkage with α-globin gene deletions belong: Hb Evanston (α14,Trp → Arg), characterized by instability of the α-globin-chain, resulting in α thalassemia and α thalassemia-like phenotype,[7] Hb Hasharon (α47,Asp → His), resulting in mild α-globin-chain instability, α thalassemia deletion with thalassemia-like phenotype; Hb G Philadelphia (α68,Asn → Lys), Hb Mahidol (α74,Asp → His), and Hb J. Tongariki (α115,Ala → Asp) exhibit α thalassemia deletion resulting in thalassemia-like phenotype; Hb Nigeria (α81,Ser → Cys) α thalassemia gene linkage resulting in thalassemia-like phenotype; Hb J Capetown (α92,Arg → Gln) α thalassemia gene linkage, erythrocytosis.

To the unstable α-globin-chain variants belong:[7] Hb Fort Worth (α27,Glu → Gly), Hb Ann Arbor (α80,Leu → Arg), Hb Suan Dok (α109,Leu → Arg), Hb Petah Tikva (α110,Ala → Asp), and Hb Quong Sze (α125,Leu → Pro).

Extended α-globin-chain variants (resulting from mutations of the translation termination codon). To this group belong: Hb Constant Spring (α term → Gln), Hb Icaria (α term → Lys), Hb Koya Dora (α term → Ser), and Hb Seal Rock (α term → → Glu).[7]

The above described α-globin variants might have arisen either by point mutations in existing α thalassemia chromosomes, or through genetic recombination between a chromosome carrying the mutant allele with a normal deleted chromosome.

2. Non-Deletion α Thalassemias

This group of α thalassemias may result from different abnormal mechanisms: defective transcription including initiation and termination of transcription, nonsense mutations, splice site mutations, unstable α-globin-chain, etc.

An α^+ thalassemia has been described with deletion of two nucleotides (-2 and -3) preceding the AUG codon. The mechanism of decreased α-globin-chain synthesis is probably caused by deficiency of translation of an α^+ mRNA thalassemic gene due to deletion of the two preceding nucleotides at the initiation site of translation.[22]

α Thalassemia caused by a polyadenylation signal mutation at the 3′ flanking region AATAAA → AATAAG of α-2 globin gene resulting in an α-2 globin gene defective expression. The α-1 globin gene in this case was completely inactive because of frameshift mutation at codon 14. This kind of an α thalassemia was described in a Saudi Arabia population.[23]

Mutation of the initiation codon (αATG → ACG) in Sardinian population causes unstable α-globin-chain resulting in α thalassemia.

B. β THALASSEMIAS

The best known β thalassemias are characterized by deficient β-globin-chain synthesis in erythroid cells and structurally abnormal β-globin-chains.[24,25] Control of the globin gene expression occurs principally at the level of transcription. Critical for proper and efficient transcription are the three upstream regulating regions of DNA, most proximal to the initiation site (-28 to -30 bp) the TATA box. This region probably serves to localize the precise site of the transcription initiation site (the so-called CAP site), influencing simultaneously the efficiency of transcription. The next control region CCAAT is localized about -80 bp upstream of the initiation site. This is probably the region where RNA polymerase II will be bound with the DNA strand. Mutation of this site also impedes the transcription process. Finally, a distant regulatory element, PuCPuCCC (Pu = purine), is located about -100 and more bp upstream of the initiation site of transcription. Its mutation reduces the amount of β-globin mRNA at twofold or more. This region is repeated once upstream of the β-globin gene, but not upstream of the poorly expressed δ-globin gene.[24] The distant regulatory region resembles those of herpes virus thymidine kinase and the SV 40 early region, necessary for their optimal transcription.

Enhancer elements have been described independently of the upstream regulating regions in several eukaryotic genes. Enhancer usually denotes sequences that increase the efficiency of transcription of a linked gene, independently in relation to the position of the gene (upstream many hundreds or even thousands bp to the initiation site, within the gene sequences—usually within the introns, and even downstream of the termination of transcription. The best known are immunoglobulin gene enhancers which amplify the gene transcription. Enhancer sequences in the case of globin genes have not been proved, although there is a distinct difference between the α- and β-globin genes. Transcription of the β-globin gene in heterologous cells, for example, can be achieved by viral gene products (e.g., E1A adenovirus gene product supplied in "*trans*" makes the presence of TATA box superfluous for β-globin-chain transcription. Contrary to this, transcription of α-globin gene in heterologous cells (e.g., HeLa cells) occurs without the viral enhancers.

The next problem in globin genes transcription is processing of the primary transcript containing exons that are translated, and introns that are not translated. The introns must be precisely excised for obtaining final mRNAs. There are signals at the primary transcript that enable correct excision of introns. Splice junction sites near 5′ or donor sites reveal the dinucleotide GT, whereas 3′ or acceptor site reveals AG. Besides, donor sites GC have been also described. The donor sites are often surrounded by (C or A) AGGT, (G or A) AGT, and the acceptor site is often preceded by a stretch of pyrimidines. Finally, the junctions reveal complementarity with U1 RNA sequences, i.e., small nuclear RNAs that participate in correct RNA splicing. For correct splicing local directionality and conformation of the precursor mRNA also may be important. Preliminary data suggest that the initial event in splicing is formation of a spliceosome,[26] composed of RNAs juxtaposed for the second step in splicing, and snRNP complexed with several snRNAs U1 to U6 which enables endonucleolytic cleavage of RNA at the proper donor site.

1. Simple β Thalassemia

Some years ago the qualification of α and β thalassemias was restricted to the entire α- or β-globin genes. The development of modern diagnostic methods on the molecular level (molecular cloning, DNA sequencing, functional analysis of mutant genes) has enabled researchers to distinguish about 30 different mutations leading to β thalassemias. In general, all these mutations include single base replacements, small deletions or insertions within or upstream of the β-globin gene sequences.

2. β Thalassemias Caused by Transcription Mutants

Transcription mutants include base replacements at the promoter region (at the distal element and TATA box), nonsense mutations with premature termination of transcription that yields incomplete peptide chains, frameshift mutations that make it impossible to synthesize a normal peptide chain because from the point of this mutation all amino acid residues until termination are inappropriately incorporated into the peptide chain, defective splicing (at the donor respectively acceptor sites, generating new acceptor or donor splice sites, exposing cryptic splice sites), and 3′ end processing disturbances.

β thalassemias caused by mutations within upstream regulatory elements. Six β^+ thalassemia gene mutations with single base substitutions within the transcription initiation region have been described in various positions upstream of the initiation (cap) site. Two of them affect the distant regulatory element position −88 CACCC → CATCC in Afro-American populations, and position −87 ACCCT → ACGCT in Italians. In both cases the expression of β-globin genes is reduced to 20 and 30% of the norm. The mild nature of these mutations is probably caused by duplication of normal distant regulatory element within position −80 to −100 which remains intact. (The mutated bases are underlined.)

Three mutations within the upstream regulatory region affect the TATA box. One at position −29 CATAA → CATGA in Afro-American populations, and two at position −28 CATAAA → CATACA in Kurdish-Jewish populations, and CATAAA → CATAGA in Chinese populations. Clinical symptoms in these β thalassemia patients are more severe than that in the first group, but milder than that in the Mediterranean population. They synthesize only about 20 to 30% of β-globin-chains as compared to the normal population. Blood transfusions are only rarely required. The TATA box mutants cause diminution of the amount of the synthesized β-globin-chains. Initiation of transcription usually occurs at the normal cap site.

The remaining mutation in the preceding cap sequences does not concern the typical upstream regulating elements (TATA box and distant regulatory element), but nevertheless it impedes the normal β-globin gene transcription. There is mutation at 5′ position −31 CATAA → CGTAA in Japanese populations. This mutation was the first in which the regulatory role of the 5′ upstream gene sequences was recognized as controlling efficiency of the gene's transcription. What is more, the case of β thalassemia caused by mutation at position −31, beyond the typical upstream regulating elements, suggests that the other upstream sequences of genes may be also important in regulation of their expression.

3. β^0 Thalassemias Caused by Nonsense Mutations

This mutation leads to premature termination of β-globin mRNA translation which results in absence of normal β-globin chains. Here are: β15, TGG → TAG in Indian populations; β17, AAG → TAG in Chinese populations and β39, CAG → TAG in Mediterranean populations.

4. β^0 Thalassemias Caused by Frameshift Mutations

These mutations result from deletion or insertion of one or more nucleotides that alters all coding triplets beginning from the mutated one until the end of mRNA. To these mutations belong:

- β6, GAG → → GG (deletion of A) in Mediterranean populations;
- β8, AAG → G (deletion of AA) in Turkish populations;
- β8-9, AAG → AAGG (insertion of G) in Indian populations;
- β16, GGC → GG (deletion of C) in Indian populations;
- β41-42, TTCT deleted in Chinese and Indian populations;
- β44,TCT → → TT (deletion of C) in Kurdish-Jewish populations;
- β71-72, TTT → TTTA (insertion of A) in Chinese populations.

Insertion of AC after codon 146 in β-globin gene causes frameshift mutation with thalassemia-like symptoms (without β^0 thalassemia) resulting in high oxygen affinity of this hemoglobin Tak, occurring in Southeast Asian populations.

5. β Thalassemias Caused by Splicing Disturbances

Nucleotide replacements or deletions either at the donor or acceptor site results in inappropriate splicing which yields altered mRNAs and consequently there is usually absence of normal β-globin-chains.

Changes at the donor site of the first intervening sequence: position 143 (intervening sequence—IVS-1 nucleotide n.1) GT → AT, causes a splicing defect resulting in β^0 thalassemia in Mediterranean populations; position 143 (IVS-1 n.1) G → T, causes a splice junction defect resulting in β^0 thalassemia in Indian populations; position 147 (IVS-1 n.5) G → C, causes a splicing defect resulting in β^+ thalassemia in Indian and Chinese populations; position 147 (IVS-1 n.5) G → T, causes a splicing defect resulting in β^0 thalassemia in Mediterranean populations; position 148 (IVS-1 n.6) T → C, causes a splicing defect resulting in β^+ thalassemia in Mediterranean populations.

Donor site at the second intervening sequence position 496 (IVS-2 n.1) G → A, causes a splice junction defect resulting in β^0 thalassemia in Mediterranean populations.

Acceptor site of the first intervening sequence: position 252, replacement G → A causes a splicing defect resulting in β^+ thalassemia in Mediterranean populations.

Acceptor site of the second intervening sequence: position 1344, replacement A → G causes an acceptor splice site mutation by generation of an abnormal splicing site resulting in β^0 thalassemia in Afro-American populations.

Generation of new abnormal donor splicing sites. Position 1149 (IVS-2) replacement C → T creates an abnormal splicing site resulting in β^0 thalassemia in Chinese populations; position 1200 (IVS-2) replacement T → G creates an abnormal splicing site resulting in β^+ thalassemia in Indian populations; position 1240 (IVS-2) replacement C → G creates an abnormal splicing site resulting in β^+ thalassemia in Mediterranean populations.

a. Enhanced Cryptic Splicing Sites

Silent substitution β24, GGT → GGA causes an anomalous splice site resulting in β^+ thalassemia in Afro-American populations; Hb E, β26, GAG → AAG (Glu → Lys) causes an abnormal splice site resulting in unstable hemoglobin with microcytosis in heterozygotes and microcytic anemia in homozygotes of African populations; Hb Knossos, β27,GCC → TCC (Ala → Ser) causes an anomalous splice site resulting in β^+ thalassemia in Greek populations.

b. Disturbances of 3′ End Processing

Sequence change at the 3′ end AATAAA → AACAAA causes disturbances in RNA processing resulting in β^+ thalassemia in black populations.[24]

Non-deletion forms of β thalassemias are represented in Table 1. As seen in this table numerous kinds of mutations affecting the transcription process can lead to thalassemias. The range of clinical manifestations is very large, from very mild forms practically without clinical symptoms to very severe forms causing death in early life. Most of the mild forms usually belong to the β^+ thalassemias with diminished amount of β-globin mRNA, whereas most of the severe thalassemias are β^0 thalassemia forms without β-globin mRNA.

This review of nondeletion forms of thalassemias has called attention to the role of intervening sequences in correct transcription. Replacements of single nucleotides, either at the donor or acceptor sites, may cause severe disturbances in normal splicing and in consequence absence of normal β-globin mRNA. These observations were precisely what indicated the obligatory presence of these sequences in normal processing of mRNA. Particularly important are the donor and acceptor sites. Therefore every replacement of these nucleotides causes inactivation of the splicing sites necessary for production of normal mRNAs. The study of β thalassemias revealed the importance for splicing of not only sequences directly engaged in the splicing process, but also the neighboring sequences and even sequences within the introns. Several β thalassemia genes exhibit nucleotide replacements within introns which exert deleterious effects on splicing, generating new donor splice sites within introns that compete with or retard normal processing,[24] e.g., premature termination of translation because the additional splice site within the intron removes a part of the mRNA.

Single base substitutions may lead to activation of cryptic donor splice sites. Usually only a fraction of transcripts are processed from the mutated cryptic donor site. Such a minor use of the mutated pathway is seen in the case of Hb E or Hb Knossos genes (with substitutions at positions 26 and 27, respectively). These mutations of the intervening sequences (IVS-1) cryptic donor sites demonstrate how sequence changes in coding sequences rather than in intervening sequences influence RNA processing. Therefore, processing of mRNA must be seen as a dynamic process with many different correct splicing factors.

Mutations may occur also at the 3′ end of eukaryotic mRNAs within the sequence of nucleotides AAUAAA which serves both as an endonucleolytic and polyadenylation signal. Mutation of AATAAA into AATAAG in American blacks caused a 900 nucleotides elongation of the transcript,[24] until a new AATAAA sequence was present on downstream 3′ flanking segment with subsequent polyadenylation. This result suggests that the sequence AATAAA at the 3′ end of eukaryotic genes is rather an endonucleolytic signal than a polyadenylation signal.

C. DELETION FORMS OF β THALASSEMIAS

Numerous β thalassemias caused by various deletions of the β globin gene cluster have been described and are presented in Figure 5. The deletions may comprise solely β-, δ- or γ-globin genes and also the whole β-globin gene cluster from ε to β genes. A very interesting paradoxical case was described (in a Dutch man): although the β-globin gene was intact (the deletion terminated 2.5 kb upstream of the β-globin gene) expression of this gene was absent.[24] This gene, cloned, revealed normal expression. It is supposed that abnormal DNA sequences brought into the vicinity of the β-globin gene inhibited its normal expression.

The different mutations of the β-globin gene cluster demonstrate various molecular

TABLE 1
Disorders Caused by Particular Amino Acid Substitutions (Mutations) of the β-Globin-Chain

Distal element	TATA box	Position 5	Position 6	Position 8	Position 9	Position 14
β^+ Thal	**β^+ Thal**	**Hb slighlty**	**Sickle cell disease**	**β^0 Thal**	**Incr O_2 aff**	**Hb unstable**
−88 C → T	−31 A → G	**unstable**	6 Glu → Val	Turkish	Porto Alegre	Sögn
−87 C → G	−29 A → G	Hb Warwickshire	C Ziguinchor	frameshift	Ser → Cys	Leu → Arg
	−28 A → C	Pro → Arg	6 Glu → Val	AAG → G (del)		Saki
	−28 A → G		58 Pro → Arg	Indian frameshift		Leu → Pro
			C Harlem	AAG → AAGG		
			6 Glu → Val	(ins)		
			73 Asp → Asn			
			S Travis			
			6 Glu → Val			
			142 Ala → Val			
			hom: m hemolysis (intracellular crystallization)			
			Hb C			
			C Glu → Lys			
			m hemolysis			
			Hb unstable			
			Leiden			
			Glu deleted			
			β^0 Thal			
			frameshift			
			Mediterranean			

TABLE 1 (continued)
Disorders Caused by Particular Amino Acid Substitutions (Mutations) of the β-Globin-Chain

Position 15	Position 16	Position 17	Positions 17-18	Position 20	Position 21	Position 23	Position 24	Position 26
Hb unstable Belfast Trp → Arg	**β^0 Thal** Indian frameshift GGC → GG (del)	**β^0 Thal** Chinese n mut Lys → Term	**m anemia** **unstable, h O_2 aff** Lyon six base del	**erythrocytosis** **h O_2 aff** Olympia Val → Met	**m anemia** **decr O_2 aff** Connecticut Asp → Gly	**erythrocytosis** **unstable** **incr O_2 aff** Strasburg Val → Asp Palmerston North Val → Phe	**hemolytic anemia** **unstable** Savannah Gly → Val Riverdale Bronx Gly → Arg Moscva Gly → Asp	**microcytosis** **spl j def** **unstable** Hb E Glu → Lys
β^0 Thal Indian nonsense mutation Trp → Term						**hemolytic anemia** **incr O_2 aff** Freiburg Val del	**β^+ Thal** Gly → Gly	**microcytic anemia** **unstable** Henri Mondor Glu → Val
						unstable **incr O_2 aff** Miyashiro Val → Gly		

Position 27	Position 28	Position 29	Position 30	IVs-1	Position 31	Position 32	Position 34	Position 35
hemolytic anemia **unstable** Volga Ala → Asp	**hemolytic anemia** **unstable** Genova Leu → Pro	**hemolytic anemia** **unstable** Lufkin Gly → Asp	**unstable** Tacoma Arg → Ser	**β^0 Thal** **spl j def** Mediterranean pos 143 G → A Indian pos 143 G → T Mediterranean pos 147 G → T	**hemolytic anemia** **unstable** Yokohama Leu → Pro	**hemolytic anemia** **unstable** Abraham Lincoln Perth Leu → Pro	**erythrocytosis** **incr O_2 aff** Pitie-Salpetriere Val → Phe $\alpha_1\beta_2$ contact	**hemolytic anemia** **unstable** **incr O_2 aff** Philly Try → Phe
β^+ Thal Knossos Ala → Ser spl j def	**hemolytic anemia** **cyanosis meth** **incr O_2 aff** **unstable** St Louis Leu → Gln			**β^+ Thal** **spl j def** Indian Chinese pos 147 G → C Mediterranean pos 148 T → C pos 252 G → A		**hemolytic anemia** **unstable** **decr O_2 aff** Castilla Leu → Arg		

TABLE 1 (continued)
Disorders Caused by Particular Amino Acid Substitutions (Mutations) of the β-Globin-Chain

Position 36	Position 37	Position 38	Position 39	Position 40	Position 41	Positions 41-42	Position 42	Position 44
Erythrocytosis s unstable **incr O_2 aff** Linköping Pro → Thr	**Incr O_2 aff** Hirose Tryp → Ser **decr O_2 aff** Rotschild Trp → Arg	**hemolytic anemia** **decr O_2 aff** Hazebrouk Thr → Pro	**unstable** Vaasa Gln → Glu **$β^0$ Thal** Mediterranean n mut Gln → Term	**incr O_2 aff** Athens-Georgia Arg → Lys $α_1β_2$ contact Austin Arg → Ser	**hemolysis drug related (?)** **unstable** Mequon Phe → Tyr	**$β^0$ Thal** **frameshift term codon at pos 59** Chinese Indian TTCT- 4 base del	**hemolytic anemia** **unstable** **heme contact** **decr O_2 aff** Hammersmith Phe → Ser **heme contact** Louisville Bucuresti Phe → Leu	**hemolytic anemia** **unstable** **decr O_2 aff** Niteroi Phe-Glu-Ser deleted **$β^0$ Thal** **frameshift term codon at pos 60** Kurdish-Jewish TCT-TT del C

Position 45	Position 48	Position 51	Position 56	Position 57	Position 60	Position 61	Position 62	Position 63
hemolytic anemia **unstable** **decr O_2 aff** Cheverly Phe → Ser	**hemolytic anemia** **unstable** **decr O_2 aff** Okaloosa Leu → Arg	**unstable** Willamette Pro → Arg	**unstable** **hemolytic anemia** Tochigi Gly-Asn-Pro-Lys deleted	**unstable** G Ferrara Asn → Lys	**unstable** Collingwood Val → Ala	**decr O_2 aff** Bologna Lys → Met	**hemolytic anemia** **unstable** **incr O_2 aff** Duarte Ala → Pro	**hemolytic anemia** **cyanosis** **meth unstable** (distal His Substit.) M Saskatoon His → Tyr **hemolytic anemia** **unstable** **incr O_2 aff** Zürich His → Arg Bicêtre His → Pro

TABLE 1 (continued)
Disorders Caused by Particular Amino Acid Substitutions (Mutations) of the β-Globin-Chain

Position 64	Position 66	Position 67	Position 68	Position 70	Position 71	Position 73	Position 74	Positions 74-75
unstable **incr O_2 aff** J Calabria Gly → Asp	**hemolytic anemia** **unstable** I-Toulouse Lys → Glu	**cyanosis meth** **decr O_2 aff** **unstable** M. Milwaukee Val → Glu heme contact **hemolytic anemia** **unstable** Bristol Val → Asp Sydney Val → Ala	**hemolytic anemia** **unstable** Mizuko Leu → Pro **erythrocytosis** **h O_2 aff** Brisbane Great Lakes Leu → His	**hemolytic anemia** **unstable** **decr O_2 aff** heme contact Ala → Asp	**hemolytic anemia** **unstable** heme contact Christchurch Phe → Ser **$β^0$ Thal** Chinese frameshift TTT-TTTA ins A	**decr O_2 aff** Mobile Asp → Val Vancouver Asp → Tyr	**hemolytic anemia** **unstable** Bushwick Gly → Val Shepherds Bush Gly → Asp	**hemolytic anemia** **unstable** Atlanta Leu → Pro **unstable** **incr O_2 aff** Pasadena Leu → Arg **hemolytic anemia** **unstable** St Antoine Gly-Leu deleted **$β^+$ Thal-like** Vicksburg Leu del

Position 79	Position 81	Position 82	Position 83	Position 85	Position 87	Position 88	Position 89	Position 91
incr O_2 aff G Hsi-Tsou Asp → Gly	**chronic hemolysis** **unstable** **incr O_2 aff** Baylor Leu → Arg	**erythrocytosis** **incr O_2 aff** 2,3-DPG binding site Rahere Lys → Thr Helsinki Lys → Met	**s unstable** Ta-Li Gly → Cys	**hemolytic anemia** **unstable** **incr O_2 aff** Buenos Aires Phe → Ser	**hemolytic anemia** **unstable** **incr O_2 aff** Tours Thr deleted heme loss	**hemolytic anemia** **unstable** **heme contact** Santa Ana Leu → Pro Boras Leu → Arg	**erythrocytosis** **h O_2 aff** Creteil Ser → Asn Vanderbil Ser → Arg	**hemolytic anemia** **unstable** heme contact site Sabine Leu → Pro Caribbean Leu → Arg

TABLE 1 (continued)
Disorders Caused by Particular Amino Acid Substitutions (Mutations) of the β-Globin-Chain

Positions 91-95	Position 92	Position 93	Position 94	Position 97	Position 98	Position 99	Position 100	Position 101
hemolytic anemia **unstable** Gun-Hill Leu-His-Cys-Asp-Lys del	**cyanosis** **meth** prox His substit M Hyde Park His → Tyr	**unstable** **incr O_2 aff** Okazaki Cys → Arg	**erythrocytosis** **incr O_2 aff** Barcelona Asp → His	**erythrocytosis** **h O_2 aff** Malmö His → Gln Wood His → Leu	**hemolytic anemia** **unstable** **incr O_2 aff** Köln Val → Met Nottingham Val → Gly Djelfa Val → Ala heme contact site	**erythrocytosis** **incr O_2 aff** $\alpha_1\beta_2$ contact Ypsilanti Asp → Tyr Yakima Asp → His Kempsey Asp → Asn Radcliffe Asp → Ala Hôtel Dieu Asp → Gly Chemilly Asp → Val	**erythrocytosis** **incr O_2 aff** Brigham Pro → Leu	**hemolytic anemia** **unstable** Rush Glue → Gln
	hemolytic anemia **unstable** Istanbul His → Gln St. Etienne Newcastle His → Pro		**incr O_2 aff** Bunbury Asp → Asn					**incr O_2 aff** British Columbia Glu → Lys
	hemolytic anemia **unstable** **incr O_2 aff** Mozhaisk His → Arg							**erythrocytosis** **incr O_2 aff** Alberta Glu → Gly Potomac Glu → Asp

Position 102	Position 103	IVS-2	Position 106	Position 107	Position 108	Position 109	Position 111	Position 112
cyanosis	**erthrocytosis**	**β^0 Thal**	**hemolytic**	**hemolytic**	**hemolysis**	**erythrocytosis**	**hemolytic**	**Thal-like**
low O_2 aff	**incr O_2 aff**	**spl j def**	**anemia**	**anemia**	**decr O_2 aff**	**h O_2 aff**	**anemia**	**unstable**
$\alpha_1\beta_2$	Heathrow	pos 496	**unstable**	**unstable**	Yoshizuka	San Diego	**unstable**	Indianapolis
contact	Phe → Leu	G → A	**incr O_2 aff**	Burke	Asn → Asp	Val → Met	Peterborough	Cys → Arg
Kansas		Mediterranean	heme contact	Gly → Arg			Val → Phe	
Asn → Thr		pos 1149	site		decr O_2 aff			
Beth Israel		C → T	Southhampton		Presbyterian			
Asn → Ser		Chinese	Casper		Asn → Lys			
		pos 1344	Leu → Pro					
low O_2 aff		A → G						
St. Mandé		Afro-American	**cyanosis**					
Asn → Tyr			**hemolysis**					
		β^+ Thal	**unstable**					
		pos 1200	**meth**					
		T → G	Tübingen					
		Italian	Leu → Gln					
		pos 1240						
		C → G						
		Mediterranean						
		acceptor						
		splice site						

TABLE 1 (continued)
Disorders Caused by Particular Amino Acid Substitutions (Mutations) of the β-Globin-Chain

Position 113	Position 115	Position 117	Position 120	Position 124	Position 126	Position 127	Position 128	Position 129
unstable New York Val → Glu	**hemolytic anemia** **unstable** Madrid Ala → Pro	**hemolytic anemia** **unstable** Saitama His → Pro	**unstable** Jianghua Lys → Ile	**erythrocytosis** **incr O_2 aff** Ty Gard Pro → Gln	**unstable** Hofu Val → Glu	**heat super-stable** Hacettepe Motown Gln → Glu	**hemolytic anemia** **unstable** J Guantanamo Ala → Asp	**unstable** **incr O_2 aff** Crete Ala → Pro

Position 130	Position 131	Position 132	Position 134	Position 135	Position 136	Position 138	Position 140	Position 141
hemolytic anemia **unstable** Wien Tyr → Asp	**unstable** **hemolytic anemia** Shelby Gln → Lys	**Thal-like** K Woolwich Lys → Gln	**microcytic anemia** **unstable** North Shore Caracas Val → Glu	**hemolytic anemia** **incr O_2 aff** **unstable** Altdorf Ala → Pro	**decr O_2 aff** Hope Gly → Asp	**unstable** Brockton Ala → Pro	**erythrocytosis** **incr O_2 aff** St Jacques Ala → Thr	**hemolytic anemia** **unstable** Olmsted Leu → Arg Coventry Leu deleted

Position 142	Position 143	Position 144	Position 145	Position 146	3′ flanking region
erythrocytosis **h O_2 aff** Ohio Ala → Asp	**erythrocytosis** **h O_2 aff** Little Rock His → Gln Syracuse His → Pro	**erythrocytosis** **h O_2 aff** Andrew Minneapolis Lys → Asn	**erythrocytosis** **h O_2 aff** Rainier Tyr → Cys alkali resistant Bethesda Tyr → His Osler Fort Gordon Nancy Tyr → Asp	**erythrocytosis** **h O_2 aff** Hiroshima His → Asp York His → Pro Cowtown His → Leu	**β^+ Thal** pos 1584 T → C AATAAA → AACAA Poly A signal mutation
unstable **hemolytic** **anemia** **h O_2 aff** Toyoake Ala → Pro	**hemolytic** **anemia** **unstable** **h O_2 aff** Saverne frameshift CAC-CC del A chain extended by 10 residues		**Chronic** **hemolysis** **unstable** chain extended by 11 residues Cranston frameshift AG ins after codon 144	**Thal-like** **h O_2 aff** Tak frameshift AC inserted after 146 chain extended by 11 residues	
			erythrocytosis **h O_2 aff** Mc Kees Rocks Tyr → Term		

Note: Thal—thalassemia; del—deletion; ins—insertion; incr O_2 aff—increased O_2 affinity; decr O_2 aff—decreased O_2 affinity; hom—homozygote; het—heterozygote; m—mild; s—slight; h—high; meth—methemoglobin; spl j def—splicing (junction) defect; n mut—nonsense mutation; IVS-1,IVS-2 intervening (intron) sequences; pos—position; substit—substitution; prox—proximal.

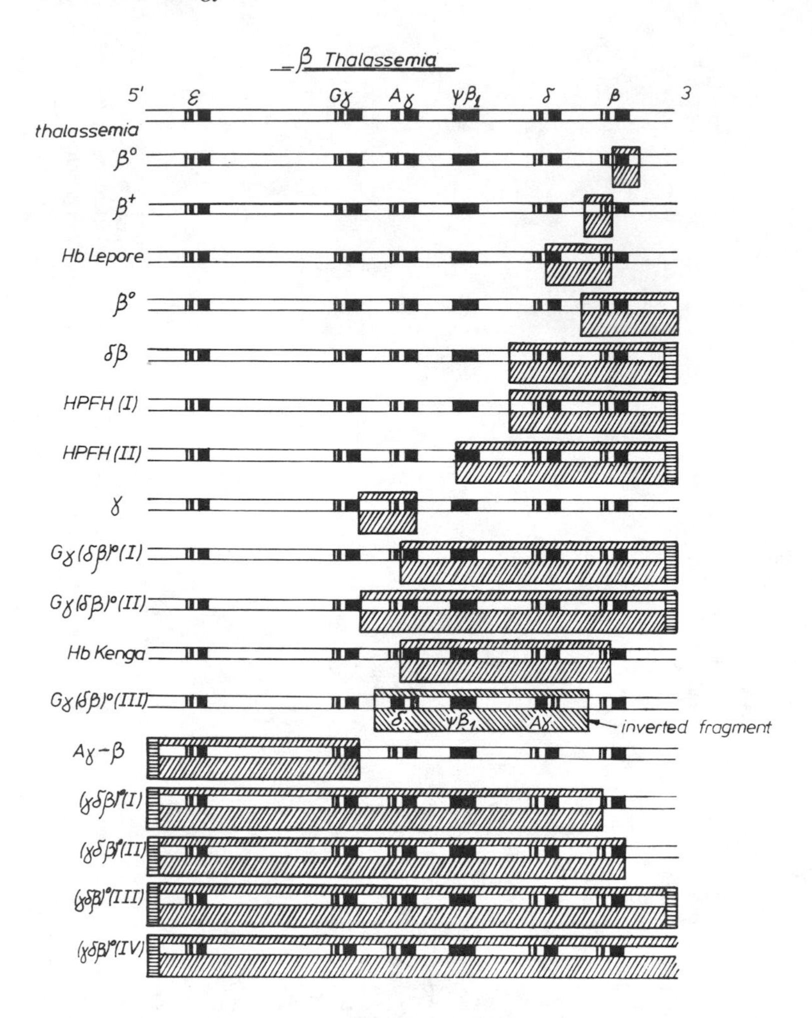

FIGURE 5. Various types of β thalassemia.

abnormalities producing clinically indistinguishable syndromes. Usually, β thalassemias demonstrate the presence of small, underhemoglobinized erythrocytes (microcytes) with various degrees of anemia. A compensatory increase of Hb F and Hb A_2 is observed. In general, the β^+ thalassemias demonstrate milder clinical symptoms, but this is not a rule, because known are also β^+ thalassemias with severe clinical manifestations. In the silent carrier type of β^+ thalassemias (heterozygotes of mild forms) the clinical findings are normal, without increase of Hb F or Hb A_2 levels. The mild hematologic changes constitute the syndrome of thalassemia minor. Heterozygotes may also show the more severe syndrome of thalassemia intermedia with jaundice, microcytic anemia and splenomegaly. Homozygous β^0 thalassemias and the severe forms of β^+ thalassemias demonstrate clinical symptoms of Cooley's anemia with severe microcytic anemia (erythrocytes are more resistant to hypotonic solutions) with hepato-splenomegaly and bone deformities, particularly that of the skull because of expanded bone marrow volume caused by inefficient erythropoesis. Most of the pathological symptoms can be alleviated by intensive transfusion therapy. However, the therapy often results in massive pathological accumulation of iron in the tissues.

δβ thalassemias reveal clinical symptoms similar to those of mild to moderately severe forms of β thalassemia. The most important difference concerns hemoglobin composition. In heterozygotes Hb A is decreased or absent, the level of Hb F is increased and the level of Hb A_2 is slightly decreased. In homozygotes Hb A and Hb A_2 are absent, and Hb F is the sole hemoglobin in these patients.

1. $\gamma\delta\beta$ Thalassemias

This disorder is fully expressed at birth. In heterozygous newborns there is hemolytic anemia with microcytosis of erythrocytes. In heterozygous adults there is a mild microcytic anemia. Homozygotes are probably incompatible, for even intrauterine survival.

2. δ Thalassemias

Because of the small amount of δ-globin-chains in normal individuals, further decrease or even absence of δ-globin-chains does not play a role. Therefore, no clinical manifestations either in δ^+ or δ^0 thalassemias are present. In heterozygotes the amount of Hb A_2 is decreased, and in homozygotes Hb A_2 is absent.

D. SYNDROME OF α/β THALASSEMIAS

Because of frequent occurrence of both α and β thalassemias in the same region, various combinations of these two thalassemias may occur in the population. Clinical manifestations in these cases depend on either α or β thalassemia defects—whichever is more severe. Usually there predominates a hypochromic, microcytic anemia.

Sometimes the presence of both α and β thalassemias in the same individual has a beneficial effect on the clinical picture of these patients. This results from the pathogenesis of clinical manifestations in thalassemias, where the decreased amount of the affected chain does not play so much of the main pathological role, as does the surplus of the normal chain which uncombined results in intramedullary precipitation causing ineffective erythropoesis with intramedullary breakdown of hemoglobin and defective maturation of erythrocytes.

E. STRUCTURAL HEMOGLOBIN VARIANTS CAUSING MANIFESTATIONS SIMILAR TO β THALASSEMIAS

A typical example of a variant with linkage to β^+ thalassemia mutation is represented by Hb Vicksburg (β75,Leu = 0). Because of the linkage, the thalassemia gene is expressed together with the Vicksburg variant. To this group belong also δ/β fusion variants: Hb Lepore Hollandia, Hb Lepore Baltimore, and Hb Lepore Boston (discussed earlier).

Further group variants with abnormal splice junctions are represented by: Hb E (β26, Glu → Lys), Hb Henri Mondor (β26, Glu → Val), and Hb Knossos (β27, Ala → Ser). The most interesting and most frequent is Hb E with an unstable β-globin-chain associated with splicing and biosynthetic defects, which altogether contribute to the thalassemia-like expression.

Highly unstable β-globin-chain variant in Hb Indianapolis (β112, Cys → Arg) causes severe clinical β thalassemia symptoms. The β-globin-chain of this variant degrades immediately after its synthesis.

Hb K Woolwich (β132, Lys → Gln), and Hb North Shore Caracas (β134, Val → Glu) cause thalassemia-like symptoms by an unknown molecular defect.

Interaction of β^+ thalassemia with abnormal variants of the β-globin-chain. A typical example is represented by β^+ thalassemia with concomitant Hb S. In this case Hb S synthesis increases to a ratio of 30:70 (the decrease of Hb A synthesis is caused by diminished synthesis of β-globin-chains), whereas in heterozygotes Hb A/Hb S ratio is 60:40. In combination with Hb S/β^0 thalassemia the only synthesized hemoglobin is Hb S, not distinguishable from homozygous Hb SS patients.

Another picture represents combination of Hb E/β^0 thalassemia with the synthesis of equal amounts of Hb E and Hb F. This result is explained by the β thalassemia-like expression of Hb E and compensatory increase of Hb F synthesis. Combination of α thalassemia with Hb S or Hb C exhibits diminished expression of Hb S or Hb C, respectively. Combination of Hb E/α thalassemia (particularly in severe forms of α thalassemia) causes diminution of Hb E synthesis to 50% of that synthesized in Hb A/Hb E patients. This is explained by a lower affinity of α globin subunits to β^S-globin subunits.

DNA translocation may inactivate the β-globin gene which may result in γβ thalassemia. It is supposed that location of a normally inactive locus in the vicinity of active gene may cause its inactivation.[27]

Noteworthy is the great accumulation of various kinds of thalassemias in geographical regions with malaria endemicity. Epidemiological studies provide strong support for the hypothesis that similar to the frequency of the sickle cell anemia/malaria relationship; the frequencies of α,β, and other kinds of thalassemias are also the result of malarial selection.[28]

VI. HEREDITARY PERSISTENCE OF FETAL HEMOGLOBIN (HPFH)

HPFH represents persistent synthesis of fetal hemoglobin in higher than normal levels in post-fetal life. HPFH syndrome occurs mainly in specific deletions of the β-globin gene cluster with normal size and hemoglobin content of erythrocytes, and usually with a normal balance between the synthesis of α and non-α-globin-chains. Clinical manifestations are mild or even absent. Among hematological features distinguishing various types of HPFH is the percentage of Hb F level and the relative presence of particular non-α-globin-chains. Genetically different gene deletions have been characterized in HPFH alleles and several point mutations at the 5′ promoter regions.[7] In several types of HPFH the molecular defect is unknown.

Most of the deletions in HPFH terminates in the $^{A}\gamma$-δ intergenic region. This leads to the suggestion that the δ gene preceding sequences play an inhibitory role in expression of γ-globin genes simultaneously with activation of the δ and β globin genes in the postnatal period of life (i.e., during switching from Hb F to Hb A). Critical deletion for HPFH production was concluded to be the removal of sequences between 4.1 to 2.5 bp upstream to the 5′ end of the δ globin gene. Alas, several types of HPFH do not exhibit deletion at this γ-δ-globin gene region.

In the $^{G}\gamma\beta^{+}$ type of HPFH a mutation C → G was discovered in position −202 bp 5′ to the $^{G}\gamma$ cap site, which creates a nucleotide sequence CGCC similar to the enhancer sequence CPuCCCC in the 5′ flanking region of the Herpes simplex or SV 40 virus. This leads to the suggestion that this mutation creates an "up-promoter" permanently activating $^{G}\gamma$-globin gene which results in persistent production of increased amounts of $^{G}\gamma$-globin-chains for production of this type of HPFH.

Similar replacements of single nucleotides in this region have been described in the Greek type of HPFH (substitution G → A at position −117 bp of the $^{A}\gamma$-globin gene);[30,31] and in Atlanta type of HPFH (substitution C → T at position −158 bp to the $^{G}\gamma$-globin gene) creating "up-promoters" for the increased synthesis of $^{G}\gamma$-globin-chains. The Greek type of HPFH exhibits the molecular canon $^{G}\gamma\downarrow\downarrow(^{A}\gamma\beta)^{+}$. Heterozygotes reveal 10 to 20% of Hb F synthesis. The mutant allele also may code for normal β-globin-chains. There are no hematologic abnormalities. Similar to this type of HPFH are the Atlanta type in blacks with molecular canon $^{A}\gamma^{0}(^{G}\gamma\beta)^{+}$ with lower Hb F production; the β^{+} African type of HPFH with nucleotide replacement C → G at position −202 bp to $^{G}\gamma$ gene and with molecular canon $^{A}\gamma^{0}(^{G}\gamma\beta)^{+}$, and the British type of HPFH with nucleotide replacement T → C at position −198 bp to $^{G}\gamma$ gene and with molecular canon $^{G}\gamma\downarrow\downarrow(^{A}\gamma\beta)^{+}$. In all these HPFH types the mutant allele is able to synthesize both δ and β-globin-chains.

The clinical picture of HPFH is very different, dependent on the type of mutation. The common African types of HPFH may be caused by different δ/β deletions not comprising the ψβ-globin gene, and the other comprising the ψβ-globin gene. The molecular canon is $(^{G}\gamma^{A}\gamma)^{+}(\delta\beta)^{0}$. Heterozygotes synthesize 20 to 30% of Hb F, homozygotes 100% of Hb F. Homozygotes exhibit mild thalassemia-like symptoms, usually hypochromic, microcytic anemia. Similar clinical symptoms also are seen in the Italian and Indian types of HPFH (comprising deletions of the ψβ-globin gene.

There are also other HPFH syndromes: the Swiss and Seattle types $(^{G}\gamma^{A}\gamma\beta)^{+}$ and the Chinese type $^{G}\gamma\downarrow\downarrow(^{A}\gamma\beta)^{+}$ with unknown molecular defect. Except for increased Hb F synthesis there are no serious clinical symptoms.

Similar to δβ thalassemias where fetal hemoglobin may be permanently synthesized, in α thalassemias embryonic ζ-globin-chains may be synthesized. Embryonic ζ-globin-chains were synthesized predominantly in Hb H disease (hemoglobin composed of four β-globin-chains in absence of α-globin-chains). The results suggest that the deletion of two α-globin genes on the same chromosome is accompanied by the continued expression of embryonic ζ-globin genes in adult individuals.[32]

VII. SOMATIC MUTATIONS OF GLOBIN GENES

Practically all the known globin gene mutants are inherited, i.e., they are present in germ cell lines. Therefore, the mutant alleles are present in all diploid cells. Somatic mutations occur in non-germinal (somatic) cell lines and abnormalities caused by these mutations appear only in limited, terminally differentiated cells. Somatic mutations are of interest as tests for mutagenic action of the environment. Globin genes are particularly useful for mutagenic testing because the normal globin-chains do not have isoleucine residues in their sequences. Therefore, the presence of isoleucine in globin-chains is the convincing record of the acquired somatic mutation. Such determinations demonstrated a linear increase of mutation frequency with age, and was significantly higher in Marshall Island inhabitants exposed to gamma radiation from nuclear fallout.[7]

Mutant forms of globin chains have been described in neoplastic blood diseases.[33]

VIII. GENERAL CONCLUSIONS

As seen from the presented material of globin mutations, all the known mechanisms of gene action may be involved causing disturbances of structure and function of the synthesized protein (in this case globin chains), independently of disturbances in gene expression. These disturbances cause various disorders: hemolytic anemias because of instability of the hemoglobin molecule and hemolytic crises with all pathological manifestations of hemolysis (e.g., sickle cell anemia in anoxemic conditions), erythrocytosis because of increased oxygen affinity of the abnormal hemoglobin molecule, decreased oxygen affinity of the hemoglobin molecule causing cyanotic disorders, various types of gene expression disturbances resulting in different thalassemias, etc. The severity of disorders caused by globin gene mutations also depend on the kind of mutations.

According to the genetic rule that all genes may undergo mutation at the same rate we must conclude that an enormous number and variety of diseases may arise because of mutations of the enormous number (probably some hundreds of thousands) of molecules present in humans. Wetherall and Clegg in their excellent paper write: ''Other human genetic disorders will probably show similar molecular diversity: there is no reason to believe that thalassemia is unique in this respect.''[21] I would like to extend this by adding that this pertains not only to thalassemias and other diseases caused by abnormal hemoglobin structure and synthesis, but also to all other protein molecules in humans.

REFERENCES

1. **Pauling, L., Itano, H., Singer, S. J., and Wells, I. C.,** Sickle cell anemia, a molecular disease, *Science,* 110, 543, 1949.
2. **Ingram, V. M.,** Gene mutations in human haemoglobin: the chemical difference between normal and sickle cell haemoglobin, *Nature (London),* 180, 326, 1957.
3. **Maxwell, C. R. and Rabinovitz, M.,** Evidence for an inhibitor in the control of globin synthesis by hemin in reticulocyte lysates, *Biochem. Biophys. Res. Commun.,* 35, 79, 1969.
4. **Farrel, P. J., Balkow, K., Hunt, T., Jackson, R. J., and Trachsel, H.,** Phosphorylation of initiation factor eIF-2 and the control of reticulocyte protein synthesis, *Cell,* 11, 187, 1977.
5. **Spritz, R. A., deRiel, J. K., Forget, B. G., and Weissman, S. M.,** Complete nucleotide sequence of the human δ-globin gene, *Cell,* 21, 639, 1980.
6. **Weatherall, D. J., Higgs, D. H., Wood, W. G., and Clegg, J. B.,** Genetic disorders of human haemoglobinas models for analysing gene regulation, *Philos. Trans. R. Soc. London, Ser. B.,* 307, 247, 1984.
7. **Honig, G. R. and Adams, III, J. G.,** *Human hemoglobin genetics,* Springer Verlag, New York, 1986.
8. **Pirastu, M., del Senno, L., Conconi, F., Vullo, C., Kan, Y. W., and Ferrara,** β^o thalassaemia caused by β^{39} nonsense mutation, *Nature (London),* 307, 76, 1984.
9. **Boissel, J. P., Kasper, Th. J., Shah, S. C., Malone, J. I., and Bunn, H. F.,** Amino-terminal processing of proteins: hemoglobin South Florida, a variant with retention of initiator methionine and Nα-acetylation, *Proc. Natl. Acad. Sci. U.S.A.,* 82, 8448, 1985.
10. **Perutz, M. F., Muirhead, H., Cox, J. M., and Goaman, L. C. G.,** Three-dimensional Fourier synthesis of horse oxyhaemoglobin at 2.8 Å resolution: the atomic model, *Nature (London),* 219, 131, 1968.
11. **Perutz, M. F.,** Stereochemistry of cooperative effects in haemoglobin, *Nature (London),* 228, 726, 1970.
12. **Perutz, M. F. and Lehmann, H.,** Molecular pathology of human haemoglobin, *Nature (London),* 219, 902, 1968.
13. **Perutz, M. F.,** Nature of haem-haem interaction, *Nature,* 237, 495, 1972.
14. **Morimoto, H., Lehmann, H., and Perutz, M. F.,** Molecular pathology of human haemoglobin: stereochemical interpretation of abnormal oxygen affinities, *Nature (London),* 232, 408, 1971.
15. **Dean, J. and Schechter, A. N.,** Sickle-cell anemia: molecular and cellular bases of therapeutic approaches, *N. Engl. J. Med.,* 299, 745, 1978; 299, 804, 1978; 299, 863, 1978.
16. **Platt, O. S., Falcone, J. F., and Lux, S. E.,** Molecular defect in the sickle erythrocyte skeleton. Abnormal spectrin binding to sickle inside-out vesicles, *J. Clin. Invest.,* 75, 266, 1985.
17. **Makowski, L. and Magadoff-Fairchild, B.,** Polymorphism of sickle cell hemoglobin aggregates: structural basis for limited growth, *Science,* 234, 1228, 1986.
18. **Williamson, D., Whisson, M. E., Whineray, M., Wells, R. M., Brennan, S. O., and Carrel, R. W.,** Polycythaemia associated with a new haemoglobin variant: haemoglobin Palmerston North, β23 (B5) Val → Phe, *N. Z. Med. J.,* 24, 98, 585, 1985.
19. **Taketa, F., Matteson, K. J., Chen, J. Y., and Libnoch, J. A.,** Methemoglobin reduction in red cells: effect of high oxygen affinity hemoglobin, *Blood,* 55, 116, 1980.
20. **Nienhuis, A. W., Anagnou, N. P., and Ley, T. J.,** Advances in thalassemia research, *Blood,* 63, 738, 1984.
21. **Weatherall, D. J. and Clegg, J. B.,** Thalassemia revisited, *Cell,* 29, 7, 1982.
22. **Morle, F., Lopez, B., Henni, T., and Godet, J.,** α-thalassaemia associated with the deletion of two nucleotides at position −2 and −3 preceding the AUG codon, *EMBO J.,* 4, 1245, 1985.
23. **Higgs, D. R., Goodbourn, S. E. Y., Lamb, J., Clegg, J. B., Weatherall, D. J., and Proudfoot, N. J.,** α-Thalassaemia caused by a polyadeynlation signal mutation, *Nature (London),* 306, 398, 1983.
24. **Orkin, H. S. and Kazazian, Jr., H. H.,** The mutation and polymorphism of the human β-globin gene and its surrounding DNA, *Ann. Rev. Genet.,* 18, 131, 1984.
25. **Spritz, R. A., Forget, B. G.,** The thalassemias: molecular mechanisms of human genetic disease, *Am. J. Hum. Genet.,* 35, 333, 1983.
26. **Sharp, Ph. A.,** Splicing of messenger RNA precursors, *Science,* 235, 766, 1987.
27. **Kioussis, D., Vanin, E., deLange, T., Flavell, R. A., and Grosveld, F. G.,** β-Globin gene inactivation by DNA translocation in γβ-thalassaemia, *Nature (London),* 306, 662, 1983.
28. **Flint, J., Hill, A. V. S., Bowden, D. K., Oppenheimer, S. J., Sill, P. R., Serjeantson, S. W., Koiri, J. B., Bhatia, K., Alpers, M. P., Boyce, A. J., Weatherall, D. J., and Clegg, J. B.,** High frequencies of α-thalassaemia are the result of natural selection by malaria, *Nature (London),* 321, 744, 1986.
29. **Collins, F. S., Stoeckert, C. J., Serjeant, G. R., Forget, B. G., and Weissman, S. M.,** $^G\gamma\beta^+$ hereditary persistence of fetal hemoglobin: Cosmid cloning and identification of a specific mutation 5′ to the $^G\gamma$ gene, *Proc. Natl. Acad. Sci. U.S.A.,* 81, 4894, 1984.
30. **Gelinas, R., Endlich, B., Pfeiffer, C., and Yagi, M., Stamatoyannopoulos, G.,** G to A substitution in the distal CCAAT box of the $^A\gamma$-globin gene in Green hereditary persistence of fetal haemoglobin, *Nature (London),* 313, 323, 1985.

31. **Collins, F. S., Metherall, J. E., Yamakawa, M., Pan, J., Weissman, S. M., and Forget, B. G.,** A point mutation in the $^{A}\gamma$-globin gene promoter in Greek hereditary persistence of fetal haemoglobin, *Nature (London),* 313, 325, 1985.
32. **Chung, S. W., Wong, S. C., Clarké, B. J., Patterson, M., Walker, W. H. C., and Chui, D. H. K.,** Human embryonic ζ-globin chains in adult patients with α-thalassemias, *Proc. Natl. Acad. Sci. U.S.A.,* 81, 6188, 1984.
33. **Anagnou, N. P., Ley, T. J., Chesbro, B., Wright, G., Kitchens, C., Liebhaber, S., Nienhuis, A. W., and Deisseroth, A. B.,** Acquired α-thalassemia in preleukemia is due to decreased expression of all four α-globin genes, *Proc. Natl. Acad. Sci. U.S.A.,* 80, 6051, 1983.

Chapter 10

THE MOLECULAR ASPECTS OF INFLAMMATION

I. INTRODUCTION

Inflammation is one of the oldest terms in medicine. For a long time it was a synonym for disease. Studies of Virchow (1871) on alterations within the connective tissue, those of Cohnheim (1873 to 1877) on vascular disturbances, and those of Matchnikoff (1882) on the phagocytic function of macrophages brought significant progress in contemporary ideas on inflammation. Inflammation may progress to loss of function. Locally inflammation is characterized by necrosis of cells, vasodilatation, edema, and accumulation of leukocytes. The inflammatory reaction belongs to the mechanisms of defense against bacteria and other damaging agents. The inflammatory reaction is induced not only by cells, but also by potent humoral factors released from cells, and produced by the leakage of plasma proteins into the inflammatory focus. Classically inflammation becomes initiated by terminal blood vessels (arterioles, capillaries, and venules) which causes vasodilatation and leakage of plasma proteins resulting from increased vascular permeability. However, it was after introducing the modern molecular concepts of inflammation that the consistent general description of this process could be outlined and its essential characters distinguished, as follows:

1. Inflammation is essentially a protective mechanism which becomes harmful to the organism after transgressing its physiological barriers.
2. Initiation and course of inflammation results from release of several locally acting mediators.
3. The course of inflammation depends on mutual interaction of inflammation mediators.

In general, the acute local inflammation is a response of vascularized tissues to injury. The main manifestations of inflammation are: reddening (rubor), warmth (calor), tumor (tissue swelling caused by exudation of fluid), and dolor (pain caused by irritation of sensory nerve endings). As a case of acute process of generalized inflammation of the whole organism the infectious diseases with fever may be considered as its main manifestation. On this case alone I will discuss the beneficial role of inflammation.

Without going into the definite arguments, fever has been considered as one of the protective mechanisms of infections, although from the beginning of medical history inflammation and fever have been observed as among the most troublesome disorders occurring among humans and all higher level living organisms. However, evolutionarily speaking, if these manifestations should be harmful to the living organisms, they would be eliminated by selection processes.

In poikilothermic vertebrates, whose thermal homeostasis is maintained by behavioral adaptations, e.g., the desert iguana, when infected with *Aermonas hydrophila* its organism reacts with an elevated body temperature. Animals permitted to maintain this temperature demonstrated a lower rate of mortality, as compared to those with abrogated temperature (by administration of aspirin) which showed an increased rate of mortality. It was not proved that the elevated temperature directly inhibited replication of the macroorganisms, however the increased body temperature caused an enhancement of the early inflammatory response with accumulation of leukocytes to the site of infection, although the motility of leukocytes did not change and phagocytosis was only slightly increased, and production of antibodies was unaltered.

In adult mice infected with various viruses (herpes simplex, coxsakie B viruses, rabies) hyperthermia during the incubation period caused a decreased mortality rate, while hyperthermia administered after the onset of clinical manifestations had no effect on the course of the disease. Investigations were performed on neonatal dogs which cannot regulate their body temperature. Adult dogs contaminated with canine herpes virus usually demonstrated a mild disease, but in newborns often a severe and fatal disease occurred. Infected newborns maintained under hyperthermic conditions demonstrated increased survival and reduced pathology as compared to newborns maintained under normal temperature. Newborn piglets infected with gastroenteritis virus and maintained under hyperthermia also demonstrated less severe morbid symptoms.

Resistance to bacterial infections (pneumococci, staphylococci) in infected animals maintained at various temperatures exhibited a direct correlation between temperature and survival. Bacterial strains used in these experiments were sensitive to the elevated temperature and the beneficial effects of temperature probably depended on direct inhibition of their growth, though infection of rabbits with *Pasteurella multocida* (growth of this pathogen was not influenced by elevated temperature *in vitro*) revealed increased resistance of infected rabbits maintained at elevated temperature.

Before the era of antibiotics, some human diseases were treated with fever (mainly malaria infection). The beneficial effects of this treatment was probably caused by the increased immune response, similar to that of vaccinia therapy for melanoma. Local hyperthermia also was used for cancer therapy.

Polymorphonuclear leukocytes are the most important cells participating in the early stages of infectious diseases. some authors described beneficial effects of elevated temperature on bactericidal function of these cells, while others did not. Recent experiments with *E. coli*, *Salmonella typhimurium* and *Listeria monocytogenes* incubated at 40°C with polymorphonuclear leukocytes exhibited an increased bactericidal effect of these cells, but not in the case of *Staphylococcus aureus*. The phagocytic activity of human monocytes was not influenced by elevated temperature although macrophages produce most of the endogenous pyrogens, responsible for induction of febrile reaction. This molecule produced by macrophages has been identified as interleukin-1 (IL-1) activating T-lymphocytes. Another lymphokine, the leukocyte migration factor, is produced by leukocytes at higher levels at 38.5°C than at 37°C, which results in leukocyte migration being greatly inhibited at higher temperatures. Rabbits challenged with viruses at higher housing temperatures reveal higher levels of this anti-viral agent than controls. The anti-viral activity of human interferons increases at temperatures above 40°C, besides, temperatures above 38°C inhibit multiplication of Daudi cells (a B-cell line) and potentiate the generation of alloreactive cytotoxic T-lymphocytes. This observation seems to suggest that temperature may drive the immune reaction to an invading pathogen from antibody-mediated to cell-mediated immune reaction.

These findings about the role of elevated temperature on cells participating in inflammation have been confirmed by *in vitro* studies on elevated temperature on proliferation of lymphocytes. These experiments also confirmed the effect of hyperthermia on T-lymphocyte mitogen responsiveness, and increased proliferation of cytotoxic lymphocytes. Further experiments revealed that T-cell proliferation exposed to IL-1 and IL-2 increased at temperatures at 39°C as compared to that of 37°C, whereas B-cell response remained constant. *In vivo* experiments with treatment of animals with sheep red blood cells (SRBC) as high temperatures revealed augmentation of T-helper cells, with neither B-cells or T-suppressor cells having been influenced. In other experiments about the influence of elevated temperature, the T-antigen dependent reaction was significantly elevated (SRBC as T-dependent antigen) while the reaction to T-independent antigen did not change. It is thought that hyperthermia acts mainly on T-lymphocytes resulting in an increase of the production of helper substances for B-lymphocyte reactions. Simultaneously, activation of these cells reduces their generation time.

To summarize, fever may influence host-parasite interaction in many aspects;[1] it may inhibit replication of the pathogen; it may increase polymorphonuclear leukocyte activity and function, influencing (by many specific reactions) the course and severity of inflammation; it may potentiate the anti-viral activities of different interferons; it may switch the immune reaction to the most adequate one for combating pathogens; and finally, through activation of specific metabolic pathways, it may activate the cell-mediated immune reaction.

However, in certain infections the main cause of damage is not the pathogen per se but the host reaction against it. A typical example represents lymphocytic choriomeningitis virus infection in mice. In newborn mice, that do not react by immune reactions against this pathogen (because of immaturity of the immunological system), infection with this parasite is harmless, but in adult mice which react by a strong immune response, this infection usually causes severe morbid manifestations.

The beneficial activities of inflammation are not limited to combating penetrating parasites but also to many other functions, especially repair of tissues damaged by the inflammation process. The repair mechanisms are induced by proliferation of cells activated by several growth factors like platelet derived growth factor (PDGF), epidermal growth factor (EGF) and others which will be discussed later.

For many years inflammation was considered among anatomo-pathological criteria, but since then several potent inflammation mediators have been discovered. At first, because the known inflammation manifestations (rubor, calor, tumor) relate to vascular phenomena, mainly vasodilatation, factors influencing vascular activities have been taken into account. The most important mechanism elucidating inflammation was the discovery of very potent local mediators. Some of them derive from plasma proteins, the other from the damaged cells.

To the most potent inflammatory mediators belong:

- Histamine
- Serotonin
- Kinins and kininogens
- Components of the complement system
- Prostaglandins
- Leukotrienes
- Acetyl glyceryl ether phosphorylcholine (acetylated alkyl phosphoglycerides)

II. HISTAMINE

The role of histamine as a local inflammation mediator has been discussed for a long time since Dale and Laidlow discovered that the effects of injected histamine resemble typical manifestations of anaphylaxis. In 1940 Rocho de Silva demonstrated histamine release under the effect of bacterial toxins and proteolysis. This suggested that histamine may be one of the main mediators at the early inflammation stages. This suggestion was strongly supported when Jancso demonstrated that granulopexis could be caused by histamine. Granulopexis involves histamine release under the effect of mechanical irritation. Histamine activated topically venous endothelium for the trapping of colloidal particles which is not observed in a nonactivated endothelium. The effect is similar to that induced by a local histamine injection. The neutralizing effect of antihistamine drugs confirmed the specific involvement of histamine in this phenomenon. Electron microscope studies demonstrated that inflammation mediators (histamine, bradykinin, serotonin) induce ''opening'' of vascular endothelium so that colloidal particles can penetrate the connective tissue resulting in its typical inflammatory reaction. Tissue penetration by collidal particles is facilitated by hyaluronidase, a permeation factor released by some bacteria. The enzyme depolymerase resolves the compact connective tissue, thus contributing to a rapid spread of the inflammatory process.

Histamine is synthesized in mast cells by histidine decarboxylation. In mast cells histamine is complexed with heparinoids. Histamine is released from mast cells by a degranulation process. Disintegration of mast cells causes release of histamine together wtih heparin and serotonin. Therefore, in anaphylactic shock a decrease of blood coagulability occurs. In rats and guinea pigs histamine is not complexed with heparin and is released without heparin, and therefore no reduction in blood coagulability occurs. In rabbits the main source of histamine are blood platelets. The human blood platelets contain only minimal amounts of histamine. Histamine release plays an early transient role in the inflammatory process by inducing increased capillary permeability and the leakage of plasma proteins.

Histamine release from mast cells is still not fully clear. There are several possibilities, including one that suggests the liberation of histamine by proteases which disintegrate protein and cell membranes binding histamine during antigen-antibody reaction in anaphylaxis.

III. SEROTONIN

Serotonin is synthesized by the enterochromaphin cells of the alimentary tract and by serotoninergic neurons. Serotonin induces a local vascular dilatation due to peripheral inhibition of vasoconstriction, and slightly increases vascular permeability. In inflammation by a combination of arteriolar dilatation and venular constriction, leakage of plasma may occur resulting in edema. In man, serotonin is released mainly by the mast cells. When injected into a rat footpad, it induces edema. Serotonin (but not histamine) probably is the main mediator of anaphylactic shock in the guinea pig. In turn, it probably plays an insignificant role in this reaction in humans.

IV. BRADYKININ AND KININOGENS

An important humoral system in inflammation is represented by bradykinin and kininogens. Bradykinin, a linear nonapeptide (M_r 1060), is suggested as the most powerful pain-producing agent. It is exceptionally potent in producing arterial and venular dilatation, acting directly on their smooth muscles. In similar ways to histamine and serotonin, but much more strongly, it induces separation of intercellular junctions of the venular endothelium, resulting in contraction and swelling of this tissue into the lumen of the vessels, pulling endothelial cells apart from each other and exposing subendothelial tissue.

Bradykinin is a prototype of several low molecular peptides, which are released from inactive plasma proteins (mainly α globulin) as substrates called kininogens by a system of enzymes called kallikreins. Initially, it was supposed that kininogens were the precursors of kinins, but several other function of kininogens have since been discovered, namely (1) its role in blood coagulation (the necessity of intact kininogen molecules—prior to the liberation of kinins—in early stages of intrinsic blood coagulation); (2) the inhibitory role of kininogens as potent inhibitors of cysteine proteinases (cathepsins, papain, calpain); and (3) one type of kininogen is the major acute phase reactant in inflammation in the rat and its synthesis increases during inflammation.[2] In this way kininogens connect the kinin-generating pathway, the blood coagulation cascade, the inhibitor defense system and the acute phase reaction of inflammation.

In mammals, three types of kininogens, differing in size, structure and function have been identified. The largest, high molecular weight kininogen (HMW-kininogen), has an M_r of 88,000 to 114,000, depending on its species origin. The low molecular weight kininogen (LMW-kininogen) has an M_r of 55,000 to 68,000, and the third (T-kininogen) with a molecular weight of 68,000, seems to be unique to the rat. Kininogens are synthesized mainly in the liver. They are typical secretory proteins (the translation products reveal a signal sequence), and undergo posttranslational modifications (glycosylation of hydroxy and acceptor groups prior to their secretion into plasma).

Kininogens are digested by kallikrein-like proteases. Proteolytic cleavage of two peptide bonds releases kinins and kinin-free kininogen molecules. Depending on the kind of kininogen and the cleavage specificity of the nuclease, four types of kinins can be generated: the nonapeptide bradykinin, the decapeptide kallidin (named also Lys-bradykinin), and undecapeptides Met-Lys-bradykinin or Ile-Ser-bradykinin. They have the same amino acid residues at their carboxy-termini (Arg-Pro-Pro-Gly-Phe-Ser-Pro-Phe-Arg) and differ at their amino termini.

Deficiency of HMW-kininogen (but not LMW-kininogen) causes prolongation of the partial thromboplastin time (time specifying blood coagulability after addition of surface active caolin to the blood). HMW-kininogen occurs in plasma complexed with prekallikrein or factor XI and after blood vessel trauma the kininogen-proenzyme complex avidly binds with the exposed subendothelial surface (contact phase) alongside factor XII. This causes optimal position of HMW-kininogen and factor XI to the factor XII, enabling activation of the clotting factors and the endogenous coagulation pathway via factor IX. This results in the triggering of endogenous coagulation to rapid blood clotting via factor IX.

In 1983 the complete amino acid sequences of HMW-kininogen and LMW-kininogen of bovine species had been deduced from sequence analysis of their cloned cDNAs.[2] The large heavy chain segment of LMW-kininogen preceding bradykinin revealed an analogical structure to the human α_2-cysteine proteinase inhibitor, present in the plasma along with the α_1-cysteine proteinase inhibitor, the precursor of which is the HMW-kininogen. It is accepted that both α_1 and α_2 cysteine proteinase inhibitors are identical to the HMW- and LMW-kininogens. Thus, the kininogens together with α_2-macroglobulin are the main cysteine proteinase inhibitors of plasma.

To summarize, the kininogens exhibit three basic functions:

1. They are substrates for kallikrein-like enzymes releasing kinins.
2. They bind coagulation proenzymes on the contact phase enabling rapid activation of the intrinsic coagulation pathway.
3. They inhibit cysteine proteinases forming inactive enzyme-inhibitor complexes.[2]

The different kininogens have a heavy chain of 50 to 60 kDa performed at the amino-terminus, the kinin segment at the core part of the molecule and the preformed light chain of variable length at the carboxy terminus. The kinin section is identical in HMW- and LMW-kininogens, but the light chains diverge considerably. The HMW-kininogen light chain has a unique histidine-rich part facilitating binding of the molecule to subendothelial surface and binding sites for prekallekrein and factor XI. The LMW-kininogens reveal the canonical bradykinin sequence preceded by the dipeptides Met-Lys or Met-Arg. LMW-kininogen lacks sites critical for contact activation (through factor XII) and proenzymes (prekallikrein) binding.

Several models of the molecular architecture of kininogens have been constructed. One of them predicts six domains for the HMW-kininogen and five domains for the LMW-kininogen.[2] According to this model two of the three heavy chain domains encoded by a triple set of exons, demonstrate reactive sites for proteinase inhibition, followed by the kinin domain. Distal from the kinin domain, the HMW-kininogen has the histidine-rich domain, and the last domain exhibits recipient structures for prekallikrein and factor XI.

The human kininogen gene consists of 11 exons (about 27 kb). The 5′ segments are identical in HMW- and LMW-kininogens, encoding the signal peptide, heavy chain and bradykinin, but completely different at the 3′ end. Both, the HMW- and LMW-kininogens are encoded by a single gene, with different polyadenylation sites and alternative splicing which generates two distinct types of mRNA for HMW-kininogen and LMW-kininogen. It is thought that this property of the kininogen gene enables feedback effects depending on physiological demands.

T-kininogen (probably unique to the rat) is not susceptible to proteolytic cleavage by kallikrein, but is hydrolyzed by large quantities of trypsin or cathepsin D. The T-kininogens have a preceding methionin residue, whereas an arginine residue precedes bradykinin in the K-kininogen sequence (K—after kallikrein). Furthermore, as compared to K-kininogen, T-kininogens exhibit two consecutive amino acid deletions in the regions preceding T-kininogen. In spite of a considerable homology of T- and K-kininogen, their mRNAs are expressed differently in response to induction of acute inflammation. The expression of T-kininogen mRNAs increased about 10-fold to 13-fold whereas expression of HMW- and LMW-K-kininogens mRNAs was unchanged. The total LMW-kininogen mRNAs comprised about 1% of liver mRNA to maximum induction which represents a major portion of the liver mRNA in the acute phase of inflammation in the rat. Further investigations revealed that the major acute phase protein mRNA is identical to the corresponding region in T-kininogen mRNA, and both T-kininogen and LMW-K-kininogen form the low molecular weight fraction of cysteine proteinase inhibitors in rat plasma, while HMW-K-kininogen contributes to the high molecular weight fraction. Mammals other than the rat, and possibly other than rodents, lack T-kininogen.

V. THE COMPLEMENT SYSTEM

The complement system comprises a group of several different plasma proteins that are activated during humoral immunologic reactions producing a variety of inflammatory mediators (see Figure 1). The plasma proteins of the complement system undergo sequential activation initiating an inflammatory reaction causing lysis of offending bacteria and infected cells. There are several inflammatory mediators that induce the following events of inflammation: increased vascular permeability, attraction of polymorphonuclear and mononuclear leukocytes, phagocytosis of immune complexes by promoting their adherence to blood phagocytic cells, damage of cell membranes by induction of osmotic lysis, and cell death.

The complement system can be activated by the classical or alternative (named also properdin) pathways. The particular proteins of the classical pathway are: C1q (M_r 400,000), C1r (M_r 400,000), C1r (M_r 168,000), C1s (M_r 90,000), C4 (M_r 204,000), C2 (M_r 117,000), C3 (M_r 191,000), C5 (M_r 185,000), C6 (M_r 95,000), C7 (M_r 120,000), C8 (M_r 150,000), and C9 (M_r 79,000); that of the alternative (properdin) pathway are: properdin (M_r 220,000), factor B (M_r 93,000), and factor D (M_r 25,000). Altogether, these proteins constitute about 10% of the serum globulins. Almost all are glycoproteins. Some of these proteins reveal homologies, e.g., C2 and factor B, both of which bear active sites for C3 cleaving. During activation, fragments of 40,000 and 35,000 are cleaved from C2 and factor B, respectively, with the major fragments of each protein participating in a Mg^{2+}-dependent complex cleaving C3, and subsequently C5.

The C1 component of complement is a 750,000 Da glycoprotein that requires Ca^{2+} for activation. Physiologically, C1 comprises two weakly interacting subunits, C1q with binding sites for activators, and C1r with enzymatic potential. Induced by immune complexes, C1 becomes activated by limited proteolysis and conformational changes. Bound to immune complexes, the interaction between C1q and C1r increases. C1 is controlled by $C\bar{1}$-inhibitor that blocks its enzymatic activity. $C\bar{1}$ half-life is 13s in the presence of $C\bar{1}$-inhibitor.

C1q is distinguished from other plasma proteins by the presence of hydroxylated amino acids, repeating glycines, and disaccharide residues, which form a collagen-like structure of this protein. C1q has several binding sites for immunoglobulins. C1q is a serum glyco-

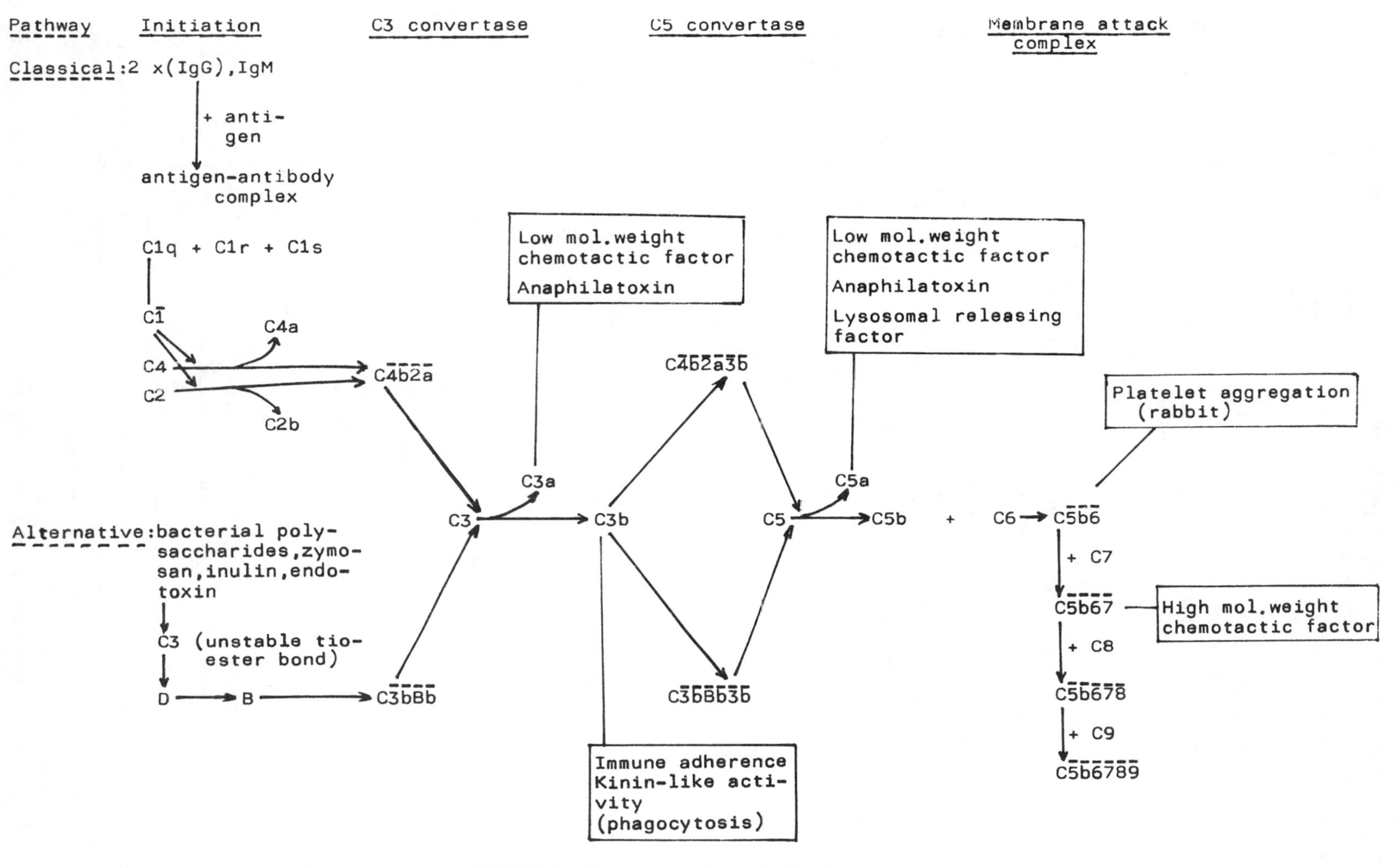

FIGURE 1. Complement pathways (explanation in text).

protein of 450,000 Da containing 18 polypeptide chains (6A, 6B, 6C) each of 226 amino acid residues, containing an N-terminal collagen-like domain and a C-terminal globular domain. Clr is a precursor of a protease. Cls is also a precursor of a single polypeptide chain. Its proteolysis yields two chains of M_r 36,000 and M_r 77,000. The active site is in the light chain.

C4 is composed of three disulfide-bonded chains of M_r 93,00, M_r 78,000, and M_r 33,000. Its proteolysis yields a 6,000 Da fragment C4b from the heavy chain. C4b becomes inactivated by an enzyme present in serum, probably the C3b inactivator. Formation of one or more disulfide bridges between the seven SH groups of C2 is probably responsible for the tenfold increase in activity and resistance to temperature-dependent decay exhibited by $C\overline{42}$.

C3 is a glycoprotein composed of two polypeptide chains (M_r 115,000 and M_r 75,000) linked by disulfide bridges. Both C3 and C4 molecules are serine proteases that contain an intrachain thioester bond essential for their hemolytic activity. Cleavage of C3 by trypsin, or by enzymes of the complement system $C\overline{42}$ or C3B yields a 7,900 Da fragment, C3a, from the heavy chain; the light chain C3b remains unchanged. Further inactivation of C3b by C3b inhibitor causes its cleavage into C3c and C3d. C5 is composed of two chains of the same size and cleaved by trypsin or an enzyme of the complement system $C\overline{423}$, it yields a 15,000 Da fragment C5a from the heavy chain. The molecular structures of C6, C7, C8, and C9 are not yet definitively known.

Properdin (M_r 220,000) consists of four chains (M_r 55,000). During activation a 30,000 Da fragment is released. Properdin factor B consists of a single peptide chain (M_r 93,000). Cleavage by trypsin or during activation yields two fragments (M_r 60,000 and M_r 35,000). The precursor of factor D during activation yields a slightly lower fragment of M_r 23,500, an active serum protease.

A. COMPLEMENT ACTIVATION

Limited proteolysis of inactive precursor molecules yields minor fragments with biological activity as enzymes or mediators to other biological reactions. These enzymes or mediators activate further reactions until the final membrane attack complex is formed. The classical pathway is activated by antigen-antibody complexes. Interaction of minimum two IgG molecules or one IgM molecule with antigen changes the conformation of the second constant domain (Fc) of immunoglobulin which enables binding with Clq. Next binding with Clr and Cls (a proesterase) occurs, which activates the complex to the esterase $C\overline{1}$ enzyme. This complex cleaves C4 into C4b and C4a, and C2 into C2a and C2b. This cleavage causes transiently covalent binding of C4b with the antigen. Next C4b binds with active C2a and forms a bimolecular enzyme $C\overline{4b2\bar{a}}$ (also named C3 convertase) which cleaves component C3 into C3a and C3b. C3b binds covalently to antigen and forms C4b2a3b (C5 conversase) which cleaves C5 into C5a and C5b initiating generation of the membrane attack complex $C\overline{56789}$ resulting in cytolysis.

Activation of the complement cascade causes release of several active substances. A vasoactive peptide is released from C2 by the action of $C\overline{1}$ in the presence of C4 and is similar to the cleavage products C3a and C5a anaphylatoxins that release histamine from mast cells. In addition, C3a is a low molecular weight chemotactic factor. C3b is attached to the cell membrane and exhibits immune adherent action which facilitates phagocytosis of foreign substances (bacteria and other substances) it exerts also a kinin-like activity. C5a has the same properties as C3a, and in addition it releases a "lysosomal releasing factor" which enables release of lysosomal enzymes without cell disintegration. C6 is responsible for blood platelet aggregation. During formation of the membrane attack complex a high molecular chemotactic factor is released.

C4 has a very important role in activation of the complement cascade by the classical pathway because of its specific interaction with seven or eight proteins. Small changes in the amino acid sequence may cause different biological activities of this protein. The best example is trypsin, chymotrypsin or elastase in which changes in two amino acid residues alters the specificity of these proteases from that for a basic amino acid residue to that for a hydrophobic amino acid residue and for specificity to many different amino acid residues.[3] C4 is activated by $C\overline{1}$ bound to Fc fragment of the antibody. C4 forms a covalent bond with antibody or cell surface by a reactive acyl group released from an intra-chain thioester bond. The acyl group of C4 diffusing slowly will react with water unless that target is close by (this denotes that the C4 molecular surface determines the role of hydrolysis by $C\overline{1}$). After binding C4, C2 activated by $C\overline{1}$, interacts non-covalently with $C\overline{4}$ and forms the active protease $C\overline{42}$ which activates through hydrolysis C3. The surface conformation of both C4 and C2 determines the rate of decay of $C\overline{42}$. Complement activation is limited by the short half-life of C3 convertase. Another limiting factor of complement activation is enzyme H that interacts with C4 and after binding with cofactor C4bp degrades $C\overline{4}$.[3]

Dissolution of immune complexes and uptake by phagocytic cells depends on the binding of $C\overline{3}$ and $C\overline{4}$ and on the presence of receptors for $C\overline{3}$ and $C\overline{4}$ and their breakdown products on macrophages. The quantity of bound $C\overline{3}$ degraded products depends on the surface of C4. Therefore, every change of amino acid residues of this molecule influences the reaction of C4 with antibody, cell surface molecules, $C\overline{1}$, C2, and C3, enzyme H and cofactor C4bp that degrades $C\overline{4}$ and $C\overline{4}$ receptors.

Another central role in regulation of the complement cascade plays the maintenance of active $C\overline{3}$ regulated by protein factor H (enzyme H) and C4-binding protein (C4bp) in plasma, and C3b/C4b receptor (CR1), decay accelerating factor (DAF), and gp 45 to 70 on cells.[4] In this aspect in human disease: paroxysmal nocturnal hemoglobinuria serves as an experiment of nature to study the cytolytic function of complement and its regulation.[5]

CR1 receptor was first isolated as a human erythrocyte membrane glycoprotein that accelerated the decay of alternative pathway convertase $\overline{C3bBb}$ and then as receptor for C3b/C4b on polymorphonuclear leukocytes, lymphocytes, monocytes, and small part of T-lymphocytes. It mediates phagocytosis of larger particles (erythrocytes and bacteria). After binding of these particles it becomes cross-linked and associated with clathrin-coated endocytic pits and coated vesicles upon internalization. *In vivo*, erythrocyte CR1 binds C3-bearing immune complexes and brings them to the liver. CR1 does not accelerate decay of enzyme complexes, but accelerates decay of convertase complexes in the fluid phase or on other targets.

C3b inactivator (I) is a plasma serine protease that specifically cleaves the α-chain of C3b or C4b. The cleavage reaction yields inactive C3b and C4b. Cofactors for this reaction are H, C4bp, and CR1. The next cleavage yields inactive C3c and C3d, or C4c, and C4d.

Decay accelerating factor (DAF) is a single-chain intrinsic membrane glycoprotein, present in erythrocytes, blood platelets and leukocytes. DAF has binding specificity for C3bBb or C4b2a accelerating decay of both the classical and alternative C3 convertases. DAF also binds with high affinity to C3b/C4b, but only when these molecules are present in membranes of the same cells on which DAF is located (so-called intrinsic functional activity). Finally, DAF prevents assemble of convertase rather than dissociating already formed enzyme. DAF accelerates the release (decay or dissociation) of C2a of Bb from the complexes, regulating in this way activation of complement components.[5]

Factor H is a single-chain plasma glycoprotein. It binds the fluid phase and covalently bound C3b. Factor H is also a cofactor for C3b inactivator (I) which mediates cleavage of C3b and decay-acceleration of the alternative pathway convertases. Three polymorphic forms of factor H have been identified.

C4-binding protein (C4bp) is a plasma glycoprotein that binds C4b in both the fluid and covalently bound phases and is a cofactor for C3b inactivator and accelerates decay of the classical pathway C3 convertase. Two phenotypes have been identified.

Gp 45-70 is a protein molecule present on peripheral leukocytes and platelets. It modulates complement activation on cell surfaces, with preferential activity for C3b or enzymes containing C3b. Gp 45-70 may also influence some functional activities of CR1.

1. Inhibitors of the Complement Cascade

One of the most potent inhibitors is the $C\overline{1}$ inhibitor that inhibits hydrolytic activity of $C\overline{1}$, active contact factor (XII), plasmin and kallikrein (plasmin degrades $C\overline{1}$ inhibitor). The $C\overline{3}$ inhibitor inhibits $C\overline{3}$ activity, and the $C\overline{6}$ inhibitor inactivates $C\overline{6}$-cell complex. The $C\overline{8}$ inhibitor inhibits complex formation of C1-7 with C8, and it inhibits hemolysis. The C4 inhibitor inhibits activation of C4. Some of these inhibitors probably are not yet sufficiently isolated specific inhibitors.

Complement activation is associated with release of the following biologically active factors:

- Immune adherence factors attracting immune complexes, and foreign particles with stimulation of phagocytosis;
- Chemotactic factors attracting leukocytes to the site of inflammation;
- Permeability factors inducing increased vascular permeability, and degranulation of mast cells with histamin release.

B. DEFICIENCY OF THE PARTICULAR COMPONENTS OF THE COMPLEMENT

C1q deficiency has been described in patients with sex-linked agammaglobulinemia. It is supposed that because C1q is synthesized by lymphoid cells and these cells are impaired in these patients it is probable that deficiency of C1q depends on general impairment of these cells.

Unusual genetically determined forms of C1q have been described in patients with unusual susceptibility to complex related diseases. In homozygotes two variant forms of C1q were present, one with reduced molecular mass of approximately 160,000 Da, and the second with increased molecular mass of each chain of about 16,000 Da, which makes in total 800,000 Da (normal molecular mass of C1q is 460,000 Da). A higher than normal molecular mass of C1q was described in human colostrum.[16]

In a family with deficient hemolytic complement function, C1q-like material has been described. All suffered from glomerulonephritis during childhood. Renal membranous glomerulopathy was confirmed by renal biopsy. Two members of this family died at the age of 23 and 20 years because of systemic lupus-like diseases. The C1q-like material revealed a molecular mass of approximately 65,000 Da that did not bind to C1r and C1s.[17]

C1r deficiency has been described in a patient with glomerulonephritis, and in a family composed of a brother and sister. The brother exhibited a systemic lupus erythematosus-like disease, and the sister exhibited recurrent infections diseases. Three other family members died in early infancy. The affected individuals lacked detectable total hemolytic complement activity.

C1s deficiency has been described in a six-year-old girl with seropositive systemic lupus erythematosus.

C4 Deficiency—A family of 15 members exhibited one patient (seven years old) with lupus erythematosus who lacked detectable serum C4. The defect was transmitted as an autosomal recessive trait. Eight of the 15 family members were heterozygotes.[6] Cold urticaria (urticaria developing upon exposure to cold) has been described in a patient with deficiency of C4 and elevated IgM.[18] Polymorphism of C4 was found in a Japanese family.[7] The variant was defined as C4B, however, hemolytic assay revealed that this variant C4B1 was hemolytically inactive, while C4B was active.

C2 deficiency has been described in several patients. In all of them the hemolytic plasma assay demonstrated decreased or negative values. C2-deficient serum was inactive in bactericidal assays, but not in patients in whom the alternative pathway of complement activation was probably used. C2 deficiency was inherited as an autosomal recessive trait. The homozygous individuals suffered from lupus erythematosus, polymyositis, glomerulonephritis or Schönlein-Henoch purpura.

C3 Deficiency—Homozygous patients suffer from recurrent infections with both pyogenic and Gram-positive bacteria resulting in severe furunculosis, pneumonia, meningitis, and septic arthritis. Typical tests (total hemolytic complement, bactericidal, chemotactic, and opsonic) were negative, immune adherence was only one third normal. It was impossible to induce the alternative pathway. All these abnormalities have been corrected by addition of C3 to the serum. In addition, a reduced amount of leukocytes was present at the area of inflammation which suggests that C3 may be responsible for leukocytes recruitment at the inflammatory focus. C3 deficiency was inherited as an autosomal recessive trait.

C5 deficiency resulting is seropositive lupus erythematosus, has been described with arthritis, hemolytic anemia, and mebranous glomerulonephritis. The total hemolytic complement test was negative, chemotactic activity reduced to 15 to 30% of normal. This deficiency was inherited as an autosomal recessive trait.

C6 deficiency was observed in an individual without major disturbances. Generation of chemotactic factor was normal, the bactericidal activity of homozygous deficient serum was absent. In humans aggregation of blood platelets was normal, and in C6-deficient rabbits it was absent. Genetic polymorphism of the sixth and seventh components of human complement was observed in a Japanese family. Three common and four rare alleles were found at locus C6.

C7 deficiency was described in patients with Raynaud's syndrome, sclerodactyly, and teleangiectasia. The complement hemolytic test was reduced to 0.4 to 3% of normal. Bactericidal activity was reduced or absent. Three common C7 allotypes were observed: C71, C72, and C74. Close linkage between loci C6 and C7 was confirmed.[8]

C8 deficiency was described in a case with prolonged gonococcal septicemia. C8 was less than 0.01% of the normal value. Bactericidal activity was normal, no hemolytic complement was found.

Polymorphism of the human C8 has been described.[9] C8 is composed of three covalently bound α-γ-chains. Nomenclature of the following C8 polymorphisms has been proposed: C81 for α-γ-chain deficiency with the alleles: C81A, C81B, C81A1, and C81QO (the silent allele); and C82 for the β chain deficiency with the alleles C82A, C82B, C82A1, and C82QO (the silent allele).

The ninth component of complement and pore-forming protein (perforin 1) from cytotoxic T-lymphocytes suggests structural and functional homologies which imply that these two killer molecules act by pore formation during cell lysis.[10]

Clinical manifestations of complement components are strikingly diverse. Increased susceptibility to infections has been observed in patients with deficiency of C3, C5, and C3b inactivator. Lack of C3b inactivator causes permanent activation of the C3 component which causes its consumption resulting in its deficiency and in recurrent infections, urticaria, angioedema, and positive Coombs test for C3. The positive Coombs test is probably caused by C3b deposition on erythrocytes and urticaria by increased anaphylatoxin release.

C1 Inhibitor Deficiency: Hereditary Angioedema—The C1 inhibitor is an α_2 neuraminoglycoprotein (M_r 105,000). It has a broad spectrum control of several blood cascades: complement, intrinsic blood coagulation by inactivation of contact factor XII, kininogenesis, and fibrinolysis (C1 inhibitor is degraded by plasmin). The disorder was described by Osler (1888). The disease is inherited as autosomal dominant. Recurrent, acute, circumscribed, transient edema of the skin is usually the most common manifestation in this disorder. Edema is characteristic in the gastrointestinal tract leading to severe abdominal pain often with watery diarrhea. Clinical signs of peritonitis are unusual. Improvement usually occurs within 48 h. Submucosal edema of the upper respiratory tract may occur which, in concert with simultaneous laryngeal edema, may cause death by asphyxiation. Attacks of angioedema may occur at different time intervals—from once weekly to a few attacks during the entire lifetime. The genetic basis of disfunctional C1 inhibitor is a structural defect at the locus for C1 inhibitor. The absence of C1 inhibitor causes spontaneous activation of C1, which in turn activates C4 and C2. Release of a kinin-like substance from C2 increases vascular permeability, however recent investigations suggest rather that absence of C1 inhibitor permits activation of kallikrein, which cleaves kininogen resulting in kinin synthesis (probably bradykinin).[11]

A special case of inflammation seen in the presence of autoantibodies against the C1-inhibitor. In this situation typical manifestations of angioedema occur caused by inactivation by antibodies of this pivotal inhibitor that inactivates proteins of the complement cascade, kinin, fibrinolytic, and contact phase systems.[13] (Besides, antibodies are known as specific pathogens in myasthenia gravis, Graves' disease, autoimmune hemolytic anemia, acquired hemophilia [antibodies against factor VIII] and glomerulonephritis).

Paroxysmal nocturnal hemoglobinuria is a disorder which enables insight into the cytolytic function of complement and its regulation.[5] The primary cause of increased hemolysis by the complement cascade is deficiency of a membrane complement regulatory protein, decay accelerating factor (DAF), and its cofactor CR1 (required for inactivation of erythrocyte-bound C3b). It is suggested that the protein primarily responsible for preventing complement activation on human erythroctyes of patients suffering from paroxysmal nocturnal hemoglobinuria is DAF (CR1 was present in these erythrocytes). The consequence of DAF deficiency is an impaired function of CR1.[19] In addition, increased hemolysis may be caused by binding of the C5bb678 complex resulting in quantitative abnormality, affecting C9 binding and/or polymerization within the C5b678-9 complex.[6]

The complement cleavage product C5a is a potent stimulant of inflammatory processes. As an anaphylatoxin it acts similarly to C3a and C4a by inducing contraction of smooth muscles, increasing vascular permeability and releasing histamine from mast cells and basophilic leukocytes. In addition, C5a, by stimulation of locomotion of mononuclear and polymorphonuclear leukocytes to inflammatory foci (chemikinesis), and directing their migration (chemotaxis), causes their accumulation with release of free radicals and tissue-digesting enzymes which probably play a major role in the tissue damage accompanying inflammation.[12]

Activation of the terminal complement pathway C5b-9 in the vicinity of cell membranes leads to insertion of this complex into the lipid bilayer of cell membranes which alters transmembrane ion fluxes and may lead to cell swelling and lysis. Lysis occurs after formation of several transmembrane channels. In contrast to erythrocytes, nucleated cells are able to eliminate complement lesions and repair their plasma membranes. Therefore, nucleated cells are more susceptible to agents that impair protein synthesis or cause release of lipids that transform into potent inflammatory agents like prostaglandins, leukotrienes, and especially free radicals. A special case in this aspect is seen in glomerular epithelial cell injury, caused by formation of stabile terminal C5-9 and C5-7 complexes in rat membranous nephropathy.[14] A beneficial role in prevention of cell damage caused by the terminal complement complexes may exhibit antibodies against them, as has been documented in protection of polymorphonuclear leukocytes.[15]

VI. PROSTAGLANDINS

The prostaglandins are a complex group of oxygenated fatty acids made up of the most potent natural bioregulators and pathogens. They are synthesized at each event of membrane damage that releases free fatty acids, mainly arachidonic acid. the free arachidonic acid (released by inflammatory or immunological stimuli, calcium ionophores, ultraviolet light, melittin of bee venom, tumor promoting agents) reacts with prostaglandin cyclooxygenase which oxygenates arachidonic acid to endoperoxide (PGG_2). PGG_2 is then converted to a different active product, the nature of which determined by the enzyme present in the tissue (e.g., platelets form primarily thromboxane A_2, whereas the aorta forms prostacyclin) (see Figure 2).

The precise mechanism of the release of the prostaglandin precursors is unclear. It is generally accepted that arachidonic acid is released from cell membranes after their injury, liberated by phospholipase A_2, but there also are some arguments that phosphatidyl inositol-specific phospholipase C, yielding diacylglycerides and arachidonic acid, is the main source of arachidonic acid. PGG_2 is converted to PGH_2 which is a substrate for thromboxane synthetase, PGI_2 synthetase, PGD_2 isomerase, and PGE_2 isomerase.

Among the different prostaglandins, PGE seems to be the most important mediator of inflammation. It causes inflammation with fever and pain. An argument for its role in inflammation is that these symptoms disappear after administration of inhibitors of the first step in prostaglandins synthesis (i.e., aspirin or indomethacin), However, edema, the primary manifestation of inflammation caused by increased vasodilatation and vascular permeability, may result from the action of PGE, but more potent in this aspect are other inflammatory mediators such as histamine and bradykinin. However, PGE potentiates the edema caused by histamine or bradykinin probably by its vasodilatatory action resulting in greater perfusion of the inflammatory focus.

Pain occurring during inflammation is partially caused by the action of PGE_1 (more potent) and PGE_2 (less potent). Physiologically, pain occurs only after administration of very high doses of PGE, thus it is suggested that although similar to bradykinin in producing edema, prostaglandins only strengthen the pain caused by bradykinin. The pain-producing activity of prostaglandins is suggested by the antagonistic beneficial action of cyclooxygenase inhibitors (aspirin and other) which inhibit the generation of prostaglandins.

PGE_1 and in lesser degree PGE_2 are potent fever-producing agents. It is supposed that the classical pyrogens act through mediation of these prostaglandins.

Blood clotting disturbances belong to the main manifestations in inflammation. A large role in this aspect is played by the antagonistic activity of thromboxane A_2 (inducing blood platelet aggregation and vasoconstriction), and prostacyclin (inhibiting aggregation of platelets and inducing vasodilatation).

Thromboxane A_2 is the most potent prostaglandin-related mediator of platelet aggregation. Thromboxane A_2 and prostacyclin (PGI_2) have a half-life of only seconds before being converted into inactive forms.

Blood platelet aggregation occurring after injury of vascular intima plays an important role in hemostasis. Damage of the intima, where prostacyclin is synthesized, potentiates the action of platelet aggregation caused by thromboxane A_2 (due to the absence of its natural inhibitor prostacyclin).

VII. LEUKOTRIENES

At present, it seems that the most potent inflammatory mediators are leukotrienes synthesized by conversion of arachidonic acid by the action of 5-lipoxygenase into leukotrienes (see Figure 2). Leukotrienes have a long history. They represent a group of compounds containing a conjugated triene structure, from which their name is derived. Primarily, in 1940 Kellaway and Threthewie described the so-called ''slowing reacting substance'' (SRS) in immediate-type hypersensitivity that generates a spasmogen which contracts smooth muscles more slowly than does histamine.

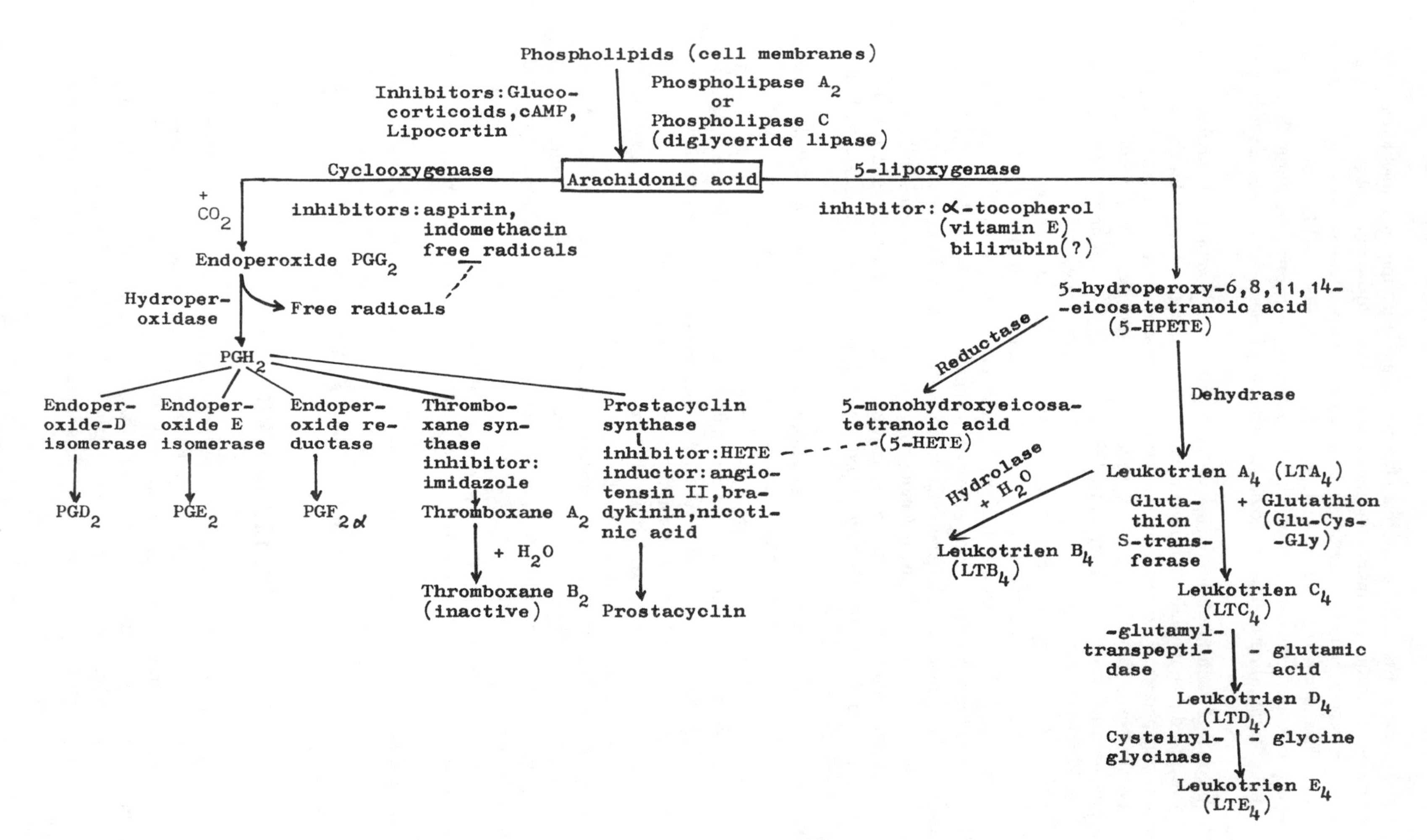

FIGURE 2. Arachidonic acid derivatives, cyclooxygenase pathway (prostaglandins), 5-lipooxygenase pathway (leukotriens) (explanation in text).

The biological activity of leukotrienes: 5-HETE after incorporation into membrane lipids modulates the motility and probably the glucose transport of neutrophils; LTB_4 is the most potent chemotactic factor on human neutrophils and eosinophils (comparable with that of fragments of the fifth component of complement); LTC_4, LTD_4, and LTE_4 exhibit vasoactive activity (more potent than that of histamine). SRS-A (the spasmogenic activity of sputum of asthmatics) is recognized as a combination of LTC_4, LTD_4, and LTE_4. The spasmogenic activity of these compounds exceeds that of histamine for LTC_4 and LTE_4 by 2 logs (EC_{50} 1×10^{-8} *M*), and for LTD_4 by 3 logs (EC_{50} 3×10^{-9}).[20]

LTC_4, LTD_4 and LTE_4 are the main mediators of the respiratory tract during anaphylaxis in which LTE_4, histamine and thromboxane play a minor role. On human bronchi LTC_4 and LTD_4 are spasmogens more than 1000 times as potent as histamine and 500 more potent than $PGF_{2\alpha}$. LTC_4 and LTD_4 exhibit a flare in the human skin at 0.1 to 1.0 nmol per site, much more persistent than that caused by histamine, suggesting some effects on arterioles and deeper veins.

Three types of interactions of the different oxidative metabolites of arachidonic acid have been stated: antagonistic or suportive for biosynthesis of a specific product (e.g., suppression of SRS-A generation by PGE_1 and by PGI_2 and enhancement by $PGF_{2\alpha}$); synergistic with or antagonistic action to the target cells (e.g., PGD_2 acts synergistically with LTB_4 on chemotaxis of human neutrophils); augmentative in generation of secondary metabolites (e.g., production of thromboxane in response on the action of leukotriene constituents of SRS-A).[20]

VIII. ROLE OF SUPEROXIDES IN INFLAMMATION

Polymorphonuclear leukocytes and macrophages consume large amounts of molecular oxygen during phagocytosis which is transformed by membranous NADPH oxidase to superoxide $O_2^{\cdot -}$). A substantial portion of this $O_2^{\cdot -}$ injures the tissue under inflammatory stimuli. The superoxide $O_2^{\cdot -}$ only has low oxidizing potential, but its conversion to hydroxyl radical OH· generates a potent oxidant reacting with any organic compound. Superoxide dismutase (SOD) dismutes superoxide $O_2^{\cdot -}$ to H_2O_2 and O_2 being in this way a potent anti-inflammatory factor. In experiments SOD by its conversion of $O_2^{\cdot -}$ to H_2O_2 demonstrated anti-inflammatory activity, while catalase by its conversion of H_2O_2 to O_2 did not. SOD derivatives showed anti-inflammatory activity in animal models of inflammation documented by a decrease in leukocyte infiltration in the inflamed focus—superoxide ($O_2^{\cdot -}$) itself, and not OH· derived from it, was assumed to participate in inflammation.[21]

Oxidizing radicals derived from the prostaglandin biosynthetic pathway have been identified (see Figure 2), released from the hydroperoxy group during the peroxidatic reduction. This oxidizing moiety was capable of deactivating certain enzymes of the prostaglandin biosynthetic pathway. Thus, in *in vitro* experiments on arachidonic acid metabolism, addition of compounds with reducing groups (epinephrine, phenol, and hydroquinone) was necessary to maintain cyclooxygenase activity. These reducing compounds served as scavengers of oxidants released from PGG_2.[21]

Free radicals released by the action of prostaglandin hydroperoxidase from PGG_2 exhibit destructive effects on cells, whereas other peroxidases (glutathione peroxidase, catalase) reduce hydroperoxides to less toxic compounds. Therefore, prostaglandin hydroperoxidase may be responsible for some cellular damage characteristic for inflammation. Low concentration of hydroperoxides is necessary for the start of cyclooxygenase activity. The reducing agents stimulate the prostaglandin pathway, probably by protecting against oxidative deactivation, because higher concentrations of free radicals inhibit this pathway probably by lowering the level necessary to initiate cyclooxygenase activity.[21]

It is noteworthy that recently bilirubin has been discovered to serve as an antioxidant.

In model experiments bilirubin prevented lipid peroxidation as efficiently as alpha-tocopherol. Thus, bilirubin at the physiological level is not a toxic end-product of heme, but rather may serve as scavenger removing highly reactive oxygen radicals from circulation that may injure several tissues.[23]

The immunoregulatory activity of leukotrienes. Leukotrienes actively synthesized by leukocytes during immune and nonimmune inflammations exert several immunoregulatory functions.[22] LTB_4 inhibits proliferation of human peripheral blood mononuclear leukocytes, proliferation of $T4^+$ (helper/inducer lymphocytes), proliferation of human B cells, and production of interferons by human $T8^+$ (suppressor/cytotoxic) lymphocytes. In turn, LTB_4 causes enhancement of suppressor function of human T-cells, proliferation and marker expression of $T8^+$ lymphocytes, enhancement of cytotoxicity and target binding by human NK (natural killer cells) and NC (naturally cytostatic cells), enhancement of γ-interferon production by mouse T-cells and human T4+ cells, enhancement of interleukin-2 production by human T4+ cells and enhancement of interleukin-1 production by human monocytes. LTA_4 causes enhancement of cytotoxicity of human NK- and NC-cells.

Inflammatory mediators, derivatives of arachidonic acid, (prostaglandins, leukotrienes, and free radicals) may be generated as consequences of cell membrane by complement activation.[24]

A significant increase of cysteinyl leukotrienes was observed after thermal or mechanical trauma in the anesthesized rat. Cysteinyl leukotrienes were rapidly eliminated from the blood plasma into bile. It is assumed that the increase of leukotrienes in trauma may be responsible for respiratory and circulatory events occurring in trauma and in general pathophysiology of trauma.[25]

An important group of proteins, known collectively as lipocortins, represent a potent inhibitor of phospholipase A with anti-inflammatory activity. In the clinic, steroids are widely used as anti-inflammatory drugs. Dissecting the mechanism of steroid action a family of anti-inflammatory proteins, called lipocortins, has been discovered independently by several groups. These proteins control the enzyme phospholipase A_2, inhibiting the release of arachidonic acid, precursor of inflammatory mediators, derivatives of arachidonic acid. Lipocortin-like proteins have been discovered in several cells (monocytes, neutrophils, and renal medullary cell preparations). The active form is a protein of M_r 40,000. Lipocortins mimic some effects of steroids and mediate anti-inflammatory activity in different *in vivo* models. Recently cloning of human lipocortin complementary DNA and its expression in *E. coli* confirmed that lipocortin is a potent inhibitor of phospholipase A_2 activity.[26]

Atypical generation of arachidonic derivatives was observed in celiac patients. Celiac or gluten-sensitive enteropathy is a disorder characterized by small intestinal mucosal injury caused by dietary exposure to wheat gluten or similar proteins in rye, barley, and triticale. There is growing evidence that this disorder may be immunologically mediated. Diagnostic jejunal biopsies of mucosa from several patients has been analyzed for the presence of arachidonic acid derivatives. The predominant arachidonic acid metabolite generated was 15-hydroeicosatetraenoic acid (15-HETE). Patients of normal diet (containing gluten) generated more 15-HETE than a group of control patients on a gluten-free diet. It is supposed that 15-HETE may be responsible for many of the biological effects of celiac disease.[27]

Congenital deficiency of thromboxane and prostacyclin has been observed in a 23-year-old woman. Clinically a bleeding disorder characterized by a mildly prolonged bleeding time and defective platelet-release reaction was caused by congenital cyclooxygenase deficiency. Their platelets did not aggregate, but after addition of synthetic PGG_2, they did aggregate. Thrombin added to her platelet-rich plasma did not generate thromboxane B_2. A biopsy vein specimen did not generate prostacyclin. Thrombotic manifestations have not been observed.[28]

IX. ACETYLGLYCERYL ETHER PHOSPHORYLCHOLINE (AGEPC)

This inflammatory mediator was first described as a ''platelet-activating factor'' derived from antigen-stimulated IgE-sensitized rabbit basophils. It induces release of stored granular constituents (e.g., serotonin) from washed platelets. Its main physiological function is contraction of smooth visceral muscles (by an independent mechanism of muscarinic, histaminic, or LTC receptors), activation of polymorphonuclear leukocytes stimulating chemotaxis, granule secretion, aggregation, and production of superoxide anions. Injected intravenously, AGEPC exhibits cardiovascular, pulmonary, and hematologic changes similar to those seen in anaphylactic shock. Intracutaneous injection in humans causes erythema, edema, and transient severe burning pain. Compared with exogenous histamine AGEPC was 1000 to 10,000 times more potent in inducing an acute transient increase in vascular permeability. The reaction induced by AGEPC was not abolished by anti-histaminic drugs. In a prolonged phase of increased vascular permeability AGEPC exhibited 100- to 1,000-fold more potent cutaneous vasoconstriction than histamine. These data show that AGEPC is the most potent known inducer of increased vascular permeability.[29]

Up to this time it was accepted that the two main components of acute inflammation, i.e., edema and leukocytic infiltration, are generated by two separate mechanisms, edema by the action of histamine and histamine-like substances, and leukocytic infiltration by plasma or tissue-generated chemostatic substances (C5a, LTB_4, etc.). The discovery that AGEPC can induce simultaneously leukocyte chemotaxis and degranulation suggests that the prolonged permeability effects are leukocyte-dependent by the same mechanism and not by two separate mechanisms.[29]

Recent investigations have shown that the microvasculature acts as more than a passive stage for the cellular and humoral manifestations in acute inflammation. Endothelial cells perform a variety of metabolic functions (combined with tissue injury), they synthesize prostacyclin and probably other arachidonate derivatives, they influence smooth muscles, leukocytes and platelets' functions, metabolism of vasoactive substances (bradykinin, angiotensin, histamine, and serotonin), they produce granulopoetin-like material and bind chemotactic compounds. The potential role of AGEPC in modulating these and other vascular functions must be elaborated.

X. ACUTE PHASE PROTEINS

In the acute phase of tissue injury, caused by inflammation, infection or malignant neoplasia, a group of so-called ''acute phase proteins'' is synthesized. The first isolated acute phase protein (in O. T. Avery's laboratory at the Rockefeller Institute) was the C-reactive protein (CRP—the name denotes that it precipitates pneumococcal somatic C-polysaccharide). Many circulating protease inhibitors, complement components, coagulation factors, and transport proteins have been shown to behave as acute phase proteins. Among these proteins the amount of C-reactive protein increases after tissue injury or inflammation 100- to 1000-fold over the resting state. Human CRP is a plasma protein, composed of five identical 21,000 Da subunits forming a non-glycosylated cyclic pentamer. The CRP is synthesized mainly in the liver. The *in vivo* rate of catabolism of CRP is rapid and not significantly different in normal animals, and animals with acute inflammation. This can be interpreted that the acute phase increase of CRP must be caused by increased rate of its sythnesis. The increased in CRP concentration exceeds by several orders of magnitude the increase for other acute phase serum proteins, haptoglobin, fibrinogen, α_1 antitrypsin, α_1 antichemotrypsin, acid glycoprotein, C3 (the third complement component), and ceruloplasmin.[30]

Induction of acute phase proteins seems to be dependent upon interleukin-1 (IL-1), the product of mononuclear phagocytes. Addition of IL-1 to murine hepatocytes in primary culture mimes many, if not all, the positive and negative changes in specific gene expression characteristic of the acute phase reaction in response to inflammation or tissue injury. One of the two major histocompatibility complex (MHC) class III genes encodes the positive acute phase protein, complement factor B. Purified IL-1 increased the expression of factor B and other positive acute-phase proteins in human hepatoma cells, and decreased synthesis of albumin, the negative acute-phase reactant. These results helped to establish viewpoints on selective and independent mechanisms for the regulation of genes engaged in molecular control of the inflammatory response.[31]

The general nature of the acute phase response, its stimulation by many forms of cell injury, and its occurrence in all homoiothermic species suggests that it has a beneficial function. In all species CRP binds to phosphorylcholine residues in the presence of Ca^{2+} A normal human plasma protein, known as serum amyloid P component because its pentagonal molecule is laid down in amyloid deposits, closely resembles CRP in structure (about 70% homology of amino acid sequences). A severe complication of sustained acute phase response, due to persistent infection or inflammation, is the deposition of serum amyloid A protein (SAA), an extracellular accumulation of protein fibrils consisting of peptide subunits arranged in anti-parallel β-pleated sheets. The protein is derived from a serum precursor (SAA). SAA is an apolipoprotein of high-density lipoprotein (HDL) synthesized during the acute phase. The trigger molecule initiating CRP-synthesis is interleukin-1 (IL-1), produced by macrophages. IL-1 also activates T-lymphocytes and is a pyrogen. Its synthesis may be induced by direct stimulation of macrophages (e.g., by bacterial lipopolisaccharide), or indirect mediators of inflammation released from damaged tissue, e.g., prostaglandins. CRP binds preferentially to damaged cell membranes activating the complement cascade. The roles of CRP and SAP (serum amyloid P component) are unknown. SAP occurs in glomerular basement membrane and in association with elastic fiber microfibrils throughout the body. Aggregated SAP selectively binds fibronectin from whole serum.[32]

Familial amyloidotic polyneuropathy (FAP)—FAP is a disease of autosomal dominant inheritance, characterized by extracellular deposition of amyloid fibrils and progressive disorder of the peripheral nerves. A variant prealbumin containing a Val → Met substitution at position 30 (in exon) is considered to lead to extracellular deposition of amyloid fibrils in Japanese, Portuguese, and Swedish types of FAP. In addition six other substitutions have been found in introns of the prealbumin gene.[33,34]

XI. COAGULATION IN INFLAMMATION

Coagulation ceased to belong to the unique domain of hematologists because there has been a deserved and growing interest in blood clotting factors which participate in inflammation and carcinogenesis.

Activation of the blood clotting cascade involves not only clot formation, but also fibrinolysis and kininogenesis. Activation of these pathways occurs by contact activation, represented in Figure 3.

Activation of the blood clotting[36] cascade occurs by the formation of several active complexes which by interaction form the prothrombinase complex that cleaves fibrinogen to fibrin, stabilized by active factor XIII.

The initiation activation complex of the intrinsic pathway includes: factor XII (Hageman factor), factor XI, prekallikrein, high molecular weight kininogen (HMW-kininogen), and the injured surface of the intima of a blood vessel. At first, factor XII undergoes limited cleavage which displays the active sites of protein enabling it to interact with the other components of the activation complex, and resulting in cleavage of factor XI, which in this way becomes activated (see Figure 4).

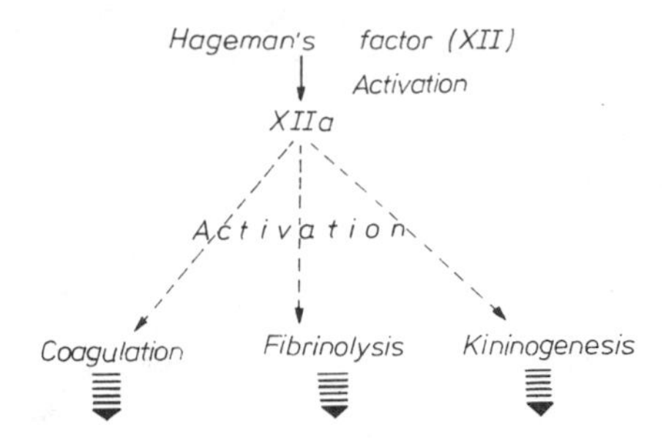

FIGURE 3. Interaction of coagulation, fibrinolysis, and kininogenesis.

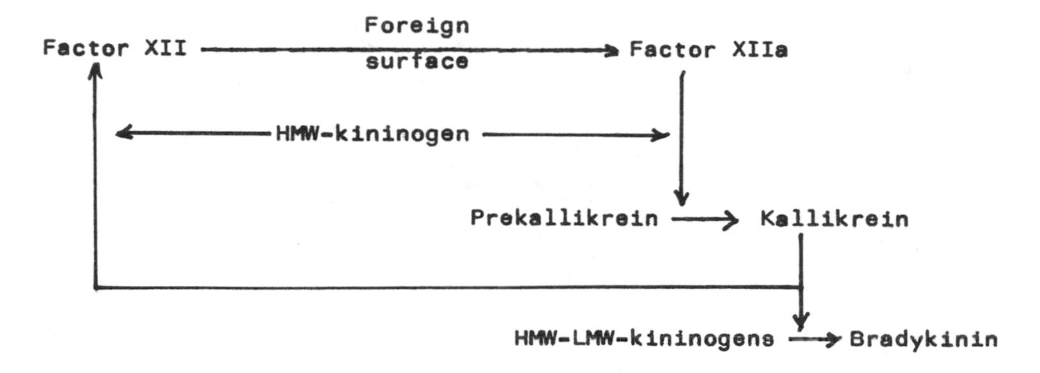

FIGURE 4. Formation of the initiation complex in coagulation.

The active XI complex includes activated factor IX, factor VIII, phospholipid and Ca^{2+}. The active complex X is composed of activated factor X, factor V, phospholipid and Ca^{2+}. The activated complex X represent the enzyme "prothrombinase" which cleaves prothrombin to thrombin. Finally, thrombin cleaves fibrinogen to fibrin which forms a network, the most obvious feature of the blood clot, and stabilized by the activated factor XIII.

The extrinsic pathway of blood coagulation takes place in the tissue where a complex composed of factor VII, a tissue factor, phospholipid and Ca^{2+} is formed. The activated complex activates factor X complex which leads to the common reaction of the intrinsic and extrinsic pathways (see Figure 5).

The terms surface and phospholipid most probably denote specific membrane receptors on the cellular constituents of the blood coagulation system or on subcellular debris. Complex formation serves to accelerate activation of the particular coagulation factors and decreases the action of natural coagulation inhibitors.

Thrombin cleaves different substrates, including fibrinogen, factor V, factor VIII, factor XIII, and prothrombin. In addition, thrombin binds with platelets, and induces their aggregation resulting in the release of several active substances.

The cellular components associated with the traumatized vessel, crucial for *in vivo* hemostasis, are the platelets. Binding sites on platelets have been described for fibrinogen, factor V and Va, factor Xa, thrombin and v. Willebrand factor. Binding with v. Willebrand factor enables binding of platelets to the subendothelium of blood vessels.

Factor XII is a single polypeptide chain (M_r 80,000). Surface bound factor XII is many times more susceptible to cleavage by kallikrein than nonsurface bound. Cleavage generates α XIIa, a two-chain structure that activates factor XI. Further cleavage yields esterase and amidase activity activating mainly prekallikrein and, only poorly, factor XI. The remainder

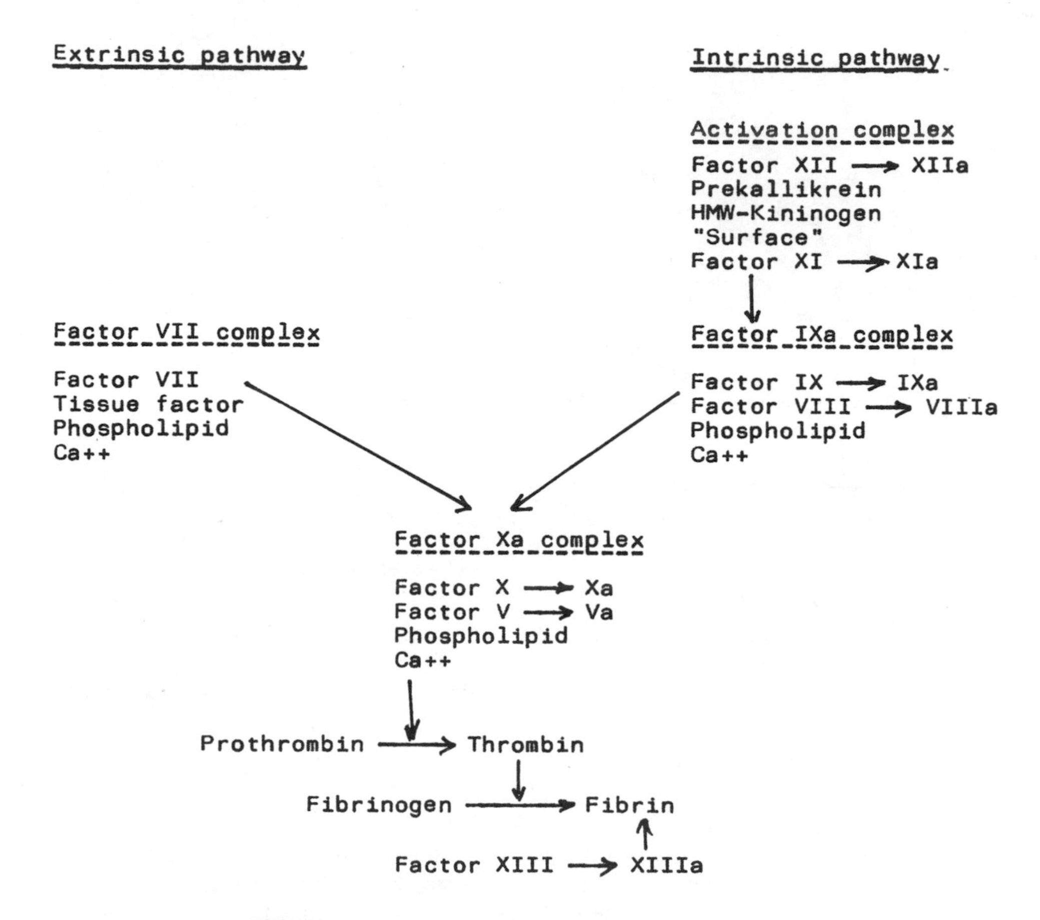

FIGURE 5. Intrinsic and extrinsic pathways of coagulation.

of the molecule (52,000 Da heavy chain) remains bound to the surface. In the presence of kaolin, HMW-kininogen and Russell's viper venom inhibitor II, the intact factor XII activates factor XI and prekallikrein at the same rate.

Factor XI is a symmetrical double molecule, consisting of two chains of 83,000 Da, each containing a serine site. Activation by XIIa cleaves each of the chains, inducing reactivity in the active site serines. XIa has also some esterase activity toward low molecular weight substrates. Physiologically XIa activates factor IX, but also has some activity toward factor XII.

High molecular weight kininogen (HMW-kininogen) cleaved by kallikrein yields bradykinin and participates in contact phase coagulation (M_r 110,000). The remainder consists of a heavy and a light chain, linked by disulfide bonds. The pro-coagulant feature is bound with the light chain.

Kallikrein is a serine protease generated by factor XIIa proteolytic activity on prekallikrein. Plasma prekallikrein circulates in a 1:1 complex with HMW-kininogen and participates in surface activation of factor XII. Kallikrein cleaves kininogen, plasminogen, and factor IX.

To the vitamin K-dependent factors belong: factor IX, factor VII, factor X, prothrombin, protein C, protein S, protein Z, and protein M. These proenzymes undergo posttranslational modification in microsomes prior to secretion into the plasma. The modification involves vitamin K, causing carboxylation of specific glutamic acids in their N-terminal regions. Carboxylation of glutamic acid in the γ position causes formation of γ-carboxyglutamic acid. This modification enables binding of Ca^{2+}. After binding metals these factors can bind vesicles composed of acidic phospholipids. Binding of metals and phospholipid is essential for these factors to be utilized as enzymes and substrates in the generation of procoagulant activity. Each of these K-dependent factors contains 10 to 12 γ-carboxyglutamic acids. The proteins C, S, and Z have been discovered in patients who had congenital deficiencies in these blood coagulation factors.

Factor IX is a single chain proenzyme (M_r 55,000), activated by factor XIa or by factor VIIa. Activation of factor IX requires calcium and two bond cleavages. The active-serine site is localized in the C-terminal 27,000 Da peptide.

Factor VII represents the tissue factor for the extrinsic coagulation pathway, able to activate factor IX. Also the uncleaved form of factor VII is biologically active, but the activity of factor VII can be stimulated through cleavage by other proteases which causes approximately a 100-fold increase of activity. The active-site serine is localized in the C-terminal peptide. The activity is destroyed by further cleavage.

Factor X is a two-chain molecule, composed of a heavy chain of 37,000 Da and a light chain of 17,000 Da bound by disulfide bonds. The light chain contains 12 γ-carboxyglutamic acid residues. Factor X is activated by cleavage by factor VII or IXa, yielding factor Xa.

Prothrombin (M_r 72,000) is activated by cleavage by factor Xa into active α-thrombin, composed of two disulfide-linked chains, an A chain of 57,000 Da and a B chain of 31,000 Da. The active-site serine is localized within the B chain. Prothrombin is also cleaved by thrombin, resulting in its inactivation.

Protein C (M_r 45,000) is activated through cleavage by thrombin. Unlike the other K-dependent proteins, activated protein C is an anticoagulant enzyme that inactivates both factor VIIIa and factor Va.

A. PROTEIN COFACTORS FOR VITAMIN K-DEPENDENT ENZYMES

Each of the vitamin K-dependent enzymes is activated up to 100,000-fold by the formation of a complex composed of a metal ion, phospholipid and a cofactor protein. Such a cofactor for factor X activation is represented by factor V, and for factor IX activation factor VIII. The increase of activity of factors X and IX by the cofactors suggests that these cofactors serve as regulators in thrombosis and hemostasis.

Factor V is an asymmetrical, high molecular weight (M_r 330,000), single-chain protein. Proteolysis of factor V by thrombin results in several peptides (D—94,000, F—71,000, D_1—92,000, G—31,000 and E—74,000 Da). The combination of D and E peptides is sufficient for Va activity. Va is inactivated by protein C.

Factor VIII is composed of multiple polypeptide chains. It is activated by cleavage by thrombin and inactivated by protein C. It serves as cofactor in factor IX activation. A tissue factor has been isolated, its M_r is approximately 43,000.

Complex formation is very important in the action of vitamin K-dependent proteins. The well-characterized prothrombin fragment 1 with Ca^{2+} concentrations sufficiently high to accommodate metal ion-dependent transitions will bind to lipid vesicles containing negatively charged phospholipid. The same factor V and Va binds tightly to phospholipid vesicles and to platelets. The binding to platelets is independent of the platelets' activation. Factor V binds to approximately 800 sites on the platelet. Factor Va binds to bovine platelets with about 1000-fold increase in affinity in its interaction with acidic phospholipid vescicles.

B. COAGULATION COMPLEXES

Although the particular coagulation factors may result in a blood clot, their concerted action in the form of protein complexes causes a very rapid acceleration of the reaction.

The prothrombinase complex composed of prothrombin, enzyme factor Xa, cofactor protein factor V (Va), calcium ions and phospholipids has been the most studied in this aspect. Accepting the relative reaction rate of factor Xa as 1.0, the reaction rate of the whole complex increases to 278,000, and considering factor V instead of Va the reaction rate increases only to 695. This results from condensation in a limited microenvironment which leads to acceleration dependent on concentration of enzyme and substrate.

Similar acceleration of the enzymatic reaction occurs in factor IXa activity. The IXa complex that activates factor X is composed of factor IXa, cofactor VIIIa, calcium ions and a phospholipid surface (possibly a platelet analogue) and the substrate factor X. Factor IXa alone can catalyze the factor X activation, but in complex with factor VIIIa a 10^5-fold increase of the reaction may occur. A similar acceleration occurs which phospholipid is replaced by platelets.

In the case of 1 + 2 − 12 the contact phase of coagulation is defined by the activated partial thromboplastin time, genetic deficiencies of factors participating in the formation of the contact activation complex, and the kind of deficiencies. It usually results in mild bleeding even without and disturbances. The contact activation complex which activates factor XI includes factor XII, prekallikrein, HMW-kininogen, factor XI, and a negatively charged surface. The surface is usually provided by the platelets. In laboratory test the surface may be provided by kaolin, glass, stearic acid, or other high molecular weight or water insoluble polyanions. The active factor XI activates factor IX to IXa by clearing it. The course of the reaction is as follows: when an appropriate surface is available these components may bind to the surface which changes conformation of factor XII resulting in its activation. The activated factor XII catalyzes limited proteolytic activation of the also immobilized HMW kininogen-prekallikrein complex, as well as the HMW-kininogen-bound factor XI. Alternatively, the intrinsic activity of prekallikrein may cleave factor XII, which bound to the surface is a very susceptible substrate to kallikrein. Once the reaction is initiated, factor XIIa can activate factor XI by its cleavage. The kallikrein products participate in kinin generation and plasminogen (fibrinolysis) activation. The primary activated fragment of factor XII is a poor factor XI activator, but it is a trigger that, through prekallikrein activation to kallikrein, accelerates activation factor XI within the initiation complex (see Figure 4).

Inhibitors of the blood clotting system, α_2 macroglobulin, α_1 antitrypsin, the complement $C\bar{1}$ inhibitor, and antithrombin III are among the inhibitors. Antithrombin III (M_r 65,000) is the universal regulator of the blood coagulation process. Antithrombin III inhibits all the known enzyme products of the blood clotting cascade except factor VII or VIIa and activated protein C. The inhibitory reaction may be increased by addition of heparin (reduced about 1000-fold). This acceleration of that inhibitory reaction by heparin can be abolished by the released platelet factor 4.

C. REGULATION OF ANTICOAGULANT ACTIVITY

The anticoagulant activity is induced by contact with the cell surface, namely, (1) cell surface heparin that accelerates inactivation of the coagulation proteases by antithrombin III; (2) thrombomodulin that alters macromolecular specificity of thrombin decreasing its ability to catalyze clot formation, simultaneously converting thrombin into a protein C activator. Active protein C (a serine protease) inactivates factors Va and VIIIa, essential for the function of factors IXa and Xa, simultaneously active protein C stimulates fibrinolysis by neutralization of plasminogen activator inhibitor. Rapid activation of protein C occurs on the surface of the endothelial cells by a complex formed by the integral membrane protein thrombomodulin and thrombin. Complexed thrombin with thrombomodulin increases 1000-fold the activation of protein C. Ca^{2+} inhibits activation of protein C, but is necessary for activation of protein C by complex thrombomodulin x thrombin. The thrombomodulin x thrombin complex in the presence of Ca^{2+} activates protein C and directly inhibits clot formation by thrombin (fibrinogen → fibrin). Protein C is a vitamin K-dependent protein which is similar to all vitamin K-dependent proteins in that their activity requires negatively charged phospholipids on the cell surface. This requirement is common both for the pro-coagulant and anticoagulant activity. Thus, the mechanisms for these two activities must differ. Binding of active protein C with the cell surface requires the presence of another vitamin K-dependent protein, namely protein S. Protein S circulates both free and bound to an acute phase protein C4bp (C4b binding protein). C4bp and protein S are in equilibrium, thus an increased level of C4bp causes a decrease of free protein S, which predisposes to thrombosis. Active protein C is inactivated by a slow-acting protease inhibitor and a plasminogen activator inhibitor which inactivates both protein C and plasminogen activator. Both, protein C and protein S are vitamin K-dependent proteins, posttranslationally modified by the vitamin K carboxylase. The modified domain is responsible for the Ca^{2+}-dependent membrane interactions. Thrombomodulin is a glycoprotein with a short cytosolic domain at the C-terminus, a membrane spanning domain and a region above the membrane composed of several growth factor domains (Figure 6).

XII. ROLE OF ANTICOAGULANT ACTIVITY IN INFLAMMATION

Inflammatory agents such as endotoxin may initiate activation of tissue factor on monocytes and endothelial cells that triggers the coagulation cascade. Simultaneous induction of anticoagulant activity prevents deep venous thrombosis or disseminated intravascular coagulation. The interaction between inflammation and coagulation may be caused by the action of interleukin-1 (IL-1) that initates tissue factor formation on endothelial cells and synthesis of leukocyte-binding sites on the cell surface. In this way the endothelial cells become deprived of natural anticoagulant properties. Under the influence of endotoxin or tumor necrosis factor these cells become deprived of the thrombomodulin function and the capacity to bind protein S and stimulate factor Va inactivation, which, in severe infections, may cause thrombosis and disseminated intravascular coagulation.[38]

A. CONGENITAL DEFICIENCY OF PROTEIN C ANTICOAGULANT PATHWAY

A severe thrombotic tendency has been described in all homozygotes with protein C deficiency, leading sometimes to purpura fulminans, a fatal disease involving uncontrolled coagulation particularly in the microvasculature of the skin. Many families with heterozygous protein C deficiency and recurrent thrombosis have been described.[38]

One of the most severe disorders caused by blood clotting disturbances is disseminated intravascular coagulation (DIC). Activation of the blood clotting cascade occurs through limited digestion which enables almost immediate activation of blood clotting, however, it may be dangerous because of depletion of coagulation factors. Depletion of coagulation factors occurs because of simultaneous activation of coagulation and fibrinolysis. Several inflammation products or agents that cause activation of blood coagulation by atypical blood coagulation inducers are very important in this aspect.

To such inducers belong:

- Bacterial endotoxins that may directly transform fibrinogen to fibrin
- Tissue thromboplastins entering circulation from the placenta, surgically operated tissues (mainly lungs), lysed erythrocytes, etc.
- Antigen-antibody complexes or aggregated gamma-globulins, activating Hageman's factor (XII) and inducing coagulation, fibrinolysis, kininogenesis, and complement activation

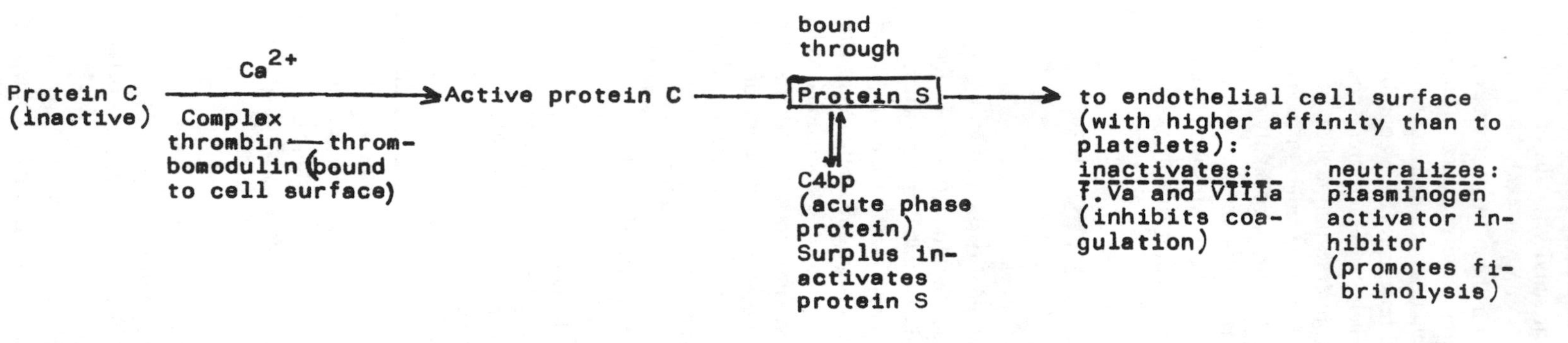

FIGURE 6. Protein C pathway. Inflammation: endotoxin or other inflammatory agents trigger the coagulation cascade; simultaneous induction (by surplus of thrombin) of coagulation and anticoagulant pathways prevents thrombosis; surplus of C4bp inactivates protein S that leads to disseminated intravascular coagulation and thrombosis.

- Colloidal particles: saturated fatty acids, chylomicrons, adhered or aggregated platelets
- Viruses activating complement
- Proteolytic enzymes entering circulation (e.g., trypsin in pancreatic necrosis)

The consequences of DIC may be severe bleeding because the blood components that should be available for clot formation already have been depleted. Alternatively, the clotting abnormalities may result in thrombosis with blocking of several blood vessels.

The dominant role of coagulation in inflammation, especially septic, is stressed by the fact that a large dose of plasma antithrombin, the universal inhibitor of coagulation, may prevent a lethal diffuse intravascular coagulation induced by a lethal dose of tissue thromboplastins.

B. GENERALIZED SHWARTZMAN REACTION

The typical Shwartzman reaction is composed of intravenous injection of a sublethal dose of bacterial endotoxin (or other toxins) followed within 24 h by an additional administration of the same or other toxin or, vice versa, at first subcutaneous injection followed by an intravenous injection of endotoxin. This results in severe, often lethal intravascular coagulation and fibrinolysis (DIC) with glomerular capillary thrombosis leading to bilateral cortical necrosis of the kidneys. The majority of studies suggested that the generalized Shwartzman reaction is caused by the endotoxins in generating intravascular coagulation by directly activating platelets, resulting in aggregation and degranulation. Recently, a hypothesis has been put forth that the generalized Shwartzman reaction may be due to endothelial cell desquamation leading to exposure of the basement membrane, platelet aggregation, and the consumption of coagulation factors.[37] The hypothesis was based on the presence of endothelial cells in the circulation of the endotoxin-treated animals. In addition, damage of the endothelium causes prostacyclin deficit because of its synthesis in these cells—therefore supplementation of exogenous prostacyclin exhibits beneficial results through inhibition of platelets aggregation and fibrin deposition in the glomerular capillaries.

The local Shwartzman reaction is caused by local skin reactivity caused by intradermal injection (the preparatory phase of the reaction) of endotoxin, followed 18 to 24 h later by an intravenous injection (the provoking phase of the reaction) of the same or other endotoxin. Some hours later hemorrhagic necrosis of the skin develops at the site of the intradermal injection. Precisely because the second reaction may be caused by various toxic agents it is thought that a common mediator must be responsible for both the preparatory and provoking reactions in the local Shwartzman reaction, and this is thought to be IL-1, the major immunoregulatory molecule produced by macrophages. IL-1 is responsible for the nonspecific host response by producing fever, neutrophil release from the bone marrow, T-cell proliferation, acute phase protein synthesis in the liver, fibroblast proliferation, and leukocyte adhesion to endothelial cells.[39]

XIII. THE ROLE OF FIBRINOLYSIS IN INFLAMMATION

Plasminogen, the precursor-enzyme which degrades fibrin, is activated physiologically by the active Hageman factor (XII). Plasminogen also can be activated by urokinase, synthesized by the kidneys, trypsin, and thrombin.

In pathological conditions plasminogen is activated mainly be bacterial products (streptokinase, etc.).

Plasmin is able to digest fibrinogen, fibrin, coagulation factors II, V, VIII and C1̄ inhibitor. The natural inhibitor of plasmin is hepatic antiplasmin.

The role of fibrinolysis in inflammation is emphasized by the fact that trasylol, an inhibitor of serine group enzymes, was found in the clinic to have a widespread application in prevention of inflammation.

XIV. THE ROLE OF CELLS IN INFLAMMATION

In the first phase of inflammation there is a transient increase in vascular permeability. The second phase is more prolonged and consists of accumulation of leukocytes by chemoattractants, their adhesion to vessel walls, migration of white blood cells into the inflammatory area, phagocytosis, and fibrin deposition accompanied by the second phase of vascular permeability. The leukocytes exhibit activities closely connected with the pathogenesis of inflammatory reactions (chemotaxis, phagocytosis, release of hydrolytic enzymes, and reactive oxygen molecules).

A. HUMAN NEUTROPHILS

The inflammatory reaction of human neutrophils consists of two phases: in the first phase, the neutrophils responding to chemotactic agents accumulate in the area of the foreign intruder, and in the second they try to eliminate the intruder by secreting lysosomal enzymes and superoxide anions. The initiation of chemotaxis of neutrophils involves dramatic transient cytosolic acidification by sodium propionate, which activates the released lysosomal enzymes and seems to be a second messenger for the initiation of chemotaxis.[40]

B. THE LYSOSOMAL ENZYMES

The accumulated neutrophils after internalization of the foreign intruders undergo degradation which releases the bulk of lysosomal enzymes into the inflammatory area. Lysosomal enzymes include about 40 acid hydrolases (cathepsins, hydrolyzing proteases, acid ribonucleases, phosphatases, enzymes, decomposing mucopolysaccharides, etc.). These enzymes are enveloped in lysosomal membranes which prevent their release in noninjured cells. Injured cells release large amounts of these enzymes after degradation which removes cell detritus, but they also are able to injure normal cells in the inflammatory area. In some instances, mainly in chronic inflammations caused by immunological disturbances, nondegraded cells also are able to release lysosomal enzymes which injure these cells. The release of lysosomal enzymes from nondegraded cells is mediated by the lysosomal releasing factor of the complement component C5a.

Human neutrophils, after binding of lipid or peptide chemotactic factors, release several enzymes, among others the plasminogen activator. The plasminogen activator bound to a specific receptor on the surface on monocytes remains active, providing a mechanism concentrating proteolytic activity on the cell surface, which suggests participation in processes requiring direct, local proteolysis.[41]

The rapid stimulation of neutrophils by chemotactic factors causes their aggregation, adherence, lysosomal degranulation, production of superoxide anions, phosphatidyl inositol turnover, internal calcium release and influx of extracellular calcium, glucose flux, and membrane depolarization.[42]

The antimicrobial system of neutrophils depends on the production of reactive oxygen intermediates and by bactericidal factors present in these cells. the bactericidal factors kill Gram-positive and Gram-negative virulent bacteria. The bactericidal factors have been localized on the membranes of azurophilic granules as a highly potent, broad spectrum, rapidly acting protein(s) effective in a physiologic medium.[43]

Besides bactericidal factors, neutrophils reveal an enzymatic activity that causes detoxification of bacterial lipopolisaccharides (endotoxins). It has been demonstrated that acyloxyacyl hydrolysis of bacterial lipopolysaccharides reduces their toxicity. It is suggested that, during invasion of Gram-negative bacteria, acyloxyacyl hydrolysis may be the defense mechanism that reduces the toxicity of lipopolysaccharides, preserving their potentially beneficial inflammatory and immune stimuli (deacylated lipopolysaccharides stimulate B lymphocytes at a somewhat decreased level).[44]

An important factor in inflammation is vascular permeability. The chemical mediators of edema in man are little known. The typical inflammatory mediators histamine, serotonin, and bradykinin insufficiently explain the chemical mechanisms of edema. More important seem to be the derivatives of complement activation. Incubation of blood plasma with zymosan (activating the alternative pathway of complement) causes rapid generation of C5a which increases vascular permeability and together with the vasodilatator, prostaglandin, induces edema, independent of histamine release. C5a demonstrates chemoattraction of polymorphonuclear leukocytes. Further experiments demonstrated that, in the absence of polymorphonuclear leukocytes, zymosan was unable to induce edema. The chemotactic factors induce only a little edema, but mixed with PGE_2 in the presence of polymorphonuclear leukocytes the induction of edema was greatly enhanced. These results suggest that polymorphonuclear leukocytes are essential for the permeability and edema in inflammation.[45]

Chronic granulatious disease may be inherited as an autosomal recessive or X-linked disorder. Phagocytosis by neutrophils, macrophages, and eosinophils is accompanied by a microbicide from non-mitochondrial respiration that is important for the killing and digestion of bacteria. This oxidase activity is based on an electron transport chain with an unusual b-type cytochrome b_{-245}. The molecular defect of the X-linked inherited disease is the absence of cytochrome b_{-245} which is present in the autosomal inherited disease. In the autosomal inherited granulomatous disease the metabolic disturbance is caused by the absence or malfunction of a proximal component of the electron transport chain in the form of a selective lack of protein phosphorylation (M_r 44,000), which may be an important component of this oxidase [46]. In the X-linked granulomatous disease the absence of two subunits of apoprotein closely linked and associated with the heme of the cytochrome b_{-245} was found, whereas both subunits were present in the autosomal recessive form of this disease.[47]

C. MACROPHAGES (MONONUCLEAR PHAGOCYTES)

Human peripheral blood monocytes secrete over 50 proteins, regulated internally during their maturation and also by extracellular agents. The development of macrophages is stimulated by four colony stimulating factors that influence the development of bone marrow precursor cells from multipotential stem cells to mature macrophages or neutrophils.[48] These substances are:

1. Multi-colony stimulating factor of M_r 19,000 to 30,000 (generated in T-cells), that acts on pluripotent stem cells;
2. Granulo/macrophage-colony stimulating factor of M_r 23,000 (generated in macrophages, fibroblasts, and endothelium), that acts on a bipotential stem cell to produce mononuclear phagocytes and granulocytes;
3. Granulocyte-colony stimulating factor of M_r 25,000 (generated in macrophages), that causes granulocyte proliferation (in higher concentration also monocyte proliferation;
4. Macrophage-colony stimulating factor of M_r 70,000 (generated in fibroblasts), that is exclusively a mononuclear phagocyte growth factor.

Unique receptors have been identified for each colony stimulating factor. These are not all the factors that act on the level of early myeloid stem cell. A specific factor is present during inflammation, so called "factor increasing monocytopoesis" not related to other known factors.[48]

Interferon gamma is the only known factor that induces activating effects on oxidative metabolism producing the reactive oxygen metabolites killing bacteria. Monocytes activated by interferon gamma can lyse tumor cells. In addition to an increase of reactive oxygen intermediates, interferon gamma increases the affinity of Fc receptors and an increase of MHC class II molecules and an alteration in the synthesis of complement components. A factor which can block the induction of interferon gamma or reverse preexisting activation was isolated from some murine tumors, named "macrophage deactivation factor".[48]

In macrophages, the released arachidonic acid is metabolized either through the cyclooxygenase pathway yielding PGE_2 thromboxan A_2, and lipoxygenase pathway yielding potent inflammatory mediators. Exogenous prostaglandins may inhibit proliferation of macrophage progenitor cells. PGE_2 stimulates an increase of phagocytosis and the number of Fc receptors, it augments endotoxin-induced collagenase production, which may cause degradation of the connective tissue observed in inflammation. Prostaglandins inhibit spreading, adherence, and migration of macrophages.[49] Human macrophages (monocytes) have a receptor for ingestion of activators of the alternative pathway of complement activation, which also functions in the absence of plasma protein, distinct from the receptor for FcIgG and the C3b fragment of component C3 of the complement. Such an activator is soluble β-glucan that binds with a specific receptor on macrophages and activates the alternative complement pathway inducing phagocytosis and leukotriene generation.[50]

Human mast cell granules contain different enzymes chemotactic factors, heparin proteoglycan, and histamine. The heparin proteoglycan seems to be the store of preformed mediators and must be solubilized for their release. About one third of heparin glycosaminoglycans have antithrombin III activity, which inhibits the alternative complement pathway. In this way the heparin released by C5a stops the production of additional activating peptides. The presence of acid hydrolases (β-hexosaminidase, β-glucuronidase and arylsulfatase) suggests that these secretory granules are modified primary lysosomes. The neutral proteases of the mast cell granule are tryptic, and probably digest proteoglycan core peptides, collagen, and elastin. Their presence clears the products of damaged tissues and facilitates their repair.

Release of mast cell granules occurs after binding of IgE to its receptor on the surface of mast cells which initiates a transmembranal signal for solubilization and movement of the swollen granules to the perigranular membrane-plasmolemma. After fusion extracellular release of the granule constituents occurs. The IgE membrane receptor is composed of an α-subunit of M_r 50,000 and β-subunit of M_r 30,000. Binding of IgE to the receptor activates the adenylate cyclase which generates cAMP from ATP. cAMP activates cytoplasmic cAMP-dependent protein kinase holoenzymes. In addition, IgE-receptor interaction causes sequential action of two methyltransferases which catalyze methylation of phosphatidylethanolamine to form phosphatidylcholine, which in turn activates the calcium channel and increases calcium influx enabling secretion (mediators release).[20] In consequence of their secretion, the arachidonic cyclooxygenase and lipoxygenase pathways become activated with release of prostaglandins, especially PGE_2 active in the inflammation process, as well as leukotrienes, especially the glutathione-containing leukotriene LTC_4, the major mediator of smooth muscle contraction released by mast cells after interaction with cell-bound IgE.[51]

D. EOSINOPHILS

For many years clinicians have been tempted to relate increase of the number of eosinophils with certain illnesses, particularly in asthma, allergic diseases, and helminthic parasites infection. None of these proved to be absolutely correct, and there is not a unified explanation about the role of eosinophils in inflammation. It is difficult to explain the sudden disappearance of the eosinophils from the blood in pyogenic infections. Modern histological examinations revealed that eosinophil material is much more widespread than was accepted earlier.

Recent investigations have discovered several proteins in the eosinophil granules. At first, eosinophil cationic protein (ECP) was discovered, then eosinophil major basic protein (MBP). The role of these proteins in inflammation is unclear. The Charcot-Leyden crystal protein is composed of lysophospholipase (derived from the plasma membrane of eosinophils), arylphosphatase B and phospholipase D. Eosinophils, after stimulation, secrete stored granule constituents and new compounds, synthesized during stimulation. To the most important belong: LTC_4 and products of oxygen metabolism including superoxides, oxygen radicals, and hydrogen peroxide (its capacity for the latter exceeds even that of neutrophils). Factors that induce secretion of eosinophils are IgE and C3. Like other inflammatory cells, eosinophils have receptors for IgE and C3.[52] Activated eosinophils produce sulfidopeptide leukotrienes particularly after contact with large particles coated with IgE.[53]

Eosinophils leave the bone marrow still immature after their final cell division. Their maturation probably occurs in the spleen under the influence of maturation factors. Mononuclear cells from human blood (T lymphocytes and monocytes) produce activating factors for eosinophil-dependent killing of antibody coated schistosomula of *Schistosoma mansoni*. The activating factor is about 40,000 Da, heat-resistant, and trypsin-sensitive. Interleukin-2 (IL-2) has a possible role in eosinophil proliferation.

XV. THE ROLE OF PLATELETS IN INFLAMMATION

Platelets contain a multitude of enzymes (catalase, hyaluronidase, esterase, β-glucuronidase, nucleotidase, tryptase, tyrosinase, pyrophosphatase). Bacteria, viruses, and antigen-antibody complexes cause degranulation of platelets with release of inflammatory mediators: histamine, PGE_2 and β-lysine.

A. PLATELET DERIVED GROWTH FACTOR (PDGF)

PDGF from human platelets is a cationic heterodimer glycoprotein, of M_r about 30,000, composed of chains A and B. Expression of PDGF gene may be regulated by specific factors released locally by activated cells. B-chain transcripts have been demonstrated in several cells (transformed cells by phorbol esters, human tumors, and developing placenta). It is suggested that in different cells inactive precursor forms of PDGF may exist or active PDGF remains associated with cells in which it is synthesized. The half-life of PDGF in the blood is extremely short. Binding of PDGF to a specific matrix may prolong its mitotic activity. Clearance of secreted PDGF may be mediated by plasma proteins, e.g., α_2-macroglobulin inhibits PDGF binding with its receptor.

1. The Biological Effects of PDGF

Interacting with responsive cells, PDGF causes release of prostacyclin (PGI_2) and PGE_2, which may result in bone resorption. These prostaglandins are potent anti-atherosclerotic agents as vasodilatatory anti-platelet agents. PDGF increases both expression of LDL receptors and uptake of LDL, which suggests that such mitrogens as PDGF may be involved in the LDL pathway because of a requirement of cholesterol by proliferating cells. PDGF specifically stimulates collagen V formation and type III versus type IV collagen. Accumulation of connective tissue matrix around the proliferating cells may cause various diseases manifested with fibrosis and cartilage destruction (rheumatoid arthritis, liver cirrhosis, atherosclerosis). PDGF is a chemoattractant for fibroblasts and smooth muscle cells which may be essential in the repair processes and for monocytes and neutrophils which may do not respond mitogenically, but release their specific granular constituents. High affinity cell-surface receptors have been described only on connective tissue cells (fibroblasts, vascular smooth muscles, chondrocytes, glial cells). PDGF also stimulates the production and release of somatomedin-C-like substances by cultured human fibroblasts. PDGF also is a potent vasoconstrictor.[54]

The main source of PDGF are α granules of the blood platelets after exposure to thrombin, collagen, and ADP. Deposition of new platelets is necessary for continuous release of PDGF. PDGF may be secreted by circulating cells (platelets, monocyte/macrophages engaged in wound healing, atherosclerosis, and inflammation) fibrotic disorders of lung, liver, and kidney, and by resident cells (megakaryocytes, endothelial cells, smooth muscle cells, human placental cells, cells transformed by simian sarcoma viruses). To the agents inducing PDGF secretion belong: thrombin, collagen, adherence factor Xa of the coagulation cascade, and others. Cells producing PDGF are capable of also producing other growth factors, for example platelets synthesize FGF (fibroblast growth factor), EGF (epidermal growth factor), TGF-β (transforming growth factor); monocyte/macrophages synthesize FGF and IL-1; smooth muscles from rat pup synthesize IGF (insulin-like growth factor). A measurable amount of PDGF is absent in normal blood. Rapid appearance of PDGF in the blood is probably caused by its clearing from the plasma compartment. The increase of PDGF after thrombin stimulation probably results by conversion of preformed precursor into active PDGF.[54]

Wound repair is well-understood by the appearance and disappearance of various cellular constituents. After deposition of platelets and blood coagulation, the next cells are granulocytes, monocytes, and lymphocytes, and then fibroblasts which secrete collagen with the consequence of forming numerous blood capillaries. Platelets, monocyte/macrophages and injured endothelial cells secrete PDGF together with other growth factors. Platelet-derived PDGF seems to be important for chemotactic properties attracting leukocytes and fibroblasts, while macrophage-derived PDGF is needed for the continuing process of fibrinogenesis. An important role is played by platelet PDGF after removal of arterial endothelium, in this case platelets immediately adhere to the exposed subendothelium. PDGF is secreted which initiates chemotaxis followed by migration and proliferation of these cells. Macrophage activation is associated with several tissue reactions, such as pulmonary fibrosis, liver cirrhosis, and inflammatory arthritis. In these disorders lung and liver fibroblasts or synovial cells produce large amounts of connective tissue.

XVI. THE INFLAMMATORY MEDIATORS OF THE IMMUNE SYSTEM

An important source of inflammatory mediators is represented by the immune system. Fractionation of cells engaged in the immune reactions led to the discovery in cytolytic T- and natural killer cells (NK) of pore-forming, lytic molecules (named perforin or cytolysin—similar to the lytic pore-forming C9 component of the complement system). In addition, in the cytotoxic T-cells and NK-cells the presence of various serine esterases have been discovered, which present another similarity with the complement system. However, there are also some dissimilarities: antibodies against perforin-mediated lysis do not inhibit T-cell mediated lysis; why are cytolytic cells not lysed by the pore-forming protein it produces?; no perforin activity was found in cytotoxic T-cells. The main difference is that in target cells the lytic process caused by cytolytic T-cells initiates in the nuclear and not in the cytoplasmic membrane with rapid DNA degradation. Precisely, this observation led to the hypothesis that under the cytolytic T-cell attack an autodestructive mechanism becomes triggered which causes a kind of ''suicide'' of the target cell.[55] It is supposed that in this case an inducer on the effector T-cell reaches a target-cell acceptor molecule, which activation causes disintegration of the target-cell by unknown mechanism.[55]

The induced-suicide model of cytolytic T-cell action is supported by a similar process seen in glucocorticoid-mediated cytolysis of immature thymocytes in which the same characteristic degradation of chromosomal DNA occurs.

A. TUMOR NECROSIS FACTOR (CACHECTIN)

Invasive diseases frequently cause metabolic disturbances manifested as a wasting of the body (cachexia) or shock. The metabolic disturbances are mediated principally by the immune system (reticuloendothelial cells and lymphocytes) which in response to a variety of stimuli secretes different cytokines: interleukin-1, lymphotoxin, gamma-interferon and tumor necrosis factor, also called cachectin. These substances are able to alter host metabolism.

The tumor necrosis factor is of special interest because it is involved in inducing shock, cachexia, and necrosis of tumor cells. The dramatic effect of anti-tumor activity on experimental tumors (hemorrhagic necrosis and regression) have not shown similar effects in cancer patients. Cachectin is a polypeptide hormone composed of subunits of molecular mass 17,000 arranged in dimeric, trimeric, or pentameric form, depending on species and method of isolation. It is a moderately hydrophobic protein containing intrachain disulphide bridges. Cachectin is secreted by macrophages stimulated by bacterial lipopolysaccharides (LPS) or other bacterial endotoxins. Cachectrin released into blood binds to specific receptors, mainly in liver, skin, kidneys, lungs, and gastrointestinal tract. The cachectin gene is located on chromosome 6, closely linked to the lymphotoxin gene. There exists considerable homology between cachectin and lymphotoxin genes, suggesting coordinate control of their expression. Isolated peritoneal macrophages contain detectable amounts of cachectin mRNA, but synthesis of cachectin occurs only after LPS induction. Glucocorticoids in nanomolar concentrations strongly inhibit cachectin gene expression, but once cachectin translation is initiated, dexamethasone is unable to arrest cachectin production. Cachectin administered to animals causes anorexia with weight loss. It seems that cachectin plays a major role in the pathogenesis of Gram-negative endotoxin induced toxemia (manifested by fever, metabolic acidosis, diarrhea, hypotension, and disseminated intravascular coagulation and fibrinolysis). Cachectin is a potent pyrogen, acting directly on the hypothalamic thermoregulatory centers.[56]

Several activities of tumor necrosis factor mimic the effects of lymphotoxin, including tumor hemorrhagic necrosis, fever, shock, and activation of neutrophils. Both are induced by LPS and share the same receptors. Macrophages induced by LPS secrete cachectin, IL-1, and granulocyte colony-stimulating factor. Several biological effects ascribed originally to IL-1, are now known to also be elicited by cachectin. Cachectin is a strong inducer of IL-1, being more potent in tumor necrotizing activity and expression of major histocompatibility complex (MHC) antigen in endothelial cells, whereas IL-1 is produced by a larger amount of cells, and activates T-cells, whereas cachectin does not.[57]

The tumor killing mechanism by both cachectin and lymphotoxin is not well understood: involvement of prostaglandins, proteases (lysosomal enzymes), free radicals, and DNA fragmentation have been suggested. The tumor necrotizing activity is potentiated by interferons, some metabolic inhibitors, and fever, whereas glucocorticoids and inhibitors of phospholipase block it. Cachectin increases procoagulant activity of the blood and enhances endothelial cell adhesiveness for inflammatory cells. The mechanism of cachexia caused by cachectin probably lies in lipoprotein lipase inhibition, the enzyme involved in lipid storage in adipose tissue. Lipoprotein lipase also may be inhibited by IL-1 and interferons. Cachectin also has anti-parasitic activity—it activates eosinophils to kill schistosomes in the presence of antibodies. In addition, there are elevated levels of cachectin in the blood of patients with visceral leishmaniasis and malaria. Finally, cachectin exhibits selective cytotoxicity for cells infected by RNA and DNA viruses.[57]

Summarizing: the tumor necrosis factor belongs to a polypeptide mediator system. LPS, or other microbial products induce synthesis of cachectin, IL-1, and granulocyte colony-stimulating factor in macrophages. In this way the macrophage exhibits a central position in inducing T-cells (by IL-1), granulocyte proliferation in bone marrow (by granulocyte colony-stimulating factor), and synthesis of cachectin stimulating IL-1 production by itself. Subsequently, T-cells synthesize IL-2 (a T-cell growth factor), granulocyte-macrophage colony stimulating factor also called IL-3 (in bone marrow), B-cell growth factor IL-4, and gamma-interferon that stimulates macrophages. Similar activity also has T-cells producing lymphotoxin.[57]

Another mechanism of cell killing is represented by cell-mediated killing by immune cells, which is an important defense barrier against proliferation of tumor cells, virus-infected cells, and foreign cells. An efficient way to destroy target cells is to produce pores in their membranes. Pore formation also serves as a killing mechanism by several bacterial toxins and parasites. For example. *Entamoeba histolytica* following contact with the target cell releases a pore-forming protein that forms a channel in the target cell membrane causing a lethal loss of water, electrolytes and macromolecules. Target-cells lysed by cytotoxic T-cells, or natural killer cells, show circular lesions on their surface that closely resemble the lesions produced by the membrane attack complex of complement. The cytotoxic T-cells and natural killer cells contain electron-dense secretory granules on the trans side of the Golgi apparatus. The isolated granules lyse tumor cells producing circular lesions on the target-cell membranes forming transmembrane channels. The purified lytic protein named perforin or pore-forming protein lyses a variety of cells. Its membranolytic activity is Ca^{2+} dependent (Ca^{2+} probably induces conformational changes enabling binding with lipid-binding domains). Cytotoxic T-cells killing target cells may be divided into four stages:

1. Receptor-mediated binding that enables interaction of the effector cell with the target cell
2. Polarization of Golgi apparatus granules and other cytoskeletal elements toward the target cell
3. Release of perforin into the intercellular space
4. Ca^{2+}-dependent assembly of membrane lesions with channel formation in the membrane of the target cell, resulting in leakage of water, salts, nucleotides, and proteins across the cell membrane of the target cell[58]

The presented model is in accordance with the observations that during cell killing, the Golgi apparatus, the microtubule organizer center and the granules of the cytotoxic T-cell reorient toward the target cell. Similar observations were made in the case of human eosinophils, the granules of which damage cell membranes by forming functional pores in the lipid bilayer of target cells. The secreted protein of eosinophils lyses parasites and degranulates mast cells at submicromolar concentrations. The described cell-killing mechanism resembles in many details the cell membrane damage produced by the humoral immune response, which involves interaction of antibodies with antigens, and the activation of the complement cascade resulting in pore formation and cytolysis. Polymerized C9 component of the complement pathway forms functional channels with properties similar to those of the lymphocyte perforin. C9 and perforin also may polymerize in solution, but once polymerized they cannot insert into membrane bilayers. This form of inactivation prevents spontaneous injuring of other cells. Serine proteases participate in cell mediated killing of cytotoxic T-cells and of natural killer cells. Serum proteases also are engaged in cell killing by *Entamoeba histolytica*, and in activation of complement cascade. The presence of transmembrane tunnels may also explain the early disintegration of the target-cell nucleus and fragmentation of target DNA, as an alternative mechanism for the earlier described "suicide" of target cells under the action of cytotoxic T-cells.

B. CHRONIC INFLAMMATION

The acute phase of inflammation involves a series of events mediated particularly by the macrophage-derived monokine, IL-1, resulting in increase of the synthesis of acute phase proteins, fever, leukocytosis, T-cell and B-cell activation, release of IL-2, activation of tissue cells (fibroblasts, chondrocytes, synoviocytes). Serine proteases, thrombin, and plasminogen activator (urokinase), simultaneously become activated. Thrombin stimulates proliferation of fibroblasts by interaction with specific cell surface components. Plasminogen activator, after conversion to plasmin, modified cell morphology and interactions of cells with their substratum. Many oncogenically transformed cells release plasminogen activator resulting in altered morphology and probably the invasive character of these cells. The interaction of various cells participating in inflammatory reactions must be correlated on a molecular basis. One of the correlating agents may be the protease nexin. Nexin released into the culture medium, becomes linked to thrombin and urokinase, and mediates most of the specific binding of these proteases to normal foreskin cells.[59] There is interesting evidence that several acute phase proteins may be deposited on newly formed elastic fibers in healing wounds, suggesting that these proteins are engaged in repair and resolution following inflammation. Therefore, the absence of an acute phase response seems to be responsible for the chronic course of several diseases.[60]

A very persuasive argument for such interpretation is the observation that hereditary deficiencies of early complement components usually are associated with the development of rheumatic diseases like systemic lupus erythematosus, while terminal component deficiency predisposes to recurrent neisserial infection. However, recently there have been described cases of rheumatic diseases in patients with C6 and other complement terminal component deficiencies.[61] Nevertheless, these observations suggest that some forms of chronic inflammation may result from specific defects in the presence of normal molecular mediators of inflammation.

Accumulation of lymphocytes in an inflamed joint space is correlated with the presence of ferritin and hemosiderin. In sites of inflammation polymorphonuclear cells release lactoferrin and macrophages release apoferritin. Lactoferrin exhibits high binding affinity for iron exceeding that of transferrin by several hundred times which causes a fall of the iron level in serum in acute inflammations. Lymphocytes have membrane receptors for ferritin, lactoferrin, and transferrin-bound iron which suggests that iron binding proteins and iron-proteins receptors may cause lymphoid cell migration and activation. Transferrin plays an important role in lymphocyte proliferation. T-lymphocytes proliferate *in vitro* only in the presence of transferrin 20 to 80% saturated with iron. Excessive iron may inhibit B-cell function which is evident in thalassemic patients which secrete abnormally low level of immunoglobulins. *In vitro* culture of B-cells with iron-chelating agents inhibits both proliferation and antibody synthesis.[62] In contrast to transferrin, the iron-storage protein, ferritin, inhibits T-cell proliferation. It is suggested that lactoferrin may regulate myeloid cell differentiation, modulation of macrophage-mediated cytotoxicity, and control of antibody production. However, these results are controversial and must be confirmed by further investigations.

Iron is a component of several enzymes in all living cells, available only in its ferric form. Microorganisms need large amounts of iron, therefore in acute infections a decrease of iron level in serum occurs, and in chronic infections the decrease of serum level of iron causes hypochromic anemia. Because of the great requirement of microorganisms for iron, the decrease of iron level in acute inflammations plays a kind of anti-bacterial mechanism.

B-cells also may be activated by α_2-macroglobulin-protease. This protease inhibitor is synthesized by hepatocytes and macrophages; B-cell activation may be blocked by trasylol, but not by soya bean trypsin inhibitor.

Chronic inflammation usually is characterized by proliferation of the connective tissue cells and matrix, which involves collagens fibronectin and other glycoproteins such as laminin. Recent studies have revealed that in the pathogenesis of chronic inflammatory diseases of joints and other tissues enhanced production of prostaglandin E_2 and collagenase by human synovial cells and fibroblasts may be induced by cachectin. In addition, cachectin may also stimulate the osteoclast's activity with bone resorption. Cachectin stimulates procoagulant activity by the vascular endothelial cells, and inhibits expression of thrombomodulin which disturbs protein C inhibitory activity on coagulation, resulting in an endotoxemia. Chronic inflammation usually causes changes in the amount and type of collagen synthesis with collagen type III the most prevalent. Interaction disturbances of the connective tissue protein fibronectin with fibrin and collagens is an important component in chronic inflammatory processes. Particularly in the rheumatoid joint, fibronectin complexes with hyaluronic acid influence the interaction of several synovial fluid components with the synovial cells (fibronectin-collagen complexes protect collagen from proteolysis).[60]

The next pathogenic agent of chronic inflammation causing tissue damage are free radicals. The pathway of free radical generation is $O_2^{\cdot -}$, OH^- and H_2O_2 formation. The most injurious is the OH^- radical. Its formation is most probably caused by reduction of iron salts. Because ferritin and lactoferrin, present in synovial fluid, bind iron in an unusual form there is free iron available for OH^- production.

Thus, the macrophage has the central position in chronic inflammation inducing tissue damage by proteases of the acute phase proteins the synthesis of which is induced by IL-1 secreted by the activated macrophages. IL-1 causes also connective tissue proliferation and cellular activation, and B-cell stimulation which are stimulated simultaneously by polyclonal B-cell activator generated by enzyme release from macrophages bound to α_2-macroglobulin. The inflammatory damage of tissues causes generation of free radicals which additionally cause damage to these tissues. Finally, iron bound to lactoferrin and ferritin activates B-cells because of the presence of receptors for these compounds on their cell membranes.[60]

The pathogenesis of chronic inflammation is associated with the accumulation of phagocytic leukocytes. Thus, the appearance of polymorphonuclear leukocytes and mononuclear leukocytes is characteristic for this kind of inflammation. Release of lysosomal enzymes from these cells due to specific stimuli or cell death, contributes to tissue damage and necrosis. The putative mediator of leukocyte activation seems to be leukotriene B_4, which is supported by the fact that inhibition of arachidonate lipoxygenase prevents tissue damage in chronic inflammation.[63]

Among the different factors which contribute to the pathogenesis of rheumatoid arthritis (genetic predisposition, impaired immunological control of virus infection, autoantibody formation), there are some clinical findings that speak for nervous system involvement: (1) symmetric distribution of synovitis; (2) in hemiplegic patients, joints in the paretic side develop later inflammatory changes; (3) exacerbation of the disease is often preceded by psychological trauma. The neuropeptide substance P may be the mediator for this interaction, since it is released into the joint tissues from primary sensory nerve fibers, and it stimulates prostaglandin E_2 and collagenase release from synoviocytes. Substance P is potent proinflammatory mediator which increases the severity of adjuvant-induced arthritis in rats.[64]

C. ANGIOGENESIS

Formation of new capillary blood vessels (angiogenesis) accompanies several normal and pathological processes in the body. The growth of blood vessels requires both cell proliferation and locomotion, dependent on several angiogenic factors. Among the first isolated was the fibroblast growth factor (FGF), isolated from brain, and the endothelial cell growth factor (ECGF), isolated from hypothalamus. The discovery that ECGF revealed a distinct affinity to heparin made it possible to isolate from chondrosarcoma a homogeneous, cationic M_r 18,000, endothelial cell mitogen. Soon two heparin-binding growth factors, basic FGF of a 146 amino acid residues polypeptide, and acidic FGF of a 140 amino acid residues polypeptide, had been determined.[65]

Receptors for FGF have been identified on capillary endothelial and aortic endothelial for acid FGF, and bovine epithelial lens cells for basic FGF. Both classes of heparin-binding growth factor stimulate endothelial cell proliferation *in vitro*, and are angiogenic *in vivo*. They induce the formation of highly vascularized granulation tissue. Heparin-binding endothelial growth factors are present in almost all normal tissues, yet endothelial proliferation in these tissues is extremely low. The question arises, how are these potent growth factors maintained in a functionally inactive state? One possibility is that they are sequestered within their cells of origin and are not secreted because of a lack of adequate stimuli. Peculiarly, they secrete PDGF. Thus, FGF may be released under specific circumstances, physiologically during ovulation, and pathologically as repair-mediating proteins.[65]

Angiogenin (M_r 14,400) isolated from human adenocarcinoma, is a potent stimulator of angiogenesis. It cleaves 28S and 18S rRNAs. Although both angiogenin and FGF are potent angiogenic stimulators, they differ in many aspects. Angiogenin does not exhibit mitogen activity for vascular endothelial cells. Thus, angiogenin and FGF must act by different mechanisms.

Transforming growth factors (TGF) are polypeptides that alter the phenotype of some normal cells to transformed cells. These factors are also angiogenic *in vivo*. TGF-α is synthesized by transformed cells, it has 30% homology to EGF and binds to EGF receptors. Both EGF and TGF-α stimulate microvascular endothelial cell proliferation. TGF-β was isolated from tumors and normal cells (kidneys, blood platelets, and other cells). TGF-β stimulates proliferation of macrophages, fibroblasts, collagen production and capillary formation. Paradoxically, TGF-β inhibits proliferation of vascular endothelial cells *in vitro*.[65]

Angiogenic activity also reveals some other factors, such as low molecular weight angiogenic factors that have been only partially purified; chemotactic factors isolated from wound fluid that stimulate directional locomotion, but not proliferation of endothelial cells; prostaglandins E_1 and E_2, and a non-dialyzable angiogenic factor isolated from mixed cultures of T-lymphocytes.[65]

The role of heparin in angiogenesis is unclear. There are some observations that prove angiogenic activity of heparin, e.g., mast cells and mast cell-derived heparin stimulate locomotion of capillary endothelial cells *in vitro*; heparin increases angiogenesis induced by tumor in the chick embryo; protamine, that binds avidly to heparin, inhibits the ability of mast cells and heparin to stimulate endothelial cell locomotion, and inhibits angiogenesis associated with embryogenesis, inflammation, and some immune reactions.[65]

1. Inhibition of Angiogenesis

Angiostatic steroids are compounds that in the presence of heparin or heparin fragments inhibit angiogenesis. These steroids lack glucocorticoid and mineralocorticoid activity. The angiostatic function of these steroids seems to be bound with specific structural configuration of the pregnane nucleus.[66] The discovery of the inhibitory function to angiogenesis of the described steroids in combination with heparin may be of great importance for the therapy of some tumors that can proliferate only when associated with the formation of new vessels.

D. PAIN

Pain belongs to a basic sensory abnormality associated with inflammation. Pain develops when nerve fiber terminals of polymodal nociceptors become sensitized by mediators of inflammation. The pain-producing inflammatory mediators are bradykinin, prostaglandins, PGE_1 and PGE_2, and leukotrienes particularly LTB_4. Prostaglandins produce hyperalgesia at low concentrations without evoking pain. Pain becomes evoked by synergistic action of bradykinin and prostaglandins. To the pain mediators also belong derivatives of the lipoxygenase pathway, particularly leukotriene B_4. The action of LTB_4 requires additionally polymorphonuclear leukocytes that recognize specifically, and respond functionally to LTB_4. Thus, the hyperalgesia produced by LTB_4 can be distinguished from that of hyperalgesia elicited by PGE_2 and bradykinin by its dependence of polymorphonuclear leukocytes and independence of the cyclooxygenation pathway of arachidonic acid.[67]

The discovery of opioid receptors gave new insight into the pain producing mechanism, followed by the discovery of endogenous opioids. At least 18 active opioid peptides have been identified, all of which are derived from three large polypeptide precursors. Posttranslational proteolytic processing of these inactive precursors liberates a mixture of active peptides: the β-endorphin-like compounds, the enkephalins and the dynorphins, which bind with different affinity to the three opiate receptors: μ, δ and κ. The opiate receptor antagonist naloxone binds with high affinity only to the μ receptors, and to the other receptors with 10 to 15 times lower affinity. The opiate system cannot be the sole analogesic system because pain may persist also in the presence of naloxone. The opiates are released from the pituitary and adrenal medulla. Acupuncture causes an increase of β-endorphin in the cerebrospinal fluid. Chronic pain seems to be associated with a decreased level of a dynorphin-like function, without change in β-endorphin or enkephalin levels. Congenital insensitivity to pain has been described in patients with very high levels of β-endorphins with no naloxone effect in one patient and the exact opposite in another.[68]

Our understanding of the natural neurochemical mechanisms of pain has increased since the discovery of a central nervous system substrate, the normal function of which seems to be pain inhibition. This substrate includes cells of medial brain stem and fibers descending from them to the spinal dorsal horns. In this system the nociceptive signals become modulated. These brain areas and pathways from them to spinal sensory circuits can be activated by focal stimulation. The analgesic function seems to be mediated by endogenous opioids. The physiological stimulus triggering pain suppression is stress. The natural stressors are fighting, sexual arousal, food deprivation, thermal stress, and others. Experimentally various opioid systems may be activated by stress. The opioid-sensitive form of stress analgesia can be determined by naloxone. By the use of naloxone it was possible to distinguish opioid-dependent and opioid-independent forms of stress analgesia which confirmed the separateness of their neurochemical substrates. Different agonists (serotonin, norephinephrine, dopamine), antagonists and depletors (α-fluoromethylhistidine for histamine and mast cell degranulator compound 48/80) were examined in an attempt to alter non-opioid stress analgesia. The histamine depletor reduced this kind of analgesia and compound 48/80 did not. This result suggests that neuronal stores of histamine may be responsible for non-opioid analgesia (antihistamines H_1 but not H_2 reduced stress analgesia).[69]

XVII. SPECIAL KINDS OF INFLAMMATION

A. BURNS

Burns belong to the major health problems throughout the world. In spite of the great progress in treatment of burns by infusion of large quantities of salt-containing-fluids, and protein to restore physiological conditions of blood circulation because of increased microvascular permeability, and generalized cell membrane defects resulting in intracellular swelling, several other problems remained unresolved. Immediately after the burn an increase of the osmotic pressure occurs in burned tissue and adjacent non-burned tissues, probably because of release of osmotically active cellular compounds. Several mediators are released from burn tissue: prostaglandins, kinins, serotonin, histamine, free oxygen radicals, and lipid peroxides. The edema in non-burned tissues is caused principally by severe hypoproteinemia, and not by the increased vascular permeability. This is particularly evident in lung edema being seen only after severe smoke inhalation. Post-burn lung dysfunction is known to be the major cause of death in burns. Tracheobronchial lesions in burns are caused by water-soluble gases from burning plastics or rubber: ammonia, sulfur dioxide, and chlorine which when reacting with water yield strong acids and alkalis that may induce bronchospasm, ulceration of mucous membranes, and edema. Lipid-soluble compounds (nitrous oxide, phosgene, hydrogen chloride, and toxic aldehydes) damage the cell membrane and impair the ciliary clearance of bacteria. In addition, alveolar macrophages become activated which release chemotactic factors increasing inflammation.[70]

Burn wounds and the lungs are the most frequent sites of infection, particularly by virulent strains of *Pseudomonas aeruginosa* which release exotoxin A that impairs protein synthesis, also relative frequent are *Staphylococcus aureus*, and fungal infections. Alterations of the cellular and humoral immune reactions predispose to these infections. Impaired phagocyte and decreased neutrophil chemotaxis and their bactericidal activity often leads to sepsis. There are some suggestions that fragments of collagen released from the burned tissue may inhibit phagocytosis. The normal lymphocyte function is also severely affected, probably because of disturbances in release of proliferation and differentiation factors (IL-1, IL-2), and abnormal thymopoetin derivative thymopentin (a fragment of thymopoetin with five amino acid residues). Abnormal fibronectin derivatives and abnormal gamma globulin have been described. Early excision of burn wounds may be beneficial through the removal of the source of these abnormalities and prevent postburn disease.[70] Moreover, several metabolic disturbances occur in burned patients, such as increased protein catabolism, ureagenesis, lipolysis, and gluconeogenesis caused by an excess of counterregulatory hormones (catecholamines, glucagon, glucocorticoids).

Healing of burns is under the control of the release of several growth factors:

1. The angiogenesis factor, secreted by hypoxic macrophages (a neovascularization factor by chemoattraction of mesothelial and endothelial cells that migrate to the wound edge and forms new blood vessels);
2. The macrophage-derived growth factor which stimulates fibroblast mitosis, and deposition of fibronectin and glycosoaminoglycan—the rate of its release is oxygen-dependent; thus, an increase of oxygen promotes healing;
3. Platelet-derived growth factor which has properties of both angiogenesis factor and macrophage-derived growth factor:
4. The epidermal growth factor (EGF) that controls reepithelialization of the superficial wound;
5. Mast cells present in burn wounds in increasing numbers which synthesize mucopolysaccharides, histamine (which results in vasodilation), and chondroitin sulfate A. These substances become deposited on collagen fibers. Myofibroblasts with the same function as fibroblasts contract, and ground substances become deposited on the collagen fibers resulting in a firm, hyperemic scar.[70]

Recent investigations have revealed that epidermal growth factor stimulating proliferation of keratinocytes accelerates epidermal regeneration and rapid healing of burn wounds. The same effect was achieved by administration of TGF-α and vaccinia growth factor which possess substantial sequence homology with EGF.[71]

XVIII. MOLECULAR ASPECTS OF INFECTIOUS DISEASES

A. THE ROLE OF MEMBRANE RECEPTORS

Adherence of bacteria to living tissues initiates the first step in infectious diseases. It is well-known that bacteria adhere to specific surface structures of cells which determines their organospecificity. Since the discovery that polysaccharides are the surface antigens of bacteria, interaction of cell surface structures with bacterial polysaccharides plays a decisive role in their adherence to cells initiating in this way the first step of infectious diseases. The importance of this interaction has been best observed in piglets resistant to infection by a virulent strain of *E. coli* that do not adhere to intestinal epithelial cells, in contrast to piglets in which this strain of *E. coli* adheres to the intestinal cells and which react with severe intestinal infection.[72]

It has been demonstrated in *in vitro* experiments that microbes present in the oral cavity reveal their preferential regions. Also, *Neisseria gonorrhoeae* in culture adheres exclusively to human cells and not to cells of any other species. Recent investigations have been devoted to the molecular basis of this bacterial adherence. The best-known evidence has been achieved in the case of various microbes of the digestive tract where the particular microbes bind with surface glycoprotiens forming glycoconjugates present at the cell surface. In this case, sugars play the role of bacterial receptors. This has been seen *in vitro* where different strains of *E. coli* and other bacteria of the digestive tract adhere to red blood cells resulting in their agglutination. However, after addition of mannose to the test tube bacteria bind to mannose, and not to the red blood cells preventing their agglutination. Further observations revealed that microbes sensitive to mannose exhibit on their surface lectins in the form of long filaments that enable binding of these bacteria to sugar residues on the cell surfaces. The binding can be inhibited by periodate which degrades (oxidizes) some sugars, or by Concanavalin A which binds specifically with mannose residues. Some strains of *E. coli* also bind to mannose derivatives on the surface of yeast. Thus, it is very probable that different glycoproteins play the role of receptors for several bacteria.[72]

There are strains of *E. coli* which cause acute pyelonephritis which do not adhere to cell surface of the digestive tract, but to the cells of the urinary tract. In this case the *E. coli* strain binds to galactose participating in the formation of membranal glycolipids, common to the determinant of blood group P. Therefore, red blood cells of individuals who do not have blood group P are not agglutinated by this strain of bacteria and their urinary tract cells do not adhere them. What is more, these individuals are relatively insensitive to urinary infections by this kind of bacteria. Other sugars may also specifically bind different strains of bacteria; for example, *Vibrio cholerae* binds with fucose and mycoplasm as with sialic acid.

Investigations upon receptors for streptococci present in the oral cavity revealed another substance binding them with the cell surface. Electronmicroscope examination exhibited on the surface of these bacteria, filaments synthesized by them and composed of lipoteichoic acid. These filaments are responsible for binding bacteria with the cell surface. In *in vitro* experiments, the lipoteichoic acid added to the culture medium inhibited adherence of streptococci to epithelial cells, but not adherence of other bacteria, for example, *E. coli*. The cellular component that seems to be cellular receptor for streptococci is fibronectin.[72]

The discovery that during infection the particular bacteria become attached to cell surface receptors in the form of various sugars of lipoproteins creates a new possibility in preventing infection. In experiments in which mannose sensitive *E. coli*, isolated from patients with pyelonephritis, were administered into the urinary bladder of mice, about 70% become diseased, but when administered together with methylmannoside only 25% become diseased. Methylmannoside administered together with other bacteria did not exhibit diminution of infection.[72] These and other experiments are promising indicating that the discovery of receptors of pathogenic bacteria may well be efficacious in preventing different infectious diseases.

B. THE ROLE OF BACTERIAL TOXINS

1. The Diphtheria Toxin

Several bacteria act on eukaryotic cells by producing toxins that interfere with their vital functions. Before the molecular era the mechanism of their action could not be elucidated. We now know that diphtheria toxin and pseudomonas toxin A produce their effects in eukaryotic cells by inactivating elogantion factor 2 (EF-2), that is essential for protein synthesis. EF-2, is a protein (M_r about 93,000) that catalyzes translocation of the growing polypeptide chain from the aminoacyl-tRNA site to the peptidyl-tRNA site on the ribosome. In this way each time an amino acid is added during chain elongation. Simultaneously, the mRNA template is shifted by a distance of one codon. The energy for this reaction derives from hydrolysis of GTP. Inactivation of EF-2 is caused by the transfer of a molecule of ADP-ribose from NAD^+ to a modified amino acid on the factor 2 (3-carboxyamido-3-[trimethylammonio]propyl) histidine, known as diphthamide. The modification of histidine residue in EF-2 is effectuated by at least three enzymes. One of these enzymes, ADP-ribosyltransferase, with a specificity for the diphthamide residue, is similar to fragment A of diphtheria toxin—in the part that reveals the mono(ADP-ribosyl) transferase activity. Since ADP-ribosylation inactivates EF-2, the endogenous enzyme ADP-ribosyltransferase may be a normal enzyme that regulates protein synthesis in eukaryotes, and this is supported by the fact that covalent modification of EF-2 is reversible in the presence of microbial fragment A and NAD.[73]

The diphtheria toxin (M_r 60,000) is composed of two polypeptide chains, fragment A (M_r 21,000) and fragment B (M_r 39,000), disulphide bonded. At first, the toxin binds with the cell surface through the fragment B, enabling entry into the cell by fragment A. Fragment A then adopts the active conformation, enabling the rebosylation reaction in which ADP-ribose of NAD becomes transferred to EF-2. Since ADP-ribosylated EF-2 is inactive, the protein synthesis stops and the cell dies.

Fragment A of diphtheria toxin is highly specific for the single amino acid residue histidine within the EF-2 molecule. The receptor on the cell surface which binds fragment B of the toxin is unknown, but it probably is a glycoprotein. The uptake of diphtheria toxin is greatly increased by lowering of pH. When cells are incubated with ammonium chloride the toxin is less active, although its binding with the cell surface is normal. The entry of fragment A is enabled by channel forming of fragment B transversing the membrane. This is possible because fragment B has a hydrophobic domain that dissolves the membrane lipids and a hydrophilic domain which can interact with the polar head group of the cell membranes.[74,75]

There are some proofs that diphtheria toxin may pass through the cell membrane in undegraded form by a mechanism resembling pinocytosis or phagocytosis. It becomes evident that diphtheria toxin penetrates into cells by specific entry sites. Therefore, in mice which do not have such entry sites the diphtheria toxin does not cause any morbid manifestations. This explains the species resistance of rodents (mice and rats) to the diphtheria infection in which the toxic dose of the diphtheria toxin must be 10,000 to 100,000 times greater than that in humans. In addition, damage to the intestinal mucosa (by dextrane, DEAE-cellulose) increases the susceptibility to diphtheria toxin.

Plant derivative toxins are glycoproteins (abrin, ricin) of similar structure to diphtheria toxin; M_r of the whole toxins is about 62,000. Their action is also similar to the diphtheria toxin; their B subunits bind to gangloside G_{M1}, then fragments A cross the membrane and bind with the 60S ribosomal subunit which inhibits protein synthesis.

Shigella toxin of M_r 68,000 is composed of three active A components (A M_r 30,000, A_1 M_r 27,000 and A_2 M_r 3000) and six or seven B subunits which, similar to diphtheria toxin, bind to ganglioside G_{M1} then the A fragment binds to the 60S ribosomal subunit which inhibits protein synthesis.

2. Cholera and Related Toxins

Cholera patients have a watery diarrhea that leads to dehydration and metabolic acidosis; this is caused by intestinal infection with *Vibrio cholerae*, which adhere the small intestine and secrete the exotoxin—cholera toxin. Cholera toxin is composed of two types of subunits: a single heavy subunit (M_r 58,000) noncovalently bound to an aggregate of five to six light subunits (M_r 28,000). The cholera toxin binds with receptors on the mucosal cells and stimulates the intestinal adenyl cyclase activity resulting in an increase of cAMP that causes diarrhea and fluid loss by inhibiting uptake of sodium chloride, apart from stimulating NaCl secretion by crypt cells.[76]

The cholera receptor is a specific ganglioside G_{M1}, which neutralizes cholera toxin in equimolar proportions. Several authors have shown that both cholera toxin and its B region can create pores in synthetic lipid bilayers containing G_{M1}. It is supposed that association of cholera toxin with G_{M1} may induce conformational changes in the B subunits resulting in exposure of hidden hydrophobic B-subunit regions, that fuse with hydrophobic protein or lipid components of the plasma membrane and enables entry of the A subunit into the cell (the B subunits interacting with membrane components may form a channel through which the A subunit passes across the membrane. The activation of the A subunit after its internalization depends on NAD, undefined cellular cytosol factors and ATP in addition to the A_1 fragment (crossing the membrane subunit A becomes cleaved) and cell membrane. Cholera toxin (like diphtheria toxin) shows ADP-ribosyltransferase activity and catalyzes the reaction NAD + acceptor protein → ADP-ribose-acceptor protein + nicotinamide + H^+. The ADP-ribosylated protein by cholera toxin is guanyl-binding component of the membrane-bound adenylate cyclase. Adenyl cyclase becomes active when GTP is bound to GTP-binding protein, and inactive when GTP is hydrolyzed to GDP by GTP-ase. The cholera toxin blocks CTP-ase which stabilizes adenylate cyclase in active conformation yielding an excess of cAMP that causes the described pathological manifestations of cholera.[76]

The amount of cholera toxin bound to intestinal cells is strictly dependent on the amount of ganglioside G_{M1} in the membrane of these cells. Because the amount of G_{M1} in the cell membranes of human, pig, and ox is 0.1, 2.0, and 43.0 n*M*, respectively, per gram of fresh tissue—the amount of cholera toxin molecules bound with these tissues is about 15,000 molecules per cell in man, 120,000 in pig, and about 2,600,000 molecules per cell in ox.

The binding of cholera toxin molecules with G_{M1} abolishes the binding of physiological stimulators (principally hormones of glycoprotein structure: thyrotropin—TSH, lutropin—LH, and human chorionic gonadotropin—hCG), which causes disturbances in physiological accumulation of fluids within the intestines, in normal lipolysis in fatty acid cells, and in steroidogenesis in adrenals. Because of the similarity of subunits A of cholera toxin with α-chains of the glycoprotein hormones, competition may exist for receptors G_{M1} between the physiological hormones and cholera toxin which may cause additional hormonal disturbances in cholera infection.

3. Pertussis Toxin

The binding of lipid and peptide chemotactic factors to cell surface receptors on polymorphonuclear leukocytes initiates a series of biochemical events (increased chemotactic migration, aggregation, adherence, lysosomal degranulation, and production of superoxide anion). The rapid stimulation, by chemotactic factors of phospholipid and protein methylation, arachidonic acid release, phosphatidyl inositol turnover, internal calcium release and influx of extracellular calcium, glucose flux, and membrane depolarization suggests that these events are associated with receptor occupancy. Modulation by guanine nucleotides of the affinity of receptors for *N*-formyl-methionyl peptides and activation of GTP-ase activity suggests that the guanine nucleotide-binding protein may participate in transduction of signals. Cholera toxin-induced ADP-ribosylation of GTP causes permanent stimulation of adenylate cyclase. A bacterial toxin isolated from *Bordetella pertussis* similarly catalyzes ADP-ribosylation of regulatory protein of adenylate cyclase which suppresses the increase of Ca^{2+} by leukotriene B_4 (LTB_4), chemotactic migration and lysosomal enzyme release. Pertussis toxin selectively decreases the number of high-affinity receptors for LTB_4.[42]

Pertussis toxin may disarm the immune response evoked by LPS through interaction with G-binding proteins mediating transmembranal signals. Because pertussis toxin inhibits activation of macrophages and B-cells by LPS, it is likely that LPS is involved in activation of the immune reaction by interaction with G-binding proteins.

4. Neurotoxins: Tetanus Toxin and Botulinus Toxin

Both toxins are extremely potent neurotoxins produced by anaerobic bacteria of the *Clostridium* genus. The teatnus toxin M_r 150,000 is composed of a L (light) chain of M_r 50,000 and H (heavy) chain of M_r 100,000. The tetanus toxin causes spastic paralysis by blocking presynaptic transmitter release mainly in the central nervous system. The toxin gains access to the nervous system by uptake at the nerve terminals, with retrograde axonal transport within the motoneuron and trans-synaptic migration into the inhibitory interneuron.

The botulinus toxin is also composed of two subunits, L and H, of the same molecular weights. Botulinus toxin acts at the peripheral level by inhibiting the release of acetylcholine at the neuromuscular junction, resulting in flaccid paralysis.

Both these toxins diffuse probably from the extracellular fluid at the neuromuscular junction, and become internalized after binding to nerve endings. The only difference is that tetanus toxin is taken up by all peripheral and central neurons, whereas botulinus toxin is taken up by motoneurons. The mechanism by which these toxins block neurotransmitter release is unknown. The first problem is in which way these water-soluble proteins can pass the hydrophobic cell membrane to reach the cytoplasm. It is supposed that acidification may induce conformational rearrangement of these toxins on exposure to a hydrophobic surface that enables them to cross the membranes.

Both toxins bind to gangliosides of the G_{1b} series which are considered to be the first receptor for the toxins. The binding is reversible which allows the toxins to be bound to a toxin-specific protein receptor, a transmembrane protein that is involved in their internalization. In the nerve cells there are only a few toxin-specific protein receptor molecules available which causes preferential uptake in cells which have larger amounts of these molecules. It also is possible that binding of the toxins to the gangliosides of the B_{1b} series causes conformational rearrangements of the toxins that increase their affinity to the toxin-specific protein receptor. The double receptor model for tetanus and botulinus toxins accommodates their high neuronal selectivity because ganglioside G_{1b} are more concentrated in nerve cells than in other tissues and because the toxin-specific protein receptor is expressed preferentially by nerve cells. In this way the toxin is first bound to the ganglioside and then to the toxin specific protein receptor.[77]

Tetanus toxin injected intercellularly inhibits exocytosis in adrenal chromaffin cells which may explain adrenal disturbances in tetanus.[79]

Some strains of *Clostridium botulinum*, apart from neurotoxin, synthesize another toxin of ADP-ribosyltransferase activity, named C2 toxin. This toxin causes hypotension, increased intestinal secretion and vascular permeability, and lung hemorrhages. Botulinum C2 toxin causes ADP-ribosylation of actin which becomes nono-ADP-ribosylated. It is noteworthy that skeletal muscle actin does not become ASP-ribosylated by this toxin.[79]

C. VIRAL INFECTIONS

1. Pathogenesis of Mammalian Reovirus Infections

Reovirus infection involves three discrete steps. The outer capsid of these viruses possesses three distinct proteins of highly specialized roles: protein $\sigma 1$; protein u1C; and protein $\sigma 3$. After entry into the gastrointestinal tract, protein $\sigma 1$ (the viral hemagglutinin) interacts with receptors on the surface of immune and nonimmune lymphocytes, and neurons which determine cell and tissue tropism. The u1C protein becomes digested by the host proteases—this protein is responsible for the host's cellular immune response. It determines the capacity of viral growth on mucosal surfaces. Protein $\sigma 3$ is responsible for inhibiting cell macromolecular synthesis which is of critical value in allowing the reovirus to initiate persistent infection of cells. Localization of the different types of reovirus of different cells is determined by the $\sigma 1$ protein, encoded by S_1 gene. Dependent on the kind of mutation, reoviruses may cause a fatal acute encephalitis in mice associated with viral replication in neurons, but no in ependymal cells, whereas other mutations of the reovirus cause nonfatal acute ependymitis without involvement of neurons.[80]

Experiments performed on reoviruses illustrate different possibilities of interaction with specific cell receptors and consequently different pathology:

1. Mutations of gene S1 (encoding protein $\sigma 1$) may cause binding of viruses with nerve cells or lymphoid cells which may result in loss of neurovirulence or altered specificity in cytolytic T-cell and humoral responses.
2. Mutations of gene M2 (encoding protein u1C) influence growth at the mucosal surface, resulting in decreased capacity to grow at 1° site. Immune induction results in loss of immune stimulation of suppressor T lymphocytes; and growth in target tissue results in relative decrease in neurovirulence.
3. Mutations of gene S4 (encoding protein $\sigma 3$) results in inhibition of the host's RNA and protein synthesis that causes persistent infection.
4. Finally, multiple gene mutations encode several proteins, resulting in pleiotropic effects (temperature-sensitive mutations have the same tropism but reduced growth *in vitro* and *in vivo*.[80]

On the basis of reovirus infection there are some possible generalizations. Any virus or microbe, if it is to produce infection in an eukaryotic host, must enter through a major portal, spread (it is produces a systemic disease) and bind with a specific cell tissue site. Several host reactions are common to all types of infections:

1. The viral receptor binding component is responsible for determining cell and tissue tropism, and is a major antigen for the cellular and humoral immune reactions.
2. The viral component of the outer capsid (protein u1C) has a critical role in determining the ability of a reovirus to replicate at the primary site in the gastrointestinal tract and the ultimate nature of the immune reaction (this component provides a second site on the surface of the reovirus that can be altered, resulting in a reduced possibility to grow at the portals of entry).
3. The $\sigma 3$ outer capsid protein responsible for inhibiting cellular synthesis of RNA and protein is the site of frequent mutations resulting in decrease of capacity of the reovirus to lyse cells.

Experiments on influenza virus, poliovirus, coxsackie virus, adenovirus, and others have revealed that virulence is dependent on several viral genes and despite multiple genes involved in infection—individual genes are responsible for different stages and properieties of the infecting parasites.[80]

2. Pathogenesis of Lentivirus Infections

Lentiviruses are a subfamily of retroviruses. The name originates from the slow time course of the infections they cause in animals and humans. This group of pathogens includes: oncoviruses causing cancer in man, mammals, birds and reptiles; visna-maedi virus causing pneumonia or meningoencephalitis in sheep and goats; caprine arthritis encephalitis virus causing arthritis, pneumonia and meningoencephalitis in sheep and goats; equine infectious anemia virus causing fever and anemia in horses; and AIDS virus named also human immunodeficiency virus (HIV) causing immune deficiency, encephalopathy, and myelopathy in man, and probably several others.

The best-known and first-discovered lentiviruses were visna and maedi. Their name derives from Icelandic names for prominent manifestations (wasting and shortness of breath) of the neurological and pulmonal diseases they cause in sheep. The same virus is responsible for both maedi—the more prevalent pulmonal disease, and visna—the paralytic disease resembling multiple sclerosis. The outbreak of the disease in Iceland occurred in the 1940s, whereas the virus was introduced into Iceland by sheep imported ten years before from Germany. The Icelandic physician B. Sigurdsson discovered that the disease was caused by a filterable agent with a long incubation time and protracted course of disease and introduced the term "slow infection". The visna-maedi virus replicates at the site of entry (the lung in natural infection), and subsequently spreads via the bloodstream or other routes (e.g., the cerebrospinal fluid). The infected organism reacts by effective immune response on the extracellular virus, but it is ineffective to eradicate the infectious agent which persists in various organs, circulating in blood and body fluids. In the lungs and central nervous system, the tissue becomes destroyed because of accumulation of inflammatory cells, which is manifested in shortness of breath or partial paralysis and weight loss. Usually in the second year of infection the animals die after a protracted and progressive disease.

The following problems emerge: How does the virus persist and spread in spite of the host's immune response? What mechanism destroys the tissue? Why do these pathological events evolve so slowly?[81]

The best explanation for the persistence of the lentiviruses is the immunologically silent nature of the infection. This occurs because of the latent state in which viral antigens are not produced in sufficient quantities to evoke destruction of infected cells by immune-surveillance mechanisms. This is best evidenced by cell culture infected by the visna virus. The visna virus reproduces rapidly; thousands of copies of genomic and mRNA copies are synthesized, and the cell degenerates within 3 days. In contrast, in animals the virus replicates slowly (the number of viral RNA copies is about two orders of magnitude less than in cells in culture), and it seems that the virus expression is blocked by an unknown mechanism. The expansion of the virus through the bloodstream is caused by infection of immune cells, mainly monocytes, in which only restricted levels of viral RNA accumulate which circulate and transfer the infectious virus to other cells.[81]

Pathological changes in lentivirus infections are mostly caused by the immune and inflammatory reaction. The inflammatory reaction predominates in visna which causes demyelination because the target cell is the myelin-producing oligodendrocyte. In equine infectious anemia both lymphoproliferative processes and immune complexes occur. The erythrocytes become coated with a viral hemagglutinin, then with antivirus antibody, and complement which causes hemolysis. Circulating immune complexes and antibodies deposited in the kidney cause glomerulonephritis.

The HIV (human immunodeficiency virus) or AIDS virus also establishes persistent, non-cythopathic infections in normal human lymphocytes which causes depletion of specific kinds of lymphocytes, the helper lymphocytes, which annihilates the immune reaction. Subsequently, the patients die because of lack of immunity. Glomerulosclerosis and thrombocytopenic purpura of AIDS may be also caused by immune complexes. The AIDS virus differs from the lentiviruses in its effects on the immune system. In AIDS the immune system is deficient, whereas in lentivirus infections the immune system is usually normal or only selectively impaired. The AIDS virus binds with the receptor on the surface of the central cell of the immune system the helper T-cells and cause their destruction or dysfunction. There also are other suspected pathogenic mechanisms in AIDS. For example, viral envelope glycoprotein shed from productive cells may become a target for autoimmune response directed against the viral antigen bound to the helper T-cell which results in destruction of these cells.[81]

The most important and until now unanswered question is the control mechanism of lentivirus replication. The answer to this would probably shed much light into the mysterious problem of slow infections in man and animals.

3. The Viral Theory of Multiple Sclerosis

Several observations suggest that retroviral infection is involved in the pathogenesis of the human demyelinating disease multiple sclerosis. After the discovery that lentiviruses may cause demyelination in visna-maedi disease in sheep, and that AIDS virus has been shown to be neurotropic in man, it was recently found that multiple sclerosis is accompanied by the presence, in the cerebrospinal fluid of multiple sclerosis patients, of antibodies against human T-cell lymphotropic virus (HTLV-I), and of HTLV-I-specific RNA.[82] Finally, the presence of HTLV-I-like sequences in the central nervous system tissue from these patients have been described.[83] If the HTLV-I virus is the cause of multiple sclerosis, it is a question that probably will be answered in the near future.

4. Scrapie and Related Diseases

Scrapie is a naturally occurring slow infection in sheep which is transmitted by prions. The scrapie agent has several unusual properties that distinguish it from conventional viruses: (1) long incubation period, and (2) its ability to cause devastating degeneration of the central nervous system without inflammatory reaction. The scrapie agent is susceptible to agents which inactivate proteins, but is resistant to nucleases and ultraviolet radiation that destroy nucleic acids. Recent investigations have revealed the presence of a very small nucleic acid genome (between 0.75 to 1.6×10^6 Da) the infectivity of which may be associated with a single proteinaceous substance. The inability to detect any scrapie-specific nucleic acid in highly infectious preparations led to the suggestion that the disease is caused by a proteinaceous infectious particle, called ''prion''. The next step was the detection of a gene encoding the prion protein, but what is most puzzling is that the prion protein is a constituent of normal cells. The question arises: Is this protein associated with fibrils and other pathological structures occurring in scrapie and other neurological diseases? It has been suggested that this protein may be a component of amyloid plagues in scrapie-infected brain. These results do not explain the nature of scrapie and any additional factors which seem to be necessary to initiate the disease. The last supposition is that prion proteins may be identical in healthy and scrapie-infected sheep which acquire specific properties after interaction with nucleic acid present in scrapie-infected tissue. Final determination of the roles of the prion protein and associated nucleic acid in scrapie will probably not be possible until their structures are known.[84]

Many features of scrapie are common with human degenerative neurological diseases: Creutzfeld-Jacob disease, kuru, and Gerstmann-Straussler syndrome. A complementary DNA whose protein product seems to be the major component of scrapie has been identified and characterized in these human diseases which suggests that this group of degenerative encephalopathies also is caused by a small proteinaceous particle—the prion.[85]

D. PROTOZOAN INFECTION

Antigenic variation is the mechanism that enables some parasitic trypanosome species to evade the immunological attack of their host. The parasite's ability to survive in spite of host's immunological reaction is based on continuous alteration of the surface coat, consisting of a single glycoprotein, against which the immunological reaction is directed. By the consecutive production of a large number of different variants of the surface glycoprotein the parasite population survives despite an active host immune reaction.

Trypanosomes are unicellular flagellates, with a parasitic lifestyle, which cause African sleeping sickness in man. The subspecies *Trypanosoma brucei, T. b. gambiense, T.b. rhodiense,* and *Trypanosoma cruzi* cause Chagas disease in South America. In addition, trypanosomes cause analogous diseases in domestic animals. In nature, *T. brucei* shuttles back and forth between mammals and its tsetse fly vector. Parasites enter the fly, where they multiply, migrate to the salivary glands, and are inoculated into a susceptible host with the insect's saliva. The surface coat protects the parasite against both immune specific and nonimmune mechanisms. The protective value of the surface coat depends on three factors:

1. The parasites are enshrouded by the surface coat, which carries the antigenic determinants of intact trypanosomes.
2. The coat is made up of a single glycoprotein, and trypanosomes are able to produce a large number of different variants of this variant surface glycoprotein (VSG) which do not share antigenic determinants (some of trypanosomes are able to produce 100 and more VSG).
3. The parasite population in a chronically infected animal does not use at the same time the whole antigen repertoire, but restricts the outgrowth of parasites carrying the different VSG to a few variants. The consecutive production of antigenically different VSGs enables the trypanosome population to survive for prolonged periods (a new population arises every 7 to 10 d) despite active immune reaction.[86]

The question arises as to what way trypanosomes restrict synthesis of VSGs to one antigen at a time. This may be dependent on the specific maturation of their mRNAs differing from that in other eukaryotes. In trypanosomes and other kinetoplastida, mRNA consists of two exons transcribed from two separate genes. The 5′ noncoding exon (named mini-exon) is common to all mRNAs and is derived from the first 35 nucleotides of a separately encoded, mini-exon-derived (medRNA) transcript.

Juxtaposition of the mini-exon and the major mRNA coding sequence requires RNA splicing.[87] This observation led to the proposal of several models that can explain discontinuous transcription. This kind of transcription that produces large polycistronic RNAs that are processed by trans-splicing and simply by adding a ''mini-exon floating cap'' trypanosomes are able to produce multiple nature mRNAs from a single precursor.[87] The precise mechanism that causes synthesis of different VSGs is not clear.

1. Schistosoma

Like other parasites, the schistosomes have a complex life cycle. Infected persons excrete schistosome eggs in their feces. When the feces enters fresh water, schistosoma embryos emerge from eggs and swim to find their hosts—a snail—where the embryos develop and multiply and after 1 to 2 months the snails release tens of thousands of schistosome larvae per day. The larvae burrow into an individual's skin, enter the microcapillaries and the developing worms travel by bloodstream to the lungs, where they remain about 10 d, and then travel to the hepatic portal veins. After development and maturation into adult males and females they mate and migrate to the final destination, e.g., *Schistosoma mansoni* migrate to mesenteric veins, *Schistosoma haematobium* to the veins of the urinary bladder. Each pair produces 300 to 3000 eggs per day. The worms may live for 20 to 30 years. The eggs migrate through the wall of the intestine (*S. mansoni*) or the urinary bladder (*S. haematobium*), and the rest with the blood to the liver where they cause severe inflammatory reaction. The eggs are excreted, and in water the newly hatched embryos enter a new snail host. Ironically, the worms are living in veins and exposed to the immune attack, but they are never attacked. In the 1960s it was discovered that *Schistosoma* worms in infected animals become coated with molecules (blood group antigens) of their host. In this way *Schistosoma* coated with the blood group antigens of their host cannot be recognized as ''not-self'' and induce an immunological attack against them. Besides, it was found that *Schistosoma* synthesize substances that suppress immunological reaction e.g., tripeptide threonine-lysine-proline that inhibits macrophages and is a potent anti-inflammatory agent. Another substance is also excreted by the worms of the same type of activity as cyclosporine inhibiting T-cells.[88]

Malaria is one of the most widespread diseases in the world, mainly in tropical and subtropical areas. The parasites (*Plasmodium malariae*) live and develop in normal erythrocytes. Erythrocytes containing pathological hemoglobin, mainly heterozygotes of Hb S (homozygotes are badly injured by the pathological hemoglobin), or mild glucose-6-phosphate dehydrogenase deficiency (severe deficiency also causes a serious disease) are relatively resistant to malaria infection. Similar to other parasites, *Plasmodium falciparum* also needs receptors on the erythrocyte membrane. In experiments the erythrocyte membranal component glycophorin blocked the invasion of *Plasmodium falciparum* into erythrocytes. The same results have been obtained with *N*-acetyl-D-glucosamine, which coupled with bovine serum was 100,000 times more effective than *N*-acetyl-D-glucosamine alone. The result suggests that the binding of *Plasmodium falciparum* to erythrocytes is lectin-like and is determined by carbohydrates on glycophorin.[89] In other experiments it was proved that the invasion of *Plasmodium falciparum* is most effectively inhibited by band 3 transmembrane protein. The lactosamine chains contributed much to this activity. This result suggests that primary interaction of the parasite on band 3 protein mediates its invasion into erythrocytes.[90]

REFERENCES

1. **Ashman, R. B. and Müllbacher, A.,** Infectious disease, fever, and the immune response, *Immunol. Today,* 5, 268, 1984.
2. **Müller-Esterl, W., Iwanaga, S., and Nakanishi, S.,** Kininogens revisited, *Trends Biochem. Sci.,* 11, 336, 1986.
3. **Porter, R. R.,** Complement polymorphism, the major histocompatibility complex and associated diseases: a speculation, *Mol. Biol. Med.,* 1, 161, 1983.
4. **Holers, V. M., Cole, J. L., Lublin, D. M., Seya, T., and Atkinson, J. P.,** Human C3b- and C4b-regulatory proteins: a new multi-gene family, *Immunol. Today,* 6, 188, 1985.
5. **Rosenfeld, S. I., Jenkins, Jr., D. E., and Leddy, J. P.,** Enhanced reactive lysis of paroxysmal nocturnal hemoglobinuria erythrocytes, *J. Ex. Med.,* 164, 981, 1986.
6. **Ochs, H. D., Rosenfeld, S. I., Thomas, E. D., Giblett, E. R., Alper, C. A., Dupont, B., Schaller, J. G., Gilliland, B. C., Hansen, J. A., and Wedgwood, R. J.,** Linkage between the gene (or genes) controlling synthesis of the fourth component of complement and the major histocompatibility complex, *N. Engl. J. Med.,* 296, 470, 1977.
7. **Suzuki, K., O'Neill, G. J., and Matsumoto, H.,** A product of the C4B locus lacking hemolytic activity, *Hum. Genet.,* 73, 101, 1986.
8. **Tokunaga, K., Dewald, G., Omoto, K., and Juji, T.,** Family study on the polymorphisms of the sixth and seventh components (C6 and C7) of human complement: linkage and haplotype analyses, *J. Hum. Genet.,* 39, 414, 1986.
9. **Rittner, C., Hargesheimer, W., and Mollenhauer, E.,** Population and formal genetics of the human C81(α-γ) polymorphism, *Hum. Genet.,* 67, 166, 1984.
10. **Young, J. D., Cohn, Z. A., and Podack, E. R.,** The ninth component of complement and the pore forming protein (perforin 1) from cytotoxic T cells: structural, immunological and functional similarities, *Science,* 233, 184, 1986.
11. **Abdullah, I. H. and Greally, J.,** C1-inhibitor, biochemical properties and clinical applications, *CRC Crit. Rev. Immunol.,* 5, 317, 1985.
12. **Greer, J.,** Model structure of the inflammatory protein C5a, *Science,* 228, 1055, 1985.
13. **Jackson, J., Sim, R. B., Whelan, A., and Feighery, C.,** An IgC autoantibody which inactivates $C\bar{1}$-inhibitor, *Nature (London),* 323, 722, 1986.
14. **Cybulsky, A. V., Quigg, R. J., and Salant, D. J.,** The membrane attack complex in complement-mediated epithelial cell injury: formation and stability of C5b-9 and C5b-7 in rat membranous nephropathy, *J. Immunol.,* 137, 1511, 1986.
15. **Campbell, A. K. and Morgan, B. P.,** Monoclonal antibodies demonstrate protection of polymorphonuclear leukocytes against complement attack, *Nature (London),* 317, 164, 1985.
16. **Reid, K. B., Bentley, D. R., and Wood, K. J.,** Cloning and characterization of the complementary DNA for the B chain of normal human serum C1q, *Philos. Trans. R. Soc. London,* Ser. B, 306, 345, 1984.

17. **Hannena, H. J., Kluein-Nelemans, J. C., Hack, C. E., Eerenberg-Belmer, A. J., Mallee, C., and van Helden, H. P.,** SLE-like syndrome and functional deficiency of C1q in members of a large family, *Clin. Exp. Immuno.,* 55, 106, 1984.
18. **Stafford, C. T. and Jamieson, D. M.,** Cold urticaria associated with C4 deficiency and elevated IgM, *Ann. Allergy,* 56, 313, 1986.
19. **Pangubrn, M. K., Schreiber, R. D., and Muller-Eberhard, H. J.,** Deficiency of an erythrocyte membrane protein with complement regulatory activity in paroxysmal nocturnal hemoglobinuria, *Proc. Natl. Acad. Sci. U.S.A.,* 80, 5430, 1983.
20. **Lewis, R. A. and Austen, K. F.,** Mediation of local homeostasis and inflammation by leukotriens and other mast cell-dependent compounds, *Nature,* 293, 103, 1981.
21. **Kuehl, F. A. and Egan, R W.,** Prostaglandins, arachidonic acid and inflammation, *Science,* 210, 978, 1980.
22. **Rola-Pleszczynski, M.,** Immunoregulation by leukotrienes and other lipoxygenase metabolites, *Immunol. Today,* 6, 302, 1985.
23. **Stocker, R., Yamamoto, Y., McDonagh, A. F., Glazer, A. N., and Ames, B. N.,** Bilirubin is an antioxidant of possible physiological importance, *Science,* 235, 1043, 1987.
24. **Imagawa, D. K., Osiefchin, N. E., Paznekas, W. A., Shin, M. L., and Mayer, M. M.,** Consequences of cell membrane attack by complement: release of arachidonate and formation of inflammatory derivatives, *Proc. Natl. Acad. Sci. U.S.A.,* 80, 6647, 1983.
25. **Denzlinger, C., Rapp, S., Hagmann, W., and Keppler, D.,** Leukotrienes as mediators in tissue trauma, *Science,* 230, 330, 1985.
26. **Wallner, B. P., Mattaliano, R. J., Hession, C., Cate, R. L., Tizard, R., Sinclair, L. K., Foeller, C., Pingchang Chow, H., Browning, J. L., Ramachandran, K. L., and Pepinsky, B.,** Cloning and expression of human lipocortin, a phospholipase A_2 inhibitor with potential anti-inflammatory activity, *Nature (London),* 320, 77, 1986.
27. **Krilis, S. A., Macpherson, J. L., de Carle, D. J., Daggard, G. E., Talley, N. A., and Chesterman, C. N.,** Small bowel mucosa from celiac patients generates 15-hydrooxyeicosatetranoic acid (15-HETE) after *in vitro* challenge with gluten, *J. Immunol.,* 137, 3768, 1986.
28. **Paret, F. I., Mannuci, P. M., D'Angelo, A., Smith, J. B., Sauterin, L., and Galli, G.,** Congenital deficiency of thromboxane and prostacyclin, *Lancet.,* Apr. 26, 898, 1980.
29. **Gimbrone, Jr., M. A.,** Blood vessels and the new mediators of inflammation in *Advances in the Biology of Disease Vol. 1,* Ed. Rubin, E., Damjanov, I., Eds., Williams & Wilkins, Baltimore, 1984 113-114.
30. **Tucci, A., Goldberger, G., Whitehead, A. S., Kay, R. M., Woods, D. E., and Colten, H. R.,** Biosynthesis and postsynthetic processing of human C-reactive protein, *J. Immunol.,* 131, 2416, 1983.
31. **Perlmutter, D. H., Goldberger, G., Dinarello, Ch. A., Mizel, S. B., and Colten, H. R.,** Regulation of class III major histocompatibility complex gene products by interleukin-1, *Science,* 232, 850, 1986.
32. **Pepys, M. B.,** C-reactive protein and the acute phase response, *Nature (London),* 296, 12, 1982.
33. **Yoshioka, K., Sasaki, H., Yoshioka, N., Furuya, H., Harada, T., and Kito, S.,** Structure of the mutant prealbumin gene responsible for familial amyloidotic polyneuropathy, *Mol. Biol. Med.,* 3, 319, 1986.
34. **Maeda, S., Mita, S., Araki, S., Shimada, K.,** Structure and expression of the mutant prealbumin gene associated with familial amyloidotic polyneuropathy, *Mol. Biol. Med.,* 3, 329, 1986.
35. **Heimark, R. L., Kurachi, K., Fujikawa, K., and Davie, E. W.,** Surface activation of blood coagulation, fibrinolysis and kinin formation, *Nature,* 286, 456, 1980.
36. **Mann, K. G., Fass, D. N.,** The molecular biology of blood coagulation in *Hematology,* Vol. 2, Fairbanks, V. F., Ed., John Wiley & Sons, New York, 1983, 347.
37. **Kunkel, S. L.,** Generalized Shwartzman reaction: an enigmatic model for therapeutic agents, in *Advances in the Biology of Disease,* Vol. 1, Rubin, E., and Damjonov, I., Eds., Williams & Wilkins, Baltimore, 1984, 118.
38. **Esmon, Ch. T.,** The regulation of natural anticoagulant pathways, *Science,* 235, 1348, 1987.
39. **Beck, G., Habicht, G. S., Benach, J. L., and Miller, F.,** Interleukin-1: Common endogenous mediator in inflammation and the local Shwartzman reaction, *J. Immunol.,* 136, 3025, 1986.
40. **Yuli, I. and Oplatka, A.,** Cytosolic acidification as an early transductory signal of human neutrophil chemotaxis, *Science,* 235, 340, 1987.
41. **Heiple, J. and Ossowski, L.,** Human neutrophil plasminogen activator is localized in specific granules and is translocated to the cell surface by exocytosis, *J. Exp. Med.,* 164, 826, 1986.
42. **Goldman, D. W., Chang, F. H., Gifford, L. A., Goetzl, E. J., and Bourne, H. R.,** Pertussis toxin inhibition of chemotactic factor-induced calcium mobilization and function in human polymorphonuclear leukocytes, *J. Exp. Med.,* 162, 145, 1985.
43. **Gabay, J. E., Heiple, J. M., Cohn, Z. A., and Nathan, C. F.,** Subcellular location and properties of bactericidal factors from human neutrophils, *J. Exp. Med.,* 164, 1407, 1986.
44. **Munford, S. R. and Hall, C, L.,** Detoxification of bacterial lipopolysaccharides (endotoxins) by a human neutrophil enzyme, *Science,* 234, 203, 1986.

45. **Wedmore, C. V. and Williams, T. J.,** Control of vascular permeability by polymorphonuclear leukocytes in inflammation, *Nature,* 289, 646, 1981.
46. **Segal, A. W., Heyworth, P. G., Cockroft, S., and Barrowman, M. M.,** Stimulated neutrophils from patients with autosomal recessive chronic granulomatous disease fail to phosphorylate a M_r-44,000 protein, *Nature (London),* 316, 547, 1985.
47. **Segal, A. W.,** Absence of both cytochrome b_{-245} subunits from neutrophils in X-linked chronic granulomatous disease, *Nature (London),* 326, 88, 1987.
48. **Hogg, N,** Factor-induced differentiation and activation of macrophages, *Immunol. Today,* 7, 65, 1986.
49. **Stenson, W. F. and Parker, Ch. W.,** Prostaglandins, macrophages, and immunity, *J. Immunol.,* 125, 1, 1980.
50. **Czop, J. K. and Austen, K. F.,** Generation of leukotrienes by human monocytes upon stimulation of their β-glucan receptor during phagocytosis, *Proc. Natl. Acad. Sci. U.S.A.,* 82, 2751, 1985.
51. **Dahinden, C. A., Clancy, R. M., Gross, M., Chiller, J. M., and Hugli, T. E.,** Leukotriene C_4 production by murine mast cells: evidence of a role for extracellular leukotriene A_4, *Proc. Natl. Acad. Sci. U.S.A.,* 82, 6632, 1985.
52. **Spry, C J. F.,** Synthesis and secretion of eosinophil granule substances, *Immunol. Today,* 6, 332, 1985.
53. **Shaw, R. J., Walsh, G. M., Cromwell, O., Moqbel, R., Spry, C. J. F., and Kay, A. B.,** Activated human eosinophils generate SRS-A leukotrienes, *Nature (London),* 316, 150, 1985.
54. **Ross, R., Raines, E. W., and Bowen-Pope, D. F.,** The biology of platelet-derived growth factor, *Cell,* 46, 155, 1986.
55. **Golstein, P.,** Cytolytic T-cell melodrama, *Nature (London),* 327, 12, 1987.
56. **Beutler, B. and Cerami, A.,** Cachectin and tumour necrosis factor as two sides of the same biological coin, *Nature (London),* 320, 584, 1986.
57. **Old, L. J.,** Tumour necrosis factor, polypeptide mediator network, *Nature (London),* 326, 330, 1987.
58. **Young, J. D. and Cohn, Z. A.,** Cell-mediated killing: a common mechanism? *Cell,* 46, 641, 1986.
59. **Baker, J. B., Low, D. A., Simmer, R. L., and Cunningham, D. D.,** Protease-nexin: a cellular component that links thrombin and plasminogen activator and mediates their binding to cells, *Cell,* 21, 37, 1980.
60. **Whicher, J. and Chambers, R.,** Mechanisms in chronic inflammation, *Immunol. Today,* 5, 3, 1984.
61. **Wisnieski, J. J., Naff, G. B., Pensky, J., and Sorin, B. B.,** Terminal complement component deficiencies and rheumatic disease: development of a rheumatic syndrome and anticomplementary activity in a patient with complete C6 deficiency, *Ann. Rheum. Dis.,* 44, 716, 1985.
62. **Brock, J. H. and de Sousa, M.,** Immunoregulation by iron-binding proteins, *Immunol. Today,* 7, 30, 1986.
63. **Higgs, G. A., Mugridge, K. G., Moncada, S., and Vane, J. R.,** Inhibition of tissue damage by the arachidonate lipoxygenase inhibitor BW755C, *Proc. Natl. Acad. Sci. U.S.A.,* 81, 2890, 1984.
64. **Lotz, M., Carson, D. A., and Vaughan, J. H.,** Substances P activation of rheumatoid synoviocytes: neural pathway in pathogenesis of arthritis, *Science,* 235, 893, 1987.
65. **Folkman, J. and Klagsburn, M.,** Angiogenic factors, *Science,* 235, 442, 1987.
66. **Crum, R. Szabo, S., and Folkman, J.,** A new class of steroids inhibits angiogenesis in the presence of heparin or a heparin fragment, *Science,* 230, 1375, 1985.
67. **Levine, J. D., Lau, W., Kwiat, G., and Goetzl, E. J.,** Leukotriene B_4 produces hyperalgesia that is dependent on polymorphonuclear leukocytes, *Science,* 225, 743, 1984.
68. **Woolf, C. J. and Wall, P.D.,** Endogenous opioid peptides and pain mechanisms: a complex relationship, *Nature (London),* 306, 739, 1983.
69. **Terman, G. W., Shavit, Y., Lewis, J. W., Cannon J. T., and Liebeskind, J. C.,** Intrinsic mechanisms of pain inhibition: activation by stress, *Science,* 226, 1270, 1984.
70. **Demling, R. H.,** Burns, *N. Engl. J. Med.,* 313, 1389, 1985.
71. **Schultz, G. S., White, M., Mitchell, R., Brown, G., Lynch, J., Twardzik, D. R., and Todaro, G. J.,** Epithelial wound healing enhanced by transforming growth factor-α and vaccina growth factor, *Science,* 235, 350, 1987.
72. **Ofek, I., Sharon, N.,** Comment les bacteries adhérent aux cellules, *Recherche,* 14, 376, 1983.
73. **Clemens, M.,** Enzymes and toxins that regulate protein synthesis, *Nature (London),* 310, 727, 1984.
74. **van Heyningen, S.,** Diphtheria toxin: which route into the cell? *Nature (London),* 292, 293, 1981.
75. **Zalman, L. S. and Wisnieski, B. J.,** Mechanism of insertion of diphtheria toxin: peptide entry and pore size determinations, *Proc. Natl. Acad. Sci. U.S.A.,* 81, 3341, 1984.
76. **Homgren, J.,** Action of cholera toxin and the prevention and treatment of cholera, *Nature (London),* 292, 413, 1981.
77. **Montecucco, C.,** How do tetanus and botulinum toxins bind to neuronal membranes? *Trends in Biochemical Sciences,* Elsevier Science Publishers, Amsterdam, 1986.
78. **Penner, R., Neher, E., and Dreyer, F.,** Intracellularly injected tetanus toxin inhibits exocytosis in bovine adrenal chromaffin cells, *Nature (London),* 324, 76, 1986.

79. **Aktories, K., Bärmann, M., Ohishi, L., Tsuyama, S., Jakobs, K. H., and Haberman, E.,** Botulinum C2 toxin ADP-ribosylates actin, *Nature (London),* 322, 390, 1986.
80. **Fields, B. N. and Green, M. I.,** Genetic and molecular mechanisms of viral pathogenesis: implications for prevention and treatment, *Nature (London),* 300, 19, 1982.
81. **Haase, A.,** Pathogenesis of lentivirus infections, *Nature (London),* 322, 180, 1986.
82. **Koprowski, H., DeFreitas, E. C., Harper, M. E., Sandberg-Wollheim, M., Sheremata, W. A., Robert-Guroff, M., Saxinger, C. W., Feinberg, M. B. Wong-Staal, and Gallo, R. C.,** Multiple sclerosis and human T-cell lymphotropic retroviruses, *Nature (London),* 318, 155 1985.
83. **Hauser, S. L., Aubert, C., Burks, J. S., Kerr, C., Lyon-Caen, O., de The, G., and Brahic, M.,** Analysis of human T-lymphotropic virus sequences in multiple sclerosis tissue, *Nature (London),* 322, 176, 1986.
84. **Robertson, H. D., Branch, A. D., and Dahlberg, J. E.,** Focusing on the nature of the scrapie agent, *Cell,* 40, 725, 1985.
85. **Cheng, Y., Liao, J., Lebo, R.V., Clawon, G. A., and Smuckler, E. A.,** Human prion protein cDNA: molecular cloning, chromosomal mapping, and biological implications, *Science,* 233, 364, 1986.
86. **Bernards, A.,** Antigenic variation of trypanosomes, *B.B.A. Libr.,* 824, 1, 1984.
87. **van der Ploeg, L. H.,** Discontinuous transcription and splicing in trypanosomes, *Cell,* 47, 479, 1986.
88. **Kolata, G.,** Avoiding the Schistosome's tricks, *Science,* 227, 285, 1985.
89. **Jungery, M., Pasvol, G., Newbold, C. I., and Weatherall, D. J.,** A lectin-like receptor involved in invasion of erythrocytes by *Plasmodium falciparum, Proc. Natl. Acad. Sci. U.S.A.,* 80, 1018, 1983.
90. **Friedman, M. J., Fukuda, M., and Laine, R. A.,** Evidence for a malarial parasite interaction site on the major transmembrane protein of the human erythrocyte, *Science,* 228, 75, 1985.

Chapter 11

THE MOLECULAR ASPECTS OF IMMUNOLOGY

I. INTRODUCTION

The generation of the immunology system was probably as essential to the evolutionary development of higher organisms as that of the nervous or endocrinological systems. The early findings in this field were those of von Behring and Kitasato (1890) who discovered molecules in the blood serum of immunized animals antibody molecules that could, by means of a specific reaction, neutralize diphtheria or tetanus toxins. Soon, this led to the discovery and differentiation of humoral and cellular immunology, which demonstrated that humoral immunity was executed by synthesized antibodies, and cellular immunity by living cells (that is humoral immunology could be transferred to a nonimmunized animal from immunized animals by the blood serum, whereas cellular immunity could be transferred only by living cells). In the 1960s it was demonstrated that only lymphocytes can produce antibodies. It was discovered that the differentiation of lymphocytes producing antibodies in the chicken occurs in the bursa Fabricii. Thus these lymphocytes were described as B-cells (in mammals the maturation of B-cells, recognized by the presence of IgM on their surfaces, occurs in the bone marrow), and lymphocytes responsible for cellular immunity mature in the thymus were described as T-cells (responsible for such reactions as graft rejection, delayed hypersensitivity, etc. which are performed without production of antibodies). Soon, the molecular structure of antibodies was clarified. It became obvious that different antibodies, at the amino-terminal regions, are composed of their polypeptide chains by different amino acid residues, and that the great diversity of antigens can be recognized by adequate antibodies from the enormous repertoire of antibody specificities produced by B-cells.

Immunological reactions are initiated by the binding of antigens with the receptors present on the surface of lymphocytes. Each lymphocyte bears on its surface a great number of identical receptors and can bind with only one kind of antigen. Thus, for binding an enormous number of different antigens with a high degree of specificity, lymphocytes must be heterogeneous with regard to antigen receptors. The receptors for B-lymphocytes are immunoglobulins synthesized by themselves (each lymphocyte synthesizes only one type of immunoglobulin). The T-lymphocyte receptors recognize a cell surface target complex involving both cell major histocompatibility complex and antigen.

The immune system of vertebrates is organized in such a manner that several types of lymphoid organs are associated with it and drain defined tissues and vascular spaces. Lymph nodes accumulate fluids from all intercellular spaces, excluding the eye, brain, and gastrointestinal tract. The gastrointestinal tract has its own lymphoid organs: Peyer's patches, the appendix, tonsils, and adenoids. Antigens from the blood are accumulated in the spleen. The antigens in these lymphoid organs are retained and processed by macrophages and dendritic cells, and the processed antigen is presented to antigen-specific lymphocytes in a manner that promotes their proliferation and differentiation into effector cells. The antigen-reactive lymphocytes are enriched by their mobility in the sense that they recirculate from blood into lymphoid organs and back to blood. These lymphocytes cross regions of antigen-presenting cells on the way to specific T- and B-cell domains through entry sites which open to the unusual postcapillary venules bearing high-walled endothelium. Only recirculating B- and T-cells adhere and migrate into the lymphoid organ. The migration of the recirculating lymphocytes was named homing, and surface structures that recognize the B- and T-cells, and to which they adhere, were named homing receptors. At least two independent homing receptors, have been discovered, one for peripheral nodes and another for Peyer's patches. The majority of the recirculating immunocompetent lymphocytes are arrested in the G_0 stage of the cell cycle for days to years. The activation of these lymphocytes by cognate antigen or by mitogenic lectins causes them to enter the cell cycle, proliferate, and differentiate into memory cells and effector cells. Antigen-activated T- and B-lymphocytes lose the ability to bind to organ-specific high-walled endothelium. In the lymphoid organs small numbers of antigen-activated T-helper cells and large numbers of antigen-specific B-lymphoblasts accumulate in the germinal centers. All antigen-specific helper and killer T-cell clones lack homing receptors. The differentiation of homing receptors of B-cells includes rearrangement of antigen receptor genes for production of functional cell surface antigen receptors; the ability to carry out their function upon stimulation with cognate antigen; expression of phenotype-specific surface molecules; acquirement of products enabling their release from maturational lymphoid organs and homing receptors that allow them to enter peripheral lymphoid organs.[1] Biochemically, the lymph node homing receptor is a branched chain, ubiquinated cell-surface glycoprotein, and the lymph node receptor is a lectin.

II. THE IMMUNOGLOBULIN MOLECULE

The immunoglobulin molecule (Ig) is a tetramer, composed of two identical heavy (H) and two identical light chains (L) bound by several disulfide bonds. Each chain is composed of a constant (C) and variable (V) part. Antibodies are encoded by three unlinked families of genes: L chains lambda and kappa, and a cluster of H chains, localized in mice on chromosomes 16, 6, and 12, respectively, and in humans on chromosome 22, 2, and 14, repectively.

First, Dreyer and Bennett (1965) postulated that the V and C parts of immunoglobulins are encoded separately, and that there are many variable region genes, but only one for the constant region sequences. The discovery of restriction endonucleases and their application to gene mapping allowed Hozumi and Tonegawa to test this hypothesis in a direct fashion.[2] They cleaved DNA from embryonic cells and plasmocytoma cells with restriction endonucleases, separated the obtained fragments, and assayed for V_λ and C_λ region genes using appropriate cDNA probes. An unexpected finding was that the C and V region genes which in the germ line DNA were separately located, were brought together during lymphocyte differentiation by somatic DNA recombination. Further studies revealed that the germ line V_λ gene encodes only 97 amino acid residues of the V_λ chain, and the remaining 13 amino acid residues are encoded by a separate DNA fragment designated ''joining'' (J_λ), located in the germ line DNA a substantial distance from the V_λ gene. The J_λ fragment itself is separated from the C_λ gene by an intervening sequence of 1.15 kb. During lymphocyte differentiation the DNA undergoes somatic rearrangement which results in the joining of the V_λ with the J_λ fragment, leaving the J_λ-C_λ intervening sequence intact (Figure 1).

In its final structure, an active lambda gene is composed of an initial hydrophobic leader sequence, separated from the somatically joined V_λ/J_λ segment by a small intervening sequence named intron; another intron separates the V_λ/J_λ segment from the C_λ segment. The entire L-V-J-C genomic sequence is transcribed, and the introns are removed by ''splicing''. In mature mRNA all L-V-J-C coding sequences are continuous.

In the mouse, the kappa chain coding system consists of several V_κ segments and five, J_κ segments, but only one C_κ region. During rearrangement, one V_κ segment joins with one J_κ segment, probably by deletion of the DNA sequences separating the V_κ and J_κ segments, which results in a single transcription unit L-V-J-C.

The V_H region is encoded by three gene segments: V-D-J^H + L (leader sequence). The germ line contains a cluster of 100 to 200 different L-V_H segments, 12 D (for diversity), 4

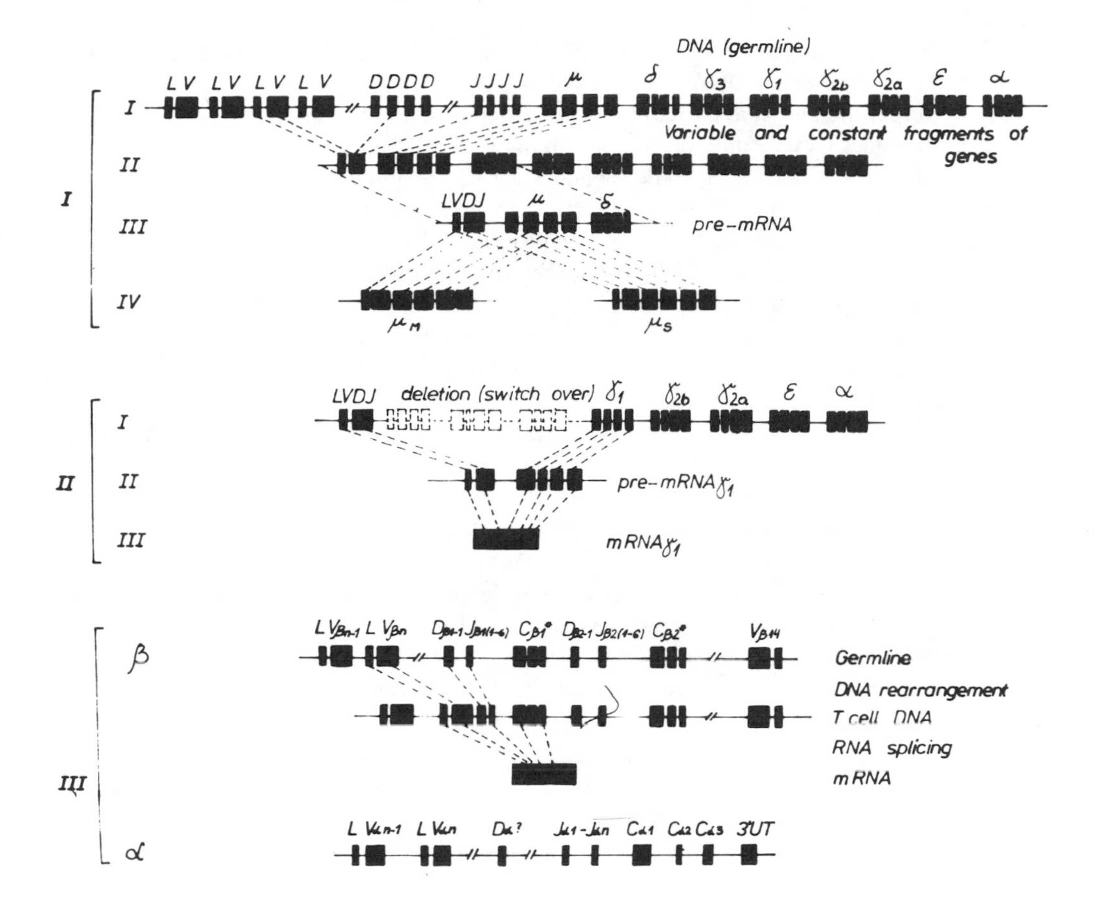

FIGURE 1. Genes of the immunoglobulin supergene family B cell: the upper line (I,I) represents the DNA germ line with separately located κ, or λ, or heavy chain immunoglobulin genes; note that the particular segments of genes (L — leader sequences, V — variable sequences, C — constant sequences, D — diversity sequences) are separately located. After rearrangement (differentiation), the particular fragments of a gene (for IgM) are joined (line I, II), and pre-mRNA is synthesized (line I, III). Processing and splicing yields two kinds of mRNA: membranal (μ_M), and secretory (μ_S) (line I, IV), mRNA of a heavy chain is composed of LVDJC fragments. The next part of the figure represents the switch over of the particular immunoglobulin classes (II, I: II, III) according to the deletion theory.

The lowest part of the Figure represents T-cell receptor for β and α chain genes which are processed similarly to immunoglobulin genes (shown on β chain). (partially redrawn after Epplen et al.[9])

J (for joining) and a cluster of 8 copies of C_H segments, one for each one of the 8 Ig classes and subclasses. The order of C_H genes in the mouse is: μ, δ, γ3, γi, γ2b, γ2a, σ and α. Each of the C_H coding sequences is composed of exons coding different domains and hinge regions of Ig constant region; there is no hinge region for C_μ and C_α.

A unique feature of the immune system is that an antibody forming cell can switch the class of Ig while maintaining the same V region, and therefore the same antigen specificity. This suggests that a single V_H region can associate sequentially with different C_H regions within one cell or cell clone. Several models have been proposed to explain this phenomenon. In the RNA-splicing model, through different modes of splicing, a putative RNA-precursor spanning the entire C_H locus would generate different heavy chain mRNAs. In the multicopy V_H insertion model, copies of particular V_H regions are inserted near each C_H gene. In the third model, one V_H gene undergoes a series of shifts to associate with different C_H genes. Up to this time none of the models has been confirmed as true.

In the deletion model, the DNA between a particular V_H gene and the first C_H gene to be expressed is excised, and successive switching occurs by deletion of the following C_H genes including intervening DNA sequences. For example, a lymphocyte clone can switch from IgM to IgG_3 deleting the C_μ and C_γ genes together with all flanking DNA sequences. The deletion model was verified on plasmocytoma cells.

Switching occurs only at specific switch sites (S) recognized by specific switching enzymes which may belong to the non-histone chromatin proteins. The deletion mechanism of switching predicts the irreversibility of this process.

Antibody-producing cells are able to synthesize a single type of both light and heavy chains that are the products of only one of the two alleles. The simplest explanation is that in an individual B-cell the rearrangement of Ig genes is restricted to one of the two homologous chromosomes. However, investigations on normal B-cells, B-cell lymphomas, showed that rearrangement may occur on both chromosomes.

Three different types of nonproductive Ig gene rearrangements are postulated, resulting from V_L-J_L or V_H-D_H-J_H joining as: out-of-frame rearrangements, rearrangement to an abnormal abortive DNA sequence, and incomplete rearrangements. Out-of-frame rearrangements join the antibody gene segments together in different translational reading frames, and can result from inaccurate joining of V_L-J_L, V_H-D or D-J_H gene segments. The V_H-D-J_H joining requires two distinct site-specific recombination events. Sometimes incomplete joining may occur, like V_H-D or D-J_H which results in a lack of Ig synthesis. Allelic exclusion may also be the result of regulatory disturbances.

In the differentiation process, two forms of IgM molecules are synthesized: membrane bound receptor IgM molecules, and IgM molecules secreted into the plasma. The two forms of IgM differ chemically, although only single C_μ copy exists. This paradox has been resolved by showing that two chemically and functionally distinct μ mRNAs can derive from a single C_μ gene by alternative modes of splicing.

The number of unique antibodies in any particular species is estimated at 10^6 to 10^8. Two opposing hypotheses have been developed to account for antibody diversity. According to the germ line theory, germ cells carry structural genes for all V_L and V_H which an individual can produce. The somatic diversification theory postulates only a few inherited genes, which underwent extensive somatic mutation to generate thousands of genes. Surprisingly, to a certain extent both theories are known to be correct.

The recent dynamic development of gene cloning and sequencing has allowed the recognition that antibody diversity arises from several sources, including: (1) multiple germ line genes encoding V, D, and J segments, (2) combinatorial joining of V, D, J segments, (3) nonprecise joining or combinatorial scrambling of V-J or V-D-J, (4) somatic mutation of the V-D-J segments after assembly, (5) combinatorial pairing of heavy and light chains.[2,3]

The variable region of an antibody presents not only a combining site, but also an antigenic profile named its idiotype, against which anti-idiotypic antibodies can by synthesized. The anti-idiotopic profile of the variable region of the given antibody molecule is not a single site, but consists of several distinct sites against which different anti-idiotypic antibody molecules can be produced. This means that an idiotype of a given antibody represents a set of different immunogenic idiotopes. Finally, the immune system of a single animal, after producing specific antibodies to an antigen, produces antibodies against antibodies made by itself. The anti-idiotopic antibodies display new idiotypic profiles, and the immune system represents a network of idiotypic interactions.[4]

The cellular reactions soon became transformed into molecular ones. This occurred because of the discovery of a great number of active compounds responsible for the proliferation and differentiation of cells participating in the immune reaction, and factors which determine its course.

The central position in the immune response has T-lymphocytes which are involved in most functions of the immune system. The T-cell receptor (TCR) plays the key role in these cells, and consists of two disulfide-bound polypeptide chains α and β, which give the T-cell specificity for both the foreign antigens and self major histocompatibility complex (MHC) determinants. The progenitor cells develop into functional immuno-competent cells in the thymus, which controls the self-MHC restricted responsiveness of mature peripheral T-cells. This denotes that during intrathymic maturation there follows elimination of "forbidden" specificities. Colonization of the fetal thymus revealed that blood-borne stem cells during their inhabitation in the thymus acquire successively several specificities (among others, also MHC specificities) which causes gradual change from blast cells to small lymphocytes. In the adult thymus all ontogenic stages of thymocyte differentiation are present simultaneously — from small numbers of immature blasts to fully mature thymocytes.

The immune response in vertebrates may be performed by B-lymphocytes which react specifically with soluble antigens, and T-lymphocytes which react with antigens present on the cell surface. The receptor molecules that mediate these two reactions exhibit similar heterodimeric structures: light (L) and heavy (H) chains of immunoglobulins serve as antigen receptors for B-cells, and α and β chains form the T-cell antigen receptors. Light and heavy chains of immunoglobulins, and alpha and beta chains of T-cell receptor molecules have variable (V) regions that recognize antigens, and constant (C) regions that are responsible for attachment to the cell membranes and probably for signal transmission. The constant part of the immunoglobulin molecule is also involved in complement activation via the classical pathway, and in the degranulation of mast cells. The population of these cells is able to bind an enormous number of different antigens in spite of the fact that individual cells are synthesizing only a single type of antigen receptor. Therefore, antigenic challenge causes clonal proliferation of cells that have receptors specific to the particular antigens.

The genes for the synthesis of immunoglobulins and T-cell receptors exhibit similar organization and undergo similar DNA rearrangements. The immunoglobulin light chains are encoded by kappa and lambda gene families, variable regions of which are encoded by V_L (leader) and J_L (joining) gene segments. The variable part of the heavy immunoglobulin chains is encoded by V_H, D_H (for diversity), and J_H gene segments. These gene segments must join to form the complete light and heavy variable part of immunoglobulins binding different antigens (Figure 1).

T-cell specific antigen receptors are encoded by three different classes of genes, alpha, beta and gamma organized in the same order as the immunoglobulin genes (Figure 1). The T-cell receptor segment comprises V_β, D_β (for diversity), and J_β gene segments which during T-cell differentiation form the V_β gene. Similarly organized are the gamma genes and also probably the alpha genes. T-cell receptor genes rearrange by mechanisms similar to those of the immunoglobulin genes. The joining boundary of each gene segment is similar as in immunoglobulins flanked by a highly conserved heptamer and a nonconserved spacer sequence of 12 or 23 nucleotides and an AT-rich nonamer. The length of the nonconserved spacer corresponds to approximately one or two turns of the DNA helix. In both immunoglobulin and T-cells the receptor recognition sequence joins at a two-turn fragment of one DNA. Besides, the joining in these gene families is imprecise, which is an additional source of diversity of the synthesized receptors.[5]

The diversity of V parts of the T-cell receptors is achieved by the same mechanisms as in B-cell antigen diversification, i.e., multiple germ line gene segments (the β gene family has 12 functional J_β gene segments (6 in each cluster, 2 D_β gene segments and, up to now, an undefined number of V_β gene segments), combinatorial joining of the gene segments, somatic mutation, and combinatorial association of the polypeptide subunits. The somatic mutational mechanism arises from the variability of sites at which the gene segments are joined, and N-region diversification is done by the random addition of nucleotides to either end of the D gene segment during the joining of the V and J gene segments.[5] Of the three functionally defined subsets of T-cells, the helper and cytotoxic cells express a heterodimer receptor composed of one alpha and one beta chain. The alpha chain is only little known.

The gamma chain of the T-cell receptor is expressed on a small subset of circulating T-cells. There is no evidence that its expression is associated with any well-defined function of a subset of T-cells, and in almost all mature T-cells they synthesize a gene that cannot be transcribed into protein. In contrast to this, early in thymic ontogeny the gamma chain gene is intensively transcribed, which leads to the suggestion that the gamma gene product may be important in the ontogeny of T-cells.

The crucial feature of the T-cell receptor is that it recognizes an antigen only when it is associated on the surface of a cell with a molecule encoded by the major histocompatibility complex (MHC). In other words, the T-cell becomes activated by self MHC associated with the antigen, but not when the antigen is its own.

The T-cell antigen receptor is associated with CD3 (T3) complex protein which seems to be identical in all T-cell clones and to be common to major subsets of T-cells. CD3 is composed of at least three molecular species (CD3γ, CD3ε, CD3σ): one nonglycosylated 20 kd molecule and two glycoprotein molecules of 20 kDa and 25 to 28 kDa, respectively. It is suggested that the hydrophobic CD3 molecule may anchor the glycosylated CD3 species and probably also the T-cell antigen receptor composed of the alpha and beta chains. The CD3 protein complex is specifically engaged in the recognition of antigens. Simonsen suggested that the CD3 complex may act as a specific inhibitor of T-cell receptors in connection with the MHC complex by a reversible inhibiting of antigen binding causing harmful interaction with crossreacting self determinants. The blocking of the CD3 receptors may be released in the case of an immunogen + self-MHC antigens that enables the binding of the immunogen with the T-cell receptor which initiates the immune reaction. The blocking mechanism seems to be very important for prevention of harmful immune reactions. The reason is that the T-cell clone, after activation by a foreign immunogen presented together with the individual MHC molecule (antigen + MHC molecule), because of a crossreaction, modifies the MHC molecule to a non-self MHC molecule which generates the immune reaction. It is thought that this mechanism may be more important for maintaining natural T-cell tolerance to self, than the complicated network of suppressor clones.[7]

The CD3 protein complex is non-covalently associated with the alpha chain of the T-cell receptor, and a mutation that prevents the appearance of the receptor heterodimer also presents the appearance of the CD3 complex. However, 3 to 8% of the peripheral lymphocytes express CD3 also in the absence of the T-cell receptor. Anti-CD3 antibodies can induce both cytotoxicity and interleukin-2 secretion, and thus seem to have the full attributes of normal T-cells.[6]

The CD3-T-cell receptor gamma chains mediate non-MHC restricted killing of target cells. The experiments revealed that apart from the classical cytotoxic T-lymphocytes which must recognize MHC products to kill target cells, there is a broadly defined category of lymphocytes that does not require prior sensitization for their cytotoxic activity (unlike classical cytotoxic cells). Such killing may be targeted by antibodies bound to the cell surface (i.e., antibody-dependent cell-mediated cytotoxicity — ADCC) or not, and described as natural killing. Most natural killing lymphocytes (NK lymphocytes) are able to react to ADCC or not.[8]

The molecular structure of all possible T-cell receptor variants seems to be far from being solved. About 95% of the T-cells (T_{helper}, $T_{cytotoxic}$, and $T_{natural\ killer}$) bear T-cell receptors αβ with one disulfide bond and the phenotype CD3(+), CD4(+), CD8(−) and CD3(+), CD4(−), CD8(+); A 5% of the T-cells (NK-like cells) bear T-cell receptors $\gamma C_1 \delta$ with one disulfide bond and the phenotype CD3(+), CD4(−), CD8(−); and another 5% of the T-cells (NK-like cells) bear T-cell receptors $\gamma C_2 \delta$ and the phenotype CD3(+), CD4(−), CD8(−). As a rule, the CD4 phenotype realizes its function on T-helper cells which are usually class II MHC restricted, and the CD8 phenotype on cytotoxic cells which are usually class I MHC restricted.[8]

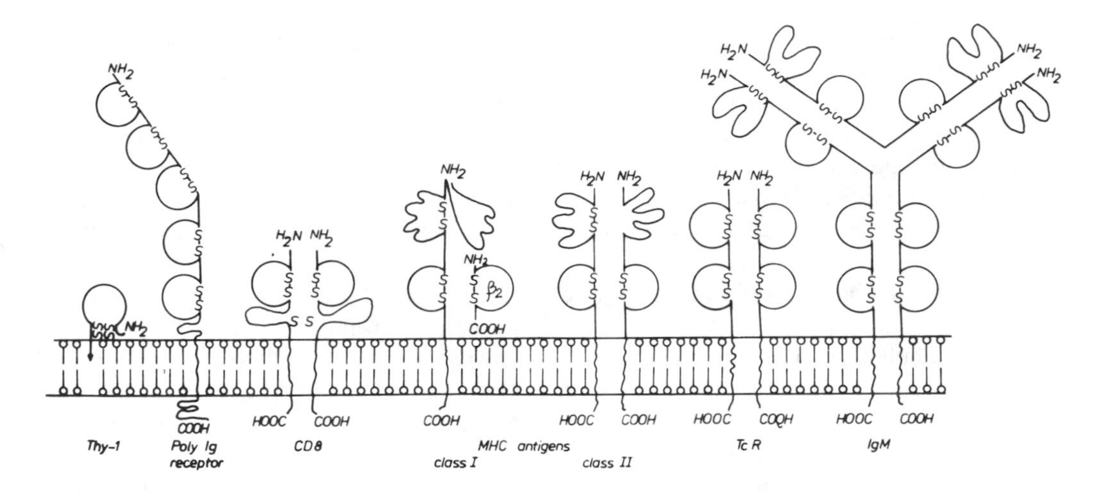

FIGURE 2. Members of the immunoglobulin gene superfamily. Single gene members include T-cell accessory molecules in class I (CD_8) and in class II (CD_4) MHC recognition and probably signal transmission (through channel formation) CD3 δ and CD3 σ. The Poly-Ig receptor is able to transport certain classes of immunoglobulins across mucosal membranes. β_2 Microglobulin is associated with class I MHC molecules. The CD8 single chain heterodimer with a V-like homology unit recognizes conserved determinants of class I molecules with a receptor domain similar to the antigen-binding domains of immunoglobulins and the T-cell receptors. Less similar sequences have been proposed for the human CD4. (CD3 δ and CD3 ε have a single V-like homology unit). It is noteworthy that all molecules cooperating with MHC are members of the immunoglobulin gene superfamily. Besides, there are several not well-defined molecules such as Thy-1 present in lymphocytes and neurons, and brain-specific molecules N-CAM and neurocytoplasmic protein 3 (NP3) the functions of which are not known. But their presence suggests that the immunoglobulin gene superfamily may play an important role in cell-cell interactions in the nervous system.

Suppressor cells which represent about 3 to 8% of the peripheral blood lymphocytes belong to the subset of T-cells that do not express the T-cell heterodimer αβ receptor.[6] The kind of polypeptide receptor molecule that these cells use is unknown.

Immunoglobulins, T-cell receptor proteins, and MHC products form a supergene family of approximate molecular structure and mechanisms of their expression.[5] This family includes Thy-1, poly-Ig receptor, CD_8, MHC class I and II antigens, T-cell receptor, and immunoglobulins (see Figure 2). Members of this superfamily are engaged in specific immune recognition. The individual subfamilies of the Ig superfamily are dispersed over the whole genome. The alpha-chain gene of the T-cell receptor resides in man and mouse on chromosome 14, the beta-chain gene resides on chromosome 7 in man and chromosome 6 in mouse, the gamma-chain genes of the T-cell receptor in man are on chromosome 7 and in mouse on chromosome 13. Human immunoglobulin heavy chains are encoded by loci on chromosome 14, gene loci for kappa chains on chromosome 2, and gene loci for lambda chains on chromosome 22. Gene loci for MHC in humans are localized on chromosome 6, and in mouse on chromosome 17. Members of the superfamily T-cell receptors exhibit similarities among each other at the level of gene organization, and primary, secondary and tertiary protein structure. Members of the Ig gene superfamily are composed of several homologous units. The exon/intron organization of T-cell receptor genes reflects closely the Ig domain structure.[9] A supergene family is a set of multigene families and single copy genes related by sequence, but not necessarily related in function. Sequence analysis of MHC class I and class II gene products and Thy-1 revealed homology to the immunoglobulin supergene family, described after the first genes of this family to be analyzed.[5]

The variability of T-cell receptors generates principally by the same gene rearrangements as in immunoglobulin genes. In rearranged immunoglobulin genes as well as in T-cell receptors (alpha, beta, gamma) short regions of inserted nucleotides have been described which were absent in the germ line. These regions play an important role in determining antigen binding specificity, and are a source of additional variability of the T-cell receptors. Another source of additional variability is the action of terminal transferase that catalyzes independently the polymerization of deoxyribonucleotide triphosphates into the 3′hydroxyl terminus of a polynucleotide on the template molecule. The activity of this enzyme is greatly expressed in cortical thymocytes and in undifferentiated bone marrow cells. Increased activities of terminal transferase are observed in peripheral blood cells in acute and chronic leukemia.[9,10]

The interaction of T-cells with their target cells is also not fully understood. T- and B-lymphocytes exhibit totally different mechanisms. B-cells bind antigens directly with surface immunoglobulin molecules synthesized by themselves, whereas T-cells recognize foreign antigen in cooperation with MHC molecules, class I or class II, present on their own membrane surfaces and the membrane surfaces of the target cells. B-cells are able to synthesize from the same genetic code different immunoglobulin molecules, for example, secretory IgM and membranal variant of IgM. Antigen presented to class II MHC-restricted T-cells is not intact. In antigen-presenting cells (mainly macrophages), antigen becomes degraded in an acidified intracellular compartment. Degraded antigens may interact physically with class II MHC molecules. This kind of antigen presentation diverges from that of antigen recognition by class I MHC-restricted T-lymphocytes (cytotoxic T-lymphocytes). These cells specifically interact with integral cell-membrane molecules (viral envelope glycoproteins, allogeneic MHC molecules, minor histocompatibility antigens, etc.). However, recent investigations demonstrated that the reacting viral protein (in influenza virus infection) was the viral nucleoprotein or its fragment that migrates to the cell surface, and not the membrane hemagglutinin molecule.[11]

The experiments with influenza antigen support the view that antigen processing is not a specialized function of a particular subset of cells, but rather an intracellular activity in all cells. This is evidenced by the fact that fibroblasts made to express class II antigens by DNA-mediated gene-transfer are able to present various protein antigens to T-cells, and that this function can be inhibited by the same inhibitors that interfere with the processing of antigens by hematopoietic-presenting cells. This clarifies the fact that class I molecules distributed on various somatic cells and virally infected cells are recognized by cytotoxic T-cells.[11] In addition it was demonstrated that there must exist two distinct mechanisms for the production of immunogenic protein fragments, and different pathways for intracellular transport for class I and class II molecules that determine the nature of recognized antigens.[11]

Biochemical studies demonstrated binding of pair reactants, i.e., the antigen with the MHC molecules, and the T-cell receptor with the antigen. Recent experiments revealed that the primary sequences of all peptides binding the same MHC molecule demonstrates a related pattern of amino acid residues (named "motif"). This motif was also detected in the MHC molecule binding the peptides, which suggests that the foreign peptide displaces an internal peptide sequence in the case of binding. The protein fragments that are recognized by T-cell receptors can be divided into amino acid residues that interact independently with the T-cell receptor (epitope) or with the amino acid residues that interact with the MHC molecule (agretope). Agretopes do not influence the specificity of the stimulated T-cell clones, but they rely on their ability to recognize the peptides as foreign.

Most surprising was the discovery that all elements of the motifs were also present in the hypervariable portion of the MHC molecule. In this aspect the hypervariable fragments of MHC molecule can be divided into two subsites: an internal subsite for T-cell receptors that interact with T-cell receptors during primary antigen selection, and another internal agretope (motif) that interacts subsequently (after primary selection by the T-cell receptor) with the MHC molecule. In this way, the foreign antigen is first recognized by the T-cell receptor and subsequently by the MHC molecule, but not earlier than after primary recognition by the T-cell receptor. If this be so, antigens not recognized by the T-cell receptors cannot be recognized by MHC molecules.[12] Also, double independent recognition of foreign antigens by reacting T-cells of two independent T-cell receptors (one for antigen and one for MHC antigen) would be reduced to one receptor with double subsequent action: the first, interaction with the T-cell receptor, and the second, with the MHC molecule receptor.

The interaction of T-cells with target cells does not consist of the simple binding of receptors alone, but interaction of several accompanying molecules as well. These molecules are: MHC class I and class II molecules, CD3, CD4, and CD8 molecules. CD3 molecules are composed of three different protein molecules: $CD3_{\gamma}$, $CD3_{\delta}$, $CD3_{\alpha}$, named also T3 complex protein. The CD3 complex, associated with the T-cell receptor subunits alpha and beta, seems to be necessary for transmission of signals to the interior of the cells. According to Simonsen's suggestion, the *N*-glycosylated glycoproteins b and c may block the access of self-antigens to the T-cell receptor, and enable access to the T-cell receptor of foreign antigen + self-MHC antigen, with the transmission of the signal to the interior of the cell, mediated by the CD3 protein. In addition, in the case of the interaction of the antigen presenting cell and the T-helper cell mediated by the MHC class II molecule, an additional protein, CD4 (in mouse, L3T4 protein), may increase the overall avidity between the reacting cells. In the case of the interaction of the T cytotoxic (killer) cell and the target cell mediated by the MHC class I molecule, an additional protein, CD8 (in mouse, Lyt 2 protein), may increase the overall avidity between the reacting cells (see Figure 3).

III. THE IMMUNE REACTION

The immune reaction results from a series of molecular and cellular interactions. The molecules involved in the immune reaction include lymphokines (secreted by lymphocytes and other cells participating in the immune reaction), immunoglobulins (secreted by B-lymphocytes) and cell surface proteins (histocompatibility antigens, receptors for the F fragment of immunoglobulins, lymphokine receptors and the T-cell receptor).

Macrophages or other antigen-presenting cells, simultaneously with the degradation of antigen to a form able to induce immune reaction, synthesize and secrete interleukin-1 (IL-1), a protein of molecular mass in the range of 12 kDa to 16 kDa. IL-1 is a potent inflammatory mediator that acts on the synthesis of acute phase protein and connective tissue proliferation and cellular activation, but also is the main factor that causes proliferation of B-cells. IL-1 does not stimulate T-cell proliferation directly; it induces synthesis of interleukin-2 (IL-2) by the activated T-helper cells which stimulates the T-cells to proliferate. In addition, IL-2 stimulates the production of another lymphokine, interferon gamma (named also immune interferon), which augments the cytotoxic activities of T-cells, macrophages, and natural killer cells.

The interferon gamma increases expression on the surface of macrophages of the MHC molecules engaged in presentation of antigen to the T-helper cells. By enhancing antigen presentation, interferon gamma amplifies the immune response to various antigens. In this way a large expansion of the antigen-specific cell population can occur. The action of IL-2 is similar to that of the polypeptide hormones. T-cells activated by IL-2 have specific receptors on their surfaces expressed by antigenically activated T-cells; in other words, activated T-cells synthesize IL-2 receptors for their own use, and it seems that the action of IL-2 is limited to T-cells. In contrast to this, IL-1 acts on a broader range of target cells, causing proliferation of the activated B-cells and antigenically activated T-cells. Physiologically, IL-1 activates the keratin-producing cells of the skin, corneal cells, and cells lining the mouth cavity.[13]

IL-1 participates also in the inflammatory reaction. It stimulates the proliferation of fibroblasts and, in abnormal cases, may contribute to fibrosis by deposition of fibrous

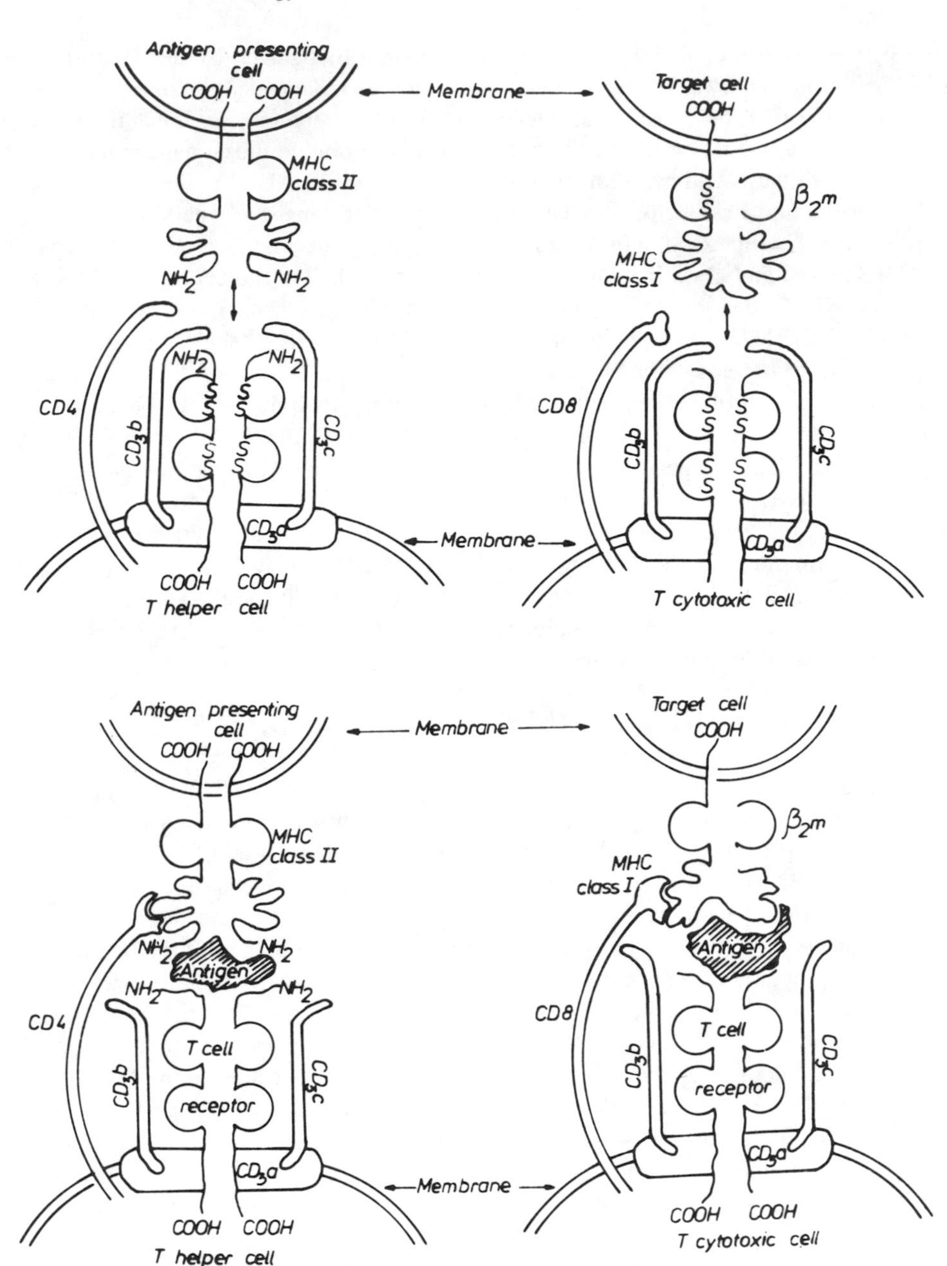

FIGURE 3. Hypothetic model of the interaction of T-cells and target cells. T-helper cells bear the CD4 molecule that recognizes MHC class II antigens, whereas T-cytotoxic cells (killer cells) bear the CD8 molecule that recognizes MHC class I antigens. The T-cell receptor recognizes the antigen every time in cooperation with the MHC molecule class II or class I. The T-cell receptor cooperates with CD_3 complex proteins. In the absence of the foreign antigen (top): CD3 b and CD3 c proteins block the T-cell receptor. In the presence of foreign antigen (bottom): In cooperation with the CD3 a protein, they display the T-cell receptor which enables interaction and transmission of signal for activation of the T-helper cell. This leads to immunoglobulin synthesis or activation of the T-cytotoxic cell, leading to the killing of the target cell. (Adapted from Simonsen[7] and Epplen et al.[9])

connective tissue. Finally, IL-1 may produce the protein-splitting enzymes collagenase and plasminogen activator. These enzymes might promote cleaning up of the debris which is necessary to healing wounds. The prolonged production of these enzymes, together with prostaglandins, especially by the synovial cells, may cause damage to joint surfaces and results in chronic arthritis.[13]

IL-1 acts also on brain cells in the temperature control region of the hypothalamus. Administered to rabbits, IL-1 causes fever. It is generally accepted that IL-1 is the endogenous pyrogen.[13]

Because of the multiple action of IL-1 it is thought that there may be several IL-1 molecules with different activities, whereas IL-2 is encoded by a single gene, and therefore only a single IL-2 molecule is acting during T-cell proliferation.

The IL-2 receptor protein (Tac antigen) binds IL-2, and has a relative molecular mass of 55 kDa. IL-2 is produced by a subset of T-helper cells induced by antigen binding simultaneously with the induction of IL-2 receptors. Induced T-cells synthesize two classes of IL-2 receptors — molecules with high affinity for IL-2 (about 10% of the IL-2 receptors), responsible for the physiological response of T-cells to IL-2; and molecules with low affinity to IL-2. Some lymphoid cells (natural killer cells and precursors of lymphokine-activated killer cells) do not synthesize IL-2 receptor protein, but instead bind and react with IL-2. These results suggest that besides the IL-2 receptor of M_r 55 kDa, there also exists another IL-2 binding protein of M_r 70 kDa. It is suggested that high affinity IL-2 receptor proteins may be composed of both M_r 55 kDa and M_r 70 kDa subunits, each of them capable of binding IL-2 with low activity.[13]

Resting B-cells stimulated with antigen produce the B-cell growth factor (BCGF) which causes proliferation of activated B-cells. Further proliferation of B-cells is stimulated by IL-1. Maturation of activated B-cells to antibody-producing plasma cells is under the control of of the B-cell differentiation factors (BCDF). The T- and B-cell growth factors represent distinct entities. Activated B-cells exhibit specific receptors for BCGF. Both BCGF and BCDF are synthesized by T-helper cells. According to some observations BCDF may be specific to the production of specific immunoglobulin classes. B-cells usually synthesize immunoglobulins of M class (IgM) first, and immunoglobulins of G class (IgG) next. It is suggested that B-cells differentiated under the influence of a specific BCDF synthesize only IgM, and IgG is synthesized by other B-cells which have matured under the influence of a specific BCDF for plasma cells secreting IgG.[14]

Human immune response seems to be realized by the induction of IL-2 synthesis induced by antigen binding. Secreted IL-2 causes IL-2 receptor expression on resting T-cells, and the synthesis is regulated by IL-2. IL-2 and adequate IL-2 receptor synthesis causes T-helper cell proliferation, probably by an autocrine mechanism. Simultaneously, T-helper cells secrete BCGF and BCDF. BCDF probably can be secreted also by other cells independently of IL-2. T-suppressor cells can interrupt this reaction at the level of IL-2 synthesis, but most probably this is done at the level of BCGF and BCDF synthesis by T-helper cells.

T-cell differentiation inside the thymus represents several unresolved problems. The thymus contains several cell populations whose exact role in T-cell maturation must be elucidated. Like all hematopoietic cells in adults, thymocytes arise from bone marrow precursors which represent the pluripotent stem cell, with low expression of Thy-1 antigen. These cells undergo differentiation to prothymocyte cells while still in the bone marrow. Next, the prothymocyte cells migrate to the thymus cortex and form early thymocytes. These early thymocyte cells are characterized by low expression of Thy-1 and Ly-1 antigens (in the mouse), and differentiate to double positive thymocytes characterized by these antigens and, in about 50% of them, the alpha/beta T-cell receptor molecules. Next, the thymocytes migrate to the thymus medulla where they mature to single positive T-cells which, at the last step of maturation, become class I or class II MHC-restricted and leave the thymus for the peripheral blood. It is generally accepted that the thymus cells become tolerant of MHC antigens expressed by bone marrow-derived cells — probably dendritic cells and epithelial cells.[15]

Recent experiments revealed that the MHC binding to antigen fragments exhibits some properties different from those of antibody-antigen binding. The antibody-antigen binding is rapid, whereas the MHC binding is slow, and once the peptide is bonded, it is released very slowly (half-life about 30 h). This may indicate that the MHC molecule and the antigen fragment must adopt adequate configuration. Experiments revealed that the antigenic peptides for T-cells have amino acid homology with the class II molecules presenting them. MHC molecules must display the homologous fragment of the protein (antigen), which takes time, and hence the slow course of the MHC binding reaction with the antigen.[16,16a]

The colony-stimulating factors (CSFs) are a family of glycoproteins that regulate the survival, proliferation, and differentiation of hemopoietic progenitor cells and the function of the mature cells. Four CSFs that control the production of granulocytes and macrophages in the mouse have been discovered; three of them have equivalents in the human hemopoietic system. All of them stimulate the formation of hemopoietic colonies in *in vitro* cultures of undifferentiated blast cells which proliferate and produce differentiated progeny. The type of CSF is indicated by the prefix of the CSF. G-CSF (CSF-β in humans) produces neutrophilic colonies; M-CSF (CSF-1 in humans) produces monocytic colonies; GM-CSF (CSF-α in humans) produces neutrophilic, eosinophilic, and monocytic colonies. CSF-α induces T-cells, endothelial cells, and macrophages; CSF-β induces macrophages; and M-CSF induces fibroblasts. The activity of the CSFs may overlap some of their functions. CSFs have specific receptors on the surface of target cells. After the binding of CSFs with their receptors, they become internalized and rapidly degraded in some cells (peritoneal macrophages), or more slowly in others (bone marrow derived macrophages). Intracellular accumulation of CSFs seems to be correlated with the biological response.[17]

Macrophages represent the major cellular component of the classical reticuloendothelial system. The initial concept of Metchnikoff that these cells are scavengers that remove unwanted material (debris) from the extracellular environment is an underestimate of the important role of these cells in modern biology. According to Unanue and Allen,[18] macrophages exhibit a unique role in the tissue response to external stimuli:

1. Macrophages can interact with foreign material (proteins and polysaccharides), engulf and submit it to intracellular metabolic changes.
2. Macrophages are secretory cells that secrete potent inflammatory mediators like proteases, complement components, and growth regulatory factors, e.g., IL-1 and arachidonic derivatives.
3. Macrophages interact with the T- and B-cells in immunological reactions.
4. Because of their location close to the microvasculature, and epithelial and mesenchymal cells, macrophages can influence vascular reactions during inflammation.
5. Macrophages exhibit surface receptors for lymphokines, and "activated macrophages" because their highly microbial and tumoricidal activities play an important role in infectious and neoplastic diseases.

The role of macrophages in the immune reaction consists of processing of antigen (partial degradation) to a form recognizable by the cellular immune system (known as antigen presentation), and release of active molecules such as interferon gamma and IL-1 with their multiple regulatory functions. In this way foreign antigens, immediately after their recognition, induce adequate immunological reactions. Antigens treated with macrophages are 1000 times more immunogeneic than antigens alone. The CD4 positive T-lymphocytes recognize proteins (antigens) only when presented by an antigen-presenting cell that bears an MHC class II molecule which limits immune recognition to the few cells that bear MHC class II molecules. The activation of CD4-positive cells causes B-cell activation and development of inflammatory reaction, and activation of CD8 positive cells to become active killer cells. The CD8 positive cell recognizes an antigenic determinant on a target cell that bears the same MHC class I antigen.[18]

For effective antigen presentation, the antigen-presenting cell (usually a macrophage) must take up the antigen from an appropriate receptor, and internalize and process it if necessary. The antigen-presenting cell must express MHC class II molecules and secrete growth-differentiating molecules. Processing means the antigen requires affinity to an MHC class II molecule by unfolding or by partial proteolysis. It selects for a portion of the molecule or epitopes that have affinity to the MHC class II molecule, creating by their association the determinant recognized by the CD4 positive cell (CD4(+)). According to Unanue and Allen,[18] a globular protein (antigen) is first taken into an acid vesicular compartment that bears Ia molecules (the endosome), and there denatured or partially fragmented. Fragments that have affinity to Ia molecules bind to it and are transported to the cell surface as "presented antigens", whereas fragments that do not have affinity to Ia molecules become totally degraded in the lysosomes. Endosomes that do not bear Ia molecules transport whole the material to the lysosomes where it will be degraded. In contrast to this, the B-cells respond to conformational determinants of antigens, whereas T-cells respond to appropriate amino acid sequences.

The CD4(+) cell recognizes the antigen on the B-cell equally as on a macrophage. Recognition then leads to B-cell activation.

Antigen presentation induces rapid production of interferon gamma, which leads to increased synthesis of mRNA for Ia molecules either in macrophages or in other cells such as epithelial, endothelial, and connective tissue cells. The next mediator secreted by macrophages is interleukin-1 (IL-1) involved in immunological cellular interactions during antigen presentation (membrane form of IL-1), and as mediator in inflammation (secreted form of IL-1). IL-1 is produced either by macrophages or by lymphoid and nonlymphoid cells. IL-1 induces synthesis of IL-2 receptors on T-cells, which allow the T-cell to respond to IL-2, i.e., the previous T-cell growth factor. IL-1 promotes growth of many cells and tissues (liver, brain, connective tissue, muscle, bone, pancreatic islets, neutrophils). It is suggested that IL-1 pathology comprises rheumatoid arthritis, osteoporosis, pulmonary fibrosis, and insulin-dependent diabetes mellitus.[18] During severe infection, macrophages secrete cachectin, (also named tumor necrosis factor).

IV. SIGNAL TRANSDUCTION IN CELLS PARTICIPATING IN THE IMMUNE REACTION

The key event in signal transduction is the hydrolysis of membrane-associated inositol phospholipid which generates products acting as second messengers in cell activation (compare Figure 3, Chapter 7). Turnover of membrane phospholipids is a characteristic event in many cell types undergoing stimulation by external ligands. Protein kinase C becomes activated when a T-cell[19] or macrophage[20] is stimulated by antigen or mitogen: the binding of the receptor and the ligand causes hydrolysis of phosphatidylinositol phosphates and production of diacylglycerol that activates protein kinase C. Mononuclear phagocytes bear about 50 surface receptors coupled to various second messengers that initiate rapid execution of macrophage function and alter the cell's potential for taking such action. In T-cells, interaction between an antigen associated with a membrane-bound MHC molecule and a corresponding cell surface receptor results in T-cell proliferation, cytolysis of target cells, and production of IL-2 or other lymphokines. These functions require activation of genes, mRNA transcription, and synthesis of new proteins. Thus signals must exist that transduce the signals originating at the cell's surface into its nucleus. The critical component in signal transduction is phosphatidylinositol 4,5-biphosphate, present in cell membranes in minute quantities. The interaction of a ligand with its cell surface receptor activates phospholipase C to hydrolyze phosphatidylinositol 4,5-biphosphate, which activates the coupling GTP-binding proteins known as regulatory proteins in signal transduction pathways, e.g., the adenylate cyclase system which uses cAMP as a second messenger. Diacylglycerol activates protein kinase C by increasing its affinity to calcium ions. The increasing intracellular concentration of Ca^{2+} may activate Ca^{2+}-dependent protein kinases, resulting in protein phosphorylation similar to protein kinase C. Protein phosphorylation by Ca^{2+} bound to calmodulin, and the action of calmodulin kinase yields substrates for DNA replication. Another mechanism may use the Na^+/H^+ exchanger, the action of which increases intracellular pH and pH-sensitive phosphorylation, resulting in DNA replication.

V. KILLER CELLS

Cell death is sometimes a pathological event, and sometimes a physiological process, regulated in the same manner as cell proliferation. There are three types of human cytotoxic lymphocytes: 1. Cytotoxic T-cell, antigen-specific, MHC-restricted; 2. Cytotoxic T-cell of broad specificity, non-MHC restricted; 3. Natural killer cell (NK) of broad specificity, non-MHC restricted.[21]

A. ANTIGEN-SPECIFIC, MHC-RESTRICTED CYTOTOXIC T-CELL

These cells have on their surface CD3/T-cell receptor complex and recognize target antigen (target cells) by the CD3/T-cell receptor. The T-cell receptor alpha and beta chains rearrange in these cells during thymic development.

Non-MHC-restricted cytotoxic T-lymphocytes also have the CD3/T-cell receptor complex on their surface. They recognize target cells by the CD3/T-cell receptor complex, non-MHC restricted. These cytotoxic cells differ from natural killer cells by recognizing target cells via the CD3/T-cell receptor, whereas natural killer cells lack on their surfaces the CD3/T-cell receptor complex.

Natural killer cells (NK) lack the T-cell receptor, alpha and beta chains, the CD3 protein complex on their surfaces, demonstrate cytotoxic activity without MHC restriction, and usually express CD16 on their surfaces. The CD16 antigen associated with the Fc receptor for IgG may also be expressed on the surface of other cells, mainly neutrophils (but not on monocytes, B-cells and eosinophils). The recognition structures of NK cells and the target antigen are unknown.

All cytotoxic activity can be mediated by antibodies (this is called antibody-dependent cellular cytotoxicity, ADCC) or by lectins independently of what kind of cytotoxic cell it is.

The terms "Natural killer cells" and "Natural killer activity" have been used variably, causing confusion and uncertainty. Reynolds and Ortaldo[22] proposed the term "NK" cells for non-MHC-restricted cytotoxicity, and the term "N-lymphocytes" for cells with lytic activity. They think that the following functions can be ascribed to N-lymphocytes:

- Control of tumor cell growth (inhibition of primary tumor development and control metastases;
- Involvement in the control of microbial and other parasitic infections (viruses, parasites, fungi, and bacteria);
- Immune regulatory properties (regulation of antibody response and cell-mediated immunity, natural suppressor cells);
- Production of cytokines;

- Control of hematopoietic stem-cell growth and differentiation;
- Involvement in allograft rejection (organ transplantation), and involvement in graft vs. host reaction;
- Involvement in pathogenesis of some forms of aplastic anemia and neutropenia, and others.

The mechanism of cell killing by cytotoxic cells was the subject of many speculations. Complement lysis as a prototype of cell lysis has been taken into consideration. Recent investigations have demonstrated that activated T-cells possess cytoplasmic granules containing a serine proteinase which may participate in T-cell mediated regulation of humoral immune functions. A T-cell derived trypsin proteinase was demonstrated in IgE immune reactions. B-cells can be activated by various serine proteases and suppressed by proteinase inhibitors. Serine proteinases can degrade basement membrane-like structures, which seems to indicate that these enzymatic activities can facilitate migration of T-cells into inflammed tissue.[23]

Recent investigations suggest that killer cells of the immune system may destroy their target cells by secreting lethal proteins directly onto them.[24] Other type of killing depends on making holes in the membranes of target cells. Antibodies in cooperation with complement components, after forming a membrane-attack complex, destroy the membranes of the target cells. Electron microscope examinations demonstrated ring-shaped pores in the target cells after treatment with cytotoxic T-cells. The pores resembled that of complement attack on the target cells. Pore-forming proteins have been demonstrated from granules of several types of killer cells. Killer cells produce lymphotoxin, which exhibits cytotoxic effects (very slow-acting, it requires 24 h to take effect). Lymphotoxin seems to act indirectly on target-cell DNA, causing its degradation and consequent death of the cell. The tumor necrosis factor secreted by activated macrophages is an another lymphokine that may destroy tumor cells.[24]

Every physiological function must be under dual control: induction by adequate stimuli and suppression through degradation of active substances or by an opposed function. Usually, activation and suppression are executed by environmental stimuli. The control of the immune system is rather inwardly organized because during the immune response, mast lympocytes interact with one another rather than from environmental stimuli. Within the immune system, the T-helper and T-suppressor cells control the activity of cytotoxic T-cells and antibody-producing B-cells. T-helper cells act directly on these effector cells, but T-suppressor cells work indirectly through inhibition of T-helper cells. The molecular mechanisms of this action are not fully explained, and the receptor and effector molecules engaged in T-helper cell inhibition are unknown.[25] According to other authors, T-suppressor cell function becomes suppressed by contrasuppressor cells, the mechanism of which is also unknown.

The fundamental problem for the immune system is to discriminate between "self" molecules that constitute its own organism and "nonself" molecules that denote foreign organisms, tissues, or substances which must be destroyed by the immune system. One theory used to explain this dogma was that during development all cells that could destroy self molecules become selectively destroyed or functionally inactivated; and the second, that tolerance is an active mechanism in which negative feedback signals (from suppressor cells) prevent self-recognition and induction of an immune reaction against molecules of the organism itself. For the antibody-producing B-cell, both mechanisms are probably correct, whereas for the T-cell, very little is known on tolerance induction. The question is how the T-cell is inhibited. Does the occupation of a great number of the T-cell receptors release negative signals to the cell interior, or does intense stimulation cause the cell to release an external feedback signal?[26,27] According to Simonsen,[7] cooperation of CD3 complex proteins with the T-cell receptor may create a situation in which self-antigen cannot be bound by the T-cell receptor, and the non-self antigen becomes bound and induces the immune reaction.

VI. INTERFERONS

Interferons have antiviral, antitumor, and immunoregulatory functions. Lymphocytes produce interferons in response to various mitogens, bacterial and viral antigens, allogeneic cells, and antisera against lymphocyte membrane determinants. Human interferon stimulates a transient two- to three-fold increase in the concentration of diacylglycerol and inositol triphosphate within 15 to 30 s of cell exposure to interferon. Human alpha and beta interferons — but not gamma interferon — stimulate the increase of diacylglycerol and triphosphate in cells with appropriate receptors on their surfaces. Both diacylglycerol increase and its antiviral effects correlate according to the dose of interferon. Thus, diacylglycerol and G proteins belong to the group of mediators of interferon reactions.[28]

Interferon gamma is a typical macrophage-activating factor producing a number of functional and biochemical modifications in these cells (increase of endocytic, biosynthetic, secretory, and effector functions). The interferon-gamma receptor on macrophages is capable of binding either natural or recombinant interferon gamma in a specific and saturable manner. Binding of interferon gamma with its receptor on macrophages initiates macrophage activation.[29]

Common biological factors discovered on the basis of their individual activity often show the same structure with another biological factor. So it was with interferon β_2, which shows the same structure as the human B-cell differentiation factor, a T-cell product that enhances human B-cell differentiation and immunoglobulin secretion in antigen stimulated B-cells without inducing cell proliferation.[30]

Defective interferon production due to insufficient viral or bactericidal activity may be the cause of recurrent infections. In two of 44 children with recurrent respiratory tract infections, and nine with recurrent stomatitis, a transient deficiency in interferon alpha was observed, but rather as a secondary, and not a primary, deficiency.[31]

VII. IMMUNE COMPLEX-INDUCED TISSUE INJURY

The immune humoral response with antibody formation to various antigens is an action by which the organism defends itself against potentially injurious factors (e.g., viral or bacterial products). Inactivation of the foreign antigen occurs by forming antigen-antibody complexes. These complexes are removed from the circulation by phagocytosis, or cleared through the kidneys. Formation of immune complexes may be injurious to the organism, as seen in the Arthus reaction and serum sickness. It is generally accepted that tissue injury in such diseases as systemic lupus erythematosus, rheumatoid arthritis, vasculitis, and some types of glomerulonephritis may be caused by the formation or deposition of immune complexes. However, many of the immune complexes present in injured tissue may be secondary to the disease, and not the cause of the disease. The most pathogenic complexes are soluble complexes effective in the activation of the complement cascade. Usually, these type of immune complexes become deposited in the tissues, and activating the complement cascade, they cause tissue injury. Particularly dangerous are immune complexes that react specifically with tissue. An example is Goodpasture's disease, in which the circulating antibody binds with antigen intrinsic to pulmonary or glomerular basement membranes. In the membranous type of glomerulonephritis, the antibody binds with subepithelial antigen, causing *in situ* immune complexes that generate this type of glomerulonephritis. In myasthenia gravis and pemphigus vulgaris also, immune complexes are formed *in situ*. A typical case of an *in situ* immune complex is the Arthus reaction in which two soluble components (antigen and antibodies) diffuse from opposite directions and form immune complexes.[32]

Once an immune complex is formed either in the blood or in the tissue, it may activate many mediator systems. The Fc part of the immunoglobulins activates the classical pathway of the complement system, generating several other mediators such as increased ph[illegible] and accumulation of leukocytes which release a series of enzymes that may co[illegible] the inflammatory processes caused by the immune complexes. Thus, neutrophils release proteases that hydrolyze basement membranes, histones and proteoglycans; activate fibrinolysis; release collagenase that hydrolizes internal elastic lamina; release basic protein bands 1,3,4 that increase vascular permeability, and basic protein band 2 that activates mast cells; activate thromboplastin; generate slow-reacting substances (leukotrienes) that increase vascular permeability and muscle contraction; activate kininogenase, with the release of kinins; generate free radicals of oxygen, causing acute tissue injury; and release surface protease that release kinin from serum alpha-globulin. Leukocytic lysosomal proteases have many tissue targets (soluble proteins of the interstitial fluid, insoluble cell matrix, such as collagens, elastin, proteoglycans, hyaluronic acid, etc.). In the lung, leukocyte proteases cause minimal inflammation with progressive emphysema, whereas complex immune damage involves intense acute inflammation, hemorrhage, and edema. In kidneys, the monocytes (macrophages) play an important role in glomerulonephritis. Depletion of macrophages from the blood prevented generation of glomerular basement membrane nephritis. Similar to neutrophils, macrophages produce tissue injury by release of several inflammatory mediators, particularly oxygen-derived free radicals.[32]

VIII. AUTOIMMUNITY

Autoimmunity comprises a congeries of various diseases from organ specific (Hashimoto's thyroiditis, pernicious anemia, Addison's disease, myasthenia gravis, autoimmune hemolytic anemia, idiopathic thrombocytopenic purpura, idiopathic leukopenia, primary biliary cirrhosis, active chronic viral hepatitis, ulcerative colitis) to non-organ-specific ones (systemic lupus erythematosus, dermatomyositis, scleroderma, rheumatoid arthritis). The organ-specific autoimmune diseases involve immune attack on a particular organ, whereas the non-organ-specific diseases involve changes which are widespread throughout the body, caused primarily by immune complexes between autoantibodies and antigens common to most cells in the body.[33]

The most important factor in generating autoantibodies may be disturbances in the normally concerted immune regulatory system.

The principal cell that reacts on immune regulatory signals is the T-helper cell. The T-helper cell is stimulated by IL-2 (former T-cell growth factor) secreted by a T-helper cell induced by IL-1 synthesized by macrophages after challenge with antigen. *In vivo* application of IL-2 enhances cytotoxic T-cell and natural killer cell responses in mice, increases transplant rejections in rats, and partly restores T-cell function in athymic mice. Little is known about the role of IL-2 in homeostasis of the immune system *in vivo*. There are two absolutely opposite theories about IL-2 and the role of T-helper cells in autoimmunity: one assumes a T-helper cell defect, and the other that IL-2 is indispensable to cause an autoimmune response.[34] In support of the former hypothesis it was observed that in spontaneous immune disorders (e.g., systemic lupus erythematosus, rheumatoid arthritis, autoimmune diabetes) a decrease of T-cell function *in vitro* has been observed (decreased proliferative responses to T-cell mitogens and IL-2, and production of IL-2). IL-2 deficiency could also impair both the induction of suppressor cells and elimination of altered host cells. However, there are some contradictive observations: the IL-2 and T-helper cell defect is not absolute; it is not certain that the decreased IL-2 production *in vitro* was also defective *in vivo* before the autoimmune responses develop. This hypothesis does not explain the absence of autoimmune reactions in other states of IL-2 hyposecretion (Nezelof syndrome, severe combined immunodeficiency syndrome, or certain viral infections) and aggravation of autoimmune diabetes after treatment with IL-2. In support of the latter hypothesis it was observed that in transfer experiments with isolated T-cells from patients with autoimmune disease (spontaneous and experimental diabetes, myasthenia gravis, rheumatoid arthritis, and cells transferred from several experimental autoimmune disorders such as autoimmune thyroiditis or encephalitis) to healthy MHC identical recipients, autoimmune responses were induced. These experiments prove that IL-2 is an important effector of autoimmunity. In addition, high concentrations of IL-2 have been observed in the synovial fluid of patients with rheumatoid arthritis.[34]

The role of an idiotype network in autoimmune processes seems to be an important clue in its elucidation.[35] The defect in autoimmune disorders seems to depend on antigen recognition. Disturbances of the complex protein receptor of T-cells in connection with MHC class I and II molecules probably may be responsible for the erroneous recognition of self-antigens as non-self with consecutive induction of autoimmune reaction which impairs function or destroys the given organ.

Typical functional disturbances with an autoimmune background is represented by Graves' disease (synonyms: Parry's disease, Basedow's disease, exophthalmic goitre, autoimmune thyrotoxicosis). Graves' disease is characterized by frequenct lymphocyte (B- and T-cells) infiltration of the thyroid, and the presence in almost all cases of thyroid stimulating antibodies. Sometimes the patients also reveal other autoimmune diseases (pernicious anemia, diabetes mellitus, myasthenia gravis, Addison's disease, idiopathic thrombocytopenic purpura, vitiligo). The presence of thyroid stimulating antibodies causes uncontrolled secretion of thyroid hormones with hyperthyroidism. Most commonly, a thyroid stimulating antibody binds another TSH receptor which is not subject to biological feedback of the normal hormone.[36]

An opposite disorder to Graves' disease is respresented by Hashimoto's thyroiditis. Hashimoto's disease is characterized by lymphocyte (T- and B-cells) infiltraiton of the thyroid, with thyroid antibodies forming immune complexes and cell-mediated immunity. These damage thyroid cells by the following mechanisms: sensitized T-lymphocytes may interact with specific antigens on the cell surface and induce T-cell-induced cytolysis; antibodies against the cell membranes may attract killer cells by the mechanisms of antibody-dependent cellular cytotoxicity (ADCC) and cause cytolysis; complement fixing antibodies may activate complement cascade which causes lysis of thyroid cells. Hypothyroidism in Hashimoto's disease may be caused by a blocking of TSH receptors by thyrotropin receptor antibodies which deprives thryoid cells of their physiological stimuli.[36]

Myasthenia gravis is an autoimmune disease of the skeletal muscle, characterized by thymic pathology which produces autoantibodies to nicotinic acetylcholine receptors that block neuromusclular transmission. The lymphoid follicles in the medulla are increased, often with germinal centers and B-lymphocytes.

IX. ALLERGY

Allergy is a disorder of the immune system, a condition of abnormally high reactivity to otherwise normal stimuli. Contrary to inflammation that represents rather a protective reaction tending to neutralize the noxious material, allergy represents an autotoxic reaction to a normally benign agent. Both involve the same cell types and mediators, mainly metabolic derivatives of arachidonic acid (prostaglandins, thromboxanes, and leukotrienes), and a phospholipid platelet-activating factor. The lipid mediators were first discovered in experimental models of anaphylaxis. Anaphylactic stimuli cause aggregation of platelets and neutrophils that release histamine, serotonin, and several enzymes, resulting in tissue injury. Antigenic exposure causes the host to synthesize various antibodies from which IgE and IgG-1 are able to bind with cells (mast cells and macrophages) in anaphylaxis, and generate typical anaphylaxis. Other types of antibodies, through antigen-antibody complexes, may activate the complement cascade with release of anaphylatoxin (components C4a, C3a and C5a may release anaphylatoxin). Recent investigations have demonstrated that the main source of anaphylatoxin is C5a, because anaphylatoxin released by C4a and C3a becomes quickly inactivated by serum carboxypeptidase.

Disturbances in the immune regulatory mechanisms are the most important factors in allergic reactions. The experiments revealed that antibody-dependent cellular cytotoxicity (ADCC), which involves inflammatory cells (macrophages, eosinophils, and platelets) rather than lymphocytes, is responsible for the defense mechanisms (killing) against helminth parasites. The antibody involved in this reaction is of the IgE class. Depletion of IgE or the use of anti Fc ϵ receptor antibody dramatically decreases the killing capacity of macrophages, eosinophils, and platelets. Recent investigations demonstrated two types of IgE receptors with different affinity on macrophages, eosinophils, platelets, and mast cells. The Fc R_1 receptor, a tetramer composed of 1 α chain (45 kDa), 1 β chain (33 kDa), and 2 γ chains (9 kDa), is present on mast cells and basophils with higher affinity range $10^9\ M^{-1}$. The Fc ϵ R_2 receptor, a trypsin-sensitive dimer composed of the α chain (45 to 50 kDa) and the β chain (25 to 33 kDa) is present on subpopulations of macrophages, monocytes, eosinophils, platelets, and T- and B-lymphocytes with a lower affinity range $10^7\ M^{-1}$ for monomers, and $10^8\ M^{-1}$ for dimers. On the basis of the discovery of two types of IgE receptors, Capron et al. suggest that mast cells and basophils may selectively bind IgE molecules, resulting in protective immunity against parasites or adverse allergic manifestations.[37]

Interaction of IgE via Fc ϵ R_2 with mononuclear phagocytes causes release of lysosomal enzymes, plasminogen activator, oxygen metabolites, IL-1, sulfidopeptide leukotrines, prostaglandins, and platelet activating factors; interaction with eosinophils causes release of oxygen metabolites, eosinophil peroxidase, and platelet activating factor; and interaction with platelets causes release of oxygen metabolites and several cytocidal mediators. It is noteworthy that platelets from thrombasthenic patients lacking IIa and IIIa glycoproteins do not express receptors for IgE, and administration of monoclonal antibodies against IIb and IIIa glycoproteins inhibit IgE binding and IgE-dependent activation of platelets.[37]

The activation of inflammatory cells through Fc ϵ R_2 receptors does not exclude binding and activation of mast cells through Fc ϵ R_1 receptors (having a higher affinity to IgE), and the release of the eosinophils chemotactic factor of anaphylaxis that enhances the expression of Fc ϵ R_2 receptors on eosinophils. These observations suggest that allergy may be caused by the interaction of IgE with the Fc ϵ R_1 receptor, which causes the release of mediators activating IgE-dependent inflammatory cells through the expression of Fc ϵ R_2 receptors, involving these cells in the pathology of allergic reactions (particularly in their inflammatory components).[37]

In spite of these conclusions, we must be aware that there also must exist yet other unknown factors that cause a wrong immune reaction resulting after exposure to an innocuous pollen in hay fever in some persons but not in others.

X. IMMUNODEFICIENCY DISORDERS

Genetically determined immune deficiencies have been known for some time, but knowledge of their nature is far from complete. Genetically determined immunologic dysfunctions represent an excellent review of the enormous number of problems in this field. The first was the discovery of autoimmunity of New Zealand mice and the absence of a thymus in the hairless mutant nude mouse, followed mutations such as beige (bg), X-linked immunodeficiency (xid), lymphoproliferation (lpr), generalized lymphoproliferative disease (gld), lipopolisaccharide responsiveness (lps), severe combined immunodeficiency (scid), motheaten (me), and the Y-linked autoimmune accelerator gene (YAA) of BXSB mice.[38] Most of these mutants demonstrated antibodies, particularly to the double-stranded DNA. The most seriously affected mice were the motheaten mutation which are difficult to maintain because of the high mortality, but fortunately, the appearance of a new mutation at the motheaten locus called ''viable motheaten'', with a threefold increase of survival enabled prolonged studies. These led to the discovery of a 25- to 50-fold increase in IgM in 6- to 7-week-old mice even though they had only one third the normal percentage of B-cells. Both the B-cells and the T-cells did not respond, either *in vivo* nor *in vitro*, to mitogens or antigens. The mice died because of intense pneumonitis characterized by macrophage aggregation. Their plasma cells contained high levels of immunoglobulins which they were unable to secrete. These mice lacked anti-red-cell antibodies, antibodies against DNA, and also antibodies against neutrophils.[38]

Another noteworthy mutation is the severe combined immunodeficiency (scid) which probably affects the development of lymphopoietic stem cells. Approximately 15% of scid mice became ''leaky'' in adults and produced structurally normal immunoglobulins, but in about 50% of ''leaky'' mice, antibodies against DNA become synthesized and often lymphomas developed.[38]

Several of the mouse models show equivalents to human diseases. For example, patients with severe immunodeficiency have been found to have adenosine deaminase deficiency (ADA deficiency, ADA-scid-deficiency). At first it was discovered that ADA-scid cells from patients with severe combined immunodeficiency disease produce defective ADA protein (adenosine deaminase, EC 3.5.4.4, that catalyzes irreversible deamination of adenosine and deoxyadenosine).[39] Then, in a cell line GM-1715 derived from an immunodeficient patient, cDNA sequence analysis revealed a point mutation (codon 101 CCC to CAG) that predicted an amino acid substitution Arg→Gln. Further analysis showed that only one dedective allele was present. Therefore, this mutation was probably responsible for the loss of function since the primary structure of this enzyme was otherwise entirely normal.[40] Some cases of partial adenosine deaminase deficiency have been described in Xhosa, and another in Kalahari San populations. In one of these cases the stability of the enzyme at 57°C was greatly decreased (the mutation caused decreased stability of the enzyme). In addition, the adenosine deaminase level in erythrocytes decreased, whereas in leukocytes it increased.[41] Adenosine deaminase is constitutively expressed in most tissues, and is particularly high in immature T-cells. The deficiency of this enzyme causes an increased level of deoxyadenosine triphosphate (or altered methylation) that has toxic effects on T maturation. Similarly, nucleoside phosphorylase deficiency leads to an increased level of deoxyadenosine triphosphate with severe immunodeficiency. Defective rearrangement of antigen receptor genes may be another cause of severe combined immune deficiency, discovered in mice (CB-17scid).[42]

As seen from the above presented cases, human immunodeficiency diseases tend to be inherited, mostly linked to X chromosome loci. Since neither the immunoglobulin genes nor the T-cell receptor genes were found defective, it has been suggested that the X-linked disease might include regulatory genes responsible for lymphocyte proliferation and differentiation. A typical example seems to be the CBA/N mice with X-linked xid mutation. These mice are incapable of producing antibodies against soluble polysaccharides because of a B-cell defect (lack of a specific subpopulation of mature B-cells?).[43] A similar defect in antibody production is presented by the X-linked Wiskott-Aldrich syndrome, accompanied by T-cell defects and disturbances in platelet function.

Severe immunodeficiency may be caused by defective MHC antigen expression. The central feature controlling the immune response expression of the mature immune system is the phenomenon of MHC restriction (see Figures 2 and 3). Immunodeficiency caused by absent or incomplete MHC class II antigens on B-cells, monocytes, Langerhans' cells and lamina propria can be corrected by administration of interferon gamma, IL-2, or injection of Epstein-Barr virus, and a variable class I expression could be corrected by interferons.

Class I antigens were absent on platelets where they appear by passive absorption. An antigen response in cases of defective MHC antigen expression was absent, but a normal reaction to mitogens was present. Patients with MHC antigen deficiency have frequent and severe viral infections, mainly enteroviruses with ultimately fatal diarrhea. Most frequently, these disorders have been observed in patients from consanguinous marriages. On the basis of genetical analysis, a possible regulatory defect was suspected. In four cases of HLA-D region antigens, hypogammaglobulinemia, malabsorption, severe diarrhea, and absence of specific antibodies have been observed, but without viral infections. In these cases, B-cells have been immature with increased expression of IgM and IgD, and the presence of MHC class I antigens only. Monocytes and endothelial cells did not express MHC class II antigens. Interferons gamma and alpha did not lead to the expression of MHC class II antigens. Although the gene was present, transcription did not occur.[44]

Absence of IL-1 receptors on T-cells was described in three patients with severe skin infections caused by fungus. The production of IL-1 was normal, but IL-1 was not taken up by T-cells. Reduced expression of the IL-1 receptor correlated with poor T-cell response in severe combined immunodeficiency disease (scid), ataxia teleangiectasia, and hypogammaglobulinemia. IL-2 expression was reduced. Defective PHA response *in vitro* was corrected by IL-2 administration.[45] Among control mechanisms that affect IL-2 synthesis, thymosin alpha-1 (the biologically active part of thymosin) plays a significant role in increasing precursor frequency of IL-2 producing cells.[44]

A. BAR LYMPHOCYTE SYNDROME

The first case was described in 1974 in an infant lacking MHC class I antigens on lymphocytes and platelets. Since then several cases have been described with remarkable heterogeneity of deficiency of MHC class I and/or class II antigens. Clinical symptoms varied from minimal to severe combined immune deficiency. Patients with MHC class II antigens deficiency demonstrate lack of these antigens on B-cells, macrophages, and Langerhans' cell surface. Most of them have normal numbers of lymphocyte and $T3^+$ cells, a normal response to mitogens and alloantigens, normal cytotoxic T-cell function, but delayed cutaneous hypersensitivity; antigen-specific T-cell proliferation and specific antibody response are lacking. Receptors for interferon gamma were present, which induces expression of MHC class I antigens, but not of MHC class II antigens. The disease is characterized by a lack of sensitization and cell-cell interactions. Family studies suggest that the disorder may be caused by abnormal regulatory gene(s) localized outside chromosome 6.[46]

Lymphocyte membrane adhesion proteins comprise three surface antigens (CR3, LFA-1, and P150,95) that share a common β chain structure which is linked noncovalently to one of the three distinct α chain types. These molecules are central to cell-cell interactions. The lymphocyte-associated-1 (LFA-1) and macrophage-1 (Mac-1) glycoproteins mediate spreading, aggregation, orientation in chemotactic gradients, antibody dependent cellular cytotoxicity, and phagocytosis of particles. Deficiency of these proteins results in multiple adhesion-related leukocyte defects, characterized by impairment of pus formation, delayed unbilical cord separation, and deficient polymorphonuclear chemotaxis. The α chain is a 155 kDa glycoprotein, equivalent to CR3 receptor of human myeloid cells. If the defect occurs at the level of isotope switching, preventing normal maturation of IgM-secreting B-cells into IgG, IgA, or IgG secretors, cells may be induced to synthesize abnormal amounts of IgM because of the absence of normal IgG feedback regulation of IgM production. Patients with X-linked hyper-IgM reveal normal lymphocyte subpopulations, normal μ chains, and poor NK activity. Some abnormalities of IgG subclasses may occur.[44,46]

IgG deficiency was observed in 7 children out of 30 with recurrent infections (four with recurrent pneumonia and sinusitis, one with a recurrent invasive *Haemophilus influenzae* type infection, one with severe pneumococcal meningitis). The immunological abnormalities were heterogeneous: two children had isolated IgG2 deficiency, two had IgG2-IgG4 deficiency, one IgG2-IgA deficiency, and one severe Ig1-Ig2 deficiency.[47]

Among various immunoglobulin deficiences are: disturbances of B-lymphocyte differentiation of pre-B-cells that produce the μ heavy-chain constant region (C_μ) without an attached heavy-chain variable region (V_H). Approximately 5% of normal pre-B-cells from adult human bone marrow produce these incomplete μ-chains. Patients with X-linked agammaglobulinemia produce exclusively this immature form (C_μ without associated V_H). This immune deficiency represents a block in differentiation of pre-B-cells (failure to express V_H genes).[48] Another immunodeficient patient revealed molecular defects in immunoglobulin-kappa chain deficiency. Both the constant kappa (C_κ) regions had a single point mutation: loss of the invariant tryptophan in one allele, and loss of an invariant cystein in the other allele, which resulted in the complete absence of kappa chains able to form a stable intradomain disulfide bond.[49]

Fibroblasts derived from an immunodeficient patient demonstrated hypersensitivity to lethal effects of DNA-damaging agents (particularly to gamma- and UV-irradiation). The results suggested that the cells might be defective in a common, late step of DNA excision repair. The experiments revealed that these cells might be defective in DNA ligase activity which causes a joining defect of Okazaki fragments during DNA replication. This defect can be detected by hypersensitive induction of sister chromatid exchanges.[50]

Defects of T-cell and B-cell differentiation may be induced also by metabolic disturbances of amino acids. Three siblings revealed central nervous system dysfunction, *candida* dermatitis, and keratoconjunctivitis. A delayed hypersensitive skin test response to *candida* was absent (also *in vitro* lymphocyte reaction to *candida*). One of them had a subnormal percentage of T-cells in peripheral blood. The first two siblings died because of central nervous system deterioration. The third child had lactic acidosis, increased excretion of β-hydroxypropionate, methylcitrate, β-methylcrotonylglycine, and β-hydroxyisovaleriate in urine typical for branched-chain amino acid catabolism by a biotin-dependent multiple carboxylase deficiency. Four days after oral administration of biotin, the excretion of the pathological metabolites in urine decreased. These findings suggest that disorders of branched amino acid catabolism may be responsible for primary immunodeficiency disease.[51]

Genetic deficiency of the C2 component of the complement is frequently associated with autoimmune diseases. This deficiency is not due to a major gene deletion or rearrangement, but the result of a regulatory defect in C2 gene expression. Linkage disequilibrium has been demonstrated between C2 deficiency and polymorphic MHC class I, class II, and complement gene products (genes located on the short arm of chromosome 6). Mononuclear phagocytes are the principal extrahepatic source of C2 production. Studies of *in vitro* cultures revealed a defect in C2 synthesis.[52]

B. ATAXIA TELEANGIECTASIA

This is a progressive fatal neuroimmunological disorder that affects approximately 1:40,000 children. The children develop normally for about 2 years, then initiate signs of degenerating cerebellar function, and at the age of 5 years, teleangiectasiae appear over the exposed bulbar conjunctiva and skin of the ear. Most of the patients have IgA and IgG2 deficiency, and different T-cell immune dysfunctions. About one of every five patients develop cancer (usually lymphoid). The most important distrubances in ataxia teleangiectasia are DNA repair/replication and chromosomal breaks. Immunologic distrubances in ataxia teleangiectasia have been discussed above.

C. CHRONIC GRANULOMATOUS DISEASE

This is a heterogenous syndrome unified by a severe predisposition to bacterial and fungal infections because of impaired ability of phagocytic leukocytes to kill certain microorganisms, and the failure to produce microbicidal oxygen metabolites. The causal biochemical defect varies from family to family, and at least six X-linked different molecular defects and three autosomal recessive forms are known.[53]

How the phagocytes destroy invading microbes has fascinated cell biologists, biochemists, and physicians for a long time. Intensive studies of the X-linked disorder, chronic granulomatous disease, has the most studied condition of this kind. Affected boys suffer from recurrent, severe bacterial and fungal infections, and develop multiple granulomas at sites of chronic inflammation. The disease is characterized by the defect in producing superoxide by the plasma membrane associated NADPH-oxidase ($NADPH + 2O_2 \rightarrow NADP^+ + 2O_2^- + H^+$). For some years it has been recognized that an unusual low b-cytochrome is absent in some but not in all cases of this chronic granulomatous disease. Modern genetic analysis has demonstrated that the gene responsible for the X-linked chronic granulomatous disease is located at a site Xp21.1, near the centromere. The components of the superoxide-generating system have been incompletely characterized. Most investigators agree that two species, an unusual low potential b-type cytochrome and a flavoprotein form part of the electron transfer system — and many other proteins (protein kinase C, various phosphorylated polypeptides — may participate in the reaction, perhaps in its activation or upon ingestion of particles or microorganisms.[54]

The heme containing protein cytochrome b-245 has been found undetectable in all 19 examined men in whom the defect was localized on the X chromosome. Female relatives (heterozygous carriers) had reduced concentrations of the cytochrome and variable proportions of cells that were unable to generate superoxide.[55] Further investigations revealed glycosylated (M_r 90 kDa) and non-glycosylated (22 kDa) polypeptides that participate in forming functional cytochrome b-complex, absent in X-linked chronic granulomatous disease.[56] Similar results were obtained by other authors, which demonstrated that cytochrome b-245 consists of an alpha chain (M_r 23 kDa) and a beta chain (M_r 76 to 92 kDa), the latter encoded by the X-linked chronic granulomatous disease gene.[57]

In the autosomal chronic granulomatous disease the cytochrome b-245 was present in both sexes, but was nonfunctional (in seven women and one man).[55] Examinations of four patients with autosomal chronic granulomatous disease revealed a lack of selective phosphorylation of a protein of a relative molecular mass (M_r 44 kDa) that was present in normal persons and in two patients with the X-linked form of this disease. The authors suggest that this molecule may be an important functional component of the NADPH-oxidase.[58]

B. AIDS — ACQUIRED IMMUNE DEFICIENCY SYNDROME

In 1981 the Centers for Disease Control in Atlanta recognized the first cases of a fatal new disease named acquired immune deficiency syndrome — AIDS. In a few years AIDS has become the most serious problem of contemporary medicine. AIDS patients die of a variety of rare infections and malignancies (particularly pneumonia caused by the protozoan *Pneumocystis carinii*, and Kaposi's sarcoma). They also suffer from other "opportunistic" infections caused by microorganisms that normally do not cause diseases. Such clinical syndromes previously have been observed only in patients with inherited immunodeficiency syndrome, and in patients whose immunity was blocked by immunosuppresive drugs administered for organ transplantation or in cancer chemotherapy.

The virus which causes the disease belongs to the human T-lymphotropic retroviruses. Human retrovirology is a relative young field of science, although the retroviruses belong to the class of viruses associated with animal cancers which were recognized at the beginning of this century. The manner by which the retrovirus may infect eukaryotic cells could not be resolved until the discovery of reverse transcriptase (Temin and Baltimore, in 1970). The discovery of the T-cell growth factor (IL-2) in Gallo's laboratory allowed the selective growth of different T-cell populations *in vitro*. In 1978 the first human retrovirus was isolated from cultured T-lymphocytes of a patient with mature T-cell malignancy, and was designated HTLV-1 (human T-lymphotropic virus). HTLV-1 was the prototype virus for over 100 isolates from many parts of the world. The human retroviruses cause two diseases which involve disturbances of the growth of CD4(+) lymphocytes (CD4 molecule interacts with MHC class II), a remarkable target molecule for the human immunodeficiency virus (HIV). The CD4(+) lymphocyte may be induced by the human T-lymphotropic virus type I (HTLV-1) to excessive proliferation (leukemia), and the HTLV-III (now HIV) to premature death (acquired immune deficiency syndrome — AIDS).[59] Recently, other types of the immunodeficiency viruses have been discovered, e.g., human immunodeficiency virus type 2.[60]

HIV attacks all cells that bear the CD4 molecule and, among others, also macrophages in brain tissue, which explains one of the common neurological complications in patients with AIDS. About 75% of AIDS patients demonstrate subacute encephalopathy with progressive dementia and cerebral atrophy. The authors suggest that blood monocytes infected with the HIV virus may cross the blood-brain barrier and infect macrophages and other cells in the brain that bear CD4 molecules, causing their degeneration.[61]

Patients with AIDS frequently reveal autoimmune phenomena. Serum autoantibody against a 25 kDa platelet protein have been demonstrated in 95% of homosexuals with immune thrombocytopenic purpura; the same symptoms occur in 80% of AIDS patients. The lupus anticoagulant (anti-phospholipid antibody) autoantibodies against erythrocytes and antinuclear antibodies have been detected in homosexual and hemophiliac patients with AIDS and AIDS-related conditions. Antibodies against CD4(+) cells (helper/inducer cells) have been detected in 60% of AIDS-patients; these antibodies did not react with CD8(+) cells (suppressor/cytotoxic cells). The authors suggest that autoantibodies in AIDS patients may contribute to CD4(+) cells destruction. The authors describe an AIDS-related serum antibody that reacts with an antigen of relative molecular mass (M_r 18 kDa) restricted to lectin-stimulated or HIV-infected CD4(+) T-cells. The antibody suppresses proliferation of CD4(+) T-cells *in vitro* and induces cytotoxicity of these cells in the presence of complement.[62]

Recent investigations revealed that the HIV genome is much more complex than that of other retroviruses. The HIV genome harbors at least four other genes whose functions are not yet known. It is thought that these genes play an important role in regulation of AIDS virus expression.[63] Much more important are genetic variations in AIDS viruses which enable the virus to escape or compromise the immune attack of an infected individual. Several reports focus on the nucleotide sequences of HIV surface glycoprotein (*env*) genes. Twelve total *env* sequences and five complete nucleotide sequences are known. Comparison of the nucleotide sequences confirms that there is considerable variation from one isolate to another. The extent of variation is not constant in all regions. There are strong patterns of interspersed variable and conserved regions, suggesting a division of the protein into regions of different function. The surface glycoproteins of all retroviruses are quite similar in organization, although highly different in sequence. They resemble receptor-binding proteins. In all retrovirus-infected cells the *env* gene is translated into a large precursor which becomes cleaved at two or three sites to yield the mature protein. The presence or absence of a receptor protein play a determining role in the host range and specificity of the infection. This interaction is highly specific, and infectivity can be reduced 10^7-fold or more in cells that lack appropriate receptors. The AIDS virus uses the CD4 protein molecule as its specific receptor. Variation observed within the *env* gene of AIDS viruses would suggest that it might be difficult to develop effective vaccines to protect against infection by these viruses. However, the information is incomplete, and more work is needed to test the structural, genetic, and immunological predictions that derive from the sequence comparisons.[64] The most important clues to be elucidated in the case of AIDS viruses are their long-lasting latency, and knowledge of factors that cause their activation with all the fatal consequences.

REFERENCES

1. **Gallatin, M., John, Zh. P. St., Siegelman, M., Reichert, R., Butcher, C., and Weissman, I. L.,** Lymphocyte homing receptors, *Cell,* 44, 673, 1986.
2. **Tonegawa, S.,** Somatic generation of antibody diversity, *Nature,* 302, 575, 1983.
3. **Horst, A.,** Molecular mechanisms of antibody synthesis, Proc. 16th FEBS Congr., Part C, 61, 1985.
4. **Jerne, N. K.,** The generative grammar of the immune system, *Science,* 232, 1057, 1985.
5. **Hood, L., Kronenberg, M., Hunkapiller, T.,** T cell antigen receptors and the immunoglobulin supergene family, *Cell,* 40, 225, 1985.
6. **Robertson, M.,** T-cell receptor: gamma gene product surfaces, *Nature,* 322, 110, 1986.
7. **Simonsen, M.,** May T3 protect us all, *Immunol. Today,* 5, 314, 1984.
8. **Reinherz, E. L.,** T-cell receptor — who needs more, *Nature,* 325, 660, 1987.
9. **Epplen, J. T., Chluba, J., Hardt, C., Hinkkanen, A., Stehnle, V., and Stockinger, H.,** Mammalian T-lymphocyte antigen receptor genes: genetic and nongenetic potential to generate variability, *Hum. Genet.,* 75, 300, 1987.
10. **Hunkapiller, T. and Hood, L.,** The growing immunoglobulin gene superfamily, *Nature,* 323, 15, 1986.
11. **Germain, R. N.,** The ins and outs of antigen processing and presentation, *Nature,* 322, 687, 1986.
12. **Schwartz, R. H.,** Fugue in T-lymphocyte recognition, *Nature,* 326, 738, 1987.
13. **Trowbridge, I.,** Interleukin-2 receptor proteins, *Nature,* 327, 461, 1987.
14. **Marx, J. L.,** Chemical signals in the immune system, *Science,* 221, 1362, 1983.
15. **Gefter, M. and Marrack, Ph.,** Development and modification of the lymphocyte repertoire, *Nature,* 321, 116, 1986.
16. **Marrack, Ph.,** New insights into antigen recognition, *Science,* 235, 1311, 1987.

16a. **Guillet, J. G., Lai, M. Z., Briner, Th. J., Buus, S., Sette, A., Grey, H. M., Smith, A., and Gefter, M. L.,** Immunological self, nonself discrimination, *Science,* 235, 865, 1987.

17. **Nicola, N. A.,** Why do hemopoietic growth factor receptors interact with each other, *Immunol. Today,* 8, 134, 1987.
18. **Unanue, E. R. and Allen, P. M.,** The basis for the immunoregulatory role of macrophages and other accessory cells, *Science,* 236, 551, 1987.
19. **Isakov, N., Scholz, W., Altman, A.,** Signal transduction and intracellular events in T-lymphocyte activation, *Immunol. Today,* 7, 271, 1986.
20. **Hamilton, Th. A., Adams, D. O.,** Molecular mechanisms of signal transduction in macrophages, *Immunol. Today,* 8, 151, 1987.
21. **Lanier, L. L., Phillips, J. H.,** Evidence for three types of human cytotoxic lymphocyte, *Immunol. Today,* 7, 132, 1986.
22. **Reynolds, C. W. and Ortaldo, J. R.,** Natural killer activity: the definition rather than a cell type, *Immunol. Today,* 8, 172, 1987.
23. **Kramer, M. D. and Simon, M. M.,** Are proteinases functional molecules of T lymphocytes?, *Immunol. Today,* 8, 140, 1987.
24. **Marx, J. L.,** How killer cells kill their targets, *Science,* 231, 1367, 1986.
25. **Mitchison, N. A.,** Suppressor T cells in the network, *Nature,* 316, 676, 1985.
26. **Taylor, R. B.,** Mechanism of T-cell tolerance, *Nature,* 307, 317, 1984.
27. **Schwartz, R.,** Induction of tolerance to self in T lymphocytes, *Nature,* 308, 690, 1984.
28. **Yap, W. H., Teo, T. S., and Tan, Y. H.,** An early event in the interferon-induced transmembrane signaling process, *Science,* 234, 355, 1986.
29. **Celada, A., Gray, P. W., Rinderknecht, E., and Schreiber, R. D.,** Evidence for a gamma-interferon receptor that regulates macrophage tumoricidal activity, *J. Exp. Med.,* 160, 55, 1984.
30. **Sehgal, P. B., May, L. T., Tamm, I., and Vilcek, J.,** Human β_2 interferon and B-cell differentiation factor BSF-2 are identical, *Science,* 235, 731, 1987.
31. **Pugliese, A., Salomone, C., Martino, S., Bigliono, A., Delpiano, A., and Tovo, P. A.,** Defective interferon-alpha production in children with recurrent respiratory tract infections. A primary or secondary deficiency, *Boll. Ist Sieroter. Milan.,* 64, 328, 1985.
32. **Johnson, K. J. and Ward, P. A.,** Newer concepts in the pathogenesis of immune complex-induced tissue injury, in *Advances in the Biology of Disease,* Rubin, E. and Damjanov, I., Eds., Williams and Wilkins, Baltimore, 1984, 104.
33. **Roitt, I. M.,** Prevailing theories in autoimmune disorders, *Triangle,* 23, 67, 1984.
34. **Krömer, G., Schauenstein, K., and Wick, G.,** Is autoimmunity a side-effect of interleukin 2 production?, *Immunol. Today,* 7, 199, 1986.
35. **Zanetti, M.,** The idiotype network in autoimmune processes, *Immunol. Today,* 6, 299, 1985.
36. **Volpe, R.,** Immunological aspects of thyroid disease, *Triangle,* 23, 95, 1984.
37. **Capron, A., Dessaint, J. P., Capron, M., Joseph, M., Ameisen, J. C., and Tonnel, A. B.,** From parasites to allergy: a second receptor for IgE, *Immunol. Today,* 7, 15, 1986.

38. **Gershwin, M. E. and Shultz, L.,** Mechanisms of genetically determined immune dysfunction, *Immunol. Today,* 6, 36, 1985.
39. **Valerio, D., Duyvesteyn, M. G. C., van Ormondt, H., Khan, P. M., and van der Eb, A. J.,** Adenosine deaminase (ADA) deficiency in cells derived from humans with severe combined immunodeficiency is due to an aberration of the ADA protein, *Nucleic Acids Res.,* 12, 1015, 1984.
40. **Bonthron, D. T., Markham, A. F., Ginsburg, D., and Orkin, S. H.,** Identification of a point mutation in the adenosine deaminase gene responsible for immunodeficiency, *J. Clin. Invest.,* 76, 894, 1985.
41. **Hart, S. L., Lane, A. B., and Jenkins, T.,** Partial adenosine deaminase deficiency: another family from Southern Africa, *Hum. Genet.,* 74, 307, 1986.
42. **Schuler, W., Weiler, I. J., Schuler, A., Phillips, R. A., Rosenberg, N., Mak, T. W., Kearney, J. F., Perry, R. P., and Bosma, M. J.,** Rearrangement of antigen receptor genes is defective in mice with severe combined immune deficiency, *Cell,* 46, 963, 1986.
43. **Darling, S. and Goodfellow, P.,** Lymphocyte development genes and immunodeficiency disease, *Nature,* 314, 318, 1985.
44. **Cunningham-Rundless, S.,** Approaches to the study of human immunoregulation, *Immunol. Today,* 6, 251, 1985.
45. **Chu, E. T., Rosenwasser, L. J., Dinarello, Ch. A., Rosen, F. S., and Geha, R. S.,** Immunodeficiency with defective T-cell response to interleukin 1, *Proc. Natl. Acad. Sci. USA,* 81, 4945, 1984.
46. **Gupta, S.,** Primary and acquired immunodeficiency disorders, *Immunol. Today,* 7, 324, 1986.
47. **Shackelford, P. G., Polmar, S. H., Mayus, J. L., Johnson, W. L., Corry, J. M., and Nahm, M. H.,** Spectrum of IgG2 subclass deficiency in children with recurrent infections: prospective study, *J. Pediatr.,* 108, 647, 1986.
48. **Schwaber, J., Molgaard, H., Orkin, S. H., Gould, H. J., and Rosen, F. S.,** Early pre-B cells from normal and X-linked agammaglobulinaemia produce C_{μ} without attached V_H region, *Nature,* 304, 355, 1983.
49. **Stavnezer-Nordgren, J., Kekish, O., and Zegers, B. J.,** Molecular defects in a human immunoglobulin kappa chain deficiency, *Science,* 230, 458, 1985.
50. **Henderson, L. M., Arlett, C. F., Harcourt, S. A., Lehmann, A. R., and Broughton, B. C.,** Cells from an immunodeficient patient (46BR) with a defect in DNA ligation are hypomutable but hypersensitive to the induction of sister chromatid exchange, *Proc. Natl. Acad. Sci. USA.,* 82, 2044, 1985.
51. **Cowan, M. J., Packman, S., Wara, D. W., and Amman, A. J.,** Multiple biotin-dependent caroxylase deficiencies associated with defects in T-cell and B-cell immunity, *Lancet,* II, 115, 1979.
52. **Cole, F. S., Whitehead, A. S., Auerbach, H. S., Lint, Th., Zeitz, H. J., Kilbridge, P., and Colten, H. R.,** The molecular basis for genetic deficiency of the second component of the human complement, *N. Engl. J. Med.,* 313, 11, 1985.
53. **Hitzig, W. H. and Seger, R. A.,** Chronic granulomatous disease, a heterogeneous syndrome, *Hum. Genet.,* 64, 207, 1983.
54. **Orkin, S. H.,** X-linked chronic granulomatous disease: from chromosomal position to the *in vivo* gene product, *Trends Genet.,* 3, 149, 1987.
55. **Segal, A. W., Cross, A. R., Garcia, R. C., Borregard, N., Valerius, N. H., Soothill, J. F., and Jones, O. T.,** Absence of cytochrome b-245 in chronic granulomatous disease, A multicenter European evaluation of its incidence and relevance, *N. Engl. J. Med.,* 308, 245, 1983.
56. **Dinauer, M. C., Orkin, S. H., Brown, R., Jesaitis, A. J., and Parkos, A.,** The glycoprotein encoded by the X-linked chronic granulomatous disease locus is component of a neturophil b complex, *Nature,* 327, 717, 1987.
57. **Teahan, C., Rowe, P., Parker, P., Totty, N., and Segal, A. W.,** The X-linked chronic granulomatous disease gene codes for the β-chain of cytochrome b_{-245}, *Nature,* 327, 720, 1987.
58. **Segal, A. W., Heyworth, P. G., Cockroft, S., and Barrowman, M. M.,** Stimulated neutrophils from patients with autosomal recessive chronic granulomatous disease fail to phosphorylate a M_r 44,000 protein, *Nature,* 316, 547, 1985.
59. **Wong-Staal, F. and Gallo, R. C.,** Human T-lymphotropic retroviruses, *Nature,* 317, 395, 1985.
60. **Guyader, M., Emerman, M., Sonigo, P., Clavel, F., Montagnier, L., and Alizon, M.,** Genome organization and transactivation of the human immunodeficiency virus type 2, *Nature,* 326, 662, 1987.
61. **Koenig, S., Genelman, H. E., Orenstein, J. M., Dal Canto, M. C., Petesh-kpour, G. H., Yungbluth, M., Janotta, F., Aksamit, A., Martin, M. A., and Fauci, A. S.,** Detection of AIDS virus in macrophages in brain tissue from AIDS patients with encephalopathy, *Science,* 233, 1089, 1986.
62. **Stricker, R. B., McHugh, Th. M., Moody, D. J., Morrow, W. J. W., Stites, D. P., Shuman, M. A., and Levy, J. A.,** An AIDS-related cytotoxic autoantibody reacts with a specific antigen on stimulated $CD4^+$ T cells, *Nature,* 327, 710, 1987.
63. **Chen, I. S.,** Regulation of AIDS virus expression, *Cell,* 47, 1, 1986.
64. **Coffin, J. M.,** Genetic variation in AIDS viruses, *Cell,* 46, 1, 1986.

Chapter 12

THE MOLECULAR ASPECTS OF CARCINOGENESIS

I. THE ORIGIN OF HUMAN CANCERS

The easiest method to obtain experimental cancers is to expose animals to physical or chemical agents that damage the DNA, particularly agents that cause point mutations. However, in most cases, repeated doses of the mutagen must be used, and not all the animals used for the experiment develop cancers. The resulting cancer appears at various time intervals from the start of the experiment. This denotes that in the development of cancers some stages must occur. It is generally accepted that the first step that initiates carcinogenesis is a mutation caused by various mutagens. The next step is characterized by the action of tumor promoters, usually nonmutagenic substances, but necessary in that the cells must pass from the normal state to the precancerous state. Under the influence of the further action of mutagens and promoters, these cells undergo the last transformation to cancer cells. The subsequent steps of promotion are necessary because repressor genes and their products may counteract the expression of the mutation, and usually more than ten cell divisions may be required following initiation to achieve their expression.[1]

The next important discovery was that infection by certain viruses may cause malignant transformation. In this case the sole infection that damages the host's DNA is insufficient to generate cancers; tumor promoters are also needed. Cancers can be produced by transplantation of certain tissues (especially embryonal tissue), or tissues implanted in sheets of plastic which deprives them of the restraining influence provided by contact with each other. Finally, cancer may also occur spontaneously in certain strains of animals.

The important question involved with the initiation of cancer is its genetic susceptibility. If the conventional dogma that cancer is usually initiated by localized somatic mutations is correct, we should expect that defects in DNA repair should have a marked effect on the incidence of the common cancers. Such an example is xeroderma pigmentosum. In xeroderma patients UV light causes skin cancers which probably result from mutations of basal skin cells. However, in spite of the repair defect that affects all tissues, the occurrence of common fatal cancers (e.g., lung cancer, digestive tract cancer) is not more frequent in these individuals than in the whole population.[1] In contrast to xeroderma pigmentosum patients, patients with Bloom's syndrome (an inherited disease characterized by high frequency of chromosomal aberrations and the cells of which do not show any increased sensitivity to various mutagens) exhibit a high incidence (about a 100-fold increase) of carcinomas, leukemias and lymphomas.[1] Fanconi's anemia and ataxia teleangiectasia associated with chromosomal instability also reveal a high incidence of leukemia.

As an evident mark of genetic disturbances in all examined cancers (mainly solid tumors such as meningiomas, gliomas, and leukemias) considerable chromosomal rearrangements have been observed, usually progressively more abnormal with time, and which cause multiple chromosomal aberrations. Unfortunately, much less is known about the earlier stages of carcinogenesis — whether these changes are primary causative factors of carcinogenesis or secondary factors produced by the carcinogeneic process. The best known example is the presence of the Philadelphia chromosome in about 90% of chronic myelogenic leukemias (translocation from chromosome 22 to 9, which creates the characteristically shortened chromosome 22). The presence of this aberration suggests that it may be primarily this kind of leukemia that allowed the precancerous stem cell to proliferate without restraint and colonize most of the bone marrow.[1] It is supposed that cases of chronic myelogenic leukemias without this aberration may be caused by another carcinogenic process.

An important impetus has been given to the modern views on carcinogenesis from the understanding that genetic damage may be one of the principal causes of this disease. This derived from the recognition of a hereditary predisposition to cancer, the presence of chromosomal aberrations in cancer cells, the connection between susceptibility to cancer and impaired ability of cells to repair damaged DNA, the relation of the mutagenic potential of substances to their carcinogenicity, and finally, the discovery of cellular genes (proto-oncogenes) that in another form as oncogenes can cause neoplastic growth.[2]

II. RETROVIRUSES

Retroviruses are a family of eukaryotic RNA viruses that replicate through a DNA intermediate. Virus replication precedes the synthesis of unintegrated DNA, and subsequently, insertion of the DNA into the cell forms the "provirus".[3] The process of retrovirus DNA integration resembles the transposition of some eukaryotic transposons (for example copia elements in *Drosophila*). The insertion is precise at the nucleotide level with respect to the viral DNA. The site of insertion in the cell genome is random or semi-random; there is no consensus recognition site for integration. Additionally, integration, similar to transposon, generates a short direct repeat sequence of specific length (usually 4, 5 or 6 bp) in the cell DNA at the site of insertion. The size of this repeat is virus-specific. After infection, viral synthesis of DNA results in a linear double-stranded DNA flanked by a retroviral long terminal repeat (LTR), namely, U3-R-U5, which may be subdivided into three regions. Following migration of the DNA to the nucleus, two circular species are formed. One circle carries a complete unique sequence and only one LTR, and the other two LTRs. Thus, there are three unintegrated molecules, precursors of the integrated DNA.[3]

Since the provirus is colinear with the unintegrated linear retroviral DNA, the outside ends of the LTRs join the cellular DNA by the use of *cis*-acting recognition sequences for the enzyme catalyzing integration. These sequences are probably analogous to the viral attachment site. The LTRs in each of the unintegrated DNA molecules are situated in different configurations. The retroviral attachment site functions in both orientations relative to the remainder of the viral DNA. The integration is partially location dependent, since the LTR-LTR junction works in either of two new positions in the viral genome. A circular LTR cannot integrate.[3] A DNA endonuclease, integrase (product of a subregion of the pol gene), plays a role in integration in that it cleaves single-stranded as well as double-stranded DNA attachment sites.[3] The 3′ ends of the pol gene are probably important for viral replication. In viruses such as ALV (avian leukosis virus) the reverse transcriptase is a α-β heterodimer, the α subunit being derived from the N terminal and from the β subunit by processing. The C terminal processed product of the ALV pol gene is a 32-kDa polypeptide with endonuclease activity, as does the reverse transcriptase itself. In other viruses reverse transcriptase may be composed of the α subunit only with processing of the precursor resulting in the formation of a 40-kDa C-terminal polypeptide with integrase activity.[3]

In contrast to previously held views, retroviral oncogenes may not be self-sufficient to produce tumors. There are hints that the spreading viral infection can recruit newly transformed cells into the tumor mass, probably avoiding one or more steps in carcinogenesis. There are suggestions that carcinogenesis by v-*abl* (Abelson leukemia virus) may depend on cooperation with celluler genes; v-*src* (Rous sarcoma virus) encodes a small protein that acts probably independently as an oncogene; several retroviruses contain two oncogenes that may act in concert to produce tumors.[4] The oncogenes of retroviruses arose by transduction of proto-oncogenes from the genomes of vertebrate cells. At least, the exact mechanism in which oncogenes transform cells remains unresolved.

III. ADENOVIRUSES

The tumorigenic action of adenoviruses require two regions, known as E1A and E1B. Each of these two regions encodes two proteins concerned with cellular transformation. The products of E1A are commonly known by the number of their amino acid residues 243 and 289. The sequence of the shorter protein is entirely contained within that of the longer, but the two proteins have independent functions. The E1B products have masses of 21 and 55 kDa, and probably three proteins in addition that represent subsets of the amino acids comprising the 55-kDa protein.[4] The exact functions of these proteins is controversial. It is certain that E1A causes indefinite growth of cells and evokes other elements of the transformed phenotype. Both products of E1A are needed for neoplastic transformation; E1A 243 is especially important for anchorage-independent growth and other aspects of the transformed phenotype; E1B products alone do not exhibit any effects on the cellular phenotype. The two regions of the andenovirus genome work together to produce the complete neoplastic phenotype.[4] The tumorigenicity varies from one adenovirus strain to another. The products of E1A (particularly the larger protein) inhibit transcription from MHC-class I genes, and thus enables the tumor to escape recognition by T-cells. E1B makes an independent contribution to tumorigenicity, although the nature of the contribution is not known.[4,5]

IV. PAPOVAVIRUSES

The genes responsible for transformation by the papovaviruses, polyoma, and SV40, are also complex. The transforming region of papovavirus encodes three proteins: large (approximately 100 kDa), middle (approximately 55 kDa), and small (approximately 22 kDa) T-antigens. The transforming region of the SV40 virus encodes a large (approximately 94 kDa) and a small (approximately 17 kDa) T-antigen. The large T-antigen of the polyoma virus causes indefinite growth, and diminishes the requirement of cells for serum growth factors; the middle T-antigen is responsible for the remaining features of the transformed phenotype; the role of the small T-antigen is unknown. The large T-antigen of the SV40 virus transforms cells, probably without cooperation from the small T-antigen. The most probable suggestion is that the small T-antigen of the SV40 virus is required to initiate transformation of resting cells, and mimics the action of some growth factors.[4,6]

The discovery of the papilloma viruses has contributed to the understanding of the genesis of human tumors, including carcinoma of the uterine cervix. Papilloma viruses can transform cells in culture. The mechanism of their tumorigenicity is unknown.[4]

V. HERPESVIRUSES

Some herpesviruses are carcinogens and the others not, and to none of them have genes responsible for tumorigenesis been assigned. Some herpesviruses can transform cells in culture. The most cogent description of transformation by herpesviruses has been achieved with noncarcinogenic herpes virus simplex and cytomegalovirus. In spite of the transformation capacity in culture, the transformed cells neither express nor retain any consistent portion of the viral DNA, but with one exception: hamster cells transformed by the DNA of Equine herpesvirus type 1 retain approximately 6% of the viral genome integrated into cellular DNA. Most of the tumorigenic herpesviruses affect lymphoid tissue, causing indefinite growth of either B- or T-lymphocytes in culture.[4] Herpesvirus saimiri naturally infects squirrel monkeys (*Saimiri sciureus*) without producing any symptoms of a disease; however, infection of other New World primates rapidly results in progressing, malignant T-cell lymphoma. A region of the viral genome (2 kbp) was identified as being required for tumorigenicity in owl monkeys. This region does not participate in replication of the virus.[7]

At present, there is no cogent evidence that this region is necessary for carcinogenesis, and that the viral product (nuclear protein EBNA) is required for replication and not for transformation.[4]

The tentative mode of action by viral oncogenes comprises:

1. Phosphorylation, with both proteins and phospholipids as substrates. The transforming protein may be a factor that causes phosphorylation, it may be the catalytic kinase itself, or may act on phosphorylation at a distance, for example, by regulating adenylate cyclase and the activity of protein kinases controlled by this system.
2. Initiation of DNA synthesis — unrestrained DNA synthesis is an inevitable component of the neoplastic phenotype.
3. Regulation of transcription by transforming proteins.[4]

VI. TRANSCRIPTION CONTROL BY ONCOGENES

Recent investigations on transcription control by oncogenes shifted from the analysis of the sequences necessary for promoter function to proteins (mainly products of oncoviruses and oncogenes) that stimulate transcription from a promoter in *trans*, named *trans*-activation. Several oncogenes synthesize proteins that are able to stimulate transcription. The most typical is the adenovirus E1A region which in E1A 289 protein is localized in the nucleus. The promoter sequence specificity for this protein has been determined. The other products of the adenovirus (E1B) lack this property; what is more, the impaired region of E1B resulting from transcription disturbances could be corrected by the E1A protein. A large number of viral and cellular promoters can be stimulated by E1A, e.g., the rat preproinsulin promoter, the human β-globin promoter, the human ϵ-globin promoter, and the SV40 early promoter become stimulated when introduced into cells with the E1A gene.[7] Promoters with strong enhancer elements usually are not stimutated by E1A; this can be interpreted to lend support to the view that these two promoters act similarly mechanistically.

A. OBSERVATIONS ON β-GLOBIN PROMOTER MUTATIONS

For example, a single base substitution at −87 in the β-globin promoter reduced the rate of transcription ten-fold but did not affect the rate of transcription when stimulated by E1A, and mutation within the TATA box of the β-globin strongly reduced the level of transcription in both cases. Other observations suggest that the E1A region may compete for RNA polymerase II, resulting in reduction of transcription in other cases.[8,9]

Stimulation of transcription from adenovirus early promoters might not be specific to the E1A proteins. For example, the early gene product of pseudorabies virus (virus of the herpesvirus family) also stimulates the promoters of the β-globin gene as well as the early gene products of herpes simplex virus. The large T-antigens of both SV40 and polyoma viruses reveal *trans*-activating activities. Cotransfection of cells with a plasmid containing the gene for the SV40 large T-antigen and a plasmid containing the SV40 late promoter results in stimulation of transcription from the late promoter. It is suggested that all papovaviruses act by *trans*-activating activity on promoters; however, this cannot be true because adenoviruses replace defective E1A mutants.[8,9]

The human T-lymphotropic viruses HTLV-I, HTLV-II and HTLV-III all encode a gene product that *trans*-activate the LTR of the homologous virus. The protein is encoded at the 3′ end of the viral coding sequences. Similar results have been observed also in experiments with bovine leukemia virus.[8]

The mechanism of E1A stimulation by *trans*-acting stimulation seems to rely on initiation of transcription. It is suggested that the rate of initiation is performed by three steps: enhancer elements that stimulate at a distance; upstream sequence elements (−50 to −200 nucleotides from the initiation site); and TATA box sequences. Mutation of any of these DNA sequences reduces the initiation rate because RNA polymerase II must recognize each of them probably by use of additional specific DNA binding factors. These factors may be specific for accurate initiation by polymerase II. The E1A protein probably stimulates a promoter by altering interaction between RNA polymerase II and one or more factors that mediate a rate-limiting step in initiation of transcription.[8]

Most of the viral promoters are capable of immortalizing, i.e., able to establish primary cells in culture. This activity may be a causal element in cellular transformation, which proves that the viral oncogene products regulate specific cellular promoters, and that certain cellular oncogenes can *trans*-activate gene expression. The heat shock protein 70 promoter is an example of such action. Transcription from this promoter is stimulated by the adenovirus E1A region. Elevation of heat shock protein 70 is also expressed in several transformed cells infected with the SV40 virus. Finally, a subset of cellular oncogenes encodes proteins that accumulate in the nucleus, for example *myc, myb, fos* and p53. The c-*myc* protein product has limited homology to adenovirus E1A protein and, similar to E1A, demonstrates biological activity to transform primary cells in complementation with the *ras* protein.[7] Both E1A and c-*myc* gene products can induce expression of heat shock protein from its promoter or from another yet unknown promoter.[8]

Of particular interest is the polyomal noncoding regulatory region which regulates both the late mRNA transcription and DNA replication. It enhances also the transcription of heterologous genes in chimeric plasmids in a manner analogous to the SV40 enhancer and the LTR region of retroviruses. Polyoma mutants grow at higher rate on differentiated cells which can differentiate in culture (e.g., neuroblastoma cells or Friend erythroleukemic cells). Sequence analysis of polyoma mutants that can grow in nonpermissive (undifferentiated) cells indicates that host specific rearrangements are necessary for viral gene expression and replication. For example, the ability of polyoma mutants to grow on embryonal carcinoma PCC4 cells always requires extensive rearrangements. Polyoma mutants that grow on embryonal cells or trophoblast cells require minor rearrangements (single bases, but sometimes also more extensive rearrangements). Rearrangements of polyoma that enable propagation in specific cell lines are correlated with specific rearrangements of the regulatory region. For example, mutants propagating on PCC4 embryonal carcinoma cells show duplication of the A domain and deletion of the B domain, whereas mutants from F9 embryonal carcinoma cells show a duplication of the B domain. These results suggest that growth of polyoma on different host cells is dependent on (or favored by) different host-specific rearrangements of the regulatory region, by specific sequences of the enhancer, which may either favor or hinder viral growth, depending on the host cell type.[10,11]

VII. GROWTH FACTORS IN CARCINOGENESIS

Cell survival requires an ability to sense and adapt to alterations in the environment. In multicellular organisms cellular proliferation and coordination requires additional complex regulatory mechanisms. In higher organisms the regulatory mechanisms involve extracellular messenger molecules, including hormones and peptide growth factors together with specific cellular receptors that bind these molecules. Dysfunction of these regulatory mechanisms results in cell death or uncontrolled cellular growth (proliferation). Consequently, the dysfunction of cellular growth factors, especially those that control normal cellular proliferation and differentiation, may be responsible for carcinogenesis. This suggestion has been strongly supported by the hypothesis of Sporn and Todaro on the autocrine mechanism of carcinogenesis. The ability of cancer cells to produce and respond to their own growth factors (autocrine secretions) has become a central concept linking oncogene and growth factor research.[12]

There are two ways in which a qualitative or quantitative alteration in a growth factor may play a primary role in carcinogenesis.[12a] First, the tumor cell itself may be the target of somatic genetic alterations, which results in autocrine secretion of growth factors; and second, the tumor lineage must not be the primary target for somatic mutation. An alteration into a different cell type could lead to production of precancerous cells. Subsequent mutation within the dividing population, together with selection would result in monoclonal proliferation reacting on the existing growth factor.

In support of the first hypothesis are the results of the analysis of bombesin-like peptides that can function as autocrine growth factors in human small-cell lung cancer.[13] Bombesin and bombesin-like peptides (e.g., gastrin-releasing peptide) demonstrate various physiological functions such as induction of gastrin cell hyperplasia, increased pancreatic DNA content *in vivo* in rats, and induction of normal human bronchial epithelial cell proliferation in culture. Human small-cell lung cancer cell lines produce and secrete bombesin-like peptides with the expression of a single class of high-affinity receptors for bombesin-like peptides. Exogenously added bombesin-like peptides stimulate the clonal growth and DNA synthesis in small-cell lung cancer cells in culture. These findings suggest that bombesin-like peptides may function as autocrine growth factors for this cancer. To test this hapothesis the authors synthesized a monoclonal antibody against the C-terminal region of the bombesin-like peptide. This antibody blocked the C-terminal region of bombesin-like peptides which inhibited the clonal growth of small-cell lung cancer in culture.

Autocrine growth also may be caused through infection by oncogene-carrying retroviruses. MH2, an avian retrovirus containing the v-*myc* and v-*mil* oncogenes, rapidly transforms chick hematopoietic cells in culture.[14] In an ES4 strain of erythroblastosis virus that also carries the two oncogenes v-*erb*A and v-*erb*B, both are required to fully induce the transformed phenotype in erythroid cells. The study on MH2 virus infection demonstrated that these two oncogenes together establish an autocrine growth system in which v-*myc* induces cell proliferation, whereas v-*mil* induces the production of chicken myelomonocytic growth factor (cMGF). MH2 efficiently induces monocytic leukemias and liver tumors, whereas deletion mutations lacking either a functional v-*mil* or v-*myc* do not.

In some cases antibodies to the relevant growth factor receptor do not inhibit cell growth. It has been suggested that in these cases the growth factor receptor interaction may occur intercellularly. The presence of chromatin receptors for growth factors (EGF, NGF, PDGF) has been detected by Rakowicz-Szulczyńska et al.[15,16] After 1 h of incubation all three growth factors were detected in the nucleus, tightly bound to chromatin. The amount of chromatin-bound growth factor increased until 49 h of incubation. The binding of EGF and NGF to isolated chromatin was inhibited by monoclonal antibodies specific to the respective growth factor. The NGF receptor was shown to be tightly bound to ''active chromatin'' (DNase II-sensitive sequences) which, after binding of NGF, became resistant to DNase II digestion.

The second possibility, although difficult to prove, can be supported by the effect of *src* infection on long-term marrow cultures which revealed an increased self-renewal of hemopoietic progenitor cells without leukemia.[17] Long-term marrow cultures from mice have been infected with a molecular recombinant of Rous sarcoma virus and murine amphitropic leukemia virus, which resulted in introduction of the *src* gene into the cultured cells and expression of its kinase funciton. A dramatic increase of stem cells (CFU-S-myeloid multipotential cells or colony-forming units-spleen) and the committed progenitor cells (GM-CFC — granulocyte/macrophage) occurred simultaneously with a decrease of mature granulocytes. The virus was absent in progenitor cells, but present in stromal cells, which suggests that the stromal cells might be responsible for transformation of the progenitor cells with the capacity for self-renewal.

Quantitative and qualitative alterations of the growth factor receptors may play a role in carcinogenesis. Increased numbers for EGF have been described in squamous carcinoma and primary brain tumors (gliomas), sometimes with amplification of the EGF gene. Amplification of the v-*erb*B-2 oncogene has been described in human adenocarcinomas *in vivo*.[18] There are two genes related to the v-*erb*B gene in the human genome — c-*erb*B-1 analogous to EGF (epithelial growth factor), and the c-*erb*B-2 gene that encodes a receptor-like protein very similar, but not indentical, to the EGF-receptor. Fresh human malignant tumors have been examined by DNA hybridization. DNA from 101 tumors demonstrated amplification of the c-*erb*B-1 gene in 5 of 63 adenocarcinomas and none of 38 other types of tumors, whereas the c-*erb*B-1/EGF-receptor gene showed amplification only in 1 of 8 squamous-cell carcinomas. The authors concluded that the protein products of the amplified c-*erb*B-2 gene might participate in the generation of adenocarcinomas, and the EGF-receptor in the generation of some squamous-cell carcinomas. Amplification, enhanced expression, and possible rearrangement of the EGF-receptor gene in primary human brain tumors have been demonstrated also by other authors.[19] EGF, through interaction with specific receptors, induces proliferation of target cells by a mechanism not well known. EGF-truncated receptor may be involved in transformation by the avian erythroblastosis virus oncogene v-*erb*B. Similar EGF receptor defects may also occur in human tumors. The authors examined eight novel brain tumors and 29 tumors studied previously, using a highly sensitive assay for EGF receptor kinase activity. They proved that some glioblastomas expressed high levels of kinase activity as compared with other tumors and with normal brain tissue. Overexpression of EGF receptors was described on the epidermoid carcinoma cell line, in various primary brain tumors, and in squamous carcinomas. In the experiments, four of ten primary brain tumors demonstrated amplification of the EGF-receptor gene. In other brain tumors amplification of the EGF-receptor gene was not detected.

A different molecular mechanism operates in adult T-cell leukemia. Type C human T-lymphotropic virus type I (HTLV-I) is generally accepted as the causative infectious agent for this kind of leukemia. Infection of human cells by this virus is usually associated with constitutive expression of large number of cellular receptors for IL-2. Leukemic cells demonstrate normal IL-2 receptor gene organization and processing IL-2 receptor mRNA. However, mitogenic stimuli that activate IL-2 receptor gene expression in normal cells inhibit transcription of IL-2 receptor gene in the HTLV-I infected leukemic cells.[20] The result of these studies speaks for disturbances in concerted regulation of IL-2 synthesis which in leukemic cells shows over-production, and IL-2 receptor synthesis which in these cells is suppressed.

Qualitative receptor abnormalities may be the cause of abnormal proliferation of cells by inappropriate or excessive stimulation. A typical example of this kind of cellular proliferation demonstrated v-*erb*B oncogene protein because of its close similarity to EGF-receptor protein.[21] Close similarity to a physiological polypeptide growth factor PDGF (platelet derived growth factor) demonstrates the putative transforming protein (p28^{sis} of simian sarcoma virus which stimulates cells in culture such as PDGF. A similar autocrine with tumorigenesis function may be expressed by other growth factors produced by transformed cells such as insulin-like growth factor (IGF), fibroblast derived growth factor, and transforming growth factors (TGFs). Besides the specificity mediated by regulation of the production of growth factors, cellular specificity can be controlled by specific receptors present only on target cells. Binding of the growth factor may alter the affinity of another receptor for its specific activity, as shown in the case of the PDGF and EGF receptor.[21]

Binding of different growth factors to their receptors can induce a cascade of biochemical reactions that cause stimulation of tyrosine-specific protein kinases, resulting in proliferation of target cells. The primary function of EGF may be to induce conformational changes of receptors, resulting in an activation step necessary for triggering a proliferative response that may reside in the receptor itself. The known intrinsic function of the EGF receptor is the ability to phosphorylate tyrosine residues. This function is specific, probably for all transforming proteins of the retrovirus family whose oncogenes are related to *src*. The tyrosine kinase activity seems to be the only functional activity associated with the oncogene retroviruses.[21] The authors suggest based on their own investigations, that the v-*erb*B oncogene encodes only the transmembrane part of the EGF receptor together with the region responsible for the tyrosine kinase activity, which suggests that the *src*-related subset to oncogenes (which includes v-*erb*B), derived probably from cellular sequences that encode growth factor receptors, and may produce transformation through expression of uncontrolled receptor functions. The absence of the EGF-binding domain might remove the control generated by ligand binding, resulting in permanent uncontrolled generation of signals (the same as produced by EGF) and causing uncontrolled proliferation of cells.[21]

VIII. AUTONOMOUS GROWTH OF CANCER

The autonomous nature of malignant cell growth has been known for many years. The best definition has been coined by Rous who stated that the cancer cell passing from one stage to another becomes more and more altered; loses more and more its attributes typical of the organ from which it derived; becomes less and less subordinated to controlling and coordinating organismic factors, which finally causes this cell of unknown origin to nourish at the cost of the host-organism; proliferates uncontrollably; migrates and invades different tissues and organs causing general wastage of the host organism, leading to death. The autonomous nature of malignant cells denotes that they require fewer exogenous growth factors for optimal growth and multiplication than do their normal counterparts. To explain this phenomenon it was suggested that malignant cells produce endogenous polypeptide growth factors which become incorporated by the same producer cell by functional external receptors, allowing phenotypic expression of the peptide by the same cell that produces it. This process has been named ''autocrine secretion''.[12]

There are now arguments that support the autocrine hypothesis. Several cultured tumor cells release stimulating growth factors to the culture medium in those cases where they have on their surfaces adequate receptors for the released growth factors. Best known in this aspect are α and β transforming growth factors (TGF-α and TGF-β). TGF-α is related to PDGF, and TGF-β to bombesin. The activity of these peptides is mediated by a distinct membrane receptor which, in turn, activates a post-receptor signalling mechanism and may lead to a mitogenic response. The signalling may be modified by oncogene expression of the growth factor or its receptor or post-receptor activity. The autocrine action of a growth factor in a cancer cell was first described in rodent cells transformed either by Moloney or Kirsten murine sarcoma viruses (Mo-MSV or Ki-MSV). The peptides released by these cells were isolated and named transforming growth factors alpha (TGF-α). The only known receptor for these peptides is the EGF receptor. Several data indicate that human cancer cells, produce and release TGF-α and have functional receptors for it. The role of TGF-α on cellular transformation was confirmed by the experiments with temperature-sensitive mutant rodent sarcoma viruses. In these experiments TGF-α was not released into the medium unless cells were grown at a temperature permissive for transformation. This suggests that oncogenes released by the sarcoma cells $p37^{mos}$ and $p21^{K\text{-}ras}$ control TGF-α at either the transcriptional or translational level. Cells infected by the Moloney, Kirsten, or Harvey (Ha) viruses also release other peptide growth factors, mainly TGF-β and PDGF-like peptides. It is noteworthy that normal rat kidney cells require for transformation the simultaneous action of all three peptides, i.e., TGF-α, TGF-β, and PDGF.[12]

The most important evidence for the autocrine hypothesis is provided by the relationship of the oncogene product of simian sarcoma virus (SSV) and PDGF. Simian sarcoma virus-infected cells release the transforming protein $p28^{sis}$, the amino acid residues sequence of which is almost identical with the N-terminal 109 amino acid residues of the B chain of the

PDGF. Production of PDGF-like peptides was described in human osteosarcoma and glioma cells, in cell lines transformed by SV40 virus, in mouse fibroblasts transformed by the Mo-MSV and Ki-MSV viruses, as well as in the human T-24 bladder cancer cell line expressing a mutated *ras* proto-oncogene. The PDGF-like substances seem to be encoded by cellular genes. The ability of several cells transformed by the simian sarcoma virus to grow in nude mice is dependent on their ability to synthesize PDGF-like peptides. The larger the release of PDGF-like peptides, the larger tumors generated, and the lower the amounts of these substances, the smaller are the tumors.[12]

Autocrine regulation of cell growth is not limited to cancer cells. Tissue injury, especially that of arterial walls, causes accumulation of platelets that secrete large amounts of PDGF that cause proliferation of smooth muscle cells characteristic of the first stage of atherosclerosis. Autocrine PDGF mechanisms are probably of critical physiological importance during the repair of tissue injury (during wound healing).

Another type of oncogenes act by a direct mechanism on growth, generating signals in such a way that changes in the membrane receptors induce specific genes in the nuclei. To these oncogenes belong Ha-*ras, myc,* and *fos*. The Ha-*ras* gene product p21 is localized on the cytoplasmic side of the plasma membrane, and binds GTP belonging to the GTP protein regulating system among other adenyl cyclase which requires bound GTP for activity, and consequently influencing the cAMP level in the cell. Normal cellular p21 exhibits GTPase activity, which is reduced when the transforming potential of p21 is activated by mutation. The homology of the *ras* oncogene products and the G-protein regulators implicates *ras* in transduction of the signal across the plasma membrane. GTP binding activity of transforming Ha-*ras* p21 is enhanced by EGF and probably also by TGF-α.[12]

In contrast to *ras* p21 proteins, peptides encoded by *fos, myc,* and *myb* oncogenes are in the nucleus, and are expressed differentially during cell proliferation and differentiation. These oncogenes act directly on cellular growth. Cells treated with PDGF enhance transcription of both *myc* and *fos* genes. Fibroblasts transfected with cellular *myc* gene attached to a strong promoter do not require PDGF for induction of replication as do normal fibroblasts stimulated by EGF; these cells undergo mitosis when treated with EGF alone. Cells tranfected with *myc* can be stimulated to form colonies in the presence of growth factors (EGF) which denotes that expression of *myc* alters cellular sensitivity to growth factors.[12]

IX. NEGATIVE AUTOCRINE GROWTH FACTORS

The old physiological dogma that every action must be balanced by counteraction is true also in case of growth factors. The transforming growth factor β (TGF-β) was first known to stimulate the growth of non-neoplastic fibroblasts. However, later it was recognized also as a growth inhibiting factor for many cells. TGF-β (or a related peptide) is released by confluent monkey kidney cells. These cells exhibit receptors for TGF-β, and respond with inhibition of growth. This result suggests that TGF-β may inhibit the cell cycle and maintain the resting state. The smallest amounts of TGF-β may inhibit growth of several cell types, both neoplastic or non-neoplastic, as either fibroblasts or epithelial cells. However, the action of TGF-β is not unequivocal; in the presence of PDGF it may stimulate proliferation of cells, whereas in the presence of EGF it may inhibit their proliferation.[22] The case of TGF-β demonstrates that autocrine regulation of cell growth may be positive through promotion of cell proliferation, and negative through inhibition of cell proliferation.[12,23]

The active TGF-β is composed of two identical protein subunits, each of 112 amino acid residues. Cloning of TGF-β gene revealed that the 112 amino acid part derives from a larger inactive protein containing 391 amino acid residues, which becomes split to the 112 TGF-β subunits. Defective splitting of the inactive form to the active one may be one of the causes of carcinogenesis.[24,25] Another insufficient function of TGF-β may result from lack of detectable TGF-β membrane receptors.[24] The reason that TGF-β induces growth of fibroblasts may be dependent on its action through *sis* oncogene activation that codes for one form of PDGF which stimulates growth of fibroblasts; it may also be by the additional activation of cellular oncogenes *fos* and *myc*.

The tumor necrosis factor (TNF) has a similar property of dual action on cell proliferation, which kills some tumor cells and is attractive as a possible anti-cancer agent. TNF is also a potent stimulatory factor for proliferation of certain types of normal cells, mainly fibroblasts. In the case of TNF a feedback system operates, whereby TNF stimulates growth of fibroblasts, and simultaneously, production of interferon β, which inhibits cell division and may be an element of feedback regulation that controls cell growth.[24] There are also observations that PDGF stimulates interferon production in fibroblasts, and that TGF-β does the same.[24]

The anti-growth and anti-tumor effects of interferons may be caused by their ability to counteract activation of oncogenes. There are observations that under the influence of interferon cells transformed with the *ras* oncogene can revert to a normal cell probably through inhibition of *ras* oncogene expression. The same anti-proliferative effects were observed after administration of β-interferon that decreased expression of *myc* oncogene.[24]

The growth-inhibitory results of interferons, TGF-β and tumor necrosis factor initiated search for chromosome(s) with gene(s)-inhibiting potential. The preliminary results suggest that such genes may lie on chromosome 11. The experiments were performed on human-human cell hybrids which lose chromosomes very slowly. The loss of chromosome 11 showed sometimes tumor-forming ability in hybrids between HeLa cells, and reintroduction of this chromosome suppressed the tumor-forming ability. This correlation is of specific interest because patients with the hereditary growth disorder (Beckwith-Wiedemann syndrome) associated with chromosome 11 alterations reveal a high susceptability to any of the three following cancers: Wilm's tumor of the kidney, liver cancer (hepatoblastoma), and muscle cell cancer (rhabdomyosarcoma). It is supposed that in all these cancers a particular mutant gene occurs on both chromosomes 11. In other experiments, chromosome 15 seems to be involved in tumor suppression because alterations of this chromosome were associated with increased tumorigenicity.[24]

Loss of differentiation is also an important feature of neoplastic cells. Another group of authors suggest that loss of a differentiation factor may cause neoplastic proliferation of cells. Normally, proliferation of cells becomes stopped by differentiation; therefore, loss of differentiation may cause uncontrolled proliferation of cells. Experiments performed with DNA from quiescent cells revealed that it was strongly inhibitory for HeLa cells. On the basis of this observation, trials with differentiation-enhancing agents have been undertaken. Such an agent is Ara-C (a drug already used for chemotherapy) which enhances tumor cell differentiation. Ara-C used in lower concentrations enhances differentiation of preleukemic cells, resulting in inhibition of proliferation, whereas the same drug in higher concentrations demonstrates direct cell toxicity. Trials have been also performed with another drug, hexamethylene biacetamide (not yet used in cancer chemotherapy), which induces the differentiation of mouse erythroleukemia cells to mature red cells. But the most promising in this aspect seems to be the granulocyte-colony-stimulating factor which induces proliferation and differentiation of neutrophils.[25]

On the basis of the differentiation experiments it is suggested that complete transformation of normal cells to malignancy requires a two-step mechanism: the first step is ''immortalization'', i.e., the cells are capable of dividing indefinitely in culture; this can be achieved either by carcinogenic chemicals or certain oncogenes, e.g., the *myc* oncogene. The second step consists of the ability to form malignant tumors in animals; this step can be achieved by another group of oncogenes of the *ras* prototype.[24]

Proteins necessary for the survival, proliferation, and differentiation of hematopoietic cells are produced by a variety of cell types. The biological activity of these factors is measured by their ability to stimulate hemopoietic progenitor cells to form colonies in soft agar; therefore the name colony-stimulating factors (CSF). CSFs are classified by the types of mature blood cells present in the culture. The best known are macrophage-CSF (M-CSF or CSF-1) and granulocyte-CSF (G-CSF), which stimulate proliferation of progenitor cells to macrophages or granulocytes, respectively; and granulocyte-macrophage-CSF (GM-CSF or CSF-2), which stimulates both progenitor cells to macrophages and granulocytes.[26] To the typical colony-stimulating factors belong interleukins (IL-1, IL-2, IL-3, or multi-CSF). The human T-cell, GM-CSF, was found to be 60% homologous with the GM-CSF cloned from murine lung mRNA.[26] The naturally HTLV-transformed T-lymphoblast cell line produces several CSFs, among them the neutrophil migration inhibitory factor (NIF-T) which inhibits migration of human neutrophils *in vitro*. The highly purified NIF-T exhibits also GM-CSF activity. This is caused by partial homology of the NH_2-terminal sequence which exactly corresponds with the sequence of GM-CSF.[26]

The discovery of naturally existing factors inhibiting proliferation of cells creates new bases for understanding the neoplastic processes that carcinogenesis may generate not only by the action of various carcinogens but also because of the loss of the inhibitory factors. In this aspect, particularly important seem to be the emerging possibilities of cancer therapy with such factors as tumor necrosis factor (TNF), transforming growth factor β (TGF-β), and interferons. For successful therapy of cancer with these factors, further investigations are needed, but in the cheerless situation of contemporary cancer therapy, this seems to be the first gleam of hope.

X. THE ACTION OF CELLULAR ONCOGENES

At present, about 30 oncogens originating from the cellular genome, and 10 or more present in the genomes of DNA tumor viruses are known. These 40 different oncogenes exhibit various functional properties that can be classified into smaller groups. On the basis of the nuclear or cytoplasmic localization of their gene products (not genes), we distinguish nuclear viral oncogens (SV40 large T-oncogene, polyoma large T-oncogene, and adenovirus E1A) and nuclear cellular oncogenes (*myc, myb,* N-*myc,* p53, *ski* and *fos*); and cytoplasmic viral oncogenes (polyoma middle T-oncogene) and cytoplasmic cellular oncogenes (*ras, src, erb*B, *neu, ros, fms, fes/fps, yes, mil/raf, mos,* and *abl*).[27] The most important feature of all oncogenes is their immortalizing ability (the ability to convert a tissue culture of limited growth to a culture that can be passaged without limits). The nuclear oncogenes (*myc,* N-*myc, myb,* and E1A) exhibit some structural homology, whereas some others (p53, polyoma large T-oncogene, and SV large T-oncogene) exhibit some functional similarities. Among the cytoplasmic oncogenes, only *ras, src,* and middle T-oncogene of polyoma have been studied in detail. The cytoplasmic oncogenes generally exhibit weak immortalizing activity. Nuclear oncogenes collaborate with cytoplasmic oncogenes in malignant transformation of normal cells. Nuclear, but not particularly cytoplasmic oncogenes are able to induce full transformation of cells in culture. An exception is presented by the SV40 large T-oncogene (*sis* oncogene) which by its single action is able to induce nuclear functions such as immortalization, and cytoplasmic functions such as anchorage independence. It is of interest that the large T-antigen may be found both in the nucleus and at the plasma membrane.[27]

These results suggest that nuclear and cytoplasmic oncogene activities are more complementary than additive. However, certain oncogenes carrying single oncogenes are able to induce tumors *in vivo*. A typical example is represented by the *ras* oncogene that, acting alone, converts embryonic fibroblasts to a tumorigenic state. This may be explained by environmental changes of cells.[27]

Nonviral carcinogenesis involves several stages of tumor progression, and some of these stages may require activation of multiple oncogenes.[28,29] At present only few such tumors have been described.

XI. THE MECHANISM OF NUCLEAR ONCOGENE ACTIVATION

Most of the nuclear oncogenes arose from altered normal proto-oncogenes, generated by processes that have lead to deregulation in the level of their encoded proteins. The greatest evidence for is the activating mechanism of *myc* expression resulting from chromosomal translocations which deprives the *myc* gene of its normal transcriptional promoter-enhancer regulators and replaces them with sequences from the immunoglobulin genes. Another mechanism of *myc* gene activation may result from the action of a retrovirus that integrates its genome nearby in the chromosome, providing promoter-enhancer segments that override normal regulation.[27,28]

These findings demonstrate that oncogenes of both cellular and viral origin may induce expression of other cellular genes. This may disturb the activity or specificity of transcription of cellular genes whose products are critical to cell growth and differentiation. The normal cell genome possesses multiple nuclear proto-oncogenes (*myc,* N-*myc, myb, fos,* p53, *ski*) the role in cell transformation of which is not fully understood. There are known oncogenes (*myc* and E1A) that can influence transcription as trans-activators of transcription and simultaneously cause immortalization of cells, the most typical feature of malignant transformation.[27]

Very interesting is the mechanism of p53 activation. This protein has a lifetime of 30 min. The SV40 virus can prolong the lifetime by complexing with the large T-antigen which increases its metabolic stability by a factor of 50. The p53 proto-oncogene can also be activated by fusion with a strong promoter which increases transcription and consequently the protein level. There are also some experimental data demonstrating amplification of *myc,* N-*myc* and *myb* genes which leads to an increase of encoded proteins.[27,28]

XII. ACTIVATION OF CYTOPLASMIC ONCOGENES

The *ras, src, erb*B, *abl, fes/fps,* and *neu* oncogenes become activated by mutations that affect the structure of their encoded proteins, with the exception of the *mos* oncogene which acquires oncogenic properties when linked to a constitutive promoter. Spontaneous point mutations have been observed in the case of *ras* oncogene which encodes p21 protein (mutations at positions 12, 13, or 61). Similar point mutations occur in the Rous sarcoma virus *src*-encoded pp60 protein ($pp60^{src}$), resulting in strong transforming activity. Interestingly, strong overexpression of the normal protein causes only partial transformation. An extreme example is represented by the *neu* gene in which overexpression does not cause transformation, whereas low levels of a structurally altered protein induces strong transformation.[27,28]

Structural alterations of the protein termini usually activate the cloned normal genes. The *abl* gene belongs to the cytoplasmic group being altered in human carcinomas because of chromosomal translocations. In contrast to the *myc* gene, the expression of which in Burkitt's lymphoma becomes deregulated because of chromosomal translocations (the protein-encoding region is intact), the *abl* protein in chronic myeologenous leukemia is altered, its amino terminus is lost, and is replaced by a protein sequence encoded by the foreign partner gene that participates in the translocation. Overexpression of the normal *abl* gene usually does not cause pathological effects on the cellular phenotype.[27,28]

The cytoplasmic proto-oncogene products are synthesized in relatively constant amounts. Their expression may be dependent on growth and differentiation that never cause transformation of cells. It seems that cytoplasmic proto-oncogenes are stimulated by upstream agonists only. To the direct-acting activation mechanisms belong guanosine triphosphatase activity (*ras* proteins), internalization and kinase-C phosphorylation of receptor proteins (*erb*B), and others not yet completely explained (*src* and *abl* oncogenes). The best known activation mechanism is the *ras* mechanism, encoded by Harvey or Kirsten viruses. The p21 proteins are membrane bound which bind guanosine diphosphate (GDP) or triphosphate (GTP). Binding of GTP causes excitation of the *ras* proteins, which transduces signals from various cell-surface receptors to adenyl cyclase, thereby causing an increase of the cAMP level. This, in turn, activates tyrosine kinases, using their bound adenosine triphosphate (ATP). Hydrolysis of GTP to GDP by an intrinsic activity of the G proteins causes inactivation of the *ras* proteins and puts a stop to transmembrane signals to adenyl cyclase, and a stop to tyrosine kinase activity. The p21 protein encoded by Ha-*ras* gene exhibits GTP-binding and hydrolysis of GTP to GDP. The p21 protein encoded by an oncogenic allele with amino acid substition on position 12 binds GTP very effectively, but its GTP hydrolytic property becomes reduced about ten times. The inability to hydrolyze GTP causes permanent activation which leads to bladder cancer.[27]

Another regulatory mechanism is used by the $pp60^{src}$ protein encoded by the Rous sarcoma virus, which also activates tyrosine kinase. Its regulatory mechanism is only poorly understood. In polyoma-infected cells the *src* protein forms complexes with the middle T-antigen encoded by the polyoma virus. The complexed $pp50^{src}$ exhibits tyrosine kinase activity (30 to 50 times higher than that of free protein). It seems that the middle T-polyoma antigen may be the regulatory subunit responsible for reversible $pp60^{src}$ protein activation. These results demonstrate that proto-oncogenes can be activated by various mechanisms: loss of transcriptional control (*myc* gene, because of chromosomal translocations); loss of GTPase activity (*ras* oncogene); and acquirement of a foreign, physiologically unresponsive element (*src* oncogene). However, these results do not explain in which way they generate malignant transformation of cells. An important indication may result from the activity of the *src* protein which involves phosphorylation and degradation of the membrane constituent-phosphatidyl inositol. The *src* protein and the related *ras* oncogene encode tyrosine kinase associated with a lipid kinase, that can convert phosphatidyl to mono- and diphosphorylated forms; there, in turn, may yield potent ''second messengers'' diacylglycerol and inositol triphosphate that activate protein kinase C and mobilize intracellular calcium with several pleiotropic changes associated with the oncogene action.[27,30-33]

XIII. THE AUTONOMY OF CANCER CELLS TOWARDS GROWTH FACTORS

Curiously, tumor cells exhibit a decreased dependence on growth factors, in contrast to normal cells. Tumors can acquire autonomy that permits proliferation even in the absence of growth factors. According to Weinberg, four different mechanisms may contribute to the autonomy of tumor cells: (1) the autocrine mechanism, (2) receptor alteration, (3) transducer alteration, and (4) deregulation of the nuclear oncogenes. The autocrine mechanism depends on the ability of tumor cells to produce growth factors that are adsorbed by appropriate membranal or even nuclear growth factor receptors, creating an uncontrolled autostimulation. An example of direct autocrine stimulation is represented by the oncogene *sis,* which is an altered, deregulated normal cellular gene encoding platelet-derived growth factor (PDGF).[34,35] In this case it happens that PDGF, because of its altered structure, became an active oncogene that generates a positive feedback loop that continually stimulates the cell with growth stimuli. The receptor alteration can stimulate cells with growth stimuli also without interaction with its normal counterpart. In this way the growth receptor acts as an oncogene. A typical case of such an oncogene is represented by the altered EGF receptor, a portion of which revealed homology with the *erb*B oncogene (present in the avian erythroblastosis virus). A similar EGF receptor structure is represented by the *neu* oncogene (detected in rat neuroblastomas and glioblastomas), and also related to the *erb*B oncogene. Finally, the *fms* oncogene (detected in feline sarcoma virus) encodes an altered version of the mononuclear phagocyte (CSF-1) growth factor receptor. The third manner of autocrine autonomous stimulation represents the alteration of transducing stimuli. A typical example is represented by the *ras* p21 proteins (one Ha-*ras*, two K-*ras*, and two N-*ras*) which interact with several growth factor receptors.[27]

Platelet-derived growth factor (PDGF) belongs to those growth factors with a particularly broad spectrum of activities. Secretion of PDGF may be induced in platelets by various agents such as thrombin, collagen, adherence with simultaneous production of FGF, EGF, and TGF-β; in monocyte/macrophages it is induced by endotoxin; concanavalin A; 12-*O*-tetradecanoylphorbol-13-acetate (TPA); foreign agents; adherence with simultaneous production of FGF and IL-1; (PDGF participates in wound healing, atherosclerosis; fibrotic disorders of the lung, liver, and kidney; inflammation; and arthritis); in endothelium it is induced by thrombin, factor Xa, injury with simultaneous production of IL-1; cells transformed by the simian sarcoma virus produce multiple subcutaneous fibrosarcomas and astrocytomas; a particularly wide variety of transforming agents induce production of PDGF in cells transformed by the simian sarcoma virus (SSV); PDGF-responsive cells cause TGF-α production in blastomas and sarcomas by autocrine stimulation; PDGF-nonresponsive cells cause TGF-α production in erythroleukemic cells, bladder carcinoma, and hepatoma by paracrine stimulation in myelofibrosis.[34]

SSV was the first acute transforming retrovirus isolated from a primate. Intramuscular injection of SSV into marmorsets induces multiple fibrosarcomas, whereas intracranial injection induces astrocytomas. All tissues transformed by PDGF express receptors in culture. PDGF activity is, and must be, mediated by binding with its receptor. Transformation of cells by SSV involves binding of v-*sis* encoding PDGF-like proteins which bind with the PDGF receptor-activating processes responsive to PDGF. Thus, transformation of cells by SSV is mediated by the production of PDGF-like material, and PDGF may be responsible for the consequences of acute cell transformation, at least by SSV.[34] Also, in a human osteosarcoma cell line, coexpression of PDGF-like growth factor and PDGF receptors have been detected, which support the autocrine receptor activation in human neoplastic disease.[36]

Epidermal growth factor (EGF) induces a variety of cellular respones, including stimulation of DNA synthesis and cell division. The EGF receptor is a tyrosine-specific protein kinase with autophosphorylating activity. A 300 amino acid residues portion of the receptor demonstrates sequences of transforming proteins from the *src* family that demonstrates tyrosine kinase activity.[37] *Erb*B protein demonstrates a truncated part of the EGF receptor with complete homology of that part of the receptor that lacks detectable tyrosine kinase activity. Three tyrosine residues near the C terminus are phosphorylated. In intact cells EGF stimulates phosphorylation of several sites, particularly tyrosine 14 from the C-terminal end.[38] The EGF receptor regulates the expression of its own receptor.[39,40] The EGF receptor has close homology with the avian erythroblastosis virus transforming gene v-*erb*B, and another oncogene belonging to v-*erb*B, namely, the *neu* oncogene isolated from rat neuroblastomas. The chromosomal location of the neu gene is close to v-*erb*B gene. In contrast to v-*erb*B, which encodes a 68 kDa truncated EGF receptor, the *neu* oncogene product is a 185 kDa cell surface antigen.[41]

XIV. CHROMOSOMAL BREAKS IN THE ACTIVATION OF ONCOGENES

A particular interest in elucidation of carcinogenesis evoked the detection of oncogenes on the breakpoints of chromosomes. There are several heritable fragile sites in the human genome, especially in chromosomes that undergo rearrangements during development. About 17 fragile sites related to cancer, 15 not related to cancer, and 17 non-fragile regions related to human malignancy close to bands involved in chromosome rearrangements have been described in the human genome.[42]

At the fragile sites there frequently are located oncogenes. Translocations of chromosomes with concomitant oncogenes may cause increased expression of oncogenes with the consequence of increased proliferation and sometimes neoplastic transformation. A typical example is Burkitt's lymphoma. Translocation of the same segment of the human chromosome 8q24 → qter to chromosome 14, 2, or 22 occurs in Burkitt's lymphoma. At this site on chromosome 8 is located the *myc* oncogene either in a silent form or expressed at very low levels. Chromosomes 14, 2, and 22 are known to carry the genes for immunoglobulin heavy, kappa, and lambda chains, respectively. During translocation the c-*myc* oncogene remains on the 8q+ chromosome, and the translocated regions coding immunoglobulin chains translocate from their chromosomes to chromosome 8, resulting in an increased expression of the c-*myc* oncogene.[43] The rearrangements in T-cell neoplasmas usually carry characteristic chromosomal rearrangements and inversions of chromosome 14 at band q11, which is the locus for the alpha chain of the T-cell receptor. The genes for the variable regions of this receptor remain on chromosomes 14, whereas the gene for the constant region translocates to chromosome 11.[44] In both Burkitt's lymphoma and T-cell leukemias disregulation of c-*myc* expression causes malignant transformation.

A total of 93 bands have been detected to be involved in chromosome rearrangements in human cancer. Of the 26 cellular oncogenes, 19 are localized in cancer-associated bands. Cytogenetic methods allow us to go this far: the clustering of oncogene sites and cancer break-points is statistically highly significant. (Fisher's exact test P — 0.0000012).[45] These cellular oncogenes, localized on the particular chromosomes, are chromosome 1p36:*src*-2 and *fgr*; chromosome 1p11:N-*ras*; chromosome 1q23:*ski*; chromosome 3p25:*raf*-1; chromosome 5q34:*fms*; chromosome 6q21:*ros*; chromosome 6q23:*myb*; chromosome 7p11:*erb*B; chromosome 7q22:*met*; chromosome 8q22:*mos*; chromosome 8q24:*myc*; chromosome 9q34:*abl*; chromosome 11p15:h-*ras*-1; chromosome 11q23:*ets*-1; chromosome 17q11:*erb*A; chromosome 17q21:*neu*; chromosome 18q21:*yes*-1; chromosome 21q22:*ets*-2.[45]

Several nonrandom chromosomal aberrations (heritable fragile sites) have been observed particularly in leukemias and lymphomas. Chromosome 6p23 contains a fragile site that is frequently the breakpoint in the chromosomal translocation in acute nonlymphocytic leukemia with an increased number of bone-marrow basophils, where translocation occurs between chromosomes 6 and 9 — translocation t(6;9) (p23;q34) with breakage at band 23 of chromosome 6, and q34 of chromosome 9. Acute myloblastic leukemia with maturation and a fragile site at chromosome 8q22 frequently demonstrates translocation t(87;21) (q22;23). In acute monoblastic leukemia, translocation t(9;11) (p21;q23) occurs at the breakpoint of chromosome 9p21. In both acute myeloid and lymphoid leukemias, and in lymphomas there frequently occur breakpoints at chromosome 11q13 and 11q23, which cause various translocations or deletions. Acute myelomonocytic leukemia with abnormal eosinophils exhibits deletion and more frequently inversion at the breakpoint of chromosome 16q22. In solid tumors and small-cell lung carcinoma there often occur deletions of chromosome 3 del(3) (p14p23) and translocations in Ewing's sarcoma t(11;22) (q23;q24;q12), involving a breakpoint at the chromosome fragile site 11q23. Several other chromosomal aberrations (rearrangements, numerical changes) occurring because of the presence of fragile sites have been described on chromosomes 2, 7, 12, 17, and 20.[46]

Fifty-one fragile sites of homologous chromosomes in lymphocytes from humans and great apes have been observed in culture when deprived of thymidine. The manifestation of these sites was increased by caffeine that inhibits DNA repair in replicating cells. Twenty fragile sites have been correlated with breakpoints observed in human malignant cells.[47]

The mechanisms of immunoglobulin and T-cell receptor production and their implications for c-*myc* activation reflect the differentiation of these cells during their maturation when rearrangements of their DNA must occur. The first step of B-cell differentiation brings a large number of transcriptionally inactive variable-region gene segments (V segment) into the transcriptionally active site of a constant-region gene segment (C segment). (For details see Chapter 11.) In this way the promoter falls under the influence of an enhancer upstream of the C segment, which causes permanent transcriptional activity at this site and synthesis of immunoglobulins. The transcriptional activity is combined with the rearranged V segment. This precisely prompted the view that c-*myc* genes bound to this site on the V segment, when translocated, become induced under the influence of the enhancer that physiologically induces transcription of immunoglobulin genes. However, the analysis of the nucleotide sequences ruled out this hypothesis because the translocated c-*myc* gene is in the opposite transcriptional orientation to the immunoglobulin locus. Soon, four alternative explanations for c-*myc* activation had been proposed:[48]

1. Translocations occur in tumors between chromosome 8 (which contains the c-*myc* locus) and chromosome 14 (which contains the immunoglobulin heavy chain locus), or chromosomes 2 and 22 (which contain the light chain loci). Recent investigations have determined that the enhancer is tissue specific but indifferent to gene orientation, and thus could activate the c-*myc* oncogene in spite of its orientation. However, translocation moves the immunoglobulin enhancer to chromosome 8 where it cannot induce the translocated oncogene on chromosome 14.
2. During translocation the cellular *myc* oncogene can be broken with loss of the entire 5′ exon which contains regulatory elements of the c-*myc* oncogene. The truncated oncogene may be transcribed by a viral gene producing a protein corresponding to the last two *myc* exons. Several authors postulated that the loss of transcriptional or translational control may be crucial for the effects of c-*myc* in tumorigenesis.
3. Further investigations demonstrated that the translocated c-*myc* differs from its normal counterpart: the 5′ half of the translocated gene is littered with mutations. C. Croce and his colleagues demonstrated that daughter cells retaining the intact human chromosome 8 produce no human *myc* transcripts, whereas those retaining the human chromosome 14 containing the translocated gene produce high levels of them.
4. Finally, it seems that the c-*myc* producing immortalizing proteins cause immortalization of cells, and only the action of a further oncogene or oncogenes will contribute to other aspects of malignant transformation.[48]

Further investigations revealed that during tumorigenesis, disturbances of regulation of activation of c-*myc* are the most important, i.e., changes of c-*myc* expression during the growth cycle of tumorigenic and non-tumorigenic cells *in vitro,* and interaction of c-*myc* with the regulatory mechanisms operating on the immunoglobulin locus in Burkitt's lymphoma cells. In cells in which the c-*myc* gene remains on chromosome 8, it remains silent, whereas translocated to an immunoglobulin locus it becomes constitutively active. Thus, deregulation must be associated with the translocated gene or its flanking DNA sequences. In addition to that the deregulation of c-*myc* gene expression is caused by an alteration of c-*myc* gene regulatory sequences.[49]

XV. *ras* PROTEINS

Most of the known human transforming genes belong to the *ras* gene family which acquires malignant properties on recombination with retroviral sequences. Three human *ras* oncogenes are known: Ha-*ras* (Harvey), K-*ras* (Kirsten) and N-*ras*. All of them synthesize the p21 protein, composed of 189 amino acid residues. The human *ras* genes acquire malignant properties by single gene mutations of the 12th or 61st amino acid residue. Mutated Ha-*ras* locus has been detected in T24 bladder carcinoma, whereas K-*ras* genes have been found in several other carcinomas and sarcomas. It is generally accepted that activation of the *ras* genes by specific mutations may cause development of certain human cancers. The *ras* proteins cause malignant transformation because of their homology with two sarcoma virus oncogenes. The normal function of the *ras* p21 protein is transduction of growth factor signals from the cell surface to the rest of the cell. The mechanism of the transforming ability of *ras* proteins by binding GTP with loss of the GTP hydrolytic function has been described earlier.[27]

Activated *ras* proteins have been detected in several human tumors. The best documented is the T24 bladder carcinoma. Investigations have demonstrated that in about 10% of randomly selected urinary tract tumors there is direct evidence as well that oncogene activation are the result of a somatic event within the tumor cell population.[50]

Activation of *ras* oncogenes was detected in colorectal cancers.[51] In more than a third of all examined colorectal cancers, mutations at the codon 12 of the K-*ras* oncogene were detected. These mutations usually preceded the development of malignancy, but without correlation between the presence of mutant oncogenes and the degree of invasiveness of the tumors.[52] Examination of mutations in K-*ras* p21 protein showed mutation at position 12 (substitution glycine → arginine) was present only in cancer cells of lung tumor, and not in normal tissue of the same patient.[53] K-*ras* activation was also detected in a serous cystadenocarcinoma of the ovary, but not in normal tissue of the same patient.[54] Activation of N-*ras* oncogene was observed in acute myeloblastic leukemia.[55] A point mutation has been detected at codon 13 of the N-*ras* oncogene in myelodysplastic syndrome.[56]

In chronic myelogenous leukemia (CML), expression of the c-*abl* oncogene was studied. Contrary to non-CML cells, which contained 7.4 and 6.8 kb *abl*-related transcripts, the CML cells contained a predominant novel 8.2 kb *abl*-related transcript. Besides, the levels of *abl*-related mRNAs were eight times higher at the blast crisis of CML as compared with CML cells during the chronic phase or with non-CML cells.[57] In a similar observation a 8.5 kb RNA transcript was present containing both v-*abl* and *bcr* (breakpoint cluster region) encoding a 210-kDa protein (p210) which may be the transforming protein in the case of CML. A 190-kDa phosphoprotein is probably encoded by the *bcr* region.[58]

In the Philadelphia chromosome (Ph^1) of CML, frequently the c-*abl* oncogene becomes translocated to *bcr* on chromosome 22, resulting in expression of a chimeric *bcr-abl* mRNA which encodes the p210*bcr-abl* tyrosine kinase. Ph^1 positive ALL cells (acute myelogenous leukemia) express unique *abl*-derived 185 kDa and 180 kDa tyrosine kinases that are distinct from the *bcr-abl*-derived p210 protein of CML. The synthesis of the 185/180 kDa proteins correlates with the expression of a novel 6.5 kb mRNA molecule. In this case, similar genetic translocations in two different leukemias (ALL and CLL) resulted in expression of distinct c-*abl*-derived proteins.[59]

In a case of pre-B-cell leukemia and in four patients with follicular lymphoma, both carrying translocation of chromosomes 14 and 18, the joining sequences have been studied. In each case, the involved segment of chromosome 18 had recombined with immunoglobulin heavy-chain joining segment (J_H) on chromosome 14 (D, diversity regions, are rearranged with the J_H segments). Special signal-like sequences have been detected on chromosome 18 near the breakpoints. Chromosome translocation (14;18) results from a mistake during the process of VDJ joining at the pre-B-cell stage of differentiation. The putative recombinase joins separated DNA segments of two different chromosomes instead of joining separated segments of the same chromosome.[60]

B-cell neoplasias resulting from *myc* gene deregulation may be caused by various translocations. The best elucidated are Burkitt's lymphoma (100% of cases) with t(8;2), t(8;22), or t(8;14). Mechanisms possibly implicated in Burkitt's lymphoma pathogenesis are chronic endemic malaria or Epstein-Barr virus infection; and follicular lymphoma (about 60% of cases) with t(8;14). To the same group belong multiple myeloma and B-chronic lymphatic leukemia in which translocations have not yet been diagnosed.[61]

Disregulation of c-*myc* expression was detected in over 95% of murine plasmocytomas and rat plasmocytomas. Infection with a murine retrovirus (J-3) containing an avian v-*myc* oncogene rapidly induces plasmocytomas in pristane-primed BALB/cAn mice, which denotes that avian v-*myc* infection replaces in these cases chromosomal translocation of c-*myc* activation, resulting in plasmacytogenesis.[62]

In human neuroblastoma cell lines a domain of DNA-designated N-*myc* is amplified 20- to 140-fold. N-*myc* was detected in neuroblastoma tissue from 24 of 63 patients. N-*myc* was a common finding in untreated human neuroblastomas, and was highly correlated with advanced stages of the disease ($p < 0.001$).[63]

XVI. PAPILLOMA VIRUSES AND CERVICAL TUMORS

Tumors of the uterine cervix and elsewhere in the female genital tract as well as cancer of the penis seem to be caused by infection. Genital warts are caused by papilloma viruses. It is also suggested that cervical tumors may be caused by papilloma viruses. Human papilloma viruses (HPV) comprise a very heterogeneous family with 24 different types already isolated. HPV 6, 11, 16, and 18 are commonly associated with various changes of the genital tract; conversely, types 16 and 18 are predominantly associated with the tumors. In one study, HPV 16 was present in about 60% of examined cervical tumors. It is supposed that ultraviolet and ionizing radiation, or chemical carcinogens, or other promoting agents may act in concert with HPV to produce malignant tumors.[64]

XVII. HUMAN BREAST CANCER

About one third of human breast cancers require hormones for their continued growth, and endocrine ablation or anti-hormone therapy may cause regression of the tumor. In this aspect, estrogen-induced factors of breast cancer cells may replace estrogen and promote tumor growth. Several estrogen-responsive cell lines have been isolated from breast tumors that secrete growth factors, e.g., TGF-α-like activity (transforming growth factor-α) that binds to the EGF (epidermal growth factor) receptor, and IGF-I (insulin-like growth factor I). These growth factors are estrogen dependent, and therefore deprivation of estrogen may be beneficial through diminishing the growth promoting activity of these growth factors.[65]

A novel nuclear, transformed form of an estradiol receptor has been discovered in human breast cancer cells in culture. This form is induced by estradiol. It is less extractable and less exchangeable with its ligand.[66]

Recent investigations have shown in human breast cancer cells the HER-2/*neu* oncogene, a member of the *erb*B-like oncogene family which is related, but not identical to, the EGF receptor. This gene is amplified in human breast cancer cells. In 189 primary human breast cancers the HER-2/*neu* oncogene was amplified from 2- to 20-fold in 30% of the tumors. Amplification of this oncogene was an important prognostic factor. The more the oncogene was amplified the worse was the prognosis.[67] This is the third observation that amplified oncogenes in human cancers may indicate poor prognosis. The other cases are neuroblastoma in children, and lung cancer. In neuroblastoma, a rare nerve cell tumor, amplification of the N-*myc* oncogene was found, and the more copies of N-*myc*, the worse was the prognosis. The same was observed in lung small-cell cancer that 5 to 10 years ago was considered to be incurable, but now is curable in 10% of patients, using an adequate combination of chemotherapy and radiation. Also this kind of cancer amplification of the c-*myc* oncogene is of great value in prognosis. Patients with oncogene amplification survive only half to one-third as long as patients without c-*myc* amplification. No patients with c-*myc* amplification have been cured.[68]

Direct evidence that integration of retroviruses into genomes of animals and man causes alteration of chromatic structure has been documented in several papers. For example, integration of two SV40 transcriptional promoter elements induces changes of chromatin structure detected by increased DNAase I sensitivity (''transcriptionally active chromatin'') when transposed elsewhere in the viral genome. The induction of a sufficiently long stretch of DNase I chromatin leads to the appearance of a visible nucleosome-free region.[69] Integration of avian leucosis virus LTR, the major hypersensitive site within the avian c-*myc* oncogene region, is within the proviral LTR, and the major hypersensitive sites normally detected in non-infected cells 5′ to the first c-*myc* coding exon become undetectable.[70]

To summarize, there are many types of data that support views of the causal nexus of proton-oncogenes, c-oncogenes, and v-oncogenes, but also others that are confusing. Among others, there are views that only the v-*src* gene of the Rous sarcoma virus is able to transform chicken embryo fibroblasts, whereas cellular *src* (c-*src*) does not.[71] There are many variants known of the connection of leukemias and lymphomas with chromosomal aberrations, e.g., deletions of the long arm of chromosome 6 (6q-) are frequently present in lymphoblastic leukemias, non-Hodgkin lymphomas, and myeloid leukemias. The c-*myb* proto-oncogene has been mapped on chromosome 6q21-24, and should be absent in deletion of the long arm of chromosome 6. Meanwhile, it remained on band 22, and in several cases *myb* mRNA levels were increased because of c-*myb* amplification. This indicates that deletion of chromosome portions may cause functional alterations of the c-*myb* locus, resulting in pathogenesis of leukemias and lymphomas.[72] Similar events may occur in Philadelphia-positive lymphoblastic leukemias where variants of c-*abl* protein products have been detected,[73,74] and a confusing result in Philadelphia-negative chronic myeloid leukemia where the c-*abl*-related protein was present in spite of the absence of chromosomal translocation.[75] Another example, the p53 cellular tumor antigen, is highly expressed in transformed cells originating from diverse species (SV40, adenovirus, Abelson virus, and a variety of chemically transformed cells), as well as in small amounts in non-transformed cells, mainly normal embryonic cells. Performed examinations demonstrated that the p53 protein cooperates with the Ha-*ras* oncogene.[76] The results suggest that the p53-encoding gene can play a role in the conversion of normal fibroblasts into tumorigenic immortalized cells.[77-79]

According to Bishop[2] there are four defined biochemical mechanisms of proto-oncogene and oncogene-derived proteins: protein phosphorylation with either tyrosine or serine and threonine as substrates; metabolic regulation of proteins that bind GTP; control of gene expression by influencing mRNA synthesis; participation in the replication of DNA. According to the same author, malfunction of a proto-oncogene or its product may be caused by the genetic damage that causes constitutive activity of the oncogene (the level of expression does not exceed the usual maximum; the abnormality may be a surfeit of an otherwise normal gene product because of gene amplification, translocation into the vicinity of a strong transcriptional enhancer, insertion of a retroviral DNA, or transduction into a retroviral genome), and mutation that changes the protein action.

To the cellular genes that can be activated by inserted proviruses of retroviruses belong the following: c-*myc* in chickens infected by the avian leukosis virus (ALV) results in bursal lymphoma, and in the mouse infected by the mouse mammary tumor virus (MMTV) or in cats infected by the feline leukemia virus (FeLV) results in T-cell lymphomas; c-*erb*B in chickens infected with ALV results in erythroleukemia; c-*myb* in the mouse infected by the murine leukemia virus (MLV) results in lymphosarcoma; c-*mos* in mouse infected by IAP virus causes plasmocytoma; c-K-*ras* in mouse infected by F-MLV generates the myeloid cell line; c-Ha-*ras* in chickens infected by MAV virus results in nephroblastoma; IL-2 in apes infected by GaLV virus results in the generation of a T-cell lymphoma cell line; and IL-3 (multi-macrophage colony stimulating factor — multi-M-CSF) in the mouse infected by IAP virus results in myelomonocytic leukemia.[80] Cellular genes activated by the particular viruses are: int-1 and int-2 in the mouse infected by MMTV (mammary carcinoma), and int-41 in the mouse infected by MMTV (mammary and kidney carcinoma); Mlvi-1, Mlvi-2, and Mlvi-3 in the rat infected by M-MLV (T-cell lymphoma); Pim-1 in the mouse infected by M-MLV (T-cell lymphoma); Gin-1 in the mouse infected by G-MLV (T-cell lymphoma); *pvt* in the mouse/rat infected by M-MLV (T of B-cell lymphoma); Fis-1 in the mouse infected by F-MLV (lymphoma, myeloid leukemia); and *tck* in the mouse infected by M-MLV.[80]

Noteworthy are variable gene interactions that cause the same result. In the case of Burkitt's lymphoma these are:

1. Epstein-Barr infection (EBV) which causes C-cell activation and immortalization *in vitro*, production of BCGF (B-cell growth factor), and synthesis of EBNA (protein product of LT-1 region of the virus that encodes synthesis of EBNA-EBV strain in which LT-1 is deleted and does not synthesize EBNA).
2. Activated *myc* where juxtaposition to an immunoglobulin locus (by chromosomal translocation) causes constitutive *myc* expression.
3. Activated *ras* and/or *Blym* documented by transformation of NIH 3T3 fibroblasts.[81]

The action of different v-oncogenes may be similar, which may be explained by homologous active gene portions. For example, products of v-*abl*, v-*src*, v-*fps*, v-*fes*, and v-*yes* all demonstrate homologous amino acid residues of their tyrosine kinases.[82] The most important biochemical mechanism of cell-surface signal transduction and tumor promotion seems to be protein kinase C mediating oncogene signals that cause cell proliferation. Protein kinase C, discovered in 1977, is a Ca^{2+}-activated, phospholipid-dependent enzyme that is activated by diacylglycerol, one of the earliest products of signal-induced inositol phospholipid breakdown. Diacylglycerol is normally absent in cell membranes. Kinetic analysis demonstrated that the smallest amount of diacylglycerol dramatically increases the affinity of protein kinase C for Ca^{2+}, which activates the enzyme without any change in the Ca^{2+} level. For the activation of protein kinase C, phosphatidylserine is indispensible. Protein kinase C is widely distributed in the tissues and organs of mammals, is independent of calmodulin, but is inhibited by chlorpromazine, dibucaine, and trifluoperazine, which are known as calmodulin inhibitors. In many tissues its activity exceeds that of c-AMP kinase (protein kinase A). Stimulated platelets rapidly produce diacylglycerol containing arachidonic acid. This reaction is accompanied by the disappearance of phospholipids. Diacylglycerol is only transiently produced in cell membranes, probably because of its conversion back to inositol phospholipids and to its further degradation to arachidonic acid for thromboxane and prostaglandins synthesis.[83]

XVIII. TUMOR PROMOTERS

Several phorbol esters — mainly 12-*O*-tetradecanophorbol-13-acetate (TPA) — that are known as potent tumor promoters can directly activate protein kinase C. TPA can substitute diacylglycerol in the activation of protein kinase C at extremely low concentrations. TPA increases its affinity to Ca^{2+}. It has been documented that protein kinase C activation and Ca^{2+} mobilization act synergistically to evoke full physiological effects. Among others, a possible cascade reaction from protein kinase C to tyrosine kinase has been postulated.[83]

pH may have an important role in the metabolic activation of quiescent cells. Growth stimulation of fibroblasts causes a rapid increase in pH due to activation of the Na^+/H^+ exchanger in the plasma membrane (this alkalinization is needed for the initiation of DNA synthesis). Recent investigations demonstrate that the Na^+/H^+ exchanger becomes activated by protein kinase C. Further experiments have revealed that tumor-promoting phorbol ester and synthetic diacylglycerol mimic the action of growth factors in raising cytoplasmic pH.[84]

Tumor promoters (phorbol esters) can modulate the action of EGF by reducing EGF-receptor binding. EGF complexed with its receptor stimulates the tyrosine phosphorylation of the receptor through a receptor-associated kinase activity, probably due to interaction with other growth factors (PDGF, insulin) and a number of retroviruses.[85] Tyrosine phosphorylation of EGF receptors by phorbol esters is independent of calcium ions.[86]

The action of tumor promoters may be caused by severe damage of cells (disruption of the mitotic apparatus), which may cause hemizygosity and expression of recessive genes. In addition, phorbol esters generate oxygen radicals which cause chromosome breaks and increase of the gene copy number. By the oxidation of arachidonic acid, phorbol esters generate prostaglandins and leukotrienes, which may link carcinogenic processes with inflammation.[87]

XIX. CELL SURFACE MOLECULES AND TUMOR METASTASIS

The mechanisms of metastasis of malignant tumor cells from a primary site to near or distant secondary sites belong to the most important problems in carcinogenesis. Treatment of primary tumors gave some successful results; however, metastatic cancer remains one of the most unfavorable for therapy. Nicolson enumerates several tumor properties and host properties that may be significant to the metastatic process.[88] According to Nicolson, the most important is the cell surface that, on the one hand, allows movement of cancer cells, their growth potential, their mechanical and osmotic properties, their adhesion and attachment possibilities, the presence of degradative enzymes that change the cell surface and which may deprive attachment factors or expose sites to be covered; and on the other hand, the host cells that have lost their tissue and stromal properties, that have normal adhesion properties, that enables or hinders attachment of cancer cells, the presence of inflammatory reactions particularly based on specific or nonspecific immunological reactions, the presence of vascularization processes that enable adequate nutrition and oxygenation of the metastatic cells, and finally, the presence of growth factors. Nicolson considers the properties of cell surface glycoprotein to be the most important of either neoplastic or host factors.

Translocation and migration of cells needs an attractive cellular mechanism for initiating the localized breakdown of connective tissue. This is mediated by a serine enzyme that degrades plasminogen to plasmin, both existing in sufficient amounts in body fluids. Plasmin degrades several body constituents, including basement membrane glycoproteins. Plasmin also activates collagenase. The secretion of plasminogen activator initiates an enzymatic cascade that can reduce or eliminate major extracellular impediments to migration; thus antibodies to plasminogen activator inhibit human tumor metastasis.[88a]

XX. ANGIOGENESIS

The formation of new capillaries plays a crucial role in many pathological processes, particularly in neoplasms, and it is well known that efficient inhibition of angiogenesis may result in tumor regression. There are several known factors that promote angiogenesis of tumors. Recently, a basic fibroblast growth factor has been isolated from richly vascularized normal tissues (brain, pituitary, retina, adrenals, kidney, corpus luteum, placenta) and neoplastic tissue. The authors thought that this factor can stimulate the proliferation of capillary endothelial cells, acting as a self-stimulating growth factor for capillary endothelial cells.[89]

Other authors describe angiogenic factors belonging to the typical growth factors, or at least related to them. To the typical angiogenic growth factors belongs the transforming growth factor TGF-α secreted by a variety of human tumors. It is suggested that this factor, related to EGF, may contribute to tumor-induced angiogenesis to a higher degree than the epidermal growth factor (EGF).[90] Analysis of endothelial cells led to the isolation of a human endothelial cell growth factor (ECGF), probably mapped on chromosome 5, whose 4.8 kb transcript (mRNA) was present in human brain stem mRNA. ECGF seems to be related to human interleukin-1 (IL-1), which shared approximately 30% of homology with ECGF.[91] Another tumor-derived angiogenic factor was identified as a complex that contained nicotinamide as the probably active factor. Microgram quantities of commercial nicotinamide induced neovascularization in animal experiments.[92] In other experiments, tumor-promoting phorbol esters (phorbol myristate acetate) induced angiogenesis *in vitro*.[93] Finally, angiogenesis inhibition and tumor regression was achieved by heparin or heparin fragments in the presence of cortisone.[94]

XXI. THE ROLE OF HEREDITY IN CANCER DEVELOPMENT

A strong support for a genetic background to carcinogenesis is shown by recessive genetic lesions in human cancer. To the best known belong: retinoblastoma and osteosarcoma (chromosomal locus 13(q14), Wilm's tumor (nephroblastoma — chromosomal locus 11(p13), embryonal tumors of the kidney, muscle liver and adrenal gland (Beckwith-Wiedemann syndrome — chromosomal locus 11p), bladder carcinoma 11p and acoustic neuroma or meningeoma (chromosomal locus 22).[2] Two kinds of genetic events may play a role in carcinogenesis. One, because it produces an active protein that cause malignant transformation of cells; the other may play an etiological role when it is inactive or absent. It is accepted that recessive genetic damage might remove regulatory functions, which enables activation of potential oncogenes. It is accepted that malignancy can be generated easily by recessive mutations in somatic cells, than by dominant ones. Recessive mutations might be brought to light by the loss or inactivation of the unaffected gene on the homologous chromosome.

To the best explained hereditary cancers belongs retinoblastoma. Retinoblastoma occurs in hereditary, nonhereditary, and a chromosomal deletion form. The hereditary form refers to a positive family history of the tumor, or in cases of sporadic disease, where the individual is bilaterally affected (tumors develop in both eyes). The nonhereditary form has no family history and no chromosomal abnormality, with unilateral disease (in one eye). In the chromosomal deletion form the peripheral lymphocytes have a deletion of chromosomal region 13q14. The locus for the hereditary form has been assigned to chromosomal region 13q14. The retinoblastoma genotypes are in sporadic unilateral cases rb−/rb− or rb−/− (constitutional Rb+/Rb+), in hereditary bilateral rb−/rb− or −/rb− (constitutional Rb+/rb−), and the 13 deletion form rb−/rb− or −/− (constitutional Rb+/−). When both eyes develop retinoblastoma, or when a positive tumor history is present, a mutant or inactive allele of germinal origin must be assumed with a genotype Rb+/rb−, and dominant inheritance is the rule. Patients with hereditary retinoblastoma (survivors of constitutional genotype Rb+/rb−) have 2000 times more frequently osteosarcomas in the skull after radiotherapy and 500 times more frequently in the extremities than would be expected in the general population.[95]

The rise of hereditary retinoblastoma needs, according to Knudson, two mutational events of the same homologous alleles from which the one that causes predisposition to retinoblastoma in these individuals is already mutated in the germ line. In such individuals the second mutation event that inactivates the homologous locus in the unaffected chromosome would be sufficient to generate the tumor, whereas in normal individuals such a single mutation event at the retinoblastoma locus does not generate retinoblastoma. In sporadic, unilateral cases, both alleles have to be inactivated by two separate events.[96]

Retinoblastoma is a typical case in which tumor-predisposing mutations are recessive. The tumor results from two distinct genetic alterations, each causing loss of function of one of the two homologous copies at the single genetic locus Rb (for retinoblastoma) at human chromosome 13q14. These mutations may be inherited from a parent, may arise during gametogenesis, or may occur somatically. Individuals with a single mutation at this locus have a high incidence of non-ocular second tumors, most frequently osteosarcomas. Using recombinant DNA technology, the authors found that cDNA for Rb locus was expressed in various tumors, but not in retinoblastomas and osteosarcomas in which this chromosomal band is usually deleted.[97]

The possibility that recessive mutations may cause neoplastic disease is confirmed by experimental fusion of normal and cancer cells when normal cells suppress the cancer phenotype. Tumor progression arises from several steps (proliferation, invasion of adjacent tissues, metastases) which may be composed of more than one abnormality within the cancer cell. It is suggested that oncogenes may play a role in these events. There is some evidence that c-*ras* may account for the appearance of a new and more aggressive variant of tumor. Amplification of different oncogenes has been found in highly malignant tumors. There is also some evidence that two or more proto-oncogenes may contribute to various stages of tumorigenesis, and similarly DNA in retroviruses. Sometimes carcinogenesis initiated by c-*myc* and c-*ras* for the later stages requires loss of a particular chromosome.[2]

The question arises whether the separate steps in tumorigenesis may represent damage to different genes. Bishop[2] answers that probably not, because in recessive mutations both copies of a gene become damaged or deleted (e.g., inherited retinoblastoma or Wilms's tumor). Further, the experimental induction of skin carcinomas in mice represents the example when two types of genetic damage occur to a single proto-oncogene (the initial carcinogen induces benign papillomas that are heterozygous for a point mutation in one allele of a c-*ras* gene; however, with progress of malignancy the mutant allele becomes homozygous or amplified, probably because of additional changes of the c-*ras* locus). Finally, the transformation of cultured cells may require the action of at least two different oncogenes (one causes proliferation, whereas the second the other aspects of the neoplastic phenotype).

Wilms' tumor, a nephroblastoma, is associated with a deletion of chromosome 11p13. Wilms' tumor, like retinoblastoma, is a childhood tumor. It occurs in hereditary, nonhereditary, and a chromosomal deletion form (the dominantly inherited form is usually bilateral and multifocal). Since the loci for aniridia and Wilms' tumor are linked in a chromosomal region 11p13, aniridia is frequently a marker for the deletion form of Wilms' tumor. The parallels between retinoblastoma and Wilms' tumor suggest that deletion in region 11q13 may be one of two genetic events for Wilms' tumor development, and may be a recessive gene similar to the retinoblastoma gene. The structure and function of the deleted genetic material is unknown. The role of this deletion in Wilms' tumor was confirmed by introducing a normal chromosome 11 into the Wilms' tumor cell line, which completely suppressed its ability to form tumor cells in nude mice.[98]

Similarly as in retinoblastoma, in the etiology of Wilms' tumor there occurs a recessive mutation on chromosome 11.[99] Similar studies of germline and tumor genotypes from seven patients revealed that five cases were consistent with the presence on chromosome 11 of α locus in which recessive mutational events are expressed.[100] Several papers described the loss of the Ha-*ras* allele in sporadic Wilms' tumor,[101] and the somatic deletion and duplication of genes of chromosome 11 in Wilms' tumor.[102]

Malignant melanoma occurs in sporadic and hereditary forms, and in approximately 5% are familial. It is suggested that the melanoma susceptibility gene is linked with the gene of blood group Rh on chromosome 1 and inherited as an autosomal dominant trait with incomplete penetrance. The examinations revealed the loss of polymorphic restriction fragments in malignant melanoma. The authors conclude that somatic mutations resulting in homozygosity or hemizygosity are common in melanoma, and evidently not restricted to a specific chromosome.[103]

Loss of an allele of loci on the short arm of chromosome 3 have been found in 11 of 18 patients with nonhereditary renal cell carcinomas.[104]

Bilateral acoustic neurofibromatosis is a genetic defect associated with multiple tumors of neural crest origin. In two acoustic neuromas, two neurofibromas, and one meningioma from bilateral acoustic neurofibromatosis, specific loss of alleles has been detected. In the two acoustic neuromas, deletions of chromosome 22 were present. The examinations suggest a common pathogenic mechanism in bilateral acoustic neurofibromatosis.[105]

In a human glioma, the gli gene located on chromosome 12 (q13 to q14) was amplified more than 50-fold. The gli gene belongs to the group of cellular genes that are genetically altered in primary human tumors[106] (Table 1).

XXII. TUMOR SUPPRESSOR GENES

A new era seems to be arising which, instead of paying attention to the carcinoma promoting factors in carcinogenesis, is focusing on tumor suppressor genes, failure of which may be the principal cause of tumors. G. Klein[111] writes:

> Genes that can inhibit the expression of tumorigenic phenotype have been detected by fusion of normal and malignant cells, the phenotypic reversion of *in vitro* transformants, the induction of terminal differentiation of malignant cell lineages, the loss of "recessive cancer genes," the discovery of regulatory sequences in the immediate vicinity of certain oncogenes, and the inhibition of tumor growth by normal cell products.

The first malignant neoplasms of genetic origin, caused at least by the mutation of 24 recessive genes in Drosophila melanogaster, have been described by Gateff[112] but confirmation at the molecular level has been provided by the ingenious theory of Knudson (1970) that retinoblastoma arises by the loss of both alleles at the same locus (RB-1), localized on chromosome 13q14 of the human genome.[111] According to this theory, familial retinoblastoma is caused by transmission of the defective RB-1 allele through the germ line, whereas the other, normal allele is lost during somatic development (mutation, unequal mitosis resulting in loss of chromosome 13, etc. Therefore, the Knudson theory is known as the two-event paradigm in carcinogenesis. Because of the hereditary background of retinoblastoma, this kind of tumor becomes the most useful tool in carcinogenesis research.

Retinoblastoma is a relatively rare tumor of embryonic retina, occuring only in childhood in two forms: heritable and sporadic (without family history and risk to offspring). Genetic analysis of retinoblastomas revealed a cellular gene, RB-1, whose homozygous inactivation causes retinoblastoma. The product of this gene is a nuclear phosphoprotein, p 105-Rb, that forms complexes with the adenovirus E1A and large SV40 T-oncoproteins. An aberrant Rb protein found in J82 bladder carcinoma was not able to form complexes with E1A and was less stable than p 105-Rb. The defective protein was caused by a point mutation within a splice acceptor, eliminating a single exon with 35 amino acids from its encoded protein product.[113]

Analysis of other tumor types revealed similar genetic model to osteosarcoma, hepatoblastoma, Wilms' tumor, rhabdomyosarcoma, adrenal carcinoma, transitional cell bladder carcinoma, renal cell carcinoma, ductal breast carcinoma, colon carcinoma, meningioma, acoustic neuroma, pheochromocytoma, and glioblastoma.[114] Besides retinoblastoma, the best known is Wilms' tumor (nephroblastoma), caused by the deletion of the chromosome 11p13 region. In two independent Wilms' tumor cell lines a deletion between 60 and 170 kb in size was detected. Moreover, the study of individuals with Wilms' tumor exhibited aniridia, genito-urinary abnormalities in males, and mental retardation (a complex named WAGR syndrome). The WAGR syndrome was the first example of autosomal microdeletion that

TABLE 1
Retroviral Oncogenes, Their Cellular Homologs and Function[107-110]

Retroviral oncogene	Viral origin	Viral gene product	Cellular homolog	Species of origin	Human localization: Chromosomal	Human localization: Subcellular	Tumor	Activity of virally encoded protein
v-*sis*	Simian sarcoma virus	$p28^{sis}$	PDGF B-chain	Woolly monkey/cat	22q11-1ter	Cytoplasm	Sarcoma	PDGF agonist
v-*src*	Rous sarcoma virus	$pp60^{v\text{-}src}$	$pp60^{c\text{-}src}$	Chicken	20	Plasma membrane	Sarcoma	Tyrosine kinase
v-*yes*	Y73 avian sarcoma virus	$p90^{gag\text{-}yes}$		Chicken			Sarcoma	Tyrosine kinase
v-*fps*	Fujinami sarcoma virus	$p140^{gag\text{-}fps}$	$p98^{c\text{-}fps}$	Chicken	15	Cytoplasm	Sarcoma	Tyrosine kinase
v-*fes*	ST-feline sarcoma virus	$p85^{gag\text{-}fes}$	$p92^{c\text{-}fes}$	Cat	15	Cytoplasm	Sarcoma	Tyrosine kinase
v-*ros*	UR2 avian sarcoma virus	$p68^{gag\text{-}ros}$		Chicken		Cytoplasmic membranes	Sarcoma	Tyrosine kianse
v-*abl*	Abelson murine leukemia virus	$p120^{gag\text{-}abl}$	$p150^{c\text{-}abl}$	Mouse/Cat	9	Plasma membrane	B-cell lymphoma	Tyrosine kinase
v-*fgr*	GR-feline sarcoma virus	$p70^{gag\text{-}fgr}$		Cat				Tyrosine kinase
v-*erb*A	Avian erythroblastosis virus	$gp65^{erbA}$	Truncated EGF receptor	Chicken	17	Plasma membrane	Erythroleukemia	Tyrosine kinase
v-*erb*B	Avian erythroblastosis virus	$gp65^{erbB}$	Truncated EGF receptor	Chicken		Plasma membrane	Sarcoma	Tyrosine kinase
v-*mos*	Moloney murine sarcoma virus	$p37^{mos}$		Mouse	8q22	Cytoplasm	Sarcoma	Serine-threonine kinase
v-*myc*	Avian myelocytomatosis virus	$p110^{gag\text{-}myc}$	$p58^{c\text{-}myc}$	Chicken	8q24	Nuclear matrix	Carcinoma, sarcoma, myelocytoma	Regulates transcription, binds DNA, causes immortalization and transformation of cultured cells

TABLE 1 (continued)
Retroviral Oncogenes, Their Cellular Homologs and Function[107-110]

Retroviral oncogene	Viral origin	Viral gene product	Cellular homolog	Species of origin	Human localization: Chromosomal	Human localization: Subcellular	Tumor	Activity of virally encoded protein
v-Ha-*ras*	Harvey murine sarcoma virus	$p21^{v\text{-}Ha\text{-}ras}$	$p21^{c\text{-}Ha\text{-}ras}$	Rat Mouse	11	Plasma membrane	Sarcoma, Erythroleukemia	Threonine kinase, binds GTP or GDP, regulates adenylate cyclase
v-K-*ras*	Kirsten murine sarcoma virus	$p21^{v\text{-}K\text{-}ras}$	$p21^{c\text{-}K\text{-}ras}$	Rat	12	Plasma membrane	Sarcoma, Erythroleukemia	
v-N-*ras*			$p21^{c\text{-}N\text{-}ras}$	Human	1		Neuroblastoma, leukemia, sarcoma	
v-*fms*	McDonough feline sarcoma virus			Cat	5	Cytoplasm, plasma membrane	Sarcoma	Tyrosine kinase
v-*raf*/*mil*	3611 murine sarcoma virus			Mouse	3		Sarcoma	
v-*myb*	Avian myeloblastosis virus			Chicken	6	Nuclear matrix	Myeloblastic leukemia	Binds DNA, immortalization and transformation of cells
v-*fos*	FBJ osteosarcoma virus			Mouse	2	Nucleus	Sarcoma	Transforms cultured cells
v-*ski*	Avian SKV770 virus			Chicken	1	Nucleus		Transforms cultured cells
v-*rel*	Reticuloendotheliosis virus			Turkey		Cytoplasm		
N-*myc*				Human	2		Neuroblastoma	
Blym				Chicken	1		Bursal lymphoma	
v-*mam*				Mouse, human			Mammary carcinoma	
v-*neu*				Rat			Neuro-glioblastoma	

Oncogene			
E1A (243 AA)	Adenovirus	Nucleus Cytoplasm	Establishes indefinite growth, transforms cultured cells, regulates transcription (?)
E1A (289 AA)	Adenovirus	Nucleus Cytoplasm	Establishes indefinite growth, transforms cultured cells, regulates transcription
E1B (21 kDa)	Adenovirus	Nuclear envelope; endoplasmic reticulum, plasma membrane	
E1B (55 kDa)	Adenovirus	Nucleoplasm, perinuclear cytoplasm, cell-cell contacts	Binds to p53
Py-t	Polyoma (small antigen)	Cytoplasm	
Py-mT	Polyoma (middle T-antigen)	Plasma membrane	Binds and stimulates $pp60^{c\text{-}src}$
Py-T	Polyoma (T-antigen)	Nucleus	Initiates DNA synthesis, regulates transcription
SV-t	Simian virus (small antigen)	Cytoplasm	
SV-T	Simian virus (T-antigen)	Plasma membrane, nucleus	Initiates DNA synthesis, regulates transcription, binds and stabilizes p53, establishes indefinite growth, transforms cultured cells

removes more than one locus and results in the combination otherwise known as single gene dominant disorders.[115] Similarly, the RB-1 gene is also involved in several other tumors. To those belong osteosarcoma, human breast carcinoma,[116] and others.

Loss and aberrations of human chromosomes have been detected since analyzing chromosomes in solid tumors, which must have been combined with disturbances of genetic informations in those cells. The most characteristic, according to Knudson's paradigm, is loss of heterozygosity, which deprives one of the tumor suppressor genes or alleles, and in case of inherited loss of the other wild-type allele causes development of tumors. This is just the case of retinoblastoma.

Soon it was found that in several cancers loss of a single allele does not cause cancer, and that in the development of cancer was preceded by the loss of several alleles localized on several chromosomes. A typical case represents colorectal cancer in which five distinct genetic changes have been found to occur sequentially, supporting the cumulative, multistep nature of tumor progression. The early stage, also found in familial adenomatous polyposis, is a deletion in chromosome 5q. Demethylation of cytosine residues and activating mutations in K-*ras* cause transformation into benign adenomata that after losses of chromosomes 17p(17p13) and 18q, transform into malignant tumors. Since deletions seems to include genes with tumor suppressor activity, at least three such genes are implicated in colorectal cancer.[117] One of them at 17q13 is probably the p53 gene, which normally has anti-oncogene activity, and disruption of which is involved in the pathogenesis of several cancers, especially the human lung cancer.[118] Characteristic genetic changes in lung cancer are at chromosome 3p12-22, 12q14, and 17p13, as well as overexpression *myc* genes and K-*ras* and c-*raf* proto-oncogenes.[117]

The most useful diagnostic test for tumor suppressor genes is loss of heterozygosity, which can be detected by cytogenetic methods, RFLP (restriction fragment length polymorphism), molecular cloning, and other methods.[117] Chromosomal aberrations found in the particular chromosomes demonstrate that several different cancer related genes may be mutated or lost in the same tumor, the same genes may be lost or mutated in different kinds of tumors, and progression of individual tumors may be caused by mutation or deletion of several tumor suppressor genes.

Loss of heterozygosity (for a review see Reference 117) has been found in tumors combined with changes within the particular chromosomes:

1p — Melanoma, MEN-2 (multiple endocrinological neoplasia), neuroblastoma, medullary thyroid carcinoma, pheochromocytoma, ductal breast carcinoma
1q — Breast carcinoma
3p — Small cell lung cancer, renal cell carcinoma, cervical carcinoma, von Hippel-Lindau disease
5q — Familial adenomatous polyposis, sporadic colorectal carcinoma
11p — Wilms' tumor, rhabdomyosarcoma, breast carcinoma, hepatoblastoma, transitional cell bladder carcinoma
11q — MEN-1
13q — Retinoblastoma, osteosarcoma, small cell lung cancer, ductal breast cancer, stomach cancer
17p — Small cell lung cancer, colorectal carcinoma, breast cancer, osteosarcoma
18q — Colorectal carcinoma
22 — Meningioma, acoustic neuroma, pheochromocytoma.

The most important question is how tumor suppressor genes prevent carcinogenesis. According to current investigations, the key role of these genes is in the inhibition of cell proliferation and differentiation. Differentiation is a kind of tumor suppression (terminally differentiated cells do not divide). In the case of retinoblastoma, only nondifferentiated retinoblasts can transform into retinoblastoma cells, whereas differentiated retinocytes cannot. Regulation of cell proliferation needs growth factors and its receptors, signal transduction from outside the cells to their nuclei, and nuclear binding proteins that regulate transcription and post-transcriptional gene regulation. Until now, oncogenes were identified as mutant cellular homologs of retroviral transformed genes or cellular homologs of these genes that contribute to homeostasis by normal regulatory controls, which are lost in oncogenic mutant alleles.

The RB protein is known as the cell-cycle regulator and is expressed in most or all normal cell types. It acts as an inhibitor of cell-cycle progression which is released by phosphorylation in normal cells, and by competitive binding of the hyperphosphorylated form in virally infected or tumor cells, thereby interfering with its normal inhibitory role. The overphosphorylated forms are not seen in resting cells, and they appear when cells approach the G_1/S boundary.

As retinoblastoma seems to be the result of loss of an RB-gene product i.e., a 110-kDa phosphoprotein, the introduction of RB gene into retinoblastoma or osteosarcoma cells should transform these cells into normal cells. Expression of the retroviral-mediated gene transfer in such a performed experiment demonstrated suppression of the neoplastic phenotype, affected cell morphology, growth rate, soft agar colony formation, and tumorigenicity in nude mice.[119]

REFERENCES

1. **Cairns, J.,** The origin of human cancers, *Nature,* 289, 353, 1981.
2. **Bishop, J. M.,** The molecular genetics of cancer, *Science,* 235, 305, 1987.
3. **Panganiban, A. T.,** Retroviral DNA integration, *Cell,* 42, 5, 1985.
4. **Bishop, J. M.,** Viral oncogenes, *Cell,* 42, 23, 1985.
5. **Bernards, R. and van der Eb, A. J.,** Adenovirus:transformation and oncogenicity, *Biochim. Biophys. Acta.,* 783, 187, 1984.
6. **Cuzin, F.,** The polyoma virus oncogenes, *Biochim. Biophys. Acta.,* 781, 193, 1984.
7. **Desrosiers, R. C., Bakker, A., Kamine, J., Falk, L., Hunt, R. D., and King, N. W.,** A region of the Herpesvirus saimiri genome required for oncogenicity, *Science,* 228, 184, 1985.
8. **Kingston, R. E., Baldwin, A. S., and Sharp, Ph. A.,** Transcription control of oncogenes, *Cell,* 41, 3, 1985.
9. **Borelli, E., Hen, R., and Chambon, P.,** Adenovirus-2 E1A products repress cancer-induced stimulation of transcription, *Nature,* 312, 608, 1984.
10. **Amati, P.,** Polyoma regulatory region: a potential probe for mouse cell differentiation, *Cell,* 43, 561, 1985.
11. **Sassone-Corsi, P., Wildeman, A., and Chambon, P.,** A *trans*-acting factor is responsible for the simian virus 40 enhancer activity *in vitro, Nature,* 313, 458, 1985.
12. **Sporn, M. B. and Roberts, A. B.,** Autocrine growth factors and cancer, *Nature,* 313, 745, 1985.

12a. Lancet, August 9, 317, 1986.

13. **Cuttitta, F., Carney, D. N., Mulshine, J., Moody, T. W., Fedorko, J., Fischler, A., and Minna, J. D.,** Bombesin-like peptides function as autocrine growth factors in human small-cell lung cancer, *Nature,* 316, 823, 1985.
14. **Graf, Th., v. Weizsaecker, F., Grieser, S., Coll, J., Stehelin, D., Patschinsky, T., Bister, K., Bechade, C., Calothy, G., and Leutz, A.,** v-mil induces autocrine growth and enhanced tumorigenicity in v-myc-transformed avian macrophages, *Cell,* 45, 357, 1986.
15. **Rakowicz-Szulczyńska, E. M., Rodeck, U., Herlyn, M., and Koprowski, H.,** Chromatin binding of epidermal growth factor, nerve growth factor and platelet derived growth factor in cells bearing appropriate surface receptors, *Proc. Natl. Acad. Sci. U.S.A.,* 83, 3728, 1986.
16. **Rakowicz-Szulczyńska, E. M. and Koprowski, H.,** Identification of NGF receptor in chromatin of melanoma cells using monoclonal antibody to cell surface NGF receptor, *Biochem. Biophys. Res. Commun.,* 140, 174, 1986.

17. **Boettiger, D., Anderson, S., and Dexter, T. M.,** Effect of src on long-term marrow cultures: increased self-renewal of hemopoietic progenitor cells without leukemia, *Cell,* 36, 763, 1984.
18. **Yokota, J., Yamamoto, T., Toyoshima, K., Terada, M., Sugimura, T., Battifora, H., and Cline, M. J.,** Amplification of c-erbB-2 oncogene in human adenocarcinomas *in vivo, Lancet,* April 5, 765, 1986.
19. **Libermann, T. A., Nussbaum, H. R., Razon, N., Kris, R., Lax, I., Soreq, H., Whittle, N., Waterfield, M. D., Ullrich, A., and Schlesinger, J.,** Amplification, enhanced expression, and possible rearrangement of EGF receptor gene in primary human brain tumours of glial origin, *Nature,* 313, 144, 1985.
20. **Krönke, M., Leonard, W. J., Depper, J. M., and Greene, W.,** Deregulation of interleukin-2 receptor gene expression in HTLV-I-induced adult T-cell leukemia, *Science,* 228, 1215, 1985.
21. **Downward, J., Yarden, Y., Mayes, E., Scrace, G., Totty, N., Stockwell, P., Ullrich, A., Schlesinger, J., and Waterfield, M. D.,** Close similarity of epidermal growth factor receptor and v-erb-B oncogene protein sequences, *Nature,* 307, 521, 1984.
22. **Roberts, A. B., Anzano, M. A., Wakefield, L. M., Roche, N. S., Stern, D. F., and Sporn, M. B.,** Type β transforming growth factor: a bifunctional regulator of cellular growth, *Proc. Natl. Acad. Sci. U.S.A.,* 82, 119, 1985.
23. **Sporn, M. B., Roberts, A. B., Wakefield, L. M., and Assoian, R. K.,** Transforming growth factor-β: biological function and chemical structure, *Science,* 233, 532, 1986.
24. **Marx, J. L.,** The yin and yang of cell growth control, *Science,* 232, 1093, 1986.
25. **Marx, J. L.,** Human trials of new cancer therapy begin, *Science,* 236, 778, 1987.
26. **Wong, G. G., Witek, J. S., Temple, P. A., Wilkens, K. M., Leary, A. C., Luxenberg, D. P., Jones, S. S., Brown, E. L., Kay, R. M., Orr, E. C., Shoemaker, Ch., Golde, D. W., Kaufman, R. J., Hewick, R. M., Wang, E. A., and Clark, S. C.,** Human GM-CSF: molecular cloning of the complementary DNA and purification of the natural and recombinant proteins, *Science,* 228, 810, 1985.
27. **Weinberg, R. A.,** The action of oncogenes in the cytoplasma and nucleus, *Science,* 230, 770, 1985.
28. **Marx, J. L.,** Cooperation between oncogenes, *Science,* 222, 602, 1983.
29. **Assoian, R. K., Grotendorst, G. R., Miller, D. M., and Sporn, M. B.,** Cellular transformation by coordinated action of three peptide growth factors from human platelets, *Nature,* 309, 804, 1984.
30. **Marx, J. L.,** Oncogene linked to cell regulatory system, *Science,* 226, 527, 1984.
31. **Fearn, J. C. and King, A. Ch.,** EGF receptor affinity is regulated by intracellular calcium and protein kinase C, *Cell,* 40, 991, 1985.
32. **Michel, B.,** Oncogenes and inositol lipids, *Nature,* 308, 770, 1984.
33. **Marx, J. L.,** The phosphoinositides revisited, *Science,* 228, 312, 1985.
34. **Ross, R., Raines, E. W., and Bowen-Pope, D. F.,** The biology of platelet-derived growth factor, *Cell,* 46, 155, 1986.
35. **Yarden, Y., Escobedo, J. A., Kuang, W. J., Yang-Feng, T. L., Daniel, T. O., Tremble, P. M., Chen, E. Y., Ando, M. E., Harkins, R. N., Francke, U., Fried, V. A., Ullrich, A., and Williams, L. T.,** Structure of the receptor for platelet-derived growth factor helps define a family of closely related growth factor receptors, *Nature,* 323, 226, 1986.
36. **Betshotz, Ch., Westermark, B., Ek, B., and Heldin, C. H.,** Coexpression of a PDGF-like growth factor and PDGF receptors in a human osteosarcoma cell line: implications for autocrine receptor activation, *Cell,* 39, 447, 1984.
37. **Basu, M., Biswas, R., and Das, M.,** 42,000-molecular weight EGF receptor has protein kinase activity, *Nature,* 311, 477, 1984.
38. **Downward, J., Parker, P., and Waterfield, M. D.,** Autophosphorylation sites on the epidermal growth factor receptor, *Nature,* 311, 483, 1984.
39. **Clark, A. J. L., Ishii, S., Richert, N., Merlino, G. T., and Pastan, I.,** Epidermal growth factor regulates the expression of its own receptor, *Proc. Natl. Acad. Sci. U.S.A.,* 82, 8374, 1985.
40. **Hunter, T.,** The epidermal growth factor receptor gene and its product, *Nature,* 311, 414, 1984.
41. **Coussens, L., Yang-Feng, T. L., Liao, Y. C., Chen, E., Gray, A., McGrath, J., Seeburg, P. H., Libermann, T. A., Schlesinger, J., Francke, U., Levinson, A., and Ullrich, A.,** Tyrosine kinase receptor with extensive homology to EGF receptor shares chromosomal location with neu oncogene, *Science,* 230, 1132, 1985.
42. **Miro, R., Clemente, I. C., Fuster, C., and Egozue, J.,** Fragile sites, chromosome evolution, and human neoplasia, *Hum. Genet.,* 75, 345, 1987.
43. **Croce, C. M., Thierfelder, W., Erikson, J., Nishikura, K., Finan, J., Lenoir, G. M., and Nowell, P. C.,** Transcriptional activation of an unrearranged and untranslocated c-myc oncogene by translocation of a C_λ locus in Burkitt lymphoma cells, *Proc. Natl. Acad. Sci. U.S.A.,* 80, 6922, 1983.
44. **Finger, L. R., Harvey, R. C., Moore, R. C. A., Showe, L. C., and Croce, C. M.,** A common mechanism of chromosomal translocation in T- and B-cell neoplasia, *Science,* 234, 982, 1986.
45. **Heim, S., Mitelman, F.,** Nineteen of 26 cellular oncogenes precisely localized in the human genome map to one of the 83 bands involved in primary cancer-specific rearrangements, *Hum. Genet.,* 75, 70, 1987.

46. **LeBeau, M. M. and Rowley, J. D.,** Heritable fragile sites in cancer, *Nature,* 308, 607, 1984.
47. **Yunis, J. J. and Soreng, A. L.,** Constitutive fragile sites and cancer, *Science,* 226, 1199, 1984.
48. **Robertson, M.,** Paradox and paradigm: the message and meaning of myc, *Nature,* 306, 733, 1983.
49. **Robertson, M.,** Message of myc in context, *Nature,* 309, 585, 1984.
50. **Fujita, J., Yoshida, O., Yuasa, Y., Rhim, J. S., Hatanaka, M., and Aaronson, S. A.,** Ha-ras oncogenes are activated by somatic alterations in human urinary tract tumours, *Nature,* 309, 464, 1984.
51. **Bos, J. L., Fearon, E. R., Hamilton, S. R., Verlaan-de Vries, M., van Boom, J. H., van der Eb, A. J., and Vogelstein, B.,** Prevalence of ras gene mutations in human colorectal cancers, *Nature,* 327, 293, 1987.
52. **Forrester, K., Almoguera, C., Han, K., Grizzle, W. E., and Perucho, M.,** Detection of high incidence of K-ras oncogenes during human colon tumorigenesis, *Nature,* 327, 298, 1987.
53. **Santos, E., Martin-Zanca, D., Reddy, E. P., Pierotti, M. A., Della Porta, G., and Barbacid, M.,** Malignant activation of K-ras oncogene in lung carcinoma but not in normal tissue of the same patient, *Science,* 223, 661, 1984.
54. **Feig, L. A., Bast, R. C., Jr., Knapp, R. C., and Cooper, G. M.,** Somatic activation of ras^K gene in a human ovarian carcinoma, *Science,* 223, 698, 1984.
55. **Gambke, Ch., Hall, A., Moroni, Ch.,** Activation of an N-ras gene in acute myeloblastic leukemia through somatic mutation in the first exon, *Proc. Natl. Acad. Sci. U.S.A.,* 82, 879, 1985.
56. **Hirai, H., Kobayaski, Y., Mano, H., Hagiwara, K., Maru, Y., Omine, M., Mizoguchi, H., Nishida, J., and Takaku, F.,** A point mutation at codon 13 of the N-ras oncogene in myelodysplastic syndrome, *Nature,* 327, 430, 1987.
57. **Collins, S. J., Kubonishi, I., Miyoshi, I., and Groudine, M.,** Altered transcription od c-abl oncogene in K-562 and other chronic myelogenous leukemia cells, *Science,* 225, 72, 1984.
58. **Ben-Neriah, Y., Daley, G. Q., Mess-Mason, A. M., Witte, O. N., and Baltlmore, D.,** The chronic myelogenous leukemia-specific P210 protein is the product of the ber/abl gene, *Science,* 233, 212, 1986.
59. **Clark, S. S., McLaughlin, J., Crist, W. M., Champlin, R., and Witte, O.,** Unique forms of the abl tyrosine kinase distinguish Ph^1-positive CML from Ph^1-positive ALL, *Science,* 235, 85, 1987.
60. **Tsujimoto, Y., Gorham, J., Cossman, J., Jaffe, E., Croce, C. M.,** The t(14:18) chromosome translocations involved in B-cell neoplasms result from mistakes in VDJ joining, *Science,* 229, 1390, 1985.
61. **Marcu, K., Melchers, F., Morse, H. C., III, and Potter, M.,** Mechanisms in B-cell neoplasia, *Immunol. Today,* 7, 249, 1986.
62. **Potter, M., Muschinski, J. F., Muschinski, E. B., Brust, S., Wax, J. S., Wiener, F., Babonits, M., Rapp, U. R., and Morse, H. C., III,** Avian v-myc replaces chromosomal translocation in murine plasmacytomagenesis, *Science,* 235, 787, 1987.
63. **Brodeur, G. M., Seger, R. C., Schwab, M., Varmus, H. E., and Bishop, J. M.,** Amplification of N-myc in untreated human neuroblastomas correlates with advanced disease stage, *Science,* 224, 1121, 1984.
64. **Crawford, L.,** Papilloma viruses and cervical tumours, *Nature,* 310, 16, 1984.
65. **Dickson, R. B., McManaway, M. E., and Lippman, M. E.,** Estrogen-induced factors of breast cancer cells partially replace estrogen to promote tumor growth, *Science,* 232, 1540, 1986.
66. **Kasid, A., Strobl, J. S., Huff, K., Greene, G. L., and Lippman, M. E.,** A novel nuclear form of estradiol receptor in MCF-7 human breast cancer cells, *Science,* 225, 1162, 1985.
67. **Slamon, D. J., Clark, G. M., Wong, S. G., Levin, W. J., Ullrich, A., and McGuire, W. L.,** Human breast cancer: correlation of relapse and survival with amplification of the HER-2/neu oncogene, *Science,* 235, 177, 1987.
68. **Kolata, G.,** Oncogenes give breast cancer prognosis, *Science,* 235, 160, 1987.
69. **Jongstra, J., Reudelhuber, T. L., Oudet, P., Benoist, Ch., Chae, C. B., Jeltsch, J. M., Mathis, D. J., and Chambon, P.,** Induction of altered chromatin structures by simian virus 40 enhancer and promoter elements, *Nature,* 307, 708, 1984.
70. **Schubach, W. and Groudine, M.,** Alteration of c-myc chromatin structure by avian leukosis virus integration, *Nature,* 307, 702, 1984.
71. **Iba, H., Takeya, T., Cross, F. R., Hanafusa, T., and Hanafusa, H.,** Rous sarcoma virus variants that carry the cellular src gene instead of the viral src gene cannot transform chicken embryo fibroblasts, *Proc. Natl. Acad. Sci. U.S.A.,* 81, 4424, 1984.
72. **Barletta, C., Pelicci, P. G., Kenyon, L. C., Smith, S. D., and Dalla-Favera, R.,** Relationship between the c-myb locus and the 6q- chromosomal aberration in leukemias and lymphomas, *Science,* 235, 1064, 1987.
73. **Kurzrock, R., Shtalrid, M., Romero, P., Kloetzer, W. S., Talpas, M., Trujillo, J. M., Blick, M., Beran, M., and Gutterman, J. U.,** A novel c-abl protein product in Philadelphia-positive acute lymphoblastic leukemia, *Nature,* 325, 631, 1987.
74. **Chan, L. C., Karhi, K. K., Rayter, S. I., Heisterkamp, N., Eridani, S., Powless, R., Lawler, S. D., Groffen, J., Foulkes, J. G., Greaves, M. F., and Wiedemann, L. M.,** A novel abl protein expressed in Philadelphia chromosome positive acute lymphoblastic leukaemia, *Nature,* 325, 635, 1987.

75. **Morris, Ch., M., Reeve, A. E., Fitzgerald, P. H., Hollings, P. E., Beard, M. E. J., and Heaton, D. C.,** Genetic diversity correlates with clinical variation in Ph′-negative chronic myeloid leukaemia, *Nature,* 320, 281, 1986.
76. **Lane, D. P.,** Cell immortalization and transformation by the p53 gene, *Nature,* 312, 596, 1984.
77. **Parada, L. F., Land, H., Weinberg, R. A., Wolf, D., and Rotter, V.,** Cooperation between gene encoding p53 tumour antigen and ras cellular transformation, *Nature,* 312, 649, 1984.
78. **Eliayahu, D., Raz, A., Gruss, P., Givol, D., and Oren, M.,** Participation of p53 cellular tumour antigen in transformation of normal embryonic cells, *Nature,* 312, 646, 1984.
79. **Jenkins, J. R., Rudge, K., and Currie, G. A.,** Cellular immortalization by a cDNA clone encoding the transformation-associated phosphoprotein p53, *Nature,* 312, 681, 1984.
80. **Nusse, R.,** The activation of cellular oncogenes by retroviral insertion, *Trends Genet.,* 2, 244, 1986.
81. **Klein, H. and Klein, E.,** Evolution of tumours and the impact of molecular oncology, *Nature,* 315, 190, 1985.
82. **Wang, J. Y. J.,** From c-abl to v-abl, *Nature,* 304, 100, 1983.
83. **Nishizuka,** The role of protein kinase C in cell surface signal transduction and tumour promotion, *Nature,* 308, 693, 1984.
84. **Moolenaar, W. H., Tertoolen, L. G. J., and de Laat, S. W.,** Phorbol ester and diacylglycerol mimic growth factors in raising cytoplasmic pH, *Nature,* 312, 371, 1984.
85. **Friedman, B., Frackelton, A. R., Ross, A. H., Connors, J. M., Fujiki, H., Sugimura, T., and Rosner, M. R.,** Tumor promoters block tyrosine-specific phosphorylation of the epidermal growth factor receptor, *Proc. Natl. Acad. Sci. U.S.A.,* 81, 3034, 1984.
86. **Moon, S. O., Palfrey, H. C., King, A. C.,** Phorbol esters potentiate tyrosine phosphorylation of epidermal growth factor receptors in A431 membranes by a calcium-independent mechanism, *Proc. Natl. Acad. Sci. U.S.A.,* 81, 2298, 1984.
87. **Ames, B. N.,** Dietary carcinogens and anticarcinogens, *Science,* 221, 1256, 1983.
88. **Nicolson, G. L.,** Cell surface molecules and tumor metastasis, *Exp. Cell Res.,* 150, 3, 1984.
88a. **Ossowski, L. and Reich, E.,** Antibodies to plasminogen activator inhibit human tumor metastasis, *Cell,* 35, 611, 1983.
89. **Schweigerer, L., Neufeld, G., Friedman, J., Abraham, J. A., Fiddes, J. C., and Gospodarowicz, D.,** Capillary endothelial cells express basic fibroblast growth factor, a mitogen that promotes their own growth, *Nature,* 325, 257, 1987.
90. **Schreiber, A. B., Winkler, M. E., and Derynck, R.,** Transforming growth factor - α: a more potent angiogenic mediator than epidermal growth factor, *Science,* 232, 1250, 1986.
91. **Jaye, M., Howk, R., Burgess, W., Ricca, G. A., Chiu, I. M., Ravera, M. W., O'Brien, S. J., Modi, W. S., Maciag, Th., and Drohan, W. N.,** Human endothelial cell growth factor: cloning, nucleotide sequence, and chromosome localization, *Science,* 233, 541, 1986.
92. **Kull, F. C., Brent, D. A., Parikh, I., and Cuatrecasas, P.,** Chemical identification of a tumor-derived angiogenic factor, *Science,* 236, 843, 1987.
93. **Montesano, R. and Orci, L.,** Tumor-promoting phorbol esters induce angiogenesis *in vitro, Cell,* 42, 469, 1985.
94. **Folkman, J., Langer, R., Linhardt, R. J., Haudenschild, Ch., and Taylor, S.,** Angiogenesis inhibition and tumor regression caused by a heparin fragment in the presence of cortisone, *Science,* 221, 719, 1983.
95. **Murphree, A. L. and Benedict, W. F.,** Retinoblastoma: clues to human oncogenesis, *Science,* 223, 1028, 1984.
96. **Harris, H.,** Malignant tumours generated by recessive mutations, *Nature,* 323, 582, 1986.
97. **Friend, S. H., Bernards, R., Rogelj, S., Weinberg, R. A., Rapaport, J. M., Albert, D. M., and Dryja, T. P.,** A human DNA segment with properties of the gene that predisposes to retinoblastoma and osteosarcoma, *Nature,* 323, 643, 1986.
98. **Weissman, B. E., Saxon, P. J., Pasquale, S. R., Jones, G. R., Geiser, A. G., and Stanbridge, E. J.,** Introduction of a normal chromosome 11 into a Wilms' tumor cell line controls its tumorigenic expression, *Science,* 236, 175, 1987.
99. **Solomon, E.,** Recessive mutation in aetiology of Wilms' tumor, *Nature,* 309, 111, 1984.
100. **Koufos, A., Hansen, M. F., Lampkin, B. C., Workman, M. L., Copeland, N. G., Jenkins, N. A., and Cavenee, W. K.,** Loss of alleles at loci on human chromosome 11 during genesis of Wilms' tumor, *Nature,* 309, 170, 1984.
101. **Reeve, A. E., Housiaux, P. J., Gardner, R. J. M., Chewings, W. E., Grindley, R. M., and Millow, L. J.,** Loss of Harvey ras allele in sporadic Wilms' tumor, *Nature,* 309, 174, 1984.
102. **Fearon, E. R., Vogelstein, B., Feinberg, A. P.,** Somatic deletion and duplication of genes on chromosome 11 in Wilms' tumors, *Nature,* 309, 176, 1984.
103. **Dracopoli, N. C., Houghton, A. N., and Old, L. J.,** Loss of polymeric restriction fragments in malignant melanoma: implications for tumour heterogeneity, *Proc. Natl. Acad. Sci. U.S.A.,* 82, 1470, 1985.

104. **Zbar, B., Brauch, H., Talmadge, C., and Linehan, M.,** Loss of alleles of loci on the short arm of chromosome 3 in renal cell carcinoma, *Nature,* 327, 721, 1987.
105. **Seizinger, B. R., Rouleau, G., Ozelius, L. J., Lane, A. H., George-Hyslop, P. St., Huson, S., Gusella, J. F., and Martuza, R. L.,** Common pathogenetic mechanism for three tumor types in bilateral acoustic neurofibromatosis, *Science,* 236, 317, 1987.
106. **Kinzler, K. W., Bigner, S. H., Bigner, D. D., Trent, J. M., Law, M. L., O'Brien, S. J., Wong, A. J., and Vogelstein, B.,** Identification of an amplified, highly expressed gene in human glioma, *Science,* 236, 70, 1987.
107. **Heldin, C. H., Westermark, B.,** Growth factors: mechanism of action and relation to oncogenes, *Cell,* 37, 9, 1984.
108. **Willeke, K. and Schäfer, R.,** Human oncogenes, *Hum. Genet.,* 66, 132, 1984.
109. **Land, H., Parada, L. F., and Weinberg, R. A.,** Cellular oncogenes and multi-step carcinogenesis, *Science,* 222, 772, 1983.
110. **Cooper, G. M.,** Cellular transforming genes, *Science,* 218, 801, 1982.
111. **Klein, G.,** The approaching era of tumor suppressor genes, *Science,* 238, 1539, 1987.
112. **Gateff, E.,** Malignant neoplasms of genetic origin in Drosophila melanogaster, *Science,* 200, 1448, 1978.
113. **Horowitz, J. M., Yandell, D. W., Park, S.-H., Canning, S., Whyte, P., Buchkovich, K., Harlow, E., Weinberg, R., and Dryja, T. P.,** Point mutational inactivation of the retinoblastoma antioncogene, *Science,* 243, 937, 1989.
114. **Hansen, M. and Cavenee, W. K.,** Tumor suppressors: recessive mutations that lead to cancer, *Cell,* 53, 172, 1988.
115. **Francke, U.,** A gene for Wilms' tumor?, *Nature,* 343, 692, 1990.
116. **Lee, E., To, H., Shew, J.-Y., Bookstein, R., Scully, P., and Lee, W.-H.,** Inactivation of the retinoblastoma susceptibility gene in human breast cancers, *Science,* 241, 218, 1988.
117. **Sager, R.,** Tumor suppressor genes: the puzzle and promise, *Science,* 246, 1406, 1989.
118. **Takahashi, T., Nau, M. M., Chiba, I., Birrer, M. J., Rosenberg, R. K., Vinocour, M., Levitt, M., Pass, H., Gazdar, A. F., and Minna, J. D.,** p53: a frequent target for genetic abnormalities in lung cancer, *Science,* 246, 491, 1989.
119. **Huang, H.-J. S., Yee, J.-K., Shew, J.-H., Chen, P.-L., Bookstein, R., Friedman, T., Lee, E. Y.-H., and Lee, W.-H.,** Suppression of the neoplastic phenotype by replacement of the RB gene in human cancer cells, *Science,* 242, 1563, 1988.

Chapter 13

MOLECULAR BIOLOGY OF THE NERVOUS SYSTEM

I. THE GENERATION OF NERVE STIMULI

Views on the action of the nervous system have changed dramatically in recent years. At present, it is accepted that the nervous system does not work similarly to a switching station of electrical stimuli, but as a secretory organ that distributes an enormous number of neurotransmitters (peptides) generated in nerve cells via neurons to the particular organs and tissues.

Large ionic gradients are maintained across surface membranes of neurons in such a way that the intracellular fluid contains a high concentration of potassium ions, and a low concentration of sodium ions and calcium ions. The ion gradient is maintained by the action of energy-dependent ion pumps specific for Na^+, K^+ and Ca^{2+}. All vertebrate cells maintain an internally negative membrane potential (about -60 mV) since their surface membranes are specifically permeable to K^+ which causes an increased outflow of K^+ as compared to inflow of Na^+ and Ca^{2+}. Nerve cells are electrically excitable because of the presence in their surface membranes of selective ion channels — a single channel for Na^+, and several classes of K^+ and Ca^{2+} channels. These channels open and close as a function of membrane voltage, allowing rapid movement of the appropriate ions down their concentration gradient, resulting in an ionic current that passes into or out of the cell, depolarizing or hyperpolarizing the membrane.

The electrical excitability of neurons is dependent on the nature of the ion channels, the density of ion channels, and the location of ion channels in the different functional compartments of the cell. Neurons possess several compartments, enabling different kinds of signal transmission. Dendrites receive synaptic input from numerous presynaptic elements, and respond with graded or propagated changes in a different membrane potential. The cell body also receives synaptic inputs, and acts as a summing point for membrane potential changes occurring in several dendrites and cell bodies. Depolarization generates one or a series of conducted action potentials which are initiated in the cell or at the initial segment of the axon and are conducted down the axon to the nerve terminal. At the nerve terminal the action potential causes depolarization, release of a neurotransmitter into the synaptic cleft, and excitation of succeeding neurons or of effector cells, e.g., skeletal muscle. In this event Na^+, K^+, and Ca^{2+} contribute to signal processing and transmission in neurons.[1]

The neuronal ion channels have been classified on their functional properties into two groups: the ligand-gated channels, the proteins of which, located in postsynaptic membranes, bind specific neurotransmitters and open a pore to function in synaptic transmission; and voltage-gated channels with membrane proteins that sense the transmission electrical field and open ion-specific pores to produce nerve impulses and other forms of electrical excitability. In the past, ligand-gated channels have been characterized by the binding of specific pharmacological substances, whereas the voltage-gated channels were recognized by the ionic selectivity of their pores and by properties of the manner in which they open and close. According to recent experimental results the channels can be classified according to their structure.[2]

The structure of the ligand-gated channel relating to nicotinic acetylcholine receptor (AChR) was described at first. Different AChRs (from muscle and brain) are constructed from distinct subunits (four in muscles and probably only two in brain) that are closely related genetically. Each subunit is structurally similar — all have a very similar pattern of alternating hydrophilic and hydrophobic regions. In particular, there are four domains, each with about 20 consecutive hydrophobic amino acids in all of the known nicotinic acetylcholine receptor subunits from all species. The hydrophobic amino acid domains probably constitute the membrane-crossing helices.[2]

Three peptides of the ligand-gated channels of the central nervous system have been already sequenced, i.e., two subunits of the $GABA_A$ (gamma-aminobutyric acid) receptor channel, and one of the glycine receptor channel. All three subunits are very similar, with about 50% homology which indicates that about half of their amino acid residues are identical or substituted in protein evolution. There exists also about 25% of homology between amino acid residues of the AChR and $GABA_A$ and glycine receptor channels, despite the fact that the AChR channel mediates excitatory transmission whereas the $GABA_A$ receptor channel and the glycine receptor channel mediate inhibitory synaptic transmission. In addition, all have the same pattern of four putative transmembrane domains. Finally, the subunits of the AChR receptor channel and of the GABA receptor channel are closely related.[2]

The amino acid sequence analysis of the voltage-gated sodium channel, in contrast to the polymeric AChR receptor ligand-gated channel, is composed of a single polypeptide chain about four times as large as any of the AChR subunits. The polypeptide chain of the voltage-gated sodium channel contains four very similar motifs (segments), and each seems to be analogous to a single subunit of a multimeric protein. Each of these segments demonstrates the same pattern of alternating hydrophobic and hydrophilic regions that are probably five membrane-spanning helices; and, what is most interesting, the general structure of each segment demonstrates the same structure (Arg-X-X), where X is a hydrophobic amino acid residue, and is long enough to cross the membrane. This region, called S4, is probably the voltage sensor that couples the transmembrane electric field to the conformational change that opens or closes the pore of the channel.[2]

Similar to the voltage-gated sodium channel in mammals, the calcium channel in *Drosophila* demonstrates a long polypeptide with four segments, with similar patterns that probably cross the membrane, and with the characteristic S4 regions with the repeated Arg-X-X structure. All known sodium and calcium channels demonstrate about 60% of homology.[2]

More interesting seems to be the potassium channel, called, for historical reasons, the A-current channel. It is also composed of a single polypeptide chain about a third as long as the sodium and calcium channels, with only a single copy of the characteristic motif (segment) represented fourfold in these channels. Only a single part of the A-current channel sequence resembles that of the sodium/calcium channel, and only about 100 amino acid residues, surrounding an S4-like region, exhibit 50% of homology. Unexpectedly, the pattern of alternating hydrophobic and hydrophilic amino acid stretches is similar to that of the sodium/calcium channel amino acid residues. It is suggested that the potassium channel may be the ancestral type of the other more modern voltage-gated channels that have evolved from the potassium channel by gene duplication.[2]

The discussion about the character of ion channels is not finished. Using selective agonists — kainic acid, *N*-methyl-D-aspartic acid (NMDA), and quisqualic acid — as probes for individual amino acid receptor subtypes, and some antagonists, it became accepted that individual receptors are coupled to unique ion channels, especially true for the excitatory amino acid receptors (on the basis of the nicotinic acetylcholine receptor). This was suggested based on the fact that the conductance mechanism activated by NMDA is permeable to calcium and selectively blocked by magnesium, whereas responses to kainate and quisqualate behaved as if they were caused by the opening of magnesium-insensitive ion channels permeable to Na^+ and K^+ but not to Ca^{2+}. Very recent results proved that the different conductance mechanisms represent different functional states of the same ion channel.[3-5] Using a single channel recording from hippocampal and cerebellar neurons in cell culture, the authors demonstrated that in most membrane patches excitatory amino acids can activate an ion channel with multiple conductance substrates, the amplitudes of which are similar for kainate, NMDA, and quisqualate. Usually, ion channels opened by NMDA are of large conductance (50 pS — picosiemens), whereas channels opened by kainate and quisqualate are of much smaller conductance (1 to 20 pS). From the experiments it must be concluded that kainate and as also NMDA and quisqualate, can open both the small and large conductance channels. A further suggestion is that the small and large conductance openings, activated by various amino acids, reflect the activity of a single molecular complex and not the activity of two or more unique channels.[4,5] However, in patches from cerebellar neurons, both kainate and quisqualate activate a channel of conductance 1 to 2 pS (without any other activity). This may suggest a new molecular species: the small conductance channels specifically activated by kainate and quisqualate receptors, and a second additional superchannel which overlaps the conductance 10 to 50 pS, which is activated by all agonists.[3-5] There is also another possibility that the conductance substrate may be performed by a single macromolecule, suggesting that several channels may act together cooperatively.[3]

The idea of a fully open channel permeable to Ca^{2+} and closed by Mg^{2+} (with substrates that exclude these ions), as well as the supposed existence of a superchannel are noteworthy. This may support the suggestion of activation of the superchannel by protein modification, for example, its phosphorylation, that may cause a long-term potentiation by favouring activation of low conductance substrates at the expense of Mg^{2+}-sensitive substrates, usually activated by NMDA receptors.[3]

In general, it seems that the activities of ion channels are much more complicated as seen from the experiments with another neurotransmitter — glycine. Glycine added in nanomolar concentrations potentiates the activity of NMDA (about 17 times) without any action on the kainate and quisqualate responses.[6]

A. SODIUM CHANNELS

The sodium channel belongs to the first fully characterized channels of the nervous system. The selective transmembrane sodium channel enables rapid increase of sodium permeability during the action potential. Selective ion permeation is mediated by a hydrophilic pore containing a sodium-selective ion coordination site named the ion selectivity filter. Ion conductance through the channel is regulated (''gated'') by activation that controls the rate and voltage-dependent opening of the sodium channel after depolarization, and inactivation that controls the rate and voltage dependence of the subsequent closing of the channel during maintained depolarization.[1] Sodium ion movement through an opened sodium channel range from 8 to 18 pS, corresponding to more than 10^7 ions per second channel at physiological temperature and Na^+ concentration. This denotes that the sodium ions remain in the channel a very short time, and interaction with the channel-forming protein is very weak.[1] For the examination of sodium channels, different neurotoxins have been used that bind with high affinity and specificity to voltage-sensitive sodium channels and modify their properties. In the unmyelinated nerve fibers of the rabbit vagus, the density of sodium channels is $100/\mu m^2$ of membrane surface. In general, the sodium channels occupy less than 1% of the total surface area. In myelinated nerves the axonal membrane is accessible to sodium ions in the extracellular fluid only at the nodes of Ranvier, where the sodium channel density is approximately 2000 channels/μm^2.[1]

The electrical properties of the cell bodies and dendrites of vertebrate neurons are complex. Action potentials propagated down axons depolarize nerve terminals that cause activation of calcium channels and release of neurotransmitters. Purified synaptosomal fractions contain a substantial complement of sodium channels; their density is about 20 to 75 sodium channels/μm^2, and at the initial axon segment, approximately 350 to 500 channels. The distribution of sodium channels plays a critical role in determining the signal processing properties of neurons. Membrane potential changes in the dendrites are graded, and represent the sum of the contributions of multiple synaptic inputs. The summed potentials reach the cell body and are modulated by additional synaptic inputs of the soma.[1]

B. PROTEIN COMPONENTS OF THE SODIUM CHANNELS IN NEURONS

Direct chemical identification of sodium channel components *in situ* was done by specific covalent labelling of a derivative of α scorpion toxin which revealed covalent binding with two polypeptides, designated the α and β_1 subunits of the sodium channel (M_r 270 kDa and 39 kDa, respectively). In mutant neuroblastoma cells there are neurotoxin-resistant sodium channels with a lack of functional voltage sensitivity (the α subunit was not present).[1]

Indirect measurements revealed the diameter of the sodium channel protein to be about 118 Å, which indicates that the channel protein is much larger than the postulated transmembrane pore through which the sodium ion moves and which is proposed to be 3 to 5 Å at its narrowest point, the ion selectivity filter.[1] Protein subunits from the mammalian brain that revealed two subunits of 260 kDa and 38 kDa (designated α and β) represent more than 90% of the sodium channel proteins. The β subunit corresponds to two nonidentical subunits of similar molecular size. To conclude, the purified sodium channel from rat brain is composed of three subunits: α subunit of 260 kDa, β_1 subunit of 39 kDa, and β_2 subunit of 37 kDa and the complex protein 316 kDa.[1] Further analysis revealed that the sodium channel protein is a glycoprotein of the post-translational glycosylation which is needed for normal function. Finally, phosphorylation of the sodium channel protein is the most important mechanism in the long-term regulation of neuronal excitability.[1] The purified sodium channel protein is phosphorylated by the cAMP-dependent protein kinase.

C. CALCIUM CHANNELS

Recent investigations have shown that neurons have a series of different calcium channels, each with its own unique properties. The physiological role of the calcium channels is based on the pivotal role of Ca^{2+} as an intracellular second messenger. With the increase of intracellular Ca^{2+} in micromolar (m*M*) concentrations, several cellular functions become initiated. The rise of intracellular Ca^{2+} can occur by its release from intracellular storage sites associated with the endoplasmic reticulum, or the intracellular Ca^{2+} concentration can rise because of an increase in the Ca^{2+} permeability of the plasma membrane, which is normally impermeable to Ca^{2+}.

The release of Ca^{2+} from intracellular stores occurs under the influence of different stimuli (hormones, neurotransmitters) upon cell surface receptors that stimulate the breakdown of the phospholipid phosphatidylinositol biphosphate that generates diacylglycerol and inositol triphosphate (IP_3). Under the influence of IP_3, receptors inside the cell release Ca^{2+} into the cytoplasm. The influx of Ca^{2+} into the cells occurs through the opening of Ca^{2+} channels that can be opened by the action of an agonist on a receptor, e.g., nicotinic cholinergic agonists or the excitatory amino acid NMDA. Besides, several cells have channels that open and close in response to changes in membrane potential; these voltage-sensitive-calcium channels (VSCC) occur on several cells (egg, endocrine cells, muscle cells, neurons, and others).[7]

In some cells VSCC stimulations secretion of hormones, for example, the adrenal medullary cell in which VSCC cause an increase of intracellular Ca^{2+}, which triggers the release of catecholamines and peptides. The function of neurons are more versatile; they possess not only the secretory function similar to the adrenal cell, but they must also integrate and communicate information; and, what is more, the functions of different portions of the same neuron may differ radically. In some cases, neurons can fire Ca^{2+}-based action potentials (spikes). The influx of Ca^{2+} through VSCC may play a decisive role in the regulation of nerve function, e.g., the gating of several kinds of ionic channels can be controlled by Ca^{2+}. The influx of Ca^{2+} may cause important secondary changes in nerve excitability. The role of Ca^{2+} as a trigger for neurotransmitter release is among the best known. Finally, Ca^{2+}-dependent enzymes such as Ca^{2+}-calmodulin-dependent kinase, protein kinase C, and several other kinases and proteases are involved in neuronal functions.[7]

Similar to the case of Na^+ channels and nicotinic receptors that allowed the elucidation of their structure, interaction of VSCC channels with several drugs (especially used in the control of cardiovascular system) used as probes allowed the elucidation of their structure and function (adequate toxins as in the case of Na^+ channels are not available). Among the most important drugs that interact with VSCC are dihydropyridines (DHP); many of them inhibit VSCC, and are "antagonists" of VSCC. Some of them relax smooth muscles in nanomolar (n*M*) concentrations. Finally, some DHPs keep VSCC open for a prolonged time that enables influx of Ca^{2+} in greater than normal amounts; these are "agonists" of VSCC. Drugs of other classes, such as phenylalkylamines, verapamil, and others also function as VSCC antagonists. It is interesting that all these different classes of drugs bind to separate but interacting sites of VSCC. The DHP "receptor" has been isolated from skeletal muscle; it is a protein of approximately molecular mass of 210 kDa. It has VSCC activity and a substrate for cAMP-dependent kinase. It is accepted that a kinase-mediated modification of VSCC is the basis for the inotropic effects of β-adrenergic agonists in the heart. The interaction between VSCC and DHP is also voltage-dependent. DHP antagonists block more efficiently in depolarized than in polarized tissues, which is probably caused by the fact that DHP binds in an inactivated state by depolarization, rather than in an activated state of VSCC.[7]

D. DIHYDROPYRIDINE RECEPTORS IN THE BRAIN

Radiolabeled DHPs were used to study their binding affinity with various tissues (muscle cells, endocrine cells, and neuronal tissue). In the brain, a heterogeneous distribution of [^{3}H] DHP was discovered; however, VSCC in neurons did not seem to be sensitive to DHPs. VSCC in neurons can be analyzed by electrophysiological methods, but Ca^{2+} spikes observed in many neurons after such examination can also be very revealing. Unfortunately, such methods enable examinations only of invertebrate neurons of colossal size. For obtaining information about VSCC in nerve terminals, other methods must be used. For example, preparations can be monitored with radioisotopes. The most important function in VSCC neurons is to provide Ca^{2+} for triggering neurotransmitter release. Thus, depolarization-evoked release of neurotransmitters can be used for examination of VSCC in nerve terminals.[7]

Further experiments have demonstrated that DHPs are potent modulators of the voltage-dependent component of the Ca^{2+} uptake, and the initial divergencies were caused by the existence of multiple calcium channels. This was first discovered on pheochromocytoma cells which normally resemble adrenal medullary cells. They possess also DHP-sensitive VSCC. After depolarization, Ca^{2+} enters the cells and triggers catecholamine and peptide secretion which can be blocked by DHP inhibitor nitrendipine and enhanced by the DHP agonist Bay k 8644. Adrenal medullary cells and sympathetic neurons are closely related. The adrenal medullary cells transformed into cells resembling sympathetic neurons by NGF (nerve growth factor) become insensitive to DHP. A similar transformation can occur in pheochromocytoma cells. After this transformation, the cells become insensitive to DHP, and the transmitter release can be blocked by Cd^{2+} (cadmium ions are known to be potent VSCC blockers). In both, nontransformed and transformed cells, depolarization influx of Ca^{2+} remains DHP-sensitive. Thus, in NGF-differentiated cells, a dissociation in the properties of evoked transmitter release and Ca^{2+} influx seems to occur. This led to the suggestion that in NGF-transformed cells two types of VSCC must exist — DHP-sensitive-channels and DHP-resistant (insensitive) channels, the latter being linked to the transmitter release process.[7]

According to electrophysiological and pharmacological investigations, three types of

VSCC to the overall Ca^{2+} current can be distinguished. The first type (named T) gives rise to a small transient Ca^{2+} current generated by small depolarizing steps from negative holding potentials. The second type of VSCC (N) is generated by stronger depolarizations, and is responsible for the second inactivating phase of Ca^{2+} current. The third type of VSCC (L), generated by strong depolarizations, is responsible for the nonactivating component of the current.[7]

The pharmacological properties of these three types of VSCC are very interesting: The T type has an activation range (10 n*M* Ca) positive to −70mV, an inactivation range of −100 to −60 mV, an approximate single channel conductance of 9 pS, a weak Cd^{2+} block, and no DHP modulation; the N type has an activation range positive to −10 mV, an inactivation range of −100 to −49 mV, a single channel conductance of 13 pS, a strong Cd^{2+} block, and DHP modulation; the L type has an activation range positive to −10 mV, an inactivation range of −60 to −10 mV, a single channel conductance 25 pS, a strong Cd^{2+} block, and positive DHP modulation.[7]

E. THE ROLE OF CALCIUM CHANNELS IN NEUROTRANSMITTER RELEASE

The experiments were performed on chick DGR cells (chick dorsal root ganglion neurons), where types N and L of calcium channels have been found, but no T channels. The properties of the N and L channels in sympathetic neurons are similar. According to several kinds of experiments the N-type calcium channels seem to be most probably responsible for neurotransmitter release at nerve termini. Of special interest is the fact that neuronal VSCC can be modulated by biochemical modifications produced by enzymes, especially kinases. A strong support for this mechanism gives the mechanism in the β-adrenergic stimulation of the heart caused by cAMP-dependent phosphorylation of VSCC. In several vertebrate neurons, neurotransmitters can reduce the action potential, resulting from reduction of the neurotransmitter release. One such example is represented by increased K^+ permeability caused by rapid spike repolarization or by the direct blockade of a voltage-dependent Ca^{2+} influx.

In other cases, protein kinase C seems to be involved in modulation of Ca^{2+} currents, especially the types of VSCC modulated by agonists (phorbol esters).[7]

Having accepted the fact of the importance of N channels to neurotransmitter release, the question arises about the role of L and T type channels. It is supposed that together with modulation of K^+ permeability, VSCC channels, especially of the T type, may play a role in the production of rhythmic bursting behavior in neurons, whereas the L channels may be responsible for injection of large amounts of Ca^{2+} into the cytoplasm of nerve cells for modifications of neuronal activity by the activation of enzymes, especially kinases and proteases. For example, it is suggested that postsynaptic contributions to long-term potentiation at the central synapses may be dependent on the influx of Ca^{2+} into the postsynaptic neuron, which is caused by a Ca^{2+}-dependent protease, calpain, in the postsynaptic cell.[7]

Finally, it has been described that the number of DHP binding sites can vary in a number of pathological conditions (aging, cardiomyopathy, hypertension, and after chronic administration of such drugs as opiates, alcohol, neuroleptics) in which changes of L-channel function may contribute to changes in nerve excitability observed in these patients.[7]

Calcium channels may also be regulated by activated receptors. Active nicotinic receptors may directly open channels that are permeable to Ca^{2+}. Activation of these receptors leads to the opening of a pathway that can accommodate Ca^{2+} as well as other cations. Receptor-operated channels (ROC) may also be present in smooth muscles where they could participate in agonist-induced muscle contraction. Another example is represented by NMDS, the receptors of which are present in many parts of the central nervous system. Activation of these receptors leads to the opening of a channel for large amounts of Ca^{2+} and probably for other cations that can pass through this channel. Calcium entry into neurons through the NMDA-linked channels is important in the generation of long-term potentiation of central synapses, and probably in the generation of epilepsy.[7]

II. NEUROTRANSMITTERS

It is generally accepted that information generated in the brain involves communication between neurons through release of neurotransmitters at synapses. The first recognized neurotransmitters were acetylcholine, norepinephrine and serotonin; next, amino acids γ-aminobutyric acid, glutamic acid, aspartic acid, and glycine; and finally, peptides, among them: gut peptides (vasoactive intestinal polypeptide — VIP, cholecystokinin octapeptide — CCH-8, substance P, neurotensin, methionin enkephalin, leucine enkephalin, insulin, glucagon); hypothalamic releasing hormones (thyrotropin releasing hormone — TRH, luteinizing hormone-releasing hormone — LHRH, somatostatin — growth hormone release-inhibiting hormone); pituitary peptides (adrenocorticotropin — ACTH, β-endorphin, α-melanocyte-stimulating hormone — α MSH) and others such as angiotensin II, bradykinin, vasopressin, oxytocin, carnosine, and bombesin.[8] These brain peptide transmitters have been discovered mostly serendipitously and it is supposed that the total number of peptide transmitters may be 200 or more.[8]

Neurotransmitters are identified as endogenous substances characterized on the basis of their receptor effects. This is best documented by the discovery of enkephalins whose receptors were discovered first, and afterwards the enkephalins. In this case the receptor localization in central sites of pain perception suggested that opiates (e.g., morphine) may interact with the normally occurring opiate-like substances in the organism, particularly because analgesia following electrical stimulation of the brainstem could be partially reversed by the opiate antagonist naloxone, which proved that a morphine-like substance was secreted. Soon, in brain extracts a substance was identified that competes for opiate receptor binding. This substance consists of two pentapeptides, methionine enkephalin (metenkephaline) and leucine enkephalin (leuenkephalin). These substances are localized at nerve endings, which is consistent with a neurotransmitter role, and are released in a calcium-dependent fashion with brain depolarization. After recognition of the peptide character of the enkephalins, it was found that they are contained within the 91 amino acids of the β-lipotropin, isolated ten years earlier from the pituitary. Within the pituitary opiate-like activity, the 31 amino acid peptide β-endorphin and dynorphin possess high opiate activity. β-Lipotropin derives from a 31 kDa precursor which also incorporates the sequence of ACTH. β-Endorphin is present in the brain at levels of about 10% of those of enkephalin in specific neuronal systems other than enkephalin neurons.[8]

Localization experiments of neurons and opiate receptors revealed further characteristic features. For example, the neuronal patterns of met- and leu-enkephalin differ. In the globus pallidus, leu-enkephalin fibers are disposed as narrow bands surrounding axon bundles, whereas the met-enkephalin fibers demonstrate dense clusters between the leu-enkephalin bands. They are also functionally distinct. The μ receptors (with a distinct preference for morphine) are preferentially localized in layers 1 and 4 of the cerebral cortex, which are involved in integrating sensory perception. The δ receptors (with selectivity for certain enkephalin derivatives) are preferentially localized in the nucleus accumbens, olfactory tubercle, and parts of the emotion-regulating limbic system. This localization seems to correspond with the distribution of met-enkephalins which occurs in the region of the brain with more μ receptors (hippocampus and thalamus), and leu-enkephalins that occur more in the regions of the brain with more δ receptors (central amygdala). The substantia gelatinosa of spinal cord and the caudate nucleus have similar levels of μ and δ receptors, and similar numbers of met-enkephalin and leu-enkephalin neurons.[8]

The interaction of opiates with their receptors is crucial to their function. Opiates block the excitatory function of glutamic acid and acetylcholine on cerebral neurons by blocking sodium channel functioning, similar to the way morphine does. Receptor interactions of opiate agonists and antagonists are also distinguishable by their reactivity with the adenyl cyclase-regulating nucleotide — guanosine triphosphate — GTP. Similar to sodium, GTP decreases the affinity of opiate agonists but not antagonists. Opium and enkephalins, through the mediation of GTP, decrease the adenylate cyclase activity, reducing cAMP levels in neuroblastoma cells which possess specific opiate receptors. Thus, cAMP is also the second messenger for enkephalin. Similar to glutamic acid and acetylcholine, which are α-adrenergic receptors, histamine receptors and muscarinic cholinergic receptors work through sodium channel regulation, and in addition they are regulated by divalent cations, with manganese being the most effective and calcium being relatively inactive. Receptor regulation by divalent cations is combined with GTP and subsequently with the adenylate cyclase regulating system. In addition, opiate stimulation in neuroblastoma cells decreases gangliosides synthesis, whereas phospholipid, protein, and nucleic acids are not affected.[8]

A. NEUROTENSIN

Neurotensin was identified as a by-product of substance P isolation. Neurotensin lowers blood pressure through dilating blood vessels; alters pituitary hormone release; and, injected into the brain, lowers body temperature. The highest concentration of neurotensin occurs in the hypothalamus and basal ganglia of the brain, and in nerve endings. It is released by calcium-induced depolarization. Neurotensin inhibits selectivity neuronal firing in the locus coeruleus, which possesses a high density of neurotensin neurons. Neurotensin is particularly localized at regions that resemble those of enkephalin localization, which implies its role in pain perception. Neurotensin has potent analgesic effects, unrelated to that of the opiate system (it is not blocked by naloxone, the potent blocker of the opiate system). Neurotensin and enkephalins are stored in distinct neurons.[8]

B. SUBSTANCE P

Substance P occurs in 20% of dorsal root ganglia cells with some processes extending to the skin and others entering the spinal cord, and giving rise to terminals in the substantia gelatinosa. Substance P is a sensory transmitter of pain. It may also regulate axon reflexes in the skin responsible for the local vasodilation that occurs around an injured area, mediated by sensory nerves. Since substance P is a pain transmitter, the blockade of its release by opiates may account in part for opiate analgesia. In substance nigra of the brainstem, substance P release is controlled by GABA. Substance P is contained in cell bodies and in the corpus striatum and fibers that descend to terminate in the substantia nigra, which contains the highest concentration of substance P in the brain. GABA inhibits the release of substance P caused by depolarization.[8]

Substance P is associated with serotonin in pain perception. Areas that are enriched in substance P, serotonin, and enkephalins include the amygdala, periaqueductal gray, raphe nuclei, and substantia nigra. Stimulation of the nucleus raphe magnus of the brainstem, whose neurons contain serotonin, activates descending serotonin pathways to the spinal cord which generates analgesia. The presence of pathways of substance P in the limbic system may predict its role in emotional behavior.[8]

C. CHOLECYSTOKININ AND GASTRIN

Cholecystokinin (CCK) was originally isolated from the duodenum as a substance contracting the gall bladder, and afterwards it was discovered in the brain. The duodenal CCK is composed of 33 amino acid residues, and the major CCK entity in the brain is the COOH terminal octopeptide (CCK-8) and a lesser amount of the COOH-terminal tetrapeptide (CCK-

4). Gastrin is contained in the large system of the hypothalamus, which projects to the posterior pituitary gland. CCK and vasoactive intestinal peptide (VIP) are the only brain peptides with cells in the cerebral cortex. CCK is the agonist of cerebellar cortical cell firing. The most prominent group of CCK cells in the brain occurs in the periaqueductal gray area, where CCK may be involved in the pain-integrating functions. Similar to substance P, CCK occurs in sensory fibers of the dorsal root ganglia and terminals in the dorsal gray matter of the spinal cord. A great number of CCK cells together with substance P, neurotensin, and enkephalin occur in the hypothalamus, whereas the central nucleus of the amygdala has many CCK fibers, but not cells. It coexists also together with serotonin in certain neurons. CCK is found with dopamine in brainstem dopamine neurons, which project to the limbic systems. This system seems to be combined with the action of antischizophrenic drugs, suggesting its role in schizophrenia. Receptors for CCK in the brain are for the most parts for CCK-4, which have been described in specific brain areas, different from the localization of CCK-8. CCK-8 injected intraperitoneally caused satiety in previously hungry rats. CCK in the smallest doses (0.01 p*M*) infused into the lateral ventricules in sheep suppresses feeding behavior.[8]

D. VASOACTIVE INTESTINAL POLYPEPTIDE (VIP)

VIP was isolated from the gut as a substance that causes vasodilation. It is composed of 28 amino acid residues with many similarities in amino acid sequences and activities to the intestinal peptides, secretin and glucagon. VIP stimulates the conversion of glycogen to glucose, enhances lipolysis and insulin secretion, inhibits the production of gastric acid, and stimulates secretion of the pancreas and small intestine. Afterwards, the presence of VIP was also discovered in the brain, with the highest levels in the cerebral cortex. In the cerebral cortex, VIP and CCK neurons are bipolar and oriented perpendicular to the surface, which makes these peptidergic neurons ideally suited to activate and synchronize neuronal activity within the vertical columns of cerebral cortical cells.[8]

E. BRADYKININ

Bradykinin is released from the α_2 globulin fraction of the blood serum incubated with trypsin or snake venom. It contracts guinea pig ileum. Bradykinin is involved in inflammation, cardiovascular shock, hypertension, pain, and rheumatoid arthritis. Bradykinin belongs to the most potent pain-producing substances in the body. Its release in tissue damage is probably the first step in pain perception; injected near the periaqueductal gray substance of the brainstem, it has potent analgesic activity. The main group of bradykinin cells occurs in the hypothalamus, in the lateral septal area (causing hypertension), and periaqueductal gray substance (responsible for pain perception).[8]

F. ANGIOTENSIN

This active substance was discovered in 1898 in the kidney. Its protease, renin, converts angiotensinogen (a large protein molecule of the blood plasma) into decapeptide angiotensin I, which becomes converted into angiotensin II by angiotensin-converting enzyme by removing a COOH-terminal histidyl-leucine from angiotensin I. Angiotensin II is a potent vasoconstrictor that causes renal sodium retention by stimulating aldosteron secretion from the adrenal cortex. Injected into the brain, it causes drinking behavior and rise in blood pressure. Cells sensitive to angiotensin II are accumulated in the subfornical organ. These cells react with neuronal firing and drinking behavior after injection of angiotensin II. The brain has the greatest angiotensin II affinity as compared to the whole body. This explains the action of angiotensin II in the brain in spite of its very low level in the brain.[8]

G. HYPOTHALAMIC RELEASING FACTORS

The hypothalamic releasing factors are secreted mainly by cells of the median eminence of the hypothalamus, but also in lesser amounts in other brain regions, and pass through the portal capillaries to the pituitary gland where they control the synthesis and release of pituitary hormones.

Thyrotropin-releasing hormone (TRH) is a tripeptide (pyroglutamyl histidylprolinamide) synthesized enzymatically, and not on the basis of the genetic code. About 80% of TRH in the rat brain occurs outside the hypothalamus. TRH-containing fibers are present in the motor nuclei of the trigeminal, facial, and hypoglossal nerves; the ventral spinal cord; the nucleus accumbens; the lateral septal nuclei; and the bed nucleus of the stria terminalis. TRH generates behavioral excitation and anorexia in animals, and may induce mood enhancement in humans.[8]

Somatostatin is a hypothalamic, cyclic peptide of 14 amino acid residues which inhibits the release of the growth hormone somatotropin and prolactin from the pituitary gland. It is also a typical gut hormone, being localized both in neurons throughout the brain and the stomach, intestine, and pancreas. It inhibits the secretion of glucagon, insulin, and gastrin. In the brain it occurs in the amygdala, parts of the hypothalamus, the hippocampus, and the cerebral cortex, with terminals localized in all these areas and in the nucleus caudatus, nucleus accumbens, and the olfactory tubercle. Somatostatin occurs also in primary sensory neurons and dorsal root ganglion cells; this localization suggests that it may be involved in pain transmission.[8]

Luteinizing hormone-releasing hormone (LH-RH) is a hypothalamic decapeptide that stimulates the secretion of both luteinizing and follicle-stimulating hormones from the anterior pituitary gland. LH-RH is also present in the pre- and supra-chiasmatic and arcuate nuclei of the hypothalamus, and several other brain areas.[8]

Insulin, because of its relatively large molecule (composed of 86 amino acid residues) does not penetrate into the brain; insulin present in the brain is synthesized by nerve cells. Insulin concentrations in the brain are 10 to 100 times higher than that in the blood plasma. The highest brain content of insulin occurs in the hypothalamus and olfactory bulb. The neurotransmitter role of insulin is suggested by the presence of insulin receptors that have similar properties as those of peripheral insulin receptors.[8]

Glucagon is a pancreatic hormonal peptide of 29 amino acid residues. The large intestinal form has a molecular mass of 12 kDa. A dense plexus of glucagon-containing nerve fibers has been detected in the hypothalamus.[8]

Vasopressin and oxytocin are nonapeptides isolated from the hypothalamus (supraoptic and paraventricular nuclei). Vasopressin is the antidiuretic hormone, and oxytocin regulates uterine contraction and milk ejection. Vasopressin and oxytocin are stored with carrier hormones: neurophysin I which binds oxytocin, and neurophysin II which binds vasopressin. Vasopressin and oxytocin are present in several regions of the brain. Their synthesis has been traced by radioactive methods in the supraoptic and paraventricular nuclei. In spite of similar pathways, vasopressin and oxytocin neurons are physically distinct. There are some suggestions that vasopressin may play a role in behavioral reactions, mainly in learning and memory.[8]

Carnosine (β-alanylhistidine) occurs in high concentration in the primary olfactory pathway which passes from the nasal epithelium to the olfactory bulb. Carnosine is synthesized by a single enzymatic step.[8]

Bombesin is a 14 amino acid peptide isolated from the frog skin. In the gastrointestinal tract, the lung and brain bombesin-like material is synthesized. The material reacting with the brain is chemically distinct from the authentic bombesin. Bombesin stimulates secretion of gastrin and gastric acid, and enzyme secretion from the exocrine pancreas. It alters intestinal mobility and mimics cholecystokinin effects on the gall bladder.[8]

III. THE MOLECULAR BASIS OF NEUROTRANSMISSION

This idea is based upon ion channels that can be directly controlled by a cyclic nucleotide ligand. The first observations demonstrated that cGMP injected into a rod cell of the retina affected membrane conductance. Next, the presence of a cGMP-gated release of cations in rod outer-segment membrane preparations have been demonstrated. Finally, the existence of cGMP-gated conductance within the plasma membrane has been discovered.[9]

The cGMP-dependent conductance is a strange channel. The cGMP binds allosterically to open the channel. It is suggested that there are 2 to 4 cooperating sites. The channel is permeable to monovalent and divalent cations. In the case of photoreceptor signalling, light absorbed by the signal-receptor rhodopsin is transduced to activate the receptor. The active receptor catalyzes the conversion of GTP to GDP, bound to the α subunit of a G protein (transducin). In the presence of GTP, the α subunit dissociates from the $\beta\gamma$ subunit and activates the cGMP phosphodiesterase by displacing (or removing) the small inhibitory subunit γ of the phosphodiesterase molecule. The cascade may be amplified because the active receptor can activate several hundred G proteins, and each activated phosphodiesterase molecule can hydrolyze about 1000 cGMP per second. Thus cGMP can be rapidly depleted, and after dropping below 10 μM, bound cGMP dissociates from the cGMP-dependent ion channel in the plasma membrane. The channel closes to produce membrane hyperpolarization — the cell signal that may be transmitted to the brain. Guanyl cyclase returns cGMP to the resting concentration.[9]

The usual way to modify channels is to phosphorylate them with cAMP-dependent protein kinase. The cAMP cascade has five main steps: (1) the neurotransmitter binds to the cell surface receptor, (2) the bound receptor activates adenylate cyclase, (3) adenyl cyclase hydrolyses ATP to cAMP, (4) cAMP activates the protein kinase, and (5) the protein kinase phosphorylates specific sites of the channel.[10] The phosphorylated channel may open less frequently or may stay open for a longer time.

The control of the cAMP system is executed by G proteins. G proteins form a family with at least four members. Two of them are G_s (stimulating) and G_i (inhibiting). The G proteins are activated by neurotransmitters binding to other receptors. These in turn influence adenyl cyclase, either stimulating or inhibiting.[10] The question arises as to why the neuromodulatory cascades are so complicated. This is because the cascade may be used in different cells with only small changes (e.g., the specificity of the receptors), whereas the remainder of the system remains the same; each step can amplify the signal because a single receptor can activate many enzymes by using a G protein intermediate; and finally, the multiple steps provide many places where the regulatory system can interact, for example, stimulation of an enzyme by one neurotransmitter through G_s can be antagonized by a second neurotransmitter through G_i.[10]

Several observations have demonstrated that neurotransmitters working through G proteins do not necessarily work through the cAMP system. It was demonstrated that acetylcholine, which causes inhibition of the heart, works through a G protein that turns on a special set of potassium channels. Another observation revealed that calcium channels in dorsal root ganglion neurons are turned off by noradrenaline and GABA (which is usually an inhibitory neurotransmitter) acting through a G protein.[10,11]

Modulation of the neurotransmission can be executed by functional classes of receptors. A typical example is represented by glutamate receptors that possess two distinct classes of receptors classified by the artificial agonists as NMDA (*N*-methyl-D-aspartate) and non-NMDA types. Acid amino acids, particularly glutamate, are to the brain what acetylcholine is to the neuromuscular junction. Glutamate is a potent agent that causes excitatory synapses of the brain. There are probably three pharmacologically defined glutamate receptors: one responding to NMDA, and two responding to kainate and quisqualate, which probably are separate types of receptors. Amino-phosphonovalerate (APV) is a specific blocker of the NMDA receptor class. Several neurons have high density binding sites for NMDA, and are strongly excited through NMDA directly applied to their surface. This excitation can be blocked by small doses of APV; however, APV has only little effect on excitatory transmission produced by stimulating the presynaptic axons. Therefore, excitatory synaptic transmission is mediated by the non-NMDA receptor class.[12]

NMDA receptors play a crucial role in the observed response of ventrobasal thalamus neurons to natural stimulation of somatosensory afferents, but do not appear to be responsible for the short-latency excitation seen on electrical stimulation of the afferents, which is apparently mediated by excitatory amino acid receptors of the non-NMDA type. This indicates an involvement of both NMDA and non-NMDA receptors on the ventrobasal neurons to stimulation of somatosensory afferents, depending on the mode of stimulation.[13] In other words, NMDA receptors are used in the case of natural stimulation, and not when afferent pathways are stimulated synchronously with electric shock.[12]

NMDA receptors are involved in long-term potentiation in the hippocampus, a form of synaptic plasticity probably occurring in learning and memory. They contribute also to epileptiform activity. NMDA receptors participate in high-frequency synaptic transmission in the hippocampus, their involvement during low-frequency transmission being greatly suppressed by Mg^{2+}. The frequency-dependent alleviation of the blockade provides a new synaptic mechanism whereby a single neurotransmitter can transmit very different information depending on the temporal nature of the input. This mechanism demonstrates the involvement of NMDA receptors in the initiation of long-term potentiation and their contribution to epilepsy.[13,14] According to these results it seems that NMDA receptors react preferentially to low concentrations of glutamate, and non-NMDA receptors respond to higher concentrations.

Another example of different activity of the same neurotransmitter based on various receptors is represented by the histamine receptors. Histamine has been known to be widely distributed in the brain, but only recently have neuronal pathways containing histamine and its synthesizing enzyme, histidine decarboxylase, been discovered. Histamine receptors have been classified into two main types: H_1 (antagonist mepyramine) and H_2 (antagonist cimetidine) the stimulation of which leads to various physiological reactions via the intracellular messengers diacylglycerol, inositol triphosphate, and cAMP with different agonists and antagonists.[15] Recently, the presence of the third histamine receptor, H_3, suspected for a long time, has been confirmed. The H_3 receptor controls the synthesis and release of histamine. Thioperamide, the competent H_3-receptor antagonist, is effective in nM concentrations, and has a negligible effect on H_1- and H_2-receptor reactions. H_3 receptors in tissues outside the brain, bound with histamine, suppress the release of noradrenaline, the chemical transmitter mediating vascular smooth muscle contraction.[16,17]

Proteins in the synaptic area probably play an important role in nervous transmission. Synapsin I is a neuron-specific protein localized on the cytoplasmic surfaces of synaptic vesicles. This protein is the major substrate for cAMP-dependent and calcium/calmodulin-dependent protein kinases. Recent investigations[18] suggest that this protein may be involved in the regulation of neurotransmitter release from the nerve terminal. In the nerve terminal, synaptic vesicles are embedded in a cytoskeletal network, consisting in part of actin. The dephosphorylated form of synapsin has the ability to bundle F-actin, which becomes reduced by phosphorylation of synapsin I by cAMP-dependent protein kinase, and abolished by phosphorylation by calcium/calmodulin-dependent protein kinase. These results suggest that synapsin I may be involved in regulating the translocation of the synaptic vesicles to their sites of release.[19]

Cerebral edema belongs to the most severe disorders of the central nervous system. Only a little is known about mechanisms that maintain the water balance across the barrier tissues between the blood and brain. Recent investigations have demonstrated that the barrier tissues contain receptors and second messenger systems for atriopeptins, the recently identified cardiac peptides involved in peripheral water regulation. The atriopeptins are able to control the rate of cerebrospinal fluid production.[20] Because the blood-brain and blood-cerebrospinal fluid barriers are involved in normal water movements within the central nervous system, the barrier tissues may be end organs for the atriopeptins, which may be useful in the therapy of water balance disorders of the central nervous system.

IV. THE NERVE GROWTH FACTOR (NGF)

The discovery of the nerve growth factor initiated a new era in the molecular analysis of the neural development. NGF is necessary for the survival of sympathetic and sensory neurons *in vivo* and *in vitro*. In culture, it promotes neuronal differentiation, and chemotactically directs growing axons. NGF is produced by peripheral target tissues picked up by nearby axons and transported back to neuronal somata.[21] The vital role of NGF was demonstrated by administration of specific antibodies and drugs which block its uptake from nerve endings or retrograde axonal transport, resulting in death of the vast majority of the sympathetic cells.[22] The trophic role of NGF and its effect on differentiation mature nerve cells have been demonstrated in neoplastic chromaffin cells in culture which, in the presence of NGF, ceased dividing and underwent morphological changes, resulting in the formation of an intricate network of electrically excitable neurites in 2 to 3 weeks.[22] NGF, administered locally in the periphery, acts as a stimulant and attractant which regulates the size and shape of the terminal field of an axon. Finally, NGF may be necessary for neuronal maintenance in adulthood, particularly after axon injury for adjustment and repair.[21]

This raises a question about motor neurons: does there also exist a motor neuron growth factor? Experiments testing for the presence of this growth factor show that partial denervation of a muscle induces the undamaged axons to sprout and reinnervate the denervated muscle. The sprouting suggests that it represents an axonal response to soluble factors released by denervated fibres. Search for a motor neuron growth factor did not succeed. Using a panel of antisera to material of organ-cultured denervated cells, it was stated that several antisera partially prevented sprouting *in vivo*. Each blocking antiserum recognized a 56 kDa protein, and most promising of all, sera from patients with amyotrophic lateral sclerosis blocked sprouting when injected; what is more, such sera recognized a 56-kDa antigen.[21] Amyotrophic lateral sclerosis is a disease in which several spinal motoneurons die, and the surviving motoneurons sprout very poorly. Thus, sera from these patients prove that the etiology of this disease may be caused by disturbances in growth factors for motoneurons and the 56-kDa protein may be the sought after motor neuron growth factor.[21]

Another question arises about the function of NGF in the brain. In the early stages of development a great excess of neurons is developed, probably under the influence of NGF. During development, because of insufficient amount of NGF, some of these neurons die. It is supposed that NGF may be critical for normal brain development and for maintenance of its function. In the peripheral nervous system, NGF is needed for the development and maintenance of the sympathetic nerve cell that uses catecholamine neurotransmitters (norepinephrine and dopamine), and for sensory neurons that produce neuroactive peptides. More recent work has shown that in the brain NGF acts on cholinergic neurons which use the neurotransmitter acetylcholine. The first sign that NGF acts on brain tissue was the finding that after its injection into the brain of newborn rats an increase of choline acetyltransferase occurred (choline acetyltransferase is the principal enzyme that is needed for acetylcholine synthesis). Acetylcholine transferase is a marker for cholinergic neurons. The region of the brain in which effects of NGF have been observed are the septal area of the basal forebrain, the hippocampus, and the cortex.[23]

It is noteworthy that the above are the regions in which loss of cholinergic neurons occurs in patients with Alzheimer's disease. These neurons have their cell bodies in the nuclei of the forebrain, which include the medial septum and nucleus basalis, and extend axonal projections to the hippocampus and the cortex, i.e., areas that contribute to memory formation. The degeneration of the cholinergic tract may be a major contributor to loss of memory and other mental deficits.[23]

The cholinergic neurons of the corpus striatum show another NGF effect in the brain where an increase of choline acetyltransferase occurs. This is precisely the region which deteriorates in humans with Huntington's disease.[23]

The next arguments for an active role of NGF in the brain were the discovery of an active NGF gene manifested by the presence of mRNA that encodes the NGF in the hippocampus and cortex, the target areas innervated by the basal cholinergic neurons; and the discovery of specific NGF receptors indistinguishable from those on peripheral nerve cells.[23]

NGF may be needed in the brain for the development of cholinergic brain neurons during embryonal life, and later in life for maintenance of the nerve cells. Antibodies against NGF injected into rat fetuses inhibited the development of the cholinergic neurons of the basal forebrain. Once the cholinergic neurons develop, they need a continuing source of NGF; otherwise they degenerate. The severing of nerve fibers running from the forebrain to the hippocampus causes degeneration of cholinergic neurons in a short time.[23]

Cholinergic neuronal degeneration after axotomy seems to be caused by the loss of retrogradely transported neurotrophic factor, probably NGF. Adult rats with bilateral lesions of all cholinergic axons projecting from the medial septum to the dorsal hippocampus received a continuous infusion of NGF into the lateral ventricles. After two weeks of NGF treatment, cholinergic neurons and an increase of the biosynthetic enzyme acetyltransferase (350%) were present.[24] A similar result has been achieved in rats, in that their neurons within the caudate nucleus had been destroyed, which resulted in rapid degeneration of the striatonigral pathway, including projections containing the inhibitory neurotransmitter GABA and delayed transneural death of neurons in the substantia nigra pars reticulata.[25] Neuronal death was prevented by long-term intraventricular infusion of the GABA agonist muscimol. These examples show that peptide growth factors may be useful in the therapy of degenerative nerve diseases.

V. THE MOLECULAR ASPECTS OF LEARNING AND MEMORY

Two basic categories of learning — nonassociative and associative — have been distinguished. Nonassociative learning results from experience with a single event, for example, habituation (a decrease in response after usually repeated stimulation). The associative learning, resulting from the conjunction of two or more events, is usually categorized into Pavlovian and instrumental conditioning. At the most basic level, associative learning concerns the causal relations between events occurring in the organism's environment. Recent views favor an essential role for the cerebellum in both learning and memory. The lateral interpositus nucleus is responsible for the learning and memory of the conditioned eye-blink response, but both it and the cerebellum are responsible for the reflex response. Lesions of this nucleus abolish the conditioned reaction. Unilateral cerebellar lesions do not impair learning of responses on the contralateral side of the body. The essential efferent conditional reflex pathway consists of fibers that exit from the interpositus nucleus ipsilateral to the trained side of the body in the superior cerebellar peduncle, that cross to relay in the contralateral magnocellular division of the red nucleus, and that cross back to descend in the rubral pathway to act on motor neurons. An unconditioned stimulus of the dorsal accessory olive is projected through the inferior cerebellar peduncle. Thus, lesions of this area prevent acquisition, and produce normal extinction, of the behavioral conditioned reflex. The inferior olive-climbing fiber system is important in adaptation of the vestibule-ocular reflex and in recovery from motor abnormalities induced by labyrinthine lesions. Plasticity of the vestibulo-ocular reflex is an intriguing example of a learning-like change in behavior as a result of altered visual input. Studies of the cerebral cortex and hippocampus have demonstrated that the striate cortex in rats is essential for learning and memory of visual pattern discrimination. Studies in human amnesia have implicated the hippocampus in learning and memory. Lesions of the hippocampus impair learning of spatial tasks in the rat, as do frontal cortical lesions.[26]

Experimental results in animals and primates demonstrate involvement of the hippocampus in a wide range in learning and memory events. However, long-term or "permanent" memory is not stored in the hippocampus in humans and animals. The cortex was supposed to be the principal site of long-term memory processes. The cerebral cortex seems to be important for "cognitive" processes in humans and higher mammals.[26]

Mechanisms of learning are based on neural plasticity that includes all biophysical changes that affect the functional properties of neurons, particularly biochemical events within neurons and synapses. There are some hints for learning-induced alterations of ion channel conductance in mammalian systems. Recent results suggest that training of eyelid conditioning in mammals may decrease after-hyperpolarization in pyramidal neurons by decreasing a Ca^{2+}-activated outward K^+ conductance. Other persistent changes occur in hippocampal tissue in eyelid conditioning, including a prolonged learning-specific increase in glutamate receptor binding. In long-term potentiation, for example, a brief tetanic electrical stimulus to certain pathways induces an increased synaptic excitability that can persist for days or weeks. It is suggested that long-term potentiation may be caused by increased intracellular calcium, and irreversible increase of the number of receptors for glutamate in the forebrain synaptic membranes by activating a protease that degrades a specific protein, which in turn could produce long-lasting changes in synaptic chemistry and ultrastructure.[26]

The elucidation of memory and learning processes is far from being resolved. At first, localization of neuronal centers is necessary so that the next step of the biochemical processes can be studied, particularly neurotransmitters and neuropeptides. It seems to be very probable that learning and memory events must involve many alterations in neuropeptide and neurotransmitter systems, particularly in synaptic transmission, which is largely a biochemical process involving many biochemical reactions in neurons and probably also in DNA.[27]

VI. DISEASES OF THE NERVOUS SYSTEM

A. ALZHEIMER'S DISEASE

A senile disease like Alzheimer's affects about 5% of the population over 65 years of age. Symptoms of Alzheimer's disease are not the result of "aging". Autopsy shows characteristics pathologic changes in the brain. There is loss of neurons, particularly in regions essential for memory and recognition. Very characteristic are accumulations of twisted filaments and other abnormal structures within neurons. There are amorphous aggregates of amyloid adjacent to and within blood vessels, and scattered foci of cellular debris and amyloid called neuritic plaques. Loss of neurons is accompanied with reduction of neurotransmitters, mainly acetylcholine.

There are several hypotheses that try to elucidate the etiology of Alzheimer's disease. The genetic hypothesis starts with the observation that the incidence of this disease in certain families is specifically high, which suggests that a heritable defect must occur in these families. Epidemiological studies revealed that about 50% of Alzheimer's patients studied had a family of the disease, the others did not. The genetic etiology is supported by the fact that patients with Down's syndrome at the age of 40 usually develop Alzheimer's disease. The next hypothesis is based on the presence of an abnormal protein present as neurofibrillary tangles within neurons; cerebral blood vessels are surrounded and invaded by amyloid-rich plaques that replace degenerating nerve terminals. All these signs reflect on the accumulation of proteins not normally present in the brain. The infectious-agent model is based on the supposition that there may be a lentivirus involved (similar to scrapie disease in sheep or Creutzfeld-Jacob disease in humans). However, this supposition has not yet been confirmed. The toxin model is established on the presence of food toxins that may injure the brain cells (particularly aluminum salts from aluminum cans and other utensils). The blood flow model assumes that the disease may develop because of reduced blood flow, with the consequence of reduced oxygen and glucose extraction from the blood, and reduced energy generated from the oxygen and glucose. Finally, the acetylcholine model is based on the abnormally low concentration of this neurotransmitter, particularly in the hippocampus and the cortex, simultaneously with the reduced amount of acetyltransferase (the enzyme that synthesizes acetylcholine from its precursors choline and acetyl coenzyme A).[28] None of these hypotheses could be confirmed, and they represent rather conceptual models for further investigations.

Recent investigations on Alzheimer's disease and Down's syndrome using cloned complementary DNA probe for the chromosome 21 gene-encoding brain amyloid polypeptide (β amyloid protein) of Alzheimer's disease in leukocytes from three patients with sporadic Alzheimer's disease, and two patients with karyotypically normal Down's syndrome, revealed three copies of this gene. The gene for the β amyloid protein was duplicated both in the three Alzheimer disease patients and the Down's syndrome patients. It is thought that duplication of the gene for amyloid protein on chromosome 21 may be the genetic defect in Alzheimer's disease,[29] and in addition, that sequence analysis of the cerebrovascular amyloid protein in Down's syndrome and Alzheimer's disease demonstrated homology.[30] Further examinations revealed ubiquitin as a component of the paired helical filaments which constitute the distinct type of pathological neuronal fiber in Alzheimer's disease.[31] Antibodies to paired helical filaments in Alzheimer's disease react solely with the paired helical filaments, and not with neurofilaments or any other cytoskeletal protein in brain sections. This indicates that paired helical filaments do not share any normal neuronal fibrous proteins.[32]

The senile plaque of disorganized neuropil up to 150 nm across, often has an extracellular core of deposited amyloid. The senile amyloid is distinct from paired helical filaments in Alzheimer's disease, consisting of 4 to 8 nm wide filaments that are not wound together in pairs. The cores of extracellular plaques are composed of an inorganic aluminosilicate and an abundant polypeptide of relative molecular mass about 4 kDa, named A4. Immunohistochemical studies confirmed that the core A4 is related to the vascular amyloid protein. A4 is derived from a much larger precursor protein, probably a membrane receptor glycoprotein. The A4 protein is present between amino acid residues 597 and 638 of the precursor protein, probably included into the putative transmembrane protein.[33] Localization of the A4 gene on chromosome 21 suggests that the cerebral amyloid deposited in Alzheimer's disease and in Down's syndrome may be caused by aberrant catabolism of the precursor protein cell-surface receptor.[34,35]

An animal model for human Alzheimer's disease is represented by the rat with the nerve tract severed from the basal forebrain to the hippocampus. Typical symptoms in learning and memory occur. NGF (nerve growth factor) infusions into the brain prevented the development of the learning and memory deficits. Besides, grafts of basal forebrain that contain cholinergic neurons into these rats improved their learning ability, as did infusion of NGF.[23] These results suggest that deficit of peptides, particularly NGF growth factors, may also contribute to the pathogenesis of Alzheimer's disease.

B. FAMILIAL AMYLOIDOTIC POLYNEUROPATHY

Amyloid fibril protein of patients with amyloidotic polyneuropathy is known to be chemically related to transthyretin, the plasma protein that is usually called prealbumin. The familial amyloidoses represent a heterogeneous group of hereditary diseases with extracellular deposits of insoluble fibrillar protein. Most of these patients are characterized by the appearance of peripheral neuropathy, transmitted by an autosomal dominant gene. There are several types of this disease, namely, the Portuguese type, the Swedish type, and the Japanese type.

The Portuguese type of familial amyloidotic polyneuropathy is characterized by a decrease of plasma transthyretin levels; and an abnormal substitution by methionine for valine at position 30 (this abnormal transthyretin is present in small amounts in the blood plasma, which enables its diagnosis). Plasma transport of vitamin A seems to be normal in these patients.[36-38] The authors accept that the mutant transthyretin selectivity deposited in the tissues may be the amyloid characteristic for this disease. The Japanese type[39,40] and the Swedish type[41] are characterized by the same substitution, i.e., methionine for valine at position 30.

C. DEMYELINATIVE DISEASES

The main demyelinative disease is represented by multiple sclerosis (MS). Demyelinative diseases are known to be both autoimmune and virus induced diseases. These diseases are characterized by patchy inflammatory lesions, with accompanying tissue damage, randomly scattered throughout myelin-containing areas of the central nervous system. The initial lesions are perivascular, and they grow by radial enlargement and confluence. The inflammation includes edema, formation of perivascular cuffs, and infiltration with T- and B-lymphocytes and macrophages.[42] The demyelination is caused by macrophages (receptor-mediated endocytosis mediated by anti-myelin antibody). There are two main trends in the elucidation of the etiopathogenesis of demyelinization, the autoimmunization process studied in experimentally evoked autoimmune encephalomyelitis, and virus induced demyelinization caused experimentally by various virus infections. The autoimmune process seems to be combined with HLA-loci on chromosome 6. There were also the first attempts to use specific immunotherapy in experimentally evoked autoimmune encephalomyelitis, but with disputable results. Further investigations are necessary to elaborate new immunological techniques that may be more successful in the demyelinating diseases. The viral hypothesis is somewhat related to the autoimmune because of HTLV-I-like type RNA (human T-cell leukemia virus that specifically destroys T-helper lymphocytes) found in cultured cells from MS patients.[43] According to this observation there is selective loss of a subset of T-helper cells in active multiple sclerosis.[44] The discovery of antibodies against the paramyxovirus SV5 in the cerebrospinal fluid of some multiple sclerosis patients supports the theory of the viral etiology of sclerosis multiplex.[45]

Creutzfeldt-Jacob disease represents a neurodegenerative disease caused by a lentivirus related to the scrapie virus. Serum analysis using Western blots against an antiserum made from scrapie-infected hamster brain revealed positive reactions in 81% of the 31 specimens from patients with Creutzfeldt-Jacob disease, kuru, and Gerstmann-Sträusler-Scheinker syndrome.[46] Among patients who died of Creutzfeldt-Jacob disease were four who received human growth hormone isolated from human pituitaries, probably infected with the lentivirus which is the causative agent of this disease.[47]

Brain destructive processes may be induced by various metabolic disturbances. Abnormal platelet glutamate dehydrogenase activity, and activation in dominant and nondominant olivopontocerebellar atrophy have been described. A partial deficiency of glutamate dehydrogenase deficiency (40 to 50% of normal activity) was found either in patients with nondominant and in dominant forms of olivopontocerebellar atrophy. Platelet glutamate dehydrogenase was inactivated by GTP and stimulated (one- to two-fold) by ADP. In affected members the enzyme activity was normal, but was not activated by ADP. These results suggest that there may be two forms of altered enzyme, one with decreased activity and the other of normal activity, but which cannot be regulated by the physiological agent (i.e., ADP).[48]

Dysfunction of the Na^+K^+-ATPase may be the cause of homeostasis disturbances of Na^+, K^+, and Ca^{2+} in cells. The function and localization of this enzyme may be important in elucidation of the pathogenesis in epilepsy, since a possible defect in cellular mechanisms to clear extracellular K^+ may contribute to this disease.[49]

Hexosaminidase A deficiency was found in 15 adults from nine unrelated Ashkenazi families. The clinical picture varied, including spinocerebellar, various motor neurone, and cerebellar syndromes. Psychosis was found in 30% of cases. Skin fibroblasts synthesized the alpha and beta chain precursors of the same molecular mass as that synthesized by normal fibroblasts; however the amount of alpha chains was reduced, and mature chains were not detected.[50]

Brain cell destruction in Huntington's disease may be caused by naturally occurring excitatory and toxic substances such as quinolinate, a derivative of tryptophan. A subpopulation of cells positive for NADPH-diaphorase remain intact. The NADPH-diaphorase positive cells lack the receptors at which quinolate acts. A link between quinolinate and the mechanism by which it acts and Huntington's disease should be elucidated.[51]

Mental diseases represent the least known group of etiopathogenetical disorders. The list of putative causes for schizophrenia alone is long and inconclusive. There are some hints that schizophrenia may be an inherited disease, supported mainly by studies of identical twins (schizophrenia develops approximately 50% or more in both the siblings). Also, adoption studies demonstrated that children of schizophrenics adopted by non-schizophrenics developed schizophrenia more frequently than children from nonschizophrenic parents. However, against the genetic theory there is evidence that about 90% of relatives of schizophrenics do not have schizophrenia, and that the disorder is not inherited as a Mendelian trait. The viral theory suggests a virus infection as the causative agent of schizophrenia. Finally, some authors suggest that neurotransmitter disturbances may contribute to the disease. Schizophrenics are usually treated by antipsychotic drugs to control their psychotic behavior. It was thought that an excess of dopamine D_2 receptors might be caused by neuroleptic drugs. However, recent investigations have demonstrated that the increase of D_2 dopamine receptors in the brain tissue of schizophrenics is independent of antipsychotic drug administration, and may be associated with the disease process rather than with the treatment.[52]

D. MANIC-DEPRESSIVE DISORDER

More successful were investigations on the manic-depressive (named also bipolar affective) disorder.[53] The response of manic patients to lithium salts and depressed patients to tricyclic antidepressants (monoamine oxidase inhibitors) pointed to affected neuronal systems in this disease. To evaluate the genetic contribution, a study of the genetically isolated Old Order Amish population in southeastern Pennsylvania was carried out, in which it was suggested that manic-depressive disease might be localized on the short arm of chromosome 11 as a single gene. Analysis of the segregation of restriction fragment length polymorphisms made it possible to localize a dominant gene responsible for a strong predisposition to manic-depressive disease on the tip of the short arm of chromosome 11. Another study demonstrated a close linkage of the manic-depressive disease to the X-chromosome markers for color blindness and glucose-6-phosphate dehydrogenase deficiency.[54] These results provide confirmation that mental illnesses can be caused by single genetic defects. Mental diseases exhibit heterogeneity which may be caused by mutations at different loci responsible for the same mental disease.[55] In manic-depressive disease a mutation affects the catecholamine neurotransmitter system (adrenaline, noradrenaline, dopamine, and tyrosine hydroxylase). Mutation of the hydroxylase gene that encodes the rate-limiting enzyme for the synthesis of these three neurotransmitters might be responsible for causing the manic-depressive phenotype.[55]

E. PARKINSON'S DISEASE

Patients with neuronal disorders are rarely cured because of deterioration of nerve cells that cannot regenerate. A typical example is Parkinson's disease because of degeneration of the dopamine pathway of the brain, the cell bodies of which are in the substantia nigra of the brainstem with axons ascending to terminate at the nucleus caudatus and putamen, areas involved in control of motor activity. Replacement of the missing dopamine may be ameliorated by the dopamine precursor L-dihydroxyphenylalanine (L-DOPA). In spite of the amelioration of the clinical symptoms, deterioration of these cells continues. Parkinson's disease can be evoked in animal research. Unilateral destruction of the nigrostriatal dopamine pathway causes rats to rotate in their cages. Implants into the denervated nucleus caudatus of a substantia nigra from a fetal rat sprout dopamine-secreting processes from the graft, which relieve the locomotor disturbances in these animals. Adrenal tissue implanted onto the surface of the nucleus caudatus caused amelioration of patients with Parkinson's disease.[56]

The positive results in brain transplantation of adrenal tissue implicate a possibility of brain transplantations with deficit of neurotransmitters, particularly in Alzheimer's disease, where one might consider grafting into the brains some peripheral acetylcholine producing tissue from the same patient rather than septal tissue from a human fetus.[56]

REFERENCES

1. **Catterall, W. A.,** The molecular basis of neuronal excitability, *Science,* 223, 653, 1984.
2. **Stevens, Ch. F.,** Channel families in the brain, *Nature,* 328, 198, 1987.
3. **Mayer, M.,** Two channels reduced to one, *Nature,* 325, 480, 1987.
4. **Jahr, C. E. and Stevens, C. F.,** Glutamate activates single channel conductances in hippocampal neurons, *Nature,* 325, 522, 1987.
5. **Cull-Candy, S. G. and Usowicz, M. M.,** Multiple-conductance channels activated by excitatory amino acids in cerebellar neurons, *Nature,* 325, 525, 1987.
6. **Johnson, J. W. and Acher, P.,** Glycine potentiates the NMDA response in cultured mouse brain neurons, *Nature,* 325, 529, 1987.
7. **Miller, R.,** Multiple calcium channels and neuronal function, *Science,* 235, 46, 1987.
8. **Snyder, S. H.,** Brain peptides as neurotransmitters, *Science,* 209, 976, 1980.
9. **Applebury, M. L.,** Biochemical puzzles about the cyclic-GMP-dependent channel, *Nature,* 326, 546, 1987.
10. **Stevens, C. F.,** Modifying channel function, *Nature,* 319, 622, 1986.
11. **Holz, G. G., Rane, S. G., and Dunlap, K.,** GTP-binding proteins mediate transmitter inhibition of voltage-dependent calcium channels, *Nature,* 319, 670, 1986.
12. **Stevens, C. F.,** Are there two functional classes of glutamate receptors?, *Nature,* 322, 210, 1986.
13. **Salt, T. E.,** Mediation of thalamic sensory input by both NMDA receptors and non-NMDA receptors, *Nature,* 322, 263, 1986.
14. **Herron, C. E., Lester, R. A. J., Coan, E. J., and Collingridge, G. L.,** Frequency-dependent involvement of NMDA receptors in the hippocampus: a novel synaptic mechanism, *Nature,* 322, 265, 1986.
15. **Hill, S. J.,** Histamine receptors branch out, *Nature,* 327, 104, 1987.
16. **Arrang, J. M., Garbarg, M., Lancelot, J. C., Lecomte, J. M., Pollard, H., Robba, M., Schunack, W., and Schwartz, J. C.,** Highly potent and selective ligands for histamine H_3-receptors, *Nature,* 327, 117, 1987.
17. **Ishikawa, S. and Sperelakis, N.,** A novel class (H_3) of histamine receptors on perivascular nerve terminals, *Nature,* 327, 158, 1987.
18. **Bähler, M. and Greengard, P.,** Synapsin I bundles F-actin in a phosphorylation-dependent manner, *Nature,* 326, 704, 1987.
19. **Baines, A. J.,** Synapsin I and the cytoskeleton, *Nature,* 326, 646, 1987.
20. **Steardo, L. and Nathanson, J. A.,** Brain barrier tissues: end organs for atriopeptins, *Science,* 235, 470, 1987.
21. **Sanes, J. R.,** More nerve growth factors?, *Nature,* 307, 500, 1984.
22. **Calissano, P., Cattaneo, A., Biocca, S., Aloe, L., Mercanti, D., and Levi-Montalcini, R.,** The nerve growth factor, *Exp. Cell Res.,* 154, 1, 1984.

23. **Marx, J. L.,** Nerve growth factor acts in brain, *Science,* 232, 1341, 1986.
24. **Kromer, L. F.,** Nerve growth factor treatment after brain injury prevents neuronal death, *Science,* 235, 214, 1987.
25. **Saji, M. and Reis, D. J.,** Delayed transneural death of substantia nigra neurons prevented by γ-aminobutyric acid agonist, *Science,* 235, 66, 1987.
26. **Thompson, R. T.,** The neurobiology of learning and memory, *Science,* 233, 941, 1984.
27. **Black, I. G., Adler, J. E., Dreyfus, C. F., Friedman, W. F., LaGamma, E. F., and Roach, A. H.,** Biochemistry of information storage in the nervous system, *Science,* 236, 1263, 1987.
28. **Wurtman, R. J.,** Alzheimer's disease, *Sci. Am.,* 252, 62, 1985.
29. **Delabar, J.-M., Goldgaber, D., Lamour, Y., Nichole, A., Huret, J. L., de Grouchy, J., Brown, P., Gajdusek, C., and Sinet, P. M.,** β amyloid gene duplication in Alzheimer's disease and karyotypically normal Down's syndrome, *Science,* 235, 1390, 1987.
30. **Glenner, G. G. and Wong, C. W.,** Alzheimer's disease and Down's syndrome: sharing of a unique cerebrovascular amyloid fibril protein, *Biochem. Biophys. Res. Commun.,* 122, 1131, 1984.
31. **Mori, H., Kondo, J., and Ihara, Y.,** Ubiquitin is a component of paired helical filaments in Alzheimer's disease, *Science,* 235, 1641, 1987.
32. **Ihara, Y., Abraham, C., and Selkoe, D. J.,** Antibodies to paired helical filaments in Alzheimer's disease do not recognize normal brain proteins, *Nature,* 304, 727, 1983.
33. **Anderton, B. H.,** Progress in molecular pathology, *Nature,* 325, 658, 1987.
34. **Kang, J., Lemaire, H. G., Unterbeck, A., Salbaum, J. M., Masters, C. L., Grzeschik, K. H., Multhaup, G., Beyreuter, K., and Müller-Hill, B.,** The precursor of Alzheimer's disease amyloid A4 protein resembles a cell-surface receptor, *Nature,* 325, 1987.
35. **Goodfellow, P. N.,** From pathological phenotype to candidate gene for Alzheimer's disease, *Trends Genet.,* 3, 59, 1987.
36. **Saraiva, M. J., Birken, S., Costa, P. P., and Goodman, D. S.,** Amyloid fibril protein in familial amyloidotic polyneuropathy, Portuguese type. Definition of molecular abnormality in transthyretin (prealbumin), *J. Clin. Invest.,* 74, 104, 1984.
37. **Saraiva, M. J., Birken, S., and Costa, P. P.,** Family studies of the genetic abnormality in transthyretin (prealbumin) in Portuguese patients with familial amyloidotic polyneuropathy, *Ann. N.Y. Acad. Sci.,* 435, 86, 1984.
38. **Nakazato, M., Kangawa, K., Minamino, N., Tawara, S., Matsuo, H., and Araki, S.,** Identification of a prealbumin variant in the serum of a Japanese patient with familial amyloidotic polyneuropathy, *Biochem. Biophys. Res. Commun.,* 122, 712, 1984.
39. **Nakazato, M., Kangawa, K., Minamino, N., Tawara, S., Matsuo, H., and Araki, S.,** Radioimmunoassay for detecting abnormal prealbumin in the serum for diagnosis of familial amyloidotic polyneuropathy (Japanese type), *Biochem. Biophys. Res. Commun.,* 122, 719, 1984.
40. **Dwulet, F. E. and Benson, M. D.,** Primary structure of an amyloid prealbumin and its plasma precursor in a heredofamilial polyneuropathy of Swedish origin, *Proc. Natl. Acad. Sci. U.S.A.,* 81, 694, 1984.
41. **Barnes, D. M.,** Neurosciences advance in basic and clinical realms, *Science,* 234, 1324, 1986.
42. **Gonatas, N. K., Greene, M. I., and Waksman, B. H.,** Genetic and molecular aspects of demyelination, *Immunol. Today,* 7, 121, 1986.
43. **Koprowski, H., DeFreitas, E. C., Harper, M. E., Sandberg-Wollheim, M., Sheremata, W. A., Robert-Guroff, M., Saxinger, C. W., Feinberg, M. B., Wong-Staal, and Gallo, R. C.,** Multiple sclerosis and human T-cell lymphotropic retroviruses, *Nature,* 318, 154, 1985.
44. **Rose, L. M., Ginsberg, A. H., Rothstein, T. L., Ledbetter, J. A., and Clark, E. A.,** Selective loss of a subset of T helper cells in active multiple sclerosis, *Proc. Natl. Acad. Sci. U.S.A.,* 82, 7386, 1985.
45. **Goswami, K. K. A., Randall, R. E., Lange, L. S., and Russel, W. C.,** Antibodies against the paramyxovirus SV5 in the cerebrospinal fluids of some multiple sclerosis patients, *Nature,* 327, 244, 1987.
46. **Brown, P., Cokker-Vann, M., Pomerov, K., Franko, M., Asher, D. M., Gibbs, C. J., and Gajdusek, C.,** Diagnosis of Creutzfeldt disease by western blot identification of marker protein in human brain tissue, *N. Engl. J. Med.,* 314, 547, 1986.
47. **Brown, P., Gajdusek, C., Gibbs, C. J., and Asher, D. M.,** Potential epidemic of Creutzfeldt disease from human growth hormone therapy, *N. Engl. J. Med.,* 313, 728, 1985.
48. **Sorbi, S., Tonini, S., Giannini, E., Piacentini, S., Marini, P., and Amaducci, L.,** Abnormal platelet glutamate dehydrogenase activity and activation in dominant and nondominant olivopontocerebellar atrophy, *Ann. Neurol.,* 19, 239, 1986.
49. **Stahl, W. L.,** The ($Na^+ + K^+$)ATPase: function, structure, and conformations, *Ann. Neurol.,* Suppl. 16, 5121, 1984.
50. **Navon, P., Argov, Z., and Frisch, A.,** Hexosaminidase A deficiency in adults, *Am. J. Med. Genet.,* 24, 179, 1986.
51. **Koh, J., Peters, S., and Choi, D. W.,** Neurons containing NADPH-diaphorase are selectively resistant to quinolate toxicity, *Science,* 234, 73, 1986.

52. **Wong, D. P., Wagner, H. N., Jr., Tune, L. E., Dannals, R. F., Pearlson, G. D., Linke, J. M., Tamminga, C. A., Brousoulle, E. P., Revert, H. T., Wilson, A. A., Toung, J. K. Th., Malat, J., Williams, J. A., O'Tuama, L. A., Snyder, S. A., Kuhar, M. J., and Gjedde, A.,** Positron emission tomography reveals elevated D_2 dopamine receptors in drug-naive schizophrenia, *Science,* 234, 1558, 1986.
53. **Egeland, J. A., Gerhard, D. S., Pauls, D. L., Sussex, J. N., Kidd, K. K., Allen, C. R., Hostetter, A. M., and Housman, D. E.,** Bipolar affective disorders linked to DNA markers on chromosome 11, *Nature,* 325, 783, 1987.
54. **Baron, M., Risch, N., Hamburger, R., Mandel, B., Kushner, S., Newman, M., Drumer, D., and Belmaker, R. H.,** Genetic linkage between X-chromosome markers and bipolar illness, *Nature,* 326, 289, 1987.
55. **Hodgkinson, S., Sherrington, R., Gurling, H., Marchbanks, R., Reeders, S., Mallet, J., McInnis, M., Petursson, H., and Brynjolfsson, J.,** Molecular genetic evidence for heterogeneity in manic depression, *Nature,* 325, 805, 1987.
56. **Snyder, S. H.,** A cure using brain transplants? *Nature,* 326, 824, 1987.

Chapter 14

MOLECULAR DEFECTS IN THE ENDOCRINE SYSTEM

I. INTRODUCTION

Since its discovery, the functioning of the mechanisms of the endocrine system on a molecular basis has been recognized. The main interdependencies of the endocrine system with other regulatory systems in a living organism can be summarized as follows: The synthesis and release of hormones from specifically differentiated organs (e.g., thyroidea, parathyroidea, adrenals, gonads) is dependent on releasing factors (termed also releasing hormones) secreted by the hypothalamus and the brainstem (e.g., thyrotropin-releasing hormone, corticotropin-releasing hormone). Their secretion, in turn, is influenced by the level of the proper hormone. In other words, their secretion is dependent on consumption of the proper hormone in target-tissues. For example, the greater the consumption of thyroid hormones in the tissues, the lower the level of thyroid hormones in the blood; this causes an increased secretion of thyrotropin-releasing hormone, with the consequence of increased synthesis and secretion of thyroid hormones. The lower consumption of thyroid hormones in the tissues causes an increased level of these hormones in the blood and consequently higher levels of thyroid hormones in the blood because of their overproduction (as seen in the case of Graves'-Basedow's disease, induced by antibodies against thyrotropin receptors, the so called LATS — long-acting thyroid stimulator); this results in reduced levels of the thyrotropin-releasing factor. Here we see a faultless interaction between the central regulatory organs with the peripheral ones and target tissues.

II. HORMONE ACTION

The general principle of hormone action consists of their

- Synthesis in specifically differentiated organs (cells)
- Transport through the blood-stream
- Binding with specific receptors at the target-cells

A. SYNTHESIS OF POLYPEPTIDE HORMONES

The synthesis of polypeptide hormones is usually realized by the synthesis of larger molecules that become cleavaged by specific enzymes to smaller active hormone-molecules. The polypeptide hormones consist of approximately 100 small protein molecules ranging from 192 amino acid residues (the growth hormone) to 3 amino acid residues (thyrotropin-releasing hormone). The non-poly-peptide hormones are synthesized by specific enzymes, and the synthesis of their polypeptide part, termed apoprotein, is also realized by the genetic code. Usually, the apoprotein molecules of non-polypeptide hormones are larger than that of polypeptide hormones.

The expression of genes and protein synthesis consists of several steps. The first step involves changes of the embryonic DNA (not well understood yet) that produce specifically differentiated cells and nuclear proteins able to express only genes specific for the differentiated cell but not others. The next step consists of initiation of transcription by the forming of an initiation complex, and transcription, resulting in the formation of pre-mRNA which usually includes several coding and noncoding sequences (exons and introns). The next step that follows is mRNA processing, which involves splicing by removing the introns (non-coding sequences), conversion of some nucleotides (e.g., their methylation), and addition of 7-methyl-guanine "cap" at the beginning of the mRNA and poly(A) at the end. The addition of the changed nucleotides at the beginning and end of the mRNA serve to protect the synthesized mRNA from digestion by nucleases and to prolong the existence of mRNAs. The next step involves translation with post-translational processing. Usually, translation occurs on the polyribosomes by the pairing of specific carrier amino acylated transfer RNAs to the corresponding codons of the mRNA, and polymerization of amino acids bound with the tRNAs into the polypeptide chain. Usually, the beginning of the polypeptide chain possesses the hydrophobic signal, also termed leading peptide, which enables transport of the synthesized polypeptide through membranes of the rough endoplasmic reticulum. This kind of molecule is termed pre-prohormone. During passage through the membrane the signal peptide is cleaved from the proper polypeptide. Posttranslational processing results in final preparation of the proper hormone, and may include cleaving of some peptide bonds, resulting in conversion of inactive precursor proteins (prohormones) into hormone molecules, which may additionally involve glycosylation (in the case of glyco-hormones), phosphorylation, acetylation, or sulfation. Finally, there occurs folding of the processed polypeptide into its native active conformation.

Very important for normal hormone activity are regulatory mechanisms of protein synthesis which consist of short polypeptide sequences, upstream of the initiation point of transcription. This is the TATA box about 28 to 30 nucleotides from the initiation point, responsible probably for accuracy of the initiation of transcription. Approximately 80 nucleotides still further upstream are the CCAAT sequences, responsible probably for the quantitative expression of the structural genes. In several genes, described mainly in the immunoglobulin genes, but also in the genes encoding hormone molecules, there are enhancer sequences that cause a several times larger expression of the given genes. Recently, there have also been described sequences responsible for termination of transcription, but up till now these have not been as precisely specified as in the case of the upstream occurring regulatory sequences.

B. HORMONE TRANSPORT

The next step in hormone action is concerned with the transport of hormones through the blood stream. The small hormone molecules can be easily secreted through the kidneys; therefore, binding of hormone molecules is usually with specific proteins, and the molecules in the blood protect them against their excessive loss through the kidneys. The situation is similar to that of hemoglobin that must bind oxygen with high affinity in the lungs and actively release oxygen in the tissues, the hormone blood-transporting molecules, although less specialized, must bind hormones adequately for its intended function. The binding must not be so strong that it would make it impossible to release hormones to the tissues, and it must not be so weak that it would cause excessive loss of hormone molecules through the kidney. To the best known protein-transporting hormones belong: thyroxine-binding α-globulin, corticosteroid-binding globulin — transcortin, and sex-hormone-binding protein. These proteins are carriers of the thyroid and adrenal gland secretions, as well as hydrophobic secretion of the gonads. There are several endocrine disorders known to be caused by inadequate blood transport of hormones by mutated carrier proteins.

A specific kind of hormone transport occurs in brain peptides. These substances are synthesized in nerve cells, neurons included, and transported within the neurons of the central, autonomic, and peripheral nervous systems, where they probably play the role of neurotransmitters. The enormous number of brain peptides, which frequently are similar to the hormones of the endocrine system, eliminates the previous strict difference between the nervous and endocrine systems.

A large number of peptides are present in the nervous system.[1] Some of them have been described previously in other tissues. The brain peptides belong to following categories:

- Hypothalamic-releasing hormones: thyrotropin-releasing hormone, gonadotropin-releasing hormone, somatostatin, corticotropin-releasing hormone, growth hormone-releasing hormone
- Neurohypophyseal hormones: vasopressin, oxytocin, neurophysin(s)
- Pituitary peptides: Adrenocorticotropic hormone, β-endorphin, α-melanocyte-stimulating hormone, prolactin, luteinizing hormone, growth hormone, thyrotropin
- Gastrointestinal peptides: vasoactive intestinal polypeptide, cholecystokinin, gastrin, substance P, neurotensin, methenkephalin, leu-enkephalin, insulin, glucagon, bombesin, secretin, somatostatin, motilin
- Others: angiotensin II, bradykinin, carnosine, sleep peptides, calcitonin.

The processing of brain peptides consists in the liberation of active peptides from larger inactive precursors by specific proteolytic enzymes. The cleavage of most of them occurs probably within the Golgi apparatuses in the developing secretory granules. The post-translational modifications of these peptides include such modifications as glycosylation, phosphorylation, acetylation, amidation, and sulfation. These modifications usually cause a several times increase of potentiation of the modified peptides.

Disturbances in brain peptide synthesis and level may cause several neurological disorders, e.g., dopamine in Parkinson's disease, acetylcholine in Alzheimer's disease, and disturbances in the catecholamine neurotransmitter system in manic-depressive disease.

C. HORMONE RECEPTORS

Receptors are indispensable links in the hormone action which, thanks to their high affinity to extract hormones from the blood stream, bind them and introduce the hormone-receptor complex into target cells — usually cells that possess adequate receptors. Upon binding the entering hormone, the receptor undergo activation, which increases the affinity of the receptor for nuclear interphase chromosomes. Molecular cloning of the glucocorticoid, estrogen, and progesteron receptors enabled determination of their primary amino acid structures responsible for their function as regulators of gene expression. This region in glucocorticoid and estrogen receptors consists of a centrally localized DNA-binding domain rich in cysteine, lysine, and arginine, and a carboxyl-terminal region where steroid hormones interact. Studies revealed that hormone-binding to the carboxyl-terminal region unmasks the DNA binding region and permits interaction of the DNA with the receptor complex, resulting in initiation of transcription.[2]

A similar cystein-rich domain is also present in thyroid hormone receptors that binds specific DNA sequences. This has led to the hypothesis that coordination of Zn^{2+} atoms may form a specific structural configuration for DNA binding.[2]

The steroid hormones influence growth, development, and homeostasis by interaction with the intracellular receptor proteins that directly regulate the transcription of target genes. Two receptor systems have been discovered for the corticosteroids: the glucocorticoid receptor (previously thought to be common for glucocorticoids and mineralcorticoids) and the mineralocorticoid receptor. However, the mineralocorticoid receptor has *in vitro* affinity for either glucocorticoids or mineralocorticoids. There probably exists (particularly in the brain) physiological interaction between these two receptors, which may contribute to target gene interactions, and may be helpful to achieve complex physiological control.[2]

Normally, about 1900 molecules of ovomucoid mRNA can be found in each tubular gland cell nucleus after 14 d of diethylstilbestrol treatment. After 14 d of withdrawal of diethylstilbestrol from the chick, the concentration of ovomucoid mRNA decreased to 3 molecules per tubular gland cell nucleus. Readministration of a single dose of diethylstilbestrol caused an increase in the concentration of these mRNA molecules from 3 to 38 in the first 4 h and to 120 mRNA molecules within 16 h. After a second injection of diethylstilbestrol at 48 h, the concentration of these mRNA molecules increased to 620 molecules per tubular gland cell nucleus, which is approximately one third of the level found in chronically stimulated chicks with diethylstilbestrol.[3]

The receptor for thyroid hormone corresponds to the c-*erb*-A protein. It is a nuclear protein that binds to DNA, resulting in activation of transcription under the influence of thyroid hormones.[4,5] It is interesting that the viral oncogene v-*erb*-A is defective in binding thyroid hormones, but is still located in the nucleus,[5] which deprives v-*erb*-A of its natural control and may contribute to cancerogenesis.

The results of studies on the intracellular localization of unoccupied glucocorticoid receptors are somewhat confusing. It is accepted that free glucocorticoid receptors are present in the cytosol only, and after binding with glucocorticoids they pass to the cell nucleus, binding selectively to hormone regulatory elements in the vicinity of hormonally-inducible promoters. In addition, it was found that the hormone-bound glucocorticoid receptor binds also with the long terminal repeat (LTR) region of mouse mammary tumor virus (MMTV). In experiments with MMTV it was shown that the free glucocorticoid receptor binds also to the hormone regulatory elements of MMTV, which seems to indicate that the free glucocorticoid receptor can interact with hormone regulatory elements *in vitro*.[6] However, experiments *in vivo* prove that protein-DNA interactions require the presence of the hormone.[7] In contrast to this, mutant glucocorticoid receptors bind to DNA as constitutive activators of transcriptional enhancement. It is supposed that in this case the conformational change of the glucocorticoid receptor unmasks pre-existing functional domains for DNA binding and the enhancer activation region.[8] Cells deprived of ATP do not bind glucocorticoid hormones, but in spite of lacking hormone-binding activity, glucocorticoid receptors are present in the nuclei in amounts comparable to those of receptors in normal cells.[9]

The steroid hormone receptors represent specific gene regulatory proteins the activity of which is an important tool in the elucidation of gene regulation in differentiated cells. The control may be realized by interaction of *trans*-acting proteins with *cis*-acting DNA promoter elements, resulting in positive or negative effects. In the case of steroid hormone receptors the activation of transcription seems to result from binding of the hormone-receptor complex to promoter enhancer elements of the target genes. Sequence comparisons, together with mutation analysis, have defined two conserved regions (C and E, Fig. 1). The putative DNA-binding domain C is a highly conserved 66-amino acid residues region which forms two DNA-binding "fingers" involving two pairs of cysteine residues. The hormone-binding domain E is located in the C-terminal end of the steroid and thyroid hormone receptors. The hydrophobic variable region D may correspond to a hinge region between the DNA- and hormone-binding regions. Region A/B is variable in length and sequence; whether this region is necessary to transcription is not known.[10]

Region C reveals a 50 to 60% similarity in chicken progesterone receptor, human glycocorticoid receptor, and in chicken and human c-*erb*-A proteins. Comparison between the chicken and human c-*erb*-A protein sequences shows a 91% and 88% sequence identity in regions C and E, respectively. The various receptors in regions C and E demonstrate the following homologies: region C steroid receptors/c-*erb*-A — 45 to 55%; human estrogen/chicken progesterone or human glucocorticoid receptors — 62%; chicken progesterone/human glucocorticoid receptors — 91%; region E: steroid receptors/c-*erb*-A — 20%; human estrogen/chicken progesterone or human glucocorticoid receptors — 30%, chicken progesterone/human glucocorticoid receptors — 55%.[10] This denotes that steroid and thyroid hormones, which are neither structurally nor biochemically related, have receptors that have evolved from a common ancestral gene.

1. Hormone Receptor Disturbances

The important role of receptors in normal life processes, representing relatively large protein molecules, must result in a relatively high rate of mutations that may cause distur-

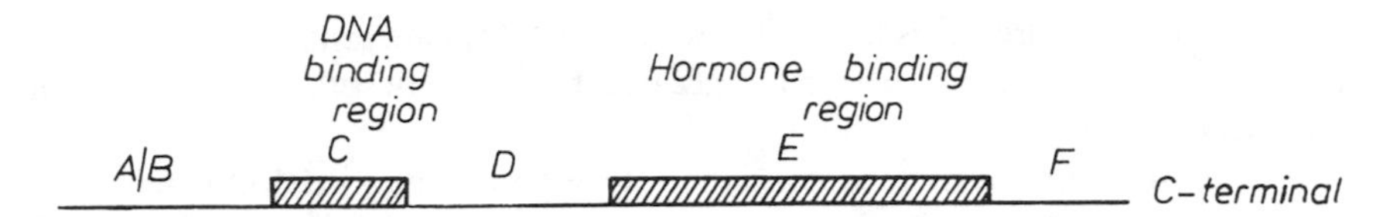

FIGURE 1. Schematic alignment of steroids (glucocorticoids and estrogen) and thyroid hormone receptors and c-*erb*-A proteins similar to v-*erb*-A protein of a gene of avian erythroblastosis virus. Segment A/B represents the N-terminal amino acid residues of the human estrogen receptor (1-185), chicken progesterone receptor (1-581), human glucocorticoid receptor (1-421), and of chicken (1-51) and human (1-102) c-*erb*-A proteins. Then follow 66 amino acid residues in steroid receptors, and c-*erb*-A proteins that form highly conserved DNA-binding regions, followed by various D segments, and finally (also highly conserved) E regions, consisting of 238 to 255 amino acid residues that form hormone-binding regions of the receptors and c-*erb*-A proteins. In total, the human estrogen receptor consists of 553 amino acid residues: the chicken progesteron receptor has 946; the human glucocorticoid receptor has 777; and the chicken c-*erb*-A protein consists of 408 amino acid residues and the human c-*erb*-A protein has 456.[10]

bances of vital functions, resulting in several disorders. Hereditary hormone-receptor disturbances can be divided into defects of intracellular and membrane receptors. In general, morbid symptoms of receptor defects are manifested by the absence of the given hormonal action, in spite of normal or, even more frequently, elevated synthesis (level) of the given hormone.

A typical case of membrane receptor deficit is represented by pseudohypoparathyroidism which is a rare hereditary disorder characterized by symptoms and signs of hypoparathyroidism with distinctive skeletal and developmental defects. The disease is due to the absence of parathyroid hormone receptors characterized by target tissue resistance to parathyroid hormone action in the target tissues, i.e., bone and kidneys. In contrast to hypoparathyroidism caused by deficient synthesis or parathyroid hormone, pseudohypoparathyroidism is characterized by excessive synthesis and secretion of parathyroid hormone. The excessive synthesis is caused by compensatory mechanisms because of the absence of feedback stimuli from target tissues that control normal hormone synthesis and secretion. Two types of pseudohypoparathyroidism are known: type I is characterized by the absence of cAMP increase in urine after parathyroid hormone administration. In these patients there occurs a decreased guanosine-triphosphate (GTP) binding protein reaction (approximately 50% reduced) which activates the cAMP system of the second messengers. The result is somewhat confusing because other hormones that use this system of transmembranal activity demonstrate only minute disturbances (TSH, glucagon, gonadotropin). Type II is characterized by normal increase of cAMP in urine. It is supposed that in this case the parathyroid hormone receptor is intact, but the cAMP receptor might be deficient.[11]

Familial nephrogenic diabetes insipidus is characterized by a decrease of number and affinity of vasopressin receptors in the kidneys. It can be discriminated from the hypophyseal form of diabetes insipidus by the total absence of reaction to administered vasopression.

Gonadotropin-resistant hypogonadism is characterized by the deficiency of the luteinizing hormone receptors, manifested by the absence of reactivity to administered luteinizing hormone. This is clinically expressed as hypogonadism, with a normal level of luteinizing hormone (LH) and follicle stimulating hormone (FSH).

Congenital TSH-resistant hypothyroidism is characterized by a defect of TSH receptors. The level of TSH in the blood is increased, and T_3 and T_4 are decreased. This is clinically expressed as cretinism.

Adrenocorticotropin (ACTH)-resistant cortisol deficit: The ACTH-receptor, up to now, has not been characterized. This is clinically expressed as hyperpigmentation, hypoglycemia, and convulsions. Cortisol level in blood and its secretion are considerably decreased.

Laron-dwarfness syndrome is characterized by the human growth hormone receptor defect with extreme high levels of growth hormone in the blood. In some cases a somatomedin defect is present.

To the group of intracellular receptor deficit disorders belong the following:[11]

Testicular feminization syndrome (complete) is characterized by the total absence of testosterone receptors. Clinically, the manifestations are: female external genitalia with underdevelopment of the labia and a blind-ending vagina, internal genitalia reduced to rudimentary anlage, gonads may be intra-abdominal along the course of the inguinal canal or in the labia, gonads similar to undescended testes, scanty or absence of axillary and pubic hair (hairless women), and female habitus with breast development. These are characterized by the presence of male sex chromatin and 46,XY karyotype; resistance to androgenic and anabolic effects of testosterone, and increased level of gonadotropin levels in the blood, usually with elevated level of testosterone.

Incomplete testicular feminization syndrome is characterized by a decreased level of testosterone-receptors or thermolabil testosterone-receptors. This is clinically less expressed than in the complete form.

Reifenstein syndrome is characterized by a decreased number of testosterone-receptors. Clinical manifestations range from moderate abnormalities (microphallus and gynecomastia) to hypospadias and complete failure of scrotal fusion. Usually, azospermia is present.

L-trijodothyronin (T_3)-receptor defect — There is a severe form, which is autosomal recessive, characterized by increased T_3, T_4 and TSH levels. This is clinically expressed as nystagmus, goiter, and backward maturation of bones. The mild form is characterized by mild clinical manifestations, but also with increased levels of T_3, T_4, and TSH.

The cortisol-receptor defect is characterized by considerably increased levels of ACTH, free cortisol in blood serum, and free cortisol in urine without symptoms of Cushing's syndrome. Clinically, there was also hypertension with hypopotassemia.

Luteinizing hormone receptor defect is characterized by an elevated level of hCG (human choriongonadotropin) with primary hypogonadism and microphallus; the homogenate from the testis showed approximately 50% decreased affinity to hCG.

Defects of insulin receptors are discussed later in this chapter.

III. MOLECULAR DISORDERS OF INSULIN AND INSULIN-DEPENDENT MECHANISMS

The mechanism of insulin action remains unresolved. Most hormones use one of the three kinds of action: the c-AMP system, the phosphatidyl-inositol/Ca^{2+} system, and ion channels, but no one of them is adequate to explain the action of insulin. Since the discovery of insulin 60 years ago, its physiological function has been very well recognized, but the molecular mechanisms remain unresolved.

Insulin binds to a cell-surface receptor, a glycoprotein, and $\alpha_2\beta_2$ heterodimer, with the α subunit outside the cell membrane, and the β subunit inside. It binds with high affinity to their receptors on the cell surface, and is subsequently internalized by receptor mediated endocytosis.[12] The insulin receptor was discovered in 1972, and is a glycoprotein with a molar mass of approximately 350 to 400 kDa. The insulin receptor exhibits insulin-dependent tyrosine kinase activity which catalyzes phosphorylation of both β subunits and exogenous peptides and proteins. The specificity of the insulin-dependent protein kinase is similar to that of the epidermal growth factor (EGF) receptor kinase and the *src* family of thyrosine-specific protein kinases.[13] Recent investigations have revealed that the transmembrane and cytoplasmic domains of the human EGF receptor are homologous to the v-*erb*-B oncogene product, and that these domains represent the corresponding cellular proto-oncogene.[13] Therefore, it is not surprising that a chimaeric receptor allows insulin to stimulate tyrosine kinase activity of the EGF-receptor, and that insulin and EGF receptors may employ similar or even identical mechanisms for signal transmission across the cell membrane.[14]

The insulin receptor becomes phosphorylated at 13 tyrosine residues in the β subunit, probably at the C terminus, at the centre and at the site near the membrane end of the molecule.[15] The replacement of insulin receptor tyrosine residues 1162 and 1163 demonstrated that tyrosine kinase-dependent phosphorylation of the insulin receptor is essential for the action of insulin.[15,16]

Insulin seems to induce the tyrosine phosphorylation of several proteins, among them probably the guanine nucleotide-binding proteins and other regulatory proteins. It stimulates GTPase activity in platelets. It is supposed that insulin inhibits adenylate cyclase, probably not via G_1 protein, but via another G protein termed G_{ins}, which may be linked to the *ras* oncogene product (in oocytes and hepatocytes, antibodies against the *ras* product prevent insulin stimulation on oocyte maturation and glycogen synthesis, respectively).[15]

Generation of a second messenger molecule is essential to all signalling cascade systems. Lacking precise knowledge of this molecule is crucial in explaining insulin action. Recently, it was suggested that the generation of a glycan containing inositol phosphate, glucosamine, and other carbohydrates may play this role.[15] The insulin mediator could be derived from the glycosyl-phosphatidylinositol structure which probably anchor proteins to the plasma membrane. This mediator has a molecular mass of 800 to 1200, and is produced 2 to 3 minutes after administration of insulin.[15]

A. DISTURBANCES OF INSULIN STRUCTURE

Insulin is synthesized as a single chain polypeptide — pre-proinsulin. The signal sequence "pre" becomes cleaved during synthesis on the rough endoplasmic reticulum, and no mutations are known that cause disturbances of removing signal sequences, because such mutations probably would be lethal.

Proinsulin is characterized by the presence of a C-peptide that joins the two A and B chains of the mature insulin molecule. Mutations occur mostly at the two critical junctions where the C-peptide is attached to the A and B insulin chains by two pairs of basic amino acids. Such defects have been recognized in families with hyperinsulinemia. The defect is inherited in an autosomal dominant pattern, and probably involves the loss of one of the basic amino acid residues that makes it impossible to cleave the proinsulin molecule at the mutation site, which results in the presence of a two-chained intermediate of proinsulin molecules secreted into the blood plasma.

B. HUMAN PROINSULIN VARIANT AT ARGININE 65

Clinically occurring glucose intolerance with abnormally high ratios of proinsulin-like material to insulin (9 to 10 as compared with normal values of approximately 0.25). The molecular weight of the proinsulin peptides is 9000 (the molecular weight of insulin is 6000). In the two-chained intermediate in which the C-peptide remains joined to the insulin A chain, the sequence is des-Arg 31, Arg 32 of proinsulin. Sequence analysis has revealed the presence of Lys 64 in the region joining the C-peptide with the insulin A chain in the proinsulin intermediate, the acetylation of which would prevent removal the C-peptide from the intermediate. Loss of Arg 65 and loss of the protective activity causing acetylation of Lys 64 make it impossible to cleave the C-peptide from the A chain of insulin. It is supposed that substitution of Arg 65 of the proinsulin molecule results in failure of cellular enzymes to cleave correctly the C-peptide from the A chain of insulin.[17] Another case of disturbances in cleavage of proinsulin to insulin was described also because of substitution of Arg → His at position 65.[18]

C. HUMAN INSULIN VARIANTS AT POSITION Phe^{B24} — Phe^{B25}

Phe^{B24} and Phe^{B25}, at the site of insulin binding to its receptor, are inactive forms of the insulin molecule. The disorder is characterized by diabetes with marked hyperinsulinemia and glucose intolerance. No evidence for elevated circulating insulin antagonists was found (normal contrainsulin hormones, insulin antibodies, insulin receptor antibodies), and the patient's insulin receptors on circulating monocytes were normal. The response to exogenous insulin (insulin tolerance test) was normal, and the biological activity of endogenous insulin reduced. Two variants were described: $Phe^{B25} \rightarrow$ Leu, and $Phe^{B24} \rightarrow$ Ser, that secreted both normal and abnormal insulin, which predicted that both alleles of the insulin gene were codominantly expressed, one normal and one abnormal.[17]

The most typical clinical symptom of abnormal insulin molecules is hyperinsulinemia with mild diabetes mellitus without the presence of insulin resistance and normal response to exogenous insulin. Because of the small molecular size the mutations of the insulin molecule are rather rare, in opposition to the insulin receptor which represents a large molecule with more frequent mutation.

D. RECEPTOR-DEPENDENT INSULIN RESISTANCE

The disorder is characterized by hyperinsulinemia because of increased resistance to endo- and exogenous insulin. Clinically the following symptoms may occur: acanthosis nigricans - caused probably by stimulation by the excessive insulin, the amount of the EGF, and EGF-receptors; hirsutism may be caused by an increased synthesis of testosterone in ovaries, which might indicate polycystic ovaries; total or partial atrophy of adipose tissue, caused by direct insulin deficit (insulin promotes lipogenesis and inhibits lipolysis). Atrophy of adipose tissue does not occur in all patients, which suggests that these disturbances may be influenced by various additional factors.[11]

Extreme insulin resistance, type A, is characterized by a marked reduction of insulin receptors on monocytes, erythrocytes, and fibroblasts; clinical symptoms are acanthosis nigricans, hirsutism, polycystic ovaries, and sometimes atrophy of adipose tissue. Type B is caused by insulin-receptor antibodies.[11]

Extreme insulin resistance, type C, is caused by the defective insulin-dependent phosphorylation of the insulin β-receptor subunit on monocytes. This disorder manifests the following clinical symptoms: acanthosis nigricans, hirsutism, polycystic ovaries, and sometimes atrophy of adipose tissue.[11]

Extreme insulin resistance, type D, is caused by the absence of insulin receptors on monocytes, erythrocytes, adipocytes, and fibroblasts. Clinically, it is manifested as acanthosis nigricans, and sometimes hirsutism and polycystic ovaries.[11]

Extreme insulin resistance, type E, is caused by the decreased affinity of insulin receptors to insulin on monocytes, erythrocytes, and fibroblasts; symptoms are teeth anomalies, acromegalic symptoms, and deformations of the skull.[11]

Leprechaunism is an inherited defect in a high affinity insulin receptor, which is characterized by a diminution of insulin receptors on fibroblasts, normal insulin-receptors with decreased insulin effects on fibroblasts, and atypical insulin-receptors on lymphoblasts with decreased pH and temperature-dependency. The clinical symptoms are acanthosis nigricans, hirsutism, polycystic ovaries, atrophy of adipose tissue, dwarfism (pre- and post-natal), characteristic leprechaun face. In response to oral glucose a marked hyperinsulinemia occurs.[11,19,20] The decreased c-AMP and c-GMP activities after insulin administration prove the postreceptor defect in leprachaunism.[21]

Lipoatrophic diabetes mellitus (Seip syndrome) is characterized by a variety of symptoms (even within the same family): a reduced number of insulin-receptors on monocytes and erythrocytes; decreased affinity of monocytes, erythrocytes, and fibroblasts to insulin; and of normal insulin-receptors to insulin on monocytes, erythrocytes, and fibroblasts. Clinically, there is acanthosis nigricans, atrophy of adipose tissue, and to a lesser degree, hirsutism and polycystic ovaries.[11]

Rabson-Mendenhall-syndrome is characterized by a reduced number of insulin-receptors on monocytes and lymphoblasts, and reduced affinity of fibroblast receptors to insulin. Clinically, there are acanthosis nigricans, hirsutism, teeth deformities, and pineal gland hyperplasia.[11]

Myotonic dystrophy is characterized by a reduced affinity of insulin-receptors to insulin on monocytes, with regulatory disturbances and normal binding on monocytes. Clinical symptoms are sometimes acanthosis nigricans, myotonia, muscle atrophy, and hypogonadism.[11]

Extreme insulin resistance in ataxia telangiectasia is characterized by progressive cerebellar ataxia, teleangiectasiae, and several abnormalities of the immune system. Usually, there exists hyperinsulinism and decreased glucose tolerance. Monocytes and cultured fibroblasts demonstrated an 80 to 85% decrease of affinity to insulin-binding. The whole plasma from ataxia teleangiectasia patients inhibited binding of insulin to its receptors on cultured human lymphocytes. It is supposed that the disturbance may be caused by the presence of an inhibitor of insulin binding.[22]

Insulin resistance occurs in several conditions with increased anti-insulin activity. To these belong Cushing's syndrome, acromegaly, pheochromocytoma, hyperglucagonemia, hyperthyroidism, malignant cachexia, insulin antibodies, and pregnancy.

1. Antireceptor Antibodies

The best known are patients with myasthenia gravia. Most of these patients have circulating antibodies to acetylcholine receptor sites that block binding of acetylcholine with their receptors and makes it impossible to activate the motor endplates, resulting in myasthenia.

Other kinds of antibodies occur in Graves's disease where antibodies against thyrotropin sites have a stimulatory effect, causing uncontrolled synthesis and secretion of thyroid hormones, and resulting in hyperthyroidism.

E. MOLECULAR GENETICS OF DIABETES MELLITUS

Recent investigations on diabetes mellitus show that diabetes is not one disorder, but several diseases, probably with various causes and mechanisms of transmission.[23] Therefore, the clinical entity, diabetes, appears to be an example of genetic heterogeneity with carbohydrate intolerance as the cardinal feature.[24] Clinically, we discern type I and type II diabetes. Type I is characterized by absolute deficiency of insulin, with early onset in lean patients, and a tendency to ketoacidosis. These patients frequently develop vasculopathy (microangiopathy and neuropathy). Type II is characterized by relative deficiency of insulin, with late onset in obese patients and no tendency to ketoacidosis. The diabetic complications are rare. Experiments performed in rats revealed two distinct 5′ flanking elements, the activity of which is restricted to the pancreatic B-cells. The combinatorial effect of multiple control elements may explain the cell-specific expression of the insulin gene.[25] A genetic analysis of the genetic basis of susceptibility in the nonobese diabetic mice revealed three different recessive loci required for insulin-dependent diabetes — one tightly linked to H-2K locus on chromosome 17, the second on chromosome 9, and the third shown only in a second backcross.[26] These results prove that transmission of genes that contribute to susceptibility to diabetes may be of a complex character.

The insulin-like growth factors (IGF-I and IGF-II, also termed intermedins). IGF-I and IGF-II are polypeptide growth factors in human plasma, chemically homologous to insulin. Similar to proinsulin, they are single chain molecules containing an NH_2-terminal B domain, a shorter C peptide, and an A domain. They also contain a 6 or 8 COOH terminal domain not present in proinsulin. IGF-I has 40% homology with insulin, and IGF-II 60%. IGF crossreacts with insulin-receptors, which activates common effector pathways; and vice versa, insulin may crossreact with IGF receptors. The difference between insulin and IGF consists in that IGF binds with specific higher molecular weight carrier proteins, which prolongs the half-life of the former. The complexed form of IGF with carrier proteins is biologically inactive. The role of the carrier proteins is to deliver active IGF molecules to the target cells.[27]

The insulin-like growth factors, insulin, relaxin (a polypeptide hormone from the corpus luteum responsible for dilatation o the symphysis pubis prior to parturition), glucagon, some pancreative polypeptides, and somatostatin form a hormone family of pancreatic hormones and homologous growth factors.[28] The concentration of IGFs in the blood plasma is regulated by growth hormone. IGFs stimulate incorporation of sulfate into proteoglycans of the cartilage, they have mitogenic activity in fibroblasts, and insulin-like activities on adipose and muscle tissue. Glucagon is a member of a large family of homologous peptides of the alimentary tract (among them, VIP — vasoactive intestinal polypeptide, and GIP — gastric inhibitory peptide, renamed glucose dependent insulinotropic peptide). Glucagon activates adenylate cyclase, and acts through a receptor distinct from those of secretin and VIP. It has a relationship to glicentin, a 100 amino acid polypeptide isolated from the intestine. Somatostatin is a small pancreatic hormone (14 amino acid residues), widely distributed in the whole body, and also in the hypothalamus. It inhibits release of insulin and glucagon.[28] The gene for IGF-II maps on the short arm of chromosome 11 that also contains the gene for insulin and proto-oncogene c-Ha-*ras*; IGF-I maps on chromosome 12 that is evolutionarily related to chromosome 11 and carries the gene for the proto-oncogene c-Ki-*ras*. The EGF gene is on chromosome 4q, in the same region as another growth factor, the T-cell growth factor. It is supposed that all these genes form the family of growth factors.[29] The IGF-I and IGF-II (70 and 67 amino acid residues, respectively) are synthesized in various tissues. They are required for normal fetal and postnatal growth and development.[30] IGF-I and IGF-II are synthesized in connective tissue or cells of mesenchymal origin. Because of their wide distribution and synthesis in various tissues and organs, it is supposed that they exert paracrine effects on nearby target cells.[31]

IV. THE PARATHYROID GLAND

Parathyroid hormone is a polypeptide of 84 amino acid residues that interacts with plasma membrane-bound receptors on bone and kidney target cells where it acts through the cAMP system. The synthetic amino terminal, 1 to 34 fragments, has full biological activity. Vitamin D must undergo hydroxylations on positions 1 and 25 to become fully active on target cells. 1,25(OH)2D vitamin is a steroid-like hormone which interacts with cytoplasmic receptors in small intestinal and bone cells where it causes synthesis of specific proteins. The integrated action of parathyroid hormone and 1,25(OH)2D maintains serum calcium within narrow limits, permitting normal neuromuscular and secretory function as well as normal bone mineralization. Abnormalities in the receptor-effector system for parathyroid hormone and 1,25(OH)2D causes disturbances characterized by resistance to the action of the hormone. A specific kind of hormone resistance (termed pseudohypoparathyroidism) is caused by the deficient action of the G protein unit that mediates activation of the cAMP system, and hereditary vitamin D-dependent rickets type II in which the abnormality in the nuclear uptake of 1,25(OH)2D may lead to impaired response to 1,15(OH)2D.[32]

Parathyroid hormone disturbances may be caused by three different mechanisms: (1) defect at the prereceptor level, (2) a defect membrane N-protein resulting in diminished second messenger production, and (3) a defect in the cytosolic response to the hormone.

Immunoreactive N-terminal parathyroid hormone, mid-C-regional parathyroid hormone, intact parathyroid hormone and bio-parathyroid hormone, vitamin D metabolites and serum calcium and phosphate, alkaline phosphatase activity, and the N-protein activity of the erythrocyte membranes have been analyzed in 24 patients receiving vitamin D metabolites (5,000 to 80,000 U/d). Eight patients with Albright's hereditary osteodystrophy revealed increased N-terminal and mid-C regional parathyroid hormone levels, and increased alkaline phosphatase activity. The levels of intact parathyroid hormone and bio-parathyroid hormone were normal. Nine patients with complete Albright's hereditary osteodystrophy revealed elevated levels of bio-, N-terminal, and mid-C-parathyroid hormone, and normal hydroxylation of vitamin D, but decreased N-protein activity of erythrocyte membranes. Seven patients with pseudohypoparathyroidism had no features of Albright's hereditary osteodystrophy and were without increased cAMP excretion in urine after exogenous parathyroid hormone administration. They had normal parathyroid hormone peptide levels and normal N-protein activity of erythrocyte membrane, but elevated 25(OH)2D and decreased 1,25(OH)2D levels. Concluding, the authors distinguished a group with normal dissociation of N-terminal and bio-parathyroid hormone, suggesting a defective N-terminal parathyroid hormone activity causing renal resistance, whereas bones responded to parathyroid hormone administration. The next group with a defective-N-protein activity of erythrocyte membrane resulted in generalized parathyroid hormone resistance, and the last group was characterized by high 25(OH)2D and low 1,25(OH)2D levels because of a defect in the cytosolic interaction of the two different second messengers for parathyroid hormone, c-AMP and Ca^{2+}.[33]

Cardiac peptides — These (atriopeptin, atrial natriuretic factor, or peptide) were discovered in 1981. By the use of modern biochemical research. It took only two years to describe the structure of a unique cardiac peptide, which made it possible to perform experiments to elucidate its physiological role. Recently, it has become clear that there are at least two types of natriuretic substances — one, produced by the hypothalamic region of the brain, that is an inhibitor of the ($Na^+ + K^+$) ATPase, the enzyme that transports sodium across membranes and may contribute to a rise in blood pressure; the other, comprised of several peptides that have been isolated from the atria of the heart. The initial observation demonstrated that atrial — but not ventricular — extracts of the heart caused a marked natriuretic and diuretic response. This natriuretic activity was associated with endocrine secretory granules of the atrial cells (the number of these granules changed during water deprivation). Further investigations demonstrated that these peptides relax also the smooth muscles of the vascular tissue, resulting in vasodilatation. It was explained that all the peptides derived from a common precursor — the atrial natriuretic sequence in its C-terminal region. All the atriopeptins require an intact disulfide bridge for biological activity.[34]

Blood levels of atrial peptides are elevated by intravenous volume overloading, salt overloading, vasoconstrictor agents, and other factors, leading to an increase of fluid volume in the bloodstream. Experiments with the injection of natriuretic peptides into the brain suggest that atriopeptin has a central role in regulating body salt and fluid levels and peripheral blood pressure.[35]

Experiments with synthetic peptides led to investigations of receptor-binding and elucidation of a second messenger molecule. It is suggested that probably cGMP could be this molecule since it interacts with atriopeptin in cultured cells, isolated tissue, and intact tissue membranes. In the kidney, atriopeptin acts at different sites and by different mechanisms. Atriopeptin-binding sites have been found in the glomerulus and in the inner medulla-papilla, especially on medullary collecting duct cells; thus, it is totally different from other diuretics. Natriuresis may result from hemodynamic effects that modulate sodium excretion in the distal nephron. There is a suggestion that atriopeptin may mediate passive flow from the interstitium into the lumen producing in this way an extreme natriuretic response. Atriopeptin acts as a modulator of aldosterone secretion, renin secretion and catecholamine release, thereby delimiting sodium uptake. Elevated plasma atriopeptin levels were found in animals with congenital hypertension, salt or water loading, and in human patients with chronic renal failure, a high salt diet, and supraventricular tachycardia.[35]

A. CONGENITAL ADRENAL HYPERPLASIA

Also known as adrenogenital syndrome, the disorder is caused by disturbances in adrenal steroid metabolism, resulting in inadequate secretion of active cortical hormones, primarily cortisol and aldosterone, and the excessive production of several precursors, e.g., deoxycorticosterone or secretion of large amounts of androgens. The type of disorder depends on the site of the disturbed steroid metabolism. The most common is the defect of specific 21-hydroxylase (also termed cytochrome P-450 C21 hydroxylase). The deficient activity of this enzyme occurs in approximately 95% of all cases of congenital adrenal hyperplasia, whereas the remaining cases are due to deficiency of 11-β-hydroxylase. Abnormal quantities of androgens are secreted by the affected adrenals, causing prenatal (congenital or classical form) or postnatal virilisation (late onset or nonclassical form). In approximately 50% of the patients with the classical form there occurs a salt wasting syndrome caused by the defective production of aldosterone.

Congenital adrenal hyperplasia is inherited as a Mendelian autosomal trait with an incidence ranging from 1 in 5,000 to 1 in 15,000 in Europe and the U.S., making its incidence of the same magnitude as phenylketonuria.[36] There are four forms of 21-hydroxylase deficiency: salt losing, simple virilising, late onset and cryptic. There exists a close genetic linkage between the salt losing syndrome, the simple virilizing form and the other forms of congenital adrenal hyperplasia and human leukocyte antigen genes (HLA-A system). These variants represent combinations of "severe" and "mild" alleles for 21-hydroxylase deficiency.[36]

Genes coding for 21 hydroxylase deficiency are located on the short arm of chromosome 6 within the class III major histocompatibility region (MHC) coding for the complement components C2, factor B, and C4. Two 21-hydroxylase genes (21A and B) exist, located at the 3′ ends of the genes for the fourth component of complement (C4).[36] Some characteristic HLA haplotypes are associated with the 21 hydroxylase classical form HLA A3 Bw47 DR7, and with the nonclassical HLA B14 DR1.[37] Recently, using a cDNA clone-encoding human adrenal cytochrome P450 specific for steroid 21-hydroxylation, it was demonstrated that 21-OH deficiency may sometimes result from the deletion of the cytochrome P450 gene.[37] Other authors have described deletions of the C4A gene associated with the haplotypes HLA B8 allele, and deletions of C4B gene associated with the B18 allele. These results show that deletions and other mutational events may frequently occur in the MHC class III region resulting in 21-hydroxylase deficiency.[37]

Using restriction enzymes EcoRI and TaqI, and a cDNA probe specific for 21 hyroxylase genes, two forms of the congenital adrenal hyperplasia have been determined: (1) the late onset form was associated with a double dose of a 14-kb fragment obtained by EcoRI and with a triple dose of a 3.2-kb fragment obtained by TaqI in patients with HLA B14 haplotype; (2) the classical form was negatively associated with the 14-kb fragment and with a 3.7-kb fragment obtained by TaqI in patients with HLA Bw47 haplotypes. The 3.2-kb fragment obtained by TaqI was negatively associated in patients with HLA B8 haplotypes.[37] In patients with the Bw47 haplotype, the molecular defect is most frequently monomorphic, with a selective loss of the 21-OH B gene present either in the salt-wasting syndrome or in the simple virilizing form. In the nonclassical form associated with HLA B14 DRI haplotype, heterogeneity could be observed with duplications extending in various degrees into adjacent C4 genes. Some patients with both types of congenital adrenal hyperplasia revealed normal hybridization patterns, suggesting point mutations of their 21 hydroxylase genes.[38] Similar results have been obtained by other authors also.[39] In other studies, gene conversion in salt-wasting congenital adrenal hyperplasia with absent complement C4B protein was observed.[40]

Clinically, the common form of congenital adrenal hyperplasia is diagnosed in the female at birth and in the male within 2 to 3 years of life. The affected female is exposed to excessive androgens from her own adrenals, which causes masculinization of external genitalia (hypertrophy of the clitoris) with normal internal female duct structures. The labioscrotal folds are bulbous, rugated, and may resemble a scrotum. This condition is termed female pseudohermaphroditism.

The biochemical basis of the adrenogenital syndrome is very heterogeneous, caused by enzyme deficiencies in steroid metabolism. 21-Hydroxylase and 11-hydroxylase deficiencies are characterized by virilization and salt wasting in 21-hydroxylase deficiency (caused by deficit of aldosterons), and hypertension in 11-hydroxylase deficiency (caused by excess of deoxycorticosterone). A mixed form may be caused by 3β-hydroxysteroid dehydrogenase deficiency (usually characterized by the salt-wasting syndrome). The nonvirilizing form may be caused by 17-hydroxylase deficiency characterized by absence of 17-ketosteroids and estrogens in the urine (17-hydroxylation is essential in the formation of C_{19} and C_{18} steroids — androgens and estrogens).

20,22-Desmolase deficiency causes lipid adrenal hyperplasia. The adrenal glands are very large, and contain large amounts of cholesterol and other lipids. 20,22-Demolase is necessary for the cleavage of the side chain of cholesterol. Therefore, practically no steroids are detectable in the urine. 17,20-Demolase deficiency was discovered in a male pseudohermaphrodite patient. The function of the adrenals seemed to be normal. There was a lack of C_{19} steroids, which prevent normal testosterone production, resulting in incomplete male development of the external male genitalia.

B. PSEUDOHYPOALDOSTERONISM

The disorder is characterized by salt-wasting, failure to thrive with high level of sodium in urine despite hyponatremia, hyperkaliemia, hyperreninemia, and increased level of aldosterone in the blood. No aldosterone binding was found in monocytes. The results prove that the cause of the disorder is a receptor defect in binding aldosterone. Investigations of several cases demonstrate that the defect may be complete or partial. The transmission of the defect is probably autosomal and recessive.[41]

REFERENCES

1. **Krieger, D. T.,** Brain peptides: what, where, and why, *Science,* 222, 975, 1983.
2. **Arriza, J. L., Weinberger, C., Cerelli, G., Glaser, T. M., Handelin, B. L., Housman, D. E., and Evans, R. M.,** Cloning of human mineralocorticoid receptor complementary DNA: structural and functional kinship with the glucocorticoid receptor, *Science,* 237, 268, 1987.
3. **Tsai, S. Y., Roop, D. R., Stein, J. P., Means, A. R., and O'Malley, B. W.,** Effect of estrogen on gene expression in the chick oviduct, regulation of the ovomuoid gene, *Biochemistry,* 17, 5773, 1978.
4. **Weinberger, C., Thompson, C. C., Ong, E. S., Lebo, R., Gruol, D. J., and Evans, R. M.,** The c-*erb*-A gene encodes a thyroid hormone receptor, *Nature,* 324, 641, 1986.
5. **Sap, J., Munoz, A., Damm, K., Goldberg, Y., Ghysdael, J., Leutz, A., Beug, H., and Vennström, B.,** The c-erb-A protein is a high-affinity receptor for thyroid hormone, *Nature,* 324, 635, 1986.
6. **Willmann, T. and Beato, M.,** Steroid-free glucocorticoid receptor binds specifically to mouse mammary tumour virus DNA, *Nature,* 324, 688, 1986.
7. **Becker, P. B., Gloss, B., Schmid, W., Stähle, U., and Schütz, G.,** *In vivo* protein-DNA interactions in glucocorticoid response element require the presence of the hormone, *Nature,* 324, 686, 1986.
8. **Godowski, P. J., Rusconi, S., Miesfeld, R., and Yamamoto, K. R.,** Glucocorticoid receptor mutants that are constitutive activators of transcriptional enhancement, *Nature,* 325, 365, 1987.
9. **Mendel, D. B., Bodwell, J. E., and Munck, A.,** Glucocorticoid receptors lacking hormone-binding activity are bound in nuclei of ATP-depleted cells, *Nature,* 324, 478, 1986.

10. **Green, S. and Chambon, P.,** A superfamily of potentially oncogenic hormone receptors, *Nature,* 324, 615, 1986.
11. **Dreyer, M. and Rüdiger, H. W.,** Erbliche Receptordefekte als Krankheitsursache, *Dtsch. Med. Wochenschr.,* 111, 465, 1986.
12. **Ebina, Y., Ellis, L., Jarnagin, K., Edery, M., Graf, L., Clauser, E., Ou, J., Masiarz, F., Kan, Y. W., Goldfine, I. D., Roth, R. A., Rutter, W. J.,** The human insulin receptor cDNA: the structural basis for hormone-activated transmembrane signalling, *Cell,* 40, 747, 1985.
13. **Ullrich, A., Bell, J. R., Chen, E. Y., Herrera, R., Petrucelli, L. M., Dull, T. J., Gray, A., Coussens, L., Liao, Y. C., Tsubokawa, M., Mason, A., Seeburg, P. H., Grunfeld, C., Rosen, O. M., Ramachandran, J.,** Human insulin receptor and its relationship to the tyrosine kinase family of oncogenes, *Nature,* 315, 756, 1985.
14. **Riedel, H., Dull, T., Schlesinger, J., Ullrich, A.,** A chimaeric receptor allows insulin to stimulate tyrosine kinase activity of epidermal growth factor receptor, *Nature,* 324, 68, 1986.
15. **Espinal, J.,** Mechanism of insulin action, *Nature,* 328, 574, 1987.
16. **Ellis, L., Clauser, E., Morgan, D. O., Edery, M., Roth, R. A., Rutter, W. J.,** Replacement of insulin receptor tyrosine residues 1162 and 1163 compromises insulin-stimulated kinase activity and uptake of 2-deoxyglucose, *Cell,* 45, 721, 1986.
17. **Rubenstein, A. H.,** Disorders of insulin structure, Syllabus of 36th Annual Postgraduate Assembly of the Endocrine Society, Dallas, 1984, 10.
18. **Shibasaki, Y., Kawakami, T., Kanazawa, Y., Akanuma, Y., and Takaku, F.,** Post-translational cleavage of proinsulin is blocked by a point mutation in familial hyperinsulinemia, *J. Clin. Invest.,* 76, 378, 1985.
19. **Elsas, L. J., Endo, F., Strumlauf, E., Elders, J., and Priest, J. H.,** Leprechaunism: an inherited defect in a high-affinity insulin receptor, *Am. J. Hum. Genet.,* 37, 73, 1985.
20. **Craig, J. W., Larner, J., Locker, E. F., and Elders, M. J.,** Mechanisms of insulin resistance in cultured fibroblasts from a patient with leprechaunism: resistance to proteolytic activation of glycogen synthase by trypsin, *Mol. Cell. Biochem.,* 66, 117, 1985.
21. **Vesely, D. L., Schedowie, H. K., Kemp, S. F., Frindik, J. P., and Elders, M. J.,** Decreased cyclic guanosine 3′5′ monophosphate and guanylate cyclase activity in leprechaunism: evidence for postreceptor defect, *Pediatr. Res.,* 20, 329, 1986.
22. **Bar, R. S., Levis, W. R., Rechler, M. M., Harrison, L. C., Siebert, C., Podskalny, J., Roth, J., and Mugeo, M.,** Extreme insulin resistance in ataxia teleangiectasia, *N. Engl. J. Med.,* 298, 1164, 1978.
23. **Craighead, J. E.,** Current views on the etiology of insulin-dependent diabetes mellitus, *N. Engl. J. Med.,* 299, 1439, 1978.
24. **Rüdiger, H. W. and Dreyer, M.,** Pathogenic mechanisms of hereditary diabetes mellitus, *Hum. Genet.,* 63, 100, 1983.
25. **Edlund, T., Walker, M. D., Barr, P. J., and Rutter, W. J.,** Cell-specific expression of the rat insulin gene: evidence for role of two distinct 5′ flanking elements, *Science,* 230, 912, 1985.
26. **Prohazka, M., Leiter, E. H., Serreze, D. V., and Coleman, D. L.,** Three recessive loci required for insulin-dependent diabetes in nonobese mice, *Science,* 237, 286, 1987.
27. **Vroede, M. A. De, Rechler, M. M., Nissley, S. P., Joshi, S., Burke, G. T., and Katsoyannis, P. G.,** Hybrid molecules containing the B-domain of insulin-like growth factor I are recognized by carrier proteins of the growth factor, *Proc. Natl. Acad. Sci. U.S.A.,* 82, 3010, 1985.
28. **Blundell, T. L. and Humbel, R. E.,** Hormone families: pancreatic hormones and homologous growth factors, *Science,* 287, 781, 1980.
29. **Brissenden, J. E., Ullrich, A., and Francke, U.,** Human chromosomal mapping of genes for insulin-like growth factors I and II and epidermal growth factor, *Nature,* 310, 781, 1984.
30. **Bell, G. I., Merryweather, J. P., Sanchez-Pescador, R., Stempien, M. M., Priestly, L., Scott, J., and Rall, L. B.,** Sequence of cDNA clone encoding human preproinsulin-like growth factor II, *Nature,* 310, 775, 1984.
31. **Han, V. K. M., D'Ercole, A. J., and Lund, P. K.,** Cellular localization of somatomedin (insulin-like growth factor) messenger RNA in the human fetus, *Science,* 236, 193, 1987.
32. **Spiegel, A. M. and Marx, S. J.,** Parathyroid hormone and vitamin D receptors, *Clin. Endocrinol. Metab.,* 12, 221, 1983.
33. **Radeke, H. H., Aufmkolk, B., Juppner, H., Krohn, H. P., Keck, E., and Hesch, R. D.,** Multiple pre- and postreceptor defects in pseudohypoparathyroidism (a multicenter study with twenty-four patients), *J. Clin. Endocrinol. Metab.,* 62, 393, 1986.
34. **Sagnella, G. A. and MayGregor, G. A.,** Cardiac peptides and the control of sodium excretion, *Nature,* 309, 666, 1984.
35. **Needleman, P.,** The expanding physiological roles of atrial natriuretic factor, *Nature,* 321, 199, 1986.

36. **Rumsby, G., Carroll, M. C., Porter, R. R., Grant, D. B., and Hjelm, M.,** Deletion of the steroid 21-hydroxylase and complement C4 genes in congenital adrenal hyperplasia, *J. Med. Genet.*, 23, 204, 1986.
37. **Mornet, E., Couillin, P., Kutten, F., Rahn, M. C., White, P. C., Cohen, D., Bone, A., Dausset, J.,** Associations between restriction fragment length polymorphisms detected with a probe for human 21-hydroxylase (21-OH) and two clinical forms of 21-OH deficiency, *Hum. Genet.*, 74, 402, 1986.
38. **Boehm, B. O., Rosak, C., Boehm, T. L. J., Kuchni, P., White, P. C., and Schöffling, K.,** Classical and late-onset forms of congenital adrenal hyperplasia caused by 21-OH deficiency reveal different alterations in the C4/21-OH gene region, *Mol. Biol. Med.*, 3, 437, 1986.
39. **Knorr, D., Albert, E. D., Bidlingmaier, F., Holler, W., and Scholz, S.,** Different gene defects in the salt wasting (SW), simple virilizing (SV), and nonclassical (NC) types of congenital adrenal hyperplasia (CAH), *Ann. N.Y. Acad. Sci.*, 458, 71, 1985.
40. **Donohoue, P. A., van Dop, C., McLean, R. H., White, P. C., Jospe, N., and Migeon, C. J.,** Gene conversion in salt-losing congenital adrenal hyperplasia with absent complement C4B protein, *J. Clin. Endocrinol. Metab.*, 62, 995, 1986.
41. **Armanini, D., Kuhnle, U., Strasser, T., Dorr, H., Butenandt, I., Weber, P. C., Stockigt, J. R., Pearce, P., and Funder, J. W.,** Aldosterone-receptor deficiency in pseudohypoaldosteronism, *N. Engl. J. Med.*, 313, 1178, 1985.

Chapter 15

MOLECULAR DISEASES OF THE BLOOD AND BLOOD-FORMING TISSUES

I. INTRODUCTION

The metabolic pathways in erythrocytes are practically limited to glucose metabolism without any possibility of replacing the disturbed metabolic reactions by another metabolic reaction. This is caused by the absence in mature erythrocytes of nuclei, mitochondria, and any other organelles. Therefore, the erythrocyte is deprived of metabolic options, and every abnormal enzyme of the glucose pathway results in accumulation of nonmetabolized precursors or absence of products of the enzymatic reaction, which in erythrocytes are easy to detect. The disturbances of glucose metabolism in erythrocytes deprives them of energetistic material (ATP), which results in the degradation of their structures with hemolysis as the consequence. Thus, hemolytic anemias usually are the consequences of these disturbances. After the detection of glucose-6-phosphate dehydrogenase and pyruvate deficiencies, erythroenzymopathies associated with hereditary hemolytic anemias have been extensively investigated. Soon, it was shown that these disorders are caused by the production of mutant enzymes. With the exception of a few enzymes that are readily available in the blood and tissues, it is difficult to obtain enough material to examine the structural and functional abnormalities of mutant enzymes. Gene cloning makes it possible to identify the DNA sequences encoding these enzyme proteins. Recently, human complementary DNA (cDNA) for aldolase, glucose-6-phosphate dehydrogenase, adenosine deaminase, and phosphoglycerate kinase have been obtained, and nucleotide sequence for some of them have been demonstrated.[1]

II. DEFICIENCIES OF MUTANT ENZYME ACTIVITIES

A. GLUCOSE-6-PHOSPHATE DEHYDROGENASE (G-6-PD) DEFICIENCY

G-6-PD deficiency is the most common disease-producing enzyme deficiency of humans (it is supposed that approximately 100 millions of humans exhibit this disorder). The enzyme activity is important to the hexose monophosphate pathway (pentose shunt) in glucose metabolism, which supplies the reduced nucleotides necessary to preserve the structure of the erythrocytes against oxidative processes. G-6-PD catalyzes the first step in this reaction, i.e., oxidation of G-6-P to 6-phosphogluconolactone by NADP which becomes reduced to NADPH. The 6-phosphogluconolactone hydrolyzes spontaneously to 6-phosphogluconate which is converted (oxidized) by 6-phosphogluconate dehydrogenase and NADP to ribulose-5-phosphate and CO_2 with the reduction of NADP to NADPH. Further steps include isomerization and intermolecular rearrangements to reach fructose-6-phosphate and xylulose-5-phosphate, normal intermediates of the Embden-Meyerhof pathway.

The deficient activity of G-6-PD may be harmless to the structure of the erythrocytes or cause the destruction of their structure manifested in hemolysis. Under the auspices of the World Health Organization, on the basis of several characteristic features (red cell blood activity — percent of normal; electrophoretic mobility — percent of normal; Michaelis constant — Km for G-6-P μM,Km for NADP μM; 2-deoxyglucose-6-phosphate utilization — percent of G-6-P; deaminoNADP utilization — percent of NADP; heat stability and pH optima), five G-6-PD variants have been distinguished: class 1 — severe enzyme deficiency associated with chronic nonspherocytic hemolytic anemia; class 2 — severe enzyme deficiency; class 3 — moderate to mild enzyme deficiency; class 4 — very mild or no enzyme deficiency; class 5 — increased enzyme activity.[2]

G-6-PD activity of class 1 is near "0" with spontaneous nonspherocytic hemolytic anemia. Variants of class 2 demonstrate usually less than 10% of normal activity, and hemolytic anemia usually occurs after administration of injurious factors (e.g., drugs — analgesics: acetanilid; sulfonamides and sulfons: sulfanilamide, sulfapyridine, thioazosulfone; antimalarial drugs: primaquine, quinacrine; antibacterial drugs: furazolidine, nitrofurantoin, nitrofurazone, chloramphenicol; others: naphthalene, trinitrotoluene, phenylhydrazine, quinine, quinidine). Class 3 is characterized by 10 to 60% normal activity, usually without clinical symptoms; only very large doses of injurious factors may cause hemolysis. Variants of class 4 reveal 60 to 100% activity, usually without clinical symptoms.

Very interesting is the observation of Yoshida[59] that the activity of G-6-PD depends also on intracellular interactions. Variants of very low activity of G-6-PD (Union — 3%; Markham — 1.5% of normal) do not cause hemolysis, whereas variants of higher activity do (Alhambra — 9 to 20%; Manchester — 25 to 30%; Tripler — 35% of normal activity). It was clarified that the variants Union and Markham, despite low activity, produce sufficient amount of reduced NADPH and glutathion, whereas the variants of Alhambra, Manchester, and Tripler, while showing greater activity, produce small amounts of NADPH and reduced glutathion.

G-6-PD may be also caused by instability of the enzyme molecules. The most severe cases of this kind of disturbances belong to the class 1 of G-6-PD deficiencies. The list of G-6-PD deficiency variants is certainly not closed, and new variants are still being described, e.g., G-6-PD Iserlohn and G-6-PD Regensburg. G-6-PD variant Iserlohn showed a reduced affinity for the inhibitor NADPH, whereas the G-6-PD variant Regensburg showed normal inhibitor constant. G-6-PD activity in leukocytes was normal in the Iserlohn variant, and decreased in the Regensburg variant, whereas both showed a decreased enzyme activity of approximately 6% of the normal. The Iserlohn variant was thermostable, whereas the Regensburg variant was thermoinstable.[3]

B. 6-PHOSPHOGLUCONOLACTONASE DEFICIENCY

6-Phosphogluconate is often regarded as a product of the glucose-6-phosphate dehydrogenase reaction. However, the δ-lactone of 6 phosphogluconate is the initial reaction product, and 6-phosphogluconate is formed only when the lactone is hydrolyzed. The course of the reaction is as follows:

$$\text{G-6-P} + \text{NADP} \rightarrow \text{6-phosphoglucono-}\delta\text{-lactone} + \text{NADPH} + H^+,$$
$$\text{6-phosphoglucono-}\delta\text{-lactone} + H_2O \rightarrow \text{6-phosphogluconate},$$
$$\text{6-phosphogluconate} + \text{NADP} \rightarrow \text{ribuloso-5-P} + \text{NADPH} + H^+ + CO_2$$

The enzyme 6-phosphogluconolactonase has been known for a long time; only because the substrate of the enzyme is very unstable, the disturbances of this enzyme were not recognized. Using an artificial substrate, γ-lactone, which is easier to prepare and may be used for the detection of 6-phosphogluconolactonase activity, it was possible to recognize a family with a deficiency of this enzyme which may interact with G-6-PD deficiency and cause moderately severe hemolytic anemia. The disorder was inherited as an autosomal dominant. The discovered interaction of the hereditary 6-phosphogluconolactonase deficiency may explain some unexplained molecular bases of glucose-6-phosphate dehydrogenase deficiency variants.[4]

C. DEFICIENCIES OF THE ACTIVITY OF ENZYMES OF THE EMBDEN-MAYERHOF PATHWAY

The second most frequent enzyme deficiency that cause hemolytic anemia is pyruvate kinase (PK) deficiency. Patients with PK deficiency demonstrate usually chronic hemolytic anemia, jaundice, slight splenomegaly, and an increased incidence of gallstones. The early onset of the disorder indicates usually a severe or moderate form of the disease. Survival of patients to adulthood is frequent and may be improved by splenectomy. The number of reticulocytes increases after splenectomy. The osmotic fragility of erythrocytes is normal. Pyruvate kinase catalyzes the conversion of phosphoenolpyruvate the pyruvate with regeneration of ATP. The PK deficiency block of glucose metabolism is at the final step of the Embden-Meyerhof pathway. PK-deficient erythrocytes demonstrate normal activity of the remaining enzymes and have a normal complement of reduced glutathione. The PK-catalyzed reaction causes regeneration of ATP from ADP and provides the pyruvate for conversion into lactate.

D. HEXOKINASE (HK) DEFICIENCY

HK catalyzes the first step of the Embden-Meyerhof pathway of glucose metabolism. Not long ago it was supposed that HK deficiency must be lethal, but recently some cases of this insufficiency have been described. The hemolytic symptoms are variable — from severe to mild. In the erythrocytes there is a diminished level of ATP because of insufficient utilization of glucose. Both glucose and fructose are converted to lactate at diminished rates. The concentrations of glucose-6-phosphate and 2,3-diphosphoglycerate may be reduced to 50% of the normal. The erythrocytes become degraded early, and therefore splenectomy may improve the severe course of the disorder. An autosomal recessive mode of transmission is probable.

E. GLUCOSE PHOSPHATE ISOMERASE (GPI) DEFICIENCY

The disorder is usually mild with a normal level of ATP. GPI catalyzes the interconversion of fructose-6-phosphate (F-6-P) and glucose-6-phosphate (G-6-P). GPI deficiency is transmitted in an autosomal recessive mode. Recently, generalized glucose phosphate isomerase deficiency has been described causing hemolytic anemia, neuromuscular symptoms (myopathy, mental retardation), and impairment of granulocytic function (decreased production of superoxide anion with reduced bactericidal activity).[5]

F. PHOSPHOFRUCTOKINASE (PFK) DEFICIENCY

PFK catalyzes irreversible conversion of fructose-6-phosphate into fructose-1,6-diphosphate using ATP which becomes converted to ADP. Two forms of PFK-deficiency are known: PFK-deficiency of erythrocytes which results in hemolytic anemia, usually mild; and PFK-deficiency with a severe myopathic syndrome, known as glycogenosis type VII (glycogen storage disease type VII). Both are transmitted in an autosomal, recessive way.

G. ALDOLASE DEFICIENCY

Aldolase deficiency has been supposed to be a possible lesion in hereditary spherocytosis. It was also found in several cases of hemolytic anemia (enzyme activities from 38% to 82% of the normal). Aldolase catalyzes interconversion of fructose-1,6-diphosphate, glyceraldehyde-3-phosphate, and dihydroacetonephosphate. Because of the multiple actions of aldolase in glucose metabolism, the pathology of aldolase deficiency is not completely elucidated. The transmission is autosomal, dominant(?).

H. TRIOSE PHOSPHATE ISOMERASE (TPI) DEFICIENCY

TPI deficiency results in severe, generalized disorder with hemolysis (congenital nonspherocytic hemolytic anemia) and progressive neuromuscular symptoms (weakness with generalized spasticity and atrophy of muscles, motor impairment, growth failure, mental retardation), and cardiac failure. Typical is an increased bacterial infectivity which results in early death (usually before 20 years of age). Triose phosphate isomerase catalyzes interconversion of dihydroxyacetone phosphate and glyceraldehyde-3-phosphate. Thus, the deficiency of this enzyme hinders utilization of half of the dihydroxy-acetone phosphate molecule because of a block in conversion of dihydroacetone phosphate into glyceraldehyde-3-phosphate, and next into α-glycerophosphate. The TPI deficiency is characterized by moderately reduced erythrocyte TPI activity and marked instability of the abnormal enzyme to heat. Culture of fetal fibroblasts enables prenatal diagnosis of this disorder.[6] The disorder is inherited in a autosomal recessive manner.

I. 2,3-DIPHOSPHOGLYCERATE MUTASE DEFICIENCY

The enzyme converts 1,3-diphosphoglycerate into 2,3-diphosphoglycerate. The deficiency of this enzyme results in congenital mild to severe nonspherocytic anemia with diminished level of 2,3-diphosphoglycerate, an increased level of fructose-1,6-diphosphate, and a slightly reduced level of ATP. The disorder is inherited in an autosomal recessive manner.

J. PHOSPHOGLYCERATE KINASE (PGK) DEFICIENCY

PGK-deficiency is associated with hemolytic anemia and metal disorders in man. The disease is X chromosome-linked. The enzyme catalyzes interconversion of 1,3-diphosphoglycerate and 3-phosphoglycerate with conversion of ADP into ATP. The deficiency of this enzyme deprives the glucose metabolism of 1 molecule of ATP in erythrocytes, and causes an increase of 2,3-diphosphoglycerate, resulting in decreased affinity of the hemoglobin to oxygen. The complete amino acid sequences of normal PGK was determined. Specific amino acid substitutions of several PGK variants associated with clinical symptoms were determined, and their functional abnormalities correlated with structural abnormalities of the enzyme.[7]

K. ATPase DEFICIENCY

Some ATPase deficiencies in nonspherocytic hemolytic anemias have been described. However, in several cases of ATPase deficiency, hemolytic anemia did not occur. Therefore, the correlation between ATPase deficiency and hemolysis must be determined by further observations. Usually, erythrocytes with ATPase deficiency demonstrate an increased concentration of sodium. An increased osmotic fragility of erythrocytes with ATPase deficiency was not observed.

III. DEFICIENCIES OF OTHER ENZYME ACTIVITIES OF THE HEXOSE MONOPHOSPHATE SHUNT PATHWAY

A. 6-PHOSPHOGLUCONATE DEHYDROGENASE (6-PGD) DEFICIENCY

In the normal course of the hexose monophosphate shunt reaction NADP is reduced to NADPH by the action of G-6-PD and 6-PGD. Next, NADPH must be reoxidized to NADP so that the hexose monophosphate shunt can operate. This reaction is combined in the erythrocytes with reduction of oxidized glutathion (GSSG) catalyzed by glutathion reductase. Thus, severe deficiencies of glutathion or any other enzymes of glutathion reduction and oxidation as well as NADP reduction and oxidation must result in hemolytic events resembling that of G-6-PD deficiency. Any defects of the hexose monophosphate oxidative denaturation of erythrocytes cause hemolytic syndromes. It would be expected that 6-PGD deficiency would cause the same disturbances as does G-6-PD deficiency. However, only a very few cases of G-6-PD deficiency have been described, and those too with only mild hemolysis.

B. GLUTATHION REDUCTASE (GSSG-R) DEFICIENCY

This disorder causes very different clinical symptoms (from mild to moderate hemolytic events). The flavine enzyme glutathion reductase reduces 1 mol of GSSG to 2 mol of GSH with simultaneous reduction of NADPH to NADP. Decreased activity of GSSG-R in erythrocytes was described. Administration of NAD and riboflavin causes an increased activity of glutathion reductase.

C. GLUTATHION PEROXIDASE DEFICIENCY

This enzyme catalyzes the degradation of peroxides (H_2O_2) with a simultaneous oxidation of glutathion to oxidized glutathion. Hemolysis occurs in cases of glutathion peroxidase deficiency mainly after administration of oxidants (sulfanilamides, nitrofurantoin, and others).

D. GLUTATHION SYNTHETASE DEFICIENCY

Hemolytic crisis may occur mainly after ingestion of oxidants. The general amount of glutathion in the blood is reduced, resulting in diminution of the activity of the glutathion peroxidase.

IV. HEMOLYTIC SYNDROMES CAUSED BY DEFICIENT ACTIVITIES OF NUCLEOTIDE METABOLISM

A. ADENYLATE KINASE DEFICIENCY

Adenylate kinase catalyzes interconversion of AMP, ADP, and ATP. Hemolytic syndromes caused by adenylate kinase deficiency are rare and resemble that of pyruvate kinase deficiency.

B. PYRIMIDINE 5′-NUCLEOTIDASE DEFICIENCY

The disorder is characterized by the presence of large quantities of uridine- and cytidine-containing nucleotides of erythrocytes, observed in no other condition, and increased amount of glutathion and severe ribose phosphate pyrophosphokinase synthetase. The erythrocytes reveal basophil stippling. With the maturation of erythrocytes, RNA becomes degraded, and because it cannot be dephosphorylated by deficient pyrimidine 5′ nucleotidase activity, it accumulates. The basophilic stippling is probably caused by accumulation of undegraded or partially degraded ribosomes. The mode of transmission is autosomal, recessive. The mechanism of hemolysis is not fully understood.

V. PAROXYSMAL NOCTURNAL HEMOGLOBINURIA

Paroxysmal nocturnal hemoglobinuria is a rare disorder characterized by spontaneous episodes of intravascular hemolysis, mainly during the night, caused probably by a relative acidosis occurring during sleep. It has been recognized that paroxysmal nocturnal hemoglobinuria is caused by the deficit of decay accelerating factor (DAF). DAF is a protein of M_r 70,000, isolated from the membranes of red blood cells, but also widely distributed on the surface membranes of platelets, neutrophils, monocytes, and B and T lymphocytes. DAF binds with high affinity to C3 convertases of the classical and alternative pathways of complement, accelerating the decay of C3 convertases, regulating in this way their activities. The highest level of DAF has been demonstrated on phagocytes and B lymphocytes, i.e., cells that most frequently interact with immune complexes resulting in activation of the complement cascade.[8,9] DAF is absent on erythrocytes of patients with paroxysmal nocturnal hemoglobinuria (which are DAF^-), and is present on erythrocytes of normal individuals (which are DAF^+). Using the criterion of DAF expression, there was no evidence of separate populations of normal and paroxysmal nocturnal hemoglobinuria type progenitor cells.[10]

VI. BLOOD CLOTTING DISORDERS

Blood clotting disturbances have been discussed in the chapter of molecular aspects of inflammation. In this chapter are discussed only particular abnormal blood clotting factors, the presence of which impairs the normal blood clotting process. Bleeding disorders belong to the most dangerous, life threatening disorders. Therefore, some of the molecular disturbances and their inheritance were recognized relatively early. Such an example is the inheritance of hemophilia A, a recessive disorder linked to chromosome X.

A. HEREDITARY DYSFIBRINOGENEMIA

The disorder is defined as the presence in the plasma of functionally abnormal fibrinogen. The first abnormal fibrinogen was recognized in 1963. Since then, inherited qualitative abnormalities of fibrinogen have been described in more than 100 families. Most of them are clinically silent, but some of them may cause various disorders like bleeding, thrombosis, or defective wound healing.[11,12]

Abnormalities of the fibrinogen molecule may impair one or more major steps involved in the conversion of fibrinogen into stabilized fibrin. These are: (1) cleavage of the fibrinopeptides by thrombin, (2) polymerization of the remaining fibrinogen molecules, (3) cross linking and stabilizing of the fibrin clot.[12]

Biochemical studies revealed that the functional defects of fibrinogen result from amino acid substitutions. The best known dysfibrinogenemias with known molecular defects are:

Fibrinogen Detroit	Arg A α 19 → Ser
Fibrinogen Lille	Asp A 7 → Asn
Fibrinogen Metz	Arg A 16 → Cys
Fibrinogen Munich I	Arg A 19 → Asn
Fibrinogen Petoskey	Arg A 16 → Hist

Each of the pathological fibrinogens is designated by the city of origin.[11]

Similar to atypical globin molecules, various functional disturbances of atypical fibrinogen may occur dependent on the site of amino acid substitution.

Abnormal fibrinopeptide release demonstrates the fibrinogens Baltimore, Bethesda, Charlottesville, Detroit, London III, Metz, Petoskey, Quebec II, Seattle, and White Marsh variants.

Impaired monomer polymerization demonstrates fibrinogens Alba/Geneva, Amsterdam, Bern I, Bern II, Bethesda III, Bondy, Boulogne, Buenos Aires I, Caracas, Chapel Hill I, Clermont-Ferrand, Cleveland I, Harlem I, Houston, Leuven, Logrono, London I, London II, Manila, Marburg, Marseille, Mitaka, Montreal I, Nagoya, Nancy, Oslo II, Paris I, Paris II, Paris III, Philadelphia, Pontoise, Quebec I, St. Louis, San Francisco, Troyes, Wiesbadan, and Zurich II variants.

Abnormal peptide release and abnormal monomer polymerization demonstrate fibrinogens Bethesda II, Chapel Hill II, Cleveland II, Copenhagen, Freiburg, Giessen I, Lille, Manchester, Mexico City, Munich I, Naples, New Orleans, New York, Stony Brook, Sydney, and Zurich I.

Impaired crosslinking demonstrates fibrinogens Oklahoma City and Tokyo variants (the list from Reference 11).

Abnormal fibrinopeptide release may consist from retarded release to total failure of release (fibrinogen Metz and Giessen). Usually, the abnormal amino acid substitution is located in the terminal end of the A-α chain near the thrombin-labile bond. The defect of fibrin polymerization is difficult to explain; it is thought that an abnormally large γ chain (present in fibrinogen Paris I) or abnormally small α chain (present in fibrinogen Chapel

Hill) may explain the defective polymerization. Fibrinogens Bethesda III and Philadelphia are catabolized at accelerated rates. Failure of crosslinking is also difficult to explain; in fibrinogen Paris I, loss of crosslinking may be caused by the mutation of C terminals of the mutant γ chain.[11]

Similar to the functional abnormalities, there also exist variable clinical symptoms: bleeding is caused by fibrinogen Schwarzach (substitution A-α 16:Arg→Cys) in which thrombin does not release fibrinopeptide A.[13] Another case of bleeding disorder demonstrates fibrinogen Kawaguchi, in which there exists defective release of fibrinopeptide A from the isolated abnormal molecule by thrombin. The abnormal fibrinogen forms a solid gel solely by the release of fibrinopeptide B upon incubation with thrombin.[14]

Thrombotic tendency may be caused by fibrinogen New York and fibrinogen Chapel Hill. Fibrinogen New York I is characterized by abnormal, nonclottable fibrinogen by thrombin (50% of the total fibrinogen in N-Y-1 and 35 to 40% in N-Y-1a). The abnormal fibrinogen polymerizes in the presence of Ca^{2+} and can be crosslinked by factor XIIIa. The release rates of fibrinopeptides A and B are slower than those of normal fibrinogen. Electrophoretically, the reduced fibrinogen demonstrates two protein bands in B-β chain region (M_r 54,000 as compared to 57,300 for the normal). There is an abnormal NH_2-terminal disulfide knot with two defective NH_2-terminal B-β chains. The abnormal B-β chain results from a deletion in the sequence from amino acid residues 9 to 72.[15]

Another case of thrombotic tendency demonstrates fibrinogen Chapel Hill. The abnormal fibrinogen is characterized by normal release of fibrinopeptides by thrombin, but defective formation of the fibrin clot. Clotting time with reptilase is significantly prolonged. The pathological fibrinogen forms extremely rigid gels, and the formed fibrin is highly resistant to plasmin. Thus, the pathology of this fibrinogen consists in forming stable clots that cannot be removed by the fibrinolytic system.[16]

B. PROTHROMBIN-THROMBIN DEFICIENCIES

1. Prothrombin Habena

This is represented by a Cuban family with congenital dysprothrombinemia manifested by umbilical bleeding after birth, followed by easy bruising and bleeding tendency. Prothrombin activity was less than 10% in one- and two-stage systems. Electrophoretic migration of the abnormal prothrombin was more anodic, and did not change after addition of Ca^{2+}. Family studies showed that the father had approximately 50% of prothrombin activity and antigen, whereas the mother had 45% of prothrombin activity and 100% of antigen, which suggests that the daughter was heterozygous for an abnormal prothrombin, and heterozygous for true prothrombin deficiency.[17]

2. Thrombin Metz

By physicochemical criteria thrombin Metz is identical to normal thrombin. It exhibits less than 4% of fibrinogen clotting activity with a decreased release of fibrinopeptides A and B, more pronounced for fibrinopeptide B which results in slow fibrin polymerization. Thrombin Metz is less than 5% as effective as normal thrombin in inducing platelet aggregation. Interaction with antithrombin III is slower than normal. It is supposed that the defect in thrombin Metz affects the catalytic site or its vicinity and the region of interaction of thrombin with antithrombin III and heparin.[18]

3. Hemophilia A — Blood Clotting Factor VIIIC

Hemophilia A is a bleeding disorder caused by deficiency or abnormality of the clotting factor VIIIC. The disorder occurs approximately in 10 to 20 males in every 100,000. Individuals with hemophilia A suffer episodes of uncontrolled bleeding. By the application of recombinant DNA methods the structure and genetics of factor VIIIC have been recog-

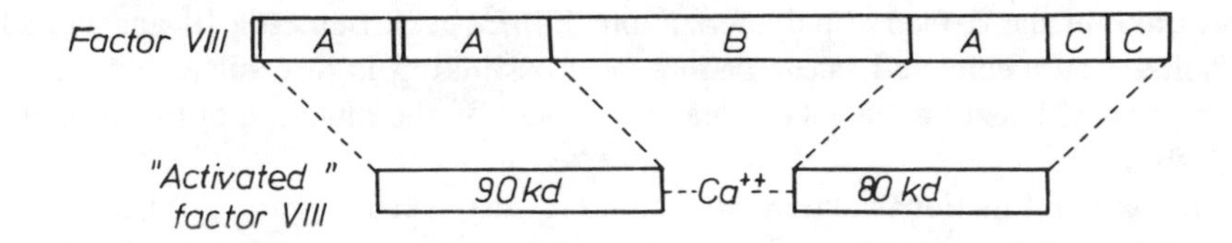

FIGURE 1. Schematic representation of the particular domains of factor VIII and activated factor VIII.

nized.[19-22] The 9-kb gene is divided into 26 exons which span 186 kb (0.1% of the entire X chromosome). The gene includes several large exons (up to 3 kb) and introns (up to 32 kb). The synthesized molecule consists of 19 amino acid residues of the signal hydrophobic peptide and 2332 amino acid residues of the proper molecule. The 330-kDa precursor molecule becomes cleaved by plasma proteases into smaller active subunits. Of particular interest are the 90-kDa N-terminal and 80-kDa (C-terminal) subfragments, which probably comprise the functional fragments of factor VIIIC.[23]

There are two distinct internal repeats within the protein molecule, a region of approximately 350 amino acid residues (A domain), repeated three times at about 30% amino acid homology, and two other domains of approximately 150 amino acid residues (C domains) (see Figure 1). The triplicated A domain reveal about 30% amino acid homology with the copper binding protein, ceruloplasmin. Between the second and third ''A'' domain the factor VIIIC contains an approximately 1000-amino acid intermediate domain ''B'' which becomes eliminated during activation of the molecule with interaction of Ca^{2+}.

Hemophilia A displays a wide range of severity caused by different mutations. In addition, about 15% of patients generate higher levels of antibodies against factor VIIIC (so called ''inhibitor'' patients). Using cloned factor VIIIC gene as a probe, several variants of the factor VIIIC have been recognized. Among others: two different nonsense mutations and two different partial deletions have been diagnosed.[24] Seven different mutations on 200 hemophilic sera (none of the same mutation) were demonstrated: 3 nonsense mutations with truncated factor VIIIC resulting in severe hemophilia, 1 amino acid substitution, and 3 deletions of several thousands of bases resulting in severe hemophilia[25]; in one family a deletion of approximately 80 kb within the factor VIIIC gene has been identified; in another family a single nucleotide change caused nonsense mutation with premature termination of factor VIIIC synthesis; and in a third family there was a small change in the size of the restriction-endonuclease fragment that correlated with the presence of the mutant gene.[26] Probably acquired variants of factor VIIIC antigen have been demonstrated in patients with renal diseases and in those with disseminated intravascular coagulation (DIC syndrome).[27]

4. Hemophilia B — Blood Clotting Factor IX

Factor IX is a serine protease (a glycoprotein of molecular mass of approximately 55 kDa). The single-chain zymogen form becomes activated by proteolytic cleavage by factor XIa, resulting in a two-chain molecule. For biological activity the protein molecule undergoes vitamin K-dependent γ-carboxylation. The synthesized factor IX protein consists of an N-terminal ''pre-pro'' sequence of the first 46 amino acid residues (from the point of peptide synthesis initiation to the processing cleavage site), followed by the ''gla'' region containing 12 γ-carboxyglutamyl residues involved in Ca^{2+} and phospholipid binding. This is followed by two EGF-like regions, and an activation peptide composed of 34 amino acid residues with Asn-linked carbohydrates and two factor XIa activation cleavage sites, and finally a "tryptic" region for the catalytic portion of the factor IX with the active center containing a serine residue.[28] Factor IX binds to the surface of endothelial cells. Phospholipid surface, particularly platelet factor 3, Ca^{2+}, and a modified form of factor VIIIC are cofactors that activate factor X by factor IXa.

The factor IX gene consists of 34 kb, and only 4% of the nucleotide sequence encodes factor IX. There exist several polymorphisms of this factor. The first recognized polymorphism was the presence of an additional Taq I restriction endonuclease cleavage site occurring precisely at the 3′ end of the fourth exon. Approximately 40% of women are heterozygous for this intragenic polymorphism. Two additional frequent restriction site polymorphisms have been described; 65 to 70% of women are heterozygous for one or more of these intragenic sites.[28]

Using cDNA probes to factor IX, mutant forms of factor IX have been described. Four of five patients with hemophilia B and circulating inhibitors (antibodies) demonstrated deletions within their Factor IX genomic structures; two of them were related, but the deletions occurred at different sites.[28] Another case with an inhibitor revealed partial deletion of the gene. Seven affected members of a family with severe hemophilia B — factor $IX_{Seattle}$ — (without inhibitors) demonstrated 10-kb deletion at the 5th and 6th exon, coding for the amino acid sequence 85 to 195 of factor IX. In the urine of these patients a small partial gene product of molecular mass of approximately 36,000 was detected.[29]

Several Factor IX mutants with a single amino acid substitution are known. Hemophilia B is a mild hereditary hemorrhagic disorder with normal level of factor IX, but its physiological function is markedly reduced. Comparison of normal factor IX and peptides from factor $IX_{Chapel\ Hill}$ revealed that the tripeptide Leu-Thr-Arg at positions 143 to 145 that is cleaved during the activation of factor IX by factor XIa, was absent, and sequence analysis revealed that Arg 145 was substituted by His. This substitution prevents cleavage by factor XIa at this peptide bond, and the activation peptide region remains associated with the light chain of factor $IX_{Chapel\ Hill}$.[30]

A point mutation has been described in a donor splice junction of exon f of the human factor IX gene resulting in severely affected patients.[31] Abnormally long factor IX molecules may also be caused by point mutations. For example: factor $IX_{Alabama}$ demonstrates an Asp → Gly point mutation at residue 47 in the epidermal growth factor region.[32] Another abnormally long factor IX with the propeptide region still attached which accumulated in plasma because of a failure in subsequent proteolysis was caused by an Arg → Gln point mutation at amino acid position-4. This abnormal factor IX lacks the signal peptide, and is normally γ-carboxylated.[32]

Defects affecting the latent catalytic site of factor IX have been described in three unrelated severe hemophilia B families: (1) Factor $IX_{Bm\ Lake\ Elsinore}$ in a patient with markedly prolonged ox brain prothrombin time, (2) Factor $IX_{Long\ Beach}$ in three patients (brothers) with moderately prolonged ox brain prothrombin time, and (3) Factor $IX_{Los\ Angeles}$ in a patient with normal ox brain prothrombin time. All three variants are indistinguishable from normal factor IX in their amino acid compositions, isoelectric points, carbohydrate distributions, and number of γ-carboxyglutamic acid residues. All of them undergo a similar pattern of cleavage by factor XIa/Ca^{2+} and by factor VIIa/Ca^{2+}; however, the cleaved factor $IX_{Bm\ Lake\ Elsinore}$ revealed only negligible activity (approximately 0.2%), the cleaved factor $IX_{Long\ Beach}$ and factor $IX_{Los\ Angeles}$ revealed significantly reduced activity (5 to 6% of normal) in activating factor VII (plus Ca^{2+} and phospholipid) ± factor VIIIC, or in binding to antithrombin-III/heparin. These results suggest that the defect in these three variant proteins must be localized near or within the latent catalytic site.[33]

Factor IX Zutphen has an extra disulfide bonded peptide. After removal of this peptide, the remaining molecule is of the normal size of factor IX. The defect is probably associated with the N-terminal portion of factor IX, where it interferes with Ca^{2+}-dependent properties in the "gla" region. Similar disturbances of the calcium-binding properties demonstrates also factor $IX_{Cambridge}$.[28]

Two families with hemophilia B have been described in which the bleeding character improved with increasing age, particularly after puberty.[28]

Congenitally low levels of all four vitamin K-dependent clotting factors have been described, probably caused by deficient activity of K-dependent carboxylase and epoxide reductase systems.[28]

5. v. Willebrand Factor

The hemostatic system must minimize blood loss after vascular injury. The initial and critical event in hemostasis is the adhesion of platelets to endothelium. The factor VIIIC molecular complex is composed of two different protein components, the antihemophilic factor VIIIC and the von Willebrand factor (vWF), which plays a major role in both platelet adhesion and fibrin formation.[34] Factor VIIIC, after activation by trace quantities of thrombin, accelerates factor X activation by factor IX, or may even form a fibrin clot. The vWF molecule is an adhesive glycoprotein, synthesized by endothelial cells and megakaryocytes. It serves as a carrier in plasma for factor VIIIC and facilitates platelet-vessel wall interactions. Some of the vWF domains bind to platelet receptor sites on glycoprotein Ib and on the glycoprotein IIb-IIIa complex, as well as to binding sites of collagen.

Whereas the VIIIC molecule is a single 220-kDa protein, vWF activity is expressed in a heterogeneous series of multimers with molecular sizes ranging from 450 to 20,000 kDa. These multimers are assembled from a single glycoprotein subunit of approximately 220 kDa. vWF accounts for 99% of the mass of the plasma factor VIII molecular complex. Blood platelet α granules contain approximately 15% of the circulating vWF. A variety of abnormalities in vWF result in von Willebrand's disease, which is a complex and heterogeneous group of hereditary bleeding disorders. In the most common type, patients have reduced vWF activity, but secrete the full range of multimers into plasma. The most common variants have a selective loss of high molecular weight multimers of vWF due to failure of assembly (type IIa), or rapid clearance due to aberrant platelet binding (type IIb).[34]

Because of the gigantic size of the vWF molecule, it is difficult to detect abnormalities of vWF in von Willebrand's disease. It is supposed that some subtypes of vWF may be the result of genetic defects at loci involved in post-translational processing, modification, or transport. Such defects may be unlinked to the vWF gene. Thus, in spite of probes for the vWF gene, such defects cannot be detected by this method. The two patients examined showed no evidence for a gross alteration at the vWF locus, which does not exclude either a small deletion or insertion, or a single nucleotide substitution in the vWF gene, or, alternatively, a defect at another locus.[34]

Two unrelated patients with "pseudo" von Willebrand's disease ("platelet type") demonstrated thrombocytopenia with prolonged bleeding time, slight cryoprecipitate-induced platelet aggregation, and lack of the high molecular weight factor VIII/vWF multimers in plasma. Isolated washed patient platelets bound more FVIII/vWF at high and low ristocetin concentrations than control platelets. The results obtained indicate that in pseudo-von Willebrand's disease the primary defect lies in the platelets and is related to a quantitative or qualitative anomaly of platelet membrane glycoprotein Ib.[35]

Eleven patients with thrombosis induced by aparaginase-prednisone-vincristine therapy of leukemia demonstrated (in 9 of the 11) a decrease in platelet counts. The level of factor VIIIC procoagulant activity, vWF, and ristocetin cofactor were similar to findings for an identically treated group who remained free of thrombosis. Qualitative examination of vWF by cross immunoelectrophoresis revealed a distinct shift to the right of the immunoprecipitin lines in each of the thrombotic patients (absent in cases without thrombosis). Two to seven months later the altered patterns had reversed to normal. This result suggests that the abnormal vWF was related to the development of thrombosis.[36]

6. Factor X Deficiency

Blood coagulation factor X mRNA encodes a single polypeptide chain containing a "pre-pro" leader sequence. Information about molecular abnormalities/activity deficiencies are very limited. In a study of 33 patients from 28 kindred families exhibiting factor X deficiency, three types of factor X deficiency could be distinguished: (1) hereditary deficiencies, (2) transient acquired deficiencies, and (3) deficiency caused by amyloidosis. Plasmas were analyzed by a one-stage assay for factor X activation by the extrinsic coagulation pathway, intrinsic coagulation pathway, and Russel's viper venom (some plasmas were analyzed by two-stage assays). The most common factor X deficiency was associated with abnormal activation by all three pathways.[37]

The acquired factor X deficiency occurs in primary amyloidosis. In a patient with primary amyloidosis with nephrotic syndrome, factor X was 5% and antithrombin III 45% of normal plasma values. During 11 months of observation no bleeding occurred. However, with progressive renal failure, antithrombin III levels increased to normal with simultaneous severe bleeding complications. This result suggests that antithrombin III deficiency may execute a protective role against bleeding in patients with severe factor X deficiency.[38]

7. Factor XI Deficiency

Factor XI (plasma thromboplastic antecedent — PTA) is a two-chain M_r 160,000 polypeptide that is synthesized in the liver. Together with factor XII, prekallikrein, and high molecular weight kininogen, factor XI becomes activated by contact with negatively charged surfaces, which results in activation of the intrinsic coagulation pathway. However, in contrast to congenital deficiencies of factor XII, prekallikrein and high molecular weight kininogen characterized by a lack of bleeding, factor XI deficiency causes excessive bleeding (usually post-operative or post-traumatic). In some cases of factor XI deficiency no bleeding occurs. The molecular mechanism of this dichotomy is not known. In a study of 78 members of 25 factor XI-deficient kindred families, deficiency of factor XI clotting activity occurred in 48 individuals (22 of whom were post-operative or post-traumatic). The remaining 26 had no clinical bleeding (many despite surgical treatment). The result obtained suggests that bleeding in cases of factor XI deficiency must be caused by an additional, unknown factor.[39]

C. MOLECULAR ASPECTS OF COAGULATION INHIBITORS

Hemorrhage resulting from abnormal coagulation factors represents one of the most dangerous clinical events. Several proteolytic systems participate in response of the blood to injury, which consists of circulating precursors of proteolytic enzymes that are activated sequentially when specific peptide bonds in these proteins are cleaved. To these systems belong the coagulation and fibrinolytic system, plasma kallikrein, and complement cascade. In the major systems controlling the activity of the circulatory proteolytic enzymes are naturally occurring inhibitory proteins, normal components of blood. At least nine plasma proteins have been recognized as protease inhibitors: α_1-antitrypsin, α_1-chymotrypsin, antithrombin III, α_2-macroglobulin, inter-α-trypsin inhibitor, α_2-plasmin inhibitor, $C\overline{1}$ inhibitor, heparin cofactor II, and protein C inhibitor. Each of these proteins is able to inhibit several proteolytic enzymes by forming complexes with the active sites of the target enzymes. The enzyme-inhibitor complexes are rapidly cleared from the blood.[40]

Four inherited deficiencies of these proteins are known: α_1-antitrypsin deficiency resulting in development of pulmonary emphysema (probably because of the lack of antileukocyte elastase activity of the lungs); $C\bar{1}$ inhibitor deficiency associated with episodic activation of the first component of complement resulting in attacks of hereditary angioneurotic edema; α_2-plasmin inhibitor and antithrombin III deficiencies cause defects in the hemostatic mechanism (antithrombin III inhibits serine proteases of coagulation factors XII, XI, X, and IX); α_2-plasmin inhibitor deficiency may cause a severe bleeding disorder because of lack of fibrinolytic inhibition.[40]

Protein C is a vitamin K-dependent plasma protein, and in contrast to other vitamin K-dependent clotting factors, activated protein C is a potent anticoagulant (see Figure 6, Chapter 10). Protein C is synthesized as a single chain polypeptide containing the light and heavy chains connected by a dipeptide of Lys-Arg. The single-chain molecule is converted enzymatically by cleavage of two more internal peptide bonds. The catalytic region near the active site serine in human protein C exhibits a high degree amino acid sequence identity with prothrombin and factors IX and X.[41]

Protein C deficiency has been found in patients with familial thrombotic complications. Newborn infants with congenital homozygous protein C deficiency develop severe thrombosis (purpura fulminans) and usually die during the neonatal period of life. Patients with systemic thrombosis demonstrate significantly decreased levels of protein C concomitant with the severity of the disseminated intravascular coagulation (DIC syndrome).[42] In a large family, two homozygotic members for protein C deficiency demonstrated activity levels of protein C of 5% and 9% of the normal. Thirteen members of this family were heterozygotes with protein C activity of 36% and 66%. The homozygotes manifested recurrent deep-vein thrombosis and pulmonary emboli.[43]

1. Antithrombin III Deficiency

Inherited deficiency of antithrombin III has been described in association with recurrent, often fatal thromboembolic disorder. The incidence of antithrombin III deficiency has been estimated at approximately 1 per 2000 in the general population (40 to 70% of the affected population reveal symptoms of thrombotic disorders at some time in their lives).[44] In two families examined by the recombinant DNA technique, deletion of antithrombin III gene was found in one family, and absence of antithrombin III gene deletion in the other family in which a point mutation or small deletion might be present (not detectable by the method used).

Antithrombin III ''Toyama'' (Arg → Cys) was described in a patient with recurrent thrombophlebitis, with normal progressive antithrombin activity but no heparin cofactor activity. The abnormal antithrombin III molecule revealed substitution of Arg 47 by Cys, which probably is an essential amino acid residue for binding with heparin[45] (heparin accelerates the inhibition of antithrombin III providing the basis for the anticoagulant activity of heparin).

Antithrombin III ''Chicago'' was described in a family with a high incidence of spontaneous thromboembolism over four generations (12 of those have experienced deep venous thromboses and/or pulmonary emboli). Seven members were examined in detail; they have had normal plasma concentrations of immunoreactive antithrombin III (mean 96%), decreased levels of progressive antithrombin activity (mean 50%), and greatly reduced amounts of plasma heparin cofactor activity (mean 42%). The defective antithrombin III exhibited a reduced ability to neutralize thrombin in the presence or absence of heparin (approximately 10 to 20%).[46]

Antithrombin III ''Vicenza'' was described in a family with thrombosis events. Normal affinity for heparin has been demonstrated, and poor inhibition of thrombin. Antithrombin "Vicenza" consists probably of a population of two different molecules; half does not form a complex with thrombin and loses its heparin affinity upon thrombin treatment.[47,48]

Antithrombin III "Budapest" was described in a thrombophilic family with congenital antithrombin III deficiency. The antithrombin III examined demonstrated decreased thrombin inactivation, reduced heparin cofactor activity, and diminished antigen concentration. On the basis of heparin affinity, antithrombin III deficiency was divided into type 1a and type 1b. Type 1a demonstrated normal antigen concentration with no functional activity, and type 1b loss of one or more of the functional properties of the antithrombin III molecule.[49]

2. Mutation of α_1-Antitrypsin into Antithrombin Pittsburg

In 1978 a 10 year old boy was described who had been hospitalized more than 50 times for posttraumatic bleeding, manifested by external bleeding, hematomas, melena, hematuria, and retrobulbar hemorrhage. He died at age 14 because of uncontrollable bleeding. Coagulation studies demonstrated prolonged bleeding, clotting and partial thromboplastin times, prothrombin and thrombin times. The addition of thrombin to platelet-rich plasma failed to produce platelet aggregation. The thrombin-inhibiting activity was associated with a variant form of α_1-antitrypsin,[40] in which the methionine 358 had been replaced by arginine, that converted the α_1-antitrypsin from its normal function as an inhibitor of elastase to that of a thrombin inhibitor. This finding indicates that the reactive center of α_1-antitrypsin is methionine 358 which acts as a bait for elastase, just as the normal center for antithrombin III is arginine which acts as a bait for thrombin. This observation demonstrated that heparin normally acts directly on antithrombin III, revealing its inherent inhibitory activity.[50]

The primary function of α_1-antitrypsin, produced by the liver, is the inhibition of neutrophil elastase, a protease capable of hydrolyzing connective tissue components. Inherited decreased levels of α_1-antitrypsin may cause early onset of emphysema (serum levels of α_1-antitrypsin are 10 to 20% of normal). Oxidants in tobacco smoke inactivate α_1-antitrypsin *(in vitro)*, which may explain the high incidence of emphysema in cigarette smokers. Oxidative inactivation is probably caused by modification of methionine residue 358 at the elastase binding site. In experimentally produced α_1-antitrypsin analogues, substitution of methionine 358 by valine made the enzyme resistant to oxidative inactivation with decreased elastase inhibiting activity; substitution of methionine 358 by arginine (the same mutation as in the α_1-antitrypsin Pittsburg) the enzyme did not inhibit elastase activity, but became an efficient thrombin inhibitor,[51] exhibiting one or more functional properties of the antithrombin III molecule.[49]

3. Fibrinolysis — Plasminogen Variants

Functional activity of plasminogen in full term newborns was approximately 18% of normal (in adults) when streptokinase was used as the plasminogen activator, and 12% of normal when urokinase was used. Proteolysis of newborn plasminogen by urokinase yielded two normal plasmin chains, but incorporation of diisopropylphosphofluoridate into the light chain of newborn plasmin was approximately 23% of that found in the light chain of adults. The results suggest that the abnormality of full-term newborn plasminogen is located in the active site of the molecule.[52]

Plasminogen "Tochigi" represents an abnormal plasminogen which lost its proteolytic activity. The performed examinations demonstrated substitution of Ala 600 → Thr (equivalent to Ala 55 in the chymotrypsin numbering system). Ala 55 is near the active site His 57. It is supposed that Thr at position 55 in plasminogen (plasmin) may perturb His 57 in such a way that proton transfers associated with the normal catalytic process cannot occur in the abnormal plasmin molecule.[53]

4. Plasminogen "Tochigi II" and "Nagoya"

Both cases demonstrated substitution of Ala 600 → Thr, with no other substitutions at the active site and substrate-binding site. The N-terminal heptapeptide sequences of the plasma light chain variants isolated from plasminogen "Tochigi II" and "Nagoya" were identical to the sequence of normal plasmin, which indicates that the same amino acid substitution was present in both "Tochigi I" and "Tochigi II" and "Nagoya" abnormal plasminogens.[54]

5. α_2-Antiplasmin Enschede[58]

This abnormal antiplasmin is associated with a severe bleeding disorder. Human α_2-antiplasmin share a significant degree of homology of amino acid residues with serine protease inhibitors (with a common inhibitory mechanism). The best known serine protease represents α_1-antitrypsin, which at activation becomes cleaved at its reactive site (a substrate-like region). The active site of α_1-antitrypsin is exposed in a strained loop that allows direct binding. The enzyme inhibition (e.g., trypsin) is executed by an irreversible complex formation in a reaction at the reactive site of the inhibitor, forming a tetrahydral intermediate that breaks down with release of an amine from the C-terminal end of the inhibitor. The mechanism which causes the serine protease inhibitor to act as an inhibitor rather than to be the substrate is unknown.

The α_2-antiplasmin Enschede becomes converted from an inhibitor to a substrate with its inactivation (by degradation). The molecular defect in α_2-antiplasmin Enschede consists of an alanine insertion between amino acid residues 353 to 357, near the active site of the enzyme. Normally, α_2-antiplasmin inhibits the action of plasmin (the clot dissolving enzyme). The defect was described in a Dutch family at Enschede. The parents proved to be heterozygotes, and the two children homozygotes for this defect. The parents revealed 50% of the normal activity, whereas the children revealed zero activity.[58]

D. MOLECULAR DEFECTS OF PLATELETS

The role of platelets in hemostasis has been recognized early in this century. The molecular bases for congenital hemorrhagic diseases have been mainly recognized in recent years. It has been documented that platelet function requires the presence of specific receptors on the platelet surface that interact with macromolecules on the blood-vessel wall or with plasma proteins, including those secreted by the platelets. These interactions include the adhesion of platelets to the endothelial tissue exposed at the cut end of the vessel, the recruitment of adjacent platelets to form a cohesive aggregate, and the generation of thrombin on the surface of platelets to form the fibrin network.

1. Defects of Platelet Adhesion to Subendothelium

Bernard-Soulier syndrome is an autosomal recessive disorder that may cause severe and even fatal hemorrhagic disease. The homozygous disease is characterized by moderate thrombocytopenia, large platelets, and a markedly prolonged bleeding time. Platelets from Bernard-Soulier syndrome revealed decreased ristocetin-induced agglutination which cannot be corrected by addition of von Willebrand factor. The platelets aggregate normally with ADP, epinephrine, and collagen. The aggregation response to thrombin is delayed and diminished. Lack of aggregation with vWF indicates absence or altered expression of the platelet-membrane receptor for vWF. The receptor defect was confirmed by experiments demonstrating that purified vWF binds to normal platelets in the presence of ristocetin, but does not bind to platelets of the Bernard-Soulier syndrome. The adhesion defect, both in Bernard-Soulier syndrome and v. Willebrand's disease, is manifested during the initial phase of platelet-subendothelium contact. The role of vWF for this binding is understood for the vWF, but not for the blood vessel component to which vWF attaches; it is collagen, but it may also be other subendothelial microfibrils.[55]

Platelet membrane glycoproteins may play an important role in platelet adhesion to subendothelium. Platelets of patients with Bernard-Soulier syndrome exhibit altered electrophoretic mobility because of a surface membrane sialic acid deficiency. Subsequent studies of different patients with Bernard-Soulier syndrome revealed deletion or severe reduction in concentration of glycoprotein Ib α and Ib β in 5 of 6 patients examined, and reduced the levels of glycoprotein Ib in the platelets of the sixth patient. Further studies suggest that glycoprotein V of molecular weight 82,000, and glycoprotein IX of molecular weight 17,000, are missing or abnormal.[56] Glycoprotein Ib is an integral membrane protein that is prominent on the cell surface, of which only a small portion penetrates the lipid bilayer of the membrane. At the internal membrane surface, glycoprotein Ib attaches to the cytoskeleton by interaction with actin-binding protein. Glycoprotein Ib is the platelet receptor for vWF. It is possible that the three glycoproteins Ib, V, and IX are derivatives of a single protein that fails to insert into the membrane. Another suggestion is the presence of a single genetic defect of an enzyme (a glycosyltransferase?) responsible for postsynthetic modification, and finally an abnormal mRNA encoding these three glycoproteins.[55] In addition, the Bernard-Soulier syndrome seems to be heterogenic, which needs further investigations.

Glycoprotein Ib is an integral platelet membrane protein of M_r 170,000 with 25,000 molecules per platelet, 50% composed of carbohydrates. It is the major contributor to the cell-surface negative charge that promotes the adhesion to the subendothelium, the site of vWF interaction.

2. von Willebrand's Disease

Willebrand's disease is characterized by the most common bleeding symptoms (epistaxis, gingival bleeding) which are similar to symptoms of primary platelet-function defects. Bleeding can be related to decreased platelet adhesion to the subendothelium, mediated by vWF. The vWF is synthesized by endothelial cells and megakaryocytes, and is present in plasma at concentration 10 μg/ml. Present in platelets are α granules subunits of M_r 200,000, circulating in multimers of 4 to 60 subunits as a flexible filament, interacting with fibrinogen and vessel-wall components, and required for initial binding to the subendothelium. Patients with severe v. Willebrand's disease (homozygotes) are deficient in factor VIIIC and vWF activity, with symptoms similar to hemophilia A. Type I von Willebrand's disease is characterized by a severe decrease of vWF in serum, and is best recognized by decreased functional assays such as ristocetin cofactor activity (the ability of plasma to agglutinate normal platelets in the presence of ristocetin) or immunoassays for the vWF antigen. Type II von Willebrand's disease is characterized by selective deficiency of the higher-molecular-weight vWF multimers. In type IIA the molecular defect is probably due to the failure of the multimer assembly of normal subunits. In type IIB the higher-molecular-multimers are depleted from the plasma because of their increased binding to platelets (and possibly also to other cells); the platelet-rich plasma from patients with type IIB demonstrates an abnormally increased reaction to ristocetin. Type IIC is characterized by changes in the size of vWF subunits in sodium dodecyl sulfate-agarose electrophoresis.[55]

Pseudo-von-Willebrand's disease is characterized by variable mild thrombocytopenia and prolonged bleeding time; increased ristocetin-induced agglutination of platelets; variable content of plasma immunoreactive vWF; and a constant deficiency of high-molecular-weight multimers. Hemorrhage is caused by the increased binding of vWF resulting in depletion of high-molecular-weight vWF multimers from plasma and resulting in decreased adhesion of platelets to subendothelium.[55]

Glanzmann's thromboasthenia is an autosomal recessive hemorrhagic disorder caused by an abnormality of the platelet-membrane glycoprotein. The clinical manifestations are similar to that of Bernard-Soulier syndrome. Glanzmann's thrombasthenia was first recognized by the failure of platelets to aggregate on a blood smear or to retract a blood clot.

There is a long bleeding time and absence of platelet aggregation by ADP in patients with normal platelet count and morphology. The abnormality may be due to a defective association of thrombasthenic platelets with contact-promoting proteins: fibrinogen, fibronectin, thrombospondin, and vWF. Fibrinogen is required for initial platelet aggregation, and next, covalently bound fibrin stabilizes the clot by polymerization. Fibrinogen is synthesized by the liver; its plasma concentration is 2.5 mg/ml. Present in platelet α granules of M_r 340,000, fibrinogen interacts with vWF, fibronectin, and thrombospondin. Fibronectin is synthesized by many cell types, including vascular endothelium; its plasma concentration is 300 μg/ml, and its serum concentration 200 μg/ml. Fibronectin is present in platelet α-granules, and is a component of fibrillar extracellular matrix. It is a dimer of M_r 450,000, and interacts with thrombospondin, fibrinogen, and collagen. Thrombospondin is synthesized by vascular endothelial cells and fibroblasts; its plasma concentration is 0.1 μg/ml, and serum concentration 20 μg/ml (from platelet secretion). Thrombospondin is a trimer of M_r 450,000, and is a major platelet α-granule protein and a component of the fibrillar matrix which interacts with fibronectin, fibrinogen, and collagen.[55]

The binding of exogenous fibrinogen to normal, nonactivated platelets is negligible, but activation of platelets by ADP or other stimulators stimulate expression of fibrinogen receptors at 40,000 sites per platelet. In Glanzmann's thrombasthenia, fibrinogen is absent or in decreased amount, and does not bind to the platelet surface. The failure of platelets to bind fibrinogen is caused by abnormal membrane glycoproteins on their surfaces. This abnormality involves deficiencies of glycoprotein IIb and IIIa. These glycoproteins promote platelet aggregation and are sites for fibrinogen action. They may be also involved in platelet-surface association with vWF, fibronectin, and thrombospondin. They function on the platelet surface as a Ca^{2+}-dependent heterodimer complex, with approximately 50,000 sites per platelet (M_r of glycoprotein IIb is 142,000, and of IIIa 99,000). Formation of the dimer glycoprotein IIb/IIIa requires micromolar Ca^{2+}, and results by direct binding of Ca^{2+} with glycoproteins. The nature of transmembrane signalling by the glycoprotein complex is unknown. During platelet activation they undergo structural transformation to become fibrinogen receptors. The initial fibrinogen binding by platelets is reversible; with time both platelet aggregation and fibrinogen binding become irreversible, and platelets secrete contact-promoting substances (fibronectin, thrombospondin, and vWF). There are known variants of Glanzmann's thrombasthenia; thrombasthenia platelets can adhere normally to exposed subendothelium fibers (reaction mediated by vWF and platelet membrane glycoprotein Ib), but the subsequent platelet to platelet interaction to form a hemostatic aggregate is defective. Patients with small amounts of glycoprotein IIb/IIIa molecules (heterozygotes) have no bleeding symptoms, but patients who have small amounts of glycoprotein IIb/IIIa and endogenous fibrinogen, and a limited ability to bind exogenous fibrinogen bleed severely. The factors that control clinical symptoms are not known. There are variants of Glanzmann's disease caused by structurally abnormal glycoproteins IIb/IIIa that prevent the formation or expression of the fibrinogen receptor.[55]

3. Congenital Afibrinogenemia

The disorder is characterized by lack of fibrinogen for platelet aggregation and fibrin formation with normal count and morphology of platelets, prolonged bleeding time, and marked plasma coagulation abnormalities. The major clinical symptoms are the same as in primary hemostatic defects of platelet adhesion and aggregation: easy bruising, epistaxis, gingival bleeding. Severe bleedings are infrequent. This is probably due to normal generation of thrombin which is a strong agonist for platelet function. Afibrinogenic platelets do not attach either to glass surfaces or to adjacent platelets.[55]

4. Defects of Platelet Secretory Granules

Platelets contain three classes of secretory granules: α-granules, dense granules, and lysosomes. The α-granules contain the nonhydrolase-secreted proteins, such as vWF, fibrinogen, fibronectin, thrombospondin, coagulation factor V, high-molecular-weight kininogen, platelet factor 4, β-thromboglobulin, platelet derived growth factor, albumin, and histidine-rich glycoprotein. The dense granules (which are inherently electron-opaque in electron micrographs) are the storage organelles for nonproteinaceous substances: calcium, pyrophosphate, serotonin, and adenine nucleotides. Congenital bleeding disorders have been described in which there is a deficiency of α-granules, dense granules, or both.

5. Gray Platelet Syndrome

Only a few clinical cases of gray platelet syndrome have been described with clinical symptoms of a primary hemostatic abnormality of platelet adhesion and aggregation (epistaxis, bruising, ecchymoses, and petechiae). The case of the disease is absence of α-granules and their secreted substance. A moderate thrombocytopenia, prolonged bleeding time, agranular platelets on routine peripheral smear, and abnormal aggregation of platelets with collagen or thrombin are present clinically. The plasma membrane of platelets is normal. The hemostatic disturbance is caused by lack of endogenous contact-promoting proteins (vWF, fibrinogen, fibronectin, and thrombospondin), and not by a deficiency of membrane receptor sites.[55]

Dense granule deficiency syndromes are caused by the absence of dense granules and their secreted material. Diminished ADP results in decreased reaction, causing adhesion and aggregation of platelets. There are normal platelet counts and morphology, variably prolonged bleeding time, and an abnormal aggregation with ADP or collagen.[55]

Abnormality of platelet arachidonic acid metabolism is caused by enzyme deficiency (cyclooxygenase or thromboxane synthetase). Decreased endoperoxide or thromboxane synthesis results in specific abnormalities of secretion-dependent aggregation. Clinically, a normal platelet count and morphology, prolonged bleeding time and a normal aggregation of platelets with ADP, collagen, and arachidonic acid are present.[55]

6. Defects of Platelet Procoagulant Activity

The syndrome is caused by deficiency of factor Va binding sites on the platelet surface, which results in decreased thrombin generation and decreased thrombin formation. Clinically, there are normal platelet counts, morphology, bleeding time, and aggregation function.[55]

Vessel-wall defects of genetic disorders of the connective tissue (mainly abnormalities of collagen, elastic fibers and fibronectin) may cause an increased vessel fragility with decreased interaction of platelets with collagen and microfibrils. Platelets may be large and collagen-induced aggregation abnormal.[55]

7. Idiopathic Thrombocytopenic Purpura

A case of this disorder was described associated with diffuse splenic histiocytosis and sphingomyelinase activity at the lower limits of the normal value. Its splenic lecithin:sphingomyelin ratio is comparable to that of 11 age-matched control subjects. It is supposed that the disorder may be due to sphingomyelinase deficiency.[57]

REFERENCES

1. **Miwa, S. and Fuji, H.,** Molecular aspects of ertyhroenzymopathies associated with hereditary hemolytic anemia, *Am. J. Hematol.*, 19, 293, 1985.
2. Treatment of Haemoglobinopathies and Allied Disorders, WHO Technical Report Series, No 509, World Health Organization, Geneva, 1972.

3. **Eber, S. W., Gahr, M., and Schroter, W.,** Glucose-6-phosphate dehydrogenase (G6PD) Iserlohn and G6PD Regensburg: two new severe enzyme defects in German families, *Blut,* 51, 109, 1985.
4. **Beutler, E., Kuhl, W., and Gelbart, T.,** 6-phosphogluconolactonase deficiency, a hereditary enzyme deficiency: possible interaction with glucose-6-phosphate dehydrogenase deficiency, *Proc. Natl. Acad. Sci. U.S.A.,* 82, 3876, 1985.
5. **Schroter, W., Eber, S. W., Bardosi, A., Gahr, M., Gabriel, M., and Sitzmann, F. C.,** Generalized glucosephosphate isomerase (GPI) deficiency causing haemolytic anaemia, neuromuscular symptoms and impairment of granulocyte function: a new syndrome due to a new stable GPI variant with diminished specific activity (GPI Homburg), *Eur. J. Pediatr.,* 144, 301, 1985.
6. **Clark, A. C. and Szobolotzky, A. M.,** Triose phosphate isomerase deficiency: prenatal diagnosis, *J. Pediatr.,* 106, 417, 1985.
7. **Yoshida, A. and Tani, K.,** Phosphoglycerate kinase abnormalities: functional, structural and genomic aspects, *Biomed. Biochim. Acta,* 42, 5263, 1983.
8. **Nicholson-Weller, A., March, J. P., Rosenfeld, S. I., and Austen, K. F.,** Affected erythrocytes of patients with paroxysmal nocturnal hemoglobinuria are deficient in the complement regulatory protein, decay accelerating factor, *Proc. Natl. Acad. Sci. U.S.A.,* 80, 5066, 1983.
9. **Kinoshita, T., Medof, M. E., Silber, R., and Nussenzweig, V.,** Distribution of decay-accelerating factor in the peripheral blood of normal individuals and patients with paroxysmal nocturnal hemoglobinuria, *J. Exp. Med.,* 162, 75, 1985.
10. **Moore, J. G., Frank, M. M., Müller-Eberhard, H. J., and Young, N. S.,** Decay-accelerating factor is present on paroxysmal nocturnal hemoglobinuria erythroid progenitors and lost during erythropoesis *in vitro, J. Exp. Med.,* 162, 1182, 1985.
11. **Bithell, T. C.,** Hereditary dysfibrinogenemia — the first 25 years, *Acta Haematol.,* 71, 145, 1984.
12. **Bithell, T. C.,** Hereditary dysfibrinogenemia, *Clin. Chem.,* 31, 509, 1985.
13. **Henschen, A., Kehl, M., and Deutsch, E.,** Novel structure elucidation strategy for genetically abnormal fibrinogens with incomplete fibrinopeptide release as applied to fibrinogen Schwarzach, *Hoppe-Seylers Z. Physiol. Chem.,* 364, 1747, 1983.
14. **Matsuda, M., Saeki, E., Kasamatsu, A., Nakamikawa, C., Manabe, S.,** Fibrinogen Kawaguchi: an abnormal fibrinogen characterized by defective release of fibrinopeptide A, *Thromb. Res.,* 37, 379, 1985.
15. **Liu, C. Y., Koehn, J. A., and Morgan, F. J.,** Characterization of fibrinogen New York 1, a dysfunctional fibrinogen with a deletion of B beta (9-72) corresponding exactly to exon 2 of the gene, *J. Biol. Chem.,* 260, 4390, 1985.
16. **Carrell, N., Gabriel, D. A., Blatt, P. M., Carr, M. E., and McDonagh, J.,** Hereditary dysfibrinogenemia in a patient with thrombotic disease, *Blood,* 62, 439, 1983.
17. **Rubio, R., Alamgro, D., Cruz, A., and Corral, J. F.,** Prothrombin Habana: a new dysfunctional molecule of human prothrombin associated with true prothrombin deficiency, *Br. J. Haematol.,* 54, 553, 1983.
18. **Rabiet, M. J., Jandrot-Perrus, M., Boissel, J. P., Elion, J., and Josso, F.,** Thrombin Metz: characterization of the dysfunctional thrombin derived from a variant of human prothrombin, *Blood,* 63, 927, 1984.
19. **Vehar, G. A., Keyt, B., Eaton, D., Rodriguez, H., O'Brien, D. P., Rotblat, F., Opperman, H., Keck, R., Wood, W. I., Harkins, R. N., Tuddenham, E. G. D., Lawn, R. M., and Capon, D. J.,** Structure of human factor VIII, *Nature,* 312, 337, 1984.
20. **Toole, J. J., Knopf, J. L., Wozney, J. M., Sultzman, L. A., Buecker, J. L., Pittman, D. D., Kaufman, R. J., Brown, E., Shoemaker, C., Orr, E. C., Amphlett, G. W., Foster, W. B., Coe, M. L., Knutson, G. J., Fass, D. N., and Hewick, R. M.,** Molecular cloning of a cDNA encoding human antihaemophilic factor, *Nature,* 312, 342, 1984.
21. **Wood, W. I., Capon, D. J., Simonsen, C. C., Eaton, D. L., Gitschier, J., Keyt, B., Seeburg, P. H., Smith, D. H., Hollingshead, P., Wion, K. L., Delwart, E., Tuddenham, E. G. D., Vehar, G. A., and Lawn, R. M.,** Expression of active human factor VIII from recombinant DNA clones, *Nature,* 312, 330, 1984.
22. **Gitschier, J., Wood, W. I., Goralka, T. M., Wion, K. L., Chen, E. Y., Eaton, D. H., Vehar, G. A., Capon, G. A., and Lawn, R. M.,** Characterization of the human factor VIII gene, *Nature,* 312, 326, 1984.
23. **Lawn, R. M.,** The molecular genetics of hemophilia: blood clotting factors VIII and IX, *Cell,* 42, 405, 1985.
24. **Gitschier, J., Wood, W. I., Tuddenham, E. G. D., Shuman, M. A., Goralka, T. M., Chen, E. Y., and Lawn, R. M.,** Detection and sequence of mutations in the factor VIII gene of haemophiliacs, *Nature,* 315, 427, 1985.
25. **Lawn, R. M. and Vehar, G. A.,** The molecular genetics of hemophilia, *Sci. Am.,* 254(March), 48, 1986.
26. **Antonorakis, S. E., Weber, P. G., Kittur, S. D., Patel, A. S., Kazazian, H., Jr., Mellis, M. A., Counts, R. B., Stamatoyannopoulos, G., Bowie, E. J., Fass, D. N., Pittman, D. D., Wozney, J. M., and Toole, J. J.,** Hemophilia A, detection of molecular defects and of carriers by DNA analysis, *N. Engl. J. Med.,* 313, 842, 1985.

27. **Weinstein, M. J., Chute, L. E., Schmitt, G. W., Hamburеger, R. H., Bauer, K. A., Troll, J. H., Janson, P., and Deykin, D.,** Abnormal factor VIII coagulant antigen in patients with renal dysfunction and in those with disseminated intravascular coagulation, *J. Clin. Invest.*, 76, 1406, 1986.
28. **Thompson, A. R.,** Structure, function, and molecular defects of factor IX, *Blood,* 67, 565, 1986.
29. **Chen, S. H., Yoshitake, S., Chance, P. F., Bray, G. L., Thompson, A. R., Scott, C. R., and Kurachi, K.,** An intragenic deletion of the factor IX gene in a family with hemophilia B, *J. Clin. Invest.*, 76, 2161, 1985.
30. **Noyes, C. M., Griffith, M. J., Roberts, H. R., and Lundblad, R. L.,** Identification of the molecular defect in factor $IX_{Chapel\ Hill}$: substitution of histidine for arginine at position 145, *Proc. Natl. Acad. Sci. U.S.A.*, 80, 4200, 1983.
31. **Rees, D. J., Rizza, C. R., and Brownlee, G. G.,** Haemophilia B caused by a point mutation in a donor splice junction of the human factor IX gene, *Nature,* 316, 643, 1985.
32. **Bentley, A. K., Rees, D. J. G., Rizza, C., and Brownlee, G. G.,** Defective propeptide factor IX caused by mutation of arginine to glutamine at position-4, *Cell,* 45, 343, 1986.
33. **Usharani, P., Warn-Cramer, B. J., Kasper, C. K., and Bajaj, S. P.,** Characterization of three abnormal factor IX variants (Bm Lake Elsinore, Long Beach, and Los Angeles) of hemophilia B. Evidence for defects affecting the latent catalytic site, *J. Clin. Invest.*, 75, 76, 1985.
34. **Ginsburg, D., Handin, R. I., Bonthron, D. T., Donlon, T. A., Bruns, G. A. P., Latt, S. A., and Orkin, S. H.,** Human von Willebrand factor (vWF) isolation of complementary DNA (cDNA) clones and chromosomal localization, *Science,* 228, 1401, 1985.
35. **Bryckaert, M. C., Pietu, G., Tobelem, C., Girma, J. P., Meyer, D., Larrieu, M. J., and Caen, J. P.,** Abnormality of glycoprotein Ib in two cases of ''pseudo''-von Willebrand's disease, *J. Lab. Clin. Med.*, 106, 393, 1985.
36. **Pui, C. H., Chesney, C. M., Weed, J., and Jackson, C. W.,** Altered von Willebrand factor molecule in children with thrombosis following asparaginase-prednisone-vincristine therapy for leukemia, *J. Clin. Oncol.*, 3, 1266, 1985.
37. **Fair, D. S. and Edgington, T. S.,** Heterogeneity of hereditary and acquired factor X deficiencies by combined immunochemical and functional analyses, *Br. J. Haematol.*, 59, 235, 1985.
38. **Quitt, M., Aghai, E., David, M., Kohan, R., Ben-Ari, Y., and Froom, P.,** Acquired factor X and antithrombin III deficiency in a patient with primary amyloidosis and nephrotic syndrome, *Scand. J. Haematol.*, 35, 155, 1985.
39. **Ragni, M. V., Sinha, D., Seaman, F., Lewis, J. H., Spero, J. A., and Walsh, P. N.,** Comparison of bleeding tendency, factor XI coagulant activity, and factor XI antigen in 25 factor XI-deficient kindreds, *Blood,* 65, 719, 1985.
40. **Harpel, P. C.,** Protease inhibitors, a precarious balance, *N. Engl. J. Med.*, 309, 725, 1983.
41. **Foster, D. and Davie, E. W.,** Characterization of a cDNA coding human protein C, *Proc. Natl. Acad. Sci. U.S.A.*, 81, 4766, 1984.
42. **Marlar, R. A.,** Protein C in thromboembolic disease, *Semin. Thromb. Hemost.*, 11, 387, 1985.
43. **Sharon, C., Tirindelli, M. C., Manucci, P. M., Tripodi, A., and Mariani, G.,** Homozygous protein C deficiency with moderately severe clinical symptoms, *Thromb. Res.*, 41, 483, 1986.
44. **Prochownik, E. V., Antonorakis, S., Bauer, K. A., Rosenberg, R. D., Fearon, E. R., Orkin, S. H.,** Molecular heterogeneity of inherity antithrombin III deficiency, *N. Engl. J. Med.*, 308, 1549, 1983.
45. **Koide, T., Odani, S., Takahashi, K., Ono, T., Sakuragawa, N.,** Antithrombin III Toyama: replacement of arginine-47 by cysteine in hereditary abnormal antithrombin III that lacks heparin-binding ability, *Proc. Natl. Acad. Sci. U.S.A.*, 81, 289, 1984.
46. **Bauer, K. A., Ashenhurst, J. B., Chediak, J., and Rosenberg, R. D.,** Antithrombin ''Chicago'': a functionally abnormal molecule with increased heparin affinity causing familial thrombophilia, *Blood,* 62, 1242, 1983.
47. **Barbui, T., Finazzi, G., Rodeghiero, F., and Dini, E.,** Immunoelectrophoretic evidence of a thrombin-induced abnormality in a new variant of hereditary dysfunctional antithrombin III (AT III ''Vicenza''), *Br. J. Haematol.*, 54, 561, 1983.
48. **Finazzi, G., Tran, T. H., Barbui, T., and Duckert, F.,** Purification of antithrombin ''Vicenza'': a molecule with normal heparin affinity and impaired reactivity to thrombin, *Br. J. Haematol.*, 59, 259, 1985.
49. **Sas, G.,** Hereditary antithrombin III deficiency: biochemical aspects, *Haematolo. Hung.*, 17, 81, 1984.
50. **Owen, M. C., Brennan, S. O., Lewis, J. H., and Carrel, R. W.,** Mutation of antitrypsin to antithrombin, $alpha_1$-antitrypsin Pittsburg (358 Met leads to Arg), a fatal bleeding disorder, *N. Engl. J. Med.*, 309, 694, 1983.
51. **Courtney, M., Jallat, S., Tessier, L. H., Benavente, A., Crystal, R. G., and Lecocq, J. P.,** Synthesis in *E. coli* of $alpha_1$-antitrypsin variants of therapeutic potential for emphysema and thrombosis, *Nature,* 313, 149, 1985.

52. **Benavente, A., Estelles, A., Aznar, J., Martinez-Sales, V., Gilabert, J., and Fornas, E.,** Dysfunctional plasminogen in full term newborn-study of active site of plasmin, *Thromb. Haemost.*, 51, 67, 1984.
53. **Miyata, T., Iwanaga, S., Sakata, Y., and Aoki, N.,** Plasminogen Tochigi: inactive plasmin resulting from replacement of alanine-600 by threonine in the active site, *Proc. Natl. Acad. Sci. U.S.A.*, 79, 6132, 1982.
54. **Miyata, T., Iwanaga, S., Sakata, Y., Aoki, N., Takamatsu, J., Kamiya, T.,** Plasminogens Tochigi II and Nagoya: two additional molecular defects with Ala-600 → Thr replacement found in plasmin light chain variants, *J. Biochem. (Tokyo)*, 96, 277, 1984.
55. **George, J. N., Nurden, A. T., and Phillips, D. R.,** Molecular defects in interactions of platelets with the vessel wall, *N. Engl. J. Med.*, 311, 1084, 1984.
56. **Nurden, A. T., Didry, D., and Rosa, J. P.,** Molecular defects of platelets in Bernard-Soulier syndrome, *Blood-Cells*, 9, 333, 1983.
57. **Hom, B. L., Belles, Q., and Oishi, N.,** Splenic histiocytosis in idiopathic thrombocytopenic purpura: a relative sphingomyelinase deficiency, *Hum. Pathol.*, 16, 1175, 1985.
58. **Holmes, W. E., Lijnen, H. R., Nelles, L., Kluft, C., Nieuwenhuis, H. K., Rijken, D. C., and Collen, D.,** α_2-antiplasmin Enschede: alanine insertion and abolition of plasmin inhibitory activity, *Science*, 238, 209, 1987.
59. **Yoshida, A.,** Hemolytic anemia and GPD deficiency, *Science*, 179, 532, 1973.

Chapter 16

MOLECULAR MECHANISMS OF METABOLIC DISORDERS

I. INTRODUCTION

Functional enzymes are required for metabolic reactions, either anabolic or catabolic. Practically, all enzymes consist of proteins which, according to the basic rule of genetics, may undergo mutations resulting almost always in disturbances of their activities and, in consequence, in metabolic disorders. An altered enzyme activity may be caused by hereditary defects such as altered molecular structure, resulting in catalytic disturbance or molecular instability that causes premature degradation of the enzyme molecule. Another cause of altered enzyme activity may be disturbances of their synthesis: defective transcription or translation, and defective transformation of synthesized proenzymes to active enzymes such as cleavage of proenzymes (e.g., excision of fragment C — joining A and B chain of the definite molecule of insulin), binding of cofactors (e.g., binding of pyridoxine to a series of proenzymes like kynureninase, cystathionine synthetase, glutamate dehydrogenase, or binding of Vitamin B_{12} to methylmalonyl-CoA isomerase), and other transformations. Finally, inappropriate substrates may irreversibly bind to enzymes and inhibit their function, or because of structural alteration they cannot be cleaved by the enzymes.

The consequences of enzyme inactivity result usually in metabolic blocks resulting in: (1) accumulation of precursors of the catalyzed reaction, (2) lack of reaction products, and (3) synthesis of inappropriate products. The accumulated precursors or inappropriate products may be toxic and cause specific disorders, or they may be harmless and metabolized on alternative metabolic pathways that do not cause any disorder. It is the same with a lack of reaction products: it may be harmful and cause severe consequences to the organism in a situation in which they are necessary for life processes but cannot be synthesized on alternative metabolic pathways (e.g., vitamins); or their presence in the organism is redundant; or they may be synthesized on other, alternative pathways.

Since 1908, when Garrod in his brilliant lecture and monograph described the first inborn errors of metabolism, the number of known metabolic disorders has continued to grow, and has increased to many hundreds, comprising practically all kinds of metabolism such as disturbances of proteins, lipoproteins and lipids, steroids, purines and pyrimidines, porphyrins and heme metals; and disorders of whole organs, tissues, cells and their organelles, the immune system, connective tissue, muscles and bones, blood and blood forming organs, cell membranes and transmembranal transport, membrane receptors, lysosomes, etc.

It is impossible to describe in this book all of the enormous number of known metabolic disorders (they are exquisitely described in other publications[1]). The aim of this book is to show known molecular mechanisms that cause different disturbances resulting from mutations of the same or various molecules, dependent on the site and kind of the molecular alterations. It points to future opportunities to recognize and deal with unresolved secrets of human pathology, especially morbidity and the course of particular diseases.

II. ATHEROSCLEROSIS

The etiology of atherosclerosis is very complex and heterogeneous. At present there are the hypotheses of cholesterol and lipoprotein disturbances, and the hypothesis of multiple recurrent damages of the intima which cause coagulation disturbances with platelet accumulation and subsequent activation, resulting in release of several active substances, particularly platelet derived growth factor (PDGF) and other growth factors.

A. THE LIPID THEORY OF ATHEROSCLEROSIS

1. Chylomicrons

Lipoprotein and cholesterol metabolism are strictly interdependent. Alimentary lipids are cleaved in the alimentary tract by the lipases of that organ system, particularly by pancreatic lipase with the participation of bile acids as emulsifying agents. In the jejunal mucosa the long fatty acids react with glycerophosphates (released from metabolized glucose), resulting in the formation of triglycerides, and alimentary cholesterol becomes esterified by the action of LCAT (lecithin:cholesterol acyltransferase). Triglycerides, cholesteryl esters, and lipoproteins form chylomicrons, the largest of the plasma lipoproteins that transport dietary fats and sterols from the small intestine into the circulation. The chylomicrons have a diameter of 750 to 12,000 Å. The lipoproteins in chylomicrons consist of apoprotein B-48, Apo E (apolipoprotein E), approximately 66% of C apolipoproteins (Apo-CI, Apo-CII, and apo-CIII), and 12% of A apolipoproteins (Apo-AI and Apo-AII).

In the circulation, the chylomicrons are the lipoproteins most rapidly catabolized by triglyceride lipase, present in the capillary endothelium, into triglycerides (utilized by adipose tissue and muscles) and chylomicron remnants. The chylomicron remnants are entrapped by chylomicron remnant receptors (mediated by Apo E and Apo B-48) on the surface of liver cells. In the liver cells, cholesterol becomes freed, and are used in part for bile acids synthesis, and in part for the synthesis of very low density lipoproteins (VLDL).

2. VLDL (Very Low Density Lipoproteins) or Prebeta-Lipoproteins

Multicellular organisms transport cholesterol in the form of cholesteryl esters with long-chain fatty acids by packaging the esters within hydrophobic cores of plasma lipoproteins.[2] The VLDL vary from 280 to 750 Å. Its composition depends on the particle size. The proportion of triglycerides and C apoproteins is greater in the larger VLDL particles; the proportions of phospholipids, Apo B and total protein are greater in smaller particles. Apo B makes up approximately 25 to 35% of the proteins of VLDL, Apo C 35 to 50%, and Apo E 7 to 12%.

VLDL transports endogenously synthesized triglycerides. The lipid droplet of VLDL is probably formed in the smooth endoplasmic reticulum of liver cells, and Apo B protein synthesis in the rough endoplasmic reticulum and carbohydrates become probably attached covalently to the apoproteins in the Golgi apparatus of the liver cells.[1] Apoproteins C are synthesized in the liver, but VLDL particles obtain their full complement of these proteins only after net transfer from circulating high density lipoproteins (HDL).

VLDLs are cleaved in the circulation into triglycerides (utilized by adipose tissue and muscles) and intermediate density lipoproteins (IDL). The IDL particles contain 85% of lipid by weight, mostly cholesteryl esters and triglycerides (25%). One part of the IDL particles is rapidly taken up by the liver (mediated by LDL receptors), where they undergo further triglyceride hydrolysis, whereas the IDL particles remaining in the blood undergo further triglyceride hydrolysis and are converted into IDL. The IDL particles contain multiple copies of Apo E and, in addition, a single copy of Apo B-100 which causes multiple copies of apo E to bind to the LDL receptors with greater affinity than LDL. After conversion of IDL to LDL, Apo E leaves the particle, and only Apo B-100 remains, resulting in the reduction of its affinity to LDL-receptors.[2]

Pathologic lipoprotein X may be present in IDL in obstructive liver disease, and in patients with familial LCAT deficiency.

3. LDL (Low Density Lipoproteins) or Beta-Lipoproteins

LDL is a product of IDL cleavage by lipoprotein lipase. It is a spherical particle with a mass of 3×10^6 Da and a diameter of 22 nm. Each LDL particle contains approximately 1500 molecules of cholesteryl ester in an oily core that is shielded from the aqueous plasma by a hydrophilic coat composed of approximately 800 phospholipid molecules, 50 unesterified cholesterol molecules, and one molecule of Apo B-100 (molecular mass 400,000). Packaging of cholesteryl esters in lipoproteins solves the problem of membrane crossing (cholesteryl esters are too hydrophobic to pass through membranes), but creates a new problem of delivery. The delivery problem is solved by lipoprotein receptors (the LDL receptor represents a typical case). The LDL particles are bound to LDL receptors which are internalized, and LDL passes through the membrane by receptor-mediated endocytosis. After binding with LDL, the LDL receptors become clustered in coated pits and internalized into the cell. The internalized lipoprotein is delivered to lysosomes where the cholesteryl esters are hydrolyzed. The liberated cholesterol is utilized for the synthesis of plasma membranes, bile acids, and steroid hormones, or stored in the plasma in the form of cytoplasmic cholesteryl ester droplets.[2] Two properties of the LDL receptor, its high affinity for LDL and its ability to cycle multiple times in and out of the cell, allow large quantities of cholesterol to be delivered to the cells, and at the same time keeps the concentration of LDL in blood low enough to avoid the buildup of atherosclerotic plaques.[2]

The intestinal apolipoprotein B-48 (Apo B-48) present in chylomicrons differs essentially from that found in VLDLs secreted by the liver. The hepatic Apo B contains 4563 amino acid residues and is termed apo B-100, whereas the intestinal Apo B has only 2152 amino acid residues and is termed apo B-48 (it should be noted that Apo B-48 contains the N-terminal end of apo B-100; (this end is 48% as large as Apo B-100)). Apo B-48, because it lacks the C-terminal LDL-binding receptor end, does not bind to LDL receptors. In the bloodstream chylomicrons and VLDLs are converted by lipoprotein lipase that removes triglycerides. The binding of chylomicron remnants with liver receptors is mediated by Apo E and apo B-48, whereas VLDLs are converted to LDLs, with Apo B-100 mediating its binding to LDL receptors.[48]

A very interesting problem is in which way are Apo B-48 and Apo B-100 produced? Are there two specific genes, or is there only a single gene that encodes both Apo B-48 and Apo B-100? The problem was solved by examination of a rare recessive disease, a-betalipoproteinemia that lacks both Apo B-48 and Apo B-100, suggesting that both are encoded by the same gene, perhaps by alternative splicing. The position at which Apo B-48 terminates lies in the middle of a huge exon of 7572 bp. However, the reverse transcription of the intestinal mRNA for Apo B-48 demonstrated that at position 6475 there was uracil instead of cytosine, which changed the codon 2153 from CAA (for glutamine) to UAA (for termination of translation). The replacement of $C \rightarrow U$ occurred during or after transcription, which suggests that there must be a highly specific enzyme in the intestine that modifies cytosine into uracil (perhaps deamination of cytosine at position 6, converting cytosine to uracil), resulting in termination of transcription.[48]

The role of this modification seems to be very important to lipoprotein catabolism. The absence of LDL-binding receptors in chylomicrons guides the chylomicron remnants to the liver, where the cholesterol constituent of these particles can be excreted (as bile acids), or bound with Apo B-100 in VLDLs which become converted into IDL and LDL and cleared by LDL receptors (as the main cholesterol source of the cellular demand).[48]

4. HDL (High Density Lipoproteins) Alpha-Lipoproteins

HDL contains more protein (45 to 55% by weight) and less lipid than the other plasma lipoproteins (phospholipids 42 to 51%, cholesterol 30 to 40%, and triglycerides 6 to 12%). HDL are, on the average, 95 to 100 Å in diameter with a molecular weight of approximately 375,000. The Apo-AI and Apo-II account for almost 90% of the total HDL protein. The remaining are Apo-C (approximately 5%) and other proteins (Apo-D and Apo-E). The central core of HDL is composed mainly of neutral lipids.

Apolipoproteins generally known by their shortened name apoproteins, comprise the lipid-free protein components of the plasma lipoproteins.

Apoprotein AI is the major protein of HDL. Its amino acid residues sequence has been determined; it is a single polypeptide chain (without cystine, cysteine, leucine, and glycoside linkages). Its molecular weight is approximately 28,331. Apo-AI activates LCAT (lecithin:cholesterol acyltransferase). Its ability to interact with and bind phospholipids is enhanced by the presence of Apo-AII.[1] 99% of Apo-AI is bound with HDL.

Apo-AII accounts for approximately 30% of Apo HDL (99% of Apo-AII is bound with HDL). Its amino acid residues sequence is known; it contains two identical chains, each of 77 amino acid residues (a single disulfide bond joins the two chains).

Apo B-100 is a 400 kDa glycoprotein which represents the sole protein bound with LDL. Apo B-100 is also present in small amounts in VLDL. Apo B-48 is present in chylomicrons. Apo B plays the main role in the transport of triglycerides from the intestine and the liver to the plasma.

Apo CI is a single polypeptide of 57 residues. The complete amino acid residues sequence has been determined. Apo CI is present in VLDL and HDL. It activates LCAT.

Apo CII is present in VLDL and HDL. Its molecular weight is approximately 8337 kDa. Apo CII is a potent activator of lipoprotein lipase of adipose tissue, but not of the hepatic triglyceride hydrolase.

Apo CIII is present in HDL and LDL. Its molecular weight is approximately 9000 kDa. It is a single polypeptide of 79 known amino acid residues. Galactose and galactosamine are attached in glycoside linkage at position 74; one or two sialic acid residues are at the end of the carbohydrate chain. Apo CIII is probably an inhibitor of lipoprotein lipase.

Apo D is present in HDL (approximately 65%) and VLDL (approximately 35%). It is a potent activator of LCAT.

Apo E is present in HDL (37%) and VLDL (27%). Its molecular weight is approximately 34 kDa. An increased level of Apo E in VLDL was described in type III of hyperlipoproteinemia and in hypothyroidism. It is an activator of LCAT.

The structural determination of apolipoprotein (a), Lp(a), discovered in 1963, turned out to be a very important phenomenon for the elucidation of pathogenesis of atherosclerosis. Lp(a) is a variant of LDL that carries one or two copies of apo(a), linked with apo B-100 by a disulfide bond. The amount of Lp(a) in plasma varies from undetectable amounts to 100 mg/dl, which means that up to 20% of LDL may bear Lp(a). High levels of Lp(a) are strongly correlated with atherosclerosis. The level of 30 mg/dl (present in 20% of the population) doubles the relative risk of coronary atherosclerosis, and elevated levels of both LDL and Lp(a) causes a five-fold increase in risk.[48]

Protein sequencing and cDNA cloning demonstrated that Apo(a) is a deformed relative of plasminogen, the precursor of the proteolytic enzyme plasmin that, when activated by tissue plasminogen activator (TPA), degrades fibrin clots. Plasminogen is proteolytically inactive until it is cleaved by TPA. Plasminogen is a protein of 791 amino acid residues that contains 5 cysteine-rich sequences (80 to 114 amino acid residues), followed by a serine protease domain. The cysteine-rich sequences have three internal disulfide bridges which cause their structure to resemble a Danish cake called kringle; therefore, these sequences are termed "kringles". Such cysteine-rich sequences (kringles) are present in several proteases of the coagulation system, including TPA and thrombin. In plasminogen they promote binding to the substrate, fibrin.

Apo(a), after the signal sequence, has 37 copies of kringle, 4 of plasminogen, followed by kringle five and the protease domain. The 36th copy of kringle four contains an extra unpaired cysteine, the probable site of disulfide linkage with Apo B-100. The protease domain in apo(a) is inactive because the arginine residue at the cleavage site for TPA is replaced by serine.[48,49]

The finding that Apo(a) is homologous to plasminogen may represent the long-sought link between lipoproteins and the clotting system. This protein, with 37 kringle copies, present in damaged intima, may adhere fibrinogen and fibrin. Furthermore, Lp(a) may inhibit proteolysis of fibrin because of a blocking activation of plasminogen by TPA. Finally, Lp(a) may compete with plasminogen for fibrin, inhibiting cleavage of fibrin clots, which may explain the clinical observation of a "fast acting" TPA inhibitor in patients with recurrent myocardial infraction.[48]

There exist some controversial opinions about the effect of Lp(a) on the thrombotic system, denying this relationship. However, because Apo(a) is encoded by a single gene, it is possible that there may be several alternative forms of Apo(a) that, by homologous recombination, add or eliminate kringles, resulting in quantitative differences in Apo(a) binding with fibrin among individuals. This is supported by the detection of a null mutation of Apo(a) caused by the absence of the unpaired 36th cysteine, which prevents the attachment to Apo B-100. It is also possible that plasminogen binds to Apo B-100, contributing to the atherogenicity of Apo B-100 itself.[48]

Classical disorders of lipoproteins are hyperlipoproteinemias. The increase in a serum lipoprotein fraction causes an increase in the individual components. For example, a rise in the chylomicron fraction causes an exogenous dietary triglyceridemia and a rise in endogenous triglyceridemia because of increase in VLDL. An increase of LDL fraction causes hypercholesterolemia. The distinction into five types of hyperlipoproteinemias is generally accepted (according to Fredrickson):

- **Type 1** — familial lipoprotein lipase deficiency (deficiency of extrahepatic triglyceride lipase). Chylomicrons are detectable in the serum after a 12-h fasting period. An increased content of exogenous triglycerides causes milky clouding of the serum. Cholesterol values are normal with the exception of a severe form of lipoprotein lipase deficiency (a very rare form).
- **Type 2a** — familial hypercholesterolemia caused by deficiency of LDL receptors. The serum remains clear. Premature atherosclerosis is typical.
- **Type 2b** — increased VLDL and LDL contents, with consequently increased values for cholesterol and endogenous triglycerides. The serum may be cloudy. Premature atherosclerosis is typical.
- **Type 3** — familial hyperlipoproteinemia is characterized by atypical lipoprotein in the beta-band of electrophoresis. There is increased content of cholesterol and triglycerides, which cause premature atherosclerosis, obesity, and hyperglycemia.
- **Type 4** — familial hypertriglyceridemia. There is an increased VLDL content, resulting in an increased triglyceride level (usually without increase in cholesterol level). In serious cases an increase of cholesterol level occurs, which may cause premature atherosclerosis. Usually, there is obesity and hyperglycemia.
- **Type 5** — familial hyperlipoproteinemia caused by increased amount of chylomicrons and VLDL, resulting in a substantial increase in endogenous and exogenous triglycerides. Usually, there is obesity and hyperglycemia. The occurrence of premature atherosclerosis is doubtful.

III. RECEPTOR-MEDIATED ENDOCYTOSIS OF LDL AND ITS FURTHER METABOLISM IN CELLS

The receptor-bound LDL becomes clustered in coated pits the cytoplasmic coat of which is composed predominantly of a single protein — clathrin. Surface proteins, being excluded from coated pits, cannot rapidly enter the cell. By the same mechanism other peptides are internalized, e.g., epidermal growth factor, insulin, asialoglycoproteins, and even lipid enveloped viruses. It was shown that receptors for several different ligands co-localize in the same coated pits.

After internalization, the coated endocytic vesicles fuse to create larger sacs of irregular contour, called endosomes or receptosomes. The pH in the endosomes falls below 6.5 because of the ATP-driven proton pumps in the membrane. At this pH, the LDL dissociates from its receptor, and the receptors find their way back to the cell surface (see Figure 1, Chapter 7). The LDL receptor makes one round trip in and out of the cell every 10 min (more than 100 trips in its 20-h lifespan).[2]

In the lysosomes, the lipoproteins become degraded and the liberated cholesterol esterified by the enzyme acyl-CoA:cholesterol acyltransferase (ACAT), resulting in cholesteryl oleate which represents deposit-cholesterol and which can be utilized by the cell after its liberation by the acid cholesterol esterase. In addition, in mammalian cells, cholesterol may be synthesized by the endogenous pathway of cholesterol synthesis from acetyl-CoA → aceto-acetyl-CoA → 3-hydroxy-3-methylglutaryl-CoA (HMG-CoA) → mevalonate → squalene → cholesterol. Crucial for the endogenous synthesis of cholesterol is the reaction of 3-hydroxy-3-methylglutaryl-CoA conversion into mevalonate catalyzed by 3-hydroxy-3-methylglutaryl-CoA reductase (HMG-CoA reductase) which becomes suppressed by free cholesterol and represents an important mechanism regulating endogenous cholesterol synthesis by negative feedback regulation of transcription;[3] this explains the reduction of enzyme molecules after cholesterol feeding or increased endogenous synthesis. The metabolic pathways of lipoproteins and cholesterol are presented in Figure 1.

The LDL receptor is a glycoprotein with two asparagine-linked (*N*-linked) oligosaccharide chains, and approximately 18 serine/threonine-linked (*O*-linked) oligosaccharide chains. The LDL receptor binds two proteins, Apo B-100 and Apo E, that are present in multiple copies in IDL and a subclass of HDL. Lipoproteins that contain Apo E bind to LDL receptors with a 20-fold higher affinity than LDL which contains only one copy of Apo B-100.

A. THE STRUCTURE AND FUNCTION OF THE LDL RECEPTOR

The LDL receptor is synthesized in the rough cytoplasmic reticulum as a precursor of molecular weight 120,000 that contains high mannose *N*-linked carbohydrate chains. The *O*-linked core sugars are added either in the endoplasmic reticulum or in a transitional zone between the endoplasmic reticulum and the Golgi apparatus. Within the next 30 min the molecular weight increases from 120,000 to 160,000, which is coincident with the conversion of the high mannose *N*-linked oligosaccharide chains to the complex endoglycosidase H-resistant form. Simultaneously, each O-linked chain is elongated by the addition of one galactose and one or two sialic acid residues. The added carbohydrates are not sufficient for the increase of molecular weight of 40,000 Da; thus it is supposed that the decrease in electrophoretic mobility is caused by a change in conformation of the LDL receptor molecule. Approximately 45 min after their synthesis, the LDL receptors appear on the cell surface (in coated pits). After binding with LDL, the coated pits invaginate to form coated endocytic vesicles. After dissociation of clathrin, the endocytic vesicles form endosomes. The decrease in pH in the endosomes below 6.5 causes dissociation of the LDL from the receptors which return to the cell surface, whereas the LDL becomes degraded in the lysosomes (after fusing of the lysosome membranes with the membranes of the endosomes).[2]

The LDL receptor is a multi-domain protein. At the extreme N terminal there is a signal hydrophobic sequence of 21 hydrophobic amino acid residues, which becomes cleaved immediately after the LDL receptor is synthesized. The mature LDL receptor (without the signal sequence) consists of 839 amino acid residues. The ligand-binding domain of the N-terminal 292 amino acid residues is composed of a sequence of 40 amino acid residues that is repeated seven times (with some variation). Each of the 40 amino acid repeats contains six disulfide bonded cysteine residues which form a cluster of negatively charged amino acids near the COOH-terminus of each repeat. The negative charges are complementary to

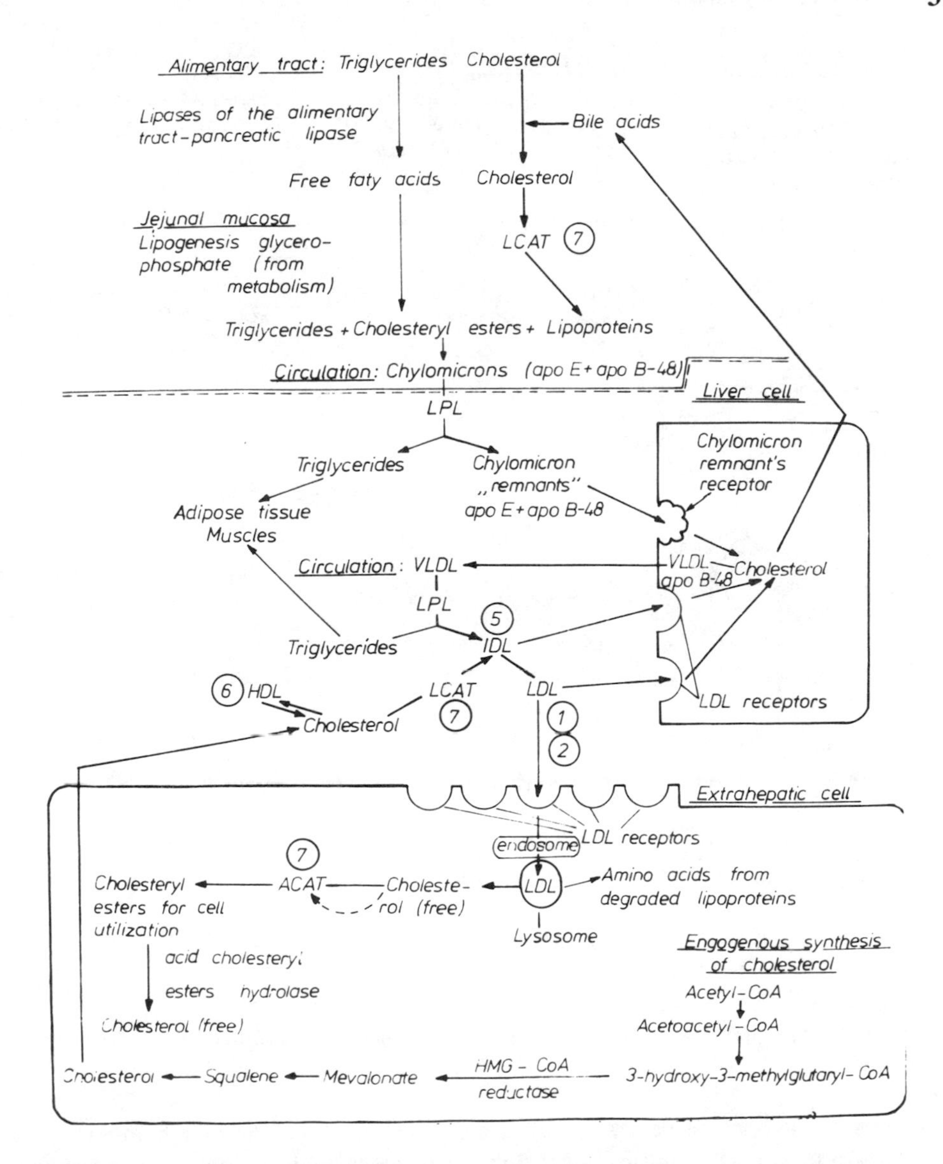

FIGURE 1. Metabolic pathways of cholesterol. (1) Familial hypercholesterolemia — LDL-receptor negative; (2) familial hypercholesterolemia — LDL-receptor positive; (3) familial hypercholesterolemia — endocytosis missing; (4a) Wolman's disease — acid cholesteryl hydrolase deficiency; (4b) cholesteryl ester storage disease — acid cholesteryl hydrolase deficiency; (5a) defective LDL-receptor synthesis (120 kDa precursor only); (5b) defective LDL-receptor synthesis (170 kDa + 40 kd → 210 kDa); (6) defective apolipoprotein E; (7) defective apolipoprotein A (apo (a)); (8) A-betalipoproteinemia — lacking apo B, VLDL, LDL, and chylomicrons; (9) familial hypo-betalipoproteinemia — low level of LDL and cholesterol; (10) Tangier disease — defective synthesis of HDL, deposits of cholesteryl esters in tissues; (11) familial lipoprotein lipase deficiency — hyperchylomicronemia, hepatosplenomegalia, xanthomatosis. Premature atherosclerosis occurs in: 1, 2, 3, 4b (benign), 5a, 5b, 6 and 7. Gene frequency for familial hypercholesterolemia: heterozygotes 1: 1,000,000; homozygotes 1: 500.

a cluster of positively charged residues of a single α-helix in Apo E, the ligand with the highest affinity to LDL receptors.[2]

The second domain of the LDL receptor, consisting of approximately 400 amino acid residues, is 35% homologous to the extracellular domain of the precursor of EGF (epidermal growth factor). The EGF precursor molecule has 1217 amino acid residues that span the

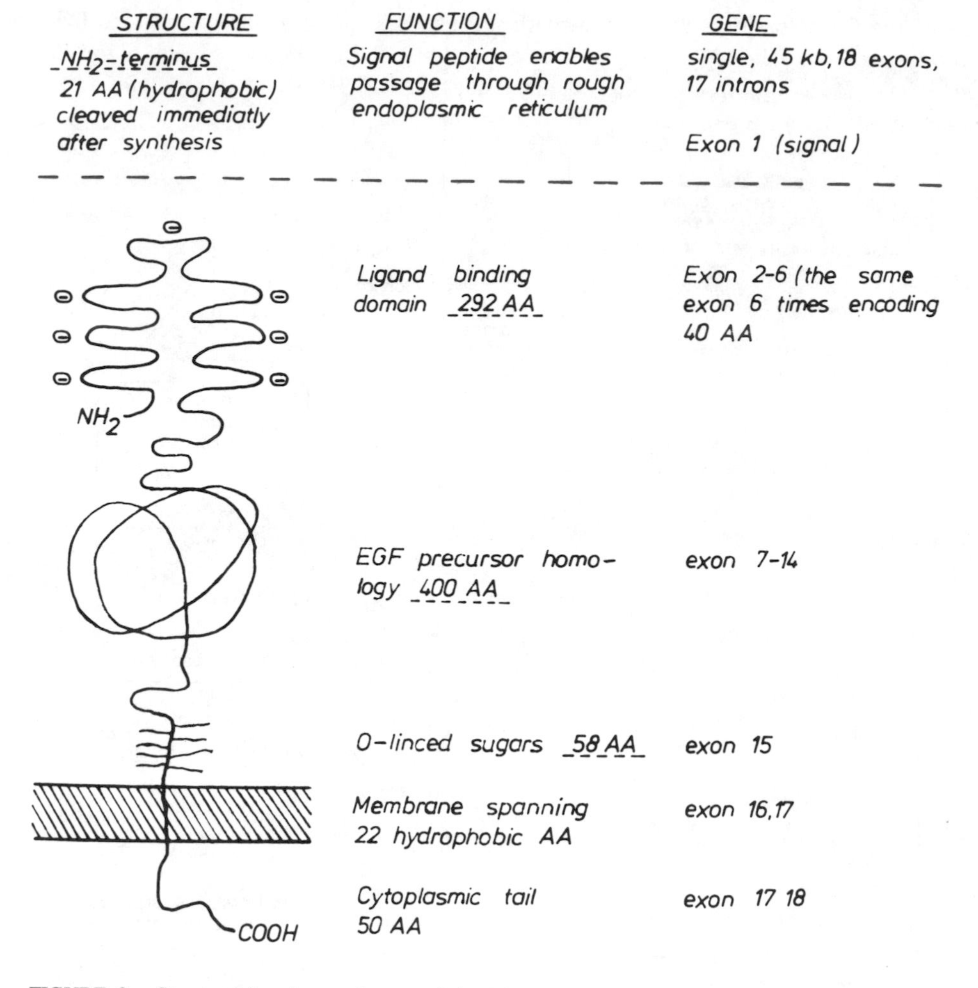

FIGURE 2. Structure function, and gene of the LDL receptor. (Redrawn from Brown, M. S. and Goldstein, J. L.[2])

plasma membrane once. Recently, it has been demonstrated that the EGF homology region of the LDL receptor is responsible for the acid-dependent ligand dissociation and recycling of this receptor.[7] The third domain consists of a stretch of 58 amino acid residues that contains 18 serine or threonine residues. This region contains the clustered *O*-linked sugar chains.[2]

The fourth domain of the LDL receptor consists of a stretch of 22 hydrophobic amino acid residues that span the plasma membrane.

The fifth domain of the LDL receptor is the COOH-terminal cytoplasmic tail. It consists of 50 amino acid residues that project into the cytoplasm. The cytoplasmic tail plays an important role in clustering in coated pits, either through interacting with clathrin or with a protein on the cytoplasmic side of the membrane.[2] The structure of the LDL receptor is shown in Figure 2.

The LDL receptor gene is a single mosaic gene on chromosome 19. It spans approximately 45 kb and consists of 18 exons, separated by 17 introns. The first intron encodes the cleaved signal sequence, and exons 2 to 6 encode the binding domain (the introns occur precisely at the ends of the repeats I, II, V, VI, and VII; repeats III, IV, and V are encoded by a single exon). Thus the binding domain is composed of a single duplicated exon. The repeat sequence is strongly homologous to a stretch of 40 amino acid residues in the middle part of the C9 component of the complement. The next eight exons encode the EGF homologous region of the LDL receptor. These exons reveal homology to several blood clotting factors (IX, X, and protein C). The *O*-linked sugar domain is encoded by exon 15, and the membrane-spanning region is encoded by parts of exons 17 and 18.[2,5,6]

The synthesis of the LDL receptor is controlled by the delivery of lipoprotein cholesterol; free cholesterol inside the cell suppresses its synthesis in the endoplasmic reticulum. Thus a surplus of free cholesterol inhibits its uptake by a lack of LDL receptors, and a lack of free cholesterol induces LDL-receptor synthesis which results in the uptake of LDL through LDL receptors and, in consequence, liberation of cholesterol for cell use.[8]

LDL receptors have been demonstrated on fibroblasts, smooth muscle cells, liver, adrenal cortex, testes, adipocytes, lymphocytes, and macrophage-monocytes. The LDL receptors bind LDL,HDL with Apo E, VLDL, chylomicron remnants, and lipoproteins. The involved ligands are Apo B-100 and Apo E. The LDL receptors play a role in the regulation of LDL blood levels and in the redistribution of cholesterol by Apo B-100 and Apo E proteins to various tissues, and deliver cholesterol for membranes and steroid hormone production.[8]

Other important lipoprotein receptors have been described: Apo E, and chylomicron remnant receptor in the liver which binds chylomicron remnants in the liver cell and HDL with Apo E (the ligand for this receptor is Apo E). The Apo E receptor enables the uptake of chylomicron remnants and cholesterol-loaded HDL with Apo E, and delivery of cholesterol to the liver for excretion (as bile acids).[4] Another receptor is the β-VLDL receptor on macrophages, which binds β-VLDL from patients with type 3 hyperlipoproteinemia, and is induced by cholesterol. These receptors may play a role in foam cell production in atherogenesis and uptake of diet-induced cholesterol (link between diet and atherosclerosis).[8]

B. MOLECULAR DEFECTS IN THE LDL RECEPTOR

The LDL receptor binds plasma LDL, a circulating cholesterol carrier, and facilitates its uptake by cells through receptor-mediated endocytosis. The defects of this receptor are the main cause of familial hypercholesterolemia (FH), which is one of the main causes of premature atherosclerosis. One in every 500 persons is heterozygous for a mutation of the LDL-receptor locus, and approximately 1 in 1,000,000 persons inherits two mutant genes as a result of marriage between two FH heterozygotes. The FH homozygotes are born with extremely high blood cholesterol levels and develop heart attacks in their early twenties, and FH heterozygotes in their forties.[9]

Studying 110 FH homozygotic patients, the authors found ten different LDL receptor mutations, separated into four classes[2] (Figure 2).

- **Class 1** — No receptors have been synthesized. This was the most common class of mutant alleles (approximately 50% of examined patients). In one case with the internalization-defective form, the LDL receptor was smaller by 10,000 Da than the size of the normal LDL receptor. The gene for this FH 274 LDL receptor demonstrated a 5 kb deletion, which eliminated the exons encoding the membrane-spanning region and the COOH-terminal cytoplasmic region of the receptor. The deletion was caused by a novel intrastrand recombination between two repetitive sequences of the Alu family that were oriented in opposite directions. In this case, only a small fraction of the synthesized receptors (in fibroblast culture) remained adhered to the cell surface; therefore they were unable to cluster in coated pits, which explains the internalization defect.[10]
- **Class 2** — The synthesized receptor is slowly transported from the endoplasmic reticulum to the Golgi apparatus. Alleles of this class synthesize inappropriate precursor molecules of molecular weight from 100,000 to 135,000 (the normal precursor has 120,000). These kinds of receptors are defective in binding sugars: the *N*-linked sugars

are not converted to the complex endoglycosidase H-resistant form, nor are the *O*-linked sugar chains elongated. These mutant receptors do not appear on the cell surface; rather, they remain in the endoplasmic reticulum where they are probably degraded. Some receptors of this class do not process sugars; only approximately one tenth of them are processed and migrate to the cell surface.[2] In this kind of mutation in hamster cells the 4-epimerase deficiency was responsible for all glycosylation defects, and could be corrected by exogenous galactose and UDP-*N*-acetylgalactosamine 4 epimerase. This result suggests that O-linked carbohydrate chains are necessary for receptor stability.[11]

- **Class 3** — Receptors are processed and reach the cell surface, but fail to bind LDL normally. In the natural form, these mutant receptors can have a normal apparent molecular weight of 160,000 or aberrant apparent molecular weights of 140,000 or 210,000. The M_r 210,000 mutant is composed of abnormal M_r 170,000 molecules (normal M_r 120,000) that after combining with the M_r 40,000 subunit, has M_r 210,000. In the case of smaller molecular weight molecules, the alterations refer also to the precursor molecule that is smaller than M_r 120,000. All mutants of this group undergo normal carbohydrate processing and reach the cell surface. The inability of binding LDL is probably caused by amino acid substitutions, deletions or duplications in the cysteine-rich LDL receptor domain, or the EGF precursor region.[2]
- **Class 4** — Receptors reach the cell surface, but fail to cluster in coated pits and do not internalize. Some of them have been elucidated in molecular detail. All involve alterations in the cytoplasmatic tail of the receptor. In one case a tryptophan codon at the third position distance from the membrane-spanning region has been converted to a nonsense codon, which results in the cytoplasmic domain being limited to two amino acid residues (instead of 50 amino acid residues).[12] The next mutation involves a duplication of four nucleotides following the codon for the sixth amino acid residue of the cytoplasmic tail, which alters the reading frame and leads to a sequence of eight random amino acid residues, followed by a stop codon.[12] The third mutation (J.D. mutation, after the patient's name) is caused by a substitution of Tyr 807 → Cys. This mutant binds LDL normally, but does not cluster in coated pits, and does not internalize, confirming that this point mutation is responsible for the internalization defect.[2,13] Deletions in the LDL receptor have been described also by other authors, e.g., a 2-kb deletion in the 3′ part of the gene.[14]

In 9 unrelated families with familial hypercholesterolemia, the gene analysis revealed two single base-substitutions, two insertions (one small and one large), and 5 large deletions. Many of the deletions occurred in Alu repetitive elements.[2]

A brilliant confirmation of the role of the LDL receptors in producing high cholesterol levels resulting in spontaneous development of coronary artery disease was given by a serendipitous finding of Watanabe (Kobe University) who found a rabbit (so-called Watanabe heritable hyperlipidemic rabbit) with a ten-fold increased blood cholesterol level. Further investigations demonstrated a mutant LDL receptor (deletion in the cysteine-rich region). The defect arises from an in-frame deletion of 12 nucleotides that eliminates 4 amino acid residues from the cysteine-rich ligand-binding domain of the LDL receptor.[15] A similar class 2 of the LDL receptor defect was detected in a patient with familial hypercholesterolemia, whose receptor also failed to be transported to the cell surface.

The discovery of the Watanabe rabbit contributed to the finding that familial hypercholesterolemic patients have a dual defect. In addition to the degrading LDL more slowly, homozygotes and heterozygotes of this disease overproduce LDL. The increase of LDL results from studies of cholesterol metabolism. LDL is not secreted directly from the liver, but is produced in the circulation from a blood-borne precursor, VLDL (see Figure 1). VLDL is a large, triglyceride-rich lipoprotein secreted by the liver. The triglycerides are removed in muscles and adipose tissue by lipoprotein lipase, and VLDL is converted to IDL. IDL is rapidly taken up by the liver LDL receptors (IDL particles contain multiple Apo E copies that determine their high affinity to LDL receptors). The remaining IDL particles in the blood undergo further triglyceride hydrolysis, being converted to LDL particles. When IDL is converted to LDL, Apo E leaves the particle, and only Apo B-100 remains. Thereafter, the affinity of apo E for the LDL receptor becomes reduced. Thus, when VLDL was administered to Watanabe rabbits, the resultant VLDL was not taken up by the liver as in normal rabbits. Rather, it remained in the circulation and was converted in increased amounts to LDL. These findings suggest that IDL is normally cleared from plasma by binding to the LDL receptors in the liver, where cholesterol can be excreted as bile acids.[2]

Normal macrophages express very few LDL receptors of the classic type present on fibroblasts and liver cells. Therefore, macrophages ingest native LDL sparingly. However, LDL modified by acetylation is not bound to the classic LDL receptor, but is recognized with high affinity by the LDL receptors on macrophages. The receptor-mediated uptake of acetyl-LDL causes massive deposition of cholesterol in the macrophages. The receptor that mediates this uptake has been named the acetyl-LDL receptor, or the scavenger cell receptor. The receptor recognizes LDL and other proteins treated with agents that alter lysine residues (e.g., succinic anhydride, maleic anhydride, and malondialdehyde) and other negatively charged compounds (e.g., sulfated glycosaminoglycans). This receptor has been partially purified; its M_r is 260,000, and is different from the classical LDL receptor of M_r 160,000. The scavenger cell receptor seems to be present only on macrophages and endothelial cells.[16] Endothelial cells have not been considered as scavenger cells, but the discovery of acetyl-LDL receptors resurrects the possibility that these cells may be scavenger cells that ingest LDL converted by chemical oxidants into a negatively charged form that binds to the acetyl-LDL receptor. When LDL accumulates in extravascular space, it might undergo oxidation with an increasing negative charge that permits binding with acetyl-LDL receptors of macrophages. This mechanism can explain the occurrence of cholesterol-loaded foam cells in atherosclerotic plaques.[16-18]

Activated platelets release a substance of M_r approximately 100,000 that blocks the uptake and degradation of acetyl-LDL by macrophages, probably as a result of competitive binding. This substance might be one of the negatively charged proteoglycans that are stored in platelets and released by thrombin, which might limit the ability of macrophages to clear lipoproteins from extracellular deposits. However, this substance might also serve as a chemotactic attractant for blood monocytes to sites of tissue injury and to activate macrophages with acetyl-LDL receptor.[16]

Another conformation of the role of cholesterol transport in the etiology of atherosclerosis originates from therapeutic results showing lowering of the plasma LDL-cholesterol levels. This can be achieved by stimulating the normal gene (in heterozygotes) to express more LDL receptors than its defective allele. This became evident from experiments that show that the production of LDL receptors can be driven by the demand of the cell for cholesterol. A surplus of free cholesterol in cells inhibits the synthesis of these receptors, whereas reduced demand for cholesterol causes accumulation of excess cholesterol in cells (the amount of the LDL receptor mRNA falls). Inasmuch as the liver is the major organ for LDL receptor expression, the increased demands for hepatic cholesterol can be achieved by inhibition of intestinal reabsorption of bile acids, and inhibition of cholesterol synthesis. The bulk of bile acids secreted by the liver is reabsorbed in the terminal ileum and reutilized in the liver. Thus the liver converts only a small amount of cholesterol into bile acids. Ingestion of resins (cholestyramine) or its physiologically equivalent procedure of ileal bypass surgery causes an increase of LDL receptors of only 15 and 20%, and a drop in plasma LDL-cholesterol levels of 15 to 20%.[2]

The second method of lowering the plasma LDL-cholesterol by increasing LDL receptor amounts is based on inhibition of endogenous cholesterol synthesis. The discovery of a potent HMG-CoA reductase inhibitor (a class of fungal metabolites called compactin, and a version developed in the USA called mevinolin) enabled more potent therapy. The administration of mevinolin results in a dual compensatory response: hepatocytes synthesize increased amounts of HMG-CoA reductase and an increased number of LDL receptors, which causes the increased amount of HMG-CoA reductase to be almost sufficient to overcome the inhibitory effect of compactin. This results in only a reduced level of cholesterol synthesis, whereas the plasma LDL level decreases because of increased LDL receptor synthesis. The fall in plasma LDL levels is compensated by the increase of LDL receptors, which causes the absolute amount of cholesterol entering the liver being the same as it was earlier. A single mevinolin treatment causes a 30% decrease in LDL-cholesterol plasma levels, and treatment together with cholestyramine causes a decrease of plasma LDL-cholesterol levels by 50 to 60%. The important point here is that this therapy lowers the plasma LDL-cholesterol level without altering cholesterol delivery.[2]

The above therapy cannot be applied to homozygotes, especially those that have totally defective LDL receptor genes, because they cannot respond with increased LDL-receptor synthesis. In these cases, removal of LDL from plasma can be achieved through extracorporeally repeated plasmapheresis, and more recently by liver transplantation. In a familial hypercholesterolemic, homozygotic, six-year-old girl after heart-liver transplantation, the plasma cholesterol level fell from 1100 mg/dl to 200—300 mg/dl, and it remained in this range for the next 13 months. The heart was transplanted because of existing severe atherosclerotic lesions. After treatment with mevinolin, her plasma cholesterol level fell to the range of 150 to 200 mg/dl, proving that the liver transplantation restored responsiveness to mevinolin which requires a normal LDL-receptor gene in order to act. Lipoprotein examination revealed that the synthesis of new LDL receptors by the transplanted liver was responsible for the dramatic drop in plasma cholesterol level.[2]

Epidemiologic studies strongly indicate a general association of high blood-cholesterol levels with heart attacks. When LDL-cholesterol levels are below 100 mg/dl (equivalent to a total plasma cholesterol level of approximately 170 mg/dl) heart attacks are rare; when LDL cholesterol levels are above 200 mg/dl (equivalent to a total plasma cholesterol level of approximately 280 mg/dl) heart attacks are frequent, especially when in addition to high plasma cholesterol the status is aggravated by smoking, hypertension, stress, diabetes mellitus, and an unknown genetic predisposition.

The normal range of LDL-cholesterol level in humans is 50 to 80 mg/dl, and reaches levels above 100 mg/dl in individuals who consume a diet rich in saturated animal fats and cholesterol.[2,19] The high level of plasma LDL is dependent on diet and heredity. The level does not rise equally in every person. In considering the genetic variability between individuals, the following possibilities must be taken into account:

1. The degree of plasma cholesterol with diet is variable (not all individuals develop hypercholesterolemia).
2. Even when the plasma cholesterol becomes elevated, the propensity for atherosclerosis varies (e.g., approximately 20% of heterozygotes of familial hypercholesterolemia escape myocardial infarction despite hypercholesterolemia from birth).
3. Genetic susceptibility to contributory risk factors is variable (some individuals are highly sensitive, whereas others, in spite of hypertension and cigarette smoking, live until the eighth or ninth decade of life).[2]

IV. LIPOPROTEIN DISTURBANCES AS CAUSES OF ATHEROSCLEROSIS

Apolipoproteins are important components of VLDL, IDL, LDL, and HDL. As such, they play an important role in normal cholesterol transport and metabolism, and may be a cause of disturbances of cholesterol metabolism leading to atherosclerosis. In addition to solubilizing lipid for transport in the blood, the plasma apolipoprotein disturbances may affect specific physiologic functions.

Apolipoprotein B (Apo B), the primary protein constituent of the cholesterol-carrying LDL, mediates cholesterol delivery to cells by interaction with specific LDL receptors. Apo B-48 is essential for the secretion of triglyceride-rich lipoproteins in the intestine, whereas Apo B-100 is necessary for secretion from the liver. Apo B secreted in the intestine differs from that secreted by the liver. The intestinal Apo B-48 has a lower molecular weight. A variable fraction of Apo B-100, secreted by the liver in VLDL, is converted to lipoproteins of higher density in the LDL which are removed more slowly from the blood. The LDL fraction is reduced in individuals with elevated VLDL. Apo B-100 is the sole protein component responsible for the receptor-mediated uptake and clearance of LDL from circulation. Apo B-100 is produced by the liver and is essential for the assembly of triglyceride-rich VLDL in the cisternae of the endoplasmic reticulum and for their secretion to the plasma. VLDL transports triglycerides to the muscles and adipose tissue, where the triglycerides are hydrolyzed by lipoprotein lipase. The resultant particle, enriched in cholesteryl ester, constitutes LDL. Plasma LDL levels are therefore determined by the balance between their rate of production from VLDL and clearance by the hepatic LDL-receptor pathway. A domain in Apo B-100, enriched in basic amino acid residues, has been identified as the binding site for the cellular uptake of cholesterol and suppression of HMG-CoA reductase by the LDL-receptor pathway (sequences 3345 to 3381).[20,21]

There is a considerable heterogeneity between LDL subfractions within individuals and characteristic differences in LDL composition amongst normal subjects and patients with hyperapobetalipoproteinemia or familial hypercholesterolemia.[22] Despite an absence of LDLs and chylomicron remnants from plasma, the rates of cholesterol synthesis and the number of LDL receptors expressed on freshly isolated cells from patients with abetalipoproteinemia are not markedly increased, suggesting that lipoproteins present in HDL (which are relatively rich in Apo E) are effective regulators of LDL-receptor activity in normal human fibroblasts.[23]

The association between mutations in plasma lipoproteins and atherosclerosis has been studied in a mutant strain of pigs. These pigs demonstrated hypercholesterolemia associated with mutations affecting structure of plasma lipoproteins. They developed atherosclerosis by 7 months of age. The affected pigs had normal LDL-receptor activity. Complex atherosclerotic lesions were formed within 2 years, and the mutant pigs died before reaching 4 years of life (normal pigs live three times longer). The main lipoprotein variant Lpb5 is Apo B (termed also Apo R). It is a 23-kDa protein present in the VLDL and HDL fractions of pig plasma.[24]

A. LIPOPROTEIN LIPASE DEFICIENCY AND REMOVAL OF APO B FROM PLASMA

Lipoprotein lipase deficiency is a condition associated with a severely impaired catabolism of plasma triglycerides. These lipoproteins are not subject to the action of lipoprotein lipase which normally degrades them to remnant particles. The metabolism of these remnants differs from those of their undegraded precursors. In normal subjects most of the Apo B-48 was removed from the blood within 15 min, and most of the Apo B-100 within 30 min. In lipoprotein lipase deficient subjects, most of the Apo B-100 remained in the blood more than 8 h; the removal of Apo B-48 was slightly more rapid. These results indicate that the removal of Apo B of both chylomicrons and large VLDL from the blood is dependent upon the hydrolysis of their components by lipoprotein lipase, and little or no Apo B-48 of chylomicrons and Apo B-100 of large VLDL is converted to LDL. The variability of conversion of VLDL to LDL may be dependent on the size and composition of the particles secreted by the liver.[25]

The lipoapoprotein E (Apo E) plays a central role in lipoprotein metabolism. It is crucial for the recognition of lipoproteins by specific lipoprotein receptors, and for cholesterol homeostasis and clinical abnormalities resulting from the occurrence of specific mutants.[26] The human hepatic receptors for LDL- and Apo E-containing lipoproteins are physiologically and genetically distinct.[27] The gene for human Apo E has been determined. It consists of 3597 nucleotides with 4 exons and 3 introns. The mature protein is composed of 299 amino acid residues.[28] By isoelectric focusing it demonstrates two major bands, E3 and E1. Normally, the Apo E1 position is occupied by sialylated derivatives of Apo E4, Apo E3, and Apo E2. A described variant occupied position E1 without neuraminidase digestion. On the basis of partial sequence analysis the variant is an example of a previously uncharacterized Apo E phenotype, E3/1. The variant contained two cysteine residues and it differs from E2 at residue 158 Arg → Cys and at position 127 Gly → Asp. The E1 variant demonstrated reduced ability for binding to LDL. It is supposed that the defective binding might be caused by cysteine at position 158. The presence of this variant was manifested by hypertriglyceridemia.[29]

Type III hyperlipoproteinemia is an inherited disturbance in plasma lipoprotein metabolism, characterized by an increase of plasma cholesterol and triglycerides, and the presence of abnormal lipoproteins. The principle abnormal lipoproteins are called β-VLDL, that differ from normal VLDL in their lipoprotein content, β electrophoretic mobility, and higher content of cholesteryl ester relative to triglycerides. β-VLDL includes two different subclasses of lipoproteins, consisting of large particles containing Apo E, and a low molecular weight form of Apo B-48. The first fraction, Iβ-VLDL appear to be of intestinal origin, the accumulation of which demonstrates impaired catabolism of chylomicron remnants; the second fraction consists of smaller, cholesterol-rich particles which contain Apo E and Apo B-100. This fraction appears to represent remnants of cholesterol-rich VLDL of hepatic origin.[30]

The primary defect of type II hyperlipoproteinemia is the presence of an abnormal form of Apo E, called Apo E2 in the β-VLDL. Apo E occurs in plasma in three major forms: Apo E3, the most common form; and Apo E2 and Apo E4 genetic variants that differ by a single amino acid substitution. Apo E3 and Apo E4 have normal binding activity, whereas Apo E2 displays defective receptor binding.[30] Three major molecular defects may predispose patients to develop type III hyperlipoproteinemia: deficiency in Apo E, a structural defect in Apo E3, and a functional defect in the liver receptor system.[31] Most patients with type III hyperlipoproteinemia have a structural defect in Apo E, associated with increased catabolism of Apo E, delayed catabolism of chylomicron remnants, and development of plasma lipoprotein abnormalities, with extensive coronary and peripheral vascular atherosclerosis.[31]

An example of abnormal Apo E is apolipoprotein E3 Leiden. In contrast with normal Apo E3, Apo E3 Leiden is defective in binding to LDL receptor, and does not contain cysteine at position 158 Arg.[32] In a study of 361 patients with hyperlipemia, the most frequent was Apo E2 isoform, not because of a higher frequency of Apo E2/2 homozygotes with type III hyperlipoproteinemia, but because of a higher frequency of heterozygotes of Apo E2. Subgrouping of hyperlipidemics into hypertriglyceridemia, hypercholesterolemia, and mixed form, the isoform Apo E2 was significantly more frequent in hypertriglyceridemia.[33] A significant difference in frequencies of isoforms Apo E3 and Apo E2 was revealed by 523 patients with myocardial infarction, compared with a control group of 1031 blood donors. All Apo E2 homozygote survivors of myocardial infarction had hyperlipoproteinemia type III. Thus, homozygotes with hyperlipoproteinemia type III have an increased risk for coronary atherosclerosis.[34]

B. THE ROLE OF APO A DEFICIENCY IN ATHEROSCLEROSIS

Epidemiological studies demonstrate that HDL deficiency is implicated in the pathogenesis of atherosclerosis, whereas increased HDL levels have been correlated with protection against coronary heart disease. It has been proved that HDL promotes cholesterol efflux from cells, and may act as a carrier of reverse cholesterol transport. This function of HDL depends mainly on Apo AI, the major protein content of HDL. Apo AI is synthesized in the liver and small intestine, and serves as a cofactor for lecithin:cholesterol acyltransferase (LCAT). The primary translation product is a preproprotein that undergoes intra- and extracellular proteolytic processing. Deficiency of Apo AI results in low levels of HDL in plasma, and may contribute to the development of premature atherosclerosis.[35] A DNA insertion of 6.5 kb in the coding region of the Apo AI gene, resulting in premature atherosclerosis, has been described.[36]

1. Tangier Disease

Tangier disease is a rare, autosomal recessive disorder characterized by the absence of HDL in plasma and an increase of cholesteryl esters stored in several cells (macrophages, Schwann cells, and intestinal smooth muscle cells). Homozygotes of this disease demonstrate hypocholesterolemia, moderate hypertriglyceridemia, and no detectable HDL on agarose electrophoresis. Tangier HDLs lack Apo AI, and contain Apo AII as their sole protein constituent. It is supposed that the decreased Apo AI level in Tangier disease may be caused by enhanced catabolism of this protein.

Patients with isoform Apo AI (pro-Apo-AI) having high density lipoproteins have been described. In contrast to normal serum in which isoprotein 4 is dominant, the Tangier serum contained isoproteins 2 and 4 in roughly equivalent amounts. The Tangier isoprotein 2 was shown to correspond to pro-Apo-AI, having 6 amino acid residues at the N terminal (Arg-His-Phe-Trp-Gln-Gln-).[37] The Tangier isoprotein 4 had normal N-terminal amino acid residues with normal HDL association, whereas the Tangier isoprotein 2 was only very little associated with HDL (less than 10%). It is thought that these cases with Tangier Apo AI have a defect in normal conversion of pro-Apo-AI to mature apo AI — either a defect in converting enzyme activity, or another structural defect. The failure of Tangier pro-Apo-AI to associate with HDL may be responsible for the HDL deficiency in Tangier patients.[38]

C. FAMILIAL APOLIPOPROTEIN AI AND APOLIPOPROTEIN CIII DEFICIENCY ASSOCIATED WITH PREMATURE ATHEROSCLEROSIS

In the plasma of two patients, LDL amounts were 35% greater than that of controls. IDL and VLDL were extremely low. Polyacrylamide gel electrophoresis demonstrated that the patients' HDL had two major HDL subclasses: HDL2b and HDL3b. The major peak in controls (HDL3a) was absent. Patients HDL contained a cholesteryl ester core, which suggests that LCAT was functional in the absence of Apo AI. The simultaneous deficiency of Apo AI and Apo AIII suggests a dual defect in lipoprotein metabolism: one in triglyceride-rich lipoproteins, and the other in HDL. The absence of Apo CIII may result in accelerated catabolism of triglyceride-rich particles, and an increased rate of LDL formation which promotes rapid uptake of Apo E-containing remnants by liver and peripheral cells.[39]

D. FAMILIAL APOLIPOPROTEIN CII DEFICIENCY

Apo CII is a peptide of 79 amino acid residues, associated with circulated triglyceride-rich lipoproteins, chylomicrons, and VLDL. It activates lipoprotein lipase which hydrolyzes the triglycerides of chylomicrons and VLDL. Individuals with Apo CII deficiency demonstrate severe hypertriglyceridemia and other lipoprotein abnormalities. The syndrome is characterized by high fasting levels of triglycerides, associated with the complete absence of Apo CII in circulating lipoproteins. Restriction enzyme analysis of DNA suggests that the defect causing Apo CII deficiency may be closely linked to the Apo CII gene.[40]

Hypertriglyceridemia may be caused by polymorphism of the Apo AII gene. Apo AII is a major protein component of HDL. Its physiological role is unclear, although Apo AII activates hepatic lipase and inhibits LCAT. It can displace Apo AI from the HDL. Apo AII has a strong negative correlation with risk of myocardial infarction.[41]

Summarizing, at present there are two hypotheses that try to elucidate the pathogenesis of atherosclerosis. The first hypothesis suggests that plasma lipids diffuse into, and are deposited in, the arterial wall. This hypothesis receives support from epidemiologic studies that relates levels of blood lipids in younger people to incidence of atherosclerosis. A strong support for this hypothesis is yielded by the incidence of atherosclerosis in familial hypercholesterolemic patients. The question arises whether besides the known mechanisms of familial hypercholesterolemia and lipoprotein disturbances, other mechanisms might play a role in atherogenesis.[2] It is generally accepted that diets high in animal fats and cholesterol may induce atherosclerosis by the mechanisms of saturation and the removal rate of LDL. Saturation of LDL receptors may occur by high dietary cholesterol intake. Once the LDL receptors are saturated, clearance of the surplus cholesterol can only be removed by nonreceptor pathways that work at a low efficiency. Saturated LDL receptors reduce the synthesis of LDL receptors, which results in hypercholesterolemia. The entry of cholesterol into the liver, mediated by the chylomicron remnant receptor, causes accumulation of cholesterol, resulting in overproduction of LDL, and atherosclerosis.[2]

The second hypothesis assumes that focal damage to the endothelium causes deposition of platelets from the blood, release of platelet derived growth factor (PDGF), and stimulation of smooth muscle cells, migration and hypertrophy. The laminated appearance of many atheromatous deposits, suggesting successive thrombus deposition and the presence of platelet antigens, supports this hypothesis.

V. THE ROLE OF PLATELETS IN ATHEROSCLEROSIS

The atherosclerosis lesions occur in the innermost layer of the affected arteries. They consist of proliferated smooth muscles surrounded by connective tissue, numerous lipid-laden macrophages and smooth muscle cells in the form of foam cells, and lymphocytes. Endothelial injury could induce endothelial cells to synthesize and release growth factors (PDGF) into the underlying artery wall. Chronic hypercholesterolemia causes monocytes attachment of macrophages to endothelium that, thanks to the acetyl-LDL receptors, can bind cholesterol and convert into foam cells.[16] The activated macrophages, together with injured or activated endothelium, and injured or activated smooth muscle cells, could also release growth factors, including PDGF. The platelet hypothesis of atherosclerosis seems to be confirmed in baboons with an indwelling catheter or with chronic hypercholesterolemia. Each of these induced platelet adherence and aggregation is followed by formation of atherosclerotic lesions. In all cases inhibition of platelet function prevented lesion formation.[42]

Another mechanism related to endothelium injury might take place. It is well known that intact endothelium secretes prostacyclin (PGI_2) that protects arteries as a potent inhibitor of platelet aggregation and activation with secretion of PDGF; whereas the injured endothelium does not secrete PGI_2, which enables platelet aggregation and activation. PDGF causes contraction of aortic strips, being more potent on a molar basis than the classic vasoconstrictor — angiotensin II.[43] Experiments with pigs that are homozygous for von Willebrand's disease are consistent with the hypothesis of platelet adhesion to the damaged area of vascular endothelium and PDGF release. The long-term effect of von Willebrand's disease is a resistance to atherosclerosis (not yet confirmed in humans).[44]

The possibility that thrombin-activated platelets may contribute to cholesterol accumulation in smooth muscle artery cells was confirmed in *in vitro* experiments in which activated platelets released cholesterol that was accumulated in cells and stored as lipid droplets (in the absence of lipoproteins and in the presence of mevinolin, the potent inhibitor of cholesterol synthesis.[45]

On the other hand, there are confusing experimental data that platelet secretory products inhibit lipoprotein metabolism by inhibiting uptake of acetyl-LDL by LDL receptors on macrophages, which prevents accumulation of cholesteryl esters in these cells.[46]

In the case of artery damage, platelets are deposited on the injured vascular endothelium, and release a number of growth factors (PDGF, EGF) and the transforming growth factor beta (TGF-β). Purified TGF-β from platelets added to growing aortic endothelial cells caused transient inhibition of their growth (cell migration and replication were inhibited during the first 24 h after wounding). The cells were blocked from entering the S phase, and the fraction of cells in G_1 was increased. Since TGF-β stimulates smooth muscle proliferation, the transient inhibition of endothelial cells may give smooth muscle cells time to fill the wounded region.[47]

VI. MOLECULAR DISORDERS OF COLLAGEN

Collagen is the major protein molecule of most connective tissues, forming the supportive framework of practically all tissues and organs. The abnormalities of these macromolecules produce an extremely diverse group of human diseases. The collagens represent a family of high molecular weight proteins that constitute the major structural component of all connective tissues. Ten different types of collagen have been identified, containing 19 genetically different polypeptide chains.

Type I collagen is the most frequent heteropolymer of molecular mass 95 kDa that is composed of two identical $\alpha1(I)$ chains, and one homologous, distinct $\alpha2(I)$ chain. This type has N- and C-terminal extensions, cleaved by vertebrate collagenase into 3/4 and 1/4 fragments (Figure 3). Type I collagen occurs in skin, bones, tendons, vessels, intervertebral discs, dentin, gingiva, muscles, and placental and fetal membranes. Type I trimer (molecular mass 95 kDa) is composed of three identical $\alpha1(I)_3$ chains (1/4 fragments) and occurs in skin, cranial bones, and inflamed gingiva with traces in fetal tendons.

Type II is a homopolymer of molecular mass 95 kDa, composed of three identical $\alpha1(II)$ chains. It occurs in cartilage and intervertebral discs.

Type III collagen is a homopolymer of molecular mass 95 kDa, composed of three identical $\alpha1(III)$ of high hydroxylysine chains, and is present in skin, blood vessel wall, gut, placental membranes, and leiomyomas.

Type IV collagen is a heteropolymer of molecular mass 160 kDa, and is composed of two distinct $\alpha1(IV)$ and $\alpha2(IV)$ chains in a triple stranded molecule. This collagen is characterized by unhelical N and C terminals with the helical portion resistant to vertebrate collagenase, persistent procollagen extensions, and disulfide-linked chains that form a diamond shaped lattice using 7S linkers. Type IV collagen is present in basement membranes, lens capsule, the placenta, glomeruli, and mouse tumor.

Type V collagen is a heteropolymer of molecular mass 95 kDa, composed of three different chains $\alpha1(V)$, $\alpha2(V)$, and $\alpha3(V)$; or (A, B, C) in heteropolymers composed of $\alpha1(V)_2$ and $\alpha2(V)$ or $\alpha1(V)$, $\alpha2(V)$ and $\alpha3(V)$; or in homopolymer $\alpha1(V)_3$, with $\alpha1(V)_2$ $\alpha2(V)$ constituting the major form. Type V collagen is present in skin and fetal membranes. Type V is a high hydroxylysine, high carbohydrate collagen.

The remaining collagens have not yet been fully characterized; they form only minority components of the connective tissues.

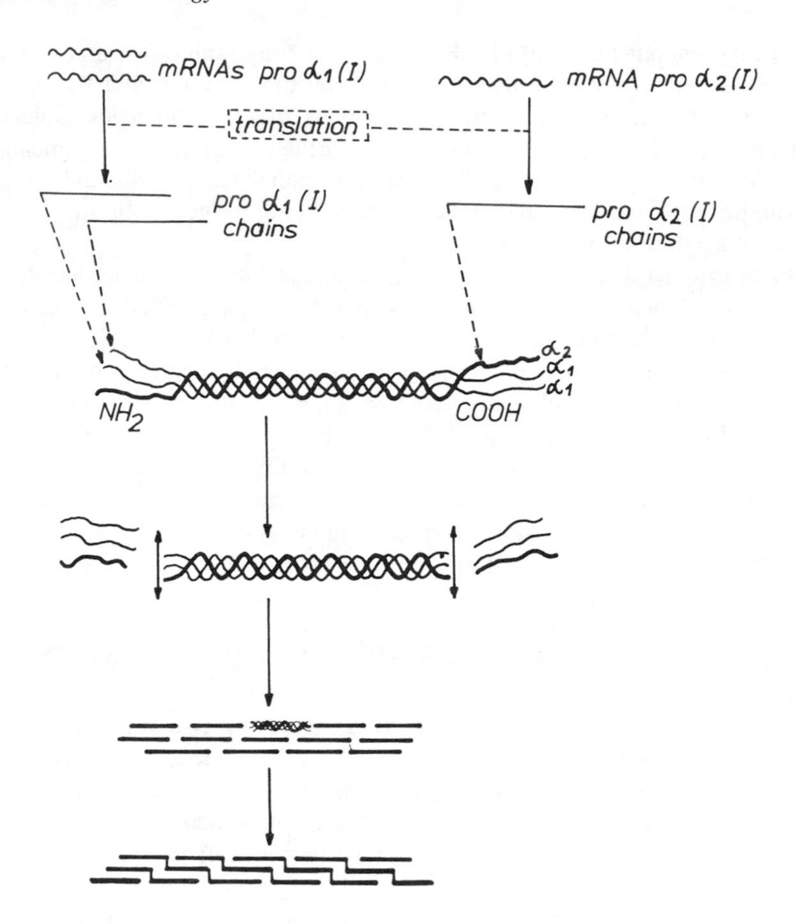

FIGURE 3. Biosynthesis of collagen (for explanation see text). (Redrawn from Prockop, D. J.[52])

Type VI collagen is a short-chain collagen, of molecular mass 30 to 70 kDa, present in the placenta and blood vessels. This collagen contains acidic and basic 40 kDa subunits with large 30 to 40 kDa globular extensions. Collagen VI forms triple helical monomers of 160 kDa, and is microfibrillar associated.

Type VII collagen of molecular weight 170 kDa has an extended chain length, and its segment with a long spacing crystallite (SLS) pattern is strongly reminiscent of the anchoring fibrils which cement basal epithelial cells to the underlying basement membrane. This type of collagen occurs in the placenta. Type VII collagen is a homopolymer of $\alpha 1(\text{VII})_3$ chains with nonhelical extensions and a central 50 kDa helix with pepsin resistance and a SLS pattern similar to the anchoring fibrils.

Type VIII collagen, a homopolymer $\alpha(\text{VIII})_3$ of molecular mass 177 and 125 kDa is present in the placenta and various cultured cell lines. It is non-disulfide bonded, with helical and globular domains, is cleaved by vertebrate collagenase into six or more fragments (no peptides) and basic 50 kDa pepsin-resistant helical protein fragments separated by 10 kDa noncollagenous fragments into 125-75 ''casettes''. Type VIII collagen is produced by endothelial cells, and has multiple interruptions in the triple helix.

Type IX and X collagens are minority collagens of cartilage which appear to be complex molecules containing multiple collagenous domains. Other collagenous components isolated from cartilage are three chains 1α, 2α, and 3α, with undefined molecular structure (collagen types after Pope and Nicholls[1]).

VII. THE STRUCTURE AND BIOSYNTHESIS OF COLLAGEN

The typical mature interstitial collagen molecule contains an association of three α chains, each of 1050 amino acid residues. Each of these chains is coiled into a left-handed helix with approximately three amino acid residues per turn. The three chains are then twisted around each other into a right-handed superhelix to form a rigid structure. The molecular formula of an α chain can be approximated as $(Gly\text{-}X\text{-}Y)_{333}$, where X is often proline, and Y often hydroxyproline, but also may consist of other amino acid residues. The presence of glycine in every third position is crucial. This small amino acid, occupying the restricted place in which the three α chains come together in the center of the triple helix, enables the triple-helical conformation, dependent on the presence of proline and hydroxyproline in the α chains. In collagen from mammals, approximately 100 of the X positions are proline, and approximately 100 of the Y positions are hydroxyproline, both rigid, cyclic amino acids which limit rotation of the polypeptide backbone, contributing to the stability of the triple helix. At both N and C terminals of each chain there are short sequences (10 to 15 amino acid residues) that do not conform to the triplet structure and cannot be involved in the triple helix. These nonhelical sequences play an important role in stabilization of the collagen fibril through covalent intermolecular crosslinking. The whole molecule forms a more or less rigid rod approximately 300 nm long and 1.5 nm in diameter. The central triple-helical fragment is resistant to most enzymes, and is cleaved by specific collagenases. The vertebrate collagenase cleaves at a single locus, yielding an N-terminal fragment which is three fourth of the molecule and a C terminal containing one fourth of the remaining triple helix.[50,51]

In the extracellular matrix the interstitial collagens aggregate into fibrils, each molecule overlapping its neighbor by approximately one fourth of its length. The supramolecular structure of the noninterstitial collagens is not so clearly defined. These collagens do not form fibrils. Many of them (types IV, VII, and VIII) have interruptions in the Gly-X-Y sequence; e.g., type IV has a large globular domain within the long collagenous domain, and a diamond-lattice network has been proposed for this collagen (four molecules crosslinked through one end of the triple helix, and two molecules linked through adjacent noncollagenous globular domains).[50]

A. BIOSYNTHESIS OF COLLAGEN

Collagen is first synthesized as a larger precursor, procollagen, which must be enzymatically cleaved at both N- and C-terminal ends to generate collagen (see Figure 3). Procollagen N-proteinase cleaves the N-terminal end, and the procollagen C-proteinase cleaves the C-terminal end. If the amino peptides are not cleaved, the protein can still form fibrils, but they are thin and irregular, and do not become adequately crosslinked. If the C-terminal propeptide is not cleaved, the large C-terminal propeptides prevent the protein assembling into fibrils.[52]

The intracellular assembly of the procollagen molecule also represents a complex process. The α-collagen chains are encoded by appropriate genes and synthesized on the rough endoplasmic reticulum. The post-translation processing include: cleavage of the signal peptides, with prolyl in the Y position being enzymatically converted to hydroxyprolyl residues, and some of the lysyl residues in the Y position being converted into hydroxylysine. The hydroxylysyl residues are substituted with galactose or glucosylgalactose. Finally, a mannose-rich oligosaccharide is added to the C-terminal propeptides. After assembly of the chains, the C-terminal propeptides associate and become disulfide linked. Post-translational modifications of prolyl and lysyl residues continue until a critical level of approximately 100 hydroxyprolyl residues per chain is reached. Then the protein folds into a triple-helical conformation. The post-translational process involves 11 separate enzymes.

It is noteworthy that ascorbic acid is an essential cofactor for the hydroxylation of both residues. Deficiency of ascorbic acid leads to the synthesis of collagen are so deficient in hydroxyproline that they cannot fold into triple-helix ure. This explains why wounds do not heal normally in scurvy, and why, s, old wounds break down.[52]

gen genes are characteristically divided into short exons of 54 bases that encode 18 amino acids with the sequence (Gly-X-Y). Most of the procollagen genes are known. Type α1(I) genes of 18 kb, 51 exons, and 70% introns are located on chromosome 17; type α2(I) genes of 39 kb, 52 exons, and 90% introns are located on chromosome 7; type α1(II) genes of 38 kb and 90% introns are located on chromosome 12; type α1(III) genes of 38 kb, 51 exons, and 90% introns are located on chromosome 2; α1(IV) genes of 51 exons are located on chromosome 7 or 17. It is supposed that type I genes are in single copies in the haploid genome. The control of type I collagen synthesis is achieved by a feedback inhibition effect of the excised propeptide extensions. Type I mRNA levels vary in different situations including: viral or chemical transformation, organ development, and cell proliferation rates. Similar data have been proposed for collagen type II, III, and IV genes.[50,52,53]

B. MUTATIONS OF COLLAGEN GENES

One of the technical problems in detecting mutations in collagen genes is the large size of the genes and their unusual exon/intron structure. The second technical problem presents the complexity of procollagen biosynthesis. A defect in one of the eleven post-translational enzymes may mimic structural mutation of the procollagen gene. Finally, several different collagens may be present in the same tissue and vital for its integrity, and it is difficult to ascribe the particular alterations to the particular procollagen genes.

The collagen diseases and abnormalities can be divided (after Pope and Nicholls[1]) into:

1. An inherited group (a) of rare single gene defects: Ehlers-Danlos syndrome — eight types or more; osteogenesis imperfecta — four main types; Marfan syndrome — three types.
2. An inherited group (1b) of common defects with genetic element: congenital aneurysm of the circle of Willis, some forms of osteoporosis, severe varicose veins, recurrent hernias, floppy mitral valve, and polycystic disease of kidneys.

A mutation that alters the structure of type I procollagen molecules has been described in osteogenesis imperfecta, Ehlers-Danlos syndrome, and related disorders. As in other molecules it becomes evident that very similar structural alterations in different regions of the protein can produce very different clinical syndromes. Mutations of the N-terminal propeptides are probably caused by primary defects in the processing of the procollagen N-proteinase, and may result in marked laxity of joints, generally classified as Ehlers-Danlos variants. Procollagen *N*-proteinase is a large, neutral metalloproteinase that requires a procollagen substrate with correct amino acid sequence and a correct three-dimensional conformation (Figure 4).[52]

C. EXAMPLES OF COLLAGEN MUTATIONS

In a patient with an imprecisely defined mutation in one allele of pro α2(I) chains, a deletion of 10 to 30 amino acid residues near the N terminal of the α2(I) chain, synthesized in fibroblasts, was detected. The deletion was approximately 100 amino acid residues, removed from the site of which the *N*-proteinase cleaves the type I procollagen molecule. The mutation caused a slippage of pro α chains in the molecule, which distorted the conformation of the cleavage site and made the procollagen resistant to the *N*-proteinase. The consequence was that half of the type I pro-*N*-collagen remained and accounted for the fact that the patient revealed dislocations of hips, knees, and other joints.[52]

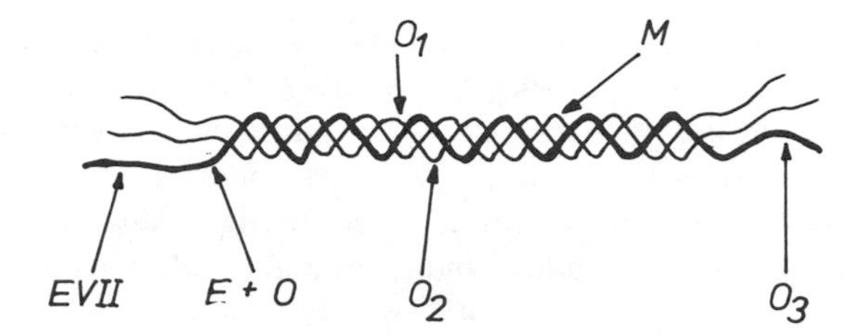

FIGURE 4. Approximate locations of mutations in the structure of type I procollagen. O_1 — procollagen $\alpha 1^S$; O_2 — procollagen $\alpha 2^S$; O_3 — procollagen $\alpha 2^{CX}$ (O — osteogenesis imperfecta); EVII — procollagen $\alpha 2^X$ (E — Ehlers-Danlos syndrome); E + O procollagen $\alpha 2^S$; M — procollagen $\alpha 2^L$ (M — Marfan syndrome). (Redrawn from Prockop, D. J.[52])

Several other variants of Ehlers-Danlos syndrome reveal mutations that change amino acid sequences in or near the cleavage site that lead to persistence of pro-*N*-collagen in tissues. The same alterations also might be caused by mutations of the procollagen *N*-proteinase. It is noteworthy that these patients generally do not have symptoms related to the bones which suggests that removal of the amino-propeptide is not critical for bone formation, as it is for the formation of normal ligaments and tendons.[52]

Several mutations in the central collagenous domain may decrease the thermal stability of the triple helix. In a variant of approximately 500 bp deletion by sporadic mutation of one allele for the pro α1(I) chain, the deletion deprived the gene of three exons with 256 bp of coding sequences, resulting in the synthesis of pro-α chains shortened by 84 amino acid residues with normal sequences on either side. This lethal mutation was explained by the so called ''protein suicide''. Since half of the pro-α1(I) chains was normal, and the shortened chains had normal sequences on either side, the chains associated and become disulfide linked. The difference between the normal and the mutated chains was so great that it prevented trimers containing these kinds of chains from folding into a normal triple helical conformation at body temperature. Therefore, the trimers containing either one or two shortened pro-α1(I) chains were degraded.[3]

The alteration in the stability of the triple helix caused by mutations of the collagenous domain of procollagen can alter the self-assembly of collagen as well as its crosslinking. In one variant of osteogenesis imperfecta, a mutation in one allele for pro-α2 chains caused a deletion of approximately 20 amino acid residues in the middle of the pro-α2 chains. The other allele of pro-α2 chains was also not functioning. In consequence, all synthesized pro-α2 chains were shorter than normal. The effect of these mutations was lethal. In another variant of osteogenesis imperfecta, a deletion of approximately 20 amino acid residues at the N-terminal end of the pro-α2(I) chain caused a moderate variant of osteogenesis imperfecta. In two other variants of osteogenesis imperfecta, a substitution that introduced a new cysteine residue into the pro-α1(I) chain caused in one of the patients an unusual intrachain disulfide bond, the presence of which resulted in a slightly decreased thermal stability of the triple helical collagen molecule.

In one atypical variant of Marfan syndrome, a longer pro-α2 chain was present, probably due to insertion of additional base sequences into an intervening sequence of the gene, thereby causing an abnormal splicing site. Two known mutations within the C-terminal propeptides, resulting in the clinical picture of osteogenesis imperfecta, indicate that structural alterations of C-terminal propeptides may cause disturbances in processing as well association of the pro-α chains. In an another case, C-terminal propeptide led to increased addition of mannose-rich carbohydrate to the C-terminal propeptide, which resulted in decreased solubility of the protein. Another variant of structural alteration was caused by deletion of four bases of the C-terminal propeptides, resulting in frameshift mutation (the last 33 amino acid residues had an abnormal sequence). This mutation was shown to prevent association of pro α2 chains with pro α1 chains. The patient with this mutation was a homozygote of moderate to severe, and progressive, osteogenesis imperfecta.[52]

The molecular defects of collagen result in four heritable diseases: osteogenesis imperfecta, Marfan syndrome, Ehlers-Danlos syndrome, and Menkes syndrome.[54]

1. Osteogenesis Imperfecta

The disorder occurs in several variants, dependent on the kind of the molecular abnormality. Molecular defects in osteogenesis imperfecta include the diminished formation of type I collagen and α1(I) mRNA, abnormal synthesis or faulty assembly of α2(I); deletion or insertion of base pairs in the gene for α1(I) or α2(I); failure to secrete type I procollagen and substitution of glycine by cysteine in the triple-helix.

Recent discoveries have brought some surprises. One surprise was that a mutation that alters the structure of one region on the type I procollagen can produce a primary disease of skin, tendons, and ligaments, whereas a mutation causing a very similar alteration in another part of the same molecule produces primary bone disease. For example, a structural mutation of the N-terminal part of the pro α2 chain results in a clinical syndrome characterized by looseness of joints, whereas mutation of the same pro α2 chain in the last third of the molecule (probably an insertion) results in the Marfan syndrome. These and several other variants represent a topological map of the protein that binds different biological functions with different regions of the same molecule. This observation explains that such diseases as osteogenesis imperfecta and Ehlers-Danlos syndrome, which initially appeared to be discrete conditions, have a common starting unity at the molecular level. And so, patients with osteogenesis imperfecta reveal some symptoms of skin, ligaments, and tendon alterations typical to Ehlers-Danlos syndrome, and vice versa.

The second surprise was the detection of "protein suicide" in both osteogenesis imperfecta and Ehlers-Danlos syndrome. It became obvious that sometimes the presence of an allele for the synthesis of a structurally abnormal protein (e.g., pro α2 chain) is more deleterious than a heterozygous null allele (absence of an allele). The principle of "protein suicide" may explain how many heterozygous gene defects, in which the amount of normal protein is reduced by no more than one-half, produce dominantly inherited diseases.

The third surprise was that mutations that shortened or lengthened the pro-α chains type I procollagen are relatively common. The reasons for the high frequency of such mutations in procollagen genes is not well understood. The shortened chains are probably caused by deletions. Procollagen genes may be particularly prone to such mutations because the coding sequences are highly repetitive for Gly-Pro-Hyp, which may predispose the genes for unequal crossover mutations during meiosis. Furthermore, the procollagen genes have a large number of intervening sequences (approximately 50) which divide the gene into short exons of 54 bp or 108 bases, which may predispose to splicing defects.[54,55]

Osteogenesis imperfecta represents a group of inherited collagen diseases. It is not a single entity, but it is a family of similar disorders with a tendency to brittle, easily fractured, collagen depleted bones. Some authors divide osteogenesis imperfecta (on a molecular basis) into types I, II, III, and atypical variant with characteristics of Ehlers-Danlos syndrome (Prockop and Kivirikko[54]).

Type I comprises nonfunctional or insufficiently functional pro α1(I) chain mutations that probably cause abnormal fibrils because of a shortened pro α1(I) chain and other unidentified mutations.

Type II is caused by a shortened pro α1(I) chain resulting in an unstable triple helix, and increased synthesis of pro α1(III) chains; pro α2(I) shortened and pro α2(I) nonfunctional chains; or an inefficiently functioning chain with unknown consequences.

Type III of pro $\alpha 1(I)$ mutations alter the structure of C-terminal propeptides, or pro $\alpha 2(I)$ mutations, likewise altering the structure of C-terminal propeptides, which in both cases results in an increase of mannose in C-terminal propeptides and increased solubility of type I procollagen.

Finally, the variant with characteristics of Ehlers-Danlos syndrome is caused by shortened pro $\alpha 2(I)$ chains caused by the resistance to procollagen *N*-proteinase resulting in persistence of intermediate pro-*N*-collagen that contains the *N*-propeptides but not the C-propeptides.

The clinical and genetic heterogeneity of osteogenesis imperfecta were established by Sillence, who distinguished 4 types of this disease, and Pope et al.[56] who modified this classification introducing the fifth type. Since type I collagen $\alpha 1(I)_2\, \alpha 2(I)$ is the predominant defective protein of bones, genes for collagen I might be faulty in this disease.

Type I is characterized by a mild disease with blue sclerae, deafness, and short stature, normal teeth and dentinogenesis, and scanty bone fractures. The disease is an inherited autosomal dominant, with variable penetrance. Two specific abnormalities of type I collagen might be associated with this type of osteogenesis imperfecta: the first, a mild disease with an amino acid substitution in the $\alpha 1(I)$ protein, and the second, a severe form with completely absent collagen $\alpha 2(I)$ chains because of a mutation in the pro $\alpha 2(I)$ collagen gene. Clinical type Ia is characterized by increased III/I collagen ratios, or by a protein defect by cysteine substitution $\alpha 1(I)$ chain (Arg or Ser $\rightarrow$ Cys).[56] Type I may be caused also by diminished pro $\alpha 1(I)$ synthesis.[56]

Cysteine substitution in osteogenesis imperfecta type I may produce lethal and mild forms of the disease. The difference was elucidated in this manner that substitution in the lethal form is at the Gly position of the Gly-X-Y repeating unit, whereas in the mild form it is at the X or Y position of the Gly-X-Y repeating unit.[57]

The molecular defect of the nonlethal variant of osteogenesis imperfecta was explained as follows: examination of cellular proteins revealed that the fibroblasts synthesized both pro $\alpha 1(I)$ and pro $\alpha 2(I)$ chains. The cellular pro $\alpha 2(I)$ chains did not become disulfide linked into dimers or trimers of pro α chains because of the mutated structure in C-terminal propeptides, which reduces their affinity to pro $\alpha 1(I)$ chains.[58]

A case of osteogenesis imperfecta type I was described where collagen did not contain $\alpha 2(I)$ chains in spite of the presence of translatable mRNA for these chains. The defect was probably caused by a structural mutation of the pro $\alpha 2(I)$ chains that prevented incorporation of these chains into triple-helical procollagen.[59]

Osteogenesis imperfecta type I represents probably several heterogeneous forms that can be clarified only after precisely recognized molecular alteration. In a study of six African families with osteogenesis imperfecta type I using three DNA probes for pro $\alpha 2(I)$ collagen, two different haplotypes were found with the mutant pro $\alpha 2(I)$ alleles.[60]

Clinical type II of oesteogenesis imperfecta is lethal *in utero* or at perinatal life. There are three characteristic types: broad bones and ribs, broad bones and thin or headed ribs, and thin or normal bones. The molecular alteration of type IIa is substitution Gly $\rightarrow$ Cys in $\alpha 1(I)$ collagen with overhydroxylated collagen and a lowered melting curve. Clinical type IIa is caused by 0.5 kb helical pro $\alpha 1(I)$ deletion with increased III/I ratios, one pro $\alpha 1(I)$ allele being shortened by 50 amino acid residues, and collagen synthesis. In the case of one normal gene, there are overhydroxylated skin and tissues collagens. Clinical type II collagen may be caused by structural pro $\alpha 2(I)$ genes; mutation in the second pro $\alpha 1(I)$ gene may cause diminished pro $\alpha 1(I)$ production with overhydroxylated three fourth portions of N terminals. Clinical type IIa caused by a 300 bp deletion in pro $\alpha 1(II)$ gene resulting in overhydroxylated pro $\alpha 1(I)$ chains.[56]

Clinical type III of osteogenesis imperfecta is characterized by severe fracture of bones *in utero* and early childhood, blue sclerae, normal teeth, progressive course, and popcorn deformities of the lower femur and upper tibia. The molecular defect is a 4 bp deletion of a pro α1(I) C-terminal peptide resulting in absence of α2(I) collagen in tissues or secreted in tissue culture, and in rapidly degraded intracellular pro α2(I) chains. Clinical type III may also be caused by substitution (of an unknown amino acid) in C-terminal propeptide, resulting in poorly secreted type I collagen and excessive mannosylation of C-terminal propeptides.[56]

Molecular analysis of a case with osteogenesis imperfecta type III revealed that the patient was heterozygous for an internal deletion of approximately 500 bp in the pro α1(I) gene. This deletion was localized between two introns of the pro α1(I) gene, and resulted in the elimination of three exons of the triple helical region. Furthermore, the termini of the rearrangement gene were located within two short inverted repeats, which suggests the formation of a DNA secondary intermediate structure serving as substrate for the deletion. Elevated type III collagen mRNA in the patient fibroblasts was also present.[61]

Clinical type IV osteogenesis imperfecta is similar to type I, with the exception of white sclerae, thin bones, normal dentinogenesis, and teeth. The molecular defect is probably a gene deletion at 5′ end of pro α2(I) gene, resulting in a short α2(I) chain deletion near the N-terminal end of the triple helix.[56]

In a patient with osteogenesis imperfecta type IV, the skin fibroblasts synthesized two populations of type I procollagen molecules — one normal, and the other of slower migrating pro α1(I) and pro α2(I) chains. The increased molecular weight of the slower migrating molecules was caused by the excessive post-translational modification (rather than peptide insertions). This alteration is consistent at the C-terminal end of the triple helical domain, which delays triple helical formation and renders all chains available for further post-translational modification N-terminal to the mutation.[62]

In a case of osteogenesis imperfecta type II, the mutant gene had undergone recombination between two nonhomologous introns, which resulted in the loss of three exons coding for 84 amino acid residues in the triplet helical domain. The deleted amino acid residues surrounded and included the methionine at the junction between the two peptides α1(I).[63]

The insertion of a single Maloney murine leukemia virus proviral copy in the first intron of the α1(I) collagen gene blocks initiation of transcription of the collagen gene and causes a recessive lethal mutation.[64] Lack of collagen I synthesis due to this insertion allowed studies about the role of collagen during mouse embryonal life. Mutant embryos developed normally to day 12 of gestation. No collagen I was present, and the other collagens, laminin and fibrobnectin, were normal. After day 12 of gestation, necrosis of mesenchymal cells occurred (mainly in the liver) resulting in sudden death caused by rupture of major blood vessels; this indicates the important role of collagen I in establishing normal blood vessels.[65]

2. Ehlers-Danlos Syndrome

The Ehlers-Danlos syndrome represents a group of inherited connective tissue disorders: fragile skin, thin tissue scars, hypermobile joints, hernias (umbilical, inguinal, femoral, subdiaphragmatic, hiatus, lumbar) and various other deformities (pectus excavatum, pectus carinatum, congenital dislocation of the hips), short stature, severe kyphoscoliosis, arterial rupture, severe premature periodontal disease.[50] All these disturbances have been classified into several Ehlers-Danlos types (after Pope and Nicholls[50]):

Type I (gravis) is characterized by soft silky hyperextensible skin, scars of forehead, elbows and knees, and rarely, aortic rupture. The molecular defect is unknown. The disease is inherited by dominant mode.

Type II (mitis) is similar to type I but less obvious clinically; bruising is common. In one family a C-terminal deletion type I collagen α1(I) chains were present. The disease is inherited as a dominant trait. The molecular defect is unknown.

Type III (benign) is characterized by normal skin with marked loose joints (found within good gymnasts and dancers). Osteoarthrosis may occur frequently in middle age. The molecular defect is unknown. The disease is inherited as a dominant trait.

Type IV of the Ehlers-Danlos syndrome is characterized by arterial fragility accompanied by a characteristic facial appearance with large eyes, thin nose and lips, and an aged appearance of the hands with prominent tendons (so-called acrogeria). The skin is thin, and venous network is prominent. Histologically, the skin appears collagen depleted. The most serious result is sudden rupture of great arteries (aorta, arteria pulmonalis, renal, splenic, femoral, or popliteal vessels) in the early years of life. The molecular defect presents a spectrum of type III collagen deficiency ranging from 0 to 50% of normal. Little or no type II procollagen is secreted into the medium of cultured skin fibroblasts from these patients. In a study of cultured fibroblasts from ten patients with type IV Ehlers-Danlos syndrome, a decreased amount of type III procollagen was recovered in the medium. The culture medium from one patient contained apparently normal amounts of type III procollagen; however, the pro $\alpha 1$(III) chains appeared as an abnormally broad electrophoretic band. Further analysis demonstrated a structural defect between amino acid residues 555 and 775 in half of the $\alpha 1$(III) chains. Most of this procollagen was susceptible to digestion with a mixture of chymotrypsin and trypsin at a temperature at which normal type III procollagen resisted digestion. The amount of type II procollagen was reduced more than fourfold. This suggests that rapid degradation of this type of collagen by collagenase resulted in a decrease of collagen in this case.[66]

Dependent on the amount of type III collagen in tissues or culture, and its abnormalities, Pope and Nicholls[50] divide type IV of the Ehlers-Danlos syndrome into following sub-types:

Type IV (a) is characterized by absence of type III collagen in tissues or culture of fibroblasts. Clinical symptoms are acrogeria, typical owl-eyes, pinched nose and lips, large arterial ruptures, early death, and autosomal recessive inheritance.

Type IV (b) is characterized by traces of type III collagen in culture of fibroblasts. Clinical symptoms are similar to type IV (a), with the exception of less frequent arterial rupture and longer life and autosomal recessive inheritance.

Type IV (c) is characterized by traces of type III collagen in culture of fibroblasts. Clinical symptoms: normal face, thin skin, varicose veins of lower extremities, and autosomal dominant inheritance.

Type IV (d) is characterized by mutant type III collagen and clinical symptoms similar to those of type IV (b).

Type IV (e) is characterized by failure of type III secretion (the synthesized collagen is retained in the cells). Clinically there is thin skin and acrogeria.

Type IV (f) is an autosomal recessive disorder with normal type III collagen in fibroblast culture with clinical symptoms similar to those of type IV (b).

Type V (a) is characterized by a deficiency of lysyl oxidase activity.

Type V (b) has no lysyl oxidase deficiency. Clinically, there is extremely severe scoliosis, deafness and fragile sclerae. The disease is inherited by an autosomal recessive order.

One mutant enzyme has been characterized as thermally labile with altered affinity for ascorbate.[67] The consequence of defective lysyl hydroxylase is lack of hydroxylysine in several connective tissues. Insufficient hydroxylation of lysyl residues was found in type I and III collagen, whereas types II, IV, and V demonstrated normal amounts of hydroxylysine. The expression of the abnormal collagen varied from one tissue to another. A complete lack of hydroxylysine was observed in skin, and less frequently in bones, tendons, lungs, and kidneys.[68]

Type VII of Ehlers-Danlos syndrome is characterized by a mutated N-terminal cleavage site for procollagen peptidase, resulting in altered processing of procollagen to collagen. Most often the disease is associated with deficient procollagen aminoprotease activity. In one case a structural mutation of the pro $\alpha 2$(I) chain was observed, resulting in incomplete cleavage of the N-terminal propeptide.[67] Clinically, there is short stature and premature severe hip dislocation. The disease is inherited by an autosomal recessive mode.

A new variant of Ehlers-Danlos syndrome type VII with a structural defect in the pro α2(I) chain has been described. The patient's skin, fascia, and bone collagens all demonstrated an abnormal additional chain pN-α2(I) running slower than the normal chain on electrophoresis. The extension was present on the N-terminal end, cleaved by human collagenase but not by pepsin, which was unable to convert this pN-α2(I) chain to the normal α2(I) chain. Skin collagen was four times more extractable, and contained fewer β-dimers and a lower concentration of crosslinking amino acids than control skin collagen. It is considered that the patient represents a spontaneous heterozygote who expressed one normal and one abnormal allele for the pro α2(I) gene.[69]

Type VIII of the Ehlers-Danlos syndrome is characterized by necrobiotic-like skin lesions, severe periodontal diseases, normally extensible skin, loose joints, and blue sclerae. The molecular defect is unknown. The disease is inherited by an autosomal recessive mode.

3. The Marfan Syndrome

The syndrome comprises a group of related disorders of patients with tall stature, long extremities (arms, legs and hands — arachnodactyly) and chest deformities (pectus excavatum). Furthermore, abnormalities of the eyes (myopia and lens dislocations) and aorta may occur. The most important are mitral and aortic valve regurgitation (from the distended aorta and a floppy valve leaflet), resulting sometimes in aortic rupture. The inheritance is probably autosomal dominant, although some variants may be autosomal recessive. There is also a variant with contractural arachnodactyly and overlapping with the Ehlers-Danlos syndrome (especially type III), which also can present aortic rupture.

The molecular defect in Marfan syndrome is only poorly understood. Increased collagen solubility has been described by several authors, as also collagen crosslinking abnormality. The most impressive collagen abnormality refers to a case of abnormal α2(I) chain in various tissues. The patient suffered lethal aortic rupture with collagen and elastin degeneration and an accumulation of mucopolysaccharide in the aortic wall. By studying skin fibroblasts, an apparent peptide insertion N-terminal to the collagenase cleavage site was found, located in cyanogen bromide peptide α2CB 3 to 5 of the mutated chain. A second possibility was thought to be a mRNA processing defect, and a third possibility was that the α2 insertion is irrelevant to the Marfan syndrome or is only one of several genes producing (in concert) the syndrome.[1] Further, precise studies (by genomic blotting and gene cloning) demonstrated that the α2(I) chain contains a 38 bp insertion on intra near the collagenase cleavage site. Genomic mapping detected no abnormalities. The relationship of this insertion on the protein abnormality is unclear, but it may serve as a marker for Marfan syndrome diagnosis.[70]

4. Cutis Laxa

Similar to other collagen disorders, cutis laxa represents a group of diseases characterized by laxity of the skin — loose skin which hangs abnormally in folds. The characteristic loose skin appears quite wrinkled and prematurely aged. In contrast to the Ehlers-Danlos syndrome, the hypertensible skin does not spring back. Easy bruising, poor wound healing, and joint hypermobility are not features of cutis laxa. Hernias and diverticula of the gastrointestinal tract and genitourinary tract are frequent. The most severe complication may be emphysema which leads to cor pulmonale and death. Dominant, recessive, and X-linked forms of cutis laxa have been observed; there are also known acquired forms. Electron microscopic examination of the skin demonstrated decreased amounts of amorphous elastin and an increase in elastin-associated microfibrils. The cultured fibroblasts did not differ from normal fibroblasts (growth rate, morphology, and total protein synthesis); however, a four- to six-fold increase was found in accumulation of a collagenous protein of M_r 140,000, distinct from collagen types I, III, IV and VIII, which was related to a cell-surface-associated glycoprotein. It is supposed that this protein may be the precursor form of pepsin-extracted type VI collagen.[71]

The molecular defect of cutis laxa consists of a decreased number of irregularly fragmented fibers; collagen fibril diameters are also irregular. The complete molecular structure of elastin molecules is not yet known. Important for proper crosslinking is lysine oxidase. In two patients with cutis laxa, lysine oxidase deficiency has been found, which results in abnormal crosslinking of collagen and elastin fibrils.[81]

5. Menkes Disease

Menkes kinky hair disease is a hereditary disease of copper metabolism in man. The disease is characterized by brittle and depigmented hair, skeletal abnormalities, severe mental retardation, and several neurological defects. It is an X-linked recessive disorder, and occurs in approximately 1 of 100,000 live births. Most of the patients die by 1 to 3 years of age. The disease is caused by maldistribution of copper. Decreased copper amounts have been demonstrated in the brain, liver, and serum in affected male infants, with subsequent decrease in the activity of essential copper-dependent enzymes (amino oxidase, cytochrome c oxidase, tyrosinase, and dopamine-β-hydroxylase). In most nonhepatic tissues the copper concentration is higher than in normal tissues, including the gut, kidney, lung, spleen, muscle, pancreas, and skin. Cultured Menkes' lymphoblasts also take up and retain more copper than normal cells, and are deficient in some copper enzymes. The intestinal transport of copper is similar in Menkes' and normal cells, but the efflux is decreased and the half-life time increased in mutant cells. The excess of copper is accumulated in a small, cysteine-rich protein — metallothionein. This protein chelates copper and other transition state elements (zinc, cadmium, and mercury) through thiolate bonds. It is supposed that overproduction of metallothionein is responsible for the increased uptake and retention of copper in Menkes' cells. The overproduction of metallothionein is probably caused by low copper concentrations on gene transcription site of metallothionein genes.[72,73]

The human genome contains multiple metallothionein-related genes, clustered on chromosome 16. Transcription of these genes is induced by heavy metals. Metallothionein-Ia and IIa genes are responsive to glucocorticoids. The absence of metallothionein sequences from the X chromosome indicates that Menkes' disease affects metallothionein genes expression by a *trans*-acting mechanism.[74]

The pathogenic effects of collagen abnormalities are not limited to bone disorders. Scleroderma fibroblasts demonstrate differences in inhibition of collagen synthesis. In normal fibroblasts collagen is inhibited in the presence of 4 μM collagen $\alpha 1$(I) by about 42%, whereas in scleroderma fibroblasts collagen synthesis was inhibited to a lesser degree (about 19%), and scleroderma fibroblasts with elevated collagen synthesis were inhibited by only 10%. This result suggests that the increased amount of collagen in scleroderma patients is caused by a defect in collagen synthesis regulatory mechanisms.[75]

In Fuchs' endothelial dystrophy corneas, the Descemet's membrane and posterior collagenous layer have been analyzed. The amino acid composition was not altered from the normal one containing the same collagen types IV and endothelial cell collagen, and only a slight discrepancy was seen in the electrophoretic mobility of some collagen chains. Fibrinogen/fibrin deposits were present only in Fuchs' endothelial dystrophy corneas. The result suggests that the appearance of 110 nm banded material in sheets and fusiform bundles characteristic of this disease is not caused by abnormal collagen molecules, but by fibrinolytic disturbances.[76]

6. Dentinogenesis Imperfecta

Hydroxyapatite crystal deposition within the collagen matrix of bone and dentin have been linked to noncollagenous proteins. Dentinogenesis imperfecta is a genetic disorder of dentin mineralization. Comparative studies of dentinogenesis imperfecta type I and type II (hereditary opalescent dentin) demonstrated the absence of the highly phosphorylated protein phosphoryn, present in normal dentin of human teeth.[77]

Collagenase plays an important role in collagen metabolism. Conversion of procollagen to collagen is necessary for proper alignment of collagen molecules to form functional fibers. This is catalyzed by at least three structurally and functionally distinct enzymes cleaving collagen types I to III. Two N-terminal proteinases cleaving type I and type II procollagen, and two C-terminal proteinases cleaving C-terminal propeptides (C-terminal propeptides of type I and III procollagen are differently affected by lysine). The regulation of this reaction process from procollagen to collagen is not well known.[78] Human skin collagenase expression is stimulated by platelet derived growth factor (PDGF) that stimulates collagenase expression through increased transcription or preferential translation of collagenase mRNA.[79] Further experiments demonstrated that collagenase is autoregulated by a protein synthesized and secreted by synovial fibroblasts (in rabbits). On the basis of this observation it is suggested that rabbit synovial fibroblasts autoregulate collagenase production, and that the level of collagenase in resting cultures of fibroblasts depends on active suppression of collagenase synthesis.[80]

Restriction fragment length polymorphism in collagen abnormalities recognition — Random variability in DNA sequences close to or within structural genes have been recognized as useful markers for detection of human genetic diseases, e.g., sickle cell hemoglobin and Huntington's chorea. Several probes have been constructed for detection of inherited abnormalities of collagen. The best known probes for detection of mutated genes encoding collagen chains (by hybridization) are: $\alpha 2$(I), $\alpha 1$(III), $\alpha 1$(II), and two polymorphisms for $\alpha 1$(I).[83] The same method was used in a case of Marfan's syndrome for explaining the question whether the Marfan syndrome can be caused by the insertion of 38 bp, as has been previously proposed. On the basis of evidence that such insertion has been found also in the normal population, it is thought that this insertion is not the potential cause of Marfan syndrome.[83]

VIII. α_1-ANTITRYPSIN DEFICIENCY

Human serum contains several protease inhibiting activities. The most potent of this group is α_1-antitrypsin which inhibits several proteases (trypsin, chymotrypsin, plasmin, kalikrein, pancreatic elastase, skin collagenase, renin, urokinase, thrombin, neutral proteases of polymorphonuclear leukocytes, and an acrosomal protease). α_1-antitrypsin is the major component of the α_1 globulin region in the electrophorogram of normal serum; therefore, its absence may be suspected by evaluation of routine serum electrophoresis pattern.

α_1-Antitrypsin is a serum glycoprotein of 394 amino acid residues, and consists of a single polypeptide chain with four carbohydrate side chains of two different types. α_1-Antitrypsin from severely deficient patients exhibits less *N*-acetyl-glucosamine, mannose, galactose and *N*-acetyl-neuraminic acid (sialic acid) as compared with normal persons.[84,85]

The synthesis of α_1-antitrypsin is controlled by an autosomal allelic system. Using starch-gel electrophoresis over 60 different phenotypes have been recognized. The system is denominated "Pi" for protease inhibitor, with a system of labelling variants based on letters of the alphabet: the electrophoretically slowest is denominated Z; the usual, M; and the faster, F. The second most common type, S, falls between M and Z. Most persons are homozygous PiM (gene frequency approximately 0.95), PiS has a gene frequency of 0.2 to 0.4, and PiZ has a gene frequency of 0.01 to 0.02. In persons of phenotype PiZ the α_1-antitrypsin activity is reduced to 10 to 15%; the phenotype MZ exhibits approximately 60% of normal activity. The loss of α_1-antitrypsin activity in ZZ and MZ individuals is related to decreased serum concentrations. Rare "silent alleles" have also been described in individuals with complete absence of α_1-antitrypsin activity.

The deficiency of α_1-antitrypsin activity results in neonatal hepatitis and chronic pulmonary disease.

Histologically, the livers of severely deficient infants reveal in almost all hepatocytes an amorphous material which appears to be a precursor of α_1-antitrypsin that has never been excreted. The basic difference between the severely deficient (PiZ) molecule and the normal (PiM) has been demonstrated in a substitution of a glutamic acid residue in PiM by lysine in the abnormal PiZ molecule, which probably inhibits attachment of sialic acid. Whether this substitution also inhibits excretion of the enzyme from the hepatocytes is not known. The abnormal molecule may aggregate, which inhibits its further processing and excretion into the circulation. Less severe types of α_1-antitrypsin deficiency have probably other substitutions that do not disturb excretion, without accumulation in liver cells. For example, PiS α_1-antitrypsin of moderately reduced activity has a substitution of glutamic acid by valine.[84] Homozygous deficiency of α_1-antitrypsin is the most inborn error in Europe.

Obstructive lung disease is caused by severe α_1-antitrypsin deficiency. Emphysema seems to be caused by the free elastolytic activity of polymorphonuclear leukocytes leading to the degradation of elastin. In animal experiments, leukocyte collagenase alone cannot produce the lesions of emphysema, whereas leukocyte elastase can do so when instilled into the trachea. Alveolar macrophages can also release an elastase which is less readily inhibited by α_1-antitrypsin than elastase from leukocytes. α_1-Antitrypsin does not inhibit papain *in vitro*, but instilled intratracheally, it prevents development of emphysematous lesions in the hamster challenged with papain. This result suggests that the main damage is being done by endogenous enzymes recruited by the exogenous irritant. Some bacterial infections (Proteus mirabilis, Pseudomonas aeruginosa) have been shown to inactivate α_1-antitrypsin, and should therefore affect both the deficient and nondeficient individuals equally.[84,85]

Because of the frequent occurrence of α_1-antitrypsin deficiency, the question arises why not all individuals with severe α_1-antitrypsin deficiency reveal lung disease? May simultaneous deficiency in leukocyte protease be protecting them? Such a variation has been found in individuals with a tendency toward a more favorable clinical course which had intermediate levels of α_1-antitrypsin and decreased levels of leukocyte protease activity. However, other authors have not confirmed this observation.[84,85]

Acquired increase in protease activity (particularly in tobacco smokers) seems to be involved in the pathogenesis of chronic obstructive lung disease because of increased numbers of leukocytes and macrophages that can be lavaged from the lungs of smokers. It was found that *in vitro* smoke condensate causes these cells to release their elastase.[85]

A. LIVER DISEASE IN CHILDREN

The first case of severe α_1-antitrypsin deficiency was described in a child with liver cirrhosis. In a 22-month screening program for α_1-antitrypsin deficiency (in Sweden), children with severe deficiency (PiZ phenotype) revealed prolonged obstructive jaundice (11%), and an additional 6% revealed some abnormality on history examination, suggesting that liver disease had been present earlier. Usually, a prolonged conjugated hyperbilirubinemia was present in these children during the first three months of age, and cleared by six to eight months of age.[84,85]

A difficult question arises in attempting to understand why only a minority of PiZ infants acquire liver disease and others do not, despite the presence of comparable amounts of amorphous material in their hepatocytes. It is assumed that there must be some other factors acting in concert with the circulating α_1-antitrypsin deficiency, or that the intracellular excess of α_1-antitrypsin produces liver lesions. On the basis of familial tendency in the occurrence of these cases, an additional heritable factor for liver disease has been suspected. However, familial occurrence of α_1-antitrypsin deficiency in which liver disease did not occur was not confirmed by other authors. Furthermore, the additional factor must not be hereditary. For example, substances from cigarette smoke could cross the placenta barrier, stimulate leukocyte elastase barrier, and overload an immature α_2-macroglobulin scavenger system, which results in an increase of elastase concentration in the reticuloendothelial system of the liver, initiating liver disease in α_1-antitrypsin deficient fetus.[84,85]

B. LIVER DISEASE IN ADULTS

The hepatocytes of homozygotes or heterozygotes of the PiZ allele exhibit similar amorphous material to be a precursor of α_1-antitrypsin, as in the neonatal livers of α_1-antitrypsin deficiency cases. In a study of 200 PiZ individuals over a ten-year period, nine cases of liver cirrhosis have been diagnosed, and six of them demonstrated liver carcinomas. In patients with hepatic carcinoma, 10 to 50% exhibited the characteristic amorphous material in hepatocytes, although usually not in the cancer cells. However, it must be concluded that in pulmonary emphysema and also in liver disease there exists a large range of individuals that escape these diseases the cause of which remains obscure.[84]

There are also some suggestions that α_1-antitrypsin deficiency may cause lesions of other organs, e.g., pancreatic fibrosis and glomerulonephritis, but there are no certain arguments confirming these suggestions.

The extent of the risk for PiMZ heterozygotes to develop chronic obstructive pulmonary disease is still a subject of considerable controversy. One of the problems in comparing the results are the different criteria used to define chronic obstructive pulmonary disease. To solve the problem patients with so-called "flaccid lung" syndrome, a subgroup of this disease, has been examined. The term "flaccid lung" refers to a loss of elasticity of lung parenchyma with a high compliance (the volume compliance was $\geqslant 1.0$ kPa^{-1}; normal lung elasticity volume lies between 0.5 and 0.8 kPa^{-1}). Flaccid lung can be found in patients with high vital capacity, in patients with spontaneous pneumothorax, and in all patients with lung emphysema. In studies on 1716 unrelated subjects, a significantly high number of PiZZ and PiZM individuals were found in the flaccid lung population than in healthy control. The relative risk for ZZ was 12.5, and for ZM 1.8. Further, the excess risk due to the deficiency is negligible compared to MM individuals, and is highly influenced or modified by other factors, probably both environmental and genetic.[86]

The number of α_1-antitrypsin variants with either normal or deficient activity is still growing. Recently, to the list of known variants a new deficiency allele $PiZ_{Augsburg}$ has been added, which was described in a 52-year-old patient suffering from respiratory disorder due to severe obstructive pulmonary disease with emphysema. He had chronic bronchitis, having been a heavy cigarette smoker for 30 years. Electrophoretically, the bands of PiM5 were located between bands PiM1 and PiM3. The variant was provisionally named $PiZ_{Augsburg}$.[87]

Necrotic panniculitis is a very rare untypical complication of α_1-antitrypsin deficiency. The patient suffered from extensive erythematous, painful plaques on the buttocks, flanks and thighs, and thromboplebitis of the right arm. His history revealed the occurrence of these lesions after minor trauma. It has been suggested that vascular injury resulted from proteolytic activity of polymorphonuclear leukocytes; in normal individuals, such activity is blocked by α_1-antitrypsin that also inhibits the proliferative response of lymphocytes to mitogenic stimulation. Finally, α_1-antitrypsin has an important inhibitory effect on activated factor XI and thrombin. In the patient described, antithrombin activity was low. In normal plasma, about 75% of the antithrombin activity is caused by α_1-antitrypsin; thus, the presence of thrombosis may be caused by a low antithrombin III activity.[88]

An extremely important case for understanding the possibilities of molecular pathology is represented by the transformation of a normal enzyme — α_1-antitrypsin Pittsburg — into an abnormal one — antithrombin — by a single substitution of methionine at position 358 by arginine (Met 358 $\rightarrow$ Arg), which yielded a mutant molecule that retained its inhibitory activity against trypsin, lost its capacity to inhibit pancreatic elastase, and displayed a 4000-fold increase in thrombin-inhibiting activity.[89,90] (See Chapter 15, Section VI.C and Section VI.C.2)

The case history is very interesting. In 1978 a 10-year-old boy had been described with a severe lifelong bleeding disorder. The boy was hospitalized more than 50 times for post-traumatic bleeding. At the age of 14 years he died of uncontrollable bleeding into the leg and abdomen. The thrombin-inhibiting activity was caused by a variant of α_1-antitrypsin that displayed a slower electrophoretic mobility than normal α_1-antitrypsin. A characteristic feature of the mutant enzyme was that heparin did not influence the activity of the mutant inhibitor, in contrast to the catalytic activity of heparin on normal antithrombin III.

The finding that a single substitution converted a normal inhibitor of elastase into that of an inhibitor of thrombin demonstrates that the reactive center of α_1-antitrypsin in methionine 358 acts as a bait for elastase, and that the normal reactive center of antithrombin III is arginine 393 which acts as a bait for thrombin. The independence of the new thrombin inhibitor for heparin control explains the bleeding disorder; it also demonstrates that heparin normally acts directly on antithrombin III, revealing its inhibitory character.[90,91]

To study the modulation of biological properties of α_1-antitrypsin the authors introduced selected sequence modifications at the reactive site by an *in vitro* mutation of a cloned α_1-antitrypsin complementary DNA. The α_1-antitrypsin analogues have been produced in *E. coli*. The first, with substitution Met $\rightarrow$ Val, was not only fully active as an elastase inhibitor, but was also resistant to oxidative inactivation (oxidants present in tobacco smoke can inactivate α_1-antitrypsin *in vitro,* which explains why cigarette smokers, because of reduced α_1-antitrypsin activity, may exhibit a high incidence of lung emphysema). Oxidative inactivation of α_1-antitrypsin is probably due to the modification of methionine residue (Met 358) at the site of the elastase binding. The second substitution (Met $\rightarrow$ Arg) does not inhibit elastase, but is an efficient thrombin inhibitor.[92] The active site of the latter is identical to that of the described α_1-antitrypsin variant (Pittsburg), which was associated with the fatal bleeding disorder.

An intriguing question of this α_1-antitrypsin mutant was that the mutant inhibitor was inducible by stress and hematoma formation. The concentration of the abnormal mutant inhibitor of the abnormal α-antitrypsin rose from approximately 0.6 to 2.0 g per liter (ten times the plasma concentration of antithrombin III). Since this inhibitory activity was four times greater than that of antithrombin III, it is evident that the patient's plasma revealed a 40-fold increase in antithrombin activity that resulted in uncontrollable bleeding. Antithrombin III is a relatively slow-acting inhibitor that allows thrombin-induced platelet aggregation and conversion of fibrinogen to fibrin. Heparin-like material on the vascular endothelium potentiates the function of antithrombin in the microcirculation. Decrease of antithrombin III (below 50%) creates a predisposition to thrombosis. The concentration of antithrombin III does not increase in response to inflammatory stress, as does α_1-antitrypsin.[89]

Cloning the human α_1-antitrypsin gene made it possible to identify the promoter and the transcription point. The cloned gene was expressed in a cell-specific way, being transcribed in human hepatoma cell line, but not in HeLa cells. The 5′ flanking region of the α_1-antitrypsin gene contains DNA sequences for efficient transcription, but not in HeLa cells. The DNA sequence activates also heterologous promoters (e.g., that of SV40), but only in one direction, suggesting that this cell-specific element does not share all the features of enhancers.[93]

Using modern techniques of synthesized specific nucleotides corresponding to the normal M and deficient Z genotypes, it is possible to recognize persons with the MM and ZZ genotypes. Using DNA from cultured amniotic cells provides a safe method for prenatal diagnosis of α_1-antitrypsin deficiency.[94]

IX. PHENYLKETONURIA

Classical phenylketonuria is an autosomal recessive human genetic disorder caused by a deficiency of hepatic phenylalanine hydroxylase (PAH — EC.1.14.16.1). The human PAH gene contains 13 exons (90 kb), and encodes for a protein of 451 amino acid residues. The genetic analysis of phenylketonuria (PKU) families indicates that mutations of the PAH gene are probably the main cause of PKU in humans. In normal subjects PAH catalyzes the rate limiting step in hydroxylation of phenylalanine to tyrosine, using tetrahydrobiopterin as cofactor. The reduced activity of PAH causes increased serum levels of phenylalanine and other normally minor metabolites. Severe PAH deficiency, known as classical phenylketonuria, causes severe mental retardation in untreated patients. By screening the newborn and by an adequate diet (restricted in phenylalanine intake), it is possible to avoid severe mental retardation. To be effective the diet must be rigidly implemented during the first decade of life.

Since the enzyme is synthesized in the liver and absent in fibroblasts, conventional methods of prenatal diagnosis by cultured aminocytes cannot be used in phenylketonuria. Therefore, prenatal diagnosis of PKU on a molecular basis was elaborated.[95,96] The investigation is begun with the cloning of cDNA for rat and human PAH, from which the complete primary structure of the human enzyme has been deduced. Cloned cDNA has been used to identify multiple restriction site polymorphisms at the human PAH locus. The procedure can be applied to prenatal diagnosis in PKU families at risk by fetal DNA analysis. It served also for the assignment of the PAH gene and the PKU locus to human chromosome 12q22 → 12q24.1.[95]

The first phenylketonuria mutation identified in the human phenylalanine hydroxylase gene was a single base substitution (GT → AT) in the cannonical 5′ splice donor site of intron.[96] Using a full-length human PAH cDNA clone, the authors obtained eight restriction fragment-length polymorphisms (RFLPs) at the human PAH locus. The RFLPs segregated in a mendelian mode, and concordantly with the mutant alleles in PKU families. Taking into account 87% of the observed heterozygosity of RFLPs in the PAH gene in the general population, it was possible to perform prenatal diagnosis in most PKU families by RFLPs analysis.[96] The authors identified 12 haplotypes of normal and PKU alleles in a Danish population, and observed a strong association among RFLP haplotypes and PKU alleles. Direct hybridization analysis using specific oligonucleotide probes demonstrated that the mutation was tightly associated with a specific RFLP haplotype among mutant alleles. The authors concluded that the detected splicing mutation is present at approximately 40% of mutant alleles and is the most prevalent phenylketonuria allele among Caucasians.[96]

A second molecular lesion was associated with the RFLP haplotype 2 mutant allele. The defect was caused by C → T transition in exon 12, resulting in the substitution of Arg → Trp at amino acid residue 408 of PAH. Direct hybridization analysis of this point mutation demonstrated that it is in linkage disequilibrium with RFLP haplotype 2 alleles of approximately 20% of the mutant PAH genes.[97]

The strong linkage disequilibrium in PKU alleles has been used to identify and characterize specific mutations causing PKU. PKU patients, homozygous for a PAH haplotype, are likely to carry the same mutation in the PAH gene. There exists a strong linkage disequilibrium among a specific splice junction mutation and a specific RFLP haplotype. This haplotype occurs only in 3% among normal (non-PKU) chromosomes, but accounts for approximately 40% of the PKU chromosomes. Using a pair of mutant and normal oligonucleotide sequences for a given mutation (e.g., splice junction mutation) the same mutation occurred on all (e.g., haplotype 3) PKU chromosomes, and never on other chromosomes (e.g., haplotype 2, characteristic for the point mutation position 408, Arg → Trp). All of the given haplotype current PKU alleles (e.g., haplotype 3) are evidently descendents of this single gene mutation. Thus, it is evident that RFLP analysis enables population studies of genetic diseases.[98]

A. HYPERPHENYLALANINEMIA DUE TO DEFICIENCY OF BIOPTERIN

The hepatic phenylalanine hydroxylating system consists of two enzymes: phenylalanine hydroxylase and dihydropteridine reductase, and the coenzyme tetrahydrobiopterin, which catalyze two reactions:

$$\text{Phenylalanine} + O_2 + \text{tetrahydrobiopterin} \rightarrow \text{tyrosine} + H_2O + \text{quinonoid dihydrobiopterin} \quad (1)$$

$$\text{DPNH (TPNH)} + H^+ + \text{quinonoid dihydrobiopterin} \rightarrow \text{tetrahydrobiopterin} + DPN^+ (TPN^+) \quad (2)$$

In a studied variant case of phenylketonuria with hyperphenylalaninemia and nephrologic disturbances (despite early dietary treatment), phenylalanine hydroxylase and dihydropteridine reductase activities were normal. Only the hydroxylation cofactor, tetrahydrobiopterin in the liver (10% of normal) was deficient. Additionally, serum and urinary levels of biopterin-like compounds were low, and the serum biopterin did not increase after administration of phenylalanine as it does in normal and phenylketonuric patients. Administration of tetrahydrobiopterin caused a transient decrease in the hyperphenylalaninemia. The phenylalanine hydroxylase activity in the diseased child was 2.3% of the normal value, which indicates that the child suffered from a deficiency of a functional hydroxylating system secondary to a defect in biosynthesis of biopterin.[99]

Another case of hyperphenylalaninemia with severe retardation in development, severe muscular hypotonia of the trunk and hypertonia of the extremities, convulsions, and frequent episodes of hyperthermia has been described.[100] Urine excretion of pterins was very low. Oral administration of L-erythrotetrahydropterin normalized the increased level of phenylalanine within 4 h, the serum tyrosine level increased briefly, and serum alanine and glutamic acid levels increased for a longer time. The described case is a new variant of hyperphenylalaninemia caused by the deficient formation of dihydroneopterin triphosphate and its pterin metabolites.[100]

X. THE LESCH-NYHAN SYNDROME

The Lesch-Nyhan syndrome is a rare disease caused by deficiency of the X-linked enzyme hypoxanthine — guanine phosphoribosyltransferase (HPRT). The disease affects males only. It is characterized by choreoathetosis (involuntary movements), spasticity with developmental and mental retardation, and self-mutilation. The deficient enzyme is known to be essential in the "salvage pathway" for recycling purine nucleotides. Therefore, the deficiency of this enzyme results biochemically in excessive production of uric acid and hyperuricemia. The reduced level of HPRT leads to intracellular loss of hypoxanthine and an increased *de novo* synthesis of purines, caused by the release of feedback inhibition. The hypoxanthine is oxidized to uric acid, which in humans builds up in the blood and urine (Figure 5). The pathological symptoms of Lesch-Nyhan syndrome develop when HPRT activity has decreased to about 1% of normal. The symptoms occur probably because of the toxic effects of abnormal levels of intermediary metabolites.[101]

HPRT is localized in the intracellular compartment, and is expressed in different degrees in all tissues. The highest activity is in the brain (particularly in basal ganglia). Human HPRT is a nonglycosylated protein of 217 amino acid residues. The N-terminal residue is acetylated. An open reading frame of 654 nucleotides has been described. Following translation, the peptide undergoes several modifications (cleavage of the N-terminal methionine

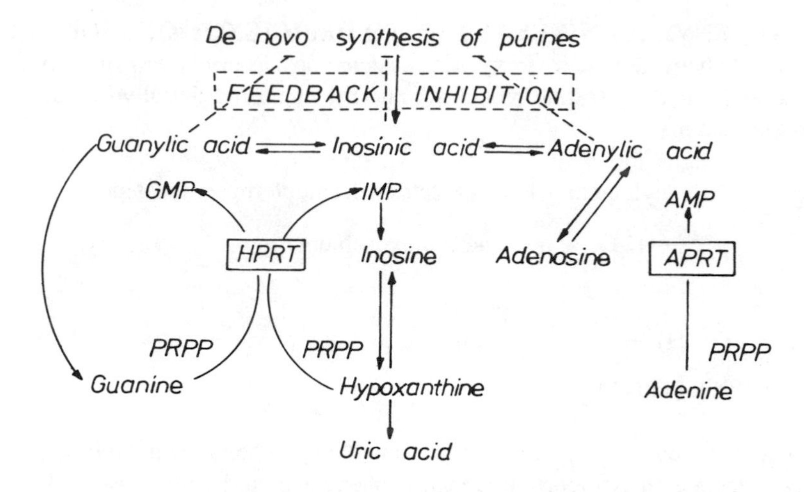

FIGURE 5. The role of hypoxanthine phosphoribosyltransferase (HPRT) in purine metabolism. HPRT catalyzes the synthesis of mononucleotides GMP, IMP by a condensation reaction with 5-phosphoribosyl-1-pyrophosphate (PRPP), liberating inorganic pyrophosphate. An analogous reaction is catalyzed by adenine phosphoribosyltransferase (APRT) in the synthesis of AMP. (5-phosphoribosyl-1-pyrophosphate serves also as substrate, together with L-glutamine in the first specific reaction of purine synthesis de novo). The deficient activity of HPRT leads to intracellular loss of hypoxanthine and in consequence to decreased levels of guanylic acid, inosinic acid, and adenylic acid, which eliminates the feedback inhibition on de novo synthesis of purines executed by these substances. (Redrawn from Hogan, B.[101])

and acetylation of the newly formed N-terminal alanine). The processed enzyme subunits aggregate to form the active native enzyme, a tetramer of four identical subunit monomers.[102,103] A comparison of the amino acid sequences of three phosphoribosyltransferases, human HPRT, adenosine triphosphate phosphoribosyltransferase from *Salmonella typhi* murium, and glutamine 5-phosphoribosyl-1-pyrophosphate (PRPP) demonstrated a minor 35-amino acid homology between HPRT and glutamine PRPP amidotransferase. However, a substantial correlation was found between all three enzymes in a region of 120 amino acid residues, suggesting a common structural feature.[103]

The entire HPRT gene is surprisingly complex. There are nine exons (from 18 to 593 bp) interrupted by eight introns (from 200 bp to 10.8 kb).

Early observations suggested that mutations responsible for HPRT deficiency might be heterogeneous. The differences from normal HPRT included Michaelis constants, sensibility to product inhibition, electrophoretic mobility, thermolability, and *in vivo* stability.[103] The deficiency of HPRT activity may be caused by abnormalities in either the catalytic function or the intracellular concentration of the enzyme. Most frequently, patients with the Lesch-Nyhan syndrome have no detectable HPRT immunoreactive protein, which indicates that enzyme deficiency in this disease is usually caused by diminished enzyme concentration.[103] A complete deficiency of hypoxanthine-guanine phosphoribosyltransferase is usually associated with the Lesch-Nyhan syndrome, whereas partial deficiency leads usually to overproduction of purines and gout.

In a study of five unrelated patients with more than 10% of normal activity (which enabled direct analysis of enzyme structure) each patient exhibited a unique structural variant. The study included: $HPRT_{Toronto}$, $HPRT_{London}$, $HPRT_{Ann\ Arbor}$, $HPRT_{Munich}$, and $HPRT_{Kinston}$.[103] In four of them structural and functional properties of these enzymes could be determined.[103]

$HPRT_{Toronto}$ was found in a patient with gout. The $HPRT_{Toronto}$ variant is characterized by a substitution at position 50. Arg → Gly, and the intracellular concentration of the enzyme was diminished, whereas the intrinsic activity (maximal velocity) and Michaelis and PRPP constants were normal.

$HPRT_{London}$ was isolated from erythrocytes and lymphoblasts of a patient with a severe form of gout. The enzyme concentration was diminished in erythrocytes and lymphoblasts with normal V_{max} and K_m for PRPP and a five-fold increased K_m for hypoxanthine, with normal isoelectric point and migration during nondenaturing polyacrylamide gel electrophoresis, and a smaller molecular weight. The $HPRT_{London}$ is characterized by a substitution at position 109, Ser → Leu.[104] The mutation in $HPRT_{London}$ is located in a predicted turn region that connects two β-pleated sheets that form the putative hypoxanthine binding site which explains its effect on binding of hypoxanthine.[103]

$HPRT_{Munich}$ represents a mutant form of human hypoxanthine-guanine phosphoribosyltransferase, isolated from a patient with gout and decreased enzyme activity. Sequence analysis of the mutant enzyme revealed a substitution at position 103, Ser → Arg. This amino acid substitution lies directly within the putative hypoxanthine-binding site of human hypoxanthine-guanine phosphoribosyltransferase, explaining its selective effect on decreased enzyme activity and binding of hypoxanthine.[105] Furthermore, the enzyme revealed normal intracellular concentration, 20-fold decreased intrinsic activity (maximal velocity), 100-fold increased Michaelis constant for hypoxanthine, and normal K_m for PRPP.[103]

The $HPRT_{Kinston}$ variant was isolated from a patient with the Lesch-Nyhan syndrome. The structural alteration in $HPRT_{Kinston}$, Asp → Asn at position 193, has dramatic effects on the Michaelis constants hypoxanthine (200-fold increase) and PRPP (also 200-fold increase). These effects on substrate binding site render the enzyme nonfunctional *in vivo*. The intracellular concentration and intrinsic activity (maximal velocity) of the enzyme were entirely normal. The amino acid substitution in $HPRT_{Kinston}$ is located outside the proposed dinucleotide fold in the area of the molecule.[103]

$HPRT_{Ann\ Arbor}$ was isolated from two male siblings with hyperuricemia and nephrolithiasis. The catalytic function of the enzyme was normal, whereas its intracellular concentration was markedly diminished (15% of normal). The intracellular degradation of $HPRT_{Ann\ Arbor}$ was substantially faster than that in normals. The Michaelis constants for hypoxanthine and PRPP, and intrinsic activity (maximal velocity) were normal. The exact structural molecular alteration of $HPRT_{Ann\ Arbor}$ is not known.[103]

In a study of 28 patients with the Lesch-Nyhan syndrome (by the method of Southern), using a near full-length HPRT cDNA as a probe, demonstrated that partial HPRT deficiencies were associated with gout, whereas the absence of activity resulted in Lesch-Nyhan syndrome. Structural alterations were identified in five patients. Gene deletions have been identified in three patients — one entire deletion and two partial deletions (one case of exons 4 through 9 deleted, and the other case of exons 7 through 9 deleted). In one case, the Southern blot analysis revealed an undefined rearrangement in the area of exons 4 through 6. None of these three deleted, and one rearranged mutant, cell lines exhibited production of a stable HPRT mRNA. Finally, an extremely labile mutant associated with a higher-molecular weight mRNA and undetectable HPRT protein demonstrated a duplication of exons 2 and 3. The duplicated exons probably had been incorporated into the mature mRNA; thus the higher-molecular weight mRNA and synthesis of an extremely labile translation product.[106]

According to Haldane's principle, new mutations at the HPRT locus must occur frequently in order for Lesch-Nyhan syndrome to be maintained in the population. Therefore, constant introduction of new mutations would be expected to result in a heterogeneous collection of genetic lesions, some of which may be novel. This rule has been confirmed in the study of 28 patients; seven among them demonstrated distinctly different mutations.[106]

2,8-Dihydroxyadenine urolithiasis (DHA) deficiency was described for the first time in 1974. Subsequently, several other cases of DHA lithiasis have been published. The authors have examined 19 Japanese families with DHA lithiasis. In 79% of them, only partial APRT deficiency has been detected. All patients with DHA lithiasis were homozygotes for defective

APRT genes, whether the deficiency was partial or complete. The disease is inherited as autosomal recessive disorder.[107]

Many mutations leading to human diseases result from single base pair (bp) alterations, and cannot be detected by the Southern method. Recently, a method has been developed to determine point mutations also. The described method of ribonuclease A cleavage, with a polyuridylic acid-paper affinity chromatography step, has been used for the identification of point mutations in 14 cases of Lesch-Nyhan syndrome in which no HPRT Southern or Northern blotting patterns had been found. Distinctive ribonuclease A cleavage patterns were identified in mRNA from 5 of the 14 cases of Lesch-Nyhan syndrome. It is supposed that the polyuridylic acid-paper affinity procedure allows HPRT mutation detection in 50% of all cases of Lesch-Nyhan syndrome, and should be used as a general method for analysis of low abundance mRNAs.[108]

Recently, two groups of authors established an experimental model of the Lesch-Nyhan syndrome in laboratory animals (mice). Embryonal stem cell lines, established in culture from peri-implantation mouse blastocysts, can colonize both the somatic and germ-cell lineages of chimaeric mice following injection into host blastocysts. Embryonal stem cells with multiple integrations of retroviral sequences can be used to introduce these sequences into the germ-line of chimaeric mice, instead of production of transgenic mice by microinjection of fertilized eggs. The newly described method has a unique possibility. Embryonal stem cells, challenged with adequate mutagens and using techniques of somatic cell genetics, can be selected *in vitro*, e.g., HPRT deficient cells that can be used for production of germ-line chimaeras, resulting in female offspring heterozygotes for HPRT-deficiency, and the generation of HPRT-deficient embryos from these families. The clonal lines carrying different HPRT alleles have given rise to germ cells in chimaeras, allowing the derivation of strains of mutant mice with the same biochemical defect as the Lesch-Nyhan patients.[101,109,110]

The same method can be used for the insertion of a wide range of mouse genes and so provide a valuable resource in studying rare recessive disorders, and also for replacing endogenous genes with specific mutations by homologous recombination.

XI. LYSOSOMAL STORAGE DISEASES

Lysosomal storage diseases represent a large group of disorders caused by several different biochemical abnormalities of lysosomal enzymes. These disturbances may occur because of:

(1) Synthesis of defective lysosomal enzymes
(2) Defective or absent activator proteins
(3) Lack of recognition material essential for targeting the newly synthesized lysosomal enzymes to the lysosome
(4) Defective transport of small molecules across the lysosomal membrane.[111]

Lysosomes are subcellular organelles containing hydrolytic enzymes that catalyze the degradation of several macromolecules. These macromolecules can be: (1) taken into the cell by endocytosis (typical example are low density lipoprotein particles — LDL); (2) macromolecules derived from the cells of structural elements by autophagy; (3) macromolecules derived from the individual's own unused secretory products in exocrine or endocrine cells.[112]

Each of the lysosomal storage diseases is due to the mutation of genes that direct the synthesis of specific lysosomal enzymes, or proteins activating lysosomal enzymes. The primary genetic defect in the lysosomal storage disease mucolipidosis III is the enzyme *N*-acetylglucosamine-1-phosphotransferase. The enzyme recognizes lysosomal enzymes (recognition function) and phosphorylation of their oligosaccharides. The correct targeting of enzymes to lysosomes depends on recognition between mannose-6-phosphate receptors and phosphomannose moieties on the oligosaccharide chains of the enzymes. Mannose residues are phosphorylated by the lysosomal phosphotransferase, which transfers *N*-acetylglucosamine-1-phosphate from UDP-*N*-acetylglucosamine to mannose residues, and a phosphodiester glucosidase that removes the *N*-acetylglucosamine residues to generate phosphomonoesters.[113,114] The lysosomal enzymes pass from cell to cell by a mechanism resembling exocytosis and endocytosis, which makes a replacement therapy potentially feasible.[112]

Until now in human pathology only the marked reduction of the lysosomal enzyme phophotransferase in mucolipidosis II (I-cell disease) and the milder defect resulting in mucolipidosis III (pseudo-Hurler polydystrophy) are known.[113]

Transport disturbances of small molecules across the lysosomal membranes have been described in nephropathic cystinosis, a fatal disease in which disulfide cystine accumulates within lysosomes because of the failure to cross lysosomal membranes. Abnormal vitamin B_{12} storage in lysosomes has also been ascribed to defective transport of free vitamin B_{12}. Finally, impaired lysosomal transport of *N*-acetylneuraminic acid (sialic acid) has been demonstrated in Salla disease, a disorder characterized by free sialic acid storage within lysosomes.[111]

The Salla disease is an autosomal recessive inherited disease, characterized by moderate to severe psychomotor retardation, spasticity, and ataxia with an early onset and slow progression. The patients excrete large amounts of free sialic acid and store 10 to 30 times normal amounts of sialic acid within several tissues. The amounts of sialic acid compounds (glycoproteins, gangliosides) are normal in cells of these patients, and there is a normal amount of acid neuraminidase activity that cleaves sialic acid from glycoconjugates in Salla disease lysosomes. The enzyme *N*-acetylneuraminate pyruvate-lyase that cleaves the neuraminic acid ring structure as the first step in sialic acid catabolism is present in Salla cells at normal activity. Further investigations demonstrated that labelled sialic acid was taken up by the lysosomes of Salla disease, but sialic acid was cleared at a much slower rate than in normal cells, which led to the conclusion that the cause of Salla disease depends on a defect of sialic acid egress from the lysosomes of these patients.[111]

The sialidosis disease is characterized by neurologic symptoms (myoclonus, epilepsy, cerebellar ataxia) with or without dysmorphic features similar to those for mucopolysaccharidoses. The disease is caused by the deficiency of α-*N*-acetyl-neuraminidase, resulting in neuronal or generalized storage of sialic-rich oligosaccharides containing glycoproteins. In some cases the defect of sialidase activity is associated with a β-galactosidase activity deficiency. For these cases the term galactosialidosis has been coined. The molecular defect of galactosialidosis is lack of an enzyme (protein) that stabilizes both sialidase and β-galactosidase, resulting in clinical symptoms of these two different diseases.[115]

The pathogenic effect of aberrant metabolism of lipids on dysfunction and degeneration of cells of the central nervous system remained for a long time unclear. Recent advances on the molecular basis of cell signaling provided an explanation in that sphingolipid metabolites, termed lysosphingolipids, may inhibit protein kinase C activity and phorbol diester binding to the enzyme. It is suggested that accumulation of galactosphingosine in Krabbe's disease and glucosphingosine in Gaucher's disease can inhibit protein kinase C continuously, with signal transduction blocked. In experiments, lysosphingolipids inhibited protein kinase C at molar percentages similar to those required for activation by phosphatidylserine or diacylglycerol. Since these substances accumulate in Krabbe disease, Gaucher disease and other sphingolipidoses progressive dysfunction of the signal transduction mechanisms vital for neural transmission, differentiation, development, and proliferation of cells occurs, which hinder these functions and may lead to cell death.[116]

A. THE LIPIDOSES

1. Ceramide Deficiency: Farber Lipogranulomatosis

This is an autosomal recessive disorder, characterized by hoarseness, painful and swollen joints, subcutaneous nodules, pulmonary infiltrations, and accumulation of lipids (ceramides and acid ceramidase) in the cytoplasm of neurons. Ceramide is the parent substance of nearly all complex sphingolipids. The biosynthesis of ceramide is initiated by the condensation of palmityl-CoA and serine to form sphingosine; next, the palmityl-CoA molecule reacts with sphingosine to form ceramide. The reaction is catalyzed by acyl CoA:sphingosine *N*-acyl transferase (EC.2.3.1.24). There are several acyl:CoA sphingosine *N*-acyl transferases that differ in fatty acid composition. Ceramide is catabolized by the ceramidase (acylsphingosine deacylase (EC.3.5.1.23)). This latter reaction occurs in the lysosomes. The deficient enzyme in Farber disease is acid ceramidase that catalyzes the reaction:

$$\text{Ceramide} + H_2O \rightleftarrows \text{sphingosine} + \text{fatty acid}$$

2. Sphingomyelin Lipidosis: Niemann-Pick Disease

This is characterized by accumulation of sphingomyelin (ceramide phosphocholine). Sphingomyelin is composed of equimolar amounts of long-chain amino alcohol sphingosine, a long chain fatty acid, and phosphocholine. The main clinical symptom is hepatosplenomegaly. The deficient enzyme is sphingomyelinase that catalyzes the reaction:

$$\text{Sphingomyelin} + H_2O \rightarrow \text{ceramide} + \text{phosphocholine}$$

The two classic types A and B represent primary lysosomal storage disorders caused by severe sphingomyelinase deficiency. Types C, D, and E are characterized by chronic neurological deterioration associated with accumulation of sphingomyelin in certain tissues. Esterification of cholesterol in types A and B Niemann-Pick cell cultures was normal, whereas all examined (20) patients with type C demonstrated a deficiency in cholesterol esterification (without evident acyl-CoA:cholesterol *O*-acyltransferase).[117]

3. Glucosyl Ceramide Lipoidosis: Gaucher Disease

The disease is an autosomal, recessive disorder characterized by hepatosplenomegaly, hypersplenism, osteoporotic erosion of long bones caused by accumulation of the sphingolipid glucocerebroside due to deficient activity of the enzyme glucocerebrosidase (EC.3.2.1.45). Glucocerebroside is composed of one molecule of sphingosine, one molecule of glucose, and a long-chain fatty acid. The deficient enzyme glucocerebrosidase catalyzes the reaction:

$$\text{Glucocerebroside} + H_2O \rightarrow \text{ceramide} + \text{glucose}$$

The most common form type I (or A) is prevalent in those people of Eastern European Jewish ethnic origin. More severe forms are termed type II (or B) and type III (or C), and are associated with central nervous system changes. The kinetic studies delineated these three distinct groups on the basis of residual activities with characteristic responses to the enzyme modifiers taurocholate (or phosphatidylserine) and glucosyl sphingosine (or *N*-hexyl glucosyl sphingosine). Type I: residual enzymes responded normally to these modifiers; type II: residual enzymes demonstrated markedly abnormal responses to these modifiers; and type III: residual enzymes revealed an intermediate response to all modifiers. Further, type II and type III enzymes had a decreased thermostability, whereas among type I enzymes three classes of thermostability were found: normal, increased, and decreased. In conclusion, these investigations indicate that type I is biochemically heterogeneous and results from at least four distinct allelic β-glucosidase mutations that alter their structure and function. The neuropathic and non-Jewish non-neuropathic phenotypes cannot be distinguished by kinetic analysis alone; the Ashkenazy type I Gaucher disease results from a unique mutation that alters a specific active site domain of the β-glucosidase.[118] Crossreacting material examinations and restriction mapping confirmed that different mutations have occurred both in the Jewish and nonJewish population.[119,120] Retrovirus mediated transfer of the human glucocerebrosidase gene to Gaucher fibroblasts demonstrated the feasibility of efficiently transferring the gene encoding glucocerebrosidase to Gaucher cells.[121]

4. Galactosylceramide Lipidosis: Globoid Cell Leukodystrophy (Krabbe's Disease)

The disease is known in humans, dogs, and mice. It is a rare, progressive, fatal disorder with clinical and pathological manifestations restricted mainly to the nervous system, particularly to the white matter and the peripheral nerves. The disease is inherited by an autosomal recessive mode. The onset of the disease is usually in the first 3 to 6 months of age. The initial symptoms are irritability which soon progresses to severe mental and motor deterioration. Sometimes the disease shows a later onset and a slower course. In all cases typical globoid cells are present. The galactosylceramide is composed of sphingosine, the amino group of which is acylated with a long-chain fatty acid (C_{14} to C_{26}), forming the ceramide molecule. The hydroxyl group of ceramide is substituted for galactose, which forms the galactosylcerebroside (or galactosylceramide) molecule. The biosynthetic process of galactosylceramide is complex. Two alternative pathways have been proposed: one through psychosine (formed from sphingosine and UDP-galactose), and the other through ceramide. Biosynthesis to sulfatide occurs through cerebroside with the "active sulfate" 3′-phosphoadenosine 5′ phosphosulfate as sulfate donor. The degradation initiates the removal of the sulfate group to convert it to galactocerebroside. This reaction is catalyzed by cerebroside sulfate sulfatase (present in arylsulfatase A). The deficiency of this enzyme causes metachromatic leukodystrophy. Galactocerebroside is degraded by the lysosomal enzyme galactocerebroside β-galactosidase to ceramide + galactose in the following reaction:

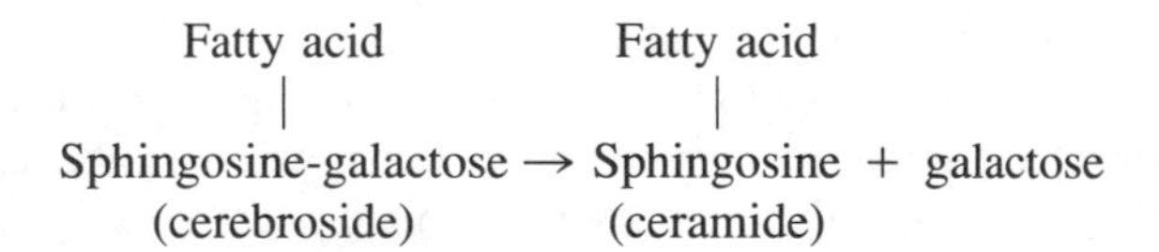

The white matter of the brain is rich in galactocerebroside and its sulfate ester, sulfatide. The gray substance contains much less of these glycolipids.

The deficient enzyme in Krabbe's disease is galactocerebroside β-galactosidase. The deficiency of this enzyme has been discovered in the brain, liver, spleen, kidney, peripheral leukocytes, and fibroblasts. Paradoxically, accumulation of galactosylceramide, the natural substrate of the missing enzyme, does not occur in the nervous system, where deterioration of the white matter occurs. It is supposed that accumulation of another metabolite galactosylsphingosine, a toxic metabolite termed psychosine, occurs in oligodendria cells, the producers of myelin, of which galactosylceramide is the major component. Therefore, devastating pathology of the white matter occurs, particularly of the rapid and total degeneration of oligodendroglia. Psychosine is absent in the brains of normal individuals, but present in the brains of patients with Krabbe disease. Psychosine is also a substrate for galactosylceramidase; thus, the deficiency of this enzyme causes accumulation of psychosine. The psychosine hypothesis for the pathogenesis of galactosylceramidase deficiency is based on the following data: deficiency of galactosylceramidase, the substrates of which are galactosylcerebroside and psychosine (galactosylsphingosine), causes accumulation of psychosine in the myelin-generating cells (oligodendroglia and Schwann cells). Psychosine is highly cytotoxic, and selectively kills the cells within which it accumulates. Since biosynthesis of galactosylceramide is restricted to myelin-generating cells, their degeneration terminates further synthesis at an early stage of myelination. Myelination continues rapidly in normal brain, and thus the amount of galactosylceramide will be lower in affected brain, even though catabolism of galactosylceramide is blocked because of the deficient enzyme activity.[122]

5. Sulfatide Lipidosis: Metachromatic Leukodystrophy

The disease comprises several closely related disorders, transmitted in an autosomal, recessive mode in which myelin degeneration is associated with the accumulation of galactosyl sulfatide and other lipids (galactosyl-3-sulfate). Two distinct forms are due to deficient activity of arylsulfatase A (termed also cerebroside sulfatase (compare with the enzyme description in the above-described Krabbe disease)). Metachromatic leukodystrophy is a lysosomal storage disorder of cerebroside sulfates. Progressive mental and motor deterioration are characteristic of this disease. The age of onset and the severity of the disease are highly variable. The most severe form is the late-infantile form which develops between the first and second year of life, and patients die usually in the first decade of life. In the late-onset forms the disease progresses more slowly, and in mild cases the disease may not be suspected during the entire lifetime. In some juvenile forms the arylsulfatase A activity is normal, and the decrease of activity is caused by the absence or deficiency of an enzyme activator. In some cases the arylsulfatase A activity was normal, but the enzyme was unstable in the lysosomes. The addition of cysteine proteinases to the culture medium of these cells protected the mutant enzyme from degradation, and normal catabolism of cerebrosides occurred.[123] Multiple sulfatase deficiency is associated with deficient activities of cerebroside sulfatase, of steroid sulfatases, and of mucopolysaccharide sulfatases.[124] Clinically, there are symptoms of metachromatic leukodystrophy and mucopolysaccharidoses.

6. Fabry's Disease

This is an inborn error of glycosphingolipid metabolism resulting from the defective activity of the lysosomal enzyme hydrolase α-galactosidase A (EC.3.2.1.22). The human active enzyme is a homodimeric protein (M_r of the subunit is 49,800) encoded by a structural gene localized on the X chromosome (q21 to q22). The consequence of the deficient enzyme is the accumulation of the major glycosphingolipid substrate; the trihexosylceramide: globotriaosyl ceramide, and related glycolipids with terminal α-galactosidic linkages. The progressive deposition, particularly in the plasma and vascular endothelium, causes ischemia and infarction. Early death occurs due to vascular disease of the heart, kidney, and/or brain. Clinically, the homozygous males have a characteristic skin lesion: angiokeratoma *corporis diffusum universale*. They have also corneal opacities, acroparestesias, and episodic pain of extremities. The glycosphingolipids involved in Fabry disease are neutral glycosphingolipids (in contrast to gangliosides that contain one or more acidic sialic groups, and sulfo-glycosphingolipids — sulfatides — that contain a sulfate ester group on the carbohydrate moiety). Neutral glycosphingolipids are synthesized by sequential enzymatic reactions. Glycosphingolipids are degraded in stepwise reactions catalyzed by specific exo-glycosidases. The primary defect in Fabry's disease is the defective activity of α-galactosidase A, which leads to the deposition of triglycosylceramide (galacto-galacto-glucosyl ceramide) and, to a lesser extent, galactobiosylceramide (galacto-galactosyl-ceramide). Recently, a cDNA clone encoding human α-galactosidase A has been found.[125]

B. THE GANGLIOSIDOSES

G_{M2}-gangliosidosis (known under the name Tay-Sachs disease) was described in 1881, and G_{M1}-gangliosidosis in 1965. At present some different subtypes of the main gangliosidoses are known. Several human genetic diseases are caused by the deficient activity of the lysosomal enzyme β-galactosidase (EC.3.2.1.23) alone, such as G_{M1} gangliosidosis and Morquio B syndrome, or in combination with a deficiency of neuraminidase (EC.3.2.1.18) in galactosialidosis. The biosynthetic pathway of G_{M1} involves the addition of five sugars in strict sequence by a multiglycosyl transferase system composed of five different glycosyl transferases; the sixth step is catalyzed by a sialyltransferase which leads to the biosynthesis of highly sialylated forms (e.g., trisialogangliosides). The five most common gangliosidosis phenotypes are: G_{M1}-gangliosidosis type 1 (infantile gangliosidosis) and G_{M1}-gangliosidosis type 2 (juvenile gangliosidosis); G_{M2}-gangliosidosis type 1 (Tay-Sachs disease); G_{M2}-gangliosidosis type 2 (Sandhoff disease), and G_{M2}-gangliosidosis type 3 (juvenile gangliosidosis).

1. Infantile G_{M1}-Gangliosidosis (Type 1)

This is a generalized gangliosidosis occurring at birth. Locomotor disturbances occur in the first 3 to 6 months. Clonic-tonic convulsions are frequent. The sensorium is dull. Respirations are labored. In the second year of life, blindness, deafness, and unresponsiveness occur. The patients usually die at the age of 2 years.

2. Juvenile G_{M1}-Gangliosidosis (Type 2)

The onset is later in life. During the first year of life no abnormalities may be seen. Locomotor ataxia occurring approximately at the age of 1 year initiates the severe clinical symptoms. Motor and mental deterioration progresses rapidly. The patients die usually between 3 and 10 years.

The enzyme defect in G_{M1}-gangliosidosis is the deficiency of degradative lysosomal acid enzyme β-galactosidase. Brain β-galactosidase activities for lactoside (ceramide-glucose-galactose) are normal. The metabolic block involves the first step in G_{M1}-ganglioside breakdown.

β-Galactosidase activity is also deficient with the keratan sulfate-like storage glycoprotein and asialofetuin as substrates. In Morquio B syndrome, the mutation does not interfere with the normal processing and intralysosomal aggregation of β-galactosidase. In infantile and juvenile G_{M1}-gangliosidosis, an 85-kDa precursor was synthesized normally, but more than 90% of the enzyme was subsequently degraded during post-translational processing. The residual 5 to 10% becomes processed with normal catalytic activity, but with reduced ability to aggregate into high molecular weight multimers.[126] In Morquio B syndrome the mutation affects the affinity and catalytic properties towards glycosoaminoglycans and ganglioside G_{M1}. The combined β-galactosidase and neuraminidase deficiency in patients with galactosialidosis is due to the deficiency of a 32-kDa "protective protein" which is normally acquired for the aggregation of β-galactosidase monomers into a high molecular weight complex with neuraminidase. Deficiency of this protein results in accelerated proteolytic degradation of β-galactosidase monomers, whereas neuraminidase remains inactive because of the 32-kDa protein is an essential subunit for this enzyme.[127,128]

The B subunit of cholera toxin, which is multivalent and binds exclusively to a specific ganglioside G_{M1}, is mitogenic for rat thymocytes. This demonstrates that endogenous plasma membrane gangliosides can mediate proliferation of lymphocytes.[129]

3. G_{M2}-Gangliosidosis (Type 1) — Tay-Sachs Disease

The onset of the disease occurs between 3 and 6 months of age with motor weakness. Motor and mental deterioration progresses rapidly. After 18 months, usually progressive deafness, blindness, convulsions, and spasticity appear, and a state of decerebrate rigidity occurs. The patients die usually from bronchopneumonia at the age of 3 years. Pathologic changes include conspicuous neuronal lipidosis of cortical, autonomic, and rectal mucosal neurons. The storage bodies contain large quantities of ganglioside G_{M2}.

4. G_{M2}-Gangliosidosis (Type 2) — Sandhoff Disease

The clinical and pathological findings are similar to those in Tay-Sachs disease.

5. G_{M2}-Gangliosidosis (Type 3) — Juvenile G_{M2} Gangliosidosis

The onset occurs between 2 and 6 years of life. Locomotor ataxia is the most frequent initial symptom, followed by progressive spasticity. Blindness occurs late. All patients deteriorate to a state of decerebrate rigidity. Death occurs between 5 and 15 years of age.

Cerebral levels of ganglioside G_{M2} are increased 100 to 300 times normal in Tay-Sachs and Sandhoff diseases. The asialoderivatives of ganglioside G_{M2} accumulate approximately 20 times the normal (much higher in Sandhoff disease than in Tay-Sachs disease). Globoside (ceramide-glucose-galactose-galactose-*N*-acetylgalactosamine) accumulates in large quantities in the viscera. In juvenile G_{M2}, gangliosidosis, cerebral levels of ganglioside G_{M2}, and asialoderivatives of ganglioside G_{M2} are markedly elevated, but not to the level seen in Tay-Sachs and Sandhoff diseases.

The deficient enzyme in G_{M2}-gangliosidoses is the lysosomal enzyme β-hexosaminidase (EC.3.2.1.30). The enzyme occurs in two major forms: hexosaminidase A (M_r 120,000) and hexosaminidase B (M_r 130,000). Hexosaminidase A is a heteropolymer consisting of the α and β subunits (54 and 28 kDa, respectively). The β subunits of hexosaminidase A and B are distinct polypeptide chains β_a (28 kDa) and β_b (27 kDa), which are joined by disulfide bonds in the native enzyme. These two chains appear to be produced by specific post-translational proteolysis of a single pre-β-chain polypeptide (63 kDa). The genes have been mapped for α chain to chromosome 15, and for β chain to chromosome 5. Mutation of the α subunit (Tay-Sachs disease) or β subunit (Sandhoff disease) causes G_{M2}-gangliosidosis characterized by storage of G_{M2} ganglioside in the brain and other tissues.[130] On the basis of DNA hybridization analysis, the α subunit seems to be intact, whereas the α subunit gene in French-Canadian patients has a 5′ deletion approximately 5 to 8 kb.[131] Cell hybridizations between fibroblasts of four variants of infantile G_{M2}-gangliosidosis (B, O, AB, and B1) revealed that B1 cells carry a mutation in the gene locus for the α subunit of β-hexosaminidase.[132]

Recently, a new type of G_{M2}-gangliosidosis has been described with β-hexosaminidase deficiency in adults. The clinical symptoms are very different. For example, one patient was demented in her 20s, whereas her mother was still active and independent at 67 years of age. The molecular defect seems to be caused by a premature intracellular proteolysis of the enzyme.[133]

C. THE MUCOPOLYSACCHARIDE STORAGE DISEASES

The mucopolysaccharidoses represent a large group of lysosomal diseases in which mucopolysaccharides accumulate in lysosomes in a progressive manner, caused by deficient activity of several lysosomal enzymes. The stored material is heterogeneous, depending on the kind of the defective enzyme. The patients have skeletal deformities including facial features, which gave the disease its former name "gargoylism".

1. Mucopolysaccharidosis Type I

This disease is caused by α-L-iduronidase deficiency. It comprises three different types: I H, Hurler syndrome; I S, Scheie syndrome; and I H/S, Hurler-Scheie type. The basic defect in Hurler, Scheie and Hurler-Scheie types is the same, i.e., deficiency of α-L-iduronidase resulting in urine excretion of dermatan sulfate and heparan sulfate. The Hurler syndrome is characterized by early clouding of corneas, severe and multiple dysostoses, mental retardation, and heart disease. Death occurs in the first 10 years of life. The Hurler syndrome is inherited by the autosomal, recessive, mucopolysaccharidosis gene I H. The Scheie syndrome is characterized by stiff joints, cloudy corneas, and aortic valve disorder.

The life-span is normal. The disease is inherited by the autosomal, recessive mucopolysaccharidosis gene I S. The Hurler-Scheie syndrome is characterized by clinical symptoms intermediate between Hurler and Scheie syndromes due to inherited mucopolysaccharidosis genes I H and I S.

2. Mucopolysaccharidosis Type II — Hunter Syndrome

This is caused by iduronate sulfatase deficiency with urinary excretion of dermatan sulfate and heparan sulfate. Clinical symptoms are milder than in mucopolysaccharidoses type I. The Hunter syndrome is inherited as an X-linked recessive disorder. There are two forms of Hunter syndrome, severe (death usually before 15 years of life) and mild (death between 30 and 60 years of life).

3. Mucopolysaccharidosis Type III A — Sanfilippo Syndrome A

This is caused by heparan sulfatase deficiency with urinary excretion of heparan sulfate. Clinically, there is severe progressive mental retardation and relatively mild somatic changes. The disease is inherited as an autosomal recessive trait. Mucopolysaccharidosis type III B, Sanfilippo syndrome B, is caused by *N*-acetyl-α-D-glucosaminidase deficiency with urine excretion of heparan sulfate. Clinically, the two forms of A and B, are indistinguishable. The Sanfilippo syndrome B is also inherited as an autosomal recessive trait.

4. Mucopolysaccharidosis Type IV — Morquio Syndrome

This disease is caused by hexosamine-6-sulfatase deficiency with urinary excretion of keratan sulfate; the syndrome seems to be caused by defective degradation of keratan sulfate. Cartilage is the most affected tissue resulting in severe bone changes (severe knockknees, pectus carinatum, short trunk, and short neck). The disease is inherited as an autosomal recessive trait. The deficient enzyme seems to hydrolyze also G_{M1} gangliosides, which explains bone deformities and storage of keratan sulfate-like material in infants with G_{M1} gangliosidosis. The patients die because of pulmonary complications and cardiorespiratory insufficiency.

5. Mucopolysaccharidosis Type VI — Maroteaux-Lamy Syndrome

This is caused by arylsulfatase B (*N*-acetylgalactosamine 4-sulfatase deficiency) with urinary excretion of dermatan sulfate. There are three distinct forms of the Maroteaux-Lamy syndrome: the severe classic form (severe bone, soft tissue, and corneal changes, survival to 20 years); intermediate form (moderately severe changes as compared to the classic form); mild form (severe bone and corneal changes). The disease is inherited as an autosomal recessive trait with probably variants of gene loci for the three variants of the Maroteaux-Lamy syndrome.

6. Mucopolysaccharidosis Type VII

This is caused by β-glucuronidase deficiency, with urinary excretion of dermatan and heparan sulfate. The disorder is inherited as an autosomal recessive trait (probably with more allelic forms). Clinical symptoms are hepatosplenomegaly, dysostosis multiplex, and mental retardation.

7. Mucopolysaccharidosis II and III

Also known as I-cell disease or inclusion cell disease, and pseudo-Hurler dystrophy, respectively, these are characterized by elevated levels of lysosomal hydrolases in body fluids and deficiency in fibroblasts. The enzymes with abnormal distribution include: α-L-iduronidase, iduronate sulfatase, β-glucuronidase, *N*-acetyl-β-hexosaminidase, arylsulfatase A, β-galactosidase, α-mannosidase, and α-L-fucosidase. The primary defect that causes this inappropriate distribution of hydrolytic enzymes is unknown.

D. GLYCOGEN STORAGE DISEASE TYPE II — POMPE'S DISEASE

Also belonging to the lysosomal storage diseases is this disease, caused by α-1,4 glucosidase (acid maltase) deficiency. There are three forms of the disease: the infantile form has its onset approximately at 2 months and death at 5 months of age; the disease is characterized by generalized glycogenosis, profound hypotonia, and cardiac failure without cyanosis in the first year of life; the heart is strikingly enlarged. The second form appears in infancy or early childhood (the patients die, usually heart failure, before 20 years of age). The adult form of the disease progresses more slowly, without severe organopathy, and with marked muscular weakness. In the liver, part of the glycogen is dispersed as in normal livers, but a large fraction is segregated in large vacuoles surrounded by a single membrane. The vacuoles are lysosomes overloaded with glycogen (because of the absence of acid maltase in the lysosomes, glycogen cannot be degraded). The disease is inherited as an autosomal, recessive trait.

E. NON-LYSOSOMAL STORAGE DISEASES

These diseases originate either because of the deficiency of enzymes essential for glycogen degradation, or abnormal molecular structure of the synthesized glycogen molecules.

1. Type I Glycogen Storage Disease — v. Gierke Disease

This is caused by glucose-6-phosphatase deficiency. The main clinical symptoms are hepatomegaly, acidosis, hypoglycemia and hyperlipidemia. The main glycogen deposits are in the liver and kidney.

2. Type III Glycogen Storage Disease — Forbes, Cori Disease

Caused by amylo-1,6-glucosidase (debrancher) deficiency, this form of glycogenosis is also termed "limit dextrinosis" because the stored glycogen resembles limit dextrin produced by degradation of glycogen by phosphorylase free of debrancher enzymes that cleaves the outer branches of glycogen (α-1,4 linkages of glucose residues); the inner branches are left intact because of amylo-1,6-glucosidase deficiency. Because of the impossibility of cleaving the remaining abnormal glycogen, hypoglycemia occurs. In this aspect type III glycogenosis is indistinguishable from the glycogenosis type I (glucose-6-phosphatase deficiency). The abnormal glycogen accumulates in liver and muscles resulting in hepatomegaly and chronic progressive myopathy. The disease is an autosomal recessive inherited malady.

3. Type IV Glycogen Storage Disease — Amylopectinosis or Andersen Disease

This is caused by α-1,4-glucan: α-1,4-glucan-6-glucosyl transferase (brancher) deficiency. The abnormal glycogen has long outer branches and resembles amylopectin. Reduced branching makes the glycogen less soluble. Clinically, there is hepatosplenomegaly with early liver cirrhosis.

4. Type V Glycogen Storage Disease — McArdle Disease

This disease is caused by muscle phosphorylase deficiency. The patients cannot perform strenuous exercise because of painful muscle cramps, following usually with myoglobinuria. The disorder is inherited as an autosomal recessive trait. Clinically, there are no disturbances. The liver phosphorylase is normal.

5. Type VI Glycogen Storage Disease — Hers Disease

This disease is caused by hepatic phosphorylase deficiency. The hepatic phosphorylase is distinct from the muscle phosphorylase which has normal activity in this disease. Clinically, there is hepatomegaly and hypoglycemia.

6. Type VII Glycogen Storage Disease — Tarui Disease

This is caused by phosphofructokinase deficiency. Clinical symptoms are the same as in muscle phosphorylase deficiency (McArdle disease). The level of phosphofructokinase in erythrocytes is markedly reduced, which causes shortening of the erythrocytes lifespan 14 to 16 days with reticulocytosis (approximately 5%).

7. Type VIII Glycogen Storage Disease — Huijing Disease

This is caused by phosphorylase kinase deficiency (phosphorylase activating enzyme deficiency). Phosphorylases require both activating and inactivating enzymes. Patients with reduced phosphorylase b kinase have reduced liver and leukocyte phosphorylase. Clinically, there is hepatomegaly and hypoglycemia.

XII. COFACTOR DEPENDENCY OF ENZYMES

Enzymes are usually synthesized in the form of inactive proenzymes which become activated by various mechanisms. One of them is selective cleavage which unmasks active sites of enzymes (this mechanism is best known in the blood coagulation cascade, where proenzymes are substrates for the active enzymes generated in the previous reaction). Another mechanism depends on the binding of apoenzymes with cofactors, usually vitamins, but also with other substances. The latter is similar to the mechanism that holds the heme ring in the hemoglobin molecule, deeply buried in a cleft termed the ''heme pocket''. The heme pocket is lined with nonpolar amino acid groups which exclude water from the vicinity of the heme. The vinyl side chains of the porphyrin ring are oriented toward the hydrophobic interior of the heme pocket, with the polar propionic acid groups toward the hydrophilic surface of the subunit. Further, the histidine residue forms a strong bond with the heme iron, and a large number (approximately 60) of weak hydrophobic bonds and van der Waals' contacts stabilize the heme group within the heme pocket. In spite of these multiple bonds, loss of single van der Waals' contacts, and especially weak hydrogen bonds, is sufficient for the heme group to be extruded from the heme pocket with simultaneous loss of function of the hemoglobin molecule.

Precisely the same situation may occur when an apoenzyme molecule undergoes mutation at the site that binds with cofactors. In this case the binding of cofactor is hindered or even made impossible, which decreases or makes impossible the activation of such enzymes. However, in these cases an increase of cofactor concentration according to the law of mass action can overcome the hindered formation of apoenzyme-cofactor complex, resulting in activation of the inactive apoenzyme. Precisely this situation occurs in many inborn errors of metabolism with deficient activation of enzymes, which can be overcome after administration of cofactors (usually vitamins) by several hundreds, or even thousands, of times the dose of the normal daily vitamin requirement. This kind of inborn error of metabolism is of special interest because administration of increased daily dose of the proper vitamin can completely neutralize the aberrant metabolism.

At first Bonner et al. (1960) demonstrated that gene mutation resulting in abnormal structure of the binding site of apoenzyme tryptophan synthase with the cofactor pyridoxal-5′-phosphate (PLP) caused its inactivation in *Neurospora crassa*. However, enrichment of the medium with pyridoxine caused normal tryptophan synthesis in *Neurospora*.[134]

The first vitamin dependency in humans was recognized in cystathioninuria. The defect is dependent on failure to cleave cystathionine to give cystine, α-ketoglutarate and ammonia, catalyzed by γ-cystathioninase deficiency. The enzyme is pyridoxal-5′-phosphate-dependent.

The pyridoxine-responsive disorders represent the most numerous group of metabolic diseases of cofactor-dependency.[2] This is explained by the participation of pyridoxine (pyridoxal-5′-phosphate) in activation of a large group of enzymes (about 50) participating in protein metabolism. To this group belong homocystinuria, cystathioninuria, xanthurenic aciduria, and ornithinemia. The primary oxaluria is probably caused by deficiency of α-ketoglutarate:glyoxylate carboligase, which is not a pyridoxal-5′-phosphate enzyme.[135,136]

Homocystinuria may be caused by deficiency of remethylation of homocysteine catalyzed by 5,10-methylenetetrahydrofolate reductase (EC.1.1.99.15) and 5-methyltetrahydrofolate:L-homocysteine methyltransferase (EC.2.1.1.13), which are less common deficiencies; or by the deficiency of cystathionine β-synthase (EC.2.2.1.13) (CS) which catalyzes the condensation of serine with homocysteine to form cystathionine — a more common deficiency.[135] The disease is characterized by abnormalities of the ocular, skeletal, and nervous systems with dislocation of lenses in almost all untreated patients. Homocystine and methionine levels in body fluids are elevated with decreased levels of cystine. The decreased metabolism leads to homocystinemia and homocystinuria. Oral administration of pyridoxine in doses ranging from 50 to 1000 times the physiological doses of 1mg/day causes in approximately 50% of patients a dramatic fall of homocystine and methionine and a rise in cystine (folate depletion in these cases must be excluded). In treated pyridoxine-responsive patients with standard oral methionine loads (4g/m[135] of body area) after adding betain (trimethylglycine) at the rate of 6g/d, the capacity to metabolize methionine increased. Betaine also increased plasma cysteine levels in patients with more severe abnormalities. The vascular complications in homocystinuria are caused by high plasma levels of homocysteine. Thus, betaine therapy may reduce this risk in patients receiving standard pyridoxine and folic acid treatment, and in whom abnormal homocysteine responses occur after the standard methionine load.[137]

Cystathioninuria is caused by γ-cystathioninase deficiency. The active enzyme contains pyridoxal-5′-phosphate as a cofactor. Mammalian γ-cystathioninase catalyzes the cleavage of cystathionine ($+H_2O$) into cysteine and α-ketobutyrate + NH_3. The same enzyme probably can catalyze also the formation of cystathionine when incubated with homoserine and homocysteine. However, the prevalent reaction is cleavage of cystathionine. Transient cystathioninuria occurs in almost all premature infants, and disappears usually at the age of 5 months. Persistent cystathioninuria may be caused either by underutilization or overproduction of cystathionine. Underutilization occurs usually in patients with genetically determined impairment of γ-cystathionase activity. Overproduction of cystathionine occurs in the case of an abnormal high amount of cystathionine β-synthase substrate:homocysteine, caused by the impairment of 5-methyltetrahydrofolate-homocysteine methyltransferase activity due to a lack of 5-methyltetrahydrofolate.[138] Cystathioninuria also may occur in nongenetic conditions like vitamin B_6 deficiency (because both cystathionine β-synthase and γ-cystathionase are pyridoxal-5′-phosphate-dependent), in thyrotoxic patients, because of a renal defect affecting transport of amino acids, and in some neural tumors.[138,139] An almost complete disappearance of amino aciduria follows in children after the administration of large doses of pyridoxine-hydrochloride daily, but not in adults.[140] Clinical symptoms of cystathioninuria are ocular abnormalities (ectopia lentis, myopia, glaucoma, optic atrophy, and retinal degeneration), skeletal abnormalities (osteoporosis, scoliosis, pectus carinatum or excavatum), mental retardation, arterial and venous thromboemboli, fair brittle hair, thin skin, fatty changes of liver, and changes in other organs.

Biochemically, the most characteristic feature in homocystinuria are abnormal high levels of plasma homocystine and methionine. For the detection of homocystine in plasma, it is necessary to precipitate protein immediately after the blood is obtained, because otherwise homocystine becomes removed with protein (homocystine forms disulfide bonds with protein). The abnormal methionine levels is probably due to enhanced rates of homocysteine methylation that affect the balance between the accumulation of methionine and homocystine. The same is observed in treatment with folic acid that shifts the methionine-homocystine balance, probably by increasing the rate of 5-methyltetrahydrofolate-dependent homocysteine methylation. Thus, administration of betaine or its metabolic precursor, choline, provides more substrate for betaine-homocysteine methyltransferase, and prevents abnormal high rates of homocysteine methylation.

The excessive dietary uptake of methionine and/or homocystine retards growth of the rat, followed by changes of a number of enzyme activities and pathological changes of various organs (pancreas, gastrointestinal tract, kidneys, spleen, thyroid, and adrenals). Simultaneous administration of glycine and arginine may alleviate these disturbances.

A. XANTHURENIC ACIDURIA

The disorder is caused by kinureninase deficiency that catalyzes the degradation of kynurenine to anthranilic acid and alanine. The enzyme is pyridoxal-5′-phosphate dependent and its deficiency results in inability to degrade kynurenine which becomes converted to xanthurenic acid, excreted in the urine. The affected pathway is that of tryptophan degradation (tryptophan is normally converted to phenylkynurenine → kynurenine → hydroxykynurenine kynureninase anthranilic acid, and alanine (anthranilic acid is the precursor of nicotinic acid). The disorder may be manifested by urticaria and bronchial asthma.

B. ORNITHINEMIA (GYRATE ATROPHY)

The deficient enzyme is ornithine δ-aminotransferase that catalyzes the reaction: ornithine + α-ketoglutarate → glutamic γ-semialdehyde + glutamate. The enzyme is pyridoxal-5′-phosphate dependent and is combined with the urea cycle, the deficiency which causes hyperammonemic syndromes. The pathway is the only known pathway for urea synthesis and the major pathway of ammonia detoxication. The disorder is manifested by convulsions appearing in the first hours of life. The pathogenesis of these convulsions is caused by deficient GABA (γ-aminobutyric acid), which is a potent neurotransmitter of inhibitory character.

C. IRON-LOADING (PYRIDOXINE-DEPENDENT) ANEMIA

The deficient enzyme is δ-aminolevulinic acid synthetase, which is pyridoxal-5′-phosphate dependent. The deficiency of this enzyme makes it impossible to synthesize porphyrins and subsequently heme. In the reticulocytes a decreased level of enzyme activity was detected that increased after administration of large doses of vitamin B_6 (pyridoxine).[135,141]

Pyridoxine dependency, its mechanism, and action of large doses of vitamin have been best studied in homocystinuria with the basic defect of (CS) cystathionine β-synthase.[135] The biochemical response to pyridoxine may be considerable, with a dramatic fall in plasma homocystine and methionine and a rise in cystine. In an attempt to correlate pyridoxine response with clinical severity, a large group of patients has been studied by Fowler.[135] Most patients of the mild/late group have been pyridoxine-dependent, whereas those of the severe early group have been pyridoxine-nondependent. To elucidate this dependency, residual CS activity in cultured fibroblasts has been measured with and without addition of the coenzyme pyridoxal-5′-phosphate. There were clear differences between the groups of cell lines from responsive and nonresponsive patients. In 26/30 of the responsive patients, a small detectable CS activity in their fibroblasts has been found, whereas only 3/16 of the nonresponsive patients revealed any detectable CS activity. Lower levels of activity, but similar differences between these groups, were found when assays were performed without pyridoxal-5′-phosphate. The important question was whether mutant CS can be stimulated by addition of pyridoxal-5′-phosphate. The results obtained confirmed the hypothesis that patients who are responsive *in vivo* have residual CS activity in fibroblasts, and conversely, that nonresponsive patients have no residual CS activity in their fibroblasts.[135] The next important question of Fowler's experiments was whether the degree to which CS activity is increased by the addition of pyridoxal-5′-phosphate is enough to explain the clinical responsiveness. It is supposed that even a few percent of CS activity control may allow significant handling in pyridoxine-treated patients, and in this way ameliorate clinical symptoms in them.[135] Restoration of enzyme activity by the *in vitro* addition of pyridoxal-5′-phosphate was elevated in homocystinuria, and particularly high in cystathioninuria (γ-cystathioninase activity) and in ornithinemia (ornithine δ-transferase activity).[135] Another explanation of the increase of CS activity may be that pyridoxal-5′-phosphate protects CS against degradation by proteolytic enzymes, as was shown in enzymes that form complexes with their coenzymes.[135] Finally, it was shown that there are CS mutants that are CRM$^+$ responsive in that they form stabilized pyridoxal-5′-phosphate complexes; and there are nonresponsive CS mutants which form unstable complexes because of low affinity to pyridoxal-5′-phosphate (CRM — cross reactive material).[135]

D. METHYLMALONIC ACIDURIA

The disorder is characterized by neonatal or infantile ketoacidosis. Methylmalonic acid is excreted in large amounts in urine in patients with vitamin B_{12} deficiency. The disorder is due to disturbances in conversion of methylmalonyl-CoA to succinyl-CoA that requires the enzyme methylmalonyl-CoA mutase, which is vitamin B_{12}-dependent. Methylmalonic aciduria is also observed in newborn infants and young children who are not vitamin B_{12}-deficient. Thus, in these children a metabolic block must exist caused by mutation of the methylmalonyl-CoA mutase.[142] Since methylmalonyl-CoA is a key intermediate in catabolism of branched-chain amino acids and fatty acids with an odd number of carbon atoms, ingestion of a high protein diet, with valine or isoleucine, causes an increase of excretion of methylmalonic acid and long-chain ketons that aggravates the pathological manifestations. Treatment of some of these children with increased doses of vitamin B_{12}, which enables binding of the enzyme with its cofactor, caused the disappearance of ketosis and urinary excretion of methylmalonic acid.[142,143] Inherited methylmalonic acidemia may not be simply considered as a state of vitamin B_{12} dependency because there are also described cases unresponsive to vitamin B_{12}.[144]

Methylmalonic aciduria also may be caused by deficient synthesis of adenosylcobalamin, the cofactor required for the activity of methylmalonyl-CoA mutase. Complementation analysis of cultured fibroblasts from patients with methylmalonic aciduria revealed five classes of defects of this mutase: defective synthesis of adenosylcobalamin synthesis A and B, and defective synthesis of both adenosylcobalamin and methylcobalamin (methylcobalamin, cobalamin C, and cobalamin D).[145] Deficiency of both cobalamin coenzymes results in methylmalonic aciduria accompanied by homocystinuria (caused by deficient activity of methionine synthase which catalyzes the conversion of homocysteine to methionine.[145]

E. BIOTIN-RESPONSIVE β-METHYLCROTONYL GLYCINURIA

The disorder is characterized by a severe metabolic acidosis and ketosis associated with β-methylcrotonylglycinuria and β-hydroxyisovaleric aciduria. The defective enzyme is probably β-methylcrotonyl-CoA carboxylase that converts β-methylcrotonyl-CoA into β-methylglutaconyl-CoA (and finally to acetyl-CoA). Thus, administration of increased doses of biotin immediately improved the clinical symptoms and caused the disappearance of the abnormal metabolites from the urine.[146] The disorder affects the normal metabolism of the branched chain amino acid — leucine.

Recently, a new complementation group of methylmalonic aciduria caused by inability to release free vitamin B_{12} from lysosomes has been described.[146] Complementation analysis was performed in cultured fibroblasts. Incorporation of the label from ^{14}C-propionate into acid-precipitable material was increased in heterokaryons formed by polyethylene glycol treatment, as compared with the culture of fibroblasts without addition of polyethylene glycol; this presents a new methylmalonic aciduria variant caused by inability to release vitamin B_{12} from lysosomes.

F. BIOTINIDASE DEFICIENCY

The disorder is caused by an inability to cleave biotin or other biotinylated peptides (biocytin) resulting from degradation of endogenous carboxylases and an inability to receive the vitamin biotin.[147] Infants in early childhood reveal biotin deficiency resulting in seizures, skin rash, alopecia, ataxia, hearing loss, developmental delay, and metabolic decompensation that can terminate in coma and death. Biochemically, the defect is characterized by accumulation of biocytin and depletion of biotin.[148] The children are normal at birth, and if treated with biotin, are symptomless, but once the clinical symptoms occur, visual impairment (acquired retinal dysplasia) and sensori-neural deafness may remain.[148] Subtle neurologic abnormalities may appear as early as two months of age.

G. THIAMINE-RESPONSIVE MEGALOBLASTIC ANEMIA

Two cases of megaloblastic anemia have been described that improved after treatment with increased doses of thiamine. In one of these cases, intermittent cerebellar ataxia with an increased level of serum alanine, and hyperalaninuria with hyperpyruvic acidemia were present. The disappearance of these symptoms after thiamine treatment seems to indicate a mutant pyruvic carboxylase that is thiamine-dependent.[149]

H. THIAMINE-RESPONSIVE MAPLE SYRUP-URINE DISEASE

Maple-syrup urine disease is an autosomal recessive disorder with the primary defect of oxidative decarboxylation of the branched chain α-keto acids derived from branched chain amino acids: leucine, isoleucine, or valine. Since 1971, when the first case was described, maple-syrup-urine disease has been classified into classical, intermediate, and intermittent types, dependent on the rapidity of onset, severity of the disease, and tolerance for dietary protein. The first described patient (a female infant with a variant form of maple-syrup-urine disease) responded to a thiamine dose 20 times the daily normal requirement.[150]

Recently, a new variant of thiamine-responsive maple-syrup-urine disease was described, caused by decreased affinity of the mutant branched chain α-keto acid dehydrogenase for α-ketoisovalerate and thiamine pyrophosphate. A system for branched chain α-keto acid dehydrogenase in disrupted human fibroblast preparations has been elaborated that demonstrated from 40 to 60% of the activity found in intact cells.[150] This system allowed a more direct measurement of enzyme activity (by elimination of pre-existing substrate pools and membrane effects). The elaborated assay permitted a study of thiamine response in normal human fibroblasts and skin culture from patients with variant forms of maple-syrup-urine disease. Using this system, the authors demonstrated a thiamine-pyrophosphate-mediated increase in the affinity of branched α-ketoisovalerate in thiamine-responsive maple-syrup-urine disease. The mutant enzyme demonstrated an elevated K_m value for thiamine pyrophosphate as compared with the enzyme from the normal or thiamine-nonresponsive patients.[150]

I. VITAMIN D-DEPENDENT SYNDROMES

In recent years the mechanism of vitamin D action has been considerably elucidated. The biological actions of vitamin D in mediating calcium homeostasis are supported by the combined presence of two dihydroxylated metabolites $1,25(OH)_2D_3$ and $24,25(OH)_2D_3$. On the basis of receptor studies for $1,25(OH)_2D_3$ and vitamin D-dependent calcium binding protein, this vitamin-endocrine system acts on several body organs: intestine, bone, kidney, colon, pancreas, chorioallantoic membrane, testes, ovary, uterus, cerebellum, placenta, pituitary gland, and parathyroids. $24,25(OH)_2D_3$ causes mobilization of calcium in bones, absorption of calcium in intestine, and reabsorption of calcium in the kidneys. It is important to distinguish vitamin D dependency from vitamin D resistancy.[151]

The clinical symptoms of vitamin D-dependent rickets type I are similar to vitamin D-deficient rickets. The onset is usually within the first year of life, with hypotonia (sometimes tetany and/or convulsions), weakness, and growth failure. There occur thickening of wrists and ankles, frontal bossing, and bone deformities with radiographic changes typical for rickets, particularly of the spine and the pelvis. Hypocalcemia is the most typical feature of vitamin D-dependent rickets with serum calcium concentration below 8.0 mg/dl. Serum immunoreactive parathyroid hormone level is increased. Intestinal calcium absorption is impaired. The serum phosphate level may be normal or low, but usually not as low as in vitamin D-resistant rickets (termed also familial hypophosphatemia). Decreased renal tubular reabsorption of phosphate and amino acids may occur and seems to be associated with the increased parathyroid hormone release. Mild hyperchloremic acidosis may occur.[150]

The vitamin D-dependent symptoms of rickets are reversible following treatment with vitamin D (doses 10 to 300 times normal). It is noteworthy that treatment of X-linked hypophosphatemia (vitamin D-resistant rickets) with a massive dose of vitamin D does not lead to complete healing of the rickets or normal growth rate. Serum 25-(OH)D_3 levels are normal and elevated in vitamin D- or 25-(OH)D_3-treated patients. The intestinal resorption of vitamin D and production of 25-(OH)D_3 are normal. The most important abnormality is a decreased serum level of 1,25$(OH)_2D_3$ level which indicates that in this kind of vitamin D-dependent rickets a partial deficiency of the renal 1α-hydroxylase must occur.[151]

The clinical symptoms of vitamin D-dependent rickets type II are similar to those in type I. The most important feature is an elevated plasma 1,25$(OH)_2D_3$ level. The onset of the disease is usually in the first year of life, with hypocalcemia, secondary hyperparathyroidism, osteomalacia, or rickets. In some families alopecia totalis may be present. There exists heterogeneity in response to vitamin D treatment. In general, almost all patients with vitamin D-dependent rickets type II reveal resistance to normalization of serum calcium and healing of bone changes, particularly families with alopecia, whereas families without alopecia responded to normal doses of vitamin D metabolites.[151] On the basis of experiments it is thought that the molecular defect in vitamin D-dependent rickets type II might be the level of 1,25$(OH)_2D_3$ receptor, or the postreceptor level. There exists also heterogeneity in the nature of the molecular defect. Patients of type II rickets may demonstrate absence of the receptor in fibroblasts, disturbances in nuclear binding of the receptor, or even have a totally (functionally) normal receptor.[150] In other studies abnormal binding, and normal binding but impaired stimulation of 24-hydroxylase activity in fibroblasts in response to treatment *in vitro* with 1,25$(OH)_2D_3$, have been found. The occurrence of alopecia in vitamin D-dependent rickets type II may be caused by receptor defects in keratinocytes similar to those seen in fibroblasts.[151]

J. HEPARIN DEPENDENCY OF LIPOPROTEIN LIPASE

Three variants of post-heparin lipoprotein lipase activity among 14 cases of hyperlipoproteinemia of type I (hyperchylomicronemia) or type V (mixed form) have been described.[152] These comprise two cases in which normal activity has been restored by the same doses of heparin which activated normal human lipoprotein lipase (variant I); one case in which only large doses of heparin caused normal activation of the enzyme (variant II: heparin-dependent lipoprotein lipase); and two cases in which even the largest doses of heparin were unable to activate the patient's blood serum to a lipolytic activity (variant II absence of heparin-induced lipoprotein lipase activity). The curves of lipolytic activities in these cases are presented in Figure 6.

Lack of lipoprotein lipase activity may be caused by: (1) lack or abnormal lipoprotein lipase molecule; (2) lack or abnormal heparin molecule; (3) pathological chylomicrons that cannot be hydrolyzed; (4) the presence of inhibitors in the patient's serum blocking the action of lipoprotein lipase or binding the heparin molecule. Possibilities (3) and (4) were excluded by cross-reactions; therefore, possibilities (1) and (2) must be taken into consid-

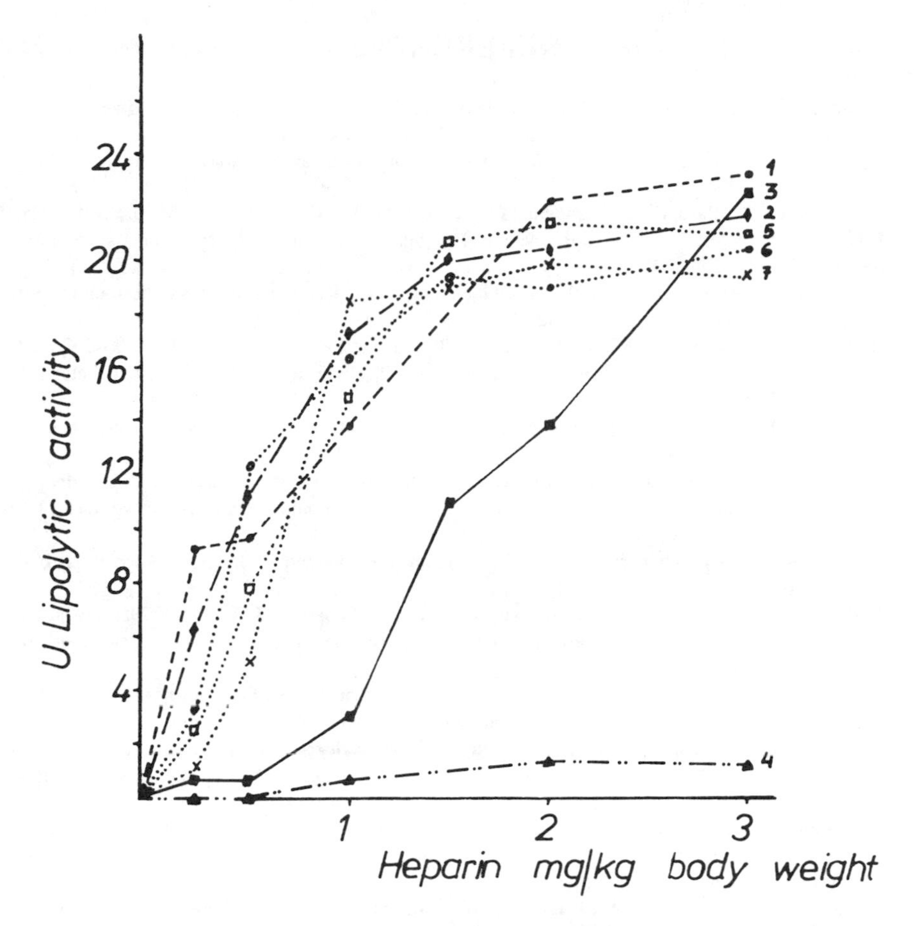

FIGURE 6. Variants of postheparin lipoprotein lipase activities in different cases of hyperlipoproteinemia. Case 1 (hyperchylomicronemia) and case 2 (mixed form of hyperlipoproteinemia) — the curves are similar to that of controls 5, 6, and 7. Case 3 — after a normal dose of heparin (1 mg/kg), lipoprotein lipase reaches only one sixth of the normal activity (reaching normal activity only after a large dose of heparin — 3 mg/kg). Case 4 — no lipolytic activity could be achieved after the largest doses of heparin.

eration. Accepting that heparin is a cofactor for the apoenzyme of lipoprotein lipase in cases 1 and 2 (Figure 6) because of the easy correction of lipolytic activity and abolishment of all clinical disturbances after a normal dose of heparin, it must be assumed that lack of activation was connected with endogenous heparin deficiency. This may be caused by deficient depolimerization of heparin because macromolecular heparin produced only approximately 13%, whereas its depolimerization products formed 59 to 66% of normal activity[153] (a similar activity as the commercial heparins used in the experiments performed).

The course of lipolytic activity in case 3 (Figure 6) suggests heparin dependency caused by an atypical apoenzyme molecular structure, resulting in defective interaction with heparin as a cofactor, which would be overcome only by excess of heparin restoring the normal function of the enzyme.

Finally, absence of the lipolytic activity even after the largest doses of heparin suggests that the molecular changes of the lipoprotein lipase might be such that the formation of an active complex with heparin was impossible, or the synthesized enzyme might immediately disintegrate, or the absence of the enzyme activity might be caused by a gene deletion (similar to the gene deletions in thalassemias).

Heparin-dependency of lipoprotein lipase indicates that enzyme cofactor-dependency may be caused either by extra- or intraorganismic cofactors.

REFERENCES

1. **Stanbury, J. B., Wyngaarden, J. B., Fredrickson, D. S., Goldstein, J. L., and Brown, M. S.,** *The Metabolic Basis of Inherited Disease,* McGraw-Hill, New York, 1983.
2. **Brown, M. S. and Goldstein, J. L.,** A receptor-mediated pathway for cholesterol homeostasis, *Science,* 232, 34, 1986.
3. **Reynolds, G. A., Basu, S. K., Osborne, T. F., Chin, D. J., Gil, G., Brown, M. S., Goldstein, J. L., Luskey, K. L.,** HMG CoA reductase: a negatively regulated gene with unusual promoter and 5′ untranslated regions, *Cell,* 38, 275, 1984.
4. **Brown, M. S., Anderson, R. G. W., and Goldstein, J. L.,** Recycling receptors: the round trip itinerary of migrant membrane proteins, *Cell,* 32, 663, 1983.
5. **Russel, D. W., Schneider, W. J., Yamamoto, T., Luskey, K. L., Brown, M. S., and Goldstein, J. L.,** Domain map of the LDL receptor: sequence homology with the epidermal growth factor precursor, *Cell,* 37, 577, 1984.
6. **Südhof, T. C., Goldstein, J. L., Brown, M. S., and Russel, D. W.,** The LDL receptor gene, a mosaic of exons shared with different proteins, *Science,* 228, 815, 1985.
7. **Davis, C. G., Goldstein, J. L., Südhof, T. C., Anderson, R. G. W., Russel, D. W., and Brown, M. S.,** Acid dependent ligand dissociation and recycling of LDL receptor mediated by growth factor homology region, *Nature,* 326, 760, 1987.
8. **Mahley, R. W. and Innerarity, T. L.,** Lipoprotein receptors and cholesterol homeostasis, *BBA,* 737, 197, 1983.
9. **Tolleshaug, H., Hobgood, K. K., Brown, M. S., and Goldstein, J. L.,** The LDL receptor locus in familial hypercholesterolemia: multiple mutations disrupt transport and processing of a membrane receptor, *Cell,* 32, 941, 1983.
10. **Lehrman, M. A., Schneider, W. J., Südhof, T. C., Brown, M. S., Goldstein, J. L., and Russel, D. W.,** Encoding transmembrane and cytoplasmic domains, *Science,* 227, 140, 1985.
11. **Kingsley, D. M., Kozarsky, K. F., Hobbie, L., and Krieger, M.,** Reversible defects in O-linked glycosylation and LDL receptor expression in a UDP-Gal/UDP-GalNAc 4-epimerase deficient mutant, *Cell,* 44, 749, 1986.
12. **Lehrman, M. A., Goldstein, J. L., Brown, M. S., Russel, D. W., and Schneider, W. J.,** Internalization-defective LDL receptors produced by genes with nonsense and frameshift mutations that truncate the cytoplasmic domain, *Cell,* 41, 735, 1985.
13. **Davis, C. G., Lehrman, M. A., Russel, D. W., Anderson, R. G. W., Brown, M. S., and Goldstein, J. L.,** The J.D. mutation in familial hypercholesterolemia: amino acid substitution in cytoplasmic domain impedes internalization of LDL receptors, *Cell,* 45, 15, 1986.
14. **Horsthemke, B., Kessling, A. M., Seed, M., Wynn, V. R., and Humphries, S. E.,** Identification of a deletion in the low density lipoprotein (LDL) receptor gene in a patient with familial hypercholesterolemia, *Hum. Genet.,* 71, 75, 1985.
15. **Yamamoto, T., Bishop, R. W., Brown, M. S., Goldstein, J. L., and Russel, D. W.,** Deletion in cystein-rich region of LDL receptor impedes transport to cell surface in WHHL rabbit, *Science,* 232, 1230, 1986.
16. **Brown, M. S. and Goldstein, J. L.,** Scavenger cell receptor shared, *Nature,* 316, 680, 1985.
17. **Steinbrecher, U. P., Parthasarathy, S., Leake, D. S., Witztum, J. L., and Steinberg, D.,** Modification of low density lipoprotein by endothelial cells involves lipid peroxidation and degradation of low density lipoprotein phospholipids, *Proc. Natl. Acad. Sci. U.S.A.,* 81, 3883, 1984.
18. **Parthasarapathy, S., Steinbrecher, U. P., Barnett, J., Witztum, J. L., and Steinberg, D.,** Essential role of phospholipase A_2 activity in endothelial cell-induced modification of low density lipoprotein, *Proc. Natl. Acad. Sci. U.S.A.,* 82, 3000, 1985.
19. **Spady, D. K. and Dietschy, J. M.,** Dietary saturated triacylglycerols suppress hepatic low density lipoprotein receptor activity in the hamster, *Proc. Natl. Acad. Sci. U.S.A.,* 82, 4526, 1985.
20. **Knott, T. J., Pease, R. J., Powell, L. M., Wallis, S. C., Rall, S. C., Innerarity, T. L., Blackhart, B., Taylor, W. H., Marcel, Y., Milne, R., Johnson, D., Fuller, M., Lusis, A. J., McCarthy, B. J., Mahley, R. W., Levy-Wilson, B., and Scott, J.,** Complete protein sequence and identification of structural domains of human apolipoprotein B, *Nature,* 323, 734, 1986.
21. **Yang, C. Y., Chen, S. H., Gianturco, S. H., Bradley, W. A., Sparrow, J. T., Tanimura, M., Li, W. H., Sparrow, D. A., DeLoof, H., Rosseneu, M., Lee, F. S., Gu, Z. W., Gotto, A. M., and Chan, L.,** Sequence, structure, receptor-binding domains and internal repeats of human apolipoprotein B-100, *Nature,* 323, 738, 1986.
22. **Teng, B., Thompson, G. R., Sniderman, A. D., Forte, T. M., Krauss, R. M., and Kwiterowich, P. O., Jr.,** Composition and distribution of low density lipoprotein fractions in hyperapobetalipoproteinemia, normolipidemia, and familial hypercholesterolemia, *Proc. Natl. Acad. Sci. U.S.A.,* 80, 6662, 1983.

23. **Illingworth, D. R., Alam, N. A., Sundberg, E. F., Hagemenas, F. C., and Layman, D. L.,** Regulation of low density lipoprotein receptors by plasma lipoproteins from patients with abetalipoproteinemia, *Proc. Natl. Acad. Sci. U.S.A.,* 80, 3475, 1983.
24. **Rapacz, J., Hasler-Rapacz, J., Taylor, K. M., Checovich, W. J., and Attie, A. D.,** Lipoprotein mutations in pigs are associated with elevated plasma cholesterol and atherosclerosis, *Science,* 234, 1573, 1986.
25. **Stalenhoef, A. F. H., Malloy, M. J., Kane, J. P., and Havel, R. J.,** Metabolism of apolipoproteins B-48 and B-100 of triglyceride-rich lipoproteins in normal and lipoprotein lipase-deficient humans, *Proc. Natl. Acad. Sci. U.S.A.,* 81, 1839, 1984.
26. **Weisgraber, K. H., Innerarity, T. L., Rall, S. C., Jr., and Mahley, R. W.,** Apolipoprotein E: receptor binding properties, *Adv. Exp. Med. Biol.,* 183, 159, 1985.
27. **Hoeg, J. M., Demosky, S. J., Jr., Gregg, R. E., Schaefer, E. J., and Brewer, H. B., Jr.,** Distinct hepatic receptors for low density lipoprotein and apolipoprotein E in humans, *Science,* 227, 759, 1985.
28. **Paik, Y. K., Chang, D. J., Reardon, C. A., Davies, G. E., Mahley, R. W., and Taylor, J. M.,** Nucleotide sequence and structure of the human apolipoprotein E gene, *Proc. Natl. Acad. Sci. U.S.A.,* 82, 3445, 1985.
29. **Weisgraber, K. H., Rall, S. C., Jr., Innerarity, T. L., Mahley, R. W., Kuusi, T., and Ehnholm, C.,** A novel electrophoretic variant of human apolipoprotein E, identification and characterization of apolipoprotein E1, *J. Clin. Invest.,* 73, 1024, 1984.
30. **Ehnaholm, C., Mahley, R. W., Chapepell, D. A., Weisgraber, K. H., Ludwig, E., and Witztum, J. L.,** Role of apolipoprotein E in the lipolytic conversion of β-very low density lipoproteins to low density lipoproteins in type III hyperlipoproteinemia, *Proc. Natl. Acad. Sci. U.S.A.,* 81, 5566, 1984.
31. **Brewer, H. B., Jr., Zech, L. A., Gregg, R. E., Schwartz, D., and Schaefer, E. J.,** NIH conference, Type III hyperlipoproteinemia: diagnosis, molecular defects, pathology, and treatment, *Ann. Intern. Med.,* 98, 623, 1983.
32. **Havekes, L., de Wit, E., Leuven, J. G., Klasen, E., Utermann, G., Weber, W., and Beisiegel, U.,** Apolipoprotein E3-Leiden. A new variant of human apolipoprotein E associated with familial type III hyperlipoproteinemia, *Hum. Genet.,* 73, 157, 1986.
33. **Utermann, G., Kindermann, I., Kaffarnik, H., and Steinmetz, A.,** Apolipoprotein E phenotypes and hyperlipidemia, *Hum. Genet.,* 65, 232, 1984.
34. **Utermann, G., Hardewig, A., and Zimmer, F.,** Apolipoprotein E phenotypes in patients with myocardial infarction, *Hum. Genet.,* 65, 237, 1984.
35. **Karathanasis, S. K., Zannis, V. I., and Breslow, J. L.,** Isolation and characterization of the human apolipoprotein A-I gene, *Proc. Natl. Acad. Sci. U.S.A.,* 80, 6147, 1983.
36. **Karathanasis, S. K., Zannis, V. I., and Breslow, J. L.,** A DNA insertion in the apolipoprotein A-I gene of patients with premature atherosclerosis, *Nature,* 305, 823, 1983.
37. **Schmitz, G., Assmann, G., Rall, S. C., Jr., and Mahley, R. W.,** Tangier disease: defective recombination of a specific Tangier apolipoprotein A-I isoform (pro-apo A-I) with high density lipoproteins, *Proc. Natl. Acad. Sci. U.S.A.,* 80, 6081, 1983.
38. **Schmitz, G., Assmann, G., Rall, S. C., Jr., and Mahley, R. W.,** Tangier disease: a disorder of intracellular membrane traffic, *Proc. Natl. Acad. Sci. U.S.A.,* 80, 6081, 1983.
39. **Forte, T. M., Nichols, A. V., Krauss, R. M., and Norum, R. A.,** Familial apolipoprotein AI (apo AI) and apolipoprotein CIII (apo CIII) deficiency, subclass distribution, composition, and morphology of lipoproteins in a disorder associated with premature atherosclerosis, *J. Clin. Invest.,* 74, 1601, 1984.
40. **Humphries, S. E., Williams, L., Myklebost, O., Stalenhoef, A. F. H., Demacker, P. N. M., Baggio, G., Crepaldi, G., Galton, D. J., and Williamson, R.,** Familial apolipoprotein CII deficiency: a preliminary analysis of the gene defect in two families, *Hum. Genet.,* 65, 151, 1984.
41. **Ferns, G. A. A., Shelley, C. S., Stocks, J., Rees, A., Paul, H., Baralle, F., and Galton, D. J.,** A DNA polymorphism of the apoprotein AII gene in hyertriglyceridemia, *Hum. Genet.,* 74, 302, 1986.
42. **Ross, R., Raines, E. W., and Bowen-Pope, D. F.,** The biology of platelet-derived growth factor, *Cell,* 46, 155, 1986.
43. **Berk, B. C., Alexander, R. W., Brock, T. A., Gimbrone, M. A., and Webb, R. C.,** Vasoconstriction: a new activity for platelet-derived growth factor, *Science,* 232, 87, 1986.
44. **George, J. N., Nurden, A. T., and Phillips, D. R.,** Molecular defects in interactions of platelets with vessel wall, *N. Engl. J. Med.,* 311, 1084, 1984.
45. **Kruth, H. S.,** Platelet-mediated cholesterol accumulation in cultured aortic smooth muscle, *Science,* 227, 1243, 1985.
46. **Phillips, D. R., Arnold, K., and Innerarity, T. L.,** Platelet secretory products inhibit lipoprotein metabolism in macrophages, *Nature,* 316, 746, 1985.
47. **Heimark, R. L., Twardzik, D. R., and Schwartz, S. M.,** Inhibition of endothelial regeneration by type-beta transforming factor from platelets, *Science,* 233, 1078, 1986.
48. **Brown, M. S. and Goldstein, J. L.,** Teaching old dogmas new tricks, *Nature,* 330, 113, 1987.

49. **McLean, J. W., Tomlinson, J. E., Kuang, W. J., Eaton, D. L., Chen, E. Y., Fless, G. M., Scanu, A. M., and Lawn, R. M.,** cDNA sequence of human apolipoprotein (a) is homologous to plasminogen, *Nature,* 300, 132, 1987.
50. **Pope, F. M. and Nicholls, A. C.,** Molecular abnormalities of collagen proteins and genes, in *Molecular Medicine,* Vol. 1, Malcolm, A. B. B., Ed., IRL Press, Oxford, 1984, 95.
51. **Prockop, D. J., Kivirikko, K. I., Tuderman, L., and Guzman, N. A.,** The biosynthesis of collagen and its disorders, *N. Engl. J. Med., Part I,* 301, 13, 1979; *Part II,* 301, 77, 1979.
52. **Prockop, D. J.,** Mutations in collagen genes, *J. Clin. Invest.,* 75, 783, 1985.
53. **Solomon, E., Hiorns, L. R., Kurkinen, M., Barlow, D., and Hogan, B. L. M.,** Chromosomal assignments of the genes coding for human types II, III, and IV collagen: a dispersed gene family, *Proc. Natl. Acad. Sci. U.S.A.,* 82, 3330, 1985.
54. **Prockop, D. J. and Kivirikko, K. I.,** Heritable diseases of collagen, *N. Engl. J. Med.,* 311, 376, 1984.
55. **Prockop, D. J.,** Osteogenesis imperfecta: Phenotypic heterogeneity, protein suicide, short and long collagen, *Amer. J. Hum. Genet.,* 36, 499, 1984.
56. **Pope, F. M., Nicholls, McPheat, J., Talmud, P., and Owen, R.,** *J. Med. Genet.,* 22, 466, 1985.
57. **Steinmann, B., Nicholls, A., and Pope, F. M,** Clinical variability of osteogenesis imperfecta reflecting molecular heterogeneity: cysteine substitutions in the α 1(I) collagen chain producing lethal and mild forms, *J. Biol. Chem.,* 261, 8958, 1986.
58. **Deak, S. B., Nicholls, A., Pope, F. M., and Prockop, D. J.,** The molecular defect in a nonlethal variant of osteogenesis imperfecta, *J. Biol. Chem.,* 258, 15192, 1983.
59. **Chu, M. L., Rowe, D., Nicholls, A., Pope, F. M., and Prockop, D. J.,** Presence of translatable mRNA for pro α2(I) chains in fibroblasts from a patient with osteogenesis imperfecta whose type I collagen does not contain α 2(I) chains, *Coll. Relat. Res.,* 4, 389, 1984.
60. **Wallis, G., Beighton, P., Boyd, C., and Mathew, C. G.,** Mutations linked to the pro α2(I) collagen gene are responsible for several cases of osteogenesis imperfecta type I, *J. Med. Genet.,* 23, 411, 1986.
61. **Chu, M. L., Gargiulo, V., Williams, C. J., and Ramirez, F.,** Multiexon deletion an osteogenesis imperfecta variant with increased type III collagen mRNA, *J. Biol. Chem.,* 260, 691, 1985.
62. **Wenstrup, R. J., Hunter, A. G. W., and Byers, P. H.,** Osteogenesis imperfecta type IV: evidence of abnormal triple helical structure of type I collagen, *Hum. Genet.,* 74, 47, 1986.
63. **Barsh, G. S., Roush, C. L., Bonadio, J., and Byers, P. H.,** Intron-mediated recombination may cause a deletion in an α 1 type collagen chain in a lethal form of osteogenesis imperfecta, *Proc. Natl. Acad. Sci. U.S.A.,* 82, 2870, 1985.
64. **Hartung, S., Jaenisch, R., and Breindl, M.,** Retrovirus insertion inactivates mouse α1(I) collagen gene by blocking initiation of transcription, *Nature,* 320, 365, 1986.
65. **Löhler, J., Timpl, R., and Jaenisch, R.,** Embryonic lethal mutation in mouse collagen I gene causes rupture of blood vessels and is associated with erythropoietic and mesenchymal cell death, *Cell,* 38, 597, 1984.
66. **Stolle, C. A., Pyeritz, R. E., Myers, J. C., and Prockop, D. J.,** Synthesis of an altered type III procollagen in a patient with type IV Ehlers-Danlos syndrome. A structural change in the α1(III) chain which makes the protein more susceptible to proteinases, *J. Biol. Chem.,* 260, 1937, 1985.
67. **Pinnel, S. R.,** Molecular defects in the Ehlers-Danlos syndrome, *J. Invest. Dermatol.,* 79 (Suppl. 1), 90, 1982.
68. **Ihme, A., Krieg, T., Nerlich, A., Feldmann, U., Rauterberg, J., Glanville, R. W., Edel, G., and Müller, P. K.,** Ehlers-Danlos syndrome type IV: collagen type specificity of defective lysysl hydroxylation in various tissues, *J. Invest. Dermatol.,* 83, 161, 1984.
69. **Eyre, D. R., Shapiro, F. D., and Aldridge, J. F.,** A heterozygous collagen defect in a variant of the Ehlers-Danlos syndrome type VII: evidence for a deleted amino-telopeptide domain in the pro α2(I) chain, *J. Biol. Chem.,* 260, 11322, 1985.
70. **Henke, E., Leader, M., Tajima, S., Pinnel, S., and Kaufman, R.,** A 38 base pair insertion in the pro α2(I) collagen gene of a patient with Marfan syndrome, *J. Cell. Biochem.,* 27, 169, 1985.
71. **Crawford, S. W., Featherstone, J. A., Holbrook, K., Yong, S. L., Bornstein, P., and Sage, H.,** Characterization of a type VI collagen-related M_r 140,000 protein from cutis laxa fibroblasts in culture, *Biochem. J.,* 227, 491, 1985.
72. **Leone, A., Paviakis, G. N., and Hamer, D. H.,** Menkes' disease: abnormal metallothionein gene regulation in response to copper, *Cell,* 40, 301, 1985.
73. **Karin, M.,** Metallothioneins: proteins in search of function, *Cell,* 41, 9, 1985.
74. **Schmidt, C. J., Hamer, D. H., and McBride, O. W.,** Chromosomal location of human metallothionein genes: implications for Menkes' disease, *Science,* 224, 1104, 1984.
75. **Perlish, J. S., Timpl, R., and Fleischmajer, R.,** Collagen synthesis regulation by the aminopropeptide of procollagen I in normal and scleroderma fibroblasts, *Arthritis Rheum.,* 28, 647, 1985.
76. **Kenney, M. C., Labermeier, U., Hinds, D., and Waring, G. O.,** Characterization of the Descemet's membrane/posterior collagenous layer isolated from Fuchs' endothelial dystrophy corneas, *Exp. Eye Res.,* 39, 267, 1984.

77. **Takagi, Y., Veis, A., and Sauk, J. J.,** Relation of mineralization defects in collagen matrices to non-collagenous protein components. Identification of a molecular defect in dentinogenesis imperfecta, *Clin. Orthop.,* 176, 282, 1983.
78. **Peltonen, L., Halila, R., and Ryhanen, L.,** Enzymes converting procollagens to collagens, *J. Cell. Biochem.,* 28, 15, 1985.
79. **Bauer, R. A., Cooper, T. W., Huang, J. S., Altman, J., and Deuel, T. F.,** Stimulation *in vitro* human skin collagenase expression by platelet-derived growth factor, *Proc. Natl. Acad. Sci. U.S.A.,* 82, 4132, 1985.
80. **Mensing, H., Krieg, T., Meigel, W., and Braun-Falco, O.,** Cutis laxa, Klassifikation, Klinik und molekulare Defekte, *Hautarzt,* 35, 506, 1984.
81. **Brinckerhoff, C. E., Benoit, M. C., and Culp, W. J.,** Autoregulation of collagenase production by a protein synthesized and secreted by synovial fibroblasts: cellular mechanism for control of collagen degradation, *Proc. Natl. Acad. Sci. U.S.A.,* 82, 1916, 1985.
82. **Pope, F. M. and Nicholls, A.,** Molecular abnormalities of collagen in human disease, *Arch. Dis. Child.,* 62, 523, 1987.
83. **Dalgleish, R., Willianms, G., and Hawkins, J. R.,** Length polymorphism in the pro $\alpha 2(I)$ collagen gene: an alternative explanation in a case of Marfan syndrome, *Hum. Genet.,* 73, 91, 1986.
84. **Morse, J. O.,** Alpha$_1$-antitrypsin deficiency, *N. Engl. J. Med., Part I,* 299, 1945, 1978; *Part II,* 299, 1099, 1978.
85. **Tetley, T. D.,** Pulmonary emphysema, in *Molecular Medicine,* Vol. 1, Malcolm, A. D. B., Ed., IRL Press, Oxford, 39, 1984.
86. **Klasen, E. C., Biemond, I., and Laros, C. D.,** α_1-antitrypsin deficiency and the flaccid lung syndrome, *Clin. Genet.,* 29, 211, 1986.
87. **Weidinger, S., Jahn, W., Cujnik, F., and Schwarzfischer, F.,** Alpha$_1$-antitrypsin: evidence for a fifth PiM subtype and a new deficiency allele PiZ$_{Augsburg}$, *Hum. Genet.,* 71, 27, 1985.
88. **Viraben, R., Massip, P., Dicostanzo, B., and Mathieu, C.,** Necrotic panniculitis with α_1-antitrypsin deficiency, *J. Amer. Acad. Dermatol.,* 14, 684, 1986.
89. **Harpel, P. C.,** Protease inhibitors — a precarious balance, *N. Engl. J. Med.,* 309, 725, 1983.
90. **Owen, M. C., Brennan, S. O., Lewis, J. H., and Carrel, R. W.,** Mutation of antitrypsin to antithrombin, alpha$_1$-antitrypsin Pittsburgh (358 Met leads to Arg), a fatal bleeding disorder, *N. Engl. J. Med.,* 309, 694, 1983.
91. **Rosenberg, S, Barr, P. J., Najarian, R. C., and Hallewell, R. A.,** Synthesis in yeast of a functional oxidation-resistant mutant of α_1-antitrypsin, *Nature,* 312, 77, 1984.
92. **Courtney, M., Jallat, S., Tessier, L. H., Benavente, A., Crystal, R. G., and Lecocq, J. P.,** Synthesis in *E. coli* of alpha$_1$-antitrypsin variants of therapeutic potential for emphysema and thrombosis, *Nature,* 313, 149, 1985.
93. **Ciliberto, G., Dente, L., and Cortese, R.,** Cell-specific expression of a transfected human α_1-antitrypsin gene, *Cell,* 41, 531, 1985.
94. **Kidd, V. J., Golbus, M. S., Wallace, R. B., Itakura, K., and Woo, S. L. C.,** Prenatal diagnosis of α_1-antitrypsin deficiency by direct analysis of the mutation site in the gene, *N. Engl. J. Med.,* 310, 639, 1984.
95. **Lidsky, A. S., Law, M. L., Morse, H. G., Kao, F.-T., Rabin, M., Ruddle, F. H., and Woo, S. L. C.,** Regional mapping of the phenylalanine hydroxylase gene and the phenylketonuria locus in the human genome, *Proc. Natl. Acad. Sci. U.S.A.,* 82, 6221, 1985.
96. **DiLella, A. G., Marvit, J., Lidsky, A. S., Güttler, F., and Woo, S. L. C.,** Tight linkage between a splicing mutation and a specific DNA haplotype in phenylketonuria, *Nature,* 322, 799, 1986.
97. **Kidd, K. K.,** Phenylketonuria: population genetics of a disease, *Nature,* 327, 282, 1987.
98. **DiLella, A. G., Marvit, J., Brayton, K., and Woo, S. L. C.,** An amino-acid substitution involved in phenylketonuria is in linkage disequilibrium with DNA haplotype 2, *Nature,* 327, 333, 1987.
99. **Kaufman, S., Berlow, S., Summer, G. K., Milstein, S., Schulman, J., Orloff, S., Spielberg, S., and Pueschel, S.,** Hyperphenylalaninemia due to a deficiency of biopterin, *N. Engl. J. Med.,* 299, 673, 1978.
100. **Niederwieser, A., Blau, N., Wang, M., Joller, P., Atares, M., and Cardesa-Garcia, J.,** GTP cyclohydrolase I deficiency with neopterin biopterin, dopamine, and serotonin deficiencies and muscular hypotonia, *Eur. J. Pediatr.,* 141, 208, 1984.
101. **Hogan, B.,** Lesch-Nyhan syndrome — Engineering mutant mice, *Nature,* 326, 240, 1987.
102. **Wilson, J. M., Young, A. B., and Kelley, W. N.,** Hypoxanthine-guanine phosphoribosyltransferase deficiency, *N. Engl. J. Med.,* 309, 900, 1983.
103. **Wilson, J. M. and Kelley, W. N.,** Molecular genetics of hypoxanthine-guanine phosphoribosyltransferase deficiency in man, *Arch. Intern. Med.,* 145, 1895, 1985.
104. **Wilson, J. M., Tarr, G. E., and Kelley, W. N.,** Human hypoxanthine (guanine) phosphoribosyltransferase: an amino acid substitution in a mutant form of the enzyme isolated from a patient with gout, *Proc. Natl. Acad. Sci. U.S.A.,* 80, 870, 1983.

105. **Wilson, J. M. and Kelley, W. N.,** Human hypoxanthine-guanine phosphoribosyltransferase. Structural alteration in a dysfunctional enzyme variant ($HPRT_{Munich}$) isolated from a patient with gout, *J. Biol. Chem.,* 259, 27, 1984.
106. **Yang, T. P., Patel, P. I., Chinault, A. C., Stout, J. T., Jackson, L. G., Hildebrand, B. M., and Caskey, C. T.,** Molecular evidence for new mutation at the hprt locus in Lesch-Nyhan patients, *Nature,* 310, 412, 1984.
107. **Kamatani, N., Terai, C., Kuroshima, S., Nishioka, K., and Mikanagi, K.,** Genetic and clinical studies on 19 families with adenine phosphoribosyltransferase deficiencies, *Hum. Genet.,* 75, 163, 1987.
108. **Gibbs, R. A. and Caskey, C. T.,** Identification and localization of mutations at the Lesch-Nyhan locus by ribonuclease A cleavage, *Science,* 236, 303, 1987.
109. **Hooper, M., Hardy, K., Handyside, A., Hunter, S., and Monk, M.,** HPRT-deficient (Lesch-Nyhan) mouse embryos derived from germline colonisation by cultured cells, *Nature,* 326, 292, 1987.
110. **Kuehn, M. R., Bradley, A., Robertson, E. J., and Evans, M. J.,** A potential animal model for Lesch-Nyhan syndrome through introduction of HPRT mutations into mice, *Nature,* 326, 295, 1987.
111. **Renlund, M., Tietze, F., and Gahl, W. A.,** Defective sialic acid egress from isolated fibroblast lysosomes of patients with Salla disease, *Science,* 232, 759, 1986.
112. **Watts, R. W.,** Some inherited lysosomal enzyme defects with special reference to the liver and prospects for their control, *Z. Gastroenterol. (Verh),* 16, 137, 1979.
113. **Alexander, D., Deeb, M., and Talj, F.,** Heterozygosity for phosphodiester glycosidase deficiency: a novel human mutation of lysosomal enzyme processing, *Hum. Genet.,* 73, 53, 1986.
114. **Lang, L., Takahashi, T., Tang, J., and Kornfeld, S.,** Lysosomal enzyme phosphorylation offer a biochemical rationale for two distinct defects in the uridine diphospho-*N*-acetylglucosamine: lysosomal enzyme precursor *N*-acetylglucosamine-1-phosphotransferase, *J. Clin. Invest.,* 76, 219, 1985.
115. **Harzer, K., Cantz, M., Sewell, A. C., Dhareshwar, S. S., Roggendorf, W., Heckl, R. W., Schofer, O., Thumler, R., Peiffer, J., and Schlote, W.,** Normomorphic sialidosis in two female adults with severe neurologic disease and without sialyl oligosacchariduria, *Hum. Genet.,* 74, 209, 1986.
116. **Hannun, Y. A. and Bell, R. M.,** Lysosphingolipids inhibit protein kinase C: implications for the sphingolipidoses, *Science,* 235, 670, 1987.
117. **Pentchev, P. G., Comly, M. E., Kruth, H. S., Vanier, M. T., Wenger, D. A., Patel, S., and Brady, R. O.,** A defect in cholesterol esterification in Niemann-Pick disease (type C) patients, *Proc. Natl. Acad. Sci. U.S.A.,* 82, 8247, 1985.
118. **Grabowski, C. A., Goldblatt, J., Dinur, T., Kruse, J., Svennerholm, L., Gatt, S., and Desnick, R. J.,** Genetic heterogeneity in Gaucher disease: physicokinetic and immunologic studies of the residual enzyme in cultured fibroblasts from non-neuronopathic and neuronopathic patients, *Am. J. Med. Genet.,* 21, 529, 1985.
119. **Beutler, E., Kuhl, W., and Sorge, J.,** Cross-reacting material in Gaucher disease fibroblasts, *Proc. Natl. Acad. Sci. U.S.A.,* 81, 6506, 1984.
120. **Sorge, J., Gelbart, T., West, C., Westwood, B., and Beutler, E.,** Heterogeneity in type I Gaucher disease demonstrated by restriction mapping of the gene, *Proc. Natl. Acad. Sci. U.S.A.,* 82, 5442, 1985.
121. **Choudary, P. V, Barranger, J. A., Tsuji, S., Mayor, J., LaMarca, M. E., Cepko, C. L., Mulligan, R. C., and Ginns, E. I.,** Retrovirus-mediated transfer of the human glucocerebrosidase gene to Gaucher fibroblasts, *Mol. Biol. Med.,* 3, 293, 1986.
122. **Igisu, H. and Suzuki, K.,** Progressive accumulation of toxic metabolite in a genetic leukodystrophy, *Science,* 224, 754, 1984.
123. **Figura, K., Steckel, F., Conary, J., Hasilik, A., and Shaw, E.,** Heterogeneity in late-onset metachromatic leukodystrophy. Effect of inhibitors of cysteine proteinases, *Am. J. Hum. Genet.,* 39, 371, 1986.
124. **Inui, K., Emmet, M., and Wenger, D. A.,** Immunological evidence for deficiency in an activator protein for sulfatide sulfatase in a variant form of metachromatic leukodystrophy, *Proc. Natl. Acad. Sci. U.S.A.,* 80, 3074, 1983.
125. **Calhoun, D. H., Bishop, D. F., Bernstein, H. S., Quinn, M., Hantzopoulos, P., and Desnick, R. J.,** Fabry disease: isolation of a cDNA clone encoding human α-galactosidase A, *Proc. Natl. Acad. Sci. U.S.A.,* 82, 7364, 1985.
126. **Hoogeveen, A. T., Graham-Kawashima, H., dAzzo, A., and Galjaard, H.,** Processing of human β-galactosidase in G_{M1} gangliosidosis and Morquio B syndrome, *J. Biol. Chem.,* 259, 1974, 1984.
127. **Mancini, G. M. S., Hoogeveen, A. T., Galjaard, H., Mansson, J. E., and Svennerholm, L.,** Ganglioside G_{M1} metabolism in living human fibroblasts with β-galactosidase deficiency, *Hum. Genet.,* 73, 35, 1986.
128. **dAzzo, A., Hoogeveen, A., Reuser, A. J., Robinson, D., and Galjaard, H.,** Molecular defect in combined β-galactosidase and neuraminidase deficiency in man, *Proc. Natl. Acad. Sci. U.S.A.,* 79, 4535, 1982.
129. **Spiegel, S., Fishman, P. H., and Weber, R. J.,** Direct evidence that endogenous G_{M1} ganglioside can mediate thymocyte proliferation, *Science,* 225, 1285, 1985.
130. **O'Dowd, B. F., Quan, F., Willard, H. F., Lamhonwah, A. M., Korneluk, R. G., Lowden, J. A., Gravel, R. A., and Mahuran, D. J.,** Isolation of cDNA clones coding for the β subunit of human β-hexosaminidase, *Proc. Natl. Acad. Sci. U.S.A.,* 82, 1184, 1985.

131. **Myerowitz, R. and Hogikyan, N. D.,** Different mutations in Ashkenazi Jewish and non-Jewish French Canadians with Tay-Sachs disease, *Science,* 232, 1646, 1986.
132. **Sonderfeld, S., Brendler, S., Sandhoff, K., Galjaard, H., and Hoogeveen, A. T.,** Genetic complementation in somatic cell hybrids of four variants of infantile G_{M2} gangliosidosis, *Hum. Genet.,* 71, 196, 1985.
133. **Navon, R., Argov, Z., and Frisch, A.,** Hexomsaminidase A deficiency in adults, *Am. J. Med. Genet.,* 24, 179, 1986.
133a. For a review of the lysosomal storage diseases see Stanbury, J. B., Wyngaarden, J. B., Fredrickson, D. S., Goldstein, J. L., and Brown, M. S., *The Metabolic Basis of Inherited Disease,* McGraw-Hill, New York, 1983.
134. **Bonner, D. M., Suyama, Y., and DeMoss, J. A.,** Genetic fine structure and enzyme formation, *Fed. Proc.,* 19, 926, 1960.
135. **Fowler, B.,** Recent advances in the mechanism of pyridoxine-responsive disorders, *J. Inher. Metab. Dis.,* 8 (Suppl. 1), 76, 1985.
136. **Scriver, C. R.,** Vitamin B_6 deficiency and dependency in man, *Am. J. Dis. Child.,* 113, 109, 1967.
137. **Wilcken, D. E., Dudman, N. P., and Tyrrel, P. A.,** Homocystinuria due to cystathionine β-synthase deficiency — the effects of betaine treatment in pyridoxine-responsive patients, *Metabolism,* 34, 1115, 1985.
138. **Mudd, S. H. and Levy, H. L.,** Disorders of transsulfuration in *The Metabolic Basis of Inherited Disease,* Stanbury, J. B., Wyngaarden, J. B., and Fredrickson, D. S., Eds., McGraw-Hill, New York, 1978, 458.
139. **Show, K. N., Liberman, E., Koch, R., and Donnel, G. N.,** Cystathioninuria, *Am. J. Dis. Child.,* 113, 119, 1967.
140. **Frimpter, G. W., Greenberg, A. J., Hilgartner, H., and Fuchs, F.,** Cystathioninuria Management, *Am. J. Dis. Child.,* 113, 115, 1967.
141. **Horrigan, D. D. and Harris, J. W.,** Pyridoxine-responsive anemia, analysis of sixty one cases, *Adv. Intern. Med.,* 12, 103, 1964.
142. **Rosenberg, L. E., Lilljequist, A. Ch., and Hsia, Y. E.,** Methylmalonic aciduria: metabolic block localization and vitamin B_{12} dependency, *Science,* 162, 805, 1968.
143. **Lindblad, R., Lindstrand, K., Svanberg, B., and Zetterstroem, R.,** The effect of cobamide coenzyme in methylmalonic acidemia, *Acta Paediatr. Scand.,* 58, 178, 1969.
144. **Zachello, F. and Tenconi, R.,** Methylmalonic acidemia and vitamin B_{12} dependency, *Acta Paediatr. Scand.,* 59, 88, 1970.
145. **Watkins, D. and Rosenblatt, D. S.,** Failure of lysosomal release of vitamin B_{12}: a new complementation group causing methylmalonic aciduria (cblF), *Am. J. Hum. Genet.,* 39, 404, 1986.
146. **Gompertz, D., Draffan, G. H., Watts, J. L., and Hull, D.,** Biotine-responsive β-methylcrotonylglycinuria, *Lancet,* July 3, 22, 1971.
147. **Wolf, B., Heard, G. S., Jefferson, L. G., Proud, V. K., Nance, W. E., and Weissbecker, K. A.,** Clinical findings in four children with biotinidase deficiency detected through a statewide neonatal screening program, *N. Engl. J. Med.,* 313, 16, 1985.
148. **Taitz, L. S., Leonard, J. V., and Bartlett, K.,** Long-term auditory and visual complications of biotinidase deficiency, *Early Hum. Dev.,* 11, 325, 1985.
149. **Lonsdale, D., Faulkner, W. R., Price, J. W., and Smeby, R. R.,** Intermittent cerebellar ataxia associated with hyperpyruvic acidemia, hyperalaninemia and hyperalaninuria, *Pediatrics,* 43, 1025, 1969.
150. **Chuang, D. T., Ku, L. S., and Cox, R. P.,** Thiamin-responsive maple-syrup-urine disease: decreased affinity of the mutant branched chain α-keto acid dehydrogenase for α-keto-isovalerate and thiamin pyrophosphate, *Proc. Natl. Acad. Sci. U.S.A.,* 79, 3300, 1982.
151. **Griffin, J. E.,** Syndromes of vitamin D resistance, in Syllabus of the 36th annual postgraduate assembly of the Endocrine Society, Dallas, Texas, October 15 to 19, 1984, 550.
152. **Horst, A., Paluszak, J., Zawilska, K., and Sobisz, S.,** Three variants of post-heparin lipoprotein lipase activity in idiopathic hyperlipoproteinemia, *Bull. Acad. Pol. Sci. Ser. Sci. Biol.,* 21, 199, 1973.
153. **Horner, A.,** Enzymic depolimerization of macromolecular heparin as a factor in control of lipoprotein lipase activity, *Proc. Natl. Acad. Sci. U.S.A.,* 69, 3469, 1972.

Chapter 17

PHARMACOGENETICS

I. INTRODUCTION

It is well known that individual reactions to the same agents may differ in a very broad range, from very small-size reactions or even absence of any reaction to very great reactions hazardous to life. This attribute of living beings can be best investigated by pharmacological studies. On this aspect van Wijngaarden (1926) performed excellent experiments that explained for the first time the character of this variation. He infused into anesthetized cats very slowly a digitalis solution for toxicity testing — to see how much digitalis should be injected until the heart stopped. He found that the responsiveness to digitalis was normally distributed, which denotes that the responsiveness to digitalis (as also other drugs) could be described by a normal frequency distribution curve, the Gauss curve. According to the normal frequency distribution, a few cats died after a relatively small dose, and some only after a relatively large dose, whereas the majority of cats died at a dose that was similar to the average of all required doses.

The essence of the molecular basis of pharmacogenetics is best illustrated by some examples from the animal kingdom.

In Vienna in 1849, a strain of rabbits was described that was able to thrive on belladonna leaves, whereas the leaves, because of the presence of atropine, are toxic to other strains of rabbits and, in general, to most living organisms. Several investigators tried, unsuccessfully, to explain the essence of this observation, and it was not until 1912, when Fleischmann and Metzner discovered that this effect was due to enzymatic degradation of atropine by the serum of these (Vienna) rabbits, that the puzzle was solved. But the decisive resolution was achieved only in 1938 when Bernheim and Bernheim applied biochemical methods for measuring the activity of the hydrolytic enzymes involved. They demonstrated that serum and liver from the "Vienna" rabbits contained an esterase that hydrolyzed atropine and homatropine, whereas serum and liver from other strains of rabbits did not. Atropinesterase in 1 ml of serum from 14 different rabbits was able to inactivate 0.04 to 0.1 mg of atropine per minute. In total, an adult rabbit with atropinesterase is able to inactivate several milligrams of atropine per minute (the average therapeutic dose in humans is less than 1 mg).

II. DRUG-INDUCED CONGENITAL MALFORMATIONS

The great majority of congenital malformations (teratogenesis) is due to an interplay of genetic and nongenetic factors. The most informative are studies on cortisone-induced cleft palate in mice. Cortisone acetate injected into pregnant mice of strain A (2.5 mg/day on day 11 to 14 after conception) caused cleft palate in virtually 100% of the offspring, whereas in C57Bl mice the same treatment caused cleft palate in only 19% of the offspring. Further studies on this problem revealed that there is a predisposition for cleft palate in strain A mice, since this malformation occurs in them also spontaneously (independent of any injection of cortisone), whereas in strain C57Bl there is no spontaneous occurrence of this malformation. A further explanation results from the conditions in which the two palatine shelves fail to fuse. The shelves grow like ridges from both sides in the embryonic mouth above the tongue. Factors that affect their fusion are the growth rate and alignment of the shelves (this is the critical factor) through their rotation, the width of the head, and the pressure of the tongue. The shelves grow and align more slowly in mice of the strain A than in mice of the strain C57Bl. Cortisone treatment slows these processes, and the two shelves cannot reach one another, resulting in cleft palate. In C57Bl mice, in spite of some retardation, the shelves are able to fuse before the head has reached its critical width. Cleft palate is a common malformation in humans, but it would be a mistake to assume that in humans it is caused by the teratogenic action of cortisone.

A. HISTAMINE

Mice of strain C3H were described as being particularly resistant to histamine. The dose that kills 50% of a group of C3H/Jax mice (LD_{50}) was 1523 mg/kg, whereas that of Swiss ICR mice was 230 mg/kg. The C3H/Jax mice were 6.6 times more resistant to histamine than the Swiss ICR mice. The administration of antihistaminic drugs (Neo-Antergan) increased the resistance to histamine of Swiss ICR mice to the approximate level seen in C3H/Jax mice, but did not increase the resistance of C3H/Jax mice. It might be supposed that the C3H/Jax mice possess an antihistaminic mechanism similar to that of antihistaminic drugs, absent in ICR mice. Therefore, antihistaminic drugs do not cause an increase in the histamine resistance of C3H/Jax mice, whereas in the ICR mice they do.

B. INSULIN

The best example of drug resistance is insulin resistance in mice of the KL strain. Most mice tolerate only three to eight units of insulin; obese mice with hereditary diabetes tolerate 20 units, whereas mice of the KL strain survive after 300 to 500 units. Additive effects of the products of three genes are responsible for the great insulin tolerance of KL mice to insulin. These are: (1) a high level of insulinase in the liver which degrades insulin at a relatively rapid rate; (2) KL mice have more than double the amount of glycogen in their livers than control mice; and (3) KL mice start to eat immediately after insulin injection, whereas control mice rather avoid food consumption. These three factors cause a rapid degradation of injected insulin, rapid glycogenolysis, and glucose intake with food, preventing the decrease of blood sugar level below its critical range.

C. SUCCINYLCHOLINE

In humans the most spectacular example is the prolonged action of succinylcholine in patients with atypical pseudocholinesterase (EC.3.1.1.8). In mammals, and in most vertebrates, the cholinesterases can be divided into two main classes: the acetylcholinesterases and the pseudocholinesterases, which in humans are different enzymes. Pseudocholinesterase is found in human plasma from the time of birth at a lower level than in adults. The level reaches its maximum at puberty (with lower levels in females than in males), and decreases in both sexes with advancing age. Pseudocholinesterase is present in plasma and several tissues, particularly in the liver, heart, intestinal mucosa, pancreas, and skin. The enzyme is species-specific, but not tissue-specific. The presence of atypical (mutant) pseudocholinesterase can modify the response to several drugs, but so far this has been established only for succinylcholine. Succinylcholine (suxamethonium) is a skeletal muscle relaxant which acts on the neuromuscular junction by initial depolarization of the nerve endplate and subsequent alteration in sensitivity to the transmitter substance, acetylcholine. The duration of succinylcholine action is very short due to its degradation by pseudocholinesterase to succinylmonocholine and choline (succinylcholine is composed of choline-succinic acid-choline). Succinylmonocholine (with little pharmacologic activity) is next degraded to inactive compounds (choline and succinic acid).

Injected intravenously, succinylcholine at once causes muscular paralysis by the action on the endplate. As a muscle relaxant, it is widely used in anesthesiology to facilitate intubation of the trachea, endoscopic procedures, and the reduction of dislocations and fractures. For short time muscle relaxation the usual dose of succinylcholine chloride is 40 to 50 mg, which causes muscular paralysis of approximately 2 to 4 min duration, manifested by apnea. However, a patient with atypical pseudocholinesterase may remain apneic for 30 to 50 min. Usually, there would be no danger to the patient when artificial respiration is given. It may happen that the prolonged effect of succinylcholine even passes unnoticed when surgery requires continuous and prolonged muscle relaxation. The danger exists when during long-lasting anesthesia a dose exceeding 1 gm/h is administered to a patient with atypical pseudocholinesterase. In patients with normal pseudocholinesterase, apnea disappears after some minutes, but in patients with atypical pseudocholinesterase the patient may not be able to breathe for 8 to 10 h. In such cases mechanical respiration or injection of normal pseudocholinesterase is required.

The duration of apnea after administration of succinylcholine may vary depending on the type of variant pseudocholinesterase and also on additional factors influencing muscular paralysis.

In psychiatry, muscle relaxants are used during electroshock treatment of mental diseases in order to avoid excessive muscle contractions (sometimes strong enough to cause bone fractures).

Not all cases of prolonged apnea after succinylcholine treatment are due to atypical pseudocholinesterase. After repeated treatment a dual type of muscle block may develop: the originally depolarizing block may change to a curariform nondepolarizing block. This block can be reversed by prostigmine, whereas apnea caused by atypical pseudocholinesterase is not terminated by prostigmine. Further, muscular paralysis may be caused also by a variety of disorders that disturb the electrolyte or acid-base balance, and which may alter the responsiveness of the endplate.

D. HEREDITARY METHEMOGLOBINEMIA

Atmospheric oxygen tends to oxidize all iron into the trivalent form. Therefore, hemoglobin is constantly being converted to methemoglobin, which in turn is continuously reduced to hemoglobin by enzymatic reduction. In individuals deficient in enzymatic reduction of methemoglobin, an increase of methemoglobin ranging from 5 to 10% of total hemoglobin occurs (normally less than 1% of hemoglobin is present as methemoglobin). The presence of abnormally high levels of methemoglobin is clinically noticeable by cyanosis. Methemoglobin cannot perform the primary transport function for oxygen. Thus, clinical symptoms of hypoxia exist.

Reduction of methemoglobin to hemoglobin occurs through four metabolic pathways: NADH- and NADH-dependent reducing pathways, and direct reducing pathways involving ascorbic acid and reduced glutathione. The most important is the NADH-dependent pathway. Erythrocytes from methemoglobinemic patients deficient in NADH dehydrogenase (termed also methemoglobin reductase or diaphorase I) reduce methemoglobin very slowly. The rate of reduction of methemoglobin does not depend on the activity of NADH-dehydrogenase only, but also on the availability of NADH and cytochrome b5. The NADPH-dependent pathway is coupled with the reduction of methylene blue to leukomethylene blue. Leukomethylene blue, in turn, spontaneously reduces methemoglobin to hemoglobin, with regeneration of methylene blue. The reaction requires regeneration of NADPH in the first steps of pentose phosphate pathway, catalyzed by glucose-6-phosphate dehydrogenase.

The first evidence for genetic variation of NADH dehydrogenase was achieved by electrophoretic examination. When erythrocyte hemolysates from patients with hereditary methemoglobinemia were subjected to electrophoresis in starch gell at pH 8.6 to 9.3, the normal NADH dehydrogenase band that migrates as a single band anodal to the position of hemoglobin A_2 was absence. Instead, a faint band with variable mobility was observed. The deficiency of NADH hydrogenase might arise from a decreased synthesis of the enzyme, an impaired catalytic function due to structural abnormalities, an increased lability, or a combination of these factors. There are several known variants of NADH dehydrogenase,

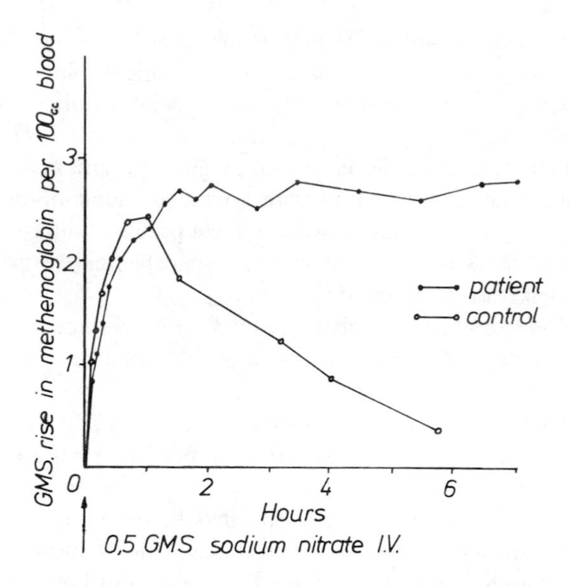

FIGURE 1. Methemoglobinemia is a patient with methemoglobin reductase deficiency after intravenous injection of 0.5 g sodium nitrate as compared with a normal subject. (redrawn from Eder, Finch and McKee, *Pharmacogenetics,* W. B. Saunders, Philadelphia, 1962).

e.g., several variants of normal catalytic activity (with increased or decreased electrophoretic mobility); variants associated with hereditary methemoglobinemia characterized by catalytic activity ranging from 0 to 62% of normal activity, mostly below 10% of the normal; and a group of methemoglobinemic patients with 0% of normal activity and no electrophoretic bands for erythrocyte NADH dehydrogenase.

The drugs nitrite and *p*-aminopropiophenone form methemoglobin in patients with hereditary methemoglobinemia in different ways — nitrite quickly and steichiometrically, and *p*-aminopropiophenone slowly and catalytically. The formation of methemoglobin caused by these drugs proceeds as it does in normal subjects. A single exposure to the methemoglobin-forming drugs causes an unusually high but a protracted increase of methemoglobin level, as illustrated in Figure 1.

REFERENCES

1. **Altland, K.,** Pseudocholinesterasen, in *Humangenetik,* Becker, P., Ed., Georg Thieme Verlag, Stuttgart, 1975, 351.
2. **Kalow, W.,** Ed., *Pharmacogenetics,* W. B. Saunders, Philadelphia, 1962.
3. **Schwartz, J. M. and Jaffé, E. R.,** Hereditary methemoglobinemia with deficiency of NADH dehydrogenase, in *The Metabolic Bases of Inherited Disease,* Stanbury, J. B., Wyngaarden, J. B., Fredrickson, D. S., Eds., 4th ed., McGraw-Hill, New York, 1978, 1464.

Chapter 18

DIAGNOSIS OF MOLECULAR DISEASES

Molecular diseases must be diagnosed by detection of abnormal molecules that cause functional disturbances of the living organisms. This can be done either by nucleic acid diagnosis which encode particular abnormal proteins (e.g., search for mutant globin genes) or by examination of protein molecules themselves (e.g., search for substitutions of particular amino acid residues within the polypeptide chain or the demonstration of abnormal physico-chemical features of the given protein molecule).

The progress in molecular DNA analysis enables detection of an increasing number of diseases by this procedure. The first step in this procedure is the isolation of DNA from any nucleated cell, usually peripheral lymphocytes, cultivated skin fibroblasts, amniotic fluid cells, chorion biopsy material, or other cells. At the second step, DNA must be cut by restriction endonucleases to obtain DNA fragments. Restriction endonucleases are enzymes that cut DNA in a sequence-specific manner. Restriction endonucleases are widespread in microorganisms, protecting them against incorporation of foreign DNA. Several hundreds of them are known, named after the organism from which they have been isolated. Each restriction enzyme cleaves DNA only at a specific DNA sequence, which is termed the recognition site. The recognition sites for the most common restriction enzymes are presented in Figure 1.

The cleavage products have flush (blunt) or staggered (sticky) ends. The Eco RI enzyme will cut DNA at each site where sequence GAATTC occurs. SV40 has only one Eco RI recognition site on its single chromosome: the *E. coli* chromosome has approximately 1000 recognition sites, and the human genome has approximately one million recognition sites. Thus the enzyme Eco RI cuts human DNA into approximately a million fragments of variable length which can be distinguished on the basis of size by electrophoresis on agarose. The shorter fragments move more quickly, the longer more slowly. Any alteration in the nucleotide sequence within the recognition site results in failure of the enzyme to cut at that site, resulting in the production of elongate fragments. Furthermore, altered recognition sites may be appropriate for other restriction enzymes, creating new recognition sites for these enzymes. In such cases, additional, atypical restriction fragments occur.[1]

The classical restriction fragment length polymorphism (RFLP) analysis comprises the following steps: isolation of DNA, cleavage by restriction enzyme, agarose electrophoresis, blotting (after Southern) of the electrophoretically obtained DNA fragments onto nitrocellulose filter with a highly radioactive DNA probe (e.g., ^{32}P β-globulin DNA probe), and finally, autoradiography, which visualizes the particular bands of DNA restriction fragments. The mutation that causes loss or gain of recognition site can be visible directly by changes in pattern fragments (e.g., sickle cell anemia). Similarly, large deletions or formation of neighboring (linked) endonuclease recognition sites that alter the size of DNA fragments can be directly detected (e.g., deletion thalassemias).

DNA polymorphisms occur approximately 1/100 bp, mostly at random within the DNA that does not code for proteins, and may be regarded as a completely neutral mutation. These polymorphisms are inherited by the mendelian mode, and are used in population studies in all living organisms, including forensic medicine (paternity identification). Developing chromosome isolation (fluorescence-activated cell sorter) makes it possible to map any phenotype by linkage inheritance studies (using translocations, or chromosome fragments, or *in situ* hybridization).

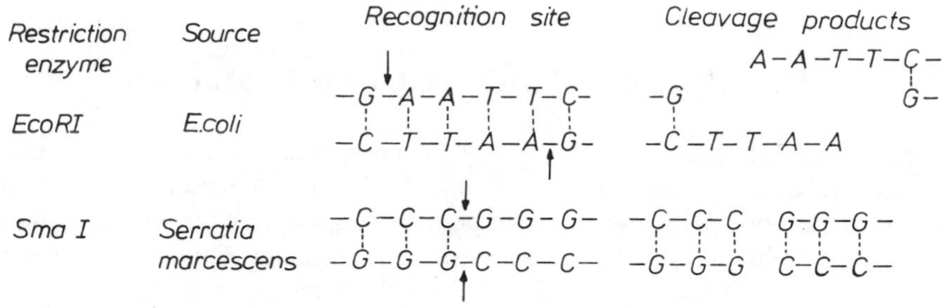

Restriction enzyme	Source	Recognition site	Cleavage products
EcoRI	E.coli	-G-A-A-T-T-C- -C-T-T-A-A-G-	-G -C-T-T-A-A A-A-T-T-C- G-
Sma I	Serratia marcescens	-C-C-C-G-G-G- -G-G-G-C-C-C-	-C-C-C -G-G-G G-G-G- C-C-C-

FIGURE 1. Restriction enzymes with staggered (sticky) ends (EcoRI) and flush (blunt) ends (SmaI).

I. cDNA CLONING

cDNA cloning may also be used as a tool for gene analysis. For this purpose, mRNA is isolated from a tissue that expresses usually only the one gene or only a limited number of other genes and treated with the enzyme reverse transcriptase to produce single stranded complementary DNA (cDNA) followed by a double-stranded cDNA. Then, the cDNA is placed in an adequate vector to acquire sufficient material for analysis. Since mRNA lacks introns, the obtained cDNA also lacks introns and flanking regulatory sequences.[1]

Another kind is cloning of genomic DNA. For this purpose, DNA from nucleated cells is cut by restriction endonuclease. The obtained fragments are placed in the cut viral vector by sticky end ligation. The recombinant viruses are amplified in bacterial host cells and then screened for the appropriate cloned sequences representing genomic libraries (in the case of isolated chromosomes cloning produces chromosome-specific libraries).[1]

In the case of the amino acid sequence of the gene product being known, it is possible to construct an artificial assembly of nucleotides forming part of the gene. Such an oligonucleotide sequence can be used for identification of the complementary genomic clone within the whole genomic or chromosomal library.

Recombinant DNA techniques are widely used for identification of molecular defects in man. These methods are highly sensitive and discriminating in detecting DNA mutations and differences between species, which promulgated them as useful tools for detection of inherited and sometimes also acquired diseases.

II. DETECTION OF COMMON DISEASE ALLELES

A classical example (the first elaborated) represents sickle cell anemia, caused by point mutation within the β-globin chain. The β-globin gene probes are used for identification of restriction fragment length polymorphism (RFLP). Using the restriction enzyme Hpa I, Kan and Dozy detected a 13.0 kb RFLP which occurred about 5 kb 3′ to the β-globin locus in a San Francisco black population and demonstrated frequent association with the β-globin sickle cell (β^S), but not with the normal (β^A) alleles. In the U.S. population, 87% β^S genes are found to be associated with the 13.0 kb RFLP, whereas only 3% association was found with the β^A (linkage disequilibrium). This single observation made it possible to establish two important conclusions: (1) natural variation in DNA structure (polymorphisms) can be readily detected by RFLPs; (2) the presence of RFLP alleles can be correlated with disease alleles, which makes disease diagnosis possible.[2]

Soon an improvement of β^S-globin mutation was elaborated using a restriction enzyme the recognition site of which comprises the DNA sequence at the β^S mutation. Mst II restriction enzyme, which recognizes the sequence CCTNAGG, was used for detection of the β^S mutatant sequence CCTNTGG (A → T) which allowed precise diagnosis by DNA hybridization analysis. In the case of HbA/HbA, RFLPs were 0.2 and 1.15 kb, whereas in HbS/HbS it was 1.35 kb (in heterozygotes HbA/HbS 0.2, 1.15, and 1.35 kb fragments were present). The second mutant common allele β^C (occurring at the same β_6 codon in which G → A) does not alter recognition site of Mst II RFLP. Thus, HbC variant hemoglobin cannot be detected by Mst II RFLP.[2]

RFLP linkage disequilibrium for a common recessively inherited disease can be identified. For example, the Z allele of inherited α_1-antitrypsin deficiency has shown linkage disequilibrium with an RFLP detected by the restriction endonuclease Ava II.[2]

A. DETECTION OF A MUTANT ALLELE BY HYBRIDIZATION WITH SYNTHETIC OLIGONUCLEOTIDES NAMED ASO (ALLELE-SPECIFIC OLIGONUCLEOTIDE METHOD)

For this purpose 19-base-long oligonucleotides are synthesized: one for the normal allele (e.g., β^A-globin allele) and another for the mutant allele (e.g., β^S or β^C-globin alleles). These nonadecanucleotides are radioactively labeled and used as probes for hybridization experiments with the normal allele. Since DNA from HbA/HbA individuals forms stable duplexes (hybridizes) only with a specific β^A probe, DNA from homozygotes HbS/HbS hybridizes only with the β^S probe, and DNA from heterozygotes HbA/HbS hybridizes with both β^A and β^S probes. Using adequate oligonucleotides β^A, β^S, and β^C, chains can be diagnosed by this method.[2,3]

However, the ASO method is limited by the high ratio of nonspecific to specific hybridization targets. The sensitivity of this method has been improved by selective amplification of the examined gene. Using two priming oligonucleotides, complementary to opposite DNA strands that flank the examined gene (e.g., β^A, β^S, and β^C alleles) by successive cycles of DNA synthesis, it was possible to generate *in vitro* a 110 bp sequence that contained these alleles. The method allowed a 220,000-fold amplification of the target sequence with only approximately 100 cells as the starting material[2,4] (so-called PCR reaction — polymerase chain reaction).

The ribonuclease A method was initially used for point mutation detection within liver mRNA. Oligonucleotide primers flanking the mutation site are used for amplification of the mutation site facilitating molecular cloning and sequencing of the point mutation. This method serves for detection of the point mutation by hybridization with a normal negative-strand probe. The fragments obtained by RNase cleavage of the mutant allele are detected by gel electrophoresis and autoradiography.[2,5] Point mutation detection of alleles arising from new mutations represents an important diagnostic tool, particularly in the frequent occurrence in X-linked recessive diseases (HPRT deficiency in Lesch-Nyhan syndrome, ornithine transcarbamylase deficiency, Duchenne muscular dystrophy, etc.).

The detection of mutation within large nuclear DNA is difficult because of multiple exons. For the study of large genomic sequences, another method has been elaborated. Single mismatches within duplexes formed by the DNA probe and the sample of the examined DNA causes denaturation of the duplex that can be detected on the basis of lack of mobility in a denaturation gradient gel. Mismatched duplexes denature and become immobile eariler in the denaturation gradient electrophoresis than perfectly matched duplexes that remain intact for a longer time.[2]

B. DETECTION OF SINGLE POINT MUTATION IN TOTAL GENOMIC DNA

Single base substitutions resulting in polymorphisms linked to mutant alleles may alter a restriction enzyme recognition site, and may be detected in the total genomic DNA by DNA blots. Base substitutions that do not cause altered recognition sites can be detected by synthetic oligonucleotides used as hybridization probes when the DNA sequences surrounding the base substitution are known. In the case of multiple point mutations within a single molecule, an adequate number of oligonucleotide probes is required. For detection of heteroduplexes formed between wild-type and mutant DNA, S1 nuclease treatment can be used for their detection. Most of the mismatched duplexes are not cleaved by this enzyme. However, it is not known if S1 nuclease treatment is applicable to heteroduplexes containing single base mismatches.[5] Therefore, a procedure has been described that enables their detection by electrophoretic separation.[6]

C. GENETIC LINKAGE ANALYSIS IN GENE MAPPING

Restriction fragment analysis of the total DNA became a tool in constructing human gene maps. There are two kinds of maps: the phenotypic map, in which a single molecule (for example, an enzyme, the deficient activity of which results in an inherited disease) is mapped to a single stained chromosome and a molecular map, in which the organization and sequence of the DNA composing the genes is determined by molecular methods.[2] The map distances differ in these two methods radically. Banding techniques permit one to distinguish approximately 20 bands on particular chromosomes. As a chromosome (e.g., X chromosome) contains approximately 100,000,000 bp, the nearest chromosome band contains approximately 5,000,000 bp, whereas a normal mammalian gene seldom exceeds 50,000 bp. Thus, the best cytogeneticist's resolution is 1/100th of the particular chromosomal band. In contrast, the molecular biologist is able to determine DNA sequences of several thousand or a thousand bp long.[6]

By the classical methods of studying genetic linkage (by comparing inheritance of a disease with a protein marked within a family, or by the method of somatic cell hybridization), only a limited number of chromosomes can be mapped — rarely more than 5,000,000 bp. The molecular methods use cloned DNA sequences, and look for polymorphisms where small or single base changes occur in the DNA sequence between two homologous chromosomes in an individual.[6]

D. VARIABLE NUMBER OF TANDEM REPEAT MARKERS (VNTR)

The recognition that RFLPs may be successfully applied for clinical diagnosis of several diseases whose gene defects are unknown but in which the inheritance of the particular disorder in a family is associated with the presence of an RFLP, raised hopes, particularly in such diseases as Huntington's chorea, adult polycystic kidney disease, cystic fibrosis, and Duchenne muscular dystrophy. The usefulness of probes for linkage analysis depends on the frequency of their RFLPs or recombinational distance from the mutant gene, or both. Single-base RFLPs are detected by restriction endonuclease cleavage and DNA hybridization analysis. For example, the restriction endonuclease Taq I recognizing TCGA has proven to be useful for RFLP identification, suggesting that CpG is a highly mutable site (confirmed in new mutational events in hemophilia A and ornithine transcarbamylase genes). Jeffreys et al. were the first to isolate probes of a highly polymorphic character because of a variable number of tandem nucleic acid repeat sequences (VNTRs).[2,8]

According to Jeffreys et al.[8] DNA polymorphisms have revolutionized human genetic analysis and have found general use in prenatal diagnosis of inherited diseases, mapping of human linkage groups, indirect localization of genetic disease loci, and analysis of the role of mitotic nondisjunction and recombination in inherited cancer. Single-copy human DNA probes are used to detect RFLPs, resulting usually from small-scale changes in DNA (base substitutions) which create new, or destroy specific recognition of restriction sites. As the mean heterozygosity of human DNA is low, only a few restriction endonucleases will recognize RFLP at a given locus, although the detection can be improved by enzymes that contain the mutable CpG doublet in their recognition site. According to Jeffreys et al. the genetic analysis in man can be simplified considerably by the availability of probes for hypervariable regions of human DNA, showing multiallelic variation and correspondingly high heterozygosities. By chance several such regions have been isolated near several genes (human insulin gene, α-related globin genes, c-Ha-*ras*-1 oncogene). In each case, the variable region consists of tandem repeats of a short sequence (or minisatellite) and polymorphism results from allelic differences in the number of repeats. These sequences can be easily detected by restriction endonucleases that do not cleave the repeat unit and provide for such loci a set of stably inherited genetic markers.[9] The authors used, at first, a short minisatellite containing four tandem repeats of a 33-bp sequence in an intron of the human myoglobin gene, which showed some similarity to three other human minisatellites which share a common short "core" sequence in each repeat unit. This observation led to isolation of several other tandem repeats that appeared to be chromosomal probes with highly informative RFLPs due to VNTRs.[8]

According to Nakamura et al.,[9] a large collection of genetic markers is needed to map genes that cause human genetic diseases. Approximately 400 polymorphic DNA markers have been described; almost all have only two alleles, and thus are uninformative for determination of genetic linkage in many families. A few marker systems detect loci that respond to restriction sites by producing fragments of many different lengths due to tandem repeats of a short DNA sequence. Because most individuals may be heterozygous at such loci, these markers provide linkage information in almost all families. Nakamura et al. used 10 oligomeric sequences derived from the tandem repeat regions of the myoglobin gene, the zeta-globin pseudogene, the insulin gene, and the X-gene region of hepatitis B-virus to develop a series of single-copy probes, which revealed new, highly polymorphic genetic loci whose allele sizes reflected variation in the number of tandem repeats.[9]

The DNA analysis developed by Geffreys et al., based on hybridization of hypervariable minisatellite regions of human DNA to produce DNA fingerprints, found application in several fields of the biological sciences. Use of genetic markers become routine in forensic medicine (identification of individuals in rape and paternity testing).[2] DNA fingerprints are also used for establishing relationships in the animal kingdom, particularly for identification of paternity in domestic species used for inseminations (e.g., in cattle).[10]

E. DNA ANALYSIS IN NEOPLASIA

DNA analysis is important to elucidate chromosomal rearrangements and point mutations resulting in activation of oncogenes. At first, it was shown that c-*myc* transcription may be activated in association with B-cell chromosomal translocations in Burkitt's lymphoma, and T-cell leukemias with chromosomal rearrangements (11:14;8:14) that involve c-*myc* and the α-chain of the T-cell receptor. Similarly, the inherited form of retinoblastoma maps to chromosome 13q21-13q13, and it is thought that loss of suppressor sequences is involved in the somatic transition to neoplasia.[2]

F. DNA ANALYSIS IN INFECTIOUS DISEASES

Recombinant DNA analysis of infectious diseases is particularly useful in viral infections enabling through *in situ* hybridization of tissues or cultured cells an early diagnosis (herpes virus, HIV — human immunodeficiency virus causing AIDS, cytomegalovirus, Epstein-Barr virus, and others).[2]

G. ANTI-SENSE RNA AS MOLECULAR TEST FOR GENETIC ANALYSIS

The introduction of "anti-sense" RNA (i.e., RNA complementary to a particular RNA) into cells inhibits expression of the "sense" mRNA transcript. With developing recombinant technology, it often became difficult to ascribe a given function to a cloned sequence. The use of anti-sense RNA is based on the observation that an mRNA is not translated *in vitro* if it is hybridized to a complementary polynucleotide. The same occurs *in vivo* when the mRNA becomes hybridized with its complementary "anti-sense" sequence of RNA. The experiments performed confirmed the usefulness of anti-sense RNA in defining cellular function of known, as well as unidentified, genes.[11]

The number of cloned human DNA sequences is still growing, and comprises some hundreds of DNA sequences that can be used for diagnosis of genetic diseases.[12-14]

III. METHODS OF DETECTING GENE MUTATIONS (SUMMARY)

A. RESTRICTION FRAGMENT LENGTH POLYMORPHISMS (RFLP)

RFLPs are based on specific DNA recognition sites of the particular restriction enzymes. RFLPs are not the result of the change that caused the lesion but a pre-existing variation in the population at large. Therefore, RFLPs that link to the disease in any family will depend upon which one existed on the ancestral allele in which the mutation arose and from whom the family is descended. Further, there is a chance that the same RFLP that characterizes the ancestral changed allele will be introduced into the family on the same chromosome as the allele lack the lesion. Thus, the diagnosis must be based on family studies.[8] RFLP analysis of β^S-globin gene by Hpa I (loss of a recognition site):

- HbA/HbA RFLPs 7.6 or 7.0 kb
- HbS/HbS RFLPs 1.3 kb
- HbA/HbS RFLPs 7.6 or 7.0 and 1.3 kb
- RFLP analysis of β^S-globin gene by Mst II (recognition site at codon β^6)
- HbA/HbA RFLPs 0.2 and 1.15 kb
- HbS/BbS RFLPs 1.35 kb
- HbA/HbS RFLPs 0.2, 1.15 and 1.35 kb
- HbC not recognizable

B. cDNA CLONING

mRNA is isolated from tissue that expresses generally only one gene and a limited number of other genes. mRNA is treated with reverse transcriptase to produce cDNA which is cloned to receive sufficient material for analysis.

1. Cloning of Genomic DNA

DNA from nucleated cells is cut by a restriction endonuclease, the fragments of which are placed into the cut viral vector by sticky end ligation. Cultures of bacteria containing cloned sequences represent genomic libraries.

C. ALLELE-SPECIFIC OLIGODEOXYNUCLEOTIDE (ASO) RECOGNITION

This is done by hybridization analysis between normal (wild-type) and synthetic non-adecanucleotide of the mutant allele (e.g., β^A, β^S, β^C-globin). Hybridization occurs only with a complementary probe (normal or mutant).

D. AMPLIFICATION OF THE MUTATED CODON (PCR REACTION)

Priming with two oligonucleotides, complementary to opposite strands that flank the examined gene with subsequent hybridization analysis (amplification 220,000-fold).

E. RIBONUCLEASE A SENSITIVITY METHOD

After amplification of the mutant allele, hybridization with negative-strand RNA probe, and, next, treatment with RNase A that cleaves the mutant allele. Fragments are detected by electrophoresis (mutant fragments migrate more quickly than the normal wild-type mRNA). Duplex melting method. Duplexes are formed between normal (wild-type) probe and sample of the examined DNA. Mismatched duplexes denature and become immobile earlier in denaturation gradient electrophoresis than perfectly matched duplexes that remain intact for a longer time and migrate in denaturation gradient electrophoresis.

F. VNTR (VARIABLE NUMBER TANDEM REPEATS) SYSTEM

Most of the DNA sequence variations are caused by base-pair (bp) changes, resulting in change in the length of a restriction fragment. Such a single marker locus has two alleles with a single cleavage site (present or absent) and the chance that a parent has two different alleles at the same locus is less than 50%. The genetic analysis can be highly improved by the analysis of hypervariable regions of human DNA exhibiting multiallelic variation and high heterozygosity. Several highly variable regions have been detected near several genes (insulin, globin genes, X-gene region of hepatitis B virus, c-Ha-*ras*-1 oncogene and others) which can be detected using any restriction endonuclease that does not cleave the repeat unit and may serve as inherited genetic markers. Each variable region contains tandem repeats of a short sequence which polymorphism results from allelic differences in the number of repeats. Genotyping with marker systems requires several probes and/or digestion with several restriction endonucleases.

G. ANTI-SENSE RNA

Introduction of "anti-sense" RNA into cells inhibits expression of the corresponding "sense" mRNA which can be used for evaluation of cellular function of the investigated genes by their inhibition.

REFERENCES

1. **Connor, J. M. and Ferguson-Smith, M. A.,** *Essential Medical Genetics,* Blackwell Scientific, Oxford, 1984.
2. **Caskey, Th.,** Disease diagnosis by recombinant DNA methods, *Science,* 236, 1223, 1987.
3. **Conner, B. J., Reyes, A. A., Morin, C., Itakura, K., Teplitz, R. L., and Wallace, R. B.,** Detection of sickle cell β^S-globin allele by hybridization with synthetic oligonucleotides, *Proc. Natl. Acad. Sci. U.S.A.,* 80, 278, 1983.
4. **Saiki, R., Scharf, S., Faloona, F., Mullis, K. B., Horn, G. T., Erlich, H. A., and Arnheim, N.,** Enzymatic amplification of β-globin genomic sequences and restriction site analysis for diagnosis of sickle cell anemia, *Science,* 230, 1350, 1985.
5. **Myers, R. M., Lumelsky, N., Lerman, L. S., and Maniatis, T.,** Detection of single base substitutions in total genomic DNA, *Nature,* 313, 495, 1985.
6. **Williamson, B.,** The cloning revolution meets human genetics, *Nature,* 293, 10, 1981.
7. **Newmark, P.,** Molecular diagnostic medicine, *Nature,* 307, 11, 1984.
8. **Jeffreys, A. J., Wilson, V., and Thein, S. L.,** Hypervariable "minisatellite" regions in human DNA, *Nature,* 314, 67, 1985.
9. **Nakamura, Y., Leppert, M., O'Connell, P., Wolff, R., Holm, T., Culver, M., Martin, C., Fujimoto, E., Hoff, M., Kumllin, E., and White, R.,** Variable number of tandem repeat (VNTR) markers for human gene mapping, *Science,* 235, 1616, 1987.
10. **Hill, W. G.,** DNA fingerprints applied to animal and bird populations, *Nature,* 327, 98, 1987.
11. **Weintraub, H., Izant, J. G., and Harland, R. M.,** Anti-sense RNA as a molecular tool for genetic analysis, *Trends Genet.,* 1, 22, 1985.
12. **Schmidtke, J. and Cooper, D. N.,** A list of cloned human DNA sequences, *Hum. Genet.,* 65, 19, 1983.
13. **Cooper, D. N. and Schmidtke, J.,** DNA restriction fragment length polymorphisms and heterozygosity in the human genome, *Hum. Genet.,* 66, 1, 1984.
14. **Cooper, D. N. and Schmidtke, J.,** Diagnosis of genetic disease using recombinant DNA Supplement, *Hum. Genet.,* 77, 66, 1987.

INDEX

A

D

E

F

G

H

I

M

N

O

P

R

S

T

U

V

W

X

Z